FOURTH EDITION

The Dynamic Universe

An Introduction to Astronomy

The first image made with the new Keck 10-m telescope, on the summit of Mauna Kea. The Keck Telescope will consist of 36 hexagonal mirrors, each about 2 meters in diameter. Nine of the mirrors were in place when this photograph was obtained in November, 1990. The object shown is a spiral galaxy; the colors were assigned by computer processing techniques. (California Association for Research in Astronomy)

FOURTH EDITION

The Dynamic Universe

An Introduction to Astronomy

Theodore P. Snow
University of Colorado at Boulder

WEST PUBLISHING COMPANY
St. Paul   New York   San Francisco   Los Angeles

Composition: Parkwood Composition Service
Copyediting: Carol Danielson
Illustrations: Alex Teshin Associates

Cover Photo: The Large Magellanic Cloud, pictured on the cover, is a neighbor galaxy to our own Milky Way. This photograph reveals some of the fascinating components of the LMC, including its barlike central region (the yellow-white area running from lower center towards the right and slightly upward) and the brilliant regions of ionized gas, which show up as red, due to emission from hydrogen gas that has been heated by the intense radiation from hot stars. In the center of the Tarantula nebula, the prominent ionized region at the left, lies a very dense cluster of young, luminous stars which can only be distinguished with extremely fine images, too fine to be made by telescopes looking up through the Earth's atmosphere. The small boxed image, showing the inner portion of this cluster of stars, was obtained by the *Hubble Space Telescope*. (Main photo: © R.J. Dufour, Rice University. Inset photo: Nasa/Space Telescope Science Institute.)

Back Cover Photo: One of the twenty-seven radio dishes of the *Very Large Array* near Soccoro, New Mexico. (National Radio Astronomy Observatory.)

Library of Congress Cataloging-in-Publication Data

Snow, Theodore P. (Theodore Peck)
    The dynamic universe : an introduction to astronomy / Theodore P. Snow.—4th ed.
      p.   cm.
    Includes bibliographical references and index.
    ISBN 0-314-77104-2 (hard)
    1. Astronomy.   I. Title.
QB43.2.S66   1991
520—dc20   90-25660
 CIP

Dedication

For Connie

CONTENTS

SECTION VII:
LIFE IN THE UNIVERSE

Chapter 32: The Chances of Companionship

GUEST EDITORIAL:

APPENDICES

To study astronomy is, in a sense, the most human thing we can do. What distinguishes us from lower creatures, if not our curiosity, our compulsion to explore and discover? And what exemplifies this compulsion better than the study of the universe?

We probe the heavens (and the Earth) by all possible means, and we do it for no other reason than to learn whatever there is to be known. Astronomy has produced many useful byproducts, of course, and could be (and often is) justified solely on that basis. That is not the real reason for astronomy, however.

This textbook represents an attempt by an astronomer to share both the knowledge and the intellectual gratification of our science. There is considerable beauty in the universe for the eye and mind to behold. Just as it is visually stimulating to gaze at a great glowing nebula or a colorful moon, it is pleasing to the intellect to grasp a new understanding of one of the grand themes of the cosmos. It is hoped that the reader of this book will gain by doing both.

This textbook is intended for the student who has not chosen science as his or her major area of study, but who needs an appreciation of science as a vital aspect of preparation for a career. It is as important for such a student to gain some perspective on the general nature of science as it is to learn a great deal of specific information about a particular discipline in the sciences. For that reason, this text stresses the philosophy and outlook of the scientist as well as the knowledge we have gathered about the physical universe we live in.

It is probably as important for the student to understand *how* we know what we know as it is to understand *what* we know. In this era of instantaneous communication and universal access to information, we need more than ever to be able to discriminate among competing hypotheses, to be able to judge the reasonableness of ideas that are advanced. This text in astronomy is written with the underlying theme that to know the workings of science is one of the most important tools we have for meeting the challenges of our technological society.

This edition, like its predecessors, covers the entire scope of astronomy, from the most nearby objects to the most distant, from the smallest to the largest. Many questions are left unanswered, because there is so much we do not know, and because it is important for students to be aware of that. Not knowing the answers, but knowing how to pursue them, is the essence of science.

This fourth edition has some new features, and of course reflects all the exciting new developments in astronomy that have occurred since the third edition was published. Particularly noteworthy are the major revision of the section on the solar system and the addition of Mathematical Methods inserts at the ends of the chapters.

Now that the age of *Voyager* is over, and eight of the nine planets have been visited by probes from Earth, a reorganization of the chapters on the solar system is timely. The *Voyager* spacecraft have largely revolutionized our knowledge of the outer planets, turning what was a collection of mysterious, faraway objects into a set of related bodies that seem a little closer to us, in mind if not in place. New themes and new understanding have emerged. In the new solar system section (Chapters 8 through 16) we now start with a new general planetary science chapter, in which an overview of the system is given, along with explanations of the major physical processes that affect the planets. With this as background, the student then explores the planets individually, but with some reorganization as compared to previous editions (for example, the Moon and Mercury are now discussed in the same chapter, as are Jupiter and Saturn). One result of this reorganization is the reduction, by one chapter, of the overall length of the book. The section ends, as in the past, with a chapter summarizing the formation of the system, but now this chapter includes a unified discussion of rings and moons, which are known today to be general phenomena associated with the giant planets. It is hoped that this revised section will provide the student with a more modern, cohesive picture of our system of planets, and of the modern planetary scientist's approach to its study.

The other major change in this edition is the addition of some quantitative material. The Mathematical Methods inserts are designed to supplement the text for those instructors and students who wish to work with the numerical representations that form the essence of astrophysics. These brief sections are set apart from the main text so that they will not disrupt the smooth flow of the chapters, but are readily at hand for those who want to study them. In each Mathematical Methods insert a single concept is presented and explored, and can form the basis for new problems and questions if the instructor so desires.

Of course the *Hubble Space Telescope* receives a great deal of attention in this edition, with many references throughout, as well as a Special Report section in Chapter 7 which presents the latest results available (as of November, 1990). While it is widely known that the *HST* was launched earlier this year and that subsequently optical problems were discovered, the reader may be surprised to learn how productive the instrument still is expected to be, and how much impact it will have on astronomy, even in its crippled state.

Other significant updates in this edition include up-to-date descriptions of new planetary probes such as *Magellan*, which is mapping Venus as we go to press, and *Galileo*, which has embarked on its 5-year journey to Jupiter. Several stunning images of details on the surface of Venus are included, revealing the fact that the clouded planet may be more Earth-like geologically than previously suspected.

In addition to these changes in the written content of the book, the new edition also incorporates many new illustrations. As always, we have obtained new photographs and data where available, but in this edition we also have replaced many of the drawings as well. Now virtually all of the original line drawings from the two-color versions of the book (editions 1 and 2) have been replaced with full color drawings. Several entirely new illustrations have been created as well, helping to make this the most visually clear and stimulating edition to date.

Comprehensive factual data are incorporated into the Appendices, which have been updated for new information (as provided by the *Voyager* planetary encounters, for example). In addition, several Appendices have been expanded, including those on stellar data, interstellar molecules, major telescopes, and groups and clusters of galaxies.

The arrangement of the text remains traditional, with an introductory section on the background of astronomy, both in history and in basic physics; a section on the solar system, dealing with the planets as individuals before discussing interplanetary bodies and then the formation of the entire system; one on stars and their lives and deaths; a section on the structure and evolution of our galaxy; a set of chapters on extragalactic astronomy and the universe as a whole; and a final, brief section on the possibilities that life may exist elsewhere. At the beginning of each of these sections is an introduction that leads the student into the material, and at the end of each is a Guest Editorial in which a leading scientist in the field shares his or her thoughts on current problems or controversies and future directions for research.

The book is designed so that the sequence of sections may be easily altered. For example, if it is desired to teach the sections on stars, the galaxy, and the universe before discussing the solar system, one need only skip directly from Chapter 7 to Chapter 17 and then go on to the end before returning to Chapter 8, where the solar system studies begin. The chapter on the Sun leads into the section on stars, so that skipping or delaying the solar system discussion will not prevent the student from learning about the nearest and best-understood star. The overview and summary chapters on the solar system (Chapters 8 and 16) include enough information on the Sun that the discussion of the system as a whole and its formation is complete as it stands.

The well-received Astronomical Insights have been carried over into this edition, with a substantial number of new ones included. These inserts, placed within the chapters, describe people, discoveries, or current controversies or new hypotheses related to the subject matter of the text. They are meant to enhance the students' enjoyment of the material, or add understanding of complex topics, but above all they are designed to increase understanding of the scientific process.

Supplemental materials for this text include an updated version of the *Study Guide*, authored by Jeffrey O. Bennett and the undersigned (both University of Colorado), and a revised edition of the *Instructor's Manual* by Stephen J. Shawl (University of Kansas). As before, the *Instructor's Manual* contains helpful discussions of strategies in teaching, provides a large number of exam questions (with answers), and gives complete answers to all the review questions from the main text. The *Study Guide*, intended to help the student get maximum benefit from the text, contains brief chapter summaries, lists of key words and phrases, self-tests, and complete bibliographies of articles on relevant topics, taken from a wide assortment of magazines and journals. In addition to the *Study Guide* and the *Instructor's Manual*, another aide to teaching is offered to large adopters of the text: a set of transparencies for use with overhead projectors, including many in color, showing a number of useful diagrams and illustrations from the text.

At every step during the preparation of this text, vital assistance was provided by a number of people, whose help is acknowledged with gratitude (with apologies to anyone inadvertently omitted). The most important guidance and support was provided by my wife, Connie; by the West Publishing Company editor, Denise Simon; and by the production editor, Ann Rudrud of West. The new and improved drawings were prepared by Larissa A. Worth of Alexander Teshin Associates. A very important contribution to the overall quality and accuracy of the book was made by Steve Shawl, who scrutinized the galley proofs and, as he has during the preparation of earlier editions, made many useful suggestions regarding content as well.

Among my colleagues at the University of Colorado and elsewhere, several have helped by either reviewing sections of the text, providing new figures, or updating data for tables and Appendices. Particularly generous in this connection were A. Stern, J. Lunine (University of Arizona), M. Geller (Harvard-Smithsonian Center for Astrophysics), and I. R. Little-Marenin (Wellesley College). Additional assistance was provided by J. M. Shull, L. Esposito, R. A. McCray, J. A. Tyson (Bell Laboratories), J. O. Burns (New Mexico State University), B. A. Goldberg (JPL), K. van Aachen (JPL), D. Malin (Anglo-Australian Observatory), P. Smiley (National Radio Astronomy Observatory), Marc Sher (William and Mary), S. V. Forgue (Ocala, Florida), G. D. Nelson (NASA and the University of Washington), R. Marcialis (University of Arizona), and E. Chaisson (Space Telescope Science Institute).

Reviewers of the revised manuscript were:

W. J. Cocke
University of Arizona

Neil Comins
University of Maine

Roger A. Freedman
University of California, Santa Barbara

J. Russell Harkay
Keene State College, New Hampshire

Terry Jay Jones
University of Minnesota

Jonathan I. Lunine
University of Arizona

Stephen J. Ratcliff
Middlebury College, Vermont

J. Scott Shaw
The University of Georgia, Athens

Alex G. Smith
University of Florida

S. Alan Stern
University of Colorado at Boulder

David W. Wingert
Georgia State University

Arthur Young
San Diego State University

A special debt is owed to those who wrote the Guest Editorials, for adding their thoughts and visions to my own less elegant discussions. Much of the excitement of astronomy lies in the pursuit of new revelations beyond the scope of current knowledge, and the essays written by leaders in this pursuit help immeasurably to impart this excitement to the reader. Those contributing new essays to this edition are France Cordova (Pennsylvania State University), Jonathon Lunine (University of Arizona), Stephen Strom and Suzan Edwards (Five Colleges Astronomy Department; Strom is located at the University of Massachusetts and Edwards is on the faculty at Smith College), and Margaret Geller (Harvard-Smithsonian Center for Astrophysics), who largely rewrote her article, originally included in the previous edition. Three essays written for earlier editions are also included here; their authors are Roger Culver (Colorado State University), Richard Larson (Yale University), and Robert Bakker (University of Colorado).

For all of these people, and to the students whose responses to my teaching philosophies have also helped to shape this book, I am grateful. With their continued input, I trust that this book will continue to evolve, as does our understanding of the dynamic universe.

Theodore P. Snow November, 1990

THE NIGHTTIME SKY AND HISTORICAL ASTRONOMY

Our study of astronomy begins with an overview of the breadth, relevance, and excitement of the science that encompasses all of the universe. We begin, in Chapter 1, with a broad description of the nighttime sky and of human beings' fascination with its splendor. Next, in Chapter 2, we examine more closely the many phenomena that can be seen and appreciated with only our eyes as observing equipment. This arms us with all the knowledge our ancestors had as they attempted to develop a successful picture of the cosmos and their place in it. In Chapter 2 we go well beyond simply describing the sky; we also uncover the explanations for the observed phenomena. Thus we start with knowledge that took human beings millenia to develop.

We then trace the historical development of astronomy, beginning with the earliest recorded astronomical observations and hypotheses, and following the ofttimes slow growth of science through the ages. The basis of our modern science lies in the Mediterranean region of Europe; therefore we emphasize developments there. We also explore the genesis of astronomy in other parts of the world, where sophisticated knowledge ultimately either led to dead ends or merged with the mainstream of development.

In Chapter 4 we discuss the groundwork for modern science that was laid during the Renaissance, when fresh ideas emerged in astronomy, as in all forms of human endeavor. We learn to appreciate the awesome breakthroughs made by giants such as Copernicus, Brahe, Kepler, and Galileo, who led the way toward a correct understanding of the universe and the place of people and the planet we inhabit.

CHAPTER 1
The Essence of Astronomy

A Moonlit view of Kitt Peak National Observatory. (National Optical Astronomy Observatories)

The oldest of all sciences is perhaps also the most beautiful. No artificial light show can rival the splendor of the heavens on a clear night, and few intellectual concepts can compare with the beauty of our modern understanding of the cosmos.

Today we study astronomy for a variety of reasons, some technological, some practical; but no one loses sight of the underlying majesty, of the human instinct for intellectual satisfaction. To study astronomy is to ask the grandest questions possible, and to find hints at their answers is to satisfy one of humankind's most deeply ingrained yearnings.

In this text we explore astronomy in the modern context, which is highly technical and sophisticated, but we will endeavor to retain the sense of wonder and beauty that has motivated the science since the beginning. Although some may argue that astronomy has little practical use, we will see that its origins are rooted in very practical requirements for methods of keeping time and maintaining calendars. In this chapter we begin our study by defining astronomy and introducing some simple terminology that will assist us in later chapters.

WHAT IS ASTRONOMY?

Astronomy is the science in which we consider the entire universe as our subject. It is the science in which we derive the properties of celestial objects and from these properties deduce the laws by which the universe operates. It is the science of everything.

Technically, we might say that astronomy is the science of everything except the Earth, or that it is the study of everything beyond the Earth's atmosphere, since the Earth and its atmosphere fall into the purview of other disciplines such as geophysics and atmospheric

science. We will find, however, that the study of astronomy necessarily includes an examination of the properties and the evolution of the Earth and its atmosphere.

In the modern sense, astronomy is probably more aptly called **astrophysics.** Ever since the time of Sir Isaac Newton (the late seventeenth century), the universe has been explored through the application of the laws of physics—most of them derived from earthly experiments and observations—to celestial phenomena. Other scientific disciplines enter into our discussions as well: To study the planets, for example, we must know something of geology and geophysics; to analyze molecules in space, we must understand the principles of chemistry.

Astronomical observations were made and records kept at least as early as the time of the most ancient recorded history (Fig. 1.1), and we believe that the skies were studied and their cyclic motions pondered long before that. In the earliest times, astronomy had a practical motivation: Knowledge of motions in the heavens made it possible to predict and plan for certain significant events, such as the changing of the seasons.

Along with the practical came the whimsical and the spiritual. In ancient times, events in the heavens were thought to exert some influence over the lives of people on Earth, and many early astronomers practiced what today we call astrology. Many a monarch retained the services of an astronomer, not only to foretell the seasons, but also to provide advice on strategies for war, love, politics, and business. In many cultures, religion and astronomy were intimately linked, so astronomers attained the importance that major religious leaders have today.

The rich and diverse Greek civilization that flourished for many centuries before and around the time of Christ developed several sciences—astronomy included—beyond the mystical and the spiritual. The Greeks had many preconceived notions about the nature of the universe, but they also made many rational advances in understanding the heavens, setting the stage for the development of modern science many centuries later. Following the era known as the Dark Ages, astronomy (like many other disciplines) experienced the pangs of rebirth during the Renaissance, becoming a rational and methodical science. During the Renaissance, battles were yet to be fought between religion and science, but the course was set. The work of pioneers such as Copernicus,

Figure 1.1 An ancient astronomical site. *Stonehenge, a monument in England, was built according to astronomical alignments in prehistoric times. (C. D. McLoughlin)*

Galileo, Kepler, and especially Newton placed astronomy on a firm physical basis.

Modern astronomy still has a practical aspect, but few who pursue the science do so for that reason. Today we study astronomy primarily for the sake of expanding our knowledge; it is research of the purest sort. Even so, many practical and concrete benefits derive from astronomy; witness, for example, the many technological spin-offs from the space program or the multitude of new physical and chemical processes, applicable here on Earth, that have been discovered through astronomical observation.

No matter how analytical we may be in modern astronomy, we never lose sight of the same basic human feelings that inspired our ancestors. The modern astronomer, who may spend observing time using a large telescope and a variety of complex electronic instruments (Fig. 1.2), still treasures the moments spent outside the telescope building, simply watching the skies with the same tools as the ancients used.

Figure 1.2 A large modern telescope. *This large reflecting telescope at the Cerro Tololo Inter-American Observatory, located in Chile and operated by the U.S. government, has a diameter of 4 meters, or roughly 13 feet. (National Optical Astronomy Observatories)*

ASTRONOMICAL INSIGHT 1.1

The Philosophy of Science

Science progresses by hypothesis and test, by trial and error. A scientist is willing to revise a theory in the light of new information, rather than adopting a hypothesis and then attempting to ignore or rationalize data that contradict a preconceived notion.

If there is anything that a scientist adopts as an article of faith, it is that no theory is above the possibility of modification—that our knowledge always has room to grow. This willingness to change does not mean that scientific theories are wrong or invalid. A scientific theory should be regarded simply as the best explanation available that is consistent with the known facts. A theory is modified or replaced when more facts become known or when a better explanation is found.

A theory usually starts as a hypothesis, an initial suggestion made to explain some phenomenon that is observed. As new information is gathered, the hypothesis develops into a theory, a framework allowing for prediction and test. Many theories fall by the wayside quickly because new information contradicts them, while others persist and become widely accepted because they are consistent with new data that become available. One feature of any theory is that it leads to specific predictions about what will happen in new situations. Part of the job of testing a new theory is to make predictions from it and then make observations or perform experiments to test those predictions.

One of the most important goals of this text is to impart to the student an understanding of how scientific progress is made—of how theories are developed, tested, and revised. For this reason, many examples of discoveries are described, some in Astronomical Insights such as this one, and others in the main text. These descriptions will help illustrate how scientists develop, test, and modify theories in the light of new information. They will show that even the best scientists make errors, but that the common thread is their willingness to discard incorrect hypotheses when conflicting evidence is found or a better explanation appears. The reader will see that scientists do not always know what to look for, but that our knowledge of the universe grows because scientists seek the best explanation for what is observed, rather than trying to modify what is observed to conform to existing theories. The hope is that these descriptions will help the student develop a sense of how science works, and a better perspective on new theories and controversies that may be encountered beyond the classroom.

A TYPICAL NIGHT OUTDOORS

Let us imagine that we are sitting outdoors on a fine, clear night, and that we are far from city lights and other distractions. This is easy to do and highly recommended; by simply getting out into the countryside, we can see many of the beautiful objects that we will be studying in more detail as we use this text.

The most obvious objects in the sky, assuming that the Moon is not in a bright phase, are the stars. They appear in profusion, scattered across the heavens, displaying a wide range of brightnesses and subtle variations in color. They twinkle, giving the appearance of vitality. Here and there we may see concentrations of stars in a cluster (Fig. 1.3), or possibly a dimly glowing gas cloud (Fig. 1.4).

If it is a moonless night, we see a broad, diffuse band of light across the sky. This is the Milky Way, our own galaxy of stars, seen edgewise from an interior position (Fig. 1.5). The Milky Way consists of billions of stars intermixed with patchy clouds of interstellar gas and dust.

We may also see a few bright, steady objects that do not appear to twinkle. These are the planets, and as many as five of them may be visible on a given night, distributed along a great arc through the sky. Careful observation over several nights will reveal that the planets are all moving gradually along this arc, changing their positions relative to the fixed stars.

The most prominent object in the nighttime sky is usually the Moon (Fig. 1.6), which shines so brightly when near full that it drowns out the light of all but the brightest stars and planets. The Moon, about one-fourth the diameter of the Earth but some 240,000

Figure 1.3 A cluster of stars. *This group of stars is held together by mutual gravitational attraction. This cluster, called the Pleiades, is easily visible to the unaided eye in the autumn and winter. (Palomar Observatory, California Institute of Technology)*

Figure 1.4 A gaseous nebula. *Clouds of gas and tiny dust particles such as this are the birthplaces of stars. This is the Orion Nebula, bright enough to be seen with only a modest telescope or pair of binoculars. (© 1981 Anglo-Australian Telescope Board. Photo by D. Malin)*

Figure 1.5 The Milky Way. *This composite photograph shows the hazy band of light across the sky known as the Milky Way. It is a cross-sectional view of the disk of our galaxy, which contains roughly 100 billion stars. (Mt. Wilson and Las Campanas Observatories, Carnegie Institution of Washington)*

miles (or about 60 times the radius of the Earth) away, presents various appearances to us, depending on how much of its sunlit portion we see. The Moon is always found somewhere along the same east-west strip of the sky, called the **zodiac,** where the planets travel.

Occasionally we may see brief flashes or trails of light called **meteors.** These "shooting stars" can be spectacular events, particularly when they arrive with great frequency, as they do during a meteor shower.

Comets are occasional visitors to our sky (Fig. 1.7), as the recent passage of Comet Halley has reminded

us all. Perhaps once or twice a year a comet is found that is bright enough to be seen with the unaided eye. These largely gaseous bodies orbit the Sun, as do the Earth and other planets, but in very elongated paths that bring them close enough to the Sun to heat up and glow visibly only for brief periods of days or weeks. Some of the more spectacular, bright comets were interpreted in ancient times as harbingers of catastrophe.

There is a complex pattern of motions in the sky, some of it evident to an alert watcher in an hour or so, other aspects requiring careful observations over

Figure 1.6 The moon. *Astronauts on one of the Apollo missions made this photograph from space. The portion at the left in this image is not visible from Earth. (NASA)*

hours, days, or weeks. The most obvious motion is the steady rotation of the entire sky; objects rise in the east and set in the west, as a reflection of the Earth's rotation. (The terms **rotation** and **revolution** are sometimes interchanged in everyday conversation, but here *rotation* means the same thing as *spin*, whereas *revolution* means *orbital motion*, such as the motion of the Earth around the Sun or of the Moon around the Earth.) Another motion that can be discerned readily is that of the Moon with respect to the stars. As it orbits the Earth, the Moon moves a distance in the sky equal to its own apparent diameter in just one hour, so it is possible to see its position change with respect to the background stars in a short time. Other cyclical motions, such as those of the planets as they gradually travel along their orbits about the Sun, require more patience and care to be observed. It is noteworthy, however, that ancient astronomers noticed many of the regular patterns, some of them quite subtle, in the motions of heavenly bodies. That they did so is a testimony to the care and diligence they applied to their studies of the skies.

THE VIEW FROM EARTH

When we look at the sky, we do not see it in three dimensions, because there are no obvious clues to tell

Figure 1.7 A comet. *This is a view of Comet Kohoutek during its 1973 passage through the inner solar system. (NASA)*

us the distances to the objects we see. Long ago this fact led to the concept of the **celestial sphere** (Fig. 1.8), in which the stars and other objects in the sky are said to lie on the surface of a sphere that is centered on the Earth. Although we no longer think of this as literal truth, it may still serve as a convenient device for discussing and visualizing the heavens.

Directions and separations of objects on the celestial sphere are measured in angular units, because lacking knowledge of distances to objects, we have no easy way to determine their actual separations in true distance units such as miles or kilometers. Thus we may specify where a star is by saying how many degrees, minutes, and seconds it is from another star or from a reference direction.

We have noted that the stars rise and set with the daily rotation of the Earth. This gives us a natural basis for timekeeping, and our standard units of time are based on the Earth's rotation. The length of the day is equivalent to the rotational period of the Earth (a more specific definition is given in the next chapter).

Anyone who has traveled from one hemisphere to the other may have noticed that the stars that are visible are not the same in the Northern Hemisphere as in the Southern (Fig. 1.9). The portion of the sky that we can see depends on our latitude (our distance, in degrees, north or south of the equator). For those of us living in the Northern Hemisphere, a large region of the southern sky is beyond our view. The constellations that we see vary as we travel north or south, a fact that was well known to early astronomers, who deduced from this and other evidence that the Earth is round. One consequence is that astronomers must have telescopes in both hemispheres in order to study the entire sky.

We usually cannot observe celestial objects during daylight, so our view of the heavens at any particular time is limited to only the half of the sky that is visible at night. Because of the Earth's motion around the Sun, however, the part of the sky that is up during the night gradually changes (Fig. 1.10). Therefore, any part of the sky that is visible from a given location on the Earth can be observed at night at some time of the year; we need only wait until the appropriate time to observe a given object.

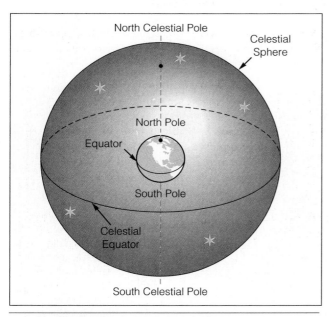

Figure 1.8 The celestial sphere. *Because we see the sky in only two dimensions, it is useful and convenient to visualize it as a sphere centered on the Earth, with the stars and other bodies set on the surface of the sphere. We measure positions of objects on the celestial sphere in angular units, because the actual distances are not directly known. (Throughout this text, we will learn about distance measurements in astronomy.)*

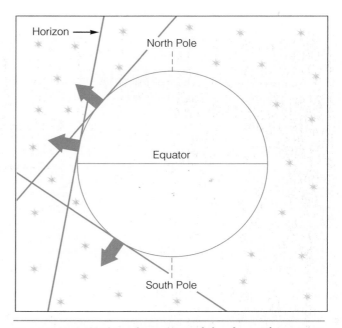

Figure 1.9 Latitude and our view of the sky. *The portion of the sky that we see depends on where we are with respect to the Earth's equator. The arrows indicate the overhead direction from three different latitudes.*

ASTRONOMICAL INSIGHT 1.2

What Is an Astronomer?

 Throughout this text we will be referring to astronomers, scientists, astrophysicists, and physicists. In a book that endeavors to summarize all that we know about the universe and its contents, we can hardly omit a description of the people who devote their time to developing this knowledge.

There are thousands of people in the United States alone who study astronomy as either a vocation or an avocation. Representative of the latter are the amateur astronomers who engage in a variety of astronomical activities, often on their own, but in many cases through local and even nationwide organizations. Telescope making, astrophotography, long-term monitoring of variable stars, public programming, and just plain stargazing are included. If you wish to join such a group, to get advice on buying a telescope, or to learn techniques such as photography of celestial objects, your best bet is to get in touch with an amateur astronomy group. Local clubs exist in most major cities, and regional associations are everywhere. These groups may be difficult to find in the telephone book, but a telescope shop or planetarium is bound to know whom to contact.

The amateur astronomers in the United States outnumber the professionals. There are some 5,000 members of the American Astronomical Society, the principal professional astronomy organization in the United States. As a rule, these people fall into a limited number of categories: those who do research and teach at colleges and universities; those who do research work at government-sponsored institutions, such as the national observatories and federal agencies like NASA; and those who perform research and related engineering functions in private industry, most often with companies involved in aerospace activities.

About 10 percent of the members of the American Astronomical Society are women. This is certainly a low percentage, but a recent study indicates that there is little discrimination in the hiring of professional astronomers. The small number of women in astronomy simply reflects the small number who choose to enter the field at the graduate level, so it is at that level that improvement must be made.

Most of the funding for research in astronomy, even for those not working directly for federal labs and observatories, comes from the government. A significant function of an astronomer on a university faculty is to write proposals, usually to the National Science Foundation or to NASA, for support of proposed research programs.

The terms *astronomer* and *astrophysicist* have come to mean much the same thing, although historically, there was a difference. An astronomer studied the skies, gathering data but doing relatively little interpretation; an astrophysicist was primarily interested in understanding the physical nature of the universe, and therefore carried out comprehensive analyses of astronomical data or did theoretical work, in both cases applying the laws of physics to phenomena in the heavens. Nearly all modern astronomers do astrophysics to varying degrees, however, ranging from the observational astronomer at one end of the spectrum to the pure theorist at the other. The two terms for people engaged in these pursuits are therefore used interchangeably today. Many modern astronomers call on the fields of engineering (for instrument development), chemistry (in studying planetary and stellar atmospheres and the interstellar medium), geophysics (in probing interior conditions in planets and other solid bodies), and sometimes even biology, but always with an underlying foundation in physics.

If you want to become a professional astronomer, you should be aware from the outset that the field is small and job opportunities are both limited and often subject to the vagaries of federal funding. If you persist, the best course is to study physics, at least through the undergraduate level, and then to plan to attend graduate school in astronomy or physics. (The latter option may make you a bit more versatile and is never a handicap when entering astronomy later.) On the bright side, demographic studies have indicated a significant shortage of astronomers for a period beginning in the early 1990s and extending through the rest of this century. Perhaps the timing will be right for you.

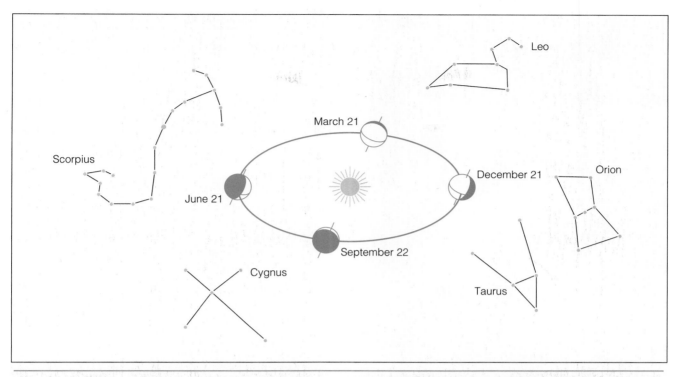

Figure 1.10 The changing view of the sky with the seasons. *As the Earth orbits the Sun, the portion of the sky that we can see at night changes. A few prominent constellations are shown here.*

FROM THE EARTH TO THE UNIVERSE: THE SCALE OF THINGS

We have been describing the appearance of the sky to the unaided eye, which has necessarily limited us to nearby objects such as the Sun, the Moon, the planets, and the stars in our part of the local galaxy. It is interesting now to expand our horizons and to try to comprehend the scale of the universe beyond this local neighborhood. Some of the distance scales we discuss are listed in Table 1.1.

Even the nearest star is much farther from the Earth than any solar system object. If we take the average Sun–Earth distance as a unit of measure (we call this the **astronomical unit** or AU), the nearest star is nearly 300,000 of these units away from us. The most distant known planet, Pluto, is only about 40 AU from the Sun, so we see that the stars are much more widely dispersed in space than the objects within our solar system.

If we could reduce the scale of the solar system, it might help us to visualize the relative distances. For example, if we let the Earth be the size of a basketball (let us say that its diameter is 1 foot), we can convert the rest of the solar system to the same scale. The Sun would be almost 110 feet in diameter, and the distance from the Sun to the Earth would be nearly 12,000 feet, or more than 2 miles. Pluto would be a tennis ball–size object roughly 90 miles away. The distance from the Sun to Alpha Centauri, the nearest star, would be over 600,000 miles, or more than twice the actual Earth–Moon distance!

Now consider our galaxy, the vast collection of stars to which our Sun belongs (Fig. 1.11). The Milky Way galaxy contains roughly 100 billion (10^{11}) stars, arranged in a huge disklike structure having a diameter of about 100,000 (10^5) light-years. (A **light-year,** the distance light travels in a year at its speed of 186,000 miles per second, is equal to about 5,900 billion miles.) The distance to Alpha Centauri is about 4 light-years, and the most distant stars easily seen with the unaided eye are several hundred light-years away. (The majority of the brightest stars in the sky are actually quite nearby, by galactic standards; see Appendix 8). Light from these stars has been traveling for hundreds of years when it reaches our eyes, and light from the far side of the

Figure 1.11 A galaxy similar to the Milky Way. *We cannot obtain an exterior view of our own galaxy, but it is thought that this one resembles ours. Note that most of the stars lie in a disk, seen face-on in this photo. (© 1980 Anglo-Australian Telescope Board. Photo by D. Malin)*

Figure 1.12 The Magellanic Clouds. *These two irregularly shaped, small galaxies lie just outside the Milky Way galaxy and orbit it, taking hundreds of millions of years for each complete orbit. (Fr. R. E. Royer)*

Table 1.1 Size Scales in the Universe

Object or Phenomenon	Size[a]
Atomic nucleus	10^{-13}cm
Atom	10^{-8}
Virus	5×10^{-6}
Interstellar dust grain	5×10^{-5}
Bacterium	10^{-4}
Human body cell	5×10^{-3}
Human	2×10^{2}
Planet Earth	1.3×10^{9}
Sun	1.4×10^{11}
Sun-Earth distance	1.5×10^{13}
Distance to nearest star	4.1×10^{18}
Milky Way galaxy	9×10^{22}
Distance to Andromeda galaxy	2×10^{24}
Local group of galaxies	5×10^{24}
Rich cluster of galaxies	10^{25}
Supercluster	10^{26}
The universe	$>10^{28}$

[a]The sizes listed are meant to be typical, to illustrate the relative scales. For round objects, the diameter is used; for irregular objects, an approximate average dimension is given.

galaxy takes about 100,000 years to reach us.

The nearest galaxies beyond the limits of the Milky Way are the Magellanic Clouds, two irregularly shaped, fuzzy patches of light visible only from the Southern Hemisphere (Fig. 1.12). The Magellanic Clouds lie between 150,000 and 210,000 light-years from the Sun, so we see that they are not very far outside the galaxy. They are considered satellites of the Milky Way, orbiting it in a time of several hundred million years. Light from the Magellanic Clouds takes roughly 200,000 years to reach us. The most distant object visible to the unaided eye is another galaxy similar to the Milky Way, called Andromeda, after its constellation, which is about 2.3 million light-years away (Fig. 1.13). When we look at the Andromeda galaxy, we are receiving light that has been traveling for more than 2 million years!

Even the distance to the Andromeda galaxy is insignificant compared to the scale of the universe itself. The Milky Way, the Magellanic Clouds, Andromeda, and a number of other galaxies all belong to a concentrated grouping, or cluster, of galaxies. Most of the other galaxies in the universe also belong to clusters (Fig. 1.14), whose diameters can be as large as tens of

ASTRONOMICAL INSIGHT 1.3

Astronomy and Astrology

There is an unfortunate tendency in modern society to confuse astrology and astronomy, or, worse yet, to consider one a legitimate alternative to the other. Behind this tendency lies an even more unfortunate misunderstanding of what science is, and the distinction between scientific theory and beliefs adopted on faith.

Astrology, the belief system based on the premise that events and configurations in the sky influence human affairs and activities on Earth, arose at a primitive stage of human development, at a time when the Earth was still thought to be a flat disk under the dome of the heavens. Although ancient Greek astronomers did much to raise the study of the heavens to a scientific level, the astrological rationale for study of the heavens persisted through the subsequent Dark Ages. Belief in the predictions of astrology began to change during the Renaissance, when the true nature of the heavens and the motions of astronomical objects were untangled, but even so, substantial interest in astrology was maintained. In fact, some of the most eminent and forward-looking astronomers of the time practiced astrology. It is unfortunately true that, even today, some people continue to profess faith in astrology.

One of the basic lessons we have learned about the universe is that it is easy to make mistakes unless we are careful to be objective and to accept only conclusions that can be verified by repeated observations or experiments, or by making predictions that can be tested. Astrology fails to meet these criteria, utterly and abysmally. Serious attempts have been made to test astrological lore by statistical analysis of people born under different signs, but no trace of a correlation has ever been found.

Unfortunately, people who accept astrology usually do not demand verification, but prefer instead to put their trust unquestioningly into the jumbled and vague predictions and aimless advice offered by astrologers. No real knowledge is to be gained by doing so; perhaps faith in astrology provides a sort of shelter from a world that is too complex for easy understanding and ready decision making.

Many phenomena defy the understanding of modern science, and many of these deserve more attention than they are getting. However, astrology is not one of them. As a phenomenon, astrology is worthy of study only by the sociological and psychological sciences, for its effects do not exist in the physical universe. Comparing astrological signs may be interesting party talk, but it is a good idea to bear in mind the difference between objective reality and subjective impressions. Failure to do so is failure to understand what science is.

---

millions of light-years. Between clusters of galaxies, space is relatively empty. (Actually, this point is controversial, as we shall see; all we can say for certain is that there are relatively few visible galaxies between clusters).

Clusters of galaxies are themselves grouped into larger conglomerates called **superclusters,** whose size scales are significant, even on the scale of the universe itself. A supercluster typically may have a diameter measured in the hundreds of millions of light-years. (*Diameter* is probably not a good word; superclusters seem to be sheetlike or filamentary structures, not rounded like clusters of galaxies.) It is difficult to imagine an organized object or collection of objects so large that it takes hundreds of millions of years for light to travel across it.

Beyond the size scale of the superclusters, we approach the scale of the universe itself. It is apparent from a variety of lines of evidence that the observable universe has an overall size scale measured in tens of billions of light-years. Light reaching us now from the farthest reaches of the universe has been traveling for many billions of years. This has the fascinating implication that we can see the universe as it was at that long-ago time when the light was emitted, a fact used by astronomers who study the origins of the universe.

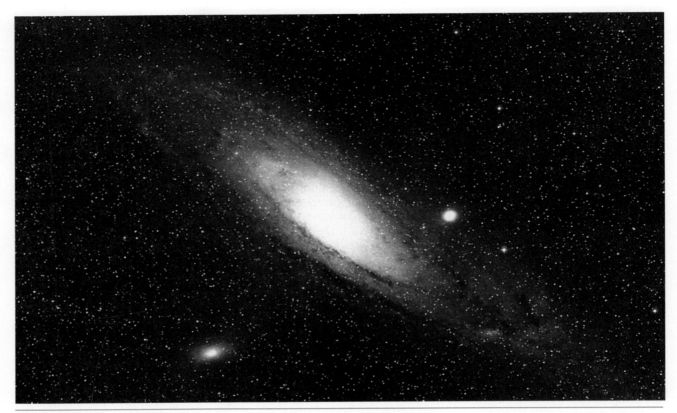

Figure 1.13 The Andromeda galaxy.   *At a distance of over 2 million light-years, this galaxy is the most distant object visible to the unaided eye. Without a telescope and a time-exposure photograph, the eye sees only an extended, fuzzy patch of light, rather than the detailed view shown here. The full extent of the galaxy, as revealed by a time exposure such as this, covers about 2½° on the sky, or 5 times the angular diameter of the Moon. (Palomar Observatory, California Institute of Technology)*

Figure 1.14 A cluster of galaxies.   *The faint, fuzzy objects in this photograph are galaxies belonging to a cluster that lies over a billion light-years from the Milky Way. Like our own, each galaxy contains billions of individual stars. (Palomar Observatory, California Institute of Technology)*

Considering the sizes and distances of objects in the universe provides us a sobering perspective on ourselves and our tiny planet. It can be quite a revelation to see how much we presume to explain about the universe and how much we think we have learned about the origins, present state, and future evolution of the cosmos and all it contains. It will be wise, however, to keep in mind that there is very much that we do not know.

PERSPECTIVE

The introduction to astronomy in this chapter prepares us for a plunge into more detailed discussions. We will begin these discussions in the next chapter by describing the nighttime sky as we see it without telescopes. We will describe in more detail the various bodies and motions observable to the unaided eye, and we will discuss the modern understanding of these phenomena.

SUMMARY

1. Astronomy is the science in which the entire universe is studied. The study of astronomy requires knowledge of several sciences, such as physics, chemistry, geology, and perhaps biology.

2. On a clear, dark night, the unaided eye can see up to five planets, the Moon, countless stars, and the occasional meteor or comet.

3. The Earth's rotation causes all objects in the sky to undergo daily motion, rising in the east and setting in the west.

4. Because we cannot directly see how far away objects are, we can only measure their positions or separations in angular units. For convenience, we visualize a celestial sphere, centered on the Earth, on which the astronomical bodies lie.

5. The portion of the sky that can be seen depends on the latitude of the observer, and the portion visible at night depends on the time of year.

6. The most distant planet in our solar system is about 40 times farther from the Sun than the Earth is, whereas the nearest star is some 300,000 times the Sun–Earth distance, or 4 light-years, away. Our galaxy is about 100,000 light-years in diameter, the nearest galaxies are 150,000 to 2.3 million light-years distant, and clusters of galaxies are typically separated by millions to tens of millions of light-years. The size of the observable universe is measured in billions of light-years.

7. The light we receive from a distant object has been traveling toward us for as long as billions of years (in the case of the most distant galaxies), so we can observe the universe as it was long ago.

REVIEW QUESTIONS

Thought Questions

1. Express in your own words the nature of scientific theory, and explain the differences between scientific theory, speculation, and dogma.

2. Discuss how you might test the validity of astrology; that is, describe an experiment or observation that would show whether predictions made by astrologers are correct or incorrect.

3. Why do you think it is best to get away from the city to view the nighttime sky?

4. We have noted that the planets appear to travel across the sky along the same strip. What does this tell us about the orbits of the planets around the Sun?

5. Explain why positions of objects in the sky are measured in angles rather than in units of linear distance such as miles or kilometers.

6. Explain why it is necessary to have observatories in both the Southern and Northern Hemispheres. At what latitude is the largest possible fraction of the sky visible?

Problems

1. Express the following numbers in scientific notation (that is, in powers-of-10 notation): 0.0045; 12,400,000,000; 120; 0.0000000000067; 187,000,000,000,000,000.

2. Perform the following calculations: $1.3 \times 10^4/6 \times 10^{-6}$; $7.48 \times 10^{12} \times 3.18 \times 10^{-8}$, 0.00000000045/1,200,000; $5,780,000,000,000 \times 14,000,000,000,000$.

3. A degree of angle contains 60 minutes of arc, and a minute of arc contains 60 seconds of arc. How many arcseconds are in a full circle (360°)? The Sun's angular diameter is 30 arcminutes. What is it in degrees and in arcseconds?

4. Suppose the scale of the solar system were changed so that the Earth was the size of a marble (with a diameter of 1 centimeter). On this scale, what would be the diameter of the Sun? What would be the Earth–Moon distance? What would be the Sun–Earth distance? How far from the Sun would Pluto be? How far away would the nearest star be? (Note: To answer these questions, you need to know the true sizes and distances involved. Most are given in the text of this chapter, but many can also be found easily in Appendix 7).

5. Using information in Appendix 7, calculate the light-travel time from the Sun to Earth, to Jupiter, and to Pluto. (You will find it easiest to use the speed of light in metric units, which is 300,000 kilometers per second.)

6. If galaxies formed 10 billion years ago, how far away must we look in order to see the universe at a time before any galaxies existed?

ADDITIONAL READINGS

We can further appreciate the essence of astronomy, as well as its beauty, by reading a wide range of books and periodicals. Many bookstores contain volumes of astronomical photographs, as well as numerous books on astronomy written for the layperson.

Periodicals that are particularly well suited to students using this text include *Sky and Telescope, Mercury,* and *Astronomy. Sky and Telescope* is especially recommended for those wishing to carry out projects such as telescope building or astrophotography; it includes monthly charts showing the positions of the stars and planets as they change throughout the year.

Culver, R. 1984. *Astrology, true or false: A scientific evaluation.* (New York: Prometheus Books)

Dean, Geoffrey. 1987. Does astrology need to be true? Parts 1 and 2. *Skeptical Inquirer* 11 (2):166, 11 (3):257.

Radner, Dasie, and Michael Radner. 1982. *Science and unreason.* Belmont, Calif.: Wadsworth, Inc. (Discusses the difference between science and pseudo-science.)

MATHEMATICAL METHODS

Powers-of-10 Notation.

In astronomy and many other sciences, it is often necessary to express and manipulate numbers that are either very large or very small. For example, the distance from the Earth to the Sun is 149,600,000 km; the diameter of an atom is approximately 0.00000001 cm; the mass of the Sun is 1,989,000,000,000,000,000,000,000,000,000,000 grams; and the mass of the proton is 0.00000000000000000000000167 gram. These and other extreme numbers arise very frequently and would be very cumbersome to work with if some sort of shorthand notation were not used.

In our numbering system, each digit in a number represents a power of 10: For example, the number 16 has a 1 in the ten's place and a 6 in the one's place, meaning that the number is equal to 1 times 10 plus 6 times 1. Similarly, 243 is equivalent to 2 times 100 plus 4 times 10 plus 3 times 1. Now note that 100 is the same as 10^2 ("ten squared"). instead of writing 243, we could have written 2.43×10^2. Similarly, since $1,000 = 10^3$, we could write 8,070 as 8.07×10^3. The power to which 10 is raised in this notation is called the *exponent*. The exponent is 2 in the number 2.43×10^2, for example.

Now consider a very large number, such as the Sun–Earth distance. The value, as noted above, is 149,600,000 km. This is the same as 1.496×10^8 km. We find the power of 10 (i.e., the exponent) by counting to the left from the one's place at the end of the number, that is, from the place where the decimal point should be. (We do not normally write in the decimal point in large numbers that have been rounded off, because this would imply that the number was more accurately known than it really is). Similarly, the mass of the Sun becomes 1.989×10^{33} grams using this notation. We see now that, for very large numbers, it is briefer and more convenient to

use the powers-of-10 notation, also called *scientific notation.*

For very small numbers, we do essentially the same thing, but we must realize that the exponent is *negative*. To see this, consider the number 0.01. This is equivalent to 1/100, or $1/10^2$. When a power of 10 is in the denominator, it can be expressed as a negative exponent; hence $0.01 = 1/100 = 1/10^2 = 10^{-2}$. In the same fashion, $0.001 = 10^{-3}$, $0.0001 = 10^{-4}$, and so on. To find the correct exponent, count to the right from the decimal point. For example, $0.00000075 = 7.5 \times 10^{-7}$. Now we see that the diameter of an atom is 1.0×10^{-8} cm, and the mass of a proton is 1.67×10^{-24} gram.

To multiply two numbers expressed in scientific notation, we multiply the lead digits and *add* the exponents. To see this, consider the product of 25 times 3,000. We first express these in scientific notation—2.5×10^1 and 3.0×10^3—then multiply the leading values and add the exponents. The result is $(2.5 \times 10^1) \times (3.0 \times 10^3) = (2.5 \times 3.0) \times 10^{1+3} = 7.5 \times 10^4$.

To divide numbers expressed in scientific notation, we divide the lead digits and *subtract* the exponents. Thus 45,000 divided by 150 is equal to $(4.5 \times 10^4)/(1.5 \times 10^2) = (4.5/1.5) \times 10^{4-2} = 3.0 \times 10^2 = 300$. To calculate the time it takes light to reach the Earth from the Sun, we divide the distance by the speed of light, or $(1.496 \times 10^8 \text{ km})/(3.00 \times 10^5 \text{ km/sec}) = 0.499 \times 10^3$ sec. We can move the decimal point one place to the right, by subtracting 1 from the exponent, so this can be rewritten 4.99×10^2 sec, or simply 499 sec (which is a little over 8 minutes).

Note that most of the astronomical numbers mentioned here contain only three or four nonzero digits. The nonzero digits are called *significant digits*. The fact that we often have only three or four (or even fewer in many cases) reflects the fact that we usually do not know the correct values to sufficient accuracy to justify adding more digits. When we multiply or divide two numbers, the result should not be credited with more significant digits than the *less accurate* of the two numbers.

CHAPTER 2

Cycles and Seasons: Motions in the Sky

A near-total lunar eclipse. (J. Kloeppel, Sommers Bausch Observatory, University of Colorado).

CHAPTER PREVIEW

Rhythms of the Cosmos
Daily Motions
    Astronomical Coordinate systems
Annual Motions: The Seasons
    Calendars
The Moon and Its Phases
Eclipses of the Sun and Moon
Planetary Motions

We view the heavens from a moving platform. The Earth spins while it travels around the Sun in a nearly circular path, and these motions create both daily and yearly cycles of celestial events as seen from the Earth's surface. The other eight planets all behave in similar fashion, and most, including the Earth, have one or more satellites orbiting them. Because we are viewing

the skies from a moving vantage point and because all celestial objects have motions of their own, our impression is that the heavens are very complex. It took human beings many centuries of thought, observation, and technological development to arrive at the simple explanation outlined in this paragraph, an explanation that most of us, in these modern times, are exposed to from early childhood.

In this chapter, we will learn about the nighttime sky as it is seen with the unaided eye. We will develop a modern understanding of the objects that can be seen without a telescope and how their simple motions create complex paths through the sky as seen from the Earth. In the process, we will develop an appreciation for the task of the ancient philosophers who strove to comprehend the workings of the universe, for the view of the heavens described here is the sum total of all the evidence available to pretechnological cultures.

Table 2.1 Periods of significant motions[a]

Motion	Period
Sidereal day	$23^h56^m4.098^s$
Mean solar day	$24^h00^m00^s$
Tropical year (equinox to equinox)	$365^d5^h48^m45^s$
Sidereal year (fixed stars)	$365^d6^h9^m10^s$
Synodic month	$29^d12^h44^m3^s$
Sidereal month	$27^d7^h43^m11^s$
Mercury: Sidereal period	87.969^d
Synodic period	115.88^d
Venus: Sidereal period	224.701^d
Synodic period	583.92^d
Mars: Sidereal period	1.88089^y
Synodic period	2.1354^y
Jupiter: Sidereal period	11.86223^y
Synodic period	1.0921^y
Saturn: Sidereal period	29.4577^y
Synodic period	1.0352^y
Uranus: Sidereal period	84.0139^y
Synodic period	1.0121^y
Neptune: Sidereal period	164.793^y
Snyodic period	1.00615^y
Pluto: Sidereal period	248.5^y
Synodic period	1.0041^y

[a]Units for lunar and planetary motions are mean solar days or tropical years.

RHYTHMS OF THE COSMOS

The observed motions of celestial bodies result from a combination of rotational and orbital motions, including the spin of the Earth and its annual movement around the Sun, as well as the individual motions of the Moon and the planets. We can best understand the overall machinery of the solar system by examining the individual parts separately.

Many of the motions in the heavens are cyclical; that is, they repeat regularly, in a well-defined period. Table 2.1 lists the periods for several motions in the solar system (unfamiliar terms in the table are explained later in the chapter).

DAILY MOTIONS

The most obvious of the many motions that affect our view of the universe are the daily cycles of all celestial objects caused by the rotation of the Earth. The Earth spins on its axis in about 24 hours, so we, on its surface, see a continuously changing view of the heavens. We see the Sun rise and set, along with the Moon, the planets, and most of the stars. Even though we understand that these daily, or **diurnal**, motions are the result of the Earth's spin, we still refer to them as though the objects in the sky themselves were moving.

The rotation of the Earth means that a person standing on the equator covers the entire circumference (about 25,000 miles) in 24 hours, yet our senses give us no feeling of motion.

In science in general, and especially in astronomy, we must always be careful to note that what we observe depends on the frame of reference in which our observations are made. Our view of the sky from a location on the Earth's surface is affected strongly by our reference frame and its own motions. Physicists have learned that there is no absolute reference frame, that in any situation we must define the frame in which we observe and make conclusions. In astronomy we can use the reference frame of the Earth, as we are doing now in discussing our view of the nighttime sky, but we use more general frames in other contexts, such as the framework established by the (approximately) fixed stars, or perhaps the reference frame of distant galaxies.

In the reference frame of the Earth, the Earth's rotation forms the basis for our timekeeping system, since the length of the day is a natural unit of time on which to base our lives—one to which, indeed, nearly all earthly species have adapted. The day is divided into 24 hours, each containing 60 minutes, and each minute consisting of 60 seconds. These divisions are based on the numbering system developed several thousand years ago, largely by the Babylonians (see Chapter 3).

Careful observation shows that, in the reference frame of the Earth, the Sun and the Moon take longer to complete their daily cycles than do the stars. Each has its own motion that carries it in the same direction as the Earth's rotation, opposite to the daily rising and setting of the stars (Fig. 2.1). The Sun and the Moon, therefore, shift to the east with respect to the stars a little each day, so it takes a little longer for each to return to the same position from our point of view. Thus, the Sun rises about 4 minutes later each day, in comparison with the stars, and the Moon rises almost an hour later each day. The planets also move, but so slowly that their diurnal motions are not easily distinguished by eye from those of the stars.

The day that forms the basis for our timekeeping system is based on the **solar day** rather than the **sidereal day** (Fig. 2.2), which is the rotation period of the Earth in the reference frame of the fixed stars. (The term *sidereal* means "with respect to the stars".) Because the Earth's orbital speed is not precisely constant, the length of the solar day varies a little throughout the year. It would be inconvenient to allow our hour,

Figure 2.1 Motions of the Earth and Moon. *The Earth and Moon spin and move along their orbits at the same time. The combination of motions affects the apparent path of the Sun and the Moon as seen by an observer on the spinning Earth. (Not to scale.)*

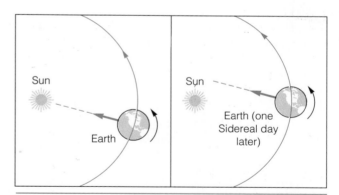

Figure 2.2 The contrast between solar and sidereal days. *The arrow indicates the overhead direction from a fixed point on the Earth. From noon one day (left), it takes one sidereal day for the arrow to point again in the same direction, as seen by a distant observer. Because the Earth has moved, however, it will be about 4 minutes later when the arrow points directly at the Sun again; hence the solar day is nearly 4 minutes longer than the sidereal day (Not to scale.)*

ASTRONOMICAL INSIGHT 2.1

Keeping in Touch with the Stars

As mentioned in the text, modern timekeeping is done with atomic clocks, which have several advantages over the use of transit telescopes to monitor sidereal time. For example, any physics laboratory can maintain its own atomic clock, and make complicated observations with it, without the need to have a transit telescope. Also, the signals from atomic clocks can be transmitted by radio from laboratory to laboratory, so that experiments requiring precise relative timing can be performed. It must be stressed, however, that the atomic clocks are used only as measures of relative time, the time from some adopted reference time. For this they are very accurate, but they cannot adequately maintain correct time relative to the Earth's rotation, because the Earth's rotation is not perfectly constant.

All methods of timekeeping are still based on the rotation of the Earth, as measured through the observations of stars. Hence, atomic clocks must occasionally be updated to account for small changes that occur in the Earth's rotation rate, when it may speed up or slow down. These changes occur for many reasons, not all of them well understood. Some appear to be random variations in the rotation rate, and others seem to be long-term, gradual changes.

One persistent change that always occurs in the same direction is a very gradual slowing of the Earth's rotation rate, probably because of internal frictional forces exerted by the tidal force of the Moon (this is discussed more fully in Chapters 5 and 9, where the effect on the Earth-Moon orbit is described). The combined effect of all the changes in the past century have increased the length of the day by about 0.0014 seconds. Given enough time, this slowing of the Earth's rotation rate reduces the number of days in a year, since each day is longer than it used to be. There is fossil evidence that the year once contained about 400 days, instead of the current 365¼.

The short-term random changes in the Earth's rotation rate can cause either increases or decreases in the length of the day. It is thought that some of these changes are caused by the flow of molten material deep in the Earth's core, but the causes of most of them are not known. Some of them are probably related to a slow shrinking of the Earth as its interior gradually cools; these minor effects act to speed up the rotation temporarily.

Over long periods of time, however, the overall trend is toward a decrease in the rotation rate and a lengthening of the day. To keep atomic clocks in synchronization with the Earth's spin, therefore, it is occasionally necessary to add some time to the clocks. Hence, every few years, a "leap second" is added to all official atomic clocks in order to bring their timekeeping into accord with the Earth's rotation, that is, with sidereal time. This is distinct from the leap year, when a full day is added every fourth year. This added day is inserted because the standard calendar has exactly 365 days in a year, whereas the actual length of the year is closer to 365¼ days long. Thus, a full day must be added every 4 years to make up for the one-quarter days that have been dropped. Interestingly, if the length of the day keeps increasing as expected, in a few thousand years the year will contain exactly 365 days, and for a while there will be no need for leap years.

You may occasionally notice small announcements in newspapers about the addition of "leap seconds" to official clocks. These small changes in official time make little practical difference in our daily lives, but now you understand that they are part of a large-scale astronomical process, with important long-term effects.

---

minute, and second to vary along with it, so the average length of the solar day, called the **mean solar day,** has been adopted as our timekeeping standard. The mean solar day is 3 minutes and 56 seconds longer than the sidereal day.

Precise timekeeping today is measured with atomic clocks but is still kept synchronized with time based on the Earth's rotation. To measure the Earth's rotation period with respect to the stars (that is, the sidereal day), astronomers refer to the **meridian,** the imaginary north-south line passing overhead at an observer's position. As the Earth rotates, each star crosses the meridian once every sidereal day. Special instruments (Fig. 2.3) are used to measure precisely the moment when a star crosses the meridian, thus measuring sidereal time. Allowance must be made to determine from this the mean solar time.

Atomic clocks measure intervals of time very accurately, because they are based on the vibration frequencies of certain kinds of atoms, which are very constant. These clocks have several advantages over the measurement of sidereal time based on observations of stars. For example, atomic clocks keep a continuous record of the passage of time, so they can be referred to whenever needed. Also, identical clocks can be placed in many locations and kept coordinated, allowing many laboratories to precisely measure the time.

We say that atomic clocks measure true physical time, but these clocks are adjusted as needed so that they stay in agreement with sidereal time. Such adjustments are needed because occasionally the Earth's rotation period changes very slightly.

Let us return now to the diurnal motions of celestial objects. We have mentioned that the stars rise and set as the Earth rotates, but in fact not all stars do this. This is because, from a given latitude on the Earth (except right at the equator), it is possible to see an area of the sky, around one of the poles, that is never obscured by the Earth. From a latitude of 30°N, for example, we can see the sky to an angular distance of 30° beyond the North Pole. Therefore, we can always see the whole part of the sky that lies within 30° of the North Pole (but we can never see the sky within 30° of the South Pole), and stars in this part of the sky circle the pole but do not set. The same thing happens around the South Pole for observers in the Southern Hemisphere (Fig. 2.4).

The point about which the stars circle, directly over the Earth's North Pole, is the **north celestial pole,** the point where the Earth's rotational pole is projected onto the celestial sphere. Similarly, the projection of the South Pole onto the celestial sphere is the **south celestial pole.** It happens that a bright star, called Polaris, lies very close to the position of the north celestial

Figure 2.3 A transit circle. *Such a device points at the meridian (the north-south line that passes directly overhead). It is used to record the times when certain reference stars pass over the meridian and is therefore helpful in measuring the sidereal day. (U.S. Naval Observatory)*

Figure 2.4 Star trails illustrating the Earth's rotation. *The circular trails are created by stars near the south celestial pole, which completed about half a circle during this all-night exposure. (© 1980 Anglo-Australian Telescope Board)*

pole and is therefore almost stationary (in the reference frame of the Earth) throughout the night. The projection of the Earth's equator onto the sky is called, naturally enough, the **celestial equator.** We now examine the usefulness of the celestial poles and the equator as reference points for measuring star positions.

Astronomical Coordinate Systems

Our discussion of diurnal motions, latitude, and the changing appearance of the sky as we move over the surface of the Earth has led us to a point where it is convenient to introduce the notion of coordinate systems for the skies. Just as it is useful to be able to describe the position of a place on the surface of the Earth, it is also important for astronomers to be able to record the position of a particular object in the sky so that they can return to it later or communicate to other astronomers where it is. The establishment of a coordinate system naturally requires us to specify the reference frame in which the coordinates are measured. Astronomers use different reference frames for different purposes, but the one most often used is based on the Earth (Fig. 2.5).

In this coordinate system, called the **equatorial coordinate system,** one reference direction is the celestial equator. To specify a star's position in this system, two coordinates are needed. One of them is the angular distance of the star north or south of the celestial equator and is called the **declination** of the star. The declination is analogous to the latitude of a position on the Earth's surface, except that here we speak of angular distance on the sky. North declinations are assigned positive values (such as $+23°14'$ for a star $23°14'$ north of the celestial equator), and south declinations are assigned negative values (the Magellanic Clouds, for example, lie at declinations near $-70°$).

The second coordinate needed to specify an object's position is called **right ascension.** The right ascension measures the angular distance of the object to the east of a fixed direction on the sky and is therefore analogous to longitude on the Earth's surface. Because the Earth is rotating, however, we cannot measure right ascension from some fixed point defined on the Earth's surface; we must instead specify a reference direction on the sky. This direction is defined by the line of intersection of the Earth's equatorial plane and the plane of its orbit. This is the direction toward the Sun at the time of the Vernal Equinox (defined in the next section). When this coordinate system was established,

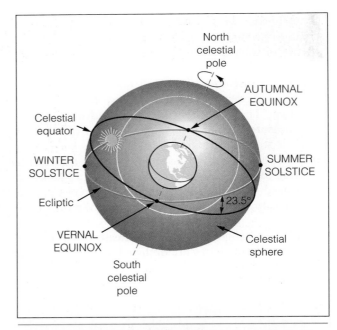

Figure 2.5 Equatorial coordinates.   *Here are illustrated the measurements of star position in declination and right ascension. Declination is the distance (in degrees, minutes, and seconds of angle) of a star north (+) or south (−) of the celestial equator. Right ascension is the distance (in hours, minutes, and seconds of time) of the star to the east of the direction of the vernal equinox, a fixed direction in space.*

the line of intersection of these planes pointed toward the constellation Aries, but because of gradual changes in the Earth's orientation it now points toward Pisces and will soon enter Aquarius; this is why popular astrology refers to the "dawning of the age of Aquarius." Nevertheless, the position toward which this line points is still called the **First Point of Aries.**

Because it takes 24 sidereal hours for a given star to make a complete circuit of the sky as the Earth rotates, its position in the east-west direction is measured in units of sidereal hours, minutes, and seconds rather than in angular units. Thus, a star located in the direction of our reference point has a right ascension of $0^h0^m0^s$, whereas a star one-quarter of the way around the sky to the east of that point has a right ascension of $6^h0^m0^s$.

There is one complication in using the equatorial coordinate system. The Earth's rotational axis slowly wobbles (Fig. 2.6) in a motion called **precession,** and this causes the celestial equator to move slowly through

ASTRONOMICAL INSIGHT 2.2

Celestial Navigation

 One of the most time-honored and well-known "practical" applications of astronomy is to find one's position on the Earth by observing the stars. Traditionally, this was the only means available to a sailor or someone exploring new territory, whereas today there are many, more modern methods (and less unexplored territory).

The essential method is quite simple, although in practice the technique involves mathematical calculations too complex to describe in detail here. The first requirement is to be able to identify one or more objects in the sky whose coordinates are known (usually from a book of tabulated positions, called an **ephemeris**). Thus, a familiarity with the constellations and the sky in general is required. Ironically, people who have learned celestial navigation are more likely to know the constellations, as well as a number of individual stars and their names, than is the typical professional astronomer.

Once a known star is recognized and its coordinates found in a catalog or ephemeris, the navigator has a reference point in the sky whose position is known. The navigator then measures the angular distance from straight overhead to that star. This measured angular distance is called the **zenith distance** of the star, the angle between the navigator's zenith (directly overhead) and the direction

to the star. Given this, as well as the sidereal time, it is possible to locate the spot on the Earth's surface where the star is directly overhead and then draw a circle about that spot (on a globe) showing all the locations where the star would have the measured zenith distance. This does not pinpoint the navigator's location, so the same procedure is carried out using a second star. Now there are two circles of possible locations, and the two intersect at only two points. Unless badly confused, the navigator normally will know which of the two intersection points is the correct location (the two points can be hundreds of miles apart). Hence, measurements of only two stars are usually sufficient, but typically a third star is measured as well to be sure there were no errors. In that case, there are three circles drawn on the map, and all three should intersect in only one point if the measurements have been made correctly.

The measurement of the zenith distance is done by using a device such as a **sextant** that allows angular separations on the sky to be measured. A sextant usually includes a pointer that can be aimed at the star and a means of giving the user a reference to the vertical direction.

In practice, the Sun and the Moon can be used for navigation, although they are both difficult to measure precisely because of their large angular diameters; also, of course, it is difficult to view the Sun directly, because of its brilliance. Furthermore, to use either, it is necessary to have rather com-

plete tables of their positions, which change throughout the year (in the case of the Sun) and throughout the month (in the case of the Moon, whose motion is sufficiently rapid to require tables showing the position hourly). The Sun is often used, however, because the horizon is easily visible during daylight, so that the **altitude,** the angular distance of the Sun above the horizon, can be readily measured in order to find the Sun's zenith distance.

Today, navigation of ships and airplanes is done by various means involving technology that has been developed in the past few decades. The well-traveled areas of the Earth are permeated with navigational radio beams that provide directional information, and modern satellite navigation systems provide very accurate data to ships and planes. In addition, devices called **gyroscopes,** which can maintain a given orientation very accurately over long periods, allow navigators to keep track of their positions over lengthy voyages without any reference to external guideposts such as the sky or a satellite navigation system.

Despite all the modern developments cited here, the art of celestial navigation is not completely lost. Military and commercial airline and ship navigators still learn how to do it and presumably could use the method if all else failed.

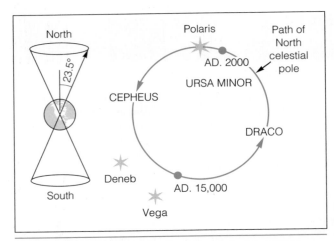

Figure 2.6 Precession. *The Earth's axis is titled 23½° away from perpendicular to the orbital plane, and it wobbles on its axis, so that an extension of the axis describes a conical pattern (left) in a time of about 26,000 years. The north celestial pole therefore follows a circular path on the sky (right).*

the sky. It takes some 26,000 years for the Earth to complete one cycle of this motion, which is exactly like the wobbling familiar to anyone who has spun a play top or gyroscope. The motion of our coordinate system is therefore very gradual, causing star positions to shift extremely slowly. Despite the small magnitude of the effect, it was noticed more than 2,000 years ago, and it must be allowed for by modern astronomers when planning observations. Typically, one of the preparations for observing is to "precess" the coordinates of all the target stars (that is, to calculate their current locations in the Earth's reference frame). Catalogs of star positions always specify the date for which they are valid so that astronomers know how much precession to allow for.

Other coordinate systems (that is, other reference frames) sometimes are used in astronomy, although equatorial coordinates are by far the most widely employed, being the most practical for use in observations. Solar system astronomers sometimes use a coordinate system based on the plane of the **ecliptic** (the plane defined by the Earth's orbit around the Sun) and the direction toward the Sun, whereas those studying the Milky Way may use coordinates based on the plane of the galaxy's disk and the direction toward its center. Ecliptic and galactic coordinates are particularly useful in studies of the structure of the solar system or of the galaxy, because an object's position immediately tells the astronomer where it lies with respect to the overall organization of the solar system or galaxy.

ANNUAL MOTIONS: THE SEASONS

We turn our attention now to celestial phenomena that are caused by the Earth's motion as it orbits the Sun. One aspect of this, the daily eastward motion of the Sun with respect to the stars, has already been mentioned in connection with the difference between the solar and the sidereal day. It must be emphasized that, in the reference frame of the fixed stars, this is only an apparent motion caused by our changing angle of view as we move with the Earth in its orbit. As the Earth moves about the Sun, it travels in a fixed plane, so the apparent annual path of the Sun through the constellations is the same each year (Fig. 2.7). This apparent path of the Sun, defined by the plane of the Earth's orbital motion about the Sun, is the **ecliptic**. The sequence of constellations through which the Sun passes is called the **zodiac**. There are twelve principal constellations of the zodiac, identified since antiquity and once thought by people to have significance for our daily lives (see the comments on astrology in Chapter 1).

Because we can see stars only in the nighttime sky, the constellations most easily visible to us at any given time are the ones at least a few hours of right ascension to the east or west of the Sun. This means that most stars, except those near the poles, cannot be observed year-round, as discussed in Chapter 1.

The orbital planes of the other planets and of the Moon are closely aligned with that of the Earth, so the planets and the Moon are always seen near the ecliptic. Hence, most of the major naked-eye objects in the solar system pass through the same sequence of constellations, the zodiac, so it is not surprising that great significance was attached to this sequence by the ancient astronomers.

Besides causing the apparent annual motions of the Sun and planets, the Earth's orbital motion has a second, and far more significant, effect on us: It creates our seasons. The Earth's spin axis is tilted with respect to the perpendicular to its orbital plane, so during the course of a year, portions of the Earth away from the equator are exposed to varying amounts of sunlight. Summer in the Northern Hemisphere occurs when the North Pole is tipped toward the Sun; winter occurs

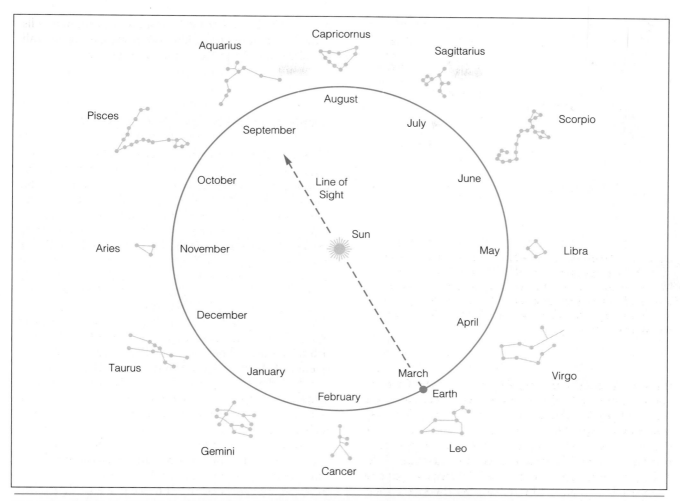

Figure 2.7 The path of the Sun through the constellations of the zodiac. *The dates refer to the position of the Earth each month. To see which constellation the Sun is in during a given month, imagine a line drawn from the Earth's position through the Sun; that line will extend to the Sun's constellation. For example, in March the Sun is in Aquarius.*

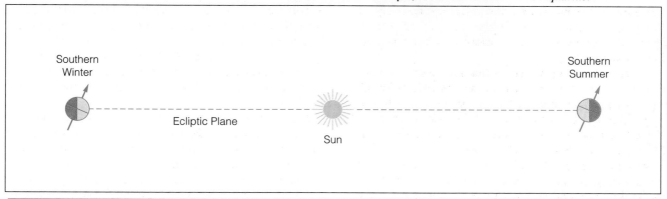

Figure 2.8 Seasons. *The Earth's tilted axis retains its orientation as the Earth orbits the Sun. Thus, at opposite points in the orbit, each hemisphere has winter and summer, depending on whether that hemisphere is tipped toward the Sun or away from it.*

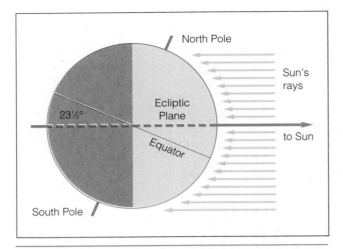

Figure 2.9 Effect of the Earth's tilted axis.   *The Sun's rays strike the ground at varying angles, depending on the latitude and the time of the year. Therefore, the solar heating, which depends on how directly the rays reach the ground, varies with the season. Here it is summer in the Northern Hemisphere; the Sun's rays are nearly perpendicular to the Earth's surface at low to moderate northern latitudes, but they strike the surface very obliquely at southern latitudes.*

during the opposite part of the Earth's orbit, as the pole, which remains fixed in orientation, is tilted away from the Sun (Fig. 2.8). The tremendous seasonal variations in climate at intermediate latitudes are caused by a combination of two effects: (1) The length of the day varies so that in summer, for example, the Sun has more time to heat the Earth's surface; and (2) the angle at which the Sun's rays strike the ground is more nearly perpendicular in the summer (Fig. 2.9), so the Sun's intensity is much greater, heating the surface more efficiently.

The Earth's axis is tilted $23\frac{1}{2}°$ from the perpendicular to the orbital plane. Therefore, during the year, the Sun, as seen from the Earth's surface, can appear directly overhead (that is, at the **zenith**) as far north and south of the equator as $23\frac{1}{2}°$, defining a region called the **tropical zone** (Fig. 2.10). For those of us who live outside the tropics, the Sun can never be directly overhead. When the North Pole is tilted most nearly in the direction toward the Sun, an occasion occurring around June 21 and called the **summer solstice,** the Sun passes directly overhead at $23\frac{1}{2}°$N latitude at local noon, but it passes to the south of the zenith for anyone at more northerly latitudes. In Boulder, Colorado, for example,

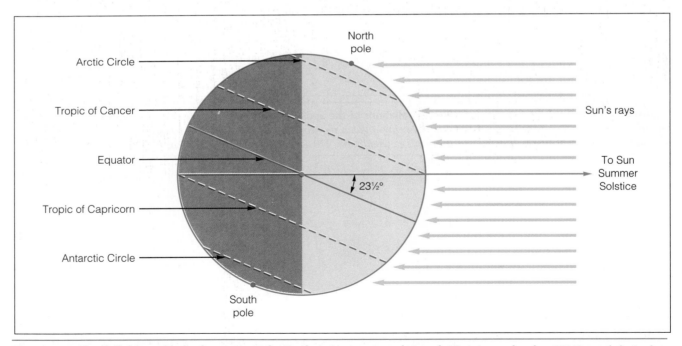

Figure 2.10 The definition of latitude zones on the Earth.   *At summer solstice, the Sun is overhead at $23\frac{1}{2}°$ north latitude, its northernmost point. This defines the Tropic of Cancer, the northern limit of the tropical zone. At the same time the entire area within $23\frac{1}{2}°$ of the North Pole is in daylight throughout the Earth's rotation, and the boundary of this region is the Arctic Circle. Similarly, the Antarctic Circle receives no sunlight at all during a complete rotation of the Earth. Six months later, the Sun is overhead at the Tropic of Capricorn.*

which is right on the fortieth parallel, the Sun on this day reaches a point 40° − 23½° = 16½° south of the zenith, or an altitude of 90° − 16½° = 73½° above the southern horizon (Fig. 2.11).

Traveling farther north, we see that at the time of the summer solstice, sunlight covers the entire north polar region to a latitude as far as 23½° south of the pole. This defines the **arctic circle** (see Fig. 2.10), and at the time of the solstice the entire circle has daylight for all 24 hours of the Earth's rotation. At the pole itself, there is constant daylight for 6 months.

During winter in the Northern Hemisphere, the Sun does not rise as high above the southern horizon at local noon as it does during the summer, because the South Pole is now tilted toward the Sun. At the **winter solstice**, when the South Pole is pointed most nearly in the direction of the Sun, the Sun's midday height above the horizon, as viewed from the Northern Hemisphere, is the lowest of the year. Returning to Boulder, we find that on this day, which usually falls near December 21, the Sun is 40° + 23½° = 63½° south of the zenith, or 90° − 63½° = 26½° above the horizon.

If we follow the Sun's motion north and south of the equator throughout the year, we find that it follows

a graceful curve as it traverses its range from +23½° (north) declination to −23½° (south) declination (Fig. 2.12). The Sun crosses the equator twice in its yearly excursion, at the times when the Earth's North Pole is pointed in a direction 90° from the Earth-Sun line. At these times, the lengths of day and night in both hemispheres are equal, and these occasions are referred to as the **vernal** (spring) and **autumnal** (fall) **equinoxes**. They occur on about March 21 and September 23, respectively. The direction to the Sun at the time of the vernal equinox coincides exactly with the direction of 0^h right ascension. (Remember, the definition of this direction is that it lies along the line of intersection of the Earth's equatorial plane and the orbital plane.)

Calendars

We have referred to the association of astronomy and timekeeping, particularly in ancient times, when astronomers were responsible for predicting major natural events such as changes of season. Today everyone knows when these things will occur and generally keeps track of dates for all purposes by the use of modern calendars (Fig. 2.13), which can be purchased well in advance for any given year. It is interesting to realize, however, that even now our calendars, and hence the annual pattern of our daily lives, are determined by the motions of bodies in the heavens.

A number of calendars were established by various ancient cultures, and all were based to some degree on the length of the year. In several cases it was deemed unlikely or unacceptable that the year should not be evenly divisible into a round number of lunar months

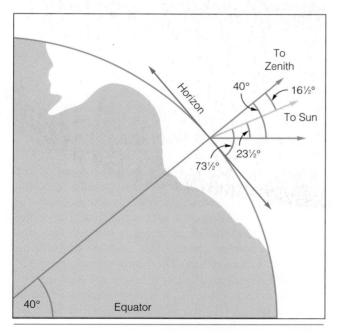

Figure 2.11 Sun's position from Boulder at the summer solstice. *Boulder, Colorado, lies on the fortieth parallel north, so when the Sun is overhead at 23½° N, as it is on the summer solstice, from Boulder it appears to be 16½° south of the overhead position, or 73½° above the southern horizon.*

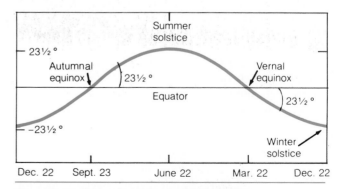

Figure 2.12 The path of the Sun through the sky. *Because of the Earth's orbital motion and the tilt of its axis, the Sun's annual path through the sky has the shape illustrated here.*

Figure 2.13 Modern calendar showing the Moon's phases. *(Sierra Club Books/Random House)*

or even days, so some of the old calendars adopted lengths of the year that did not agree with the actual period of the Earth's orbit of the Sun. All such calendars had to be adjusted every so often or fall badly out of synchronization with the seasons.

The length of the year that is important in calendar making is not the sidereal period of the Earth's orbit, as we might at first expect. The reason is that, because of precession, the seasons will gradually shift with respect to the sidereal year. Consider the length of time from one vernal equinox to the next. As the Earth travels around the Sun, the orientation of its axis does not stay precisely fixed, but instead changes just slightly because of precession. The result is that the next vernal equinox, which occurs when the Earth's axis returns to precisely the same orientation (with respect to the stars) as it had at the previous equinox, takes place just a little sooner than it would have if precession did not occur. The difference between the sidereal year and the **tropical year** (the length of time for a full cycle of seasons to occur) is just 20 minutes. Hence, calendars today are based on the tropical year. Precise values for both sidereal and tropical years are included in Table 2.1 and Appendix 2.

Until the time of Julius Caesar, most calendars were based on the assumption that the year included a whole number of days, even though it was well known that this was not the case. The errors that built up were allowed to accumulate until extra days would be inserted in order to bring the calendar back into synchronization with the seasons. Caesar commissioned the development of a new calendar, now called the **Julian calendar,** in which the length of the year was taken to be 365¼ days, only a few minutes in error. Every fourth year an extra day was inserted, a practice that we still follow. The year was divided into 12 months that did not correspond exactly to lunar months, so that at least there would be a whole number of months in a year. Caesar was also responsible for beginning the year on January 1; it had begun in March before his reform, which took place in 46 B.C.

To allow for the few minutes of error that crept into the Julian calendar because the tropical year is not precisely 365 days long, in 1582 Pope Gregory XIII ordered additional reform, with the result that in the modern calendar (called the **Gregorian calendar**), leap year is occasionally not observed. Leap year is ignored (the extra day is not inserted after February 28) on century years not divisible by 400. Thus, there will be a leap year in the year 2000, which is divisible by 400, but not in the years 2100, 2200, and 2300. With this refinement, the Gregorian calendar builds up an error of only 1 day in 3,300 years, a sufficient degree of accuracy for most people's appointment books!

THE MOON AND ITS PHASES

Except for the Sun, the Moon is the brightest object in the sky, and it, too, has its own complex motions as seen from the Earth. The Moon's motion with respect to the stars is much more rapid than that of the Sun, and it is therefore more easily noticed. It is not difficult, in fact, to observe this lunar motion during the course of an evening. The Moon moves about 12° across the sky every day, or 1° every 2 hours. Its angular diameter is about ½°, which means that the moon moves a distance in the sky about equal to its own diameter every hour.

The plane of the Moon's orbit is closely aligned with that of the Earth's orbit, so the Moon stays near the ecliptic, as seen from the Earth. It takes the Moon a little over 27 days ($27^d7^h43^m11^s.5$) to make one trip around the Earth, in the reference frame of the fixed stars. Because the Earth moves at the same time, however, it appears to us in the Earth's reference frame that the Moon takes longer than 27 days to make one complete circuit; from our point of view, the Moon takes $29^d12^h44^m2^s.8$ to complete its full cycle of phases (Fig. 2.14). The shorter orbital period is called the **sidereal period,** because it is the time required for the Moon to go around the Earth, as seen in a fixed reference frame with respect to the stars. The observed cycle of phases, which is the time required for the Moon to return to a given alignment with respect to the Sun, is called the **synodic period,** or the **lunar month.**

The difference is akin to the distinction between the solar and sidereal days discussed earlier in this chapter. In both cases, it is the Earth's motion about the Sun that lengthens the time it takes to complete a full cycle as we see it.

The Moon always keeps the same side facing the Earth, because its rotation period is equal to its orbital period (Fig. 2.15). It is a common misconception that the Moon is not rotating. It should be clear that the Moon is rotating in the reference frame of the stars, however, for if it were not, we would see all sides of the Moon as it circled the Earth. The fact that the orbital and spin periods are equal is not a coincidence

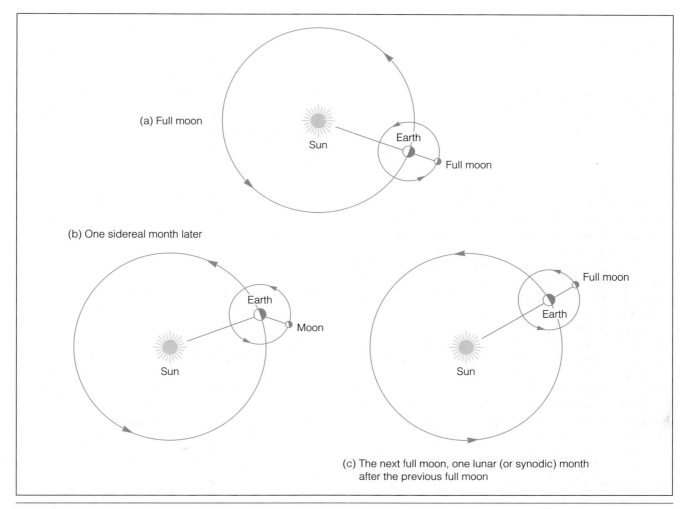

Figure 2.14 Sidereal and synodic periods of the Moon.   *Because the Earth moves in its orbit while the Moon orbits it, the Moon goes through more than one full circle (as seen by an outside observer) to go from one full Moon to the next. Hence the lunar (or synodic) month is about two days longer than the Moon's sidereal period.*

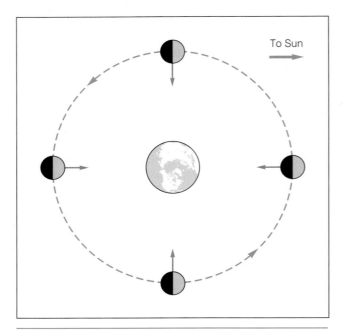

To Sun

Figure 2.15 Synchronous rotation of the Moon. *The arrow, fixed to a specific point on the Moon, illustrates that the Moon spins once during each orbit of the Earth, thereby keeping the same side facing the Earth at all times. (Not to scale.)*

but is due to the tidal forces exerted on the Moon and the Earth by their mutual gravitational pull (this is discussed in Chapters 5 and 9). The phenomenon of matching orbital and spin periods is called **synchronous rotation,** and it is common in the universe, both within the solar system and in double stars.

Since the Moon does not emit light of its own but shines instead by reflected sunlight, we easily see only those portions of its surface that are sunlit. As the Moon orbits the Earth and our viewpoint relative to the Sun's direction changes, we see varying fractions of the daylit half of the Moon. This causes the Moon's apparent shape to change drastically during the month, the sequence of shapes being referred to as the **phases** of the Moon (Fig. 2.16). The full cycle of phases is completed during one synodic period, or lunar month, of about 29½ days.

The extremes of the cycle are represented by the **full moon** (occurring when it is directly opposite the Sun, so we see its entire sunlit hemisphere) and the **new moon** (when it is between the Earth and the Sun, with its dark side facing us). The new moon cannot be ob-

served, because it is dark and is so close to the Sun. Hence, our nighttime sky is moonless at the time of the new moon.

Just as we speak of phases of the Moon, which really refer to its apparent shape as seen from the Earth, we also can speak of its **configurations,** which describe its position with respect to the Earth-Sun direction. For example, when a full moon occurs, the Moon is at **opposition** (it is in the direction opposite to that of the Sun), and a new moon occurs at **conjunction** (when the Moon lies in the same direction as the Sun). We can follow the Moon through its phases as we trace its configurations, beginning with the new moon, which takes place at conjunction (refer to Fig. 2.16). During the first week following conjunction, as the Moon moves toward **quadrature** (the position 90° from the Earth-Sun line), it appears to us to have a crescent shape that grows in thickness each night. This is called the **waxing crescent** phase; when the Moon reaches quadrature, in which we see exactly one-half of the sunlit hemisphere, it has reached the phase called **first quarter.** For the next week, as the Moon goes from quadrature toward opposition and we see more and more of the daylit side, its phase is said to be **waxing gibbous.** After the full moon, as it again approaches quadrature, the phase is **waning gibbous,** and this time at quadrature the phase is called **third quarter.** First quarter can be distinguished from third quarter easily by noting the time of night: If the Moon is already up when the Sun sets, it is first quarter, but if it does not rise until midnight, it is third quarter. After third quarter, as the Moon moves closer to the Sun, we see a diminishing slice of its sunlit side, and it is in the **waning crescent** phase.

ECLIPSES OF THE SUN AND MOON

An eclipse of the Sun or **solar eclipse,** occurs when the Moon passes directly in front of the Sun, as seen from the Earth. A **lunar eclipse,** by contrast, occurs when the Moon passes through the Earth's shadow, so that for a brief period the Moon is not directly illuminated (Fig. 2.17).

From what has been stated so far, it may seem that the Moon should pass directly in front of the Sun on each trip around the Earth, and through the Earth's shadow at each opposition, producing alternating solar and lunar eclipses at 2-week intervals. However, this

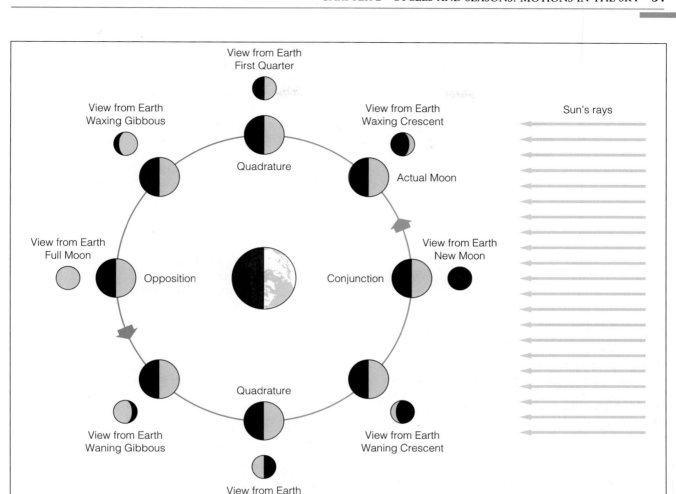

Figure 2.16 Lunar phases and configurations. *As the Moon orbits the Earth, we see varying portions of its sunlit side. The phases sketched here (outside the circle representing the Moon's orbit) show the Moon as it appears to an observer in the northern hemisphere.*

is not the case; the reason is that the Moon's orbital plane does not lie exactly in the ecliptic but is tilted by about 5° (Fig. 2.18). Therefore, the Moon usually passes just above or below the Sun as it goes through conjunction and similarly misses the Earth's shadow at opposition. The Moon passes through the ecliptic at only two points on each trip around the Earth, where the planes of the Earth's and the Moon's orbits intersect. Because the Moon's orbital plane wobbles slowly, in a precessional motion similar to that of the Earth's spin axis, the line of intersection with the Earth's orbital plane slowly moves around. The combination of

this motion, the Moon's orbital motion, and the movement of the Earth around the Sun creates a cycle of eclipses, with the same pattern recurring every 18 years (see Table 2.2 for a list of future solar eclipses). This cycle of eclipses, called the **saros**, was recognized in antiquity.

It is purely coincidental that the Moon and the Sun have nearly equal angular diameters, so that the Moon neatly blocks out the disk of the Sun during a solar eclipse (Figs. 2.19 and 2.20). The angular diameter of an object is inversely proportional to its distance, meaning that the farther away it is, the smaller it looks.

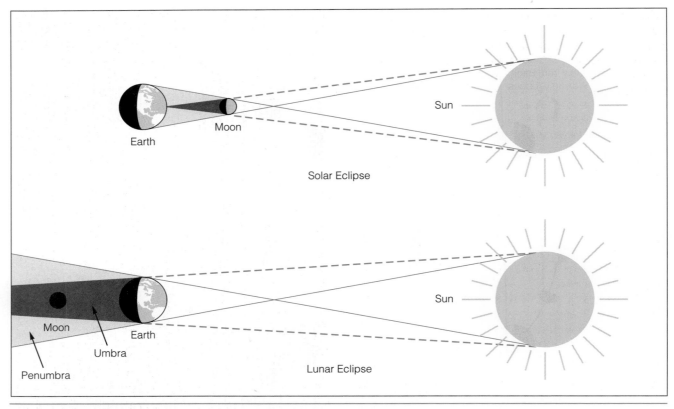

Figure 2.17 Solar and lunar eclipses. *The relative positions of the Sun, Earth, and Moon are shown at the times of solar (upper) and lunar (lower) eclipses. Only the dark portion of the shadow (the umbra) is shown in each case. (Not to scale.)*

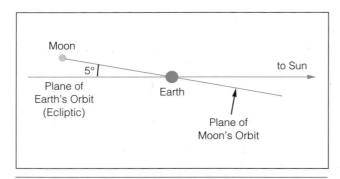

Figure 2.18 The tilt of the Moon's orbit. *This is an edge-on view illustrating the 5° angle between the plane of the Moon's orbit and the orbital plane of the Earth. Here we see the Moon at opposition line (i.e., the time of full moon), but with no lunar eclipse.*

The Sun is much larger than the Moon, but it is also much more distant. The two objects have almost exactly the same angular diameter, because the ratio of the Sun's diameter to that of the Moon just happens to be almost the same as the ratio of the Sun's distance to that of the Moon (Fig. 2.21).

If a total solar eclipse occurs at the time when the Moon is farthest from the Earth in its slightly noncircular orbit, it does not quite block all of the Sun's disk, instead leaving an outer ring of the Sun visible. This is called an **annular eclipse.** Because a total (or annular) solar eclipse requires precise alignment of the Sun and the Moon, an eclipse will appear total only along a well-defined, narrow path on the Earth's surface (Fig. 2.22). There is a wider zone outside of that, where the Moon appears to block only a portion of the Sun's disk; people in this zone see a partial solar eclipse.

During a lunar eclipse, when the Moon passes through the Earth's shadow, observers everywhere on the nighttime side of the Earth see the same portion of the Moon eclipsed. If the Moon passes through the **umbra** (the dark inner portion of the Earth's shadow), the eclipse is total, as no part of the Moon's surface is exposed to direct sunlight. If the Moon misses being fully immersed in the umbra, it passes through the **penumbra** and undergoes a partial eclipse.

Figure 2.19 A total solar eclipse. *This occurs when the Moon entirely blocks our view of the Sun. At this time we see the Sun's corona. (Fr. R. E. Royer)*

Figure 2.20 A solar eclipse sequence. *This series of photographs illustrates the Moon's progression across the disk of the Sun during a partial solar eclipse. (NASA)*

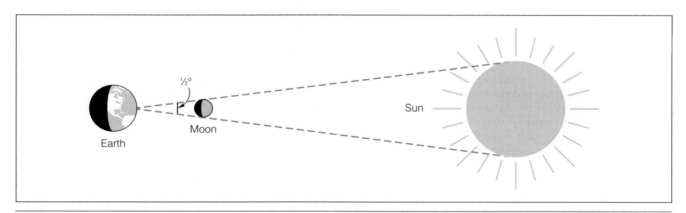

Figure 2.21 Angular diameters of the Moon and Sun. *The actual diameters and the relative distances of the Sun and Moon compensate one another so that the angular diameters are equal. (Not to scale.)*

ASTRONOMICAL INSIGHT 2.3

Chasing a Solar Eclipse

 Total eclipses of the Sun are infrequent events, and the path of totality (the region on the Earth where the eclipse appears total) can fall virtually anywhere (see Table 2.2). Total solar eclipses are important to scientists because of the opportunities they provide for study of the Sun, and they can be profoundly moving events to experience. Therefore, every total eclipse in modern times has been the cause of several major expeditions of scientists and sightseers, sometimes to quite exotic locales.

Because the angular diameters of the Sun and Moon are so nearly equal, the alignment between them must be very precise in order to produce a total solar eclipse. This means that the eclipse appears total only to observers in a very small region on the Earth. As the Moon moves across the disk of the Sun, this location travels across the Earth's surface; this is why we refer to a "path" of totality. The duration of the eclipse as seen from any single point on the Earth depends on the details of the relative distances of the Moon and the Sun and on the speed of the Moon in its orbit, so the length of time when an eclipse is total varies from one eclipse to another, and even from one point in the path of totality to another. The duration of totality is generally between 3 and 7 minutes.

During this brief time interval, the activity at the location where totality is occurring is frantic, particularly on the part of the scientific expedition members. This is the culmination of months of planning by the solar scientists, as they finally get their chance to collect data on the Sun's outer layers (which are normally invisible due to the overwhelming brightness of the solar disk). Before the eclipse begins, there is an air of tension and expectation, as instruments undergo final tests and everyone watches the weather. When totality begins, everyone is preoccupied with their observations; some scientists are too busy to look at the eclipse at all.

The sightseers who have come to watch have a better chance than the scientists to enjoy the esthetic qualities of the experience. (Those who are busy making photographs, however, may have an experience more like that of the scientists). These people may be friends or relatives of members of the scientific teams, or they may have traveled to the eclipse observation site as part of a tour organized by a travel company. Several such tours usually accompany a total solar eclipse, particularly if the locale is not too difficult to reach and the eclipse will be of substantial duration.

If the weather is clear, a great sense of anticipation builds as time for the eclipse approaches. As *first contact* occurs, when the edge of the Moon first begins to occult the Sun's disk, the partial phase of the eclipse begins. The Moon's silhouette covers more and more of the Sun, and it gradually gets darker at the observing site. As totality nears (usually more than an hour after first contact), the sky becomes much darker. At *second contact*, the beginning of totality, only bits of the Sun, called **Bailey's beads,** are seen as the last rays of sunlight pass through valleys on the edge of the Moon's disk. As the eclipse becomes total, the Sun's **corona** and **chromosphere** become visible. These tenuous outer layers of the Sun, lying above the visible surface, are so dim that normally they cannot be seen directly. The opportunity to observe these layers is one of the main reasons that eclipses are valuable to solar physicists. The corona is an extended, pale white wreath around the Sun, often showing streaks and rays directed outwards. The chromosphere, lying closer to the Sun than the corona, has a distinctly red color, and often shows great jets of gas rising above the solar surface.

During totality at the observing site it is almost as dark as nighttime. Plants and animals behave accordingly: Birds become silent, nocturnal animals may emerge from their lairs, and flowers may close up. Quite suddenly this phase ends, as *third contact* is reached, and the Sun begins to emerge from behind the Moon. Soon the world takes on a much more normal appearance to the observers, and the scientists and sightseers begin to relax. The eclipse officially ends at *last contact,* when the Moon's disk completely uncovers the Sun, ending the partial phase of the eclipse. Observers within a broad strip outside of the path of totality see only a partial eclipse which may last for two hours or more.

Even though substantial time and expense are usually required in order to experience a total solar eclipse, those who do it are uniformly enthusiastic and grateful for the opportunity. Perhaps you will have a chance to enjoy the experience on the occasion of one of the future eclipses listed in Table 2.2.

Table 2.2 Total solar eclipses of the future

Date of eclipse	Duration (min)	Location
July 11, 1991	7.1	Hawaii, Central America, Brazil
June 30, 1992	5.4	South Atlantic
November 3, 1994	4.6	South America
October 24, 1995	2.4	South Asia
March 9, 1997	2.8	Siberia, arctic regions
February 26, 1998	4.4	Central America
August 11, 1999	2.6	Central Europe, central Asia
June 21, 2001		South Atlantic, South Africa
December 4, 2002		South Africa, Indian Ocean
March 29, 2006		South Atlantic, Africa, Middle East
August 1, 2008		Greenland, North Atlantic, USSR
July 22, 2009		Indonesia, South Pacific
July 11, 2010		South Pacific
November 13, 2012		South Pacific
March 20, 2015		North Atlantic
August 21, 2017		United States, Atlantic Ocean

Figure 2.22 The Moon's shadow during a solar eclipse. *This photograph, taken from space, shows the shadow of the Moon on the Earth during a solar eclipse. The eclipse appeared total only to observers on the Earth who were located directly in the center of the shadow's path (that is, in the umbra). (NASA)*

PLANETARY MOTIONS

Just as in the case of the Sun and the Moon, careful observation reveals that the planets, too, move with respect to the background stars. It was this fact that gave the planets their generic name, since **planet** is the Greek word for "wanderer." The planets all orbit the Sun in the same direction as the Earth and, as mentioned earlier, in nearly the same plane, so they appear to move nearly in the ecliptic, through the constellations of the zodiac.

The two planets Mercury and Venus, which orbit closer to the Sun than does the Earth, are called **inferior planets** and can never appear far from the Sun in our sky (Fig. 2.23). For Mercury, the greatest angular distance from the Sun, called the **greatest elongation,** is about 28°, whereas Venus can be seen as far as 47° from the Sun. Like the Moon, the planets have specific configurations, referring to their positions with respect to the Sun-Earth line. An inferior planet is said to be at **inferior conjunction** when it lies between the Earth and the Sun (or in **transit** if the orbit happens to be aligned with that of the Earth so that the planet is seen against the disk of the Sun) and at **superior conjunction** when it is aligned with the Sun but lying on the far side.

The outer planets, or **superior planets,** can be seen in any direction with respect to the Sun, including opposition, when they are in the opposite direction from the Sun (Fig. 2.23). Conjunction for a superior planet can occur only when the planet is aligned with the Sun but on the far side of it, in analogy with a superior

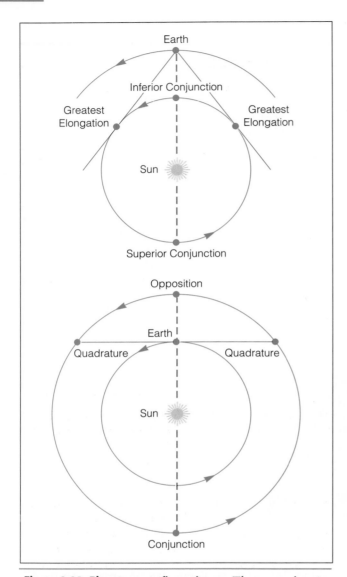

Figure 2.23 Planetary configurations. *The upper drawing illustrates the configurations (planetary positions relative to the Sun-Earth line) for an inferior planet; the lower drawing shows the configurations for a superior planet.*

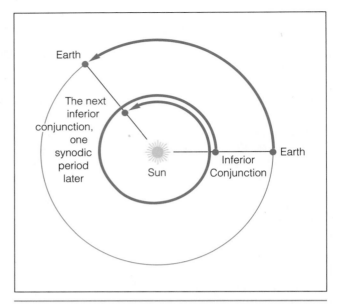

Figure 2.24 The synodic period of an inferior planet. *The inner planets travel faster than the Earth in their orbits, and therefore "lap" the Earth, much as a fast runner laps a slower runner on a track. This illustration shows approximately the situation for Mercury, which as a synodic period of about 116 days, or roughly one-third of a year.*

The time it takes the faster runner to overtake the slower one is analogous to the synodic period, whereas the time it takes to simply circle the track corresponds to the sidereal period. Table 2.1 lists the sidereal and synodic periods for the planets.

The motion of the Earth has one very important effect on planetary motions. As we go outward from the Sun, each successive planet has a slower speed in its orbit (see the discussion of Kepler's laws of planetary motion in Chapter 4). This means that the Earth, moving faster than the superior planets, periodically passes each of them (this occurs once in every synodic period). As the Earth overtakes one of the superior planets, there is an interval of time during which our line of sight to that planet sweeps backward with respect to the background stars, making it appear that the planet is moving backward. The same thing happens when we, in a rapidly moving automobile, pass a slowly moving vehicle; for a brief moment the other vehicle appears to move backward with respect to the fixed background. This apparent backward movement, called **retrograde motion,** was thought in ancient times to represent a real motion of the superior planets, rather than being recognized as a reflection of the Earth's

conjunction for an inferior planet. **Quadrature** occurs when a superior planet is seen 90° from the direction of the Sun.

Each planet has a sidereal period and a synodic period, the former being the orbital period as seen in the fixed framework of the stars and the latter being the length of time it takes the planet to pass through the full sequence of configurations, as from one conjunction or one opposition to the next (Fig. 2.24). The situation is much like that of two runners on a track.

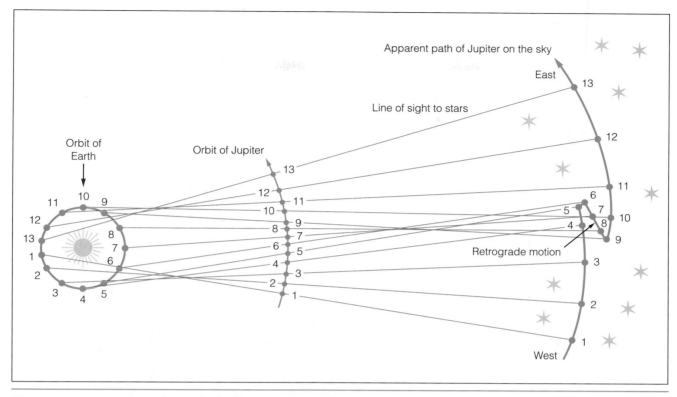

Figure 2.25 Retrograde motion.   *As the faster-moving Earth overtakes a superior planet in its orbit, the planet temporarily appears to move backward with respect to the fixed stars. This sketch illustrates the modern explanation of something that took ancient astronomers a long time to understand correctly.*

motion (Fig. 2.25). This greatly complicated many of the early models of the universe and in fact gave rise to a whole class of theories in which the planets were thought to move in small circles called epicycles, which in turn orbited the Earth (see the discussions of Hipparchus and Ptolemy in Chapter 3).

The inferior planets also undergo retrograde motion, but for a different reason. As these planets orbit the Sun, their direction of motion with respect to the fixed stars changes as seen from the Earth. Since an eastward motion with respect to the stars is the normal, or **prograde** direction, we say that the inferior planets are in retrograde motion when in the part of their orbits where they move west with respect to the stars. This occurs when they are on the far side of the Sun.

PERSPECTIVE

We have now developed a modern picture explaining

most of the astronomical phenomena visible to the naked eye. There is much left for us to learn about the universe and the laws that govern it, but already we have available to us all of the information upon which the first astronomers based their theories of cosmology. In the next two chapters, we will see how, presented with this evidence, the ancients fared.

SUMMARY

1. The Earth's rotation causes diurnal motions: the daily rising and setting of the Sun, the Moon, the stars, and the planets.

2. The solar day is a few minutes longer than the sidereal day because of the Earth's orbital motion about the Sun.

3. The most commonly used astronomical coordinate system, called the equatorial coordinate system, is based on the projection of the Earth's equator onto the celestial sphere (from which declination is measured) and on a fixed direction in the sky (from which right ascension is measured).

4. Precession is a slow drift of the Earth-based equatorial coordinate system with respect to the stars and is caused by a gradual change in the orientation of the Earth's rotation axis.

5. The orbital motion of the Earth about the Sun causes annual motions, such as the apparent motion of the Sun through the ecliptic, and it is responsible, along with the tilt of the Earth's axis, for our seasons.

6. Modern calendars are based on the tropical year, and require only very minor adjustment from time to time. Early calendars were usually based on an inaccurate length of year, and required frequent adjustment.

7. The Moon orbits the Earth, while the Earth orbits the Sun, and from the Earth we see the various phases of the Moon as it passes through different configurations.

8. Solar and lunar eclipses occur when the Moon passes directly in front of the Sun or through the Earth's shadow, respectively. The occurrence of these alignments is affected by the tilt and precession of the Moon's orbit.

9. All of the planetary orbits lie nearly in the same plane, so they are always seen along the ecliptic in various configurations with respect to the Sun-Earth line.

10. The planets go through temporary retrograde motion because of the relative speed with which they pass or are passed by the Earth.

REVIEW QUESTIONS

Thought Questions

1. Describe the difference between solar time, sidereal time, and the time kept by atomic clocks.

2. Explain why it is convenient to have the Earth divided into time zones, rather than using local solar time at every location.

3. Can you think of any way to tell, from observations or experiments, that the Earth is moving (spinning and orbiting the Sun)?

4. Explain how our seasons would be different if the Earth's axis were not tilted (that is, if the axis were perpendicular to the Earth's orbital plane). What would the seasons be like if the Earth's axis were tipped 90° instead of 23½°?

5. Use Figure 2.7 to compare the constellation where the Sun was when you were born with your sign of the zodiac, according to astrology. Explain why they are not the same.

6. Our calendar, like many ancient ones, has the year divided into months that are based on the motions of the Moon. Is it the Moon's sidereal or synodic period that is used in these calendars? Explain.

7. Explain why, on certain nights, the Moon is never visible at all.

8. If a quarter moon is already up when the Sun sets, is it first or third quarter? Explain.

9. Table 2.1 lists sidereal and synodic periods for all the planets. Explain why the synodic periods of the outer planets are not much more than 1 year.

10. Briefly summarize the types of motions that have been described in this chapter. Would all of these motions appear in the nighttime sky to a person on the planet Venus? (Ignore the fact that Venus is cloud-covered.)

Problems

1. How fast, in miles per hour, does a point on the Earth's equator travel due to the Earth's rotation?

2. The difference between a solar day and a sidereal day is 3^m56^s. Over the course of 1 year, how many days does this difference amount to? Discuss the significance of your answer.

3. What is the right ascension of a star that is directly overhead at midnight at the time of the vernal equinox? What is the right ascension of a star on the celestial equator that rises 6 hours after sunset at the time of the vernal equinox?

4. The Mauna Kea Observatory in Hawaii lies at a latitude of roughly 20°N, and the declination of the Magellanic Clouds is about −70°. Can the Magellanic Clouds be observed from Mauna Kea? Explain.

5. As noted in the text, precession causes the Sun's position in the zodiac to shift gradually, taking 26,000 years to make a full cycle through all twelve constellations. If the twelve constellations are equal in extent on the zodiac, how long does it take for the Sun to move from one constellation to the next? Discuss how this knowledge could help you determine historically when astrology was founded.

6. Suppose that you live at latitude 30°N. How far above the southern horizon would the Sun appear at noon at the time of the winter solstice? At the summer solstice?

7. What is the declination of the Sun at the time of each solstice and each equinox?

ADDITIONAL READINGS

A number of magazines, including *Mercury, Sky and Telescope, Astronomy,* and the *Griffith Observer,* contain practical information for the sky watcher, such as planetary positions and the seasonal appearance of the constellations. There are also annually published handbooks with similar data. One of the most widely used of these is the *Observer's Handbook,* by Roy L. Bishop (Toronto: Royal Astronomical Society of Canada). Some practical exercises in astronomy can be found in such books as *Astronomy: A Self-Teaching Guide,* by Dinah L. Moché (New York: Wiley, 1981).

MATHEMATICAL METHODS

The Measurement of Angles

 When we view the sky at night, we have no direct means of determining distances between objects, nor of finding how far they are from us. We see the stars, planets, the Moon, and all other celestial objects as though they were attached to a spherical shell centered on the Earth. The only thing we can measure meaningfully is the *direction* to an object.

To measure a direction, we must have a reference point or framework, and we must use angles to define or measure directions in that framework. For example, a certain star may be 30° above the horizon and 85° to the west of north. Or we might find that one star is 5° to the southeast of another.

As we have seen in this chapter, astronomers have defined a coordinate system, or framework, in which they measure positions of objects in the sky. The system chosen has the advantage that it does not refer to the horizon or the overhead direction, since these depend on where you are on the Earth, but is instead based on a framework that is fixed in the sky (except for the gradual shifting due to precession). This equatorial coordinate system is established by the celestial equator (from which the north-south position is measured) and a fixed direction along the line of intersection of the Earth's equatorial plane and the ecliptic (from which the east-west position is measured). Since we see the sky in only two dimensions, only two coordinates are needed to fully specify a position.

We may also want to consider the apparent sizes of objects seen in the sky, and again we have no way to do this in absolute terms. When we look at a planet, for example, we cannot tell how large it really is (in kilometers or miles); we can only say how big an angle it covers on the sky.

Because angular measurements are so common in astronomy, it is useful to gain some familiarity with the system we use for measuring angles. As we will discuss in the next chapter, this system was handed down to us from the ancient Babylonian culture, which flourished some 3,500 years ago. The Babylonian calendar divided the year into 360 days (even though the true length of the year was accurately known), so it became natural for the Babylonians to divide a circle into 360 equal parts. Hence, we have 360 degrees in a circle. The number 60 had special meaning to the Babylonians (explained in Chapter 3), and so each degree was divided into 60 parts, which we call *minutes of arc* or *arcminutes*. Similarly, each arcminute is subdivided into 60 equal parts, called *seconds of arc* or *arcseconds*. In a full circle, there are $360 \times 60 \times 60 = 1,296,000$ arcseconds.

The ancient astronomers, using only pointers and angular scales of fairly crude construction, were able to measure positions of stars or planets to an accuracy of a few arcminutes (for comparison, keep in mind that the angular diameter of the Moon is about 30 arcminutes). By the Renaissance, better instruments and the use of refined measuring techniques (as well as data averaging; see Chapter 4) allowed positions to be measured as accurately as one arcminute or perhaps a bit better. Now astronomers, using modern technology, are able to measure positions to much less than an *arcsecond* (about 0.01 arcsecond; this is discussed in Chapter 18), and with new instruments such as the *Hubble Space Telescope* and others that are planned, accuracies may soon reach 0.001 arcsecond. These numbers become quite remarkable when we realize that the angular diameter of a dime seen from a distance of 2 miles is about 1 arcsecond! There is more discussion of superaccurate angular measurements in Chapter 7, where a technique called *interferometry* is described.

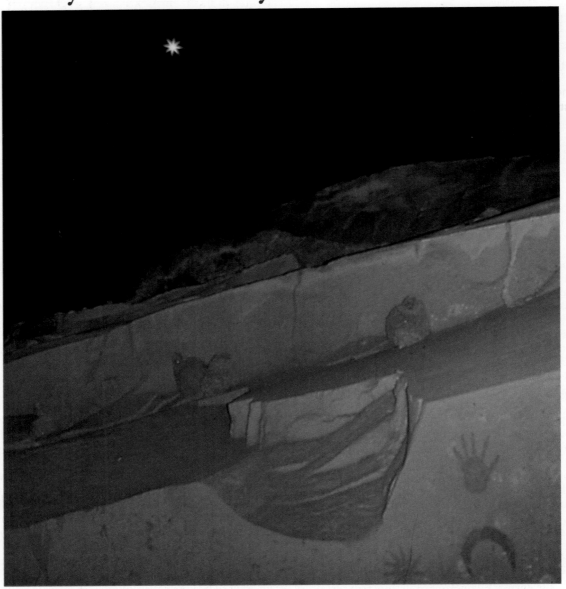

CHAPTER 3

Early Astronomy

The nighttime sky at Chaco Canyon, site of Native American astronomical petroglyphs. (J. A. Eddy)

CHAPTER PREVIEW

From the time we first became aware of our environment, we have been preoccupied with the heavens. The speculative mood that we, in these modern times, can conjure up only by disregarding our daily pressures and escaping into the countryside on a clear night to look at the stars must have dominated the nighttimes of the earliest cultures.

For the most part, we can only speculate about the astronomical knowledge of the prehistoric peoples who left no written records. We must assume that our distant ancestors in all parts of the world developed some awareness of astronomy, but we can discuss with real knowledge only those cultures that left records of their achievements. We find that parallel developments occurred in several parts of the world, including Asia, India, the Americas, and of particular importance to us, the Middle East and Mediterranean regions, wherein lie the roots of our modern astronomical science.

To develop the historical sequence leading to the present, we devote this chapter to astronomy in the vicinity of the Mediterranean, the center of the regions where the Babylonian, Egyptian, and Greek civilizations flourished (Fig. 3.1). Later in the chapter, accounts of the early Asian and American astronomies are presented, but largely for historical interest, since

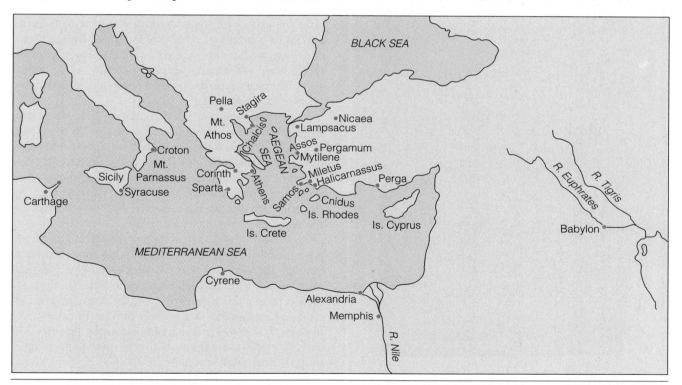

Figure 3.1 The ancient Mediterranean. *This map shows the locations of many of the sites mentioned in the text. To the right is the region of Babylonia, now Iraq. On the northern shores of the Mediterranean Sea lay most of the Greek empire, in what is now Italy, Greece, and Turkey. (C. Roman, The Astronomers, 1964 [New York: Hill and Wang])*

these cultures, by accident of geography, did not contribute to the mainstream development.

BABYLONIAN ASTRONOMY

In what is now Iraq, along the valleys of the Tigris and Euphrates rivers, there arose one of the earliest civilizations known to have left written records. The first people in this region were the Sumerians, who occupied the southern portion by about 3000 B.C. and invented a written language, cuneiform script, which was used to record astronomical data on clay tablets (Fig. 3.2).

The entire region was united around 1700 B.C. under the leadership of King Hammurabi, who established his capital at the city of Babylon (not far from the present Baghdad), lending the kingdom the name Babylonia. The clay tablets from that era reveal that the

Figure 3.2 A cuneiform clay tablet. *Thousands of tablets such as this one, containing Babylonian records of astronomical data, have been uncovered in what is now Iraq. (The Granger Collection)*

movements of the planets were given great significance. Tables were constructed for astrological purposes, showing that the Babylonians knew a great deal about the various cyclical motions of the heavens. They recognized that the Sun moves through the sky, and they divided the zodiac into twelve parts, probably because the year contains twelve lunar months (although not precisely twelve). The length of the year was known to within 4 minutes' accuracy, but the Babylonian calendar contained twelve 30-day months, creating a discrepancy that was corrected by the addition of an extra month every few years. A numbering system was developed based on powers of 60, much as our own system is based on powers of 10, and it was the Babylonians who bequeathed to us our units of angular measure (360° in a circle corresponding to the 360 days in their year, 60 minutes of arc per degree, and 60 seconds per minute).

The land between the Tigris and the Euphrates, known to the Greeks as Mesopotamia, saw several empires come and go, well into the first centuries A.D. The Assyrians dominated during the era around 600 to 800 B.C., and they preserved the clay tablets containing astronomical data from earlier times. Around 200 B.C., after the center of power had moved west to Greece, a massive amount of data was compiled by the then-occupants of Babylonia, the Chaldeans. The Chaldean tables were sufficiently accurate so that solar, lunar, and planetary motions could be predicted, as could the occurrence of eclipses.

The primary motivation throughout these times was the application of astronomical information to astrology. The Babylonians seem to have speculated very little about the nature of the universe, apparently envisioning the heavens as consisting of a dome supported by the mountains surrounding their region. No evidence has yet been uncovered for any theories of the motions of the heavenly bodies.

Some astronomical development occurred contemporaneously in Egypt, but little was accomplished there beyond the establishment of timekeeping techniques also known to the Babylonians, who used sundials to record local solar time. In Egypt, the main purpose in studying astronomy was neither philosophical nor astrological, but was practical and religious, centered on anticipating the seasons and other events of importance to agriculture and commerce.

As the Babylonian science developed and the region underwent several successive conquests, a separate civilization began to organize itself along the shores of the Mediterranean, a civilization that was to reach the greatest heights of scientific accomplishment in ancient times. The Greek empire that arose there was greatly influenced in its early astronomical development by the Babylonians. Major contributions of the early Greeks are summarized in Table 3.1.

THE EARLY GREEKS AND THE DEVELOPMENT OF SCIENTIFIC THOUGHT

The foundations of the Greek civilization were established some 5,000 years ago in the eastern Mediterranean. There arose a seafaring culture, whose earliest home was the island of Crete, where the legends were born that gave us the names of most of our constellations. This civilization prospered around 1600 B.C. during the reign of King Minos, and it was about this time that the settlement of what is now the Greek mainland began. Much of what we know about this era is gleaned from the writings of Homer in the *Iliad* and the *Odyssey* (Fig. 3.3), written later, probably around 900–800 B.C. Interspersed in these epic poems were the early Greek views of the universe, along with the names of the constellations.

The cosmology of this time consisted of the dome of the heavens, with a disklike Earth floating on water. This was similar to the Babylonian view, developed at about the same time, but the Greeks went further, hypothesizing—for what were essentially theoretical reasons—that there was an underworld comparable in scope and complexity to the heavens. Apparently these people had developed a sense of aesthetics and unity in the universe, which shows that their thinking went beyond mere description of the visible skies.

The first formal scientific thought evolved after the time of Homer's writing, and its inception is associated with the philosopher **Thales** (c. 624–547 B.C.), who lived in Miletus, on the eastern shore of the Aegean Sea. His principal contribution was the idea that rational inquiry can lead to understanding the universe, going beyond describing it. This general precept was adopted by followers of Thales, who formed the so-called Ionian school of thought, named for that part of Greece (now Turkey) surrounding Miletus. One of Thales' most singular claims to fame as an astronomer was his supposed prediction of a solar eclipse that took place in 585 B.C., a feat that some historians believe did not actually occur. Much of the Greek knowledge of the motions of the heavens at that time came from the Babylonians, and it is not certain that even they were yet able to predict eclipses accurately.

Thales did create a primitive cosmology, encompassing the idea that all the basic elements of the universe were formed from water, the primeval substance. An associate of his, **Anaximander** (c. 611–546 B.C.), altered this view and adopted the idea of a universal medium, of unknown properties, from which all material substance was made. Anaximander also developed the notion, entirely without observational sup-

Table 3.1 Notable Greek Achievements

Date	Name	Discovery or achievement
c. 900–800 B.C.	Homer	*Iliad* and *Odyssey;* summaries of legends
c. 624–547 B.C.	Thales	Rational inquiry leads to knowledge of universe
c. 611–546 B.C.	Anaximander	Universal medium; primitive cosmology
c. 570–500 B.C.	Pythagoras	Mathematical representation of universe; round Earth
c. 500–400 B.C.	Philolaus	Earth orbits central fire
c. 500–428 B.C.	Anaxagoras	Moon reflects sunlight; correct explanation of eclipses
c. 428–347 B.C.	Plato	Material world imperfect; deduce properties of universe by reason
c. 408–356 B.C.	Eudoxus	First mathematical cosmology; nested spheres
c. 384–322 B.C.	Aristotle	Concept of physical laws; proof that Earth is round
c. 310–230 B.C.	Aristarchus	Relative sizes, distances of Sun and Moon; first heliocentric theory
c. 273–? B.C.	Eratosthenes	Accurate size of Earth
c. 265–190 B.C.	Apollonius	Introduction of the epicycle
c. 200–100 B.C.	Hipparchus	Many astronomical developments; full mathematical epicyclic cosmology
c. A.D. 100–200	Ptolemy	*Almagest;* elaborate epicyclic model

port, that outside the Earth the firmament is filled with flame, contained in rotating tubes with holes that allowed the light within to be seen. Apparently it was his concept that the Sun, Moon, and stars were openings in the sides of these hoops, so that their rotation created the observed motions of the heavenly bodies.

The Pythagoreans

Pythagoras (c. 570–500 B.C.) was born on the island of Samos, off the Aegean coast not far from Miletus. His early life is obscure, but at around the age of 40,

Figure 3.3   Homer's ODYSSEY AND ILIAD. *In these two epic poems, Homer described the much more ancient legends of astronomical lore from the civilization that had flourished on the isle of Crete. (The Granger Collection)*

following several years of travel and study, he established a school at Croton, on the southern tip of Italy, in which scientific and religious teachings were intermixed. The lasting contribution of this school was the notion that all natural happenings can be described by numbers, laying the foundation for geometry and trigonometry. Pythagoras himself is thought to have been the first to assert that the Earth is round and that all heavenly bodies move in circles, ideas that never lost favor thereafter in ancient times.

Most of what we know about Pythagoras, who himself left no written records, comes from the works of his follower **Philolaus** (c. 500–400 B.C.), who added the idea that the Earth moves about a central fire (not the Sun), always hidden on the far side of the Earth. Both he and Pythagoras, for largely philosophical reasons rather than logic, believed the planetary distances to correspond to the lengths of vibrating strings that produced harmonious musical notes. This led to the concept of a celestial harmony that later had considerable influence in philosophical thinking.

There were numerous followers of the Pythagorean school who in various ways refined this view of the universe. One of particular note was **Anaxagoras** (c. 500–428 B.C.), who is credited with the realization that the Moon shines by reflected sunlight, rather than by its own power. This awareness, based on observation and deduction rather than personal philosophy, allowed him to correctly attribute solar and lunar eclipses to the passage of the Moon in front of the Sun and through the Earth's shadow, respectively.

Plato and His Followers

On the Greek mainland, a number of democratically ruled city-states arose, of which Athens became preeminent by about the fifth century B.C., dominating artistic and literary affairs. Even after the military defeat in 404 B.C., at the hands of the rival city of Sparta, Athenian culture continued to flourish. It was in this environment that **Plato** (428–347 B.C.; Fig. 3.4) reached maturity, having studied philosophical thought under the tutelage of Socrates.

After some years of traveling, Plato established an academy outside Athens, in about the year 387 B.C., where he taught his ideas of natural philosophy. His fundamental precept was that what we see of the material world is only an imperfect representation of ideal creation. The corollary of this doctrine was that we can learn more about the universe by reason than by observation, since, after all, observation can give us only an incomplete picture. Hence Plato's ideas of the universe, described in his *Republic,* were based on certain idealized assumptions that he found reasonable. One of the most important was that all motions in the universe are perfectly circular and that all astronomical bodies are spherical. Thus he adopted the Pythagorean view that the Sun, Moon, and planets move in various combinations of circular motion about the Earth. It seems that Plato thought of these objects as affixed to clear, ethereal spheres that rotated. The teachings of Plato, particularly the premise that the mysteries of the cosmos should be solved through reason rather than by direct observation, dominated much of western thinking until the Renaissance, nearly 2,000 years after his time.

A pupil of Plato's was **Eudoxus** (c. 408–356 B.C.), who made the first attempt to put Plato's view of the universe on a firm mathematical footing; he constructed a series of concentric spheres on which the Sun, Moon, and planets moved in perfect circular motions (Fig. 3.5). Each body required three or four spheres, each with its own rotation rate, to produce the composite motion necessary to duplicate the overall motion. A total of twenty-seven spheres was needed for all the known heavenly bodies. It is not clear whether Eudoxus thought of these spheres as material objects or merely as mathematical constructs.

The most renowned student of Plato was **Aristotle** (c. 384–322 B.C.) who, after the death of Plato in 347 B.C., established his own small academy across the Aegean Sea at Assos. He then moved to Pella, where he tutored Alexander, the son of King Phillip of Macedonia, leader of the Greek empire. Aristotle later created a school in Athens, where he taught his own view of the universe, one formed from Plato's mold of reason rather than observation.

Aristotle was the first to adopt physical laws and then to show why, in the context of those laws, the universe works as it does. He taught that circular motions are the only natural motions and that the center of the Earth is the center of the universe. He also believed, without any experimental or observational support, that the world is composed of four elements: earth, air, fire, and water. He demonstrated, in the context of his adopted physical laws, that both the universe and the Earth are spherical. He had three ways of proving that the Earth is spherical: (1) only at the surface of a sphere do all falling objects seek the center

Figure 3.4 Plato. *Plato's beliefs that the universe and all celestial objects were perfect and immutable had great influence on later philosophers. (The Granger Collection)*

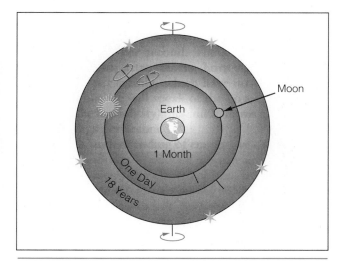

Figure 3.5 The spheres of Eudoxus. *This shows the relative orientations and directions of rotation for spheres thought to be responsible for the monthly motions of the Moon, as well as its 18-year eclipse cycle.*

by falling straight down (it was another premise of his that falling objects are following their natural inclination to reach the center of the universe); (2) the view of the constellations changes as one travels north or south; and (3) during lunar eclipses it can be seen that the shadow of the Earth is curved (Fig. 3.6). By relating his theories to observation in this manner, Aristotle broke with the tradition of Plato to some extent, although, as we have seen, he still approached the problem in the same manner, letting reason rather than observation guide the way.

Aristotle also concluded that the universe is finite in size. This led directly to his belief that the heavenly bodies can follow only circular motions, because otherwise they might encounter the edge of the universe. Another tenet of Aristotle was that the heavenly bodies were made of a fifth fundamental substance, which he called the *aether*.

The Geometric Genius of Aristarchus and Eratosthenes

After Aristotle's period, the center of Greek scientific thought moved across the Mediterranean Sea to Alexandria (the capital city established in 332 B.C. by Aristotle's former pupil, Alexander the Great), near Cairo.

The most prominent astronomer of this era, however, was born on the Mediterranean island of Samos, and it is not well known where he lived and worked, although it may have been in Alexandria. This was **Aristarchus** (c. 310–230 B.C.), the first scientist to adopt the idea that the Sun, not the Earth, is at the center of the universe. This was some 1,700 years before the time of Copernicus, who usually is given credit for this concept. The heliocentric hypothesis of Aristarchus failed to attract many followers at the time because of a lack of concrete evidence that the Earth was in motion, and a general satisfaction with the Aristotelian viewpoint, which had no generally recognized flaws.

It is interesting to consider how Aristarchus reached his revolutionary idea. By geometrical means, he was able to deduce the relative sizes and distances of the Moon and the Sun; and although his results were not very accurate, he did correctly conclude that the Sun is far larger than either the Moon or the Earth. Accepting this, he found it difficult to imagine that the Sun should orbit the lesser Earth, so he adopted the contrary point of view.

A younger contemporary of Aristarchus was **Eratosthenes** (c. 273–? B.C.), who worked in Alexandria during his long career of research in geometry and astronomy. His greatest achievement was the determination of the size of the Earth, by a method of great simplicity (Fig. 3.7). It was known that on a certain

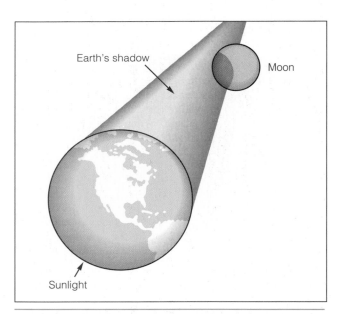

Figure 3.6 Curvature of the Earth's shadow on the Moon. *Only a spherical body can cast a circular shadow for all alignments of the Sun, Moon, and Earth.*

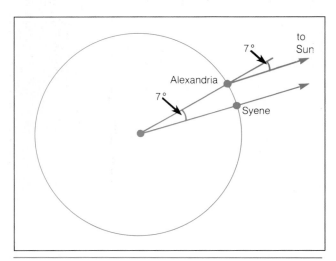

Figure 3.7 The method of Eratosthenes for measuring the size of the Earth. *Eratosthenes deduced the portion of a circle that corresponded to the distance between Alexandria and Syene. From this he was able to calculate the circumference of the Earth.*

day in the summer, sunlight penetrated all the way to the bottom of a deep well located at Syene, near present-day Khartoum. On the same day in Alexandria, the Sun's midday position was some 7° south of the zenith, indicating that the distance between the two cities corresponded to 7° of the 360° circle of the Earth's circumference (this assumes that the Sun is sufficiently far away that rays of light from it are essentially parallel when they reach the Earth). Hence the circumference was equal to 360°/7° times the distance between Alexandria and Syene, which was a little less than 5,000 stadia, a stadium being a unit of measure equal to about one-tenth of a mile. The circumference of the Earth, as calculated by Eratosthenes, was 252,000 stadia, or just under 25,000 miles—within 2 percent of the correct value!

Geocentric Cosmologies: Hipparchus and Ptolemy

By the third century B.C., the need for more precise mathematical models of the universe became apparent, as better observing techniques were developed. A mathematical concept that provided the needed precision while still preserving the precepts of Aristotle was the **epicycle** (Fig. 3.8). It was **Apollonius** (c. 265–190 B.C.) who put the epicycle on a firm geometric

footing, providing the first mathematically viable alternative to the idea of spheres developed by Eudoxus. An epicycle is a small circle on which a planet moves, whose center in turn orbits the Earth following a larger circle called a **deferent**. This construct allows a combination of motions whose speeds and circular radii can be adjusted as needed to provide a close match with the observed motions. The epicycle provided a clear advantage for explaining the retrograde motions of the superior planets, something that confounded Eudoxus' theory that the planets move on rotating spheres.

The epicyclic motions of the celestial bodies were refined further by Hipparchus (Fig. 3.9), who was active during the middle of the second century B.C. (Very little is known about his life, even the dates of his birth and death.) Hipparchus, most of whose work was done at his observatory on the island of Rhodes, was arguably

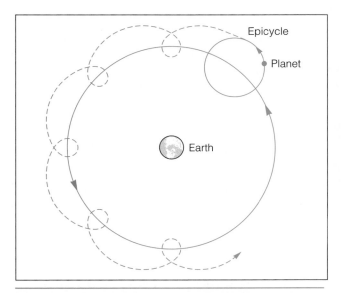

Figure 3.8 The epicycle. *It was realized that planetary motions could be represented by a combination of motions involving an epicycle, which carries a planet as it spins while orbiting the Earth.*

Figure 3.9 A fanciful rendition of Hipparchus at his observatory. *Hipparchus made important discoveries based on his observations and on his advanced methods for analyzing data. (The Bettmann Archive)*

ASTRONOMICAL INSIGHT 3.1

The Mythology of the Constellations

 As noted in the text, the writings of Homer in the *Iliad* and the *Odyssey*, which appeared around 900–800 B.C., contain descriptions of the constellations and their meanings. These descriptions were based on legends from the Minoan civilization that were already ancient in Homer's time.

Little precise information is available as to when and how the mythology of the constellations arose. However, it has been possible to deduce roughly the era that gave birth to them. Careful examination of the ancient constellations shows them to be distributed symmetrically about a point in the sky that was at the north celestial pole around 2600 B.C., so it is probable that the constellations were invented at about that time. (The pole has since shifted away from the center of these constellations because of precession.) There are few ancient constellations near the southern pole, in regions not visible from the latitude of Crete, a fact that gives independent support to the supposition that the legends arose in the Minoan culture.

It is likely that the constellations were not taken as literally as is usually supposed. Rather than being considered faithful depictions of the people and events

with which they are associated, the constellations should more properly be viewed as symbolic representations. They were probably not originally designated on the basis of their imagined resemblance to certain characters; instead, areas of the sky were dedicated in honor of prominent figures of mythology, and the familiar pictures of these figures were then fitted to the patterns of bright stars. This helps account for the lack of obvious resemblance between the star patterns and the figures and events they supposedly represent.

The constellation names were translated from the Greek of Homer to Latin when the Roman Empire rose to dominance, and most of the names with which we are familiar today are the Latin ones. Interestingly, the names of prominent stars went through another transformation, into Arabic, and today's star names are Arabic designations that usually are literal indications of the place these stars hold in their constellations. Betelgeuse, for example, the name of the prominent red star in the shoulder of Orion, is translated as "the armpit of the giant" or "the armpit of the central one." Thus, in today's standard usage, we often adopt Arabic names for stars in constellations with Latin designations, although both translations are based on Greek descriptions of the ancient Minoan legends.

Ancient and nonscientific though they are, the constellations

have a significant impact on the nomenclature of modern astronomy. The modern constellations, many of which correspond to ancient ones, refer to very specific regions of the sky with well-defined boundaries that have been agreed on by the international community of astronomers. The "official" constellations are based on those of legend, containing within them the ancient figures of mythology, but their boundaries have been extended so that every part of the sky falls within one constellation or another. A map of the constellations looks a bit like a map of the western United States, where the boundaries are generally straight lines, but the shapes are irregular. As an alternative to the Arabic names for the brightest stars, modern astronomers often use designations based on a star's rank within its constellation, with letters of the Greek alphabet used to indicate specific stars. Thus, Betelgeuse, the brightest star in Orion, is also called α Orionis.

It can be fun and informative to learn the prominent constellations, and this is taught in many general astronomy classes. It is worth keeping in mind, however, that memorizing star patterns on the sky is not the same as scientific inquiry, but should be regarded instead as a convenient method for remembering the locations of objects that are interesting to view.

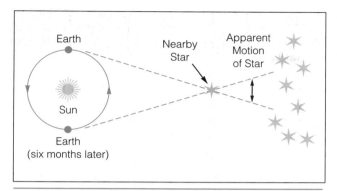

Figure 3.10 Stellar parallax. *As the Earth orbits the Sun, our line of sight toward a nearby star varies, causing the star's position (with respect to more distant stars) to change. The change of position is greatly exaggerated in this sketch; in reality, even the largest stellar parallax displacements are too small to have been measured by ancient astronomers.*

Figure 3.11 Ptolemy. *Ptolemy summarized all of the Greek knowledge of astronomy, including significant contributions of his own, in the* Almagest. *(The Granger Collection)*

the greatest astronomer of antiquity. Among his major contributions were (1) the first use of trigonometry in astronomical work (in fact, he is largely credited with its invention, although many of the concepts were developed earlier); (2) the refinement of instruments for measuring star positions, along with the first known use of a celestial coordinate system, enabling him to compile a catalogue of some 850 stars; (3) the refinement of the methods of Aristarchus for measuring the relative sizes of the Earth, Moon, and Sun; (4) the invention of the stellar magnitude system for estimating star brightnesses, a system still in use today (with minor modifications); and, perhaps most impressively, (5) the discovery of the precession of star positions, which he accomplished by comparing his observations with some that were made 160 years earlier.

Although Hipparchus did not add much that was new to the then-current models of planetary motions, he did compile an extensive series of observations of the solar motion. This led him to a refinement of the epicycle theory, namely, that the Earth was off-center in the large circle on which the Sun moved. This idea of an **eccentric,** or off-center, orbit accounted rather accurately for the observed variations of the speed of the Sun in its annual motion.

Despite the prior suggestion that the Sun was the central body, Hipparchus clung to the Earth-centered view. Apparently his major reason was that the geo-centric picture was quite capable of explaining the observations, and there was no evidence that the Earth moved. There was, in fact, an observational argument that the Earth did *not* move. Hipparchus knew that nearby stars should appear to move back and forth on the sky, as a reflection of the Earth's orbital motion (Fig. 3.10), but no such **stellar parallax** could be detected. (The shifts do occur, but they are far too small to have been measured with the crude instruments of the Greek astronomers.)

After the great work of Hipparchus, almost 300 years passed before any significant new astronomical developments occurred. Claudius Ptolemaeus (Fig. 3.11), or simply **Ptolemy,** lived in the middle of the second century A.D. Attempting to summarize all the world's knowledge of astronomy, he published a treatise now called the *Almagest,* which was based in part on the work of Hipparchus, but which also contained some new developments of his own. The thirteen books of the *Almagest* range from a summary of the observed motions of the planets to a detailed study of the motions of the Sun and Moon, and from a description of the workings of all the astronomical instruments of the day to a reproduction of Hipparchus's star catalog. Most important in his treatise, however, are the detailed models of the planetary motions. Here Ptolemy made his greatest personal contribution, for these models were so accurate in their ability to predict the planetary positions that they were used for the next 1,000 years.

To provide this accuracy, Ptolemy had to adopt and extend an idea from Hipparchus, that of the eccentric (Fig. 3.12). To satisfactorily account for all the observed irregularities in a planet's motion, Ptolemy assumed not only that the center of the deferent was off-center from the position of the Earth, but also that the rate of motion of the epicycle as it orbited the Earth was constant as seen from another off-center point, on the opposite side of the Earth from the geometrical center of the deferent. This combination of offsets, with properly chosen dimensions and rotational speeds, allowed good representation of the observed motions while preserving the hypothesis that all the heavenly motions are circular. It is interesting that other doctrines of Aristotle, such as the precise centering of all the planetary orbits on the Earth, were not strictly followed.

It seems from some of Ptolemy's writings that he may not have considered his model of the universe to represent physical reality, but instead may have viewed it as simply a mathematical tool for predicting planetary positions. This is something modern scientists do when confronted with data on phenomena for which the underlying physical principles are not clear, and in this sense, Ptolemy may be considered as one with modern astronomers. Whether or not Ptolemy believed his model to represent physical reality, it was taken literally in the centuries following his time. The notion of a geocentric universe with only circular motions dominated for some fourteen centuries afterward and was expunged only after great turmoil.

The history of Greek astronomy came to an end with the work of Ptolemy, as did most significant astronomical development in Europe and western Asia until the time of Renaissance. We will return to pick up the thread of astronomical knowledge and growth during the Dark Ages, after we digress and consider what was going on in the rest of the world.

EARLY ASTRONOMY IN ASIA

Chinese Accomplishments

One of the most ancient recorded astronomies outside the Middle East arose in China. Legends there refer to astronomical activities as far back as 2000 B.C., although the oldest writings are much more recent.

Ancient China had a tumultuous history, with periods of peace, prosperity, and cultural advance occurring throughout numerous centralized dynasties, interspersed with eras of disarray resulting from barbarian invasions and civil unrest. Historical records were sometimes lost during the transitions; thus it is not known just how far in the past the first significant astronomical activity began. Legend refers to a pair of astronomers named Hi and Ho, who, having failed to predict a solar eclipse that occurred in 2137 B.C., were put to death.

We know that timekeeping was well established in China by the fourth century B.C., since records from that era indicate that the length of the year and the lunar month were known with considerable accuracy. The principal roles of astronomers were to keep the time and to officiate over state functions, ensuring that all was done in accordance with the rules of the gods. Astrology played a major role, and the constellations

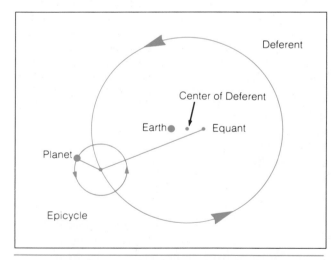

Figure 3.12 The cosmology of Ptolemy. *To account for nonuniformities in the planetary motions, Ptolemy devised a complex epicyclic scheme. The deferent, the large circle on which the epicycle moves, is not centered on the Earth and has a rotation rate that is constant as seen from another displaced point called the equant. This meant that from the Earth, the planet appeared to move faster through the sky when on one side of its deferent than when on the other. Each planet had its own deferent and epicycle, with motion and offsets adjusted to reproduce the observations.*

were identified with the emperor and with government organization (Fig. 3.13). The north polar region of the sky, for example, was thought to represent the seat of the emperor, and within a constellation, stars were given royal rank according to their brightness. A catalog of some 809 stars was constructed in the fourth century B.C. by the astronomer Shih-shen, predating by about 200 years Hipparchus' catalog.

The prevalent concept of the universe, however, was not nearly as advanced as that of the Greeks. The cosmos was thought to consist simply of the sphere of the sky with its daily rotation, and little significance was attached to the motion of the planets. The poles were considered to be the exalted regions of the sky, whereas the celestial equator and the ecliptic seem to have been given no great importance. Heaven and Earth were thought to be closely related, with events in one domain affecting those in the other; thus, irregularities in the heavens were taken as omens of calamities on Earth.

In later centuries, observations became more accurate, with the employment of simple measuring devices. By the first century A.D., accurate tables of lunar motions had been constructed and eclipses could be predicted. Apparently, precession was known to the Chinese astronomers of this era as well.

Contact with the western world developed during the first few centuries A.D., as a result of Chinese conquests to the west and the eastward diffusion of Greek culture, which reached China by way of India. From this time on, it is difficult to separate independent Chinese developments from those introduced from the West, but the records indicate that numerous important advances were made, particularly in the improvement of measuring devices.

Records of celestial events were kept continuously until medieval times, providing later generations with important data to be found nowhere else. Among the events mentioned in the Chinese records is the appearance of a "guest star" in A.D. 1054, identified in modern times with the supernova explosion that created the Crab Nebula.

Because of China's remote location as seen from Europe, astronomical developments there had little influence elsewhere, except for the maintenance of observing records during the Dark Ages in Europe, when no one else was recording what was seen in the skies.

Hindu Astronomy

In India, as in China, it is evident that some astronomical lore had developed by very early times, but our only knowledge of the oldest accomplishments is based on ancient legends and on the astronomical orientations of some buildings (Fig. 3.14). There are written records of lunar motions dating back to about 1500 B.C., and by 1100 B.C. a calendar had been established

Figure 3.13 Ancient Chinese depiction of the celestial sphere. *Places near the North Pole were assigned to nobility, with lesser beings relegated to lower latitudes. (The Granger Collection)*

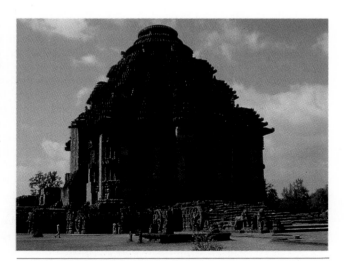

Figure 3.14 Ancient building in India. *This is one of the largest and finest surviving Hindu temples built along astronomical alignments. (J. M. Malville)*

ASTRONOMICAL INSIGHT 3.2

The Mysteries of Stonehenge

By now we are aware that some form of astronomical influence affected nearly every ancient culture. The timing of major developments seems to have been surprisingly similar in most parts of the world. The earliest records of accurate positional measurements and the establishment of calendars often date from the second millennium B.C., with sufficient sophistication for predictions of eclipses and planetary motions not developing until the last few centuries B.C.

It is most impressive, then, to consider the stone circle monuments of the British Isles, for they predate recorded astronomical achievements in the rest of the world, yet they show that their builders had substantial knowledge of astronomical skills, possibly including the prediction of eclipses.

Of the 900 or so of these monuments, by far the best known is Stonehenge, located on Salisbury Plain, about 70 miles west of London. This is a large and complex monument characterized by immense upright stones arranged in various circular patterns, in some instances with horizontal lintels connecting the vertical members.

Little is known about the builders of Stonehenge. The monument was built in three stages, the first commencing about 2800 B.C., and the most recent being completed by 2075 B.C. The sophisticated astronomical principles that guided

C. D. McLoughlin

the construction were incorporated from the beginning, even though it appears that the people inhabiting the region changed. The original Stone Age civilization gave way to a new group, called the Beaker People, who arrived from what is now the Netherlands in approximately 2500 B.C. Subsequent stages were built by this latter group, who apparently embraced the logic that had led to the original construction.

How the stones, which weigh up to 50 tons, were moved to Salisbury Plain is not known. Some came from Wales, possibly by sea, and many came from a region some 20 miles distant. The monument is now in a state of partial ruin and disarray, with some stones missing. Besides the large stones, there are a number of holes that may have once contained wooden posts to mark certain directions. Even today new parts of Stonehenge are occasionally found, in the form of postholes or distant mounds that may have been used to mark significant alignments.

There is no question that Stonehenge was laid out according to astronomical principles. There is substantial controversy, however, about the details. Despite a great

deal of publicity in the 1960s over interpretations of Stonehenge as an early computer used to predict eclipses, today there is little agreement as to what is the correct interpretation.

One axis of Stonehenge, delineated by the sight line to the Heelstone from the center of the monument, aligns with the position of sunrise at the summer solstice. The alignment is better now, however, than it was in 2800 B.C., when the first stage was built. (The reason for this is that the tilt of the Earth's axis has changed very slightly). A circle of fifty-six holes around the center of Stonehenge has also attracted a great deal of attention; some people thought it possible that the holes were used to predict eclipses, although it appears unlikely.

Whatever the correct detailed interpretation of Stonehenge, the fact remains that extensive knowledge of astronomical phenomena was available to its builders as early as 2800 B.C., more than 1,000 years before other cultures such as the Babylonians are known to have developed such a high level of sophistication. It is unfortunate that we do not know more about the remarkable people who built it.

based on the solar year. Apparently all the prominent celestial bodies were named for gods or goddesses and were thought to represent their characters.

Astronomy in India flourished during and immediately after the time of the great prince Rama, who was born in the year 961 B.C. During this period, tables of planetary motions were constructed for use in astrology, since the Hindus thought that heavenly occurrences represented earthly affairs. A very complex calendar was adopted that was based on an observed 247-year cycle of solar and lunar motions of great significance. The construction of this calendar, as well as the discovery of precession, show that the Hindu astronomers made use of positional records that must have been maintained over hundreds of years.

Sometime in the first millenium B.C., there was established cultural contact between India and the Middle Eastern civilizations of the Babylonians and the Greeks, and it is difficult to be certain of the direction in which information flowed. Some people claim that the entire basis of the Babylonian and Greek astronomical development was borrowed from India in about the eighth century B.C., whereas others argue that the developments in the two regions were quite independent. In any case, it is clear that by A.D. 500, western influence dominated Hindu astronomy, and that few new developments took place in India after that time.

ANCIENT ASTRONOMY IN THE AMERICAS

A modern appreciation of ancient astronomical developments in the New World has been slow in coming. The principal reasons for this were the nearly complete destruction of the writings of the Mayan civilization at the time of the Spanish conquest, and the lack of a written language of the North American natives.

Archeological studies have revealed that in Central and South America sophisticated cultures existed before the time of Columbus. The ancient cities of the Mayan and Incan civilizations, now in ruins, show highly complex and organized structures requiring sound knowledge of astronomical phenomena. It has been discovered that in some cases alignments of principal buildings and avenues were designed to point toward significant astronomical locations, such as the rising point of the Sun on the day of its passage through the zenith. A great deal of what we know about astronomy in these ancient cultures is read from the record of their architecture, rather than from their writings.

The center of the Mayan civilization was located in what is now Mexico and several Central American nations to the south, in a region known today as Mesoamerica. Other cultures in this region, which flourished between 500 B.C. and A.D. 1500, were the Olmec, Zapotec, and Aztec.

As in other ancient cultures, the principal motivations behind the Mesoamerican astronomy were timekeeping and astrology. The Mayas developed a very accurate calendar, and measured with great precision the motions of the celestial bodies. A particular significance was associated with the planet Venus, which inspired sacrificial rites and other ceremonies. The motions of Venus were extensively tabulated in a surviving document known as the *Dresden Codex* (Fig. 3.15).

In some locations, special astronomical observatories were constructed, usually with many significant alignments made in the directions of important celestial events. The most famous is the Caracol tower at Chichén Itzá, in the Yucatán, a building whose strange architecture defied explanation until the numerous astronomical sightlines it entails were discovered in the twentieth century (Fig. 3.16).

The Mayan cosmology apparently consisted of a layered universe, each level of which contained one type of celestial body. Above the Earth was the domain of the Moon, then the clouds, the stars, the Sun, Venus, comets, and so on. Similarly, there was thought to be an underworld consisting of nine levels, providing symmetry with the heavens, an idea parallel to that of the early Greek cosmologies described by Homer.

In South America, in the Andean territory where modern-day Peru lies, the Incan civilization flourished in the century or so before the Spanish conquest. Here, too, great attention was paid to astrological omens, and much of the astronomical effort was devoted to the construction of accurate timekeeping systems. The Incas had a rather complex calendar that consisted of 12 months, but it was otherwise quite unlike the modern calendar, not even containing the same number of days in a week that we have today. The Incas, like their Mesoamerican neighbors to the north, often laid out their temples and cities in accordance with astronomical reference points, in at least one case using markers set out over an entire valley as a framework for computing dates.

Astronomy practiced by North American natives in ancient times is much more difficult to assess, there being no written language and few records of any kind left, with the exception of scattered rock drawings and artifacts such as mounds, religious symbols, and artwork (Fig. 3.17). It seems evident, however, that the Indians of North America were intimately familiar with the skies and significant events such as the solstices that foretold the changing of the seasons. Among the few records that do exist are illustrations either carved or painted on rocks (Fig. 3.18), found in the Southwest, some of which may depict the appearance of the supernova of A.D. 1054. This was the same event observed by the Chinese, but except for the possible native American sighting, it was not mentioned in the records of any other cultures. That the petroglyphs do represent the supernova of 1054 is by no means established, however.

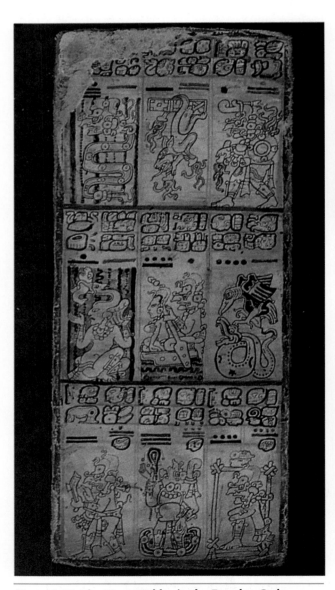

Figure 3.15 The Venus Tables in the Dresden Codex (Maya). *These tables show that the Mayans attached particular significance to the planet Venus. (Historical Pictures Service).*

Figure 3.16 The Caracol Tower at Chichen Itzá. *This is one of the most significant of the many astronomically oriented structures in Mesoamerica. (J. A. Eddy)*

Figure 3.17 Pawnee Indian sky map. *This chart, embossed on hide, appears to depict constellations of the Northern Hemisphere skies. (From Von Del Chamberlain 1982), When Stars Came Down to Earth: Cosmology of the Skidi Pawnee Indians of North America [Ballena Press: Los Altos, Calif.] Skidi Pawnee chart of the heavens, Field Museum of Natural History, photograph by Von Del Chamberlain.)*

Some structures survive that appear to have astronomical significance, including the Sun temple at Mesa Verde and other ruins in the Southwest that may have been influenced by the Mesoamerican culture to the south. In addition, there are earthen mounds of various descriptions located in many parts of the Midwest, and some seem to represent celestial events or follow astronomically significant alignments.

In the American West there are several circular monuments, called medicine wheels, marked out in rock. These were probably left behind by the Plains Indians, and they show astronomical influences to varying degrees. Among the best studied is the Bighorn medicine wheel, located high in the Bighorn Mountains of Wyoming (Fig. 3.19). This is a circular structure some 90 feet in diameter, with rock cairns around its perimeter. Lines traced through these piles of rock coincide with a number of astronomically significant directions, including a primary line pointing in the direction of the rising Sun on the morning of the summer solstice.

Unfortunately, it is impossible to be certain when these monuments were constructed, so we do not know just how long ago our predecessors on this continent developed their understanding of the heavens.

ASTRONOMY IN THE DARK AGES

Let us now pick up the main historical thread, with a brief mention of events after the time of Ptolemy. The Christian influence began to be felt throughout what had been the Greek empire, turning people away from the belief that astronomical events dominated their fates, and leading to the dissipation of efforts to improve the understanding of the heavens. The prophet Mohammed was born in the mid-seventh century A.D., and within a century, his followers had conquered most of the Mediterranean and some of southern Europe, and had burned what was left of the library at Alexandria.

The need for accurate calendars caused some revival of interest in astronomy, and the Arabs resurrected a great deal of the Greek knowledge. During their period, our modern numbering system, brought by the Arabs from India, was established, and new mathematical techniques, notably algebra, were developed. Although some new astronomical observations were made by Arab scientists, their principal contribution was to keep alive the Greek knowledge of astronomy and to pass it on to the western European civilization in the twelfth century, through Spain.

In the thirteenth century, Spanish King Alfonso X ordered the construction of new tables of planetary motions, the old tables of Ptolemy finally having begun to show inadequacies. The new tables, derived by a group of astronomers using the epicyclic model of Ptolemy, became known as the Alfonsine Tables, and were used for the next 300 years.

Meanwhile, the philosophical teachings of Aristotle, adhered to by the Islamic astronomers, were adopted

Figure 3.18 North American Native Petroglyphs. *Drawings such as this depict astronomical objects and events, reflecting substantial knowledge of the skies. (E. E. Barnard Observatory, photograph by G. Emerson)*

Figure 3.19 The Bighorn Medicine Wheel. *This prominent and well-studied North American Indian medicine wheel is located in the Bighorn Mountains of Wyoming. (U.S. Forest Service, provided by J. A. Eddy)*

by the scientists of the Christian world as well. Thus, as the sixteenth century dawned, European astronomy was based almost entirely on the teachings and observations of the ancient Greeks.

PERSPECTIVE

We have traced the development of astronomy, beginning with civilization's earliest attempts to understand the natural world. These attempts were based on fear of the unknown and the primitive belief that events in the heavens ruled the inhabitants of the Earth. This kind of astrological outlook emerged independently in many parts of the world, yet only in the vicinity of the Mediterranean, with the rise of the Greek philosophers, did there exist the intellectual stimulus to understand the universe just for the sake of understanding. Political, economic, and, most of all, religious developments then brought about a prolonged intercession, when no real progress was made. Nearly 900 years after the death of Ptolemy, the first stirrings of intellectual revolution began to be felt.

SUMMARY

1. The earliest recorded astronomical data were found in the region known as Babylonia, now Iraq.

2. The first rational inquiry and the earliest cosmological theories arose in the ancient Greek empire, on the shores of the Mediterranean.

3. From 570 B.C., Pythagoras and his followers developed the notion that the universe can be described by numbers, and adopted the belief that the Earth is spherical.

4. Plato and his pupil Aristotle stated underlying principles (which could not be tested by observation) that they believed to govern the universe; their principles included the beliefs that the Earth is the center of the universe and that all heavenly bodies are spherical.

5. In the third century B.C., Aristarchus used geometrical arguments to conclude that the Sun, not the Earth, is at the center of the universe.

6. During the same period, Eratosthenes made an accurate measurement of the size of the Earth, by comparing the directions toward the Sun from two widely separated points on the Earth's surface.

7. Hipparchus, in the second century B.C., made precise observations, applied new mathematical techniques to astronomy, compiled a large star catalog, developed a system for measuring star brightnesses, discovered precession, and developed the epicyclic theory of planetary motion.

8. Ptolemy, in the second century A.D., summarized all astronomical knowledge in the *Almagest,* and used the epicyclic theory to construct tables of planetary motion that were used throughout the Dark Ages.

9. Astronomy in ancient China and India led to significant observational achievements such as the compilation of large star catalogs and the development of accurate calendars, but did not lead to sophisticated cosmological theories.

10. Astronomy in the Americas was highly developed, but few records were left other than the astronomical alignments of buildings and monuments.

REVIEW QUESTIONS

Thought Questions

1. Discuss the relationship between observations of the sky and ancient cultural developments on Earth.

2. Astronomy was studied by ancient cultures for practical and philosophical reasons. Into which of these categories would you put the astronomical interests of the Babylonians? Of the early Greeks? Which reason for studying astronomy do you think is given by most modern astronomers?

3. As far as the records show us, astronomy played some role in the early development of virtually every culture on Earth. Choose a geographical region or society of particular interest to you, and try to find out the extent of astronomical interest and influence in that region or society.

4. What is significant about the early Greek view that there is an underworld, comparable to the heavens?

5. Discuss the relationship between the attitude of Plato toward scientific inquiry and the modern concept of scientific theory, as described in Chapter 1.

6. Even though Aristarchus was correct in asserting that the Sun, not the Earth, was the central body in the solar system, his idea found few followers in his time. Discuss the reasons for this.

7. Discuss the geocentric cosmology of Hipparchus and Ptolemy in the context of our modern view of scientific theory. Did their model conform to the requirement that a theory make predictions that can be tested? How well did it do when tested?

8. In what ways did Ptolemy's epicyclic model obey the "laws of physics" postulated by Aristotle, and in what ways did it violate those laws?

9. Explain in your own words what parallax is, and why nearby stars should undergo parallactic motion as seen from the moving Earth.

10. Why are astronomical developments in the Americas not considered to have been influential in the development of modern astronomy, despite the high level of sophistication reached by early American cultures?

Problems

1. How often did the Babylonians have to add an extra month to their calendar to compensate for their adoption of a calendar with 12 months of 30 days each?

2. How many arcseconds are in a degree? How many arcminutes are in a full circle?

3. Suppose, in the method of Eratosthenes for measuring the circumference of the Earth, that the distance between Syene and Alexandria was 3,300 km, and the zenith angle of the Sun at Alexandria was 30° on the day that the Sun was overhead at Syene. What would you find for the circumference and diameter of the Earth in this case? How do your answers compare with modern values?

4. Hipparchus discovered precession by comparing star positions that he measured with some that were measured 160 years before, and noticing changes in positions. If his measurements had an uncertainty of 20 arcminutes, how much precession had to occur per year in order for Hipparchus to be able to detect it?

5. Refer to Figure 3.10, which shows stellar parallax. If Hipparchus could not measure any angle smaller than 20 arcminutes, how close would a nearby star have to be in order to have a parallax motion that Hipparchus could have detected? (Note: You will need to use simple geometry or trigonometry to solve this.)

ADDITIONAL READINGS

Here are some readings on ancient astronomical history. Others may be found by looking up references cited in these articles and books. It is also useful to browse through journals, such as *Vistas in Astronomy* and the *Journal for the History of Astronomy*, both of which can be found in most science-oriented libraries.

Aveni, Anthony F. 1986. Archeoastronomy: Past, present and future. *Sky and Telescope* 72(5):456.

————. 1984. Native American astronomy. *Physics Today* 37(6):24.

Heath, T. L. 1969. *Greek astronomy.* New York: AMS.

Krupp, E. C. 1978. *In search of ancient astronomies.* Garden City, N.J.: Doubleday.

————. 1983. *Echoes of the ancient skies.* New York: Harper and Row.

Neugebauer, O. 1969. *The exact sciences in antiquity.* New York: Dover.

Piini, Ernest W. 1986. Ulugh Beg's forgotten observatory. *Sky and Telescope* 72(6):542.

MATHEMATICAL METHODS

The Geometry of Aristarchus

 As discussed in the text, Aristarchus believed that the Earth orbited the Sun, because he deduced that the Sun was much larger than the Earth. It is not difficult to understand how he made this deduction, with the use of simple geometry. The accompanying sketch illustrates his reasoning (see next page).

Aristarchus reasoned that when the Moon appears exactly at first quarter or at third quarter, the angle between the Earth–Moon line and the Sun–Moon line must be exactly 90°, as shown, because at these times we see exactly one-half of the Moon to be in sunlight. To deduce the other angles in the triangles shown in the sketch, Aristarchus then had to determine the angle from first-quarter Moon through the Earth to third-quarter Moon (i.e., the angle from M_1 through E to M_2 in the sketch). If this angle could be measured, then all three angles of the triangles M_1ES and M_2ES could be determined, leading in turn to the relative lengths of the sides of these triangles.

The difficult aspect of this technique is that this angle is very nearly 180°; that is, the Sun is so far away that first-quarter and third-quarter lunar phases occur almost on opposite sides of the Moon's orbit. Aristarchus attempted to measure the angle by timing how long it took the Moon to move from first to third quarter, and then from third back to first quarter, reasoning that the ratio of these times equaled the ratio of the angles M_1EM_2 to M_2EM_1.

continued on next page

MATHEMATICAL METHODS

The Geometry
of Aristarchus (continued)

Now that Aristarchus had values for all the angles in the triangles shown, he could determine the relative lengths of the sides, because the ratios of the sides of a triangle are proportional to the opposing angles (that is, if angle ESM_1 is half of angle M_1ES, then side M_1E is half of side M_1S). He found that the Sun–Earth distance (SE) is about 20 times longer than the Earth–Moon distance (EM_1 or EM_2).

The Sun and the Moon have the same angular size as seen from the Earth, so if the Sun is really 20 times farther away, it must be 20 times larger than the Moon. To determine the size of the Earth relative to that of the Sun, Aristarchus next had to determine how much larger or smaller the Earth is

compared to the Moon. To do this, he used observations made during lunar eclipses, when the Earth's shadow falls onto the Moon. The size of the Earth's shadow relative to the diameter of the Moon allowed Aristarchus to construct a scale drawing of the Sun, the Earth, and the Moon, as shown in the accompanying figure.

This reconstruction showed that the Sun had to be at least 6 or 7 times larger than the Earth, and this led Aristarchus to adopt the belief that the smaller Earth must orbit the larger Sun.

While his assumptions were quite correct, Aristarchus had rather poor observational data. The Sun–Earth distance is actually almost 400 times the Earth–Moon distance, and the angle M_1EM_2 is very nearly 180°. Rather than being 20 times larger than the Moon, the Sun is more nearly 400 times larger. The Sun is actually about 110 times larger than the Earth. Thus Aristarchus erred in the sense that he thought the Sun was only a few times larger than the Earth, but this was enough to lead him to the correct conclusion.

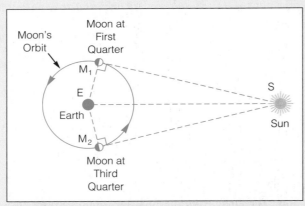

Measuring the Sun–Earth Distance. *By noting that angles SM_1E and SM_2E are right angles at the time of first- and third-quarter phases of the Moon, Aristarchus was able to reconstruct these triangles and determine the relative lengths of sides M_1E (or M_2E) and SE.*

The Relative Sizes of the Earth and the Moon. *Observations of the Earth's shadow on the Moon allowed Aristarchus to determine that the Moon is about 3/8 the size of the Earth. Combining this information with his knowledge of the relative sizes of the Moon and the Sun led Aristarchus to the conclusion that the Sun is much larger than the Earth. (Not to scale.)*

CHAPTER 4

The Renaissance

A Renaissance model of the epicylic cosmology. (© 1981 O. Gingerich, with permission of the Houghton Library, Harvard University)

The fifteenth century saw the beginnings of a re-awakening of intellectual spirit in Europe. Some important scientific studies began at the major universities, increased maritime explorations brought demands for better means of celestial navigation, and the art of printing was discovered, opening the way for widespread dissemination of information.

COPERNICUS: THE HELIOCENTRIC VIEW REVISITED

Some 19 years before the epic voyage of Columbus, Mikolaj Kopernik (Fig. 4.1) was born in Torun, in the northern part of what is now Poland. As a young man he attended the university in Kracow, where his entrancement with Latin, the universal language of scholars, led him to change his surname to Copernicus. At Kracow he nourished an abiding interest in astronomy, becoming fully acquainted with the Aristotelian view as well as the Ptolemaic model of the planetary motions.

Copernicus continued his academic pursuits in Italy, studying ecclesiastical law at Bologna and Ferrara, and medicine at Padua, before returning to Poland to assist his uncle, the bishop (and therefore chief government administrator) of a region called Ermland. During all this time, he persisted in his studies of astronomy, and by 1514 he had developed some doubts about the validity of the accepted system.

His reasons for doing so have been the subject of some uncertainty and misconceptions. It was long assumed that Copernicus was encouraged to adopt the Sun-centered view of the universe because he recognized shortcomings in the geocentric model of Ptolemy. There is, however, no evidence of widespread dissatisfaction with the Ptolemaic system, nor are there any records indicating that Copernicus himself found serious inaccuracies in it. His reasons for adopting the heliocentric viewpoint were more subtle.

The basis for his conversion was primarily philosophical. The new system, in the mind of Copernicus, presented a pleasing and unifying model of the universe and its motions. Although he was, no doubt, encouraged by the climate of change and cultural revolution that was sweeping Europe with the advent of the Renaissance, he did not adopt the heliocentric hypothesis just to be different, nor did he do it to improve the accuracy or reduce the complexity of the accepted view. The Copernican model was, in fact, no more accurate in predicting planetary motions than was the Ptolemaic system, and it was nearly as complex. Copernicus adhered to the notion of perfect circular motions and was obliged to include small epicycles in order to match the observed planetary orbits. Apparently, one of the most pleasing aspects of the Sun-centered system to Copernicus was the fact that the relative distances of the planets could be deduced (Fig. 4.2) and were found to have a certain regularity, the spacings between planets growing systematically with increasing distance from the Sun. Copernicus was also able to determine the

Figure 4.1 Nicolaus Copernicus. *The view of Copernicus that the Sun, rather than the Earth, occupied the central place in the cosmos had great influence on later scientists. (The Bettmann Archive)*

relative speeds of the planets in their orbits, finding that each planet moves more slowly than the next one closer to the Sun.

Copernicus first circulated his ideas informally sometime before 1514, in a manuscript called *Commentariolis,* and it drew increasing attention during the next several years. The Church voiced no opposition, despite the fact that the ideas expressed in the work strongly contradicted commonly accepted Church doctrine. Copernicus was reluctant to publish his findings in a more formal way, for fear of raising controversy, and was continually rechecking his calculations. Eventually a young Protestant scholar known as Rheticus encouraged him to publish his work. Having been persuaded of the validity of the heliocentric theory, Rheticus had published his own summary of the ideas of Copernicus, and then delivered the manuscript written by Copernicus to a Lutheran priest named Osiander, who assumed the responsibility for publishing it.

Fearful of strong religious opposition, Osiander added an unsigned preface in which he stated that the ideas within did not represent physical reality, but should be used only as a mathematical tool for predicting planetary motions. There is no doubt that Copernicus

himself believed in the literal truth of his heliocentric model, however. The book was titled *De Revolutionibus Orbium Coelestium (On the Revolution of the Celestial Sphere).* Copernicus died at just about the time his work was published, so he never knew of its impact upon the world.

The impact was profound, although not immediately. The many important improvements in understanding the solar system that were embodied in the new theory eventually made a strong enough impression on other scientists to win them to this view. Among the persuasive aspects were the natural ordering of the planets (the relationship between orbital period and distance from the Sun) and the simple explanation for the differing sizes of the retrograde loops that the planets go through. The extent of the backward motion observed for each planet, if it is a result of the Earth's relative motion as Copernicus believed, depends on how far away the planet is. Copernicus also was able to explain the cause of the seasons, and made the daring suggestion that the reason stellar parallax could not be observed was that the stars are simply too far away to have perceptible displacements resulting from the Earth's motion around the Sun.

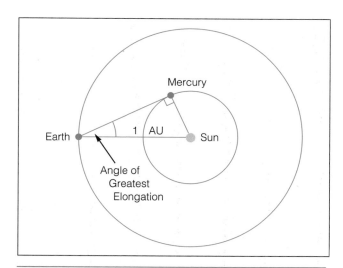

Figure 4.2 The method of Copernicus for finding relative planetary distances from the Sun. *For an inferior planet such as Mercury or Venus, Copernicus knew the angle of greatest elongation and was therefore able to reconstruct the triangle shown, which provided the Sun-Mercury (or Sun-Venus) distance relative to the Sun-Earth distance (that is, the astronomical unit). For superior planets, similar but slightly more complicated considerations provided the same information.*

TYCHO BRAHE: ADVANCED OBSERVATIONS AND A RETURN TO A STATIONARY EARTH

Although Copernicus had contributed little observationally and had not demanded a close match between theory and observation as long as general agreement was found, there was, in fact, a strong need for improved precision in astronomical measurements. The next influential character in the historical sequence filled this need; if he had not done so, further progress would have been seriously delayed, as would have the eventual acceptance of the Copernican doctrine.

Tycho Brahe (Fig. 4.3) was born in 1546, some 3 years after the death of Copernicus, in the extreme southern portion of modern Sweden (at the time the region was part of Denmark). Of noble descent, Tycho spent his youth in comfortable surroundings, and was well educated, first at Copenhagen University, then at Leipzig, where he insisted on studying mathematics and astronomy despite his family's wish that he pursue a law career. Interruptions created by war and plague caused Tycho to change universities several times, with

Figure 4.3 Tycho Brahe. *Tycho's principal contribution, a massive collection of accurate observations of planetary positions, was crucial to subsequent advances in understanding planetary motions. (The Bettmann Archive)*

most of his formal training eventually taking place at the university in Wittenberg. He developed a strong reputation as an astronomer and as an astrologer (the distinction was still scarcely recognized), and attracted the attention of the Danish King, Frederick II. In 1575 the king ceded to Tycho the island of Hveen, about 14 miles north of Copenhagen, along with enough servants and financial assistance to allow him to build and maintain his own observatory (Fig. 4.4).

Even before this time, Tycho had shown an acute interest in astronomical instruments, and with the grant to build his observatory, this interest bore fruit. He devised a variety of instruments which, although they really did not encompass any new principles, were capable of more accurate readings than any before his time. The largest was a quadrant built into a wall of the observatory (Fig. 4.5), with a 31½-foot radius. A quadrant is a quarter circle with graduated markings indicating degrees and minutes of arc around its circumference. A sight located at the center allows the observer to read the angular heights of celestial objects above the horizon. The principal improvements of Tycho's instruments over those of his contemporaries were that the angular scales were more precisely marked, and that a new sighting design allowed the scales to be read more accurately.

Tycho made two specific observational discoveries of importance, in addition to amassing a large body of data on planetary motions. One, the observation of a supernova, occurred in 1572, and the other, the observation and analysis of a bright comet, took place in 1577, soon after the observatory became active. By comparing his data with those of astronomers elsewhere, Tycho was able to determine in both cases that the new object was very distant, well beyond the orbit of the Moon (Fig. 4.6). This showed that the heavens were not perfect and immutable. The comet was particularly significant, because Tycho's analysis showed that it passed right through the crystalline spheres of the planets, demonstrating that they could not be the solid material objects supposed to exist by the ancient Greeks. Tycho's analysis of the supernova of 1572 was also quite significant, for it was in recognition of this achievement that King Frederick II granted Tycho the

Figure 4.4 Detail of Tycho's observatory at Hveen. *(Photo Researchers, Inc.)*

ASTRONOMICAL INSIGHT 4.1

The Accuracy of Angular Measurements

 We have referred often to measurements of planet and star positions without saying much about how these measurements are made, or about their accuracy. Throughout the history of astronomy, the quest for ever more accurate measurements of positions has been constant. In the earliest times, positional measurements, particularly of the Sun, were needed for timekeeping purposes. As scientists and philosophers began to probe the nature of the universe by attempting to deduce the motions of the planets, positional measurements for the Sun and planets became increasingly important. As we learned in Chapter 3, Hipparchus (c. 150 B.C.) was obliged to introduce the notion of the equant into his epicyclic model of the cosmos because positional measurements revealed errors in the model. In this chapter we have seen that Tycho Brahe's (c. 1600 A.D.) positional measurements were sufficiently accurate to allow Johannes Kepler to deduce the elliptical shape of planetary orbits, but not accurate enough to reveal stellar parallax.

The accurate measurement of positions in the sky amounts to measuring angles with the best precision possible. The earliest devices for doing this were simple sighting instruments used to determine the height of the Sun above the horizon, often by noting the angle made by the shadow of a thin object such as a peg. This angle could be measured directly, through the use of a quadrant (a device much like the protractor that you used in geometry class), or by measuring the length of the shadow and using trigonometry to deduce the direction toward the Sun. The earliest such techniques produced only rather crude results, with uncertainties of 1° or more. (Keep in mind that by our ordinary standards a degree is a rather small angle; to see this, try to draw a 1° angle, using a protractor.)

By the time of Hipparchus, around 150 B.C., somewhat better accuracies (within 20 arcminutes) were achieved, though the techniques were rather similar. The next major advance came some 1,500 years later, when metalworking technology had been developed, and it became possible to engrave angular scales on metal. Tycho in particular pushed this development, building ever-larger quandrants. A **quadrant** was a device with a pointer that could pivot so that the pointer could be aimed directly at the target object (see Fig. 4.5). The angle from the horizon to the target could then be read off of a scale engraved on a quarter-circular arc of brass. Tycho's largest quadrant was attached to a fixed wall of his observatory, but others were able to swivel around a vertical post, so that objects in any direction could be measured. The advantage of its large size was that the markings on the metal arc representing degrees and minutes would be farther apart and therefore easier to read accurately. Through the use of such devices (and subsequent averaging of multiple measurements), Tycho's observations produced positional measurements accurate to about 1 arcminute. This was sufficient to allow Kepler to determine that the orbit of Mars is an ellipse, even though he found other shapes that reproduced the observed motion of Mars to within a few arcminutes.

The human eye cannot perceive angles smaller than about 1 arcminute, so no further advance could be made until telescopes, which offer greater angular refinement, were available. Within a century following the time of Tycho Brahe, telescope technology made great advances. Measuring angular positions directly from the mounting of the telescope became possible. Most telescopes are mounted so that they can rotate on two axes, one parallel with the Earth's axis, and the other at right angles to the first. Then the position of a star in the usual two coordinates (right ascension and declination; see Chapter 2) corresponds directly to the amount by which each axis of the telescope is turned. By the beginning of the eighteenth century, positions could be measured in this way to an accuracy of better than 10 arcseconds.

To make substantial improvements over this, astronomers developed methods for measuring *relative* positions with great accuracy. Now, instead of measuring the position of a planet or star with respect to the Earth's reference frame, astronomers found that angular separations between objects in the sky could be measured more precisely. Selected stars were chosen as standards for reference in the determination of positions of others. (Today these

(continued on next page)

ASTRONOMICAL INSIGHT 4.1

The Accuracy of Angular Measurements (continued)

standards are selected and periodically revised by an international panel of astronomers.) At first the rotation axes of the telescope were used for this, as cross hairs in the eyepiece were set alternately on the target star and on one or more nearby standards, and the amount of rotation was measured (the results of multiple measurements then being averaged). This kind of technique finally allowed stellar parallax to be measured, in 1838 (by no less than three independent observers of three different stars!). The largest parallax angle, for the closest star to the Earth, is less than 1 arcsecond, so the precision of po-

sitional measurements had indeed improved dramatically from the ancient times. (To envision just how small an angle of 1 arcsecond is, try to imagine how big a dime would look as seen from a distance of two miles.)

Soon photographic techniques began to be used in astronomy to measure positions. The measurement consisted of determining the physical distance between star images on a photographic plate where the plate scale (the number of arcseconds per millimeter) is known accurately. This has the advantage that many stars could be measured through the use of only a small amount of telescope time. Modern photographic techniques are capable of measuring separations of stars to an accuracy of 0.01 arcsecond or a little better, and this seems to be about as well as can be done by measuring ordinary stellar images, which

themselves are usually about an arcsecond in diameter (because of blurring caused by the Earth's atmosphere).

Nevertheless, astronomers are developing capabilities for even more accurate positional measurements. These generally involve either placing telescopes in space, to avoid the blurring effects of the Earth's atmosphere, or making use of the complex wave properties of light, something we will discuss later (see the discussion of interferometry in Chapter 7). Today accuracies of 0.001 arcsecond appear feasible, and some scientists are beginning to think of even better precision. Throughout the rest of this text, you will learn of the importance of ever-better capabilities for measuring angles precisely, for it is one of the everlasting challenges to astronomy.

resources to build and operate his observatory. The supernova also provided evidence against the accepted doctrine, which held that the heavens were immutable.

Tycho was unable to accept the heliocentric view, primarily because he could find no evidence that the Earth was moving. He tried and failed to detect stellar parallax, which he supposed he should see with his accurate observations if the Earth really moved. Furthermore, as a strict Protestant, he found it philosophically difficult to accept a moving Earth, when the Scriptures stated that the Earth is fixed at the center of the universe.

On the other hand, he realized that the Copernican system had advantages of mathematical simplicity over the Ptolemaic model, and in the end he was ingenious enough to devise a model that satisfied all of his criteria. He imagined that the Earth was fixed, with the

Figure 4.5 The great mural quadrant at Hveen. *This instrument was used to measure the angular positions of stars and planets with respect to the horizon. Its large size made the angle markings easy to read accurately. (Photo Researchers, Inc.)*

Figure 4.6 Tycho's analysis of the comet of 1577.
Because the direction to the comet as seen from separate locations (A and B) was the same, whereas for an atmospheric object like a cloud the direction varied, Tycho showed that the comet was far beyond the Earth's atmosphere and belonged to the realm of the planets and stars.

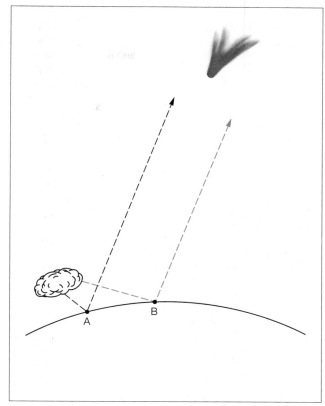

Sun orbiting it, but that all the other planets orbited the Sun (Fig. 4.7). Mathematically, this is equivalent to the Copernican system in terms of accounting for the motions of the planets as seen from the Earth. The idea never received much acceptance, however, and Tycho is remembered primarily for his fine observations.

His observational contribution consisted, in part, of the unprecedented accuracy of his data and, in part, of the completeness of his records. Until his time, it had been the general practice of astronomers to record the positions of the planets only at notable points in their travels, such as when a superior planet comes to a halt just before beginning retrograde motion. Tycho made much more systematic observations, recording planetary positions at times other than just the significant turning points in their motions. He also made

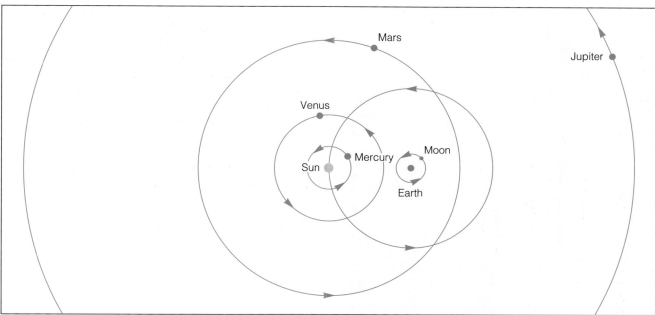

Figure 4.7 Tycho's model of the universe.   *Tycho held that the Earth was fixed, and the Sun orbited the Earth. The planets in turn orbited the Sun. This system was never worked out in mathematical detail, but it successfully preserved both the advantages of the Copernican system and the spirit of the ancient teachings. It also accounted for the lack of observed stellar parallax, since the Earth was fixed.*

multiple observations in many instances, allowing the results to be averaged together to improve their accuracy. Tycho himself did not attempt any extensive analysis of his data, but the vast collection of measurements that he gathered over the years contained the information needed to reveal the basis of the planetary motions. The task of unraveling the secrets contained in Tycho's data was left to those who followed him, particularly a young astronomer named Johannes Kepler.

KEPLER AND
THE LAWS OF PLANETARY MOTION

Until now it has been possible to follow developments in a straight sequence, but here we must begin to cover events that occurred at nearly the same time, but in different locations. To complete the thread begun with our discussion of Tycho Brahe, we now turn to Johannes Kepler (Fig. 4.8), who worked briefly at the side of Tycho and then spent many years analyzing the great wealth of observational data that Tycho had

Figure 4.8 Johannes Kepler.   *Kepler's fascination with numerical relationships led him to discover the true nature of planetary motions. (The Bettmann Archive)*

accumulated. We must keep in mind, however, that during the same period a scientist and philosopher named Galileo Galilei was at work to the south, in Italy, and that the two were aware of each other's accomplishments.

Kepler was born in 1571 in Weil, Germany, a year before Tycho's supernova was observed. A sickly youngster, he seemed headed for a career in theology, but he had difficulty conforming to the fundamentalist religious philosophy at the university he attended in Tübingen. While there he encountered a professor of astronomy who inspired in him a strong interest in the Copernican system.

Having completed his education, Kepler accepted a teaching position in the Austrian city of Graz in 1594. While there, he published his first scientific work, a book called the *Mysterium Cosmographicum,* in which he outlined several arguments favoring the Copernican theory and stated his own view that the distances of the planets from the Sun were determined by a series of regular geometric solids separating the crystalline spheres of the planets (Fig. 4.9). Kepler's reasons for adopting this notion and the Copernican system as well were based on his love of mathematical simplicity and his intuitive feeling that the universe was constructed according to a harmonious, unified plan. He devoted his life to seeking out the underlying principles of this universal harmony, and, as we shall see, this led him to a number of profound discoveries. The hypothesis put forth in the *Mysterium Cosmographicum* was, of course, not correct, but the book had a major impact anyway, especially because of its eloquent arguments in favor of the Sun-centered universe.

In 1598 a decree banning all Protestants forced Kepler out of Graz and, after first visiting extensively, in 1600 he accepted a post as chief assistant to Tycho, who had by then moved his observatory to Prague following the death of his sponsor, King Frederick II of Denmark. Tycho died the next year, and Kepler succeeded him as court mathematician to Emperor Rudolph.

Kepler's main mission, owing to both Tycho's wishes and his own, was to develop a refined understanding of the planetary motions and to upgrade the tables used to predict their positions. He set to work first on the planet Mars, for which the data were particularly extensive, and whose motions were among the most difficult to explain in the established Ptolemaic system (and, in fact, in the Copernican system, with its requirement of only circular motions). By a very complex process, Kepler was able to separate the effects of the

Earth's motion from those of Mars itself, so that he could map out the path that Mars followed with respect to the Sun.

He soon realized that the Sun must play a central role in making the planets move, and this inspired him to look for a physical link between the Sun and the planets. In this sense, he was like a modern scientist and quite unlike other astronomers of his time and before. Whereas they sought only a mathematical device for predicting motions, Kepler believed in a true physical machinery of the solar system, and he sought the underlying principles.

The data on Mars led Kepler eventually to the realization that the Sun could not be at the true center of the planet's orbit. He experimented with various geometric shapes, including epicycles, which he became convinced were contrived and not likely to represent physical reality. At one point, he was forced to reject a system he had devised because of a small error in predicting the position of Mars, an error that would have been acceptable to anyone else who had worked with the planetary motions.

By 1604, Kepler had determined that the orbit of Mars was some kind of oval, and further experimentation revealed that it was fitted precisely by a simple geometric figure called an **ellipse** (Fig. 4.10). This is a closed curve defined by a fixed total distance from two points called **foci** (singular: *focus*), and indeed Kepler found that the Sun was at the precise location of one focus of the ellipse (Fig. 4.11).

Further analysis of the motion of Mars revealed a second characteristic, having to do with the fact that the planet moves fastest in its orbit when it is nearest the Sun, and slowest when it is farthest from the Sun. Mathematically, Kepler's second discovery was that a line connecting Mars to the Sun sweeps out equal areas of space in equal intervals of time (see Fig. 4.11).

These two properties of the orbit of Mars were later generalized as Kepler's first two laws of planetary motion, applicable to all the planets. Kepler's results on the orbit of Mars were published in 1609 in a book entitled *Astronomica Nova,* or *The New Astronomy: Commentaries on the Motions of Mars.* The book received a great deal of attention.

Figure 4.9 Kepler's nested geometric solids. *Kepler expended considerable effort in exploring the possibility that the planetary distances from the Sun were determined by the spacings of nested regular geometric solids (that is, figures whose faces were regular polyhedrons; the number of gaps between possible solids of this sort was equal to the number of known planets). Kepler eventually abandoned this particular scheme, but never relented in his search for regularity in the solar system. (O. Gingerich)*

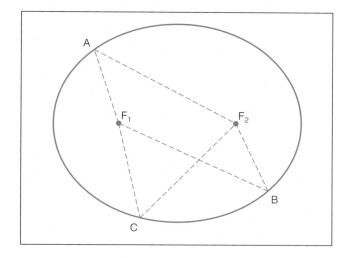

Figure 4.10 The ellipse. *From any point on this closed curve, the sum of the distances to the two foci is always the same. Thus, in this example, the distances F_1AF_2, F_1BF_2, and F_1CF_2 are all equal.*

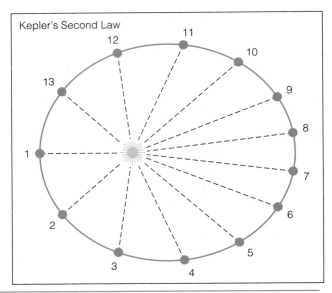

Figure 4.11 The ellipse. *At left is an exaggerated ellipse representing a planetary orbit with the Sun at one focus; at right is a similarly exaggerated ellipse with lines drawn to illustrate Kepler's second law. If the numbers represent the planet's position at equal time intervals, the areas of the triangular segments are equal.*

Kepler's first two laws of planetary motion, in general terms, are (1) The orbits of the planets are ellipses, with the Sun at one focus; and (2) for each planet, the area swept out in space by a line connecting that planet to the Sun is equal in equal intervals of time.

Most of the planets have orbits that are more nearly circular than is that of Mars, and it is indeed fortunate that Kepler chose to work with the Mars data first, or he might not have been able to discover his first two laws.

In 1612 Emperor Rudolph, who had been supporting Kepler's research, died, and Kepler moved to Linz, in northern Austria. There his work on the planetary motions continued, as he still sought the true nature of the relationship among the planets. In 1619 he published a book entitled *Harmonica Mundi*, or *The Harmony of the World*, in which is reported his discovery of a simple relationship between the orbital periods of the planets and their average distances from the Sun. Now known as Kepler's third law, or simply, the harmonic law, it states that the square of the period of a planet is proportional to the cube of the semimajor axis (which is half of the long axis of an ellipse). In other words, this says that $P^2 = a^3$, where P is the sidereal period of a planet in years, and a is the semimajor axis in terms of the average Sun–Earth distance (that is, in terms of the **astronomical unit** or *AU*; see Table 4.1).

Table 4.1 Testing Kepler's third law

Planet	*a* (AU)	*P* (years)	*a³*	*P²*
Mercury	0.387	0.241	0.058	0.058
Venus	0.723	0.615	0.378	0.378
Earth	1.000	1.000	1.000	1.000
Mars	1.524	1.881	3.533	3.538
Jupiter	5.203	11.86	140.85	140.66
Saturn	9.555	29.46	867.98	867.89
Uranus	19.22	84.01	7055.79	7057.68
Neptune	30.11	164.8	27162.32	27159.04
Pluto	39.44	248.5	61349.46	61752.25

In another major work called the *Epitome of the Copernican Astronomy*, published in parts in 1618, 1620, and 1621, Kepler presented a summary of the state of astronomy at that time, including Galileo's discoveries. In this book Kepler generalized his laws, explicitly stating that all the planets behaved similarly to Mars, something that had clearly been his belief all along.

By the time the *Epitome* was published, the Roman Catholic Church had joined the Lutheran one in taking a very intolerant stance toward what were considered heretical teachings, in contrast with the situation at the time of Copernicus, nearly 100 years before. Kepler's treatise soon found itself in the *Index of Prohibited Books,* along with *De Revolutionibus.*

In 1627, Kepler published his last significant astronomical work, a table of planetary positions based on his laws of motion, which could be used to predict accurately the planetary motions. These tables, which he called the *Rudolphine Tables* in honor of his former benefactor, were used for the next several years. The *Rudolphine Tables* represented an improvement in accuracy over any previous tables by nearly a factor of 100, a resounding and remarkable confirmation of the validity of Kepler's laws. In a very real sense, the *Rudolphine Tables* represented Kepler's life's work, since, with their publication, he completed the task set before him when he first went to work for Tycho. Kepler died in 1630, at the age of 59.

GALILEO, EXPERIMENTAL PHYSICS, AND THE TELESCOPE

Very strong contrasts can be drawn between Kepler and his great contemporary, Galileo (Fig. 4.12). Whereas Kepler was fascinated with universal harmony and therefore with the underlying principles on which the universe operates, Galileo was primarily concerned with the nature of physical phenomena and was less devoted to finding fundamental causes. Galileo wanted to know how the laws of nature operated, whereas Kepler sought the reason for the existence of the laws. Galileo's approach was level-headed and rational in the extreme. He used simple experiment and deduction in advancing his perception of the universe. He has frequently been cited as the first truly modern scientist, although others of his time probably deserve a share of that recognition.

Galileo was born in 1564 in Pisa, in northern central Italy. Although his family had intended for him a career in a trade, his obvious intellectual abilities were soon recognized, and in 1581 he began formal training at the university in Pisa. There, his reluctance to accept dogmatic ideas without question was soon noted, and he earned the nickname "The Wrangler" for his tendency toward debate. Poverty forced him to live at home for a while following his fourth year of study, and for some time he worked at a teaching post in Pisa. In 1592 he moved to the University of Padua, where he was to stay until 1610.

During his early academic years, he carried out numerous experiments in physics, discovering many simple laws, and in the process completely overturning the

Figure 4.12 Galileo Galilei. *From his numerous experiments and observations, Galileo deduced the nature of the cosmos. His methodical attacks on traditional beliefs caused him personal troubles but greatly influenced contemporary thinking. (The Granger Collection)*

accepted method of studying science. Whereas a follower of Plato and Aristotle (whose works still dominated in Galileo's time) would proceed by rational thought from standard unproven assumptions, Galileo placed great emphasis on experimental and observational confirmation. When Galileo argued a point, he did so on the basis of what could be demonstrated, although he certainly mixed in a large share of logical deduction. To him, the supreme test of a hypothesis was whether it fit the observed phenomena, not whether the observed phenomena fit a previously adopted hypothesis. In developing this approach, Galileo founded an entirely new basis for scientific inquiry, an achievement in many ways more profound than his contributions to astronomy, which were considerable.

Galileo's discoveries in physics, having to do with the motions of objects, were published in his later years, after his astronomical career had been forcibly ended by Church decree. It was, in fact, an early interest in **mechanics,** the science of the laws of motion, that lured Galileo away from a career in medicine, the subject of his first studies.

Among his most important contributions to mechanics was his elucidation of a principle called **inertia,** the tendency of a moving object to remain in motion or of an object at rest to resist being moved. Galileo was the first to clearly recognize this natural tendency and to understand that an object sliding along a surface stops only because the force of friction acts upon it. He later applied this idea to the motion of the Earth, in explaining how the atmosphere and all the objects on its surface could move along with it.

Galileo also understood the concept of **acceleration,** the rate of change of velocity, and showed that falling objects are subject to a constant acceleration, regardless of their **mass,** the total quantity of matter they contain. Supposedly this point was made in a graphic manner when Galileo is said to have dropped balls of unequal weight from the Leaning Tower of Pisa, demonstrating to a crowd of onlookers that the balls hit the ground at the same time. There is some doubt as to whether, or under what circumstances, this experiment was actually performed, although it is certain that Galileo did carry out comparable experiments, perhaps without attendant ceremony.

Galileo's other contributions to mechanics include the discovery that the oscillation period of a pendulum is approximately independent of how far it swings, which showed that it could be used as a timekeeping device. (Galileo began to build a pendulum-driven clock in his later years, but was forced to abandon the project because of failing eyesight).

Galileo's astronomical discoveries were sufficient to earn him a major place in history, even if he had never done his work in mechanics. It was his astronomical studies, along with his flair for debate and his habit of ridiculing those with whom he disagreed, that made him famous (though not universally loved) in his own time.

His first astronomical contribution was the observation of a supernova in 1604, one also seen by Kepler. Both men showed that its lack of parallax placed it in the distant realm of the stars.

In 1609 Galileo learned of the invention of the telescope, and devised an instrument of his own, which

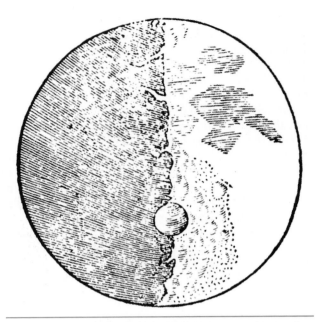

Figure 4.13 Galileo's sketch of the Moon.   *Galileo's observation of mountains, craters, and "seas" led him to believe that the Moon was not the perfect heavenly body envisioned in the ancient Greek teachings. (The Granger Collection)*

he soon put to use by systematically viewing the heavens with it. Despite the poor quality of the instrument, Galileo made a number of important discoveries almost at once from his observation tower at Padua and reported them in 1610 in a publication called *Sidereus Nuncius* ("The Starry Messenger"). Here Galileo showed that the Moon was not a smooth sphere, but instead was covered with craters and mountains (Fig. 4.13). He also reported that the Milky Way, previously thought to be a cloud, actually consisted of countless stars, contrary to the ancient teachings. Most significant of all, he showed that Jupiter was attended by four satellites, whose motions he observed long enough to establish that they orbited the parent planet (Fig. 4.14). All these discoveries, but especially the latter, violated the ancient philosophies of an idealized universe with the Earth at the center. The satellites of Jupiter showed beyond reasonable question that there were centers of motion other than the Earth.

Once Galileo had begun his astronomical observations, he became fascinated with the structure of the universe, and for a time his work in mechanics was put aside. He spent much of the next several years in pursuit of evidence for the Copernican model (Table 4.2).

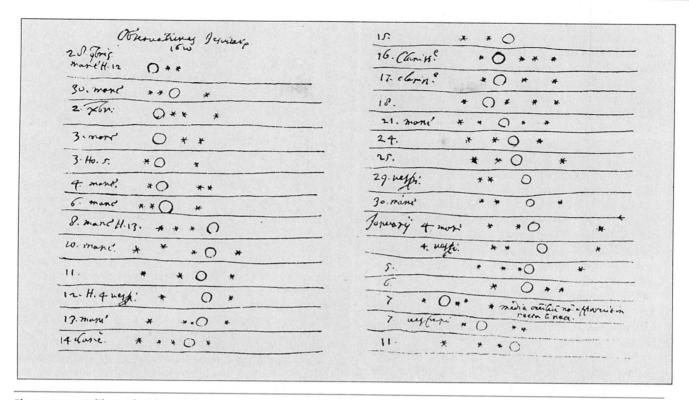

Figure 4.14 Galileo's sketches of the moons of Jupiter. *This series of drawings by Galileo of Jovian observations is often attributed to* The Starry Messenger, *but in fact was made some years later. Galileo's discovery of moons orbiting Jupiter showed that there are heavenly bodies that do not orbit the Earth. (Yerkes Observatory)*

Mostly as a result of the reputation Galileo earned with the publication of *Sidereus Nuncius,* he was able to negotiate successfully for the position of court mathematician to the Grand Duke of Tuscany, and in 1610 he moved to Florence, where he was to spend the remainder of his long career. Once established there, Galileo continued his observations and soon added new discoveries to his list. The first of these was the varying aspect of Venus as its position changed with respect to the Sun. Galileo quickly realized that this was analogous to the phases of the Moon—we see from the Earth varying portions of the sunlit side of Venus (Fig. 4.15). This had two important implications: It showed that Venus shines by reflected sunlight rather than by its own power; and it demonstrated that Venus orbits the Sun instead of the Earth, since the apparent size of Venus varies (from being smallest when Venus is fully sunlit to being largest when the dark side of Venus is facing the Earth).

At about the same time, Galileo began to study the dark spots that occasionally appeared on the face of

Table 4.2
Some of Gallileo's arguments for heliocentric theory

Discovery	Argument
Many faint stars	Difficult to reconcile with idea of stars as points attached to crystalline sphere
Craters on the Moon	Moon is not perfect, immutable heavenly body
Moons of Jupiter	A body other than Earth as center of motion
Phases of Venus	Explained only if Venus orbits the Sun, and shines by reflected sunlight
Sunspots	Spots are on solar surface, showing that the Sun is not perfect
Variable planetary sizes	Angular size variations explained by motion of planets around Sun

the Sun. These had been independently seen by a number of astronomers, and there was substantial controversy as to what they were. Some thought the spots to

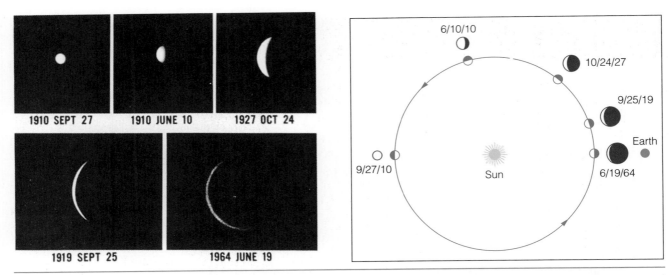

Figure 4.15 The phases of Venus.   *As Venus orbits the Sun and its position changes relative to the Sun–Earth line, its phase varies as we see differing portions of its sunlit side. In addition, its apparent size varies because of its varying distance from Earth. (Photos from NASA)*

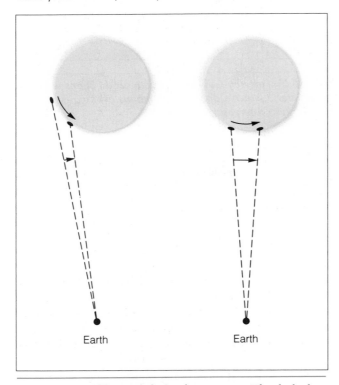

Figure 4.16 Galileo's analysis of sunspots.   *The dashed lines indicate the apparent (angular) motion of a sunspot, as seen from Earth, during two equal time intervals. The variation in observed speed of a sunspot led Galileo to argue that sunspots actually lie on the Sun's surface and were not shadows of planets passing between the Earth and the Sun.*

be small planets close to the Sun that moved across its face during their orbital motion, but Galileo showed that they must actually be spots on the surface of the Sun. He did this by noting that the spots moved very slowly when near the edge of the Sun's disk, which he attributed to foreshortening as the Sun's rotation carried them toward the Earth on the approaching edge of the Sun and away from the Earth on the receding edge, with significant sideways motion only when they moved across the central portion of the disk (Fig. 4.16). That the Sun should have blemishes on its surface, and rotate as well, were contrary to the established view.

In the years immediately following these discoveries, Galileo began to draw increasingly heavy criticism from the Church, and he made efforts to develop good relations with high-ranking officials in Rome. Nevertheless, in 1616 he was pressured into refuting the Copernican doctrine, and for several years thereafter was relatively quiet on the subject. Except for a well-publicized debate on the nature of comets, Galileo spent most of his time preparing his greatest astronomical treatise, which was finally published, after some difficulties with Church censors, in 1632. To avoid direct violation of his oath not to support the Copernican heliocentric view, Galileo wrote his book in the form of a dialogue among three characters, one of whom, named Simplicio, represented the official position of the Church; another, Salviati, that of Galileo (although this was, of course, not stated explicitly); and third, Sagredo, who was always quick to understand and

ASTRONOMICAL INSIGHT 4.2

An Excerpt from the *Dialogue on the Two Chief World Systems*

Galileo got into trouble with the Roman Catholic Church when he portrayed the official view of the nature of the universe as not only incorrect, but illogical and even silly. It might help us to see how inflammatory *Dialogue on the Two Chief World Systems* was by reading a small sample from it.

During the course of three days of discussion, the three characters in Galileo's *Dialogue* thoroughly air all the available evidence and arguments bearing on the understanding of mechanics and the structure and nature of the universe. About midway through, a discussion develops that is central to Galileo's principal point: the Sun, not the Earth, is at the center of the universe. In the following excerpt,* we see an example of Salviati's persuasive style, Simpli-

cio's dogged reluctance to give up the ideas of Aristotle, and Sagredo's ready comprehension of Salviati's arguments.

Salviati: *Now if it is true that the center of the world is the same about which the circles of the mundane bodies, that is to say, of the Planets, move, it is most certain that it is not the Earth but the Sun, rather, that is fixed in the center of the World. So that as to this first simple and general apprehension, the middle place belongs to the Sun, and the Earth is as far remote from the center as it is from that same Sun.*

Simplicio: *But from whence do you argue that not the Earth but the Sun is in the center of planetary revolutions?*

Salviati: *I infer the same from the most evident and therefore necessarily conclusive observations, of which the most potent to exclude the Earth from the said center, and to place the Sun therein, are that we see all the planets sometimes nearer and sometimes farther off from the Earth, with so great differences, that, for example, Venus when it*

is at the farthest is six times more remote than when it is nearest, and Mars rises almost eight times as high at one time as at another. See therefore whether Aristotle was somewhat mistaken in thinking that it was at all times equidistant from us.

Simplicio: *What in the next place are the tokens that their motions are about the Sun?*

Salviati: *It is shown in the three superior planets, Mars, Jupiter, and Saturn, in that we find them always nearest to the Earth when they are in opposition to the Sun and farthest off when they are towards the conjunction, and this approximation and recession imports thus much, that Mars near at hand appears sixty times greater than when it is remote. As to Venus, in the next place, and to Mercury, we are certain that they revolve about the Sun in that*

(continued on next page)

agree with Salviati's arguments. In this treatise, called *Dialogue on the Two Chief World Systems,* Galileo, through the character Salviati and at the expense of Simplicio, systematically destroyed many of the traditional astronomical teachings of the Church. The book was published in Italian, rather than in the scholarly Latin, and its contents were therefore accessible to the general populace.

Despite a lengthy preface in which Galileo disavowed any personal belief in the heliocentric doctrine, the Church reacted strongly, and within a few months Galileo was summoned before the Roman Inquisition, while existing copies of the book were heavily censored

and further publication was banned. It joined the works of Copernicus and Kepler on the *Index* (where it was to remain until the 1830s), and Galileo himself was sentenced to house arrest, eventually serving his punishment at his country home near Florence. That is where he spent the remainder of his days.

Before his life came to a close, however, Galileo published one more major work, a summary of his studies of mechanics, based largely on experiments done many years before. The book on mechanics appeared in 1638, and Galileo died in 1642, having suffered blindness the last 4 years of his life.

ASTRONOMICAL INSIGHT 4.2

An Excerpt from the *Dialogue on the Two Chief World Systems** (continued)

they never move far from it, and in that we see them sometimes above and sometimes below it, as the mutations in the figure of Venus necessarily prove. Touching the Moon, it is certain that it cannot in any way separate itself from the Earth, for the reasons that shall be more distinctly alleged hereafter.

Sagredo: I expect that I shall hear more admirable things that depend upon this annual motion of the Earth than were those dependent upon the diurnal revolution.

In this exchange, Galileo, in the guise of Salviati, advances several arguments based on observations, including the well-known phases of Venus, and the less widely quoted arguments having to do with the varying distances of the planets from the Earth. In this and other passages, Galileo went out of his way to mock the followers of Aristotle, who, according to Salviati in an earlier paragraph, "would deny all the experiences and all the observations in the world, nay, would refuse to see them, that they might not be forced to acknowledge them, and would say that the world stands as Aristotle writes and not as Nature will have it."

The Church did not take the drastic action of excommunicating Galileo for expressing his views in the *Dialogue,* but it did take severe measures, nevertheless. There the matter stood, as scientific and

religious attitudes evolved over the subsequent centuries. In a spirit of reconciliation with science and in the interest of clearing up the clouded past, the Roman Catholic Church recently reopened the case of its treatment of Galileo. After a study of the old records, the Church announced in 1983 that Galileo's name had been cleared and that the Church had erred in its treatment of him more than 300 years ago.

This recent action may seem pointless, since Galileo himself is no longer alive. It was, however, an important statement by a major church that science and religion should tolerate each other— that there is philosophical room for both. In an age when religious attacks on science are occurring anew, this position by an important worldwide religion may have some helpful effect.

PERSPECTIVE

We have seen how, in the space of less than 150 years, the science of astronomy advanced from adherence to the 1,500-year-old doctrines of Plato, Aristotle, and Ptolemy to a nearly modern view. Despite Galileo's suppression by the Church, enough questions had been raised through his work and that of Copernicus, Brahe, and Kepler, that serious consideration of the nature of the cosmos thereafter centered on the heliocentric system.

It has been interesting to examine the lives, and especially the frame of mind, of the characters who contributed to this revolution, ranging from Copernicus, whose intuition forced him to reject the established view and adopt an entirely new one; to Tycho, who philosophized little, but contributed much through his observations; to Kepler, whose search for universal harmony led him to the laws of planetary motion; and to Galileo, whose implacable common sense and brilliant deduction put it all on a firm, logical footing. The stage was now set for the missing link, the physical cause of the motions in the universe, to be discovered.

SUMMARY

1. Copernicus developed the heliocentric theory because it provided a unifying picture of the universe, not because it was more accurate or simpler than the Earth-centered theory.

2. With his new heliocentric theory, Copernicus was able to calculate the relative distances of the planets from the Sun, and he was able to explain the seasons and the lack of detectable parallax.

3. Tycho Brahe made vast improvements in the quantity and quality of astronomical observations of stars, and especially the planets, accomplishing this by making more precise and complete measurements than his predecessors.

4. Tycho was unable to detect stellar parallax, and for this and philosophical reasons, rejected the heliocentric hypothesis.

5. Kepler sought the underlying harmony among the planets and its physical basis, and in the process he discovered three laws of planetary motion.

6. Kepler's first law states that each planet orbits the

Sun in an ellipse, with the Sun at one focus.

7. Kepler's second law states that a planet moves fastest in its orbit when it is closest to the Sun, and slowest when it is farthest away, in such a manner that the area swept out in space by a line connecting the planet to the Sun is equal in equal intervals of time.

8. Kepler's third law relates the periods and semi-major axes of the planets; it states that the square of the period (in years) is equal to the cube of the semi-major axis (in AUs).

9. Galileo, by performing experiments, defined several important concepts of physics, including most notably the ideas of inertia and acceleration.

10. Galileo's use of a telescope to observe the heavens led to numerous discoveries that he used to refute the geocentric hypothesis.

11. Despite Church opposition, Galileo brought his ideas before the public in the form of a published fictional dialogue between believers of opposing points of view.

REVIEW QUESTIONS

Thought Questions

1. Summarize the reasons Copernicus adopted the heliocentric view.

2. Describe how the belief of Copernicus conformed with the teachings of Aristotle, and how it violated those teachings.

3. There is a principle adopted by most scientists that the best explanation for something is the simplest one, the one requiring the fewest unprovable assumptions. Copernicus and Hipparchus had different explanations of retrograde motion. Which explanation better satisfies the principle of simplicity? Explain.

4. Summarize the ways in which Tycho Brahe's observations were superior to those made by earlier astronomers.

5. Discuss the philosophical outlook of Johannes Kepler. In what ways was he like the followers of Plato and Aristotle, and in what ways was he like a modern scientist?

6. Why was it so fortunate that Kepler worked first on Mars as he attempted to deduce the nature of planetary motion?

7. Discuss how Kepler's third law, the harmonic law, was the most pleasing to Kepler himself, in view of his philosophical motivations.

8. Summarize the contrasts in style and philosophy between Kepler and Galileo.

9. Make a list of discoveries made by Galileo with the telescope, and explain how each was used by him to support the heliocentric view.

10. Do you think modern scientists are ever subjected to censure and oppression of their views as Galileo was? Explain.

Problems

1. Tycho's great quadrant consisted of a quarter circle with a radius of 31.5 feet, covering an angle of 90°. Marks on the quadrant corresponded to degrees and minutes of arc. How far apart were the degree marks? The minute marks?

2. Suppose that the true altitude of a certain star above the horizon was 27°14′. Suppose further that Tycho's measurements of the altitude of the star were 27°12′, 27°13′, 27°15′, 27°13′, and 27°16′. How close is the average of these values to the true value? Discuss the comparison between the average of several measurements and the accuracy of individual measurements.

3. Using a board, a couple of thumbtacks, a loop of string, and a pencil, practice drawing ellipses (stick the tacks in the board, then keep the loop of string taut as you move the pencil all the way around until you have an ellipse). Notice how the shape of the ellipse varies as the distance between the tacks (foci) is varied. What happens if the two tacks are at the same place?

4. Verify Kepler's third law by comparing P^2 and a^3 for the planets. Use data given in Appendix 7.

5. What would be the period of a planet located 3 AU from the Sun? What would be the semimajor axis of a planet whose period was 11.18 years?

6. Galileo cited the changing apparent size of Mars as an argument that Mars orbits the Sun. At its closest approach to Earth (at opposition), Mars is about 0.5 AU away, and at its farthest (conjunction), it is 2.5 AU away. How much larger does it appear from Earth at opposition than at conjunction?

ADDITIONAL READINGS

The references listed here are primarily sources of biographical and historical data on the people discussed in this chapter; many of them contain additional lists of references. Readings on the principles of physics described in this chapter can most easily be found in elementary physics texts, which are available at all levels ranging from completely nonmathematical to any degree of mathematical sophistication desired.

Beer, A., and P. Beer, eds. 1975. *Kepler. Vistas in astronomy*, Vol. 18. New York: Pergamon Press.

_____, and K. A. Strand, eds. 1975. *Copernicus. Vistas in astronomy,* Vol. 17. New York: Pergamon Press.

Drake, S. 1980. Newton's apple and Galileo's dialogue. *Scientific American* 243(2):150.

Draper, J. L. E. 1963. *Tycho Brahe: A picture of scientific inquiry in the sixteenth century.* New York: Dover.

Galileo, G. 1632. *Dialogue on the two chief world systems.* Translated by G. de Santillana. Chicago: University of Chicago Press.

Gingerich, O. 1973. Copernicus and Tycho. *Scientific American* 229(6):86.

_____ . 1973. Kepler, In *Dictionary of scientific biography,* Vol. 7, ed. C. G. Gillispie. New York: Scribner's.

_____ . 1983. The Galileo affair. *Scientific American* 247(2):132.

Gingerich, O., ed. 1975. *The nature of scientific discovery.* Washington, D.C. Smithsonian Institution Press.

Kuhn, T. 1957. *The Copernican revolution.* Cambridge, Mass.: Harvard University Press.

Whiteside, D. T. 1962. The expanding world of Newtonian research. *History of Science* 1:15.

MATHEMATICAL METHODS

The Ellipse and Conic Sections

The ellipse belongs to a family of curves called *conic sections*. These curves can be represented by mathematical equations, but their name derives from the fact that they can be formed by cutting through a hollow cone (see diagram).

If a plane cuts through a cone, the shape of the curve where the two intersect is determined by the angle of intersection. If the plane cuts through exactly perpendicular to the axis of the cone, the curve where they intersect is a circle. If the plane intersects at some other angle, but still cuts all the way through the cone, making a closed curve, this is an ellipse. If the plane cuts through at an angle equal to the angle of the cone, the result is an open-ended curve called a parabola, and if the plane cuts through at an angle closer to alignment with the axis of the ellipse, the resulting open curve is a hyperbola. The parabola represents the dividing line between a closed curve (ellipse) and an open one (hyperbola).

One of the properties of an ellipse, already mentioned in the text, is that the sum of the distances from any point on the curve to the two foci is always the same. Thus, no matter where point P is on the curve, the sum of the distances PF_1 plus PF_2 is always the same.

This property of the ellipse provides us an easy

Circle   Elipse

Parabola   Hyperbola

way to draw one: Stick two pins or tacks into a board, and then hook a loose loop of nonelastic string around them. Holding the string taut with a pen or pencil, trace out a closed curve, as shown in the accompanying figure. The string does not stretch, and this forces the distance from any point on the curve to the two foci to always be the same. You can experiment with changing the separation of the tacks (the foci) relative to the length of the string, in order to draw ellipses of different shape or eccentricity (this is discussed further in the Mathematical Methods section at the end of Chapter 8).

The ellipse is the general shape of any closed orbit—that is, an orbit between two bodies that are unable to escape each other's gravitational attraction. Thus the orbits of the planets around the Sun or of a satellite around its parent planet are all ellipses (but note that a perfect circle is possible; this is, after all, only a special kind of ellipse). It is also possible to have orbits that are not closed, whose shapes are represented by open curves. Even if the two bodies are not bound together by their mutual gravitational attraction, their gravitational attraction for each other still affects their relative motion. Then the general shape of their orbit is a hyperbola. They

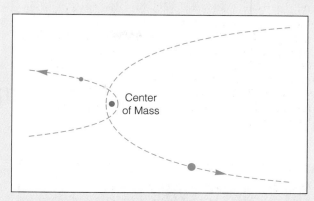

Hyperbolic orbits during an encounter.

pass near each other just once, and then they each escape, never to return (see figure).

It is also possible to have a parabolic orbit, if the relative speed of the two bodies is exactly at the boundary between trapped orbits and escape, or hyperbolic orbits. As we will discuss in Chapter 15, many comets have orbits about the Sun that are very close to being parabolic.

Astronomy on the Fringe

BY ROGER CULVER

Professor Roger Culver is a distinguished astronomer who has pursued and analyzed various claims made by adherents of "fringe science." The term fringe science refers to the many belief systems being promoted today in which people are expected to accept on faith various phenomena ranging from astrology to "New Age" concepts such as channeling, crystal power, or the reality of UFOs; none of these are based on rational, objective investigation. Dr. Culver has been a professor of astronomy at Colorado State University for more than 23 years. During that time, he has received numerous awards for his teaching, including the prestigious CSU Teacher of the Year award in 1977. He has been a Guest Observer on many occasions at the Kitt Peak National Observatory and has also been a Visiting Fellow at both the Lowell Observatory in Arizona and the Mount Stromlo Observatory in Australia. Dr. Culver has authored or co-authored several dozen publications covering a wide range of astronomical topics, including the Barnes and Noble College Outline in Astronomy *and* The Gemini Syndrome *(also published under the title* Astrology: True or False,

A Scientific Evaluation [Prometheus Books], which is a detailed analysis of the contemporary relationship between astrology and astronomy. Dr. Culver is a member of the Committee for the Scientific Investigation of Claims of the Paranormal, an organization of people who believe in rational inquiry into the basis for belief systems. In his work in this area and as a skeptic of astrology, Dr. Culver has made guest appearances on television talk shows such as "The Oprah Winfrey Show" and "Straight Talk." In this essay, Professor Culver discusses rational inquiry and fringe science.

A generation ago, a music group called the Fifth Dimension rose to the top of the popular music charts with a song that proclaimed the "dawning of the Age of Aquarius" in which "peace will guide the planets and love will steer the stars." Since then, the Age of Aquarius has outgrown its comparatively humble and obscure astrological beginnings, changed its name, and has become the "New Age" of contemporary society. Complete with an impressive array of trappings that include cosmic vibrations, crystal power, UFOs, Ramtha, and Shirley MacLaine, the New Age movement is an amalgam of medieval semiscience and modern hucksterism whose thriving existence in an era of unprecedented technological and scientific advance represents one of the more fascinating contradictions of the closing years of the twentieth century.

Scientists and astronomers would be more than happy to leave the New Age adherents to their spectral beings, astral planes, and sidereal synchronicites, were it not for many of the New Agers' brazen pretentions to scientific validity. New Age lectures and literature abound with terms such as *electomagnetic harmony, cosmic vibrations,* and *crystal energy,* which sound scientific and create the illusion of scientific credibility. Even more blatant are the claims that many of the New Age fields of endeavor, including astrology, crystal power, and telekinesis are, in fact, empirically verifiable. It is over these claims that the scientist cringes with anguish and anger. For more than 300 years, men and women of science have painstakingly chipped and carved out an experimental and observational model of the workings of the physical world, using as their philosophy of investigation a procedure called the scientific method. That method dictates that the laws and theories we employ to represent the physical world must be consistent with the experiments, observations, and measurements made on that world. Moreover, these scientific laws and theories are not considered to be the "ultimate truth" of nature, but rather are regarded as representations and approximations of that truth, and as such are eminently negotiable. For the scientist, a given law or theory is only as good as the next set of experimental results. It is this ability to adjust to new information that allows scientists to distinguish between competing theories and to embrace views of nature that describe an ever-widening variety of phenomena in the physical world. By contrast, the conflicts between the competing belief systems of various religious faiths probably can never be resolved, because of the inherent untestability of most of the tenets involved and because of the belief by the members of those religious groups that theirs is the "true faith" or the "Word

of God" and is therefore not to be tampered with in any way. It is against this backdrop that one can examine some astronomically related aspects of the New Age.

Perhaps the most ancient and enduring of the New Age art forms is astrology, the belief that "harmonies" or "synchronicities" exist between celestial configurations and human interactions, events, and personalities. Astrology's beginnings can be traced at least as far back as the Mesopotamian civilizations of the third millenium B.C. In the earlier centuries of its existence, astrology was, in certain measure, intertwined with astronomy. Indeed, the Hellenistic Greek Claudius Ptolemy wrote not only the *Almagest,* a second-century-A.D. compendium of the state of affairs in astronomical knowledge, but also a text called the *Tetrabiblos,* which was a similar compilation of the astrological knowledge of the time. Such historical entanglements were summarily terminated, however, with the Scientific Revolution of the seventeenth century. Astronomy emerged from the era of Copernicus, Tycho, Galileo, Kepler, and Newton as a full-fledged, modern science whose practitioners completely embraced the scientific method.

To this day, astrologers continue with the medieval methodology of Nostradamus and steadfastly resist any attempts to tamper with the "truth" of astrology. Thus, when astronomical science was able to demonstrate the far-flung distances of celestial objects, thereby reducing to minuscule levels the influences on human beings of "cosmic"

interactions such as gravity, tidal forces, and electromagnetic radiation, the astrologers responded by abandoning the idea of direct celestial influences in favor of "sidereal synchronicities" and "correspondences" that provide a "roadmap of life" to those who would take the time to read and interpret the map. When statistical studies indicated that the supposed astrological synchronicities between human characteristics and various celestial configurations such as Sun signs, Moon signs, and so on do not exist, the astrological response was that such tests of isolated one-on-one correlations do not constitute a valid test of astrological principles. Instead, according to astrological thought, one must look at the entire horoscope, a requirement that makes one wonder about the myriad books and newspaper columns on Sun-sign astrology and also conveniently renders any statistical test of astrology impossible to perform. The very simplest of horoscopes, for example, will include 12 astrological signs, 12 astrological houses, and 10 "luminaries" (Sun, Moon, and eight planets). If one employs just these basic components of the horoscope, one finds that it is statistically possible to generate some 10^{34} different horoscopes. By comparison, the total number of human beings who have lived on the planet is a number of the order of 10^{10} and the number of grains of sand on all the world's beaches is about 10^{28}.

Astrologers enjoy an overall predictive success rate that is statistically indistinguishable from

(continued on next page)

random guessing. In light of such predictive failure, a scientific approach to astrology would demand an extensive overhaul or even a complete dismissal of some of the traditional astrological principles and correspondences. A similar event occurred in physics early in the twentieth century, when the theories of relativity and quantum mechanics were introduced as a result of the predictive failures of the earlier, "classical" theory of physics. Instead, the astrologers have developed an impressive array of explanations whose sole purpose is to excuse or minimize the predictive failures of astrology without threatening or tampering with the basic "truths" of astrology. Astrologer Linda Goodman states in her book *Linda Goodman's Sun-Signs:*

"Alone among the sciences, astrology has spanned the centuries and made the journey intact. We shouldn't be surprised that it remains with us, unchanged by time—because astrology is truth—and truth is eternal."

One might argue that Goodman is correct, provided the word *sciences* is changed to *religions*.

Not as ancient as astrology but just as deeply ingrained in New Age beliefs are unidentified flying objects (UFOs). Unlike astrology, whose origins lie buried in prehistory, the UFO phenomenon began in June, 1947, when private pilot Kenneth Arnold, while flying to Yakima, Washington, reported seeing a formation of dazzling, disk-shaped objecs skimming the tops of the Cascade mountains at incredible speed. Arnold's account spurred a barrage of UFO sightings in the weeks that followed, and the "UFO era" was on. In the decades since Arnold's sightings, there have been thousands of reports of various "encounters" of one sort or other with mysterious objects and entities from the heavens. The New Age has eagerly embraced UFOs as extraterrestrial spacecraft piloted by beings who are visiting the Earth for a variety of purposes ranging from the performance of scientific and medical experiments on human beings, plants, and animals to helping human beings prepare for entry into the New Age. Scientists and skeptics are far less enthusiastic about the possible existence of such New Age "UFOnauts." More than 99 percent of all reported UFO sightings can be attributed to known, natural phenomena such as bright stars and planets, ball lightning, St. Elmo's fire, and meteors, or to human artifacts such as aircraft, artificial satellites, and high-altitude weather balloons.

Space exploration has shoved the UFOnauts' place of origin from nearby Venus—of Frank Scully's best-selling 1950 UFO book *Behind the Flying Saucers*—to a location that must necessarily be well outside our solar system. Undaunted, the UFOnaut believers are now certain that the UFOs come to us either from a distant star system such as zeta Reticuli or from another "dimension" or "plane of existence." Interestingly, both views preserve the "truth" of the existence of visiting extraterrestrials while at the same time placing them safely out of reach of our telescopes and space probes. Moreover, the UFOs that visit us seem able to violate the laws of motion, with their high-speed, sharp-angle turns, supersonic flight through the Earth's atmosphere without a sonic boom, and velocities faster than the speed of light. The existence of vehicles capable of such supernatural feats must be greeted with the highest degree of scientific skepticism, particularly in view of the lack of physical evidence in support of such claims. Yet the belief in UFOnauts is so strong and unshakable that in 1958 it prompted Swiss writer-psychoanalyst Carl Jung to write a treatise on the UFO phenomenon entitled "Flying Saucers: A Modern Myth of Things Seen in the Skies." Jung argues that in the post–World War II era of overwhelming science and technology, the historical human interest in and quest for salvation has changed from the more traditional mythological and religious forms to those having a technological imagery; and indeed, in the reported encounters with aliens, be they hero or villain, these beings always seem to possess the technological omnipotence of a god or goddess. If Jung's thesis is correct, then the UFO phenomenon may be little more than a technological update of a human psychological phenomenon that far predates either Kenneth Arnold or the dawning of the Age of Aquarius.

Unfortunately, there is a sinister component to the New Age as well. All too often, the "evidence" in support of New Age claims is demonstrably fraudulent. This state of affairs is totally unacceptable to a scientific method which requires scrupulously honest data, measurements, and observations to function properly. As a result, scientific investigative teams must employ among their numbers magicians, illusionists, and sleight-of-hand artists whenever they research New Age claims, so that the integrity of the work is not compromised by trickery.

As both a century and a milennium come to a close, we find that we do indeed live in a new age filled with many wonders. But along with the truly genuine wonders, there also exist many which are counterfeit, some manifestly so, some more subtly so. For all its faults and limitations, science has proved itself to be, at least for the present, the best means by which rational beings can distinguish what is genuine from what is not. The marvels of our age and its technology have also brought with them an assortment of difficult problems that promise to challenge humanity in ways unheard of in our history. If, as New Agers suggest, these new problems require new ways of thinking in order to be solved, then let them indeed by truly new ways of thinking rather than recycled medieval philosophies.

THE TOOLS OF ASTRONOMY

Before we can plunge into a full description of the universe and an understanding of it, we must first learn about the methods and tools used by astronomers to gain this understanding. One of the principal goals of this text is to describe *how* we know what we know, going well beyond a simple summary of the phenomena that are observed.

The first of the three chapters in this section is devoted to a description of some of the basic physical principles that underlie the motions and interactions of bodies large and small. We start with a discussion of Isaac Newton, whose intellectual powers unearthed many of these principles, and then we move on to the laws of physics that govern such diverse phenomena as planetary orbits, the motions of molecules in a gas, and the tides on the Earth and other bodies.

We would know nothing of the external universe were it not for the light that reaches us from faraway objects, and the next chapter describes the nature of light and the way we decipher its messages. We find that an amazing variety of information can be derived from the spectra of objects like planets and stars. Things once considered forever beyond our grasp are now routinely measured, and in Chapter 6 we learn how this is done.

The final chapter of this section is devoted to the instruments that are used to collect light from afar. Telescopes on the ground and in space are the material monuments to our quest for knowledge of the universe, and in Chapter 7 we will learn the principles of their design and the practical considerations of their construction and use. Having done this, we will, at last, be ready to explore the dynamic universe.

CHAPTER 5

Isaac Newton and the Laws of Motion

Space Shuttle astronauts demonstrating weightlessness. (NASA)

After the publication in 1632 of Galileo's *Dialogue on the Two Chief World Systems,* a 50-year period elapsed during which few major discoveries were made, but substantial progress nevertheless occurred. Telescopic observations were extended to a greater variety of objects, the surface of the Moon was mapped, star catalogs were expanded, and the planets were studied in more detail. Perhaps the greatest astronomer of the period was Christian Huygens (1629–1695), who refined the telescope and was able to detect a moon of Saturn. He also discovered the true nature of the misshapen appearance of the planet (Galileo's crude telescope had only revealed a distortion in the shape of Saturn, whereas Huygens was able to determine that the planet is encircled by a great ring system). The Earth was accurately measured by an assortment of people, and it was even discovered that it is not precisely spherical, but instead bulges at the equator.

THE LIFE OF NEWTON

In the year 1643, a few months after the death of Galileo, Isaac Newton (Fig. 5.1) was born in Woolsthorpe, England, into a climate of inquiry and scientific ferment. Newton's childhood was unremarkable, except that he showed a growing interest in mathematics and science, and at age 18 he entered college at Cambridge, receiving a bachelor's degree in early 1665. He then spent 2 years at his home in Woolsthorpe, largely because the Plague, or Black Death, made city living rather dangerous. It was during this time that he made a remarkable series of discoveries in physics, astron-

omy, optics, and mathematics, in what surely must have been one of the most intense and productive periods of individual intellectual effort in human history.

Newton had a tendency to exhaust a subject, get bored with it, and go on to new fields, so that nothing of his work was published for some time, during which some of his discoveries were repeated independently by others. Finally, after persistent urging by his friend and fellow astronomer Edmond Halley, in 1687 Newton published a massive work called *Philosophiae Naturalis Principia Mathematica,* now usually referred to as the *Principia* (Fig. 5.2). In this three-volume book, Newton established the science of mechanics (which he viewed as merely background material and relegated it to an introductory section), and applied it to the motions of the Moon and the planets, developing the law of gravitation as well. His work in optics was published separately in 1704, although it was probably written much earlier than that.

The *Principia* received great notice, particularly in England. As a result, Newton's later life was a public one, with various government positions which left less and less time for scientific discovery. With the help of younger associates, he did revise the *Principia* on two

Figure 5.1 Isaac Newton. *Newton's experiments and brilliant mathematical intuition led him to profound new understandings of physics and mathematics. (The Granger Collection)*

occasions, in 1713 and in 1726, making some improvements each time. Newton died in 1727, at the age of 84. His impact lives today, in our modern understanding of physics and mathematics. Newton's conclusions on the nature of motions and gravity are inadequate for some circumstances, as shown by Einstein, but are still quite useful for most everyday applications.

NEWTON'S LAWS OF MOTION

Inertia and the First Law

We have already discussed the concept of inertia, first developed by Galileo. Newton expanded the idea, recognizing inertia as a fundamental property of matter, very closely related to another property, the **mass** of an object.

The mass of an object reflects the amount of matter it contains, which in turn determines other properties such as weight and momentum. It is the other properties that are easily observed, not the mass, so mass can be a difficult concept. Mass and weight are particularly easy to confuse, because at the Earth's surface, a given amount of mass will always have the same weight (because the gravitational force is very nearly the same everywhere on the Earth). To see the difference, consider an astronaut who goes to the Moon. On the Moon, he weighs only one-sixth as much as he does on Earth, but his mass is the same. The relationship between mass and weight will become clearer in the next section.

Mass is usually measured in units of **grams** or **kilograms.** One gram is the mass of a cubic centimeter of water, and a kilogram, which weighs about 2.2 pounds at sea level, is the mass of 1,000 cubic centimeters (or one **liter**) of water.

Inertia and mass are closely related because the mass of a body determines how much **inertia,** or resistance to change in motion, the body has. The more massive an object is, the more inertia it has, and the more difficult it is to start it moving, to stop it, or to alter its motion once it is moving (Fig. 5.3). For example, imagine two crates of equal size, one containing books, the other pillows. Each has wheels on the bottom, minimizing friction. You may be able to move the box of pillows rather easily, but not the box of books. The box of books contains more mass, and therefore has more inertia, more resistance to being moved. Similarly, a

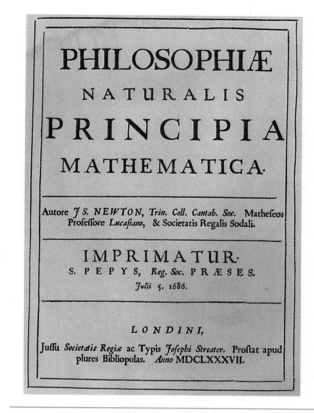

Figure 5.2 The title page from an early edition of the Principia. *This massive volume is still considered one of the greatest and most influential books ever written. (The Granger Collection)*

Figure 5.3 Inertia. *An object in motion tends to remain in motion. (H. Roger Viollet)*

massive object in motion is more difficult to turn or stop than a less massive one. Imagine, for example, the deftness required to bring a large ship to rest at a pier without destroying the pier.

Newton summarized the concept of inertia in what has become known as his first law of motion:

> A body at rest or in a state of uniform motion tends to stay at rest or in uniform motion unless an unbalanced force acts upon it.

As we have seen, this had been stated by Galileo, although he did not make it quite so general as Newton, who realized that it applies to the Earth, planets, and all celestial bodies. One very important consequence of the first law of Newton is the implication that a planet should follow a straight line, rather than a curved path around the Sun, unless some force acts upon it. We shall return to this point shortly.

Force and Acceleration

Having stated that a force is required to change an object's state of rest or uniform motion, Newton went on to determine the relationship between force and the change in motion that it produces. To understand this, we must discuss the idea of acceleration.

Acceleration is a general word for any change in the motion of an object. It is acceleration when a moving object is speeded up or slowed down, or when its direction of motion is altered (Fig. 5.4). It is also acceleration when an object at rest is put into motion. A planet orbiting the Sun is constantly undergoing acceleration; otherwise it would fly off in a straight line.

When we change the motion of an object, we change its velocity; in fact, acceleration is defined formally as the rate of change of velocity. The units in which acceleration is expressed are units of change of velocity per second. Thus, if a car accelerates from rest to 60 miles per hour in 10 seconds, its acceleration is 6 miles per hour per second. In astronomy, velocities are usually measured in centimeters per second or kilometers per second, so the units of acceleration can be either cm/sec/sec or km/sec/sec, usually written as cm/sec^2 or km/sec^2. An object falling in the Earth's gravitational field increases its downward speed by 980 cm/sec (or 32 feet/sec) for every second it falls, so we say that the acceleration due to gravity at the Earth's surface is 980 cm/sec^2 (or 32 ft/sec^2).

Since Newton's first law specifies that a body will remain in a state of rest or constant motion (i.e., con-

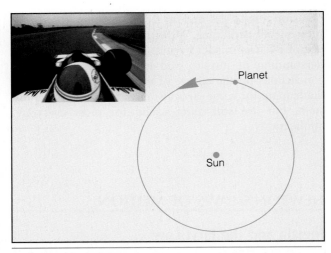

Figure 5.4 Acceleration.   *A car that is speeding up is accelerating, as is a planet that follows a curved path, even if its speed is constant. Any change of speed or direction is an acceleration. (Photo Researchers, Inc.)*

stant velocity) unless an unbalanced force acts on it, we can restate this to say that in order for an object to be accelerated, an unbalanced force must act on it. Newton's second law spells out the relationship between force, the resultant acceleration, and the mass of the object:

> An object is accelerated when an unbalanced force is applied to it, and the acceleration is proportional to the force and inversely proportional to the mass of the object.

This may be written mathematically as $a = F/m$, where a is the acceleration, F is the force, and m is the mass. More commonly it is written in the alternative (but equivalent) form $F = ma$.

We can illustrate the second law as follows: If one object has twice the mass of another and equal forces are applied to the two, the more massive one will be accelerated only half as much. Conversely, if unequal forces are applied to objects of equal mass, the one to which the greater force is applied will be accelerated to a greater speed.

Action and Reaction

Newton's third law of motion is probably more subtle than the first two, although in some circumstances it is quite obvious. It states:

> For every action there is an equal and opposite reaction.

ASTRONOMICAL INSIGHT 5.1

The Forces of Nature

One of the most difficult concepts to grasp is that of a force, particularly when there is no concrete, visible object to exert it. We think of a force as a push or a pull, and we can visualize it easily in certain circumstances, such as when a gardener pushes a wheelbarrow and it moves. It is perhaps less obvious in the case where no movement results, and even less so when no tangible agent exerts the force. In the case of gravity, the force is exerted invisibly and over great distances.

In general terms, a force exerted by a concrete object is referred to as a **mechanical force,** whereas those exerted without any such agent are **field forces.** Gravity is the most familiar force that is created by a field.

There are four basic kinds of field forces in nature, and in fact these forces form the basis for *all* forces, mechanical or field. Gravity was the first of the four to be discovered, although in "discovering" gravity, and in mathematically describing its behavior, Newton did not develop a real fundamental understanding of how or why it works. The reason that gravity was the first to be discovered is simple; no special conditions are required for gravitational forces to be exerted (*all* masses attract each other), and it operates over very great distances. After all, Newton deduced its properties by noting how it controls the motions of planets that are separated from the Sun by as

much as a billion kilometers.

The second most obvious type of field force, and the second to be discovered and described mathematically, is the **electromagnetic force.** There is an intimate relationship between electric and magnetic fields. The interaction of these fields with charged particles, first described mathematically by the Scottish physicist James Clerk Maxwell, creates forces. The electromagnetic force is actually much stronger than the force of gravity; the electromagnetic force binding an electron to a proton in the nucleus of an atom is 10^{39} times stronger than the gravitational force between the two particles. The electromagnetic force, like gravity, is inversely proportional to the square of the distance between charged particles. Therefore we might expect this force always to dominate over gravity, as it does on a subatomic scale. But it does not. The reason is that most objects in the universe, composed of vast numbers of atoms containing both electrons and protons, have little or no net electrical charge, whereas they always have mass and are therefore subject to gravitational forces. If the planets and the Sun had electrical charges in the same proportion to their masses as the electron and proton do, then electromagnetic forces, rather than gravity, would control their motions.

Besides being immensely stronger, the electromagnetic force also differs from gravity in that it can be either repulsive or attractive, depending on whether the electrical charges are the same or opposite.

The remaining two forces were not discovered until the 1930s. (Maxwell's work on electromag-

netics was done in the late 1800s.) Both of these new forces involve interactions at the subatomic level, and it was not until the development of the science of quantum mechanics that they were discovered. One of these is the **strong nuclear force,** which is responsible for holding together the protons and neutrons in the nucleus of an atom, and the other is the **weak nuclear force,** some 10^{-5} times weaker than the strong nuclear force. The strong nuclear force, in turn, is about 100 times stronger than the electromagnetic force, making it the strongest of all, but it operates only over very small distances. Within an atomic nucleus, where protons with their like electrical charges are held together despite their electromagnetic repulsion for each other, it is the strong nuclear force that acts as the glue that keeps the nucleus from flying apart. The weak nuclear force plays a more subtle role, showing its effects primarily in certain modifications of atomic nuclei during radioactive decay.

From the smallest to the largest scales, these four forces appear to be responsible for all interactions of matter. It is ironic that at the most fundamental level, the mechanism that makes the forces work is not understood. To understand the fundamental nature of forces is one of the principal goals of modern physics. One hope is to develop a mathematical framework that encompasses all of the forces, a framework first sought over 60 years ago by Albert Einstein, and referred to as a **unified field theory.** Progress has been made since Einstein's time, particularly towards unifying the electromagnetic and nuclear forces, but the ultimate goal still eludes us.

In other words, forces always occur in pairs such that when one object exerts a force on another, the second object exerts a force on the first that is equal in strength but opposite in direction (Fig. 5.5). This may sound confusing, because the accelerations of the two objects are not necessarily equal, and because in most common situations, other forces such as friction complicate the picture. Furthermore, there are many static situations in which no acceleration of either object occurs.

Figure 5.5 Action and reaction. *Some manifestations of Newton's third law are rather subtle, while others are not. Here a book exerts a force on a table, and the table exerts an equal and opposite force on the book. There is a static situation; there is no motion. When a cannon is fired, however, the shell is accelerated one way and the cannon the other. The force applied to each is the same, but the cannon has more mass than the shell and is therefore accelerated less, as Newton's second law states. (bottom: U.S. Army)*

We can visualize the third law by considering instances in which friction is not important. For example, imagine standing in a small boat and throwing overboard a heavy object, such as an anchor. The boat will move in the opposite direction from the anchor, because the anchor exerts a force on you as you throw it. The "kick" of a gun when it is fired is another example of action and reaction. When a person jumps off the ground by pushing against the Earth, both the person and the Earth are accelerated by the mutual force, but of course the immensely greater mass of the Earth prevents it from being accelerated noticeably.

The third law of motion states the principle on which a rocket works. In this case, hot, expanding gas is allowed to escape through a nozzle, creating a force on the rocket. The gas is accelerated in one direction, and the rocket is accelerated in the opposite direction (Fig. 5.6). Anyone who has inflated a balloon and then let go of it, allowing it to zoom through the air, is familiar with the operating principle of a rocket.

GRAVITATION AND ORBITS

We return now to a point raised earlier—namely, that the planets would fly off along straight lines if no force

Figure 5.6 Newton's third law applied to the launch of a rocket. *Hot gases are forced out of a nozzle (or several, as in this case), and in return they exert a force on the rocket that accelerates it. This is the first launch of the space shuttle Columbia. (NASA)*

was acting upon them. Newton's first law says that this should happen, yet obviously it does not. Newton realized that the planets must be undergoing constant acceleration toward the center of their orbits, that is, towards the Sun (Fig. 5.7). He set out to understand the nature of the force that creates this acceleration.

If you feel confused about the direction of this force, remember what you have just learned about acceleration and inertia: A planet needs no force to keep it moving, but it does require a force to keep its path curving as it travels around the Sun. What is needed is something to pull the planet inward, toward the Sun. This force can be compared to the tension in the string tied to a rock that you whirl about your head; if you suddenly cut the string, the rock would fly off in whatever direction it happened to be going at the time.

What, then, is the string that keeps the planets whirling about the Sun? Newton realized that the Sun itself must be the source of this force, and he made use of Kepler's third law, as well as observations of the Moon's orbit and falling objects at the Earth's surface, to discover its properties. He was led to formulate his law of universal gravitation:

Any two bodies in the universe are attracted to each other with a force that is proportional to the product of the masses of the two bodies and inversely proportional to the square of the distance between them.

The law of gravitation is one of the fundamental rules by which the universe operates. As we shall see in later chapters, it explains the motions of stars about each other or about the center of the galaxy, the movements of the Moon and planets, and the motions of galaxies about one another. Gravity, in fact, appears to be the dominant factor that will determine the ultimate fate of the universe. Table 5.1 shows the relative gravitational forces on a person on Earth due to various bodies in the universe.

It is useful to consider a few examples illustrating how the law of gravitation is applied. The weight of an object is simply the gravitational force between it and the Earth, separated by a distance equal to the Earth's radius (i.e., the mass of a spherical body acts as though it were concentrated in a single point at the center). If, for example, the diameter of the Earth were suddenly doubled (while its mass remained constant), our weight would decrease by a factor of 4. If the Earth were three times smaller, our weight would be $3^2 = 9$ times greater. If we climb to the top of a high mountain, our weight decreases, but not very much, because even the highest mountains are small compared to the radius

Table 5.1 Gravitational forces acting on a human standing on Earth

Source of force	Relative strength of force
Earth	1.0
Moon	3.4×10^{-6}
Sun	8.6×10^{-4}
Venus (at closest approach)	1.9×10^{-8}
Jupiter (at closest approach)	3.3×10^{-8}
Nearest star	1.4×10^{-14}
Milky Way galaxy	2.1×10^{-11}
Virgo cluster of galaxies	10^{-15}

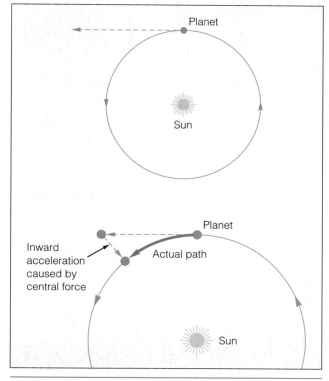

Figure 5.7 Planetary orbits and central force. *A planet would fly off in a straight line if no force attracted it toward the center of its orbit. From diagrams like the lower one shown here, Newton was able to determine the amount of acceleration required by an orbiting body to keep it in its orbit.*

of the Earth. The Earth exerts a gravitational force on an astronaut in orbit, but there is no acceleration relative to the spacecraft, so the astronaut is weightless in the local environment (Fig. 5.8).

Figure 5.8 Weightlessness. *Although astronauts orbiting the Earth are subject to the Earth's gravitational force and are "falling around" the Earth along with their spacecraft, they are weightless because they experience no gravitational acceleration relative to their surroundings. This photo shows two astronauts aboard* Skylab, *an orbiting scientific research station. (NASA)*

The law of gravitation states that the force also depends on the masses of the two objects that are attracting one another. It is easy to imagine that a person's weight would double if his or her mass doubled; similarly, the force between the Earth and the Moon depends on the masses of these two bodies. If we triple the mass of the Moon, the force is tripled; if we triple the mass of the Moon but decrease the mass of the Earth by a factor of 3 at the same time, the force remains unchanged.

Although it is useful for the purpose of illustration to work out simple thought examples as we have just done, a more quantitative approach is needed to deal with more realistic situations. As we might have expected, the law of universal gravitation can be written in the form of an equation:

$$F = \frac{Gm_1m_2}{r^2},$$

where F is the strength of the force between two objects whose masses are m_1 and m_2 and whose separation is r. The symbol G represents a constant that is required for the value of F to be expressed in normal units of force. In the system of units used most commonly by astronomers, force is measured in dynes (where 1 dyne is the force exerted by a mass of 1 gram under an acceleration of 1 centimeter per second per second).[1] A person who weighs 150 pounds (or a mass of 67 kilograms) has a weight in this system of 6.7×10^7 dynes, so we see that the dyne is a rather small unit of force. The value of the constant G is 6.67×10^{-8} when the masses are expressed in grams; the separation between them in centimeters; and the force in dynes.

The law of universal gravitation can be used to show something that Galileo had postulated long before Newton's time: namely, that the acceleration due to gravity is independent of the mass of an object. Suppose, in the previous equation, that m_1 represents the mass of the Earth, and m_2 the mass of an object falling at the Earth's surface. In that case, r is the Earth's radius (it can be shown that the Earth acts gravitationally as though all its mass were in a single point at its center, so we say that the separation between the object and the Earth is equal to the Earth's radius).

Now recall that the acceleration on a body is equal to the force applied to it, divided by its mass; hence, the acceleration of the falling object is F/m_2. If we call this acceleration g, because it is due to gravity, then we have

$$g = \frac{F}{m_2} = \frac{Gm_1}{r^2} = \frac{GM}{R^2},$$

since the symbols M and R are generally used for the mass and radius of a planetary body.

---

[1] While astronomers usually use the **centimeter-gram-second,** or **cgs,** system of units, physicists generally use the **meter-kilogram-second,** or *MKS,* system (more commonly known as the **Système Internationale,** or *SI,* system). In this system, force is measured in units of newtons, where 1 newton is defined as the force exerted by a mass of 1 kilogram under an acceleration of 1 meter per second per second. A force of 1 newton is equal to a force of 10^5 dynes.

At the surface of the Earth, the value of g is 980 centimeters per second per second, meaning that the speed of a falling object increases by 980 centimeters per second for every second of fall. In more familiar units, this corresponds to 32 feet per second per second. To find the surface gravity on some other planet or satellite, the same expression for g can be used, along with the appropriate values for the mass of the planet, M, and its radius, R (see Table 5.2). For example, the Moon has 0.273 times the radius of the Earth, and 0.0123 times the mass. Thus, the acceleration of gravity at the Moon's surface is $0.0123/0.273^2 = 0.165$ that at the surface of the Earth. Therefore, astronauts on the Moon weigh approximately one-sixth as much as they do on the Earth.

Table 5.2 Surface gravities on solar system bodies

Body	Surface gravity $(g)^a$
Earth	1.0
Sun	27.9
Moon	0.17
Mercury	0.38
Venus	0.90
Mars	0.38
Ceres (largest asteroid)	0.000167
Jupiter	2.64
Saturn	1.13
Uranus	0.89
Neptune	1.13
Pluto	0.07:

$^a g = 980$ cm/sec^2.

ENERGY, ANGULAR MOMENTUM, AND ORBITS: KEPLER'S LAWS REVISITED

Even though Newton made use of Kepler's third law in deriving the law of gravitation, the latter is in fact more fundamental. It was soon possible for Newton to show that all three of Kepler's laws follow directly from Newton's laws of motion and gravitation. Kepler's studies of planetary motions revealed the *result* of the laws of motion and gravitation, whereas Newton found the *cause* of the motions. To appreciate how this was done, we must further discuss some basic physical ideas.

An important concept in understanding not only orbital motions but also many other aspects of astrophysics is **energy.** We can define energy as the ability to do work. Energy can take on many possible forms, such as electrical energy, chemical energy, heat, and others. All forms of energy can be classified as either **kinetic energy,** the energy of motion, or **potential energy,** which is stored energy that must be released (converted to kinetic energy) if it is to do work. A speeding car has kinetic energy because of its motion, whereas a tank of gasoline has potential energy in the form of its chemical reactivity, a tendency to release large amounts of kinetic energy if ignited. Thus a car operates when this potential energy is converted to kinetic energy in its cylinders.

The units used for measuring energy can be expressed in terms of the kinetic energy of specified masses moving at specified speeds. The kinetic energy of a moving object is $\frac{1}{2}mv^2$, where m is the mass of the object

and v is its speed. In astronomy, the most commonly used unit of energy is the **erg,** a small amount of energy that is equivalent to the kinetic energy of a mass of 2 grams moving at a speed of 1 centimeter per second. (The technical definition is that an erg is the amount of energy required to move a distance of 1 centimeter against a force of 1 dyne.) A larger unit of energy often used by physicists is the **joule.** (A joule is equal to 1 newton-meter), the equivalent of a mass of 2 kilograms moving at a speed of 1 meter per second. Thus, 1 joule is equal to 10^7 ergs.

We often speak in terms of **power,** which is simply energy expended per second. The familiar **watt** is 1 joule per second; astronomers, however, tend to use units of ergs per second, which have no special name as a unit. Thus we will speak of the power (or, equivalently, of the **luminosity**) of a star in terms of its energy output in ergs per second.

Using this understanding of energy, we can now discuss orbital motions in a much more general way than we have previously done. Two objects subject to each other's gravitational attraction have kinetic energy resulting from their motions, and potential energy because they each feel a gravitational force. Just as a book on a table has potential energy that can be converted to kinetic energy if it is allowed to fall, an orbiting body also has potential energy by virtue of the gravitational force acting on it.

Newton's laws and the concepts of kinetic and potential energy can be used to show that there are many types of possible orbits when two bodies interact gravitationally. Not all are ellipses, because if one of the objects has too much kinetic energy (exceeding the potential energy caused by the gravity of the other), it will not stay in a closed orbit, but will instead follow an arcing path known as a hyperbola, and will escape after one brief encounter. Some comets have so much kinetic energy that after one trip close to the Sun, they escape forever into space, following hyperbolic paths. A rocket launched from Earth with sufficient velocity will escape into a similar orbit (Fig. 5.9).

If the kinetic energy is less than the potential energy, as it is for all the planets, then the orbit is an ellipse, as Kepler found. It is technically correct to say that a planet and the Sun orbit a common **center of mass** (Fig. 5.10), rather than saying that the planet orbits the Sun. The center of mass is a point in space between the two

bodies where their masses are essentially balanced; more specifically, it is the point where the product of mass times distance from this point is equal for the two objects. Since the Sun is so much more massive than any of the planets, the center of mass for most Sun-planet pairs is always very close to the center of the Sun, so the Sun moves very little, and we do not easily see its orbital motion. It is true, however, that the Sun's position changes as it orbits the centers of mass established by its interaction with the planets, especially the most massive ones. In a double-star system, where the two masses are more nearly equal, it is easier to see that both stars orbit a point in space between them. Thus Kepler's first law as he stated it requires a slight modification:

> Each planet has an elliptical orbit, with the center of mass between it and the Sun at one focus.

Kepler's second law also can be restated in terms of Newton's mechanics. Any object that rotates or moves around some center has **angular momentum**. This is dependent on its mass, speed, and distance from the center of motion. In the simple case of an object in circular orbit, the angular momentum is the product

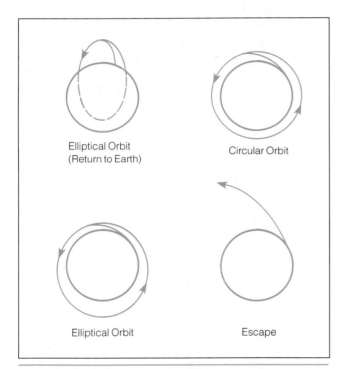

Elliptical Orbit
(Return to Earth)

Circular Orbit

Elliptical Orbit

Escape

Figure 5.9 Orbital and escape velocities. *A rocket launched with insufficient speed for circular orbit would tend to orbit the Earth's center in an ellipse, but would intersect the Earth's surface. Given the correct velocity, it will follow a circular orbit. A somewhat larger velocity will place it in an elliptical orbit that does not intersect the Earth. If given enough velocity so that its kinetic energy is greater than its gravitational potential energy, however, it will escape entirely.*

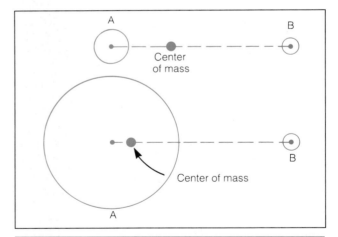

A

B

Center
of mass

B

Center of mass

A

Figure 5.10 Center of mass. *The upper sketch depicts a double star where star A has twice the mass of star B, so the center of mass, about which the two stars orbit, is one-third of the way between the centers of the two stars. The lower sketch shows a case where star A has 10 times the mass of star B, so the center of mass is very close to star A. The Sun is so much more massive than any of the planets that the center of mass for any Sun-planet pair is very near the center of the Sun, so the Sun's orbital motion is very slight.*

mvr, where *m* is its mass, *v* its speed, and *r* its distance from the center of mass.

The total amount of angular momentum in a system is always constant. Because of this, a planet in an elliptical orbit must move faster when it is close to the center than when it is farther away, so that its velocity compensates for the changes in distance (Fig. 5.11). Thus, a planet moves faster in its orbit near **perihelion** (its point of closest approach to the Sun) than at **aphelion** (its point of greatest distance from the Sun). Kepler's second law really stated that

Angular momentum in a two-body system is constant.

Kepler's third law was also revised by Newton, and this revision is especially important. Newton discovered that the relationship between period and semimajor axis depends on the masses of the two objects. Kepler could not have realized this, because the Sun is much more massive than any of the planets, so that the sum of the masses for any Sun-planet pair is always about the same. Kepler wrote his third law in the form $P^2 = a^3$, where P is the planet's period in years and a is the semimajor axis in astronomical units. Newton realized that the sum of the masses times the square of the period of a pair of bodies in orbit is proportional

to the cube of the semimajor axis. In equation form, this says

$$(m_1 + m_2)P^2 = a^3,$$

where m_1 and m_2 represent the masses of the two bodies, for example the Sun and one of the planets. The masses must be expressed in terms of the Sun's mass in this equation. If we use other units, then additional numerical constants are required for the equation to be valid. For example, in the units used by astronomers, where masses are expressed in grams, the period is in units of seconds, and the semimajor axis is expressed in centimeters, the equation becomes

$$(m_1 + m_2)P^2 = \frac{4\pi^2}{G}a^3,$$

where G is the gravitational constant (the equation has the same form in the meter-kilogram-second system of units, but the value of G is different in that system).

The practical importance of Newton's revised version of Kepler's third law is that it becomes possible to use the law to determine the masses of distant objects. In any case where the period and the semimajor axis of an orbiting object can be observed directly, the equation can be solved for the sum of the masses:

$$m_1 + m_2 = \frac{a^3}{P^2}.$$

For example, consider how Newton could have derived the mass of Saturn. Titan, the largest of Saturn's moons, has an orbital period of 15.945 days = 0.04365 years, and a semimajor axis of 1,222,000 km = 0.00816 AU. Substituting these values for P and a in the expression just cited leads to $m_1 + m_2 = 0.000285$ solar masses; in other words, the sum of the masses of Titan and Saturn is only 0.000285 times the mass of the Sun. Since Titan is very much smaller than Saturn, we can assume that its mass is negligible and that we have found the mass of Saturn. (Compare this with the value in Appendix 7 for the mass of Saturn.)

It is possible to determine the masses of orbiting stars or even orbiting galaxies by the same technique. We will find occasion to use Kepler's third law in this manner throughout our study of astronomy.

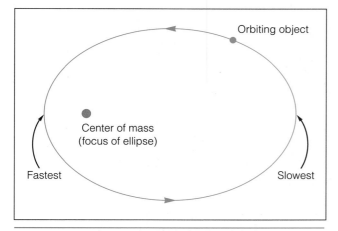

Figure 5.11 Conservation of angular momentum. *An object moves in an elliptical orbit with varying speed, and the product of its mass times its velocity times its distance from the center of mass (that is, its angular momentum) is constant (this statement is precisely correct only for a circular orbit; for an ellipse, the velocity in question is the component that is at right angles to the Sun-planet line). Kepler's second law of planetary motion is a rough statement of this fact.*

ESCAPE VELOCITY AND GAS MOTIONS

The consideration of orbital motions in terms of kinetic and potential energy leads to the concept of an **escape**

speed. If an object in a gravitational field has greater kinetic than potential energy, it will escape the gravitational field entirely. To launch a rocket into space (that is, completely free of the Earth) requires giving it enough upward speed at launch so that its kinetic energy is greater than its potential energy caused by the Earth's gravitational attraction (see Fig. 5.9). It so happens that the speed required to accomplish this is the same for an object of any mass. In equation form it is

$$v_e = \sqrt{\frac{2\,GM}{R}},$$

where v_e is the escape speed, G is the gravitational constant, and M and R are the Earth's mass and radius, respectively. For the Earth this speed is 11.2 kilometers per second, or slightly more than 40,000 kilometers per hour. Escape speeds for all the planets are given in Table 5.3.

The particles in a gas such as the Earth's atmosphere move all the time, and the speed with which they do so is related to the temperature of the gas (Fig. 5.12). In a hot gas, the atoms and molecules move more rapidly than in a cool gas. This is why a heated gas expands; the individual particles move more rapidly and therefore exert greater pressure on their surroundings. In a strict sense, temperature can be defined in terms of molecular motions (this leads directly to the concept of **absolute zero,** the temperature at which all molecular activity ceases). We can speak of the kinetic energy of individual particles in a gas, or more realistically, the average kinetic energy, which in turn is based on the average particle speed. For any gas, the temperature is proportional to average kinetic energy.

In physics and astronomy, temperatures are usually expressed in terms of the absolute scale, in which zero is absolute zero, and the degrees, equal to one one-hundredth of the difference between the freezing and boiling points of water are called **Kelvins** (K). Water freezes at 273 K and boils at 373 K. Room temperature is about 295 K. For details, see Appendix 4.

When we discuss the planets, we will be concerned with the likelihood of specific gases escaping into space, thus leaving the atmosphere devoid of those gases. We can see from this discussion that the probability that a given gas will escape a planet depends on the temperature of the atmosphere. Recall also that the kinetic energy of a moving particle depends on its mass as well as its speed; this means that in a gas where the particles have a uniform average kinetic energy (as is normally the case), the more massive particles, on average, are moving more slowly than the light particles. Therefore, the lightest gases are the most likely to escape; their higher velocities are more likely to exceed the planet's escape velocity. We will find, for example, that some of the planets (including Earth) have lost their light gases such as hydrogen and helium while retaining the

Table 5.3 Escape speeds

Body	Escape speed (km/sec)
Earth	11.2
Sun	618
Moon	2.4
Mercury	4.3
Venus	10.3
Mars	5.0
Ceres (largest asteroid)	0.6
Jupiter	60.8
Saturn	36.6
Uranus	21.2
Neptune	23.5
Pluto	1.2

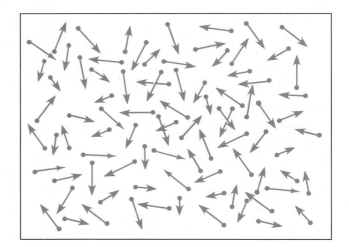

Figure 5.12 Gas motions.   *The average kinetic energy of particles in a gas is proportional to the temperature. If the gas particles, either because of high temperature or low mass, exceed a planet's escape velocity, they can escape into space. In any gas, there is a range of particle velocities, and if the average velocity for a particular gas is as much as one-sixth of the escape velocity, it will eventually escape altogether.*

ASTRONOMICAL INSIGHT 5.2

Relativity

In the text, Newton's work was characterized as being nearly modern in the sense that his laws of motion are still thought today to be adequate representations of the manner in which forces and bodies interact. Although it is true that Newtonian mechanics is sufficient for most practical applications, there are circumstances in which this is not so. By the late 1800s, some difficulties with Newton's laws were becoming recognized, and in the first decades of this century these problems led Albert Einstein to develop his theories of special and general relativity.

The circumstances in which Newton's laws fail are those in which we deal with extremely high velocities (that is, velocities that are a significant fraction of the speed of light), in which case special relativity theory must be used; or when very strong gravitational fields are involved, when general relativity theory is required. Actually, relativity is always a more accurate representation than Newton's laws, but the two theories give virtually identical results except under the circumstances just cited.

Mathematically, both special and general relativity are too complex to be presented here in any detail. It is possible, however, to provide some intuitive description of what is involved. The dilemma that led Einstein to develop special relativity had to do with the speed of light. To see

this, first let us consider a case involving lower speeds. Suppose a jet aircraft moving at 1,000 miles per hour fires a missile in the forward direction at a speed relative to the plane of 2,000 miles per hour. The speed of the missile with respect to the ground would then be the sum of the plane's speed and the speed with which the missile left the plane, that is, 3,000 miles per hour. This is the result expected from Newtonian mechanics.

Now imagine a train moving along a track, and suppose that it turns on its headlight. The light leaves the train at the normal speed of light relative to the train, and by analogy with the previous examples, we might expect that the speed of the light with respect to a bystander would be equal to the normal speed of light plus the speed of the train. For firm theoretical and experimental reasons, however, it is thought to be impossible for the normal speed of light to be exceeded. In other words, an observer on the train would actually measure the same speed for the light as would an observer standing by the track, in contradiction to the example of the plane and its missile. In Einstein's time, an actual measurement of this sort had not been made, but to him there was another reason for postulating that the speed of light should be the same for all observers: If this were not so, it would be implied that there was some fundamental difference between one frame of reference and others. There would be a "fundamental" frame in which light would have the speed required by theory, and all other frames would be somehow less

fundamental. This would imply that there was a preferred reference frame in the universe, an idea contrary to the lessons learned long before, when it was realized that the Earth was not the center of the universe.

Einstein's solution to the problem involved a surprising postulate: that the rate of passage of time itself depends on the relative speed between observers. Let us think about the train example again. If we believe that the speed of light must be the same for someone on the train as for someone standing by the track, even though the train is moving, then we are forced to accept Einstein's postulate if both observers are to measure the same speed of light. For the two observers to measure the same speed for the light despite their motion with respect to each other, it is necessary for the time interval to be different for the two observers.

If the rate of time depends on the motion of the observers with respect to each other, there are a number of interesting consequences, some of which can be (and have been) tested experimentally. For example, people who travel at high speeds through space will return to Earth younger than they would have been if they had stayed home, because time will not pass as rapidly for them. Even the sizes and masses of objects are affected: The length of an object that is moving at high speed will appear smaller to an observer at rest than to someone moving with the object, and its mass will be greater as measured by the person at rest.

(continued on next page)

ASTRONOMICAL INSIGHT 5.2
Relativity (continued)

Whereas special relativity states that the laws of physics are the same for observers moving at different *velocities*, general relativity addresses the situation for observers undergoing different *accelerations*. We have seen that a gravitational field such as the Earth's creates acceleration: A falling object increases its speed with every second as it falls. Even without falling, an object is subject to a force because of the acceleration of gravity; we are all familiar with the fact that an object at the surface of the Earth has weight. Now suppose we are somewhere out in space, inside an enclosed spacecraft, and that the spacecraft is accelerating at a rate equivalent to the gravitational acceleration at the Earth's surface. Inside the spacecraft we would "feel" this acceleration, which would create a force (in the direction opposite the spacecraft's motion) exactly identical to the weight we would have on the surface of the Earth. There is no experimental way to tell the difference between the two situations.

Now suppose that the person in the accelerating spaceship fires a gun. The bullet, being "left behind" as the ship continues to increase its speed, will follow a curving path with respect to the spacecraft, falling toward the rear. This is identical to the curving path of a bullet fired at the surface of the Earth; as the bullet travels toward its target, it drops toward the ground because of gravity. Now suppose that the "bullet" is a beam of light. In the spaceship, the beam would curve away from the direction of the ship's acceleration, just as a material bullet would. For the two situations to be fully equivalent, this implies that a beam of light on Earth also would curve downward because of gravity. Thus, Einstein predicted that a gravitational field should bend light, and this was later confirmed experimentally. Since light always follows the shortest possible path between two points, Einstein characterized the effect of a gravitational field as a "bending" of space itself. Hence, the light could travel the shortest path in the gravitational field, but this path would look like a curve to an observer not subjected to the same gravitational field.

Today we speak of **space-time**, indicating that time itself is thought of as a dimension because of the role it plays in governing the measured properties of an object; we also speak of **curved space-time** because of the curvature that occurs in the presence of a gravitational field. Throughout this text, but especially in the later chapters, where we discuss situations involving very high velocities or extreme gravitational fields, these concepts of special and general relativity will come into play.

---

heavier ones. We will speak often of this concept in Section III, where we describe the solar system and the properties of the planets.

TIDAL FORCES

A number of important astronomical phenomena can now be understood in terms of gravitational forces. One is tides, which, as we will see, occur in many situations besides our earthly oceans.

We have seen that the gravitational force resulting from a distant body decreases with distance. This means that an object subjected to the gravitational pull of such a body feels a stronger pull on the side nearest that body, and a weaker pull elsewhere. For example, the side of the Earth facing the Moon feels the strongest attraction toward the Moon, and the side opposite the Moon feels the weakest force. The Earth is therefore subjected to a **differential gravitational force,** which tends to stretch it along the line toward the Moon. Of course the Sun exerts a similar stretching force on the Earth, but it is too distant to have as strong an effect as the Moon. (However, its total gravitational force on the Earth is much greater than that of the Moon.) A **tidal force,** as differential gravitational forces are called, depends on how close one body is to the other, because the key is how rapidly the gravitational force decreases over the diameter of the body subject to the tidal force, and it decreases most rapidly at close distances. (The tidal force, in contrast with the total gravitational force, decreases as the cube, not the square, of the distance between two bodies (see the Mathematical Methods section at the end of Chapter 13). This is why the Sun, as already mentioned, exerts a greater total gravitational force on the Earth than the Moon does, even though the Moon's tidal force on the Earth is greater. Thus the Moon, not the Sun, dominates tides on the Earth.)

The Earth is more or less a rigid body, so it does not stretch very much as a result of the differential gravitational force of the Moon. The Earth's oceans, however, are fluid, and are influenced strongly by the Moon's differential gravitational force. In Fig. 5.13, consider the bold arrows indicating the strength of the Moon's gravitational force on the Earth on the near side, the center of the Earth, and on the far side. The force on the near side is greater than on the center, so the difference between these forces is a net force toward the Moon, relative to the Earth's center, and this creates a bulge of ocean water on this side. The force on the far side of the Earth is *less* than that on the Earth's center, so the difference between these two forces amounts to a net force *away* from the Moon on the Earth's far side, relative to the center of the Earth, and another bulge of ocean water is created on that side. The *differential* forces are indicated by the thin arrows in Figure 5.13. Thus, the Earth's oceans have two tidal bulges—one on the side facing the Moon, and one on the opposite side. These two bulges tend to maintain their alignment toward and away from the Moon's direction, so that as the Earth rotates, at any given location on its surface we encounter high tides twice

a day, with low tides in between. The separation between high tides is actually a bit longer than 12 hours, because the Moon moves along in its orbit while the Earth rotates.

It is interesting to consider what is happening to the Moon at the same time. It is subjected to a more intense differential gravitational force than the Earth, since the Earth is more massive than the Moon. Even though the Moon is a solid body, its shape is deformed by this force, and it has tidal bulges. The Moon is slightly elongated along the line toward the Earth. This has had drastic effects on the Moon's rotation, causing it to keep one side facing the Earth (Fig. 5.14). The original spin of the Moon is thought to have been much faster than it is today, but tidal forces have slowed the spin until the rotation period and the orbital period became equal. This situation, called **synchronous rotation,** also occurs for many other satellites in the solar system.

There are many other examples of tidal forces, both in the solar system and outside it. The satellites of the massive outer planets are subjected to severe tidal forces, and there are double-star systems in which the two stars are so close together that they are stretched into elongated shapes. We will discuss these in more detail later, along with galaxies that are affected by tidal forces, sometimes even tearing each other apart.

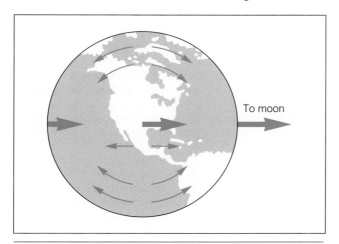

Figure 5.13 The Earth's tides. *The differential gravitational force caused by the Moon tends to stretch the Earth (large arrows). Seawater at any given point on the Earth is subjected to a combination of vertical and horizontal forces, causing it to flow toward either the side of the Earth facing the Moon or the side opposite it (curved arrows).*

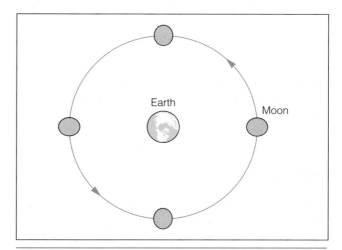

Figure 5.14 Synchronous rotation of the Moon. *The Moon is subjected to a tidal force caused by the Earth that is strong enough to slightly deform the shape of the Moon. This sketch shows how the elongated (here greatly exaggerated) Moon keeps one bulge facing the Earth at all times.*

PERSPECTIVE

We have traced the development of astronomy to the point where we can now describe, in terms of a few simple laws of physics, the motions of the bodies in the solar system. These laws also provide a basis for understanding the motions of more distant astronomical objects. The law of universal gravitation will be invoked time and time again in our study of the solar system and the rest of the universe, because so many important phenomena are explained by it.

Newton's laws of motion and gravitation have stood the test of time rather well. Einstein's theory of general relativity, as we will learn in Chapters 24 and 32, may be viewed as a more complete description of gravity and its interaction with matter. For most situations on the Earth's surface and in space, however, Newton's laws are perfectly adequate.

We are now ready to discuss other laws of physics, particularly those which govern the emission and absorption of light.

SUMMARY

1. Isaac Newton, in the late 1600s, developed the laws of motion, the law of gravitation, and calculus, and made many important contributions to our knowledge of the nature of light and telescopes.

2. Newton's first law states that an object at rest or in a state of uniform motion tends to remain in that state; this is the principle of inertia previously formulated by Galileo.

3. The second law of motion states that when an object is accelerated by a force, the amount of acceleration is equal to the force divided by the mass of the object.

4. Newton's third law states that for every force, there is an equal and opposite force.

5. The law of universal gravitation states that any two objects in the universe attract each other with a force that is proportional to the product of their masses and inversely proportional to the square of the distance between them.

6. Newton's laws, along with the concepts of kinetic and potential energy and angular momentum, can be used to explain orbital motions.

7. Kepler's third law was modified by Newton to show that the relationship between the period and semimajor axis of an orbit is dependent on the sum of the masses of the two objects; this is an important tool

for calculating the masses of distant objects.

8. Every body such as a planet or a star has an escape speed, the speed at which an object moving upward has more kinetic than potential energy and will therefore escape into space.

9. The average speed of particles in a gas is a function of the temperature of the gas and the mass of the particles; hence the probability that a given gas will escape a planetary atmosphere depends on these two quantities.

10. Differential gravitational forces are responsible for tides on the Earth and in the interiors of other planets and satellites.

REVIEW QUESTIONS

Thought Questions

1. Summarize the achievements of Newton in physics and astronomy.

2. Contrast the explanation Newton would have given for the fact that a moving object stops unless you keep pushing it with the explanation that Aristotle would have given.

3. Explain, in your own words, the difference between weight and mass.

4. Why does the law of inertia (Newton's first law) imply that there must be a force of attraction (rather than repulsion) between each planet and the Sun?

5. Explain the difference between acceleration and velocity, and between velocity and speed.

6. Explain how a book lying on a table can be stationary, even though it exerts a force on the table and the table exerts a force on it.

7. Explain why an astronaut in an orbiting spacecraft is "weightless." Is there really no force of gravity exerted on the astronaut?

8. Explain the difference between potential and kinetic energy, and between energy and power.

9. Why did Kepler not discover that his third law depends on the sum of the masses of the Sun and each planet?

10. Explain why the Earth's oceans have two high tides per day.

Problems

1. Using your own sea-level weight, compute your mass in kilograms.

2. Suppose that you are in orbit building a space station, and you push a crate that has one-third the mass that you have, so that the crate floats off in one direction and you

float away in the opposite direction. How does your acceleration compare with that of the crate?

3. Jupiter is about 5 times farther from the Sun than the Earth is, and has about 300 times the mass of the Earth. Compare the gravitational force between Jupiter and the Sun with that between the Earth and the Sun. Perform the same calculation for Saturn, which is about 10 times farther from the Sun than the Earth is, and has about 100 times the mass of the Earth.

4. How much less would you weigh on the summit of Mt. Everest, which is 29,000 feet (8.84 km) above sea level, if you weigh 130 lb at sea level? The radius of the Earth is 6,367 km.

5. What would you weigh on the surface of Titan, the large satellite of Saturn? (Use data from Appendix 7.)

6. Suppose you drop a rock from the edge of a high cliff. How fast is it falling after 3 seconds? Express your answer in feet per second and miles per hour.

7. How many ergs per second are emitted by a 100-watt light bulb?

8. Find the mass of Jupiter, in solar units, using Kepler's third law and data from Appendix 7 on the orbital period and semimajor axis of Callisto. Convert your answer from solar units to grams. How does your answer compare with the value given in Appendix 7?

9. Calculate the escape speed for Mars, and compare this with the escape speed for the Earth.

10. Convert a temperature of 55°F to the centigrade and absolute temperature scales. Convert a temperature of 77 K to the centigrade and Fahrenheit scales.

ADDITIONAL READINGS

Authoritative and thorough information on Newton, as well as many additional references, can be found in the following works.

Christianson, Gale E. 1987. Newton's Principia: A retrospective. *Sky and Telescope* 74(1):8.

Cohen, I. B. 1960. Newton in light of recent scholarship. *Isis* 51:489.

————— 1974. Newton. In *Dictionary of scientific biography,* vol. 10, ed. C. G. Gillispie. New York: Scribner's.

Drake, S. 1980. Newton's apple and Galileo's *Dialogue: Scientific American* 243(2):150.

Whiteside, D. T. 1962. The expanding world of Newtonian research. *History of Science* 1:15.

There are many books on conceptual understanding of physics in which more detailed explanations of the principles discussed in this chapter may be found. Many are used as textbooks in introductory physics courses, and should be easy to locate in a bookstore or library. A particularly good example is *Conceptual Physics,* by P. Hewitt (1981) (Little, Brown: Boston).

MATHEMATICAL METHODS

The Uses of Kepler's Third Law

Kepler's discovery of a relationship between orbital period and orbital semimajor axis allowed some properties of planetary orbits to be calculated. For example, use of this law made it possible to find the orbital period of any Sun-orbiting body, if its orbital size (i.e., semimajor axis) could be measured. Conversely, if the period were known, the orbital semimajor axis could be found.

We can illustrate each case with a simple example. In 1781, William Herschel discovered the planet Uranus. Repeated observations revealed that its distance from the Sun was about 19 AU. From Kepler's third law we find

$$P = \sqrt{a^3} = \sqrt{(19)^3} = \sqrt{6859} = 82.8 \text{ years.}$$

Thus it was possible to determine the period for the new planet without waiting for it to complete a full orbit (actually, multiple observations of the planet's position at different times allowed much more detailed calculations of its orbit, including the shape of the ellipse as well as the period). More accurate values are given in Appendix 6.

When Newton's associate Edmond Halley realized that the bright comets that were seen every 76 years must actually be the same object orbiting the Sun with a 76-year period, he was able to calculate its orbital size (semimajor axis), again by using Kepler's third law:

$$a = \sqrt[3]{P^2} = \sqrt[3]{(76)^2} = \sqrt[3]{5776} = 17.9 \text{ AU.}$$

(The symbol $\sqrt[3]{\ }$ means "cube root of", i.e., the number which, when multiplied by itself 3 times, yields the value inside the symbol. This can be found by trial and error, or by using logarithms, or by using a calculator.) Halley's comet follows an elliptical orbit that is very elongated (i.e., very eccentric); again, multiple observations of its position allowed

(continued on next page)

more detailed calculations of orbital shape to be carried out. The orbit of Halley's comet is described more fully in Chapter 15.

In the text, we presented an example of a very important and useful type of calculation that can be made using the full form of Kepler's third law as derived by Newton, which involves the sum of the masses of the two bodies. If the period and the semimajor axis are known for a pair of bodies in mutual orbit, then it is possible to calculate the sum of the masses of the two bodies:

$$m_1 + m_2 = \frac{a^3}{P^2}.$$

The example of this that was presented in the text involved a satellite orbiting Saturn, and we were able to make the simplifying assumption that the mass of the satellite was very much smaller than that of Saturn, so that we could ignore the satellite's mass. This allowed us to determine the mass of Saturn alone.

Let us now consider a case in which the two masses are comparable, so that we cannot ignore one of them. Suppose a double star is found, in which two stars orbit a common center of mass (such systems are quite common, as we will see in Chapter 18). Let us suppose that the orbital period for the double star is 10 years, and the semimajor axis of the orbit is 10 AU. Now when we solve for the sum of the masses we find

$$m_1 + m_2 = \frac{(10)^3}{(10)^2} = \frac{1000}{100} = 10 \text{ solar masses.}$$

The sum of the two masses is 10 times the mass of the Sun.

Since we are dealing with two stars here, we cannot simply decide that one of the two masses can be ignored. They might be similar to each other, or even equal. We must find further information, so that we can determine the individual masses. If we knew the ratio of the masses as well as the sum, this would allow us to solve for the individual masses. This can be done if the relative distances of the two stars from the center of mass can be found. The more massive star will be closer to the center of mass by a factor that is inversely proportional to its mass (you can think of the analogy with two people on a seesaw; the heavier one will have to sit closer to the pivot point if the two are to balance). Suppose, in our present example, that one star was twice as far from the center of mass as the other (see figure). This would mean that the closer star was twice as massive as the outer star.

If the sum of the masses is 10 solar masses, and one star is twice as massive as the other, then the in-

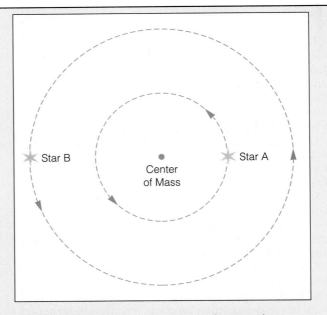

dividual masses have to be 6.66 and 3.33 solar masses (i.e., two-thirds and one-third of the total mass). It is usually helpful to write equations for the sum and the ratio of the masses, since most real cases do not involve such simple totals as 10 solar masses or fractions of one-third and two-thirds. If we designate the two stars as Star A (the one closer to the center of mass) and Star B, then we can write

$$m_A + m_B = 10$$

and

$$m_A = 2m_B.$$

Now we can substitute the second equation for m_A into the first equation:

$$2m_B + m_B = 10,$$

or

$$3m_B = 10,$$

which yields

$$m_B = \frac{10}{3} = 3.33 \text{ solar masses.}$$

Then we find m_A from

$$m_A = 10 - m_B = 10 - 3.33 = 6.66 \text{ solar masses.}$$

In practice, it is not always possible to find the ratio of the masses, usually because the orbits cannot be well enough observed to determine the relative distances of the two stars from the center of mass. Hence, astronomers often are unable to determine the two masses precisely, and are forced to use crude estimates. Even these would not be possible without the use of Kepler's third law.

CHAPTER 6

The Nature of Light

Refraction of light by water droplights. (D. Johnson, Science Source/Photo Researchers)

CHAPTER PREVIEW

Some of the tools for unlocking the secrets of the universe became available with the publication of Newton's *Principia* in the 1680s, but others had to wait 200 years or more to be discovered. The laws of motion allow astronomers to understand how the heavenly bodies move, and were of fundamental importance in unraveling the clockwork mechanism of the solar system. To understand the essential nature of a distant object, however, to learn what it is made of and what its physical state is, requires an understanding of light and how it is emitted and absorbed. The only information we can obtain on the nature of a distant object is conveyed by the light from it. Fortunately, an enormous amount of information is there, if we know how to dig it out.

THE ELECTROMAGNETIC SPECTRUM

One characteristic of light is that it acts as a wave. It is possible to think of light as passing through space like ripples on a pond (although, as we will discuss shortly, the picture is actually somewhat more complicated than that). The distance from one wavecrest to the next, called the **wavelength** (Fig. 6.1), distinguishes one color from another. Red light, for example, has a longer wavelength than blue light. It is possible to spread out the colors in order of wavelength, using a prism to obtain the traditional rainbow. Newton was the first to discover that sunlight contains all the colors, and he did so by performing experiments with a prism (Fig. 6.2). Whenever light is spread out by wavelength, the result is called a **spectrum**; more technically, a spectrum is the arrangement of light from an object according to wavelength. The science of analyzing spectra is called **spectroscopy**, and will be discussed at some length later in this chapter.

Let us consider for a moment what lies beyond red at one end of the spectrum or beyond violet at the other end. By the mid-1800s, experiments had been done to demonstrate that there is invisible radiation from the Sun at both ends of the spectrum. At long wavelengths, beyond red, is **infrared** radiation, and at short wavelengths is **ultraviolet** radiation. The spectrum continues in both directions, virtually without limit. Going toward long wavelengths, after infrared

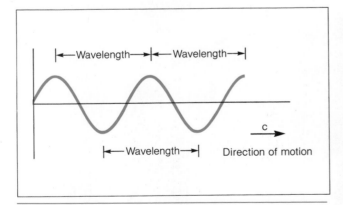

Figure 6.1 Properties of a wave. *Light can be envisioned as a wave moving through space at a constant speed, usually designated c. The distance from one wavecrest to the next is the wavelength, often denoted by the Greek letter lambda, λ. The frequency f is the number of wavecrests to pass a fixed point per second and is related to the wavelength and the speed of light by $f = c/\lambda$.*

Figure 6.2 Prism dispersing light. *The rainbow of colors seen here is a spectrum. (Runk/Schoenberger, Grant Heilman Photography)*

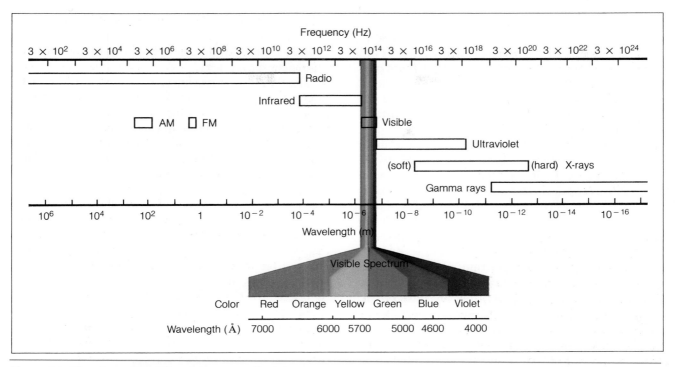

Figure 6.3 The electromagnetic spectrum. *All of the indicated forms of radiation are identical except for wavelength and frequency.*

light, are microwave and radio waves; and toward short wavelengths, after ultraviolet, come X-rays and then gamma rays (γ-rays). All these kinds of radiation are just different forms of visible and invisible light, distinguished only by their wavelengths, and together they form the **electromagnetic spectrum** (Fig. 6.3). Electromagnetic radiation is a general term for visible light, X-rays, radio waves, etc.

The reason for the name *electromagnetic radiation* is that the wave motion associated with this radiation consists of alternating electric and magnetic fields (Fig. 6.4). These fields propagate through space (or a medium) as they alternate, with the electric and magnetic fields always lying in planes that are perpendicular to each other. The speed of propagation in a vacuum is always the same (almost exactly 300,000 km/sec, or about 186,000 miles per second), but is slower in a medium such as air or glass. The electric and magnetic fields are said to be *in phase* with each other, meaning that both reach their maximum and minimum values at the same time as the wave propagates, with the relative orientation shown in Figure 6.4.

The range of wavelengths from one end of the electromagnetic spectrum to the other is immense (see

Table 6.1). Visible light has wavelengths ranging from 0.00004 to 0.00007 centimeters (cm). A special unit called the **Ångstrom** (Å) is used to measure light, defined such that 1 cm = 100,000,000 Å, or 1 Å = 0.00000001 cm = 10^{-8} cm. Thus, visible light lies

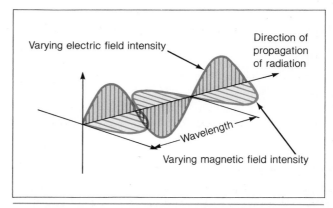

Figure 6.4 An electromagnetic wave. *This illustrates how the electric and magnetic fields oscillate in an electromagnetic wave. The directions of the fields and the direction of the wave travel always have the relative orientations shown here.*

between 4,000 Å and 7,000 Å in wavelength.

A variety of other special units are used by astronomers in referring to electromagnetic radiation of different types, but to reduce confusion, we will use only Ångstroms and centimeters. Even this is perhaps a bit complicated, but the usage is so standard that the student of introductory astronomy should be comfortable with both units.

Infrared radiation has wavelengths between about 7,000 and a few million Ångstroms—that is, between 7×10^{-5} cm and 2–3×10^{-2} cm. Microwave radiation (which includes radar wavelengths) lies roughly between 0.1 and 50 cm, with no well-defined boundary separating this region from radio, which simply includes all longer wavelengths, up to many meters or even kilometers.

At the other end of the spectrum, ultraviolet radiation is usually considered to lie between 100 and 4,000 Å, whereas X-rays are in the range of 1 to 100 Å, anything shorter than that being considered γ rays.

Thus the entire electromagnetic spectrum has a wavelength range from less than 1 Å to many kilometers—that is, from 10^{-16} to 10^6 cm. In principle, it extends even farther at both ends, but few physical processes are capable of producing such radiation. The portion of the spectrum to which the human eye responds is extremely limited compared with the full range. The eye undoubtedly has evolved to be sensitive to just this region for a combination of reasons: (1) The Sun emits most strongly in visible wavelengths; (2) the Earth's atmosphere blocks out many wavelengths of radiation, but visible light lies in one of the wavelength regions that can pass through to the ground. Therefore, it was advantageous for creatures dwelling on the Earth's surface to develop the ability to see radiation at these particular wavelengths.

The concept of **frequency** is often used as an alternative to wavelength in characterizing electromagnetic waves. The frequency is the number of waves per second that pass a fixed point, and it is determined by the wavelength and the speed with which the waves move. The speed of light, usually designated c, is constant, and the frequency of light (f) with wavelength λ is

$$f = \frac{c}{\lambda}.$$

The standard unit for measuring frequency is the **hertz** (Hz), 1 Hz being equal to 1 wave per second. The frequency of visible light with a wavelength of 5,000 Å is 6×10^{14} Hz. For a radio photon in the AM part of the spectrum, frequencies are typically around 10^6 Hz, and at the other extreme, γ-rays have frequencies around 10^{19} Hz.

Particle or Wave?

So far we have discussed light only as a wave, but there are situations in which it acts more like a stream of particles. Newton developed a corpuscular theory of radiation in which he assumed that light consisted of particles. But at the same time other experimenters, including most notably Huygens, performed experiments showing light to have definite wave characteristics.

Table 6.1 Wavelengths and frequencies of electromagnetic radiation

Type of Radiation	Wavelength Range (cm)*	Frequency Range (per second)
Gamma rays	$< 10^{-8}$	$> 3 \times 10^{19}$
X-rays	1–200×10^{-8}	1.5×10^{16}–3×10^{18}
Extreme ultraviolet	200–900×10^{-8}	3.3×10^{15}–1.5×10^{16}
Ultraviolet	900–$4,000 \times 10^{-8}$	7.5×10^{14}–3.3×10^{15}
Visible	$4,000$–$7,000 \times 10^{-8}$	4.3×10^{14}–7.5×10^{14}
Near-infrared	0.7–20×10^{-4}	1.5×10^{13}–4.3×10^{14}
Far-infrared	20–100×10^{-4}	3.0×10^{12}–1.5×10^{13}
Radio	> 0.01	$< 3 \times 10^{12}$
(Radar)	$(2$–$20)$	$(1.5$–$15 \times 10^9)$
(FM radio)	$(250$–$350)$	$(85$–$120 \times 10^6)$
(AM radio)	$(18,000$–$38,000)$	$(540$–$1,600 \times 10^3)$

*All wavelengths are given in centimeters; recall that 1×10^{-8} cm = 1 Ångstrom.

ASTRONOMICAL INSIGHT 6.1

The Perception of Light

Newton's experiments in optics are commonly thought to have marked the beginning of scientific inquiry into the nature of light, but in fact Newton was preceded in this area by many scientists and philosophers. Much of the early research, and even, to some extent, Newton's work, did not deal with the physical properties of light, but rather was aimed at understanding how the human eye senses light. The ancient Greeks, the earliest philosophers who studied the nature of light, did not recognize the distinction between light and the act of seeing it.

The Greeks studied the principles of vision in much the same way they studied astronomy: They performed very few experiments or even systematic observations, choosing instead to unravel the secrets of the universe by reason and logic. The prevalent concept, described by such notable figures as Hippocrates and Aristotle, was that the eye somehow emitted rays or beams with which it sensed things. This was analogous to the sense of touch, where a hand is extended to feel objects. It was known that objects appeared distorted or bent when viewed under water, but this was attributed to bending of the rays that came from the eye, not to effects on something coming from the object being viewed. It would have been possible for the Greeks to experiment with optics by us-

ing natural crystals or eyes removed from dead animals, but no such experimentation was done.

In a definitive summary of all that was known about vision, the Greek scientist Galen, in the second century A.D., wrote of an "animal spirit" that flowed from the brain along the optic nerve to the eye, where it was converted into a "visual spirit" in the retina, a membrane covering the rear interior portion of the eyeball. The lens, in the front of the eyeball, was thought to be responsible for sending out the beam with which the external world was perceived.

In analogy with the astronomical work of Ptolemy, the concepts of the nature of light and vision that were summarized by Galen were accepted as doctrine for some fifteen centuries to come, until the time of the Renaissance. Some of the ancient Greek ideas, such as the notion that the optic nerve was hollow so that the animal spirit could flow along it from the brain, were especially persistent. Even Leonardo da Vinci, who in the late fifteenth and early sixteenth centuries carried out pioneering work in anatomy, accepted much of the old thinking. Da Vinci took a more modern view in that he believed that light rays passing from an object to the eye play an important role, but he was unwilling to entirely give up the idea that other rays emanate outward from the eye.

In the early seventeenth century, none other than Johannes Kepler and the great French philosopher René Descartes developed an understanding of refraction and the formation of an image by a lens, and both correctly viewed vision as the process of image formation on the retina

by the lens in the front of the eyeball. Thus, when Christian Huygens and then Newton performed their experiments later in the seventeenth century, it was already well established that seeing was accomplished by sensing something that passes from the object to the eye.

Both men were more concerned with the nature of this radiation than with the properties of the eye, although Newton did discuss the perception of color, which he realized to be entirely a function of the eye. This is an important concept: The rays of light are not colored, nor, technically, is the object being viewed. It is the eye that senses wavelength differences and the brain that interprets them as colors. Light rays consist simply of alternating electric and magnetic fields.

Other advances in understanding the physical properties of light, made by scientists who followed Newton, are described in the text. Research has also been aimed at learning how the human eye perceives light. We now know that the eye has a logarithmic response, meaning that we see differences in brightness in a way that is not directly proportional to the actual intensity ratios of the observed objects. This point is explained more fully in Chapter 19, in the discussion of the stellar magnitude system.

Entire books have been written on the complex interplay between what the brain perceives and physical reality, and, of course, this interplay is important in astronomy. To a large extent, modern observational astronomy is the science of recognizing and bypassing the limitations or distortions created by our natural light-gathering system.

The fact that light bends around sharp corners, in a process called **diffraction** (which also affects water waves), is evidence for a wave nature, as is the fact that light waves can interfere with each other if their troughs and crests match up in such a way as to cancel each other out. On the other hand, it has been found that light can carry energy only in specific amounts, as though it came in individual packets, and that it can travel in a total vacuum, with no medium to transmit it.

Out of all the seemingly contradictory evidence has developed the concept of the **photon.** A photon is thought of as a particle of light that has a wavelength associated with it. The wavelength and the amount of energy contained in the photon are closely linked; in general terms, the longer the wavelength, the lower the energy. The energy can be expressed mathematically as

$$E = \frac{hc}{\lambda},$$

where h is called the Planck constant, c is the speed of light, and λ is the wavelength. It is important to understand that a photon carries a precise amount of energy, not an arbitrary or random quantity, and that when light strikes a surface, this energy arrives in discrete bundles, like bullets, rather than in a steady stream. When a photon is absorbed, this energy can be converted into other forms, such as heat.

From this discussion, it may be seen that short-wavelength radiation, such as ultraviolet light, X-rays, and γ-rays, has high energy associated with it, whereas longer-wavelength photons have low energies. It is extremely important to remember the relationship between wavelength and energy in the discussions of the emission and absorption of light later in this chapter.

Polarization

The light from a typical source, such as a light bulb or a distant star, consists of vast numbers of photons, each consisting of traveling electric and magnetic fields. Each photon has a characteristic orientation as it travels through space; that is, the planes in which its electric and magnetic fields oscillate remain fixed as the photon travels. Normally, the orientations in a stream of photons from a source are random, but in some circumstances they are not. When light is **polarized,** all the photons tend to have the same alignment of their electric and magnetic planes.

Light can become polarized in many different ways. When light from a source is reflected from a flat surface, usually one orientation is favored, and the reflected light is at least partially polarized. Perfect polarization, in which all the photons have precisely the same orientation, is very rare. Some kinds of stars are embedded in disks of gas that cause the emitted light to be partially polarized. Elongated grains of interstellar dust can partially polarize the light traveling through the galaxy from a star.

It is possible to determine whether the light from a source is polarized by observing it with a polarizing filter. This is a piece of semitransparent glass or plastic that allows only light with a certain orientation to pass through. (Polarizing sunglasses use such material to screen out reflected light, thus reducing glare.) If the filter is rotated, the greatest intensity of light will be seen through it when its orientation is the same as that of the polarized light from the source. Thus, if the brightness of a source is seen to vary when a polarizing filter is rotated, we know that the source is polarized.

CONTINUOUS RADIATION

Newton, in examining the spectrum of light obtained by passing sunlight through a prism, saw only the smooth spread of light over all the colors. Later observers, however, notably W. H. Wollaston and Joseph Fraunhofer, were able to see dark lines across the Sun's spectrum at certain fixed positions (Figs. 6.5 and 6.6). Fraunhofer noted more than 600 of these lines, and in 1817 published a catalog of many of the stronger ones. (Today these features are still called **Fraunhofer lines.**)

It must be emphasized that the position of each of these lines corresponds to a specific wavelength of light. Remember that the spectrum is a spread of light according to wavelength, and that each position in the spectrum corresponds to a certain wavelength of light. The fact that the lines are dark means that somehow the Sun emits less light at these particular wavelengths than at other wavelengths on either side.

In analyzing the light from a distant object such as a star, an astronomer either may focus attention on these lines, from which certain types of information can be derived, or may ignore the lines and concentrate instead on the overall distribution of the brightness of light as it varies with wavelength. Here we will make a distinction between **continuous radiation,** as the overall distribution is called, and spectral lines (Fig. 6.5).

Stars are glowing objects with a number of properties that can be described by a few simple laws of continuous radiation. Some of these laws were discovered in the mid-1800s through laboratory experiments by people such as Wien, Stefan, Boltzmann, and Rayleigh, and in the 1920s were placed in proper theoretical framework by Bohr and Planck.

Any object with a temperature above absolute zero emits radiation over a broad range of wavelengths, simply because it has a temperature (Fig. 6.7). Thus, not only stars, but also such commonplace objects as a human body or the wall of a room emit radiation. The continuous radiation emitted due to an object's temperature is called **thermal radiation.** (We will discuss only this type of continuous radiation here; in other chapters, such as 13 and 31, we will examine nonthermal sources of radiation.)

For all objects, thermal radiation is emitted over a broad range of wavelengths, with a peak in intensity at some particular wavelength. (Technically, in fact, at least some radiation is emitted at all wavelengths.) For the Sun, this peak is near a wavelength of 5,500 Å in the green portion of the spectrum, but to our eyes the Sun appears yellow because it emits almost as strongly over the much broader range of wavelengths that we see as yellow.

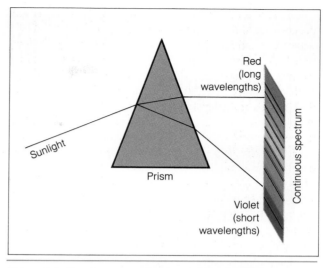

Figure 6.5 Continuous spectrum and spectral lines. *When sunlight is dispersed by a prism, the light forms a smooth rainbow of continuous radiation, gradually merging from one color into the next, with a maximum intensity in the yellow portion of the spectrum. Superimposed on this continuous spectrum are numerous spectral lines, wavelengths at which little or no light is emitted by the Sun.*

Figure 6.6 Lines in the Sun's spectrum. *This is a drawing of the solar spectrum made by Josef Fraunhofer, the first to systematically study the dark lines in the spectrum (now called Fraunhofer lines). The colors illustrate roughly the human-eye response to the wavelengths in the spectrum. Above is a graph showing the relative intensity of sunlight as a function of wavelength. (Deutsches Museum, Munich)*

ASTRONOMICAL INSIGHT 6.2

The Measurement of Wavelengths

We have seen that visible light has wavelengths that are exceedingly short by any ordinary standards, lying in the range from about 0.00004 to 0.00007 cm. You may have wondered how such small wavelengths are measured. Surprisingly, this was first done nearly 200 years ago, in 1801, in a famous experiment conducted by the English physicist Thomas Young.

Young's experiment relied on the fact that light has the properties of waves. (The experiment actually demonstrated this and is still considered one of the best arguments for a wave nature of light). When waves run into each other, they **interfere,** meaning that they interact in such a way that the magnitude of the waves is increased if the peaks and valleys match (this is **constructive interference**), or decreased if the peaks of one wave match the valleys of the other (**destructive interference**). Young showed that light waves undergo interference, thus demonstrating their wave nature, and in the process established a method for measuring their wavelengths.

Young allowed light to pass through a pair of closely spaced parallel slits (see figure), so that a pair of light beams emerged from the slits. What happens next is easiest to see if you imagine that the light all has the same wavelength. We know that the light in

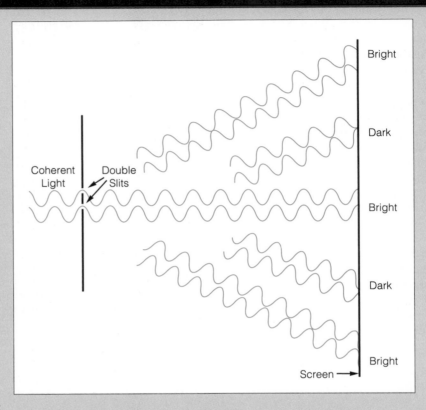

the two beams could interfere, depending on whether the troughs and valleys matched or not. If the light emerging from the slits is *in phase,* meaning that the troughs and peaks in the two beams match, then there will be constructive interference as the two beams go straight ahead from the slits. Thus, on a screen placed some distance behind the slits, a bright line is seen directly behind the slits.

Now imagine that instead of two simple beams of light going straight ahead, light actually emerges from the slits in all directions. (This is due to another wave property called **diffraction.**) Some goes straight ahead, but

some emerges in other directions, creating a diffuse glow that covers much of the screen that is placed beyond the slits. We may now view it as though there were many pairs of beams going in many directions as they emerge from the two slits. The pair that travels straight through to the screen interferes constructively as before, combining to create a bright strip on the screen directly behind the slits.

But now consider a pair of beams that emerges at an angle. One of the beams will travel a slightly longer path to the screen than the other, because of the separation between the two slits. Thus, the waves in the two beams

ASTRONOMICAL INSIGHT 6.2

are shifted with respect to each other, and the amount of shift will determine whether they interfere constructively or destructively. If one beam is shifted by half a wavelength with respect to the other, then the peaks from one will match the troughs of the other, and destructive interference occurs. There is therefore a dark strip at the location on the screen where this pair of beams would have reached it. On the other hand, if one beam is shifted with respect to the other by a distance equal to the wavelength (or a multiple of the wavelength), then the two beams will interfere constructively, creating a bright strip on the screen. The amount of shift between the two beams is related to the angle at which they emerge from the slits; as the angle changes, the shift between the beams changes, first reaching half of the wavelength, then the full wavelength, then one-and-a-half times the wavelength, then twice the wavelength, and so. Thus, interference between the two beams creates a pattern of alternating

bright and dark strips on the screen.

The key to measuring the wavelength of the light is that the separation of these bright and dark strips depends on the wavelength of the light. The longer the wavelength, the greater the shift between beams that is required in order to change from constructive to destructive interference, and the greater the separation between adjacent bright and dark strips on the screen. There are details that we have overlooked in this discussion, but you should see that measuring the wavelength of the light in this experiment essentially consists of measuring the separation of the light and dark strips on the screen.

If white light, consisting of all the colors, is used in the experiment, the bright and dark strips become broadened and less clearly separated. This is because different wavelengths of light will create the bright and dark strips at different locations on the screen. Each bright strip will now look like a rainbow because of this, as

different colors of light are brightest at different positions. Thus, a pair of slits creates a **spectrum** (actually, a series of spectra), consisting of light that is spread out according to wavelength. Instead of just a pair of slits, a series of many parallel slits can be used, with the same result.

If instead of an array of slits we use a reflecting surface that has many parallel grooves in it, then the reflected light will show the same separation of colors, because of shifts that occur in beams of light reflected from the edges and the bottoms of the grooves. This is how a **diffraction grating** works (see Fig. 7.11, page 136).

That an experiment devised as long ago as 1801 is still regarded as a modern tool for analyzing light is quite remarkable. Equally remarkable is that quantities as small as the wavelength of visible light were measured so long before there was any understanding of the processes by which light is emitted and absorbed.

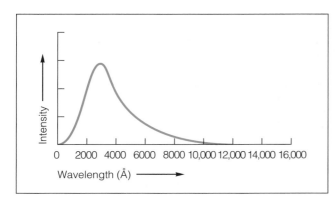

Figure 6.7 An intensity plot of a continuous spectrum. *This kind of diagram shows graphically how the brightness of a glowing object varies with wavelength. The curve shown roughly represents a star of surface temperature 10,000 K, whose continuous radiation peaks near 3,000Å in the ultraviolet.*

A simple relationship between the wavelength of maximum emission and the temperature of an object was discovered in 1893 by W. Wien and is now referred to as **Wien's law.** The law states that

The wavelength of maximum emission is inversely proportional to the absolute temperature.

That is, the hotter an object is, the shorter the wavelength of peak emission (Fig. 6.8). This explains the variety of stellar colors as being a result of a range in stellar temperatures (Fig. 6.9). A hot star emits most of its radiation at relatively short wavelengths and thus appears bluish, whereas a cool star emits most strongly at longer wavelengths and appears reddish. The Sun is intermediate in temperature and color.

When speaking of the colors of stars, we must keep in mind that a star emits light over a broad range of wavelengths, so we do not have pure red or pure blue stars. Our eyes receive light of all colors, and stars therefore are all white or pale shades of color near

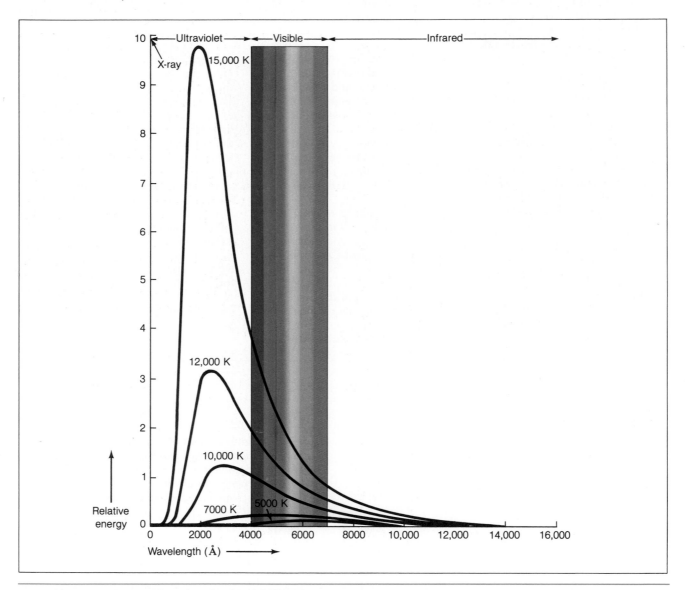

Figure 6.8 Continuous spectra for objects of different temperatures.   *This diagram illustrates Wien's law, which says that the wavelength of maximum emission is inversely proportional to the temperature (on the absolute scale). It also illustrates another property of thermal emission: The hotter of the two objects is brighter at* **all** *wavelengths.*

white. Our impression of color arises from the fact that there is a wavelength (given by Wien's law) at which a star emits more strongly than at other wavelengths. It is not as though the star emits only at that wavelength.

Wien's law applies as well to radiation beyond the visible range. The hottest stars actually emit most strongly in ultraviolet wavelengths, and very cool ones emit mostly in the infrared. The wavelength of maximum radiation is dictated by the temperature of the object. The human body, as well as all objects in the normal room-temperature environment, radiates in far-infrared wavelengths. Imagine the problems this creates for infrared astronomers, whose telescopes glow at the very wavelengths they are trying to observe! (We will describe the solution to this problem in Chapter 7.)

A second property of glowing objects, known as either **Stefan's law** or the **Stefan-Boltzmann law,** has to do with the total amount of energy emitted per second over all wavelengths and how this total energy is related to the temperature of an object. This law says:

The total energy radiated per square centimeter of surface area is proportional to the fourth power of the temperature.

This shows that the total energy emitted is very sensitive to the temperature; if we change the temperature a little, we change the energy a lot. If, for example, we double the temperature of an object (such as the electric burner on our stove), we increase the total energy per square centimeter it radiates by $2^4 = 2 \times 2 \times 2 \times 2 = 16$. If one star is 3 times hotter than another, it emits $3^4 = 3 \times 3 \times 3 \times 3 = 81$ times more total energy per square centimeter of surface area.

Notice that we have been careful to express this law in terms of the surface area. The total energy emitted by an object per second, which astronomers call the **luminosity** of an object, depends also on how much surface area it has. If we talk of stars or other spherical objects, which have a surface area of $4\pi R^2$, where R is the radius, we can say that

The total energy emitted is proportional to the fourth power of the temperature and to the square of the radius.

This can be illustrated by considering two stars, one that is twice as hot but has only half the radius of the other. The hotter star emits $2^4 = 16$ times more energy per square centimeter of surface but has only $(1/2)^2 = (1/4)$ as much surface area, hence it is $16 \times (1/4) = 4$ times brighter overall. If, on the other hand, this star were twice as hot and 3 times as large in radius as the other, it would be $2^4 \times 3^2 = 144$ times more luminous.

Both Wien's law and the Stefan-Boltzmann law were first found experimentally, much in the same manner as Kepler's discovery of the laws of planetary motion. In the case of planetary motions, it remained for Newton to find the underlying reasons for the laws, and he was able to derive them strictly on a theoretical basis. Analogously, Max Planck, the great German physicist who was active early in the twentieth century, found a theoretical understanding of thermal emission and was able to derive Wien's law and the Stefan-Boltzmann law from purely theoretical considerations. The basis

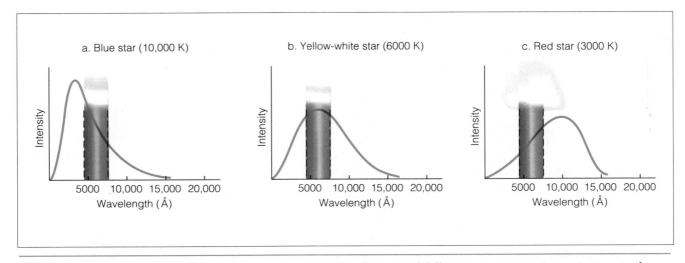

Figure 6.9 The colors of stars.   *The figure shows intensity plots for stars of different temperatures in comparison to the colors perceived by the human eye. This illustrates why stars have a range of colors as seen by the eye.*

of Planck's new understanding was the **quantum** nature of light—the fact, already discussed, that light has a particle nature and carries only discrete, fixed amounts of energy.

Finally, in addition to taking into account the temperature and size of a glowing object, we must consider the effect of its distance. So far we have discussed the energy as it is emitted at the surface, but not how bright it looks from afar. What we actually observe, of course, is affected by our distance from the object. For an object that emits in all directions the brightness varies according to the **inverse square law:**

The perceived brightness decreases as the square of the distance

as illustrated in Figure 6.10. (This should remind you of the law of gravitation.) Thus, if we double our distance from a source of radiation, it will appear $(1/2)^2$ = $(1/4)$ as bright. If we approach, reducing the distance by a factor of 2, it will appear 2^2 = 4 times brighter. As shown in later discussions of stellar properties, we must know the distances to stars before we can compare other properties having to do with their brightnesses.

THE ATOM AND SPECTROSCOPY

We turn our attention now to the spectral lines first noted by Wollaston and Fraunhofer in the spectrum of the Sun. These lines are difficult to see and are not apparent in the spectrum of light obtained with an ordinary prism. However, with the development of superior techniques for examining the spectrum, their existence became well established (see Fig. 6.6).

In the late 1850s, the German scientists R. Bunsen and G. Kirchhoff performed experiments and developed theories that made clear the importance of the Fraunhofer lines. Bunsen observed the spectra of flames created by burning various substances and found that each chemical element produced light only at specific places in the spectrum (Figs. 6.11 and 6.13). The spectrum of such a flame in this case is dark everywhere except at these specific places, indicating that the flame is emitting light only at certain wavelengths. The bright lines seen in this situation are called **emission lines** for that reason.

It was soon noticed that some of the dark lines seen in the Sun's spectrum coincide exactly in position with some of the bright lines seen by Bunsen in his laboratory experiments. Kirchhoff studied this in detail and was able to show that a number of common elements such as hydrogen, iron, sodium, and magnesium must be present in the Sun because of the coincidence in wavelengths of the lines. This was the first hint that the chemical composition of a distant object could be determined, something previously thought by astronomers to be forever beyond their ability to learn.

Kirchhoff further analyzed the absorption and emission of light and, through a series of experiments, was able to develop a set of rules describing when contin-

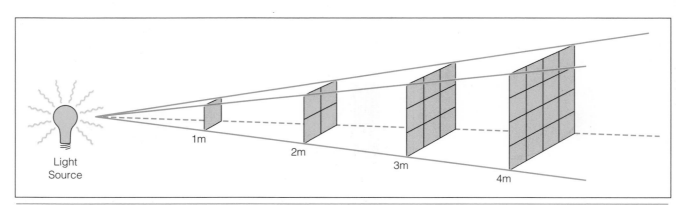

Figure 6.10 The inverse square law of light propagation. *This shows how the same total amount of radiation energy must illuminate an ever-increasing area with increasing distance from the light source. The area to be covered increases as the square of the distance; hence, the intensity of light per unit of area decreases as the square of the distance.*

uous radiation will be observed, when emission lines will be seen, and when dark **absorption lines** will form instead (Fig. 6.12):

1. A hot, dense gas or a hot solid produces a continuous spectrum (one with no lines).
2. A hot, rarefied gas produces emission (bright) lines.
3. A cool gas in front of a continuous source of light produces absorption (dark) lines.

These statements were postulated to explain why Bunsen saw only emission lines when he observed the spectra of flames, which are rarefied, hot gases, and they tell us something important about the Sun. The fact that the Sun's spectrum contains absorption lines (Fig. 6.14) implies that the outer layers of the Sun must be cooler than the interior, so that the continuous radiation is formed inside, where it is hot and dense, and absorption lines are formed as this continuous radiation passes out through the outer layers. Thus, a simple understanding of how gases in the laboratory emit and absorb light reveals something about the structure of the Sun.

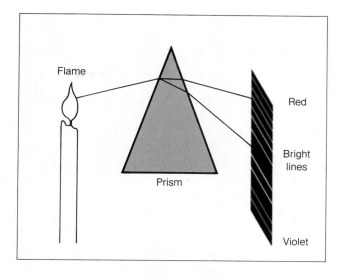

Figure 6.11 Emission lines. *The spectrum of a flame is devoid of light except at specific wavelengths where bright emission lines appear.*

Figure 6.12 Continuous, emission-Line, and absorption-line spectra. *The positions of the emission and absorption lines match because the same element emits or absorbs at the same wavelengths. Whether it emits or absorbs depends on the physical conditions, as described in Kirchhoff's laws.*

Molecular Hydrogen

Atomic Hydrogen

Sodium Lamp

Helium

Neon

Lithium

Mercury Lamp

Iron Arc

Barium

Calcium

Fraunhofer Lines

Tungsten Lamp

Fluorescent Lamp

Figure 6.13 Spectra of elements. *These are emission-line spectra for an assortment of chemical elements. Note that the pattern of lines is distinctly different from one element to the next. Experiments with such spectra, along with observations of dark lines in the Sun's spectrum, led to the realization that it is possible to identify the chemical composition of the Sun and distant stars. (Courtesy of Bausch and Lomb)*

Figure 6.14 Solar spectrum.   *In this portion of the Sun's spectrum, we see several strong absorption lines, most due to hydrogen atoms in the Sun's outer layers. (K. Gleason, Sommers Bausch Observatory, University of Colorado)*

Energy Levels and Photons

The work of Bunsen and Kirchhoff demonstrated that each element has its own unique spectrum. As we have seen, this discovery had profound implications, for it meant that astronomers could gain information about distant objects previously thought to be forever beyond their grasp. Although astronomers in the later half of the nineteenth century were able to use these astrophysical fingerprints, no real understanding of their nature and cause was developed until the early years of this century.

Spectral lines were studied by a number of scientists, and various regularities in the arrangement of the lines from a given element were noted. Although it was suspected that these regularities reflected some aspect of the structure of atoms, it was not until 1913 that the true relationship was discovered by Niels Bohr. By that time it had been established that an atom consists of a small, dense, positively charged nucleus surrounded by a cloud of negatively charged particles called electrons. Bohr found that these electrons were responsible for the absorption and emission of light, gaining energy (absorption) or losing it (emission) in the form of photons.

The key to the fixed pattern of spectral lines for each element lies in the fixed pattern of energy levels the electrons can be in. We can visualize an atom as a miniature solar system, with electrons orbiting the nucleus. Each kind of atom (each element) has its own characteristic number of electrons, and in each case the electrons have a certain set of possible orbits. It may be helpful to visualize a ladder, with each rung representing an orbit or energy level (Fig. 6.15). The spacing between the rungs of the ladder for one element, say hydrogen, are different from those of the ladder for any other element. The energy associated with each level increases the higher up the ladder, or the farther from the nucleus, the electron goes.

An electron can absorb a photon of light only if the photon carries the precise amount of energy needed to move the electron to one of the possible discrete levels

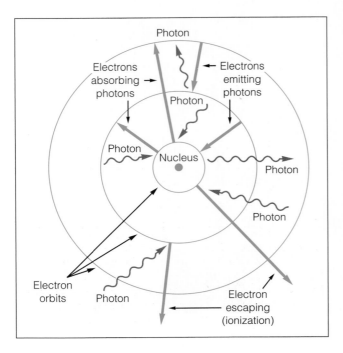

Figure 6.15 The formation of spectral lines.   *An electron must be in one of several possible orbits, each representing a different electron energy. If an electron absorbs exactly the amount of energy needed to jump to a higher (more outlying) orbit, it may do so. This is how absorption lines are formed, because the wavelength of the photon absorbed corresponds to the energy difference between the two electron orbits, and that is fixed for any given kind of atom. Conversely, when an electron drops from one orbit to a lower one, it emits a photon whose wavelength corresponds to the energy difference between orbits, which is how emission lines form. Note that it is possible for an electron to gain sufficient energy to escape from the atom altogether. This is called ionization, and is discussed in the next section.*

higher than the one it is in. Imagine an atom in space, with light streaming past it: The atom will let pass every photon that comes by, until it finds one with just the right amount of energy to boost one of its electrons to a higher level. Since the energy of a photon is closely related to its wavelength, this means that the atom will absorb only photons that have specific wavelengths, corresponding to the spacings between its energy levels. Because each kind of atom has its own unique set of energy levels, each can absorb light only in its own unique set of wavelengths. Thus Bohr was able to explain the findings of Bunsen and Kirchhoff.

Emission is the reverse process of absorption. Emission will occur whenever an electron in an upper level drops down to a lower one, and the energy of the photon emitted corresponds, as before, to the energy difference between the two levels. Since the energy levels are fixed for a given element, the emission lines and the absorption lines occur at the same wavelengths. Whether an atom produces emission lines or absorption lines depends on whether the electrons are moving downward or upward.

We can now state Kirchhoff's three rules in a different form:

1. In a hot, dense gas or a hot solid, the atoms crowd together so that their energy levels are distorted and all their lines blend together,[1] and we see a continuous spectrum.

2. In a hot, rarefied gas, the electrons tend to be in high energy states and create emission lines as they drop to lower levels.

3. In a cool gas in front of a hot, continuous source of light, the electrons tend to be in low energy levels and absorb radiation from the background continuous source, absorbing photons of specific wavelengths and forming absorption lines.

Ionization and Excitation

We have seen that electrons can move from one level to another by giving up or gaining energy. Even though each atom has many levels, there is always a limit to the range of levels an electron can have. If the electron gains too much energy, more than what is necessary

---

[1] The production of a continuous spectrum is actually a bit more complicated than this. When free electrons combine with ions, they emit at any wavelength, not just in spectral lines. See the discussion of ionization in the following section.

to reach the highest level the atom has, the electron will escape entirely (Fig. 6.16). This process is called **ionization,** and it leaves the atom with one electron short of its usual number. The atom then has a positive electrical charge and is called an **ion.** It is possible for an atom to lose more than one electron, becoming more highly ionized.

Ionization can occur in two ways. The electron can receive the energy needed for escape either by absorbing a photon with enough energy or by colliding with another atom. In any gas, the individual particles move about randomly, occasionally colliding with each other. Energy may be passed from one atom to another in such a collision, and when it is, the electrons of the atom that gained the energy may move up a few rungs in the ladder of energy levels. If the collision takes place at sufficiently high velocity, one or more of the electrons may gain enough energy to escape. The speed of motion of the atoms depends on the temperature of the gas, so the hotter the gas is, the more ionizations occur. Hence, a hot star has a higher degree of ionization in its outer layers than a cool star does.

The spectrum of an atom changes drastically when it has been ionized, because the arrangement of energy levels is altered, and because different electrons are now available to do the absorbing and emitting of photons. The spectrum of neutral helium, for example,

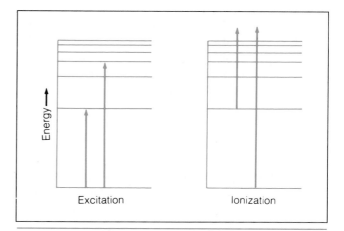

Figure 6.16 Excitation and ionization. *In the energy-level diagram at left, electrons are shown jumping up from the lowest level to higher ones. This is excitation, and it can be caused either by the absorption of photons or by collisions between atoms. On the right, electrons are shown gaining sufficient energy to escape altogether, a process called ionization.*

is quite different from that of ionized helium, so the astronomer not only can see that helium is present in the spectrum of a star, but also can tell whether it is ionized. This provides information about the temperature in the outer layers of the star, where the absorption lines are formed. By analyzing the degree of ionization of all the elements seen in the spectrum of a star, the astronomer can determine the gas temperature quite precisely.

Let us now consider an atom that is not ionized, and which collides at relatively low velocity with another atom. Its electrons may gain energy, but if it is not enough to cause ionization, the electrons may land in energy levels that are higher than those they were in before the collision (see Fig. 6.16). Ordinarily an electron will tend to stay in the lowest available energy state, and when it gets into a higher state, the atom is said to be **excited**, or in an **excited state**. When an electron does move to an excited state, it usually will drop back down almost immediately, and it will emit a photon each time it descends from the one level to a lower one. In gas in which collisions are rare, at any given time very few atoms will be in excited states. In a dense gas, however, where collisions occur often, at any moment a certain fraction of the atoms will be in excited states. It should be clear that if an electron is in an excited state, the wavelengths at which it can absorb photons are different from those at which the electrons can absorb when it is in the lowest level. The amount of energy required for an electron to move upward in the ladder of energy levels depends on which rung the electron is on to begin with; hence the wavelength of photon that it can absorb depends on which level it is in. Therefore the spectrum of an atom depends on how excited the atom is (Fig. 6.17 and Table 6.2). Observation of which lines appear in the spectrum of a star can tell the astronomer how many electrons are in excited states. This in turn reveals how

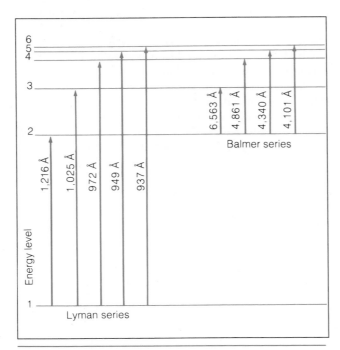

Figure 6.17 The effect of excitation on the spectrum of hydrogen. *As shown here, the absorption lines that can be formed by an atom depend on its degree of excitation. At left are the wavelengths of absorption lines originating in the lowest energy level of hydrogen, while at right are the wavelengths of absorption lines arising from an electron in the first excited level. Note that the Lyman lines are in ultraviolet wavelengths, while the Balmer lines are in the visible portion of the spectrum. This sketch shows only a few of the many energy levels of hydrogen.*

Table 6.2 Principal lines in the spectrum of hydrogen

Series	Lower Level	Line	Wavelength (Å)
Lyman (ultraviolet)	1	Alpha	1,216
		Beta	1,025
		Gamma	972
		Delta	949
		Epsilon	937
		(Limit)	(912)
Balmer (visible)	2	Alpha	6,563
		Beta	4,861
		Gamma	4,340
		Delta	4,101
		Epsilon	3,970
		(Limit)	(3,646)
Paschen (infrared)	3	Alpha	18,751
		Beta	12,818
		Gamma	10,938
		Delta	10,049
		Epsilon	9,545
		(Limit)	(8,204)
Brackett (infrared)	4	Alpha	40,512
		Beta	26,252
		Gamma	21,656
		Delta	19,445
		Epsilon	18,175
		(Limit)	(14,585)

frequently collisions are taking place in the star's outer layers, so in the end it is possible to infer the density of the gas.

From careful analysis of a star's spectrum, then, we can determine both the temperature of the outer layers of the star (from the ionization) and the density (from the excitation). Although most of the examples used in the discussion have referred to the Sun or the stars, the same considerations can be applied to the spectra of the planets.

The Doppler Effect

There is something else that spectral lines can tell us about a distant object, in addition to all the physical data we have been discussing. We can also learn how rapidly an object such as a star or planet is moving toward or away from us.

To understand this, we must visualize light from a distant object as though it were a series of concentric waves moving away from the source like ripples on a pond into which a pebble has been dropped (Fig. 6.18). If the source of light is moving, the waves will tend to get bunched up in the direction toward which the motion is occurring, and stretched out in the opposite direction. An observer sitting ahead of the moving source will see waves that are closer together than would be the case if the source were sitting still; that person will observe a shorter wavelength. An observer on the other side, with the source moving away, will see longer wavelengths.

This analogy is oversimplified in some ways, but it does illustrate an effect that really happens with light. When a source of light such as a star is approaching, all the lines in its spectrum are shifted toward wavelengths that are shorter than those observed when the light source is at rest, and if the source is moving away from the observer, all the lines are shifted toward longer wavelength. These two situations are called **blueshifts** (approach) and **redshifts** (recession), respectively, because the spectral lines are shifted toward either the blue or the red end of the spectrum.

A general term for any wavelength shift resulting from relative motion between source and observer is **Doppler shift,** in honor of the Austrian physicist who first explored the properties of such shifts. The effect does not apply just to light waves. Most of us, for example, have noticed the Doppler shift in sound waves when a source of sound passes by. The whistle on an

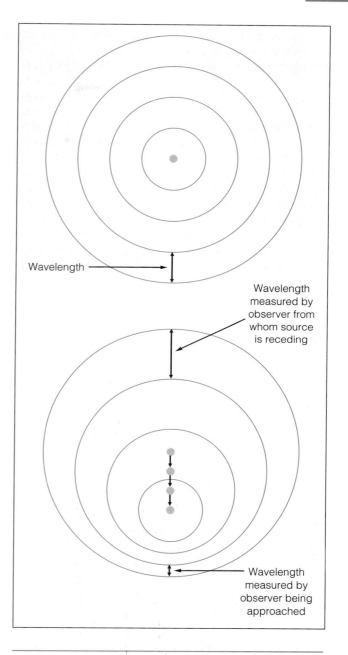

Figure 6.18 The Doppler effect. *The light waves from a stationary source (upper) remain at constant separation (that is, constant wavelength) in all directions, whereas those from a moving source get "bunched up" in the forward direction and "stretched out" in the trailing direction (lower). This causes a blueshift or a redshift for an observer who approaches or recedes from a source of light. Note that it does not matter which is moving, source or observer.*

ASTRONOMICAL INSIGHT 6.3

Measuring the Speed of Light

The enormous rate at which light travels makes its speed very difficult to measure. For all practical purposes, light travels instantaneously between points on the Earth (the light-travel time from New York to Los Angeles is only about 0.015 seconds, and under most circumstances people are not able to see things more than a few miles away, so the light-travel time is even shorter). Nevertheless, scientists have devised methods for measuring the speed of light to high accuracy.

Two of the earliest measurements were based on astronomical observations. One, first attempted by J. Bradley in 1727, makes use of the fact that the Earth's orbital motion causes light to reach the ground at a small angle. To see why, imagine running through a rainstorm where the drops are falling vertically. The faster you run, the more you have to tilt your umbrella in the forward direction, because your speed makes the drops appear to slant in toward you. The angle at which the drops strike you is determined by the ratio of your speed to that of the falling raindrops.

The same effect occurs for starlight. As the Earth moves in its orbit, photons of light from a star appear to reach us from a slightly shifted angle. The amount of displacement, called *stellar aberration*, is determined by the ratio of the Earth's orbital speed to the speed of light; hence, a measure-

ment of the displacement of star positions yields a value for the speed of light, if we know the Earth's orbital speed. This was not very accurately known in 1727, so the result of Bradley's measurement was only a crude estimate of the speed of light.

The second astronomical technique was actually suggested before Bradley's experiment but was not carried out until later. The Danish astronomer Ole Roemer pointed out that observations of the orbital periods of the satellites of Jupiter could be used to determine the speed of light, if these observations were made at different times of the year, so that the Earth's motion relative to that of Jupiter changed between observations. When the Earth is moving

nearly straight toward or away from Jupiter, the observed orbital period of one of Jupiter's satellites would either be shorter or longer than it would be when the Earth is moving perpendicular to the direction toward Jupiter (see figure). The reason for this is that the light from Jupiter's satellite would arrive a little early if the Earth is moving toward Jupiter, and a little late if the Earth is moving away from Jupiter, because of the light-travel time needed to cover the distance the Earth moved while the satellite was circling Jupiter. Again, this method depends on accurate knowledge of the Earth's orbital speed.

More modern methods are based on measurements made on the Earth. For example, in 1849,

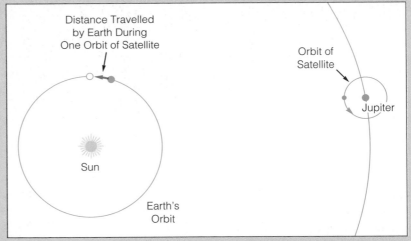

Roemer's Method for Measuring the Speed of Light. *Here we see the Earth receding from Jupiter. Each time Jupiter's moon completes an orbit, the Earth is farther away and the light-travel time correspondingly longer. Careful timing of the observed orbital period, combined with knowledge of how far the Earth moves during each orbit of Jupiter's moon, yields a value for the speed of light.*

ASTRONOMICAL INSIGHT 6.3

the French physicist Hyppolyte Fizeau devised a technique in which a beam of light was sent to a distant mirror through gaps in the edge of a rotating wheel. If the wheel turned rapidly enough, the light would pass through a gap on its way to the mirror, but would find itself blocked on the return trip, because the wheel had rotated enough so that the tooth adjacent to the gap now blocked the light (see figure). By knowing the round-trip distance to the mirror and the rate of rotation of the wheel, Fizeau was able to determine the speed of light to an accuracy of about 4 percent.

Even better accuracy was gained from a technique developed by Jean Foucault, another Frenchman. He sent a beam of light to a distant mirror that was rotating rapidly, and which reflected the light to a distant fixed mirror. From there it was reflected back to the rotating mirror (see figure). Because this mirror had turned a little by the time the beam came back from the fixed mirror, the final reflected beam was deflected at an angle to the original beam's direction. The size of the angular displacement depended on how far the rotating mirror had turned while the light beam was making its round trip to the fixed mirror. If the rotation rate of the mirror was known, then the measured angle indicated how long it took for the light to make this round trip. Combining this travel time with the measured distance between the mirrors then yielded a value for the speed of light. Foucault was able to measure this to an accuracy of about

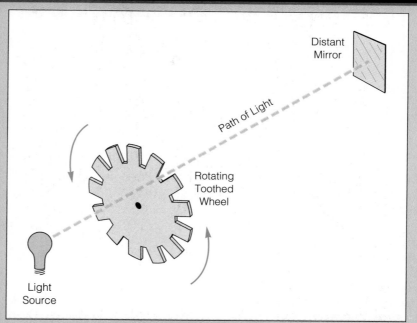

Fizeau's Method. *A beam of light passes through a gap in the toothed wheel and is reflected back from a distant mirror. The rotation speed of the wheel is adjusted so that the return beam is blocked by the tooth next to the gap; the known rotation speed of the wheel indicates the light-travel time to the mirror and back.*

Foucault's Method. *Light is sent to the rotating mirror, and from there to the fixed mirror. On its return, it is reflected in a slightly different direction, because the rotating mirror has turned a little. Measurement of the displacement angle, combined with the known rotation rate of the mirror, yields the light-travel time between the mirrors.*

ASTRONOMICAL INSIGHT 6.3

Measuring the Speed of Light (continued)

one-half of 1 percent. As in Fizeau's earlier technique, it was necessary to place the mirrors as

far apart as possible so as to maximize the measured effect, and this had the disadvantage that large distances are more difficult to measure accurately.

Today similar concepts are used, but the development of highly precise timing methods (using atomic vibrations; see discussion of timekeeping in Chapter 2)

has allowed such accurate measurement of light-travel times that laboratory-sized experiments are capable of better accuracy than the long-pathlength experiments just described. The estimated uncertainty in the modern value of the speed of light is only about one meter per second, or roughly three parts in a billion!

approaching train or the siren on an approaching emergency vehicle will suddenly change to a lower pitch at the moment the train or emergency vehicle passes by, because the wavelength received by the listener suddenly shifts to a longer one.

With the Doppler shift of light, we can determine the speed with which the source of light is approaching or receding, from the formula

$$v = \left(\frac{\Delta\lambda}{\lambda}\right)c,$$

where v is the relative velocity between source and observer, $\Delta\lambda$ is the shift in wavelength (the observed wavelength minus the rest, or laboratory, wavelength of the same line), λ is the laboratory wavelength of the line, and c is the speed of light.

Suppose that we find a star with a spectral line at a wavelength of 5,994 Å. Examination of the pattern of lines in this star's spectrum shows that this is a line of iron, which is measured in the laboratory to have a wavelength of 6,000 Å. Then the speed of the star with respect to the Earth is

$$v = \left(\frac{5,994\text{Å} - 6,000\text{Å}}{6,000\text{Å}}\right) \times 300,000 \text{ km/sec}$$

$$= \left(\frac{-6\text{Å}}{6,000\text{Å}}\right) \times 300,000 \text{ km/sec}$$

$$= -0.001 \times 300,000 \text{ km/sec}$$

$$= -300 \text{ km/sec}.$$

In this example, we find a negative velocity, because the observed wavelength is smaller than the rest wavelength. This means that the star is coming toward the Earth, and we have a blueshift. If, on the other hand,

the observed wavelength of this line had been 6,006 Å, we would have found a velocity of +300 km/sec, and the star would have been moving away from the Earth.

The Doppler shift tells us only about relative motion between the source and the observer. There is no fixed reference frame in which we can say one is moving while the other is at rest. We stress also that the Doppler shift tells us only about the part of the motion directly toward or away from the Earth, which is called the **radial velocity** (Fig. 6.19). There is no Doppler shift resulting from motion perpendicular to our line of sight. If a star is moving at some intermediate angle with

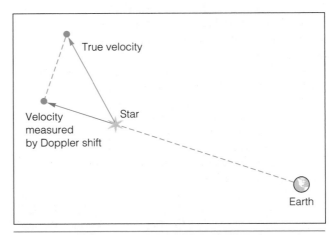

Figure 6.19 Radial velocity. *The Doppler shift only reflects motion along the line of sight, called the radial velocity. Here a star is moving at an angle to the line of sight, so that from the Doppler shift we can measure only a fraction of its true velocity.*

respect to the Earth, as is usually the case, we can determine the part of its velocity that is directed straight toward or away from us (that is, the radial velocity), but we cannot determine its true direction of motion or its speed transverse to our line of sight.

The Doppler shift is another powerful tool for astronomers. From it we learn about the motions not only of stars, but also of planets, galaxies, and any other objects in the universe that have lines in their spectra.

PERSPECTIVE

We now have learned how light is absorbed and emitted, and in doing so have begun to see how astronomers can derive great amounts of information about distant objects. We have seen how the continuous spectrum and overall intensity of light from a star depend on its temperature and size, and how the spectral lines provide information on temperature, density, and motion of a star, as well as chemical composition.

We are almost ready to get on with the job of seeing what is out there, but first we must learn how the light is captured by astronomers so that it can be analyzed.

SUMMARY

1. Visible light is just one part of the electromagnetic spectrum, which extends from the γ-rays to radio wavelengths.
2. Light has properties associated with both waves and particles, and these aspects are combined in the concept of the photon.
3. Astronomers gain information about a source of light, such as a star, from both the continuous radiation and the spectral lines.
4. The continuous radiation provides information on the temperature of a star, through the use of Wien's law. The temperature, luminosity, and radius of a star are related by the Stefan-Boltzmann law. Planck's law is a general one that embodies both Wien's law and the Stefan-Boltzmann law.
5. The observed brightness of a glowing object is inversely proportional to the square of the distance between the object and the observer.
6. Spectral lines are produced by transitions of elec-

trons between energy levels in atoms, molecules, and ions. Absorption occurs when an electron gains energy and jumps to a higher level, and emission occurs when it loses energy and drops to a lower level, the wavelength in both cases corresponding to the energy gained or lost as the electron changes levels.
7. Each chemical element has its own distinct set of spectral-line wavelengths, and therefore the composition of a distant object can be determined from its spectral lines.
8. The ionization and excitation of a gas can be inferred from its spectrum, the former yielding information on the temperature of the gas, and the latter providing data on its temperature and density.
9. Any motion along the line of sight between an observer and a source of light produces shifts in the wavelengths of the observed spectral lines, and measurement of this Doppler effect, as it is called, can be used to determine the relative speed (radial velocity) of source and observer.

REVIEW QUESTIONS

Thought Questions

1. Summarize the various forms of electromagnetic radiation, explaining their differences and similarities. Think of everyday examples of each form of radiation.
2. Explain why the human eye is sensitive to the particular wavelengths of electromagnetic radiation that we can see. What wavelengths do you think beings on a planet circling a much hotter star than the Sun would be able to see?
3. Summarize the ways in which light acts as a stream of particles, and the ways in which it acts as though it consists of waves.
4. Explain how the color of a star is related to its surface temperature.
5. Why are infrared-sensitive cameras useful devices for spotting people hiding in the dark?
6. Explain how the spectral lines formed by a given type of atom are related to the structure of that atom.
7. Explain why an atom can form emission lines under some circumstances and absorption lines under other circumstances.
8. What are the processes that can cause atoms to become ionized?
9. Explain why the Doppler effect tells us only about the component of motion directly along the line of sight between a source of light and an observer.

10. Summarize the information that can be gained about a distant object such as a star or a planet from the analysis of its spectral lines.

Problems

1. What is the wavelength of your favorite FM radio station?

2. How long does it take for a photon of light from the Sun to reach the Earth?

3. A photon in the blue portion of the visible part of the electromagnetic spectrum has a wavelength of 4,500 Å. Express this wavelength in centimeters. What is the energy of this photon in ergs?

4. Calculate the wavelengths of maximum emission for a star of surface temperature 25,000 K, a star of surface temperature 2,500 K, the Sun's corona, whose temperature is 2,000,000 K, and a human body, whose temperature is 310 K. (Note: To do this, you will need the value of the Wien constant, which is given in Appendix 2.)

5. Suppose Mars has a surface temperature 5 times greater than that of Pluto. Which of these two bodies emits more energy per second per square centimeter of surface area, and how much more does it emit?

6. If Mars has twice the diameter of Pluto, compare the relative total energy emitted per second by the two planets.

7. Pluto is approximately 25 times farther from the Sun than Mars is. Compare the intensity of sunlight reaching the two planets.

8. Suppose that in the hydrogen atom the energy difference between two energy states is 4.581×10^{-12} erg. What is the wavelength of a photon emitted when an electron drops from the upper to the lower of these two states? From Figure 6.17, can you identify which two states these are?

9. The energy needed to ionize a hydrogen atom is 2.18×10^{-11} erg. What wavelength of photon does this correspond to? How fast would an electron have to be moving to have this much kinetic energy? (The electron mass is 9.11×10^{-28} grams; recall the formula for kinetic energy in Chapter 5).

10. Suppose the strong line of hydrogen, whose rest wavelength is 6562.7, is observed in the spectrum of a star and found to have a wavelength of 6563.3 Å. What is the radial velocity of this star?

ADDITIONAL READINGS

The best sources for further explanation of the properties of light and radiation are elementary physics books of the sort used as texts in introductory courses. In addition, there are a few nontechnical books on the nature of light, such as the following works:

Bragg, W. 1959. *The universe of light*. New York: Dover.
Rublowsky, J. 1964. *Light*. New York: Basic Books.
Ruechardt, E. 1958. *Light, visible and invisible*. Ann Arbor: University of Michigan Press.

MATHEMATICAL METHODS

Modeling the Atom

 In unveiling the nature of planetary orbits, scientists first found mathematical relationships that *describe* planetary motion, and then discovered the underlying forces that *cause* these motions. Kepler found the mathematical description, and Newton discovered the underlying causes. Progress in understanding the nature of light and how atoms emit and absorb it followed a very similar course.

Starting around the turn of the century, scientists began to make strides toward determining the nature of atoms. Laboratory experiments showed that an atom has most of its mass concentrated in a positively charged nucleus, and that surrounding the nucleus is a cloud of negatively charged particles (electrons). At the same time it was discovered that atoms can emit or absorb light only with certain quantities of energy (i.e., only with certain wavelengths).

In 1885, the Swiss scientist Johan Jacob Balmer developed a mathematical description of the wavelengths emitted or absorbed by the hydrogen atom. This description, in the form of an equation, is analogous to Kepler's laws of planetary motion in that it provided no insight into the underlying reasons for this behavior of the hydrogen atom. The Balmer formula is, however, extremely useful in that it allows one to calculate the wavelengths of spectral lines due to hydrogen atoms, much as Kepler's third law allows one to calculate orbital periods or sizes for the planets.

The Balmer equation is

$$\frac{1}{\lambda} = R\left(\frac{1}{m^2} - \frac{1}{n^2}\right),$$

where λ is the wavelength, R is a numerical constant, and m and n are simple integers such as 1, 2, 3, and so on (m always has to be smaller than n). The value of R is R = 0.00109678 if the wavelength is measured in Ångstroms.

As an example, suppose m = 1 and n = 2. Substitutions into the equation yields

$$\frac{1}{\lambda} = R\left(\frac{1}{1^2} - \frac{1}{2^2}\right) = R\left(1 - \frac{1}{4}\right)$$
$$= 3R/4 = 0.000822585,$$

which leads to

$$\lambda = 1216\text{Å}.$$

This wavelength is in the ultraviolet part of the spectrum, and it corresponds to a very important and strong spectral line of hydrogen.

Now suppose that m = 3 and n = 2. Repeating the steps in the calculation above leads to

$$\lambda = 6564\text{Å},$$

which corresponds to the wavelength of the strong red line of hydrogen. Other combinations of m and n yield the wavelengths of other hydrogen lines, and the Balmer formula is capable of reproducing the entire set of spectral lines due to hydrogen, throughout the spectrum from the shortest wavelengths (ultraviolet) to the longest (radio) where this atom emits and absorbs radiation (see Table 6.2 for a list of many lines of hydrogen atoms).

The Danish physicist Nils Bohr took another step toward understanding atomic structure when he proposed that the numbers m and n in Balmer's formula correspond to energy levels in the atom. If we visualize these energy levels as rungs on a ladder, as we did in the text, then m and n represent the rungs. The lowest rung, or energy level, is designated as level 1, the next one up is 2, and so on. So in our example where we had m = 2 and n = 1, we calculated the wavelength that is emitted or absorbed when an electron moves from the lowest level to the next one, or vice versa. Our second example involved an electron moving between levels 2 and 3. Now the Balmer formula was related to a physical picture of the atom, rather than being only an equation that could reproduce the observed wavelengths at which atoms emit or absorb.

Bohr's work enabled him to determine the value of the constant R from fundamental atomic parameters. This was somewhat akin to Newton's derivation of Kepler's third law, and it was a major step toward a full understanding of why the atom has the energy level structure it has. Bohr was also able to modify the equation to describe other particles besides hydrogen atoms, although he still was successful only with species having a single electron. Thus Bohr's work provided calculated wavelengths for hydrogen atoms, for helium ions (i.e., helium atoms that have lost one of their two electrons), doubly ionized lithium (lithium that has lost two of its three electrons), and so on. The physics and mathematics become much more complicated for atoms and ions having more than one electron, and it required the full development of quantum mechanics for these spectra to be represented accurately. Nevertheless, the work of Balmer and Bohr was very useful, much as Kepler's and Newton's discoveries had been several centuries earlier.

CHAPTER 7

Telescopes on Earth and in Space

The primary mirror of the 10-m Keck Telescope, with three hexagonal mirror segments in place. (T. P. Snow)

The preceding two chapters illustrate the techniques used by astronomers to determine the properties of faraway objects. These techniques, however, require that certain measurements be made, and we have not yet discussed the instruments and methods used to make these measurements. Hence this chapter will outline the principles of telescopes and their uses.

THE NEED FOR TELESCOPES

There are several basic benefits of using telescopes. One is to collect light from a large area and bring it to a focus, so that fainter objects than the eye could see unaided can be observed. The human eye has a collecting area only a fraction of a centimeter in diameter, whereas the largest telescopes are 4 to 6 meters in diameter. The light-gathering power of a telescope depends on the area of its collecting surface; since the area of a circle of radius r is equal to πr^2, a 2-inch diameter telescope collects 4 times as much light as a 1-inch telescope, for example.

Another basic advantage of telescopes over the unaided eye is that the telescope can be equipped to record light over a long period of time, by using photographic film or an electronic detector, while the eye has no capability to store light. A long-exposure photograph taken through a telescope reveals objects too faint to be seen with the eye, even by looking through the same telescope.

By combining both large size and the capability of long exposures, the largest telescopes can detect objects more than a billion times fainter than can be seen by the unaided eye, even under the best conditions.

A third major advantage of large telescopes is that they have superior **resolution,** the ability to discern fine detail. For visible-wavelength telescopes, the Earth's atmosphere creates practical limitations on how fine the resolution can be, but for radio telescopes and optical telescopes in space, the atmosphere is not a hindrance. The resolution of a telescope is normally expressed in terms of the smallest separation between a pair of objects that can be discerned by the telescope. Thus, small resolution is good. The resolution expressed in this way is proportional to the wavelength being observed, and inversely proportional to the diameter of the telescope. For visible light (wavelength 5,500 Å, for example), a telescope 10 centimeters in diameter (about 4 inches) has a resolution of about 1 arcsecond. (This is really quite good; an arcsecond is about the angular diameter of a dime seen at a distance of 2 miles!). The human eye, with a pupil diameter of about 1 millimeter, has a resolution of roughly 100 arcseconds, or a little over 1.5 arcminutes. In principle, a large telescope such as the 5-meter Palomar telescope has a resolution far smaller than 1 arcsecond, but, in practice, the Earth's atmosphere limits the resolution that is actually achieved. Normally it is impossible to separate objects closer than about 0.5 arcsecond when looking up through the turbulent atmosphere, and often the blurring effect called "seeing" makes resolution much worse.

We can vastly improve resolution by using telescopes in pairs or groups. This technique, called **interferometry,** makes it possible to measure the precise direction toward a source of radiation by analyzing the **interference** of the waves arriving at separate telescopes. Interference refers to the fact that wavecrests and troughs can either add or cancel each other when waves overlap, depending on how the waves are shifted with respect to one another. The shift depends on the separation of the antennae and the direction of the source. Sources can be mapped with very high precision by using telescopes arranged in series along different alignments, as is done with the *Very Large Array* (see Fig. 7.23), which can achieve resolutions of better than 0.1 arcsecond. Interferometry is obviously a very complex undertaking, but it has been successfully applied to radio astronomy, and there are now preliminary plans to do the same for infrared and perhaps visible light. (The shorter the wavelength being observed, the more precision is required in interferometry. The planned *Very Large Telescope,* a European project using four large mirrors to achieve the equivalent collecting area of a single 16-meter mirror, is designed to make possible interferometry in infrared and perhaps visible light.) The more widely separated the telescopes are, the better the resolution, and radio astronomers are today engaged in *Very Long Baseline Interferometry* (or simply *VLBI*), using telescopes separated by intercontinental distances.

One technical problem must be overcome in telescope construction: The Earth rotates, and if nothing is done to compensate for this, a star quickly moves out of the field of view. To avoid this problem, telescopes are mounted so that they can be moved by a motor in the direction opposite the Earth's rotation, keeping a target object centered in the field of view. The telescope mounting usually allows the telescope to move independently in declination (north–south motion) and right ascension (east–west motion), so the drive mechanism that compensates for the Earth's rotation needs to operate only in the east–west direction. There is no need to do this, of course, for telescopes that are not attached to the Earth, such as the various observatories in space.

REFRACTORS AND REFLECTORS

We begin by discussing the principles of telescopes used for visible light, since these have the longest history and because many of the basic ideas can be used for infrared and ultraviolet telescopes as well.

The telescope used by Galileo utilized lenses to bring light to a focus and to magnify the image (Fig. 7.1). This technique was later refined by Huygens and studied by Newton, who realized it has a fundamental

shortcoming: The lens acts as a prism, bending different wavelengths of light to different degrees as it passes through the glass, thus making it impossible to bring all the light of a star to a focus at a single point. The blue light focuses closer to the lens than the longer-wavelength red light: The image of a star gets spread out into a rainbow. This problem, called **chromatic aberration,** was solved much later by the use of lenses made of combinations of different kinds of glass, so that the separation of colors caused by a layer of one kind of glass could be reversed by a layer of another kind of glass.

Telescopes that use a lens to bring light to a focus are called **refractors.** A large lens, called the *objective*, at the end of the telescope tube, bends the light rays so that they come together and form an image at a point called the **focus.** The size and curvature of the

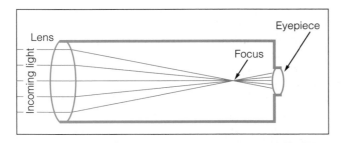

Figure 7.1 The refracting telescope.   *The path of light is bent when it passes through a surface such as that of glass. A properly shaped lens thus can bring parallel rays of light to a focus, where the image of a distant object may be examined. A second lens is used to magnify the image.*

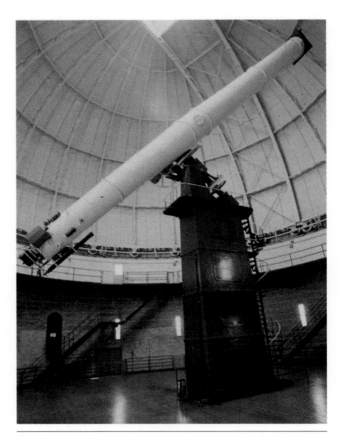

Figure 7.2 The world's largest refractor.   *This is the 40-inch refracting telescope located at the University of Chicago's Yerkes Observatory in Williams Bay, Wisconsin. (Yerkes Observatory)*

primary lens determine the focal length, that is, the distance from the primary lens to the focus point when a very distant object is observed. At this point (where the other end of the tube is located) is an eyepiece to magnify the image for the human observer.

The crucial aspect of a telescope is not its ability to magnify images, as is commonly thought, but is instead its ability to collect as much light as possible from a source. This light-gathering power is determined by the size of the principal light-collecting element, in this case the objective lens. This is what determines how sensitive the telescope will be for observing faint objects. The eyepiece does nothing but magnify the image to a scale that is easier to discern; it does not enhance the light-gathering capability of the telescope.

The largest refractors ever made are less than 4 feet in diameter. The biggest is the refractor at the Yerkes Observatory at Williams Bay, Wisconsin, with a primary lens 40-inches in diameter (Fig. 7.2). Another large refractor, with a 36-inch diameter, is located at the Lick Observatory, on Mount Hamilton, east of San Jose, California.

Newton, having discovered (but not overcome) the problem of chromatic aberration in refracting telescopes, devised an alternative type of telescope that uses a concave mirror to bring the light to a focus (Figs. 7.3, 7.4, and 7.5). This arrangement, called a **reflecting telescope,** has the advantage that all wavelengths of light reflect from a mirror at the same angle,

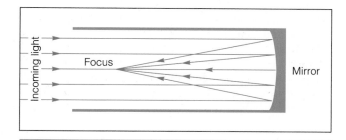

Figure 7.3 The reflecting telescope. *A properly shaped concave mirror can be used instead of a lens to bring light to a focus. Usually an additional mirror is used to reflect the image outside the telescope tube.*

Figure 7.4 A reflecting telescope built by Isaac Newton. *The image is reflected to a focus outside the telescope tube, near the top in this illustration. (The Granger Collection)*

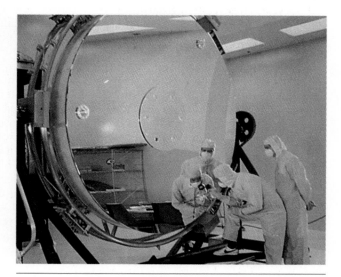

Figure 7.5 A large primary mirror. *This photo shows that the proper shape for a telescope is not very highly concave. Only the distorted reflections in this case reveal that the mirror is not flat. This is the 2.4-meter primary mirror for the Space Telescope. (Perkin Elmer Corporation)*

ASTRONOMICAL INSIGHT 7.1

A Night at the Observatory

 A typical astronomer's night at the observatory will vary quite a bit from one individual to another, and especially from one observatory to another. There is a big difference between sitting in a heated room viewing an array of electronics panels and standing in a cold dome peering into an eyepiece while recording data on a photographic plate. Yet both extremes (and various possibilities in between) are part of modern observational astronomy.

Despite apparent differences in practice, however, the principles are very much the same, regardless of where the observations take place, or with what equipment. Here is a brief outline of a typical night at a large observatory, where a modern electronic detector is to be used to record spectra of stars.

By mid-afternoon the astronomer and his or her assistants (an engineer who knows the detector and its workings, and a graduate student) are in the dome, making sure the detector is working properly. To do this, the electronics must be tested in various ways, and test spectra from special calibration lamps must be recorded. Everything in order, they eat supper in the observatory dining hall a couple of hours before sunset, then return to the dome to complete preparations for the night's work. A night assistant (an observatory employee) arrives shortly before dark and is given a summary of the plans for the night. It is the night assistant's job to control the telescope and the

dome, and this person is responsible for the safety of the observatory equipment. At large observatories, the telescope is considered so complex and valuable that a specially trained person is always on hand so that the astronomer (who may use a particular telescope only a few nights a year) never handles the controls, except for fine motions required to maintain accurate pointing during the observations.

Before it is completely dark, a number of calibration exposures are taken, using lamps and even the Moon, if it is up. These measurements will help the astronomer analyze the data later, by providing information on the characteristics of the spectrograph and the detector.

Finally, everything is in order, and if things are going smoothly, the telescope will be pointed at the first target star just as the sky gets dark enough to begin work. Observing time on large telescopes is difficult to obtain and is very valuable; not a moment is to be wasted.

A pattern is quickly established and repeated throughout the night. First, after the night assistant has pointed the telescope at the requested coordinates, the astronomer looks at the field of view to confirm which star is the correct one. (This is often done by consulting a chart prepared ahead of time.) When the star is properly positioned so that its light enters the spectrograph, the observation begins. Depending on the brightness of the star and the efficiency of the spectrograph and the detector, the exposure time may be anywhere from seconds to hours. During this time, the astronomer or the graduate student assistant continually checks to see

that the star is still properly positioned, making corrections as needed by pushing buttons on a remote control device.

If the observations are made using an instrument at the coudé focus, the astronomer and the assistants spend the entire night in an interior room in the dome building, emerging only occasionally to view the skies, give instructions to the night assistant, or go to the dining hall for lunch (usually served around midnight).

As dawn approaches and the sky begins to brighten, last-minute strategy is devised to get the most out of the remaining time. Finally, the last exposure is completed, and the night assistant is permitted to close the dome and park the telescope in its rest position (usually pointed straight overhead, to minimize stress on the gears and support system). The astronomer and the assistants make final calibration exposures before shutting down their equipment and trudging off to get some sleep. The new day for them will begin at noon or shortly thereafter.

The scenario just described depicts a typical night at a major modern observatory, such as Kitt Peak, with extensive resources and the largest telescopes. At smaller facilities, such as a typical university observatory, it is more common to find the astronomer working alone, most often using a Cassegrain focus instrument, so that he or she must spend the night out in the dome with the telescope (where it can be very cold in the winter), and getting refreshment in the form of a homemade lunch. The work pattern is similar, however, as is the desire not to waste any telescope time as the research program is carried out.

so no separation of colors occurs. A disadvantage, however, is that for us to look at the image created by a concave mirror, we need to block the incoming light, because the image forms in front of the mirror. One solution to this problem was suggested by J. Gregory, who placed a **secondary mirror** in front of the primary to reflect the image back through a hole in the primary to a point behind it, where the observer could look at it without blocking the incoming light. This design is similar to the modern **Cassegrain focus** arrangement (Fig. 7.6) and is used in nearly all large, reflecting telescopes. Newton developed a different means of de-

flecting the image outside the telescope tube; he placed a flat mirror at a 45° angle just before the focal point so that the light passed out through a hole in the side of the tube before coming to a focus. This **Newtonian focus** is used in many small- to moderate-sized reflectors, particularly of the sort often used by amateur astronomers.

Another, more complex, design is sometimes available with large telescopes (Fig. 7.6). Called a **coudé focus**, this arrangement uses up to five mirrors (including the primary) to reflect the image out of the telescope tube along one of the main axes of motion to a focus in a separate room, some distance from the telescope. The reason for going to all this trouble is to allow the use of heavy instruments in the analysis of light. In the Cassegrain or Newtonian designs, any equipment needed to record or measure the light must be attached directly to the telescope and must be light enough to move with it. That is not practical if the instrument is very massive. In the coudé design, however, the image does not move, no matter where the telescope is pointed, so large instruments can be fixed in place to analyze it.

Yet another arrangement, possible only with the largest telescopes, is called **prime focus** (Fig. 7.6). In this case, the light is allowed to focus inside the tube, and the astronomer actually sits there, in a small cage, with the proper equipment for analyzing the light (usually a simple camera in this instance). The advantage of this design is that it reduces the number of reflections to just one, thereby minimizing the loss of light and image quality that inevitably occurs at each reflection. Prime focus is used, therefore, for observations of especially faint objects.

All the focal arrangements used in reflecting telescopes have the disadvantage that some fraction of the incoming light has to be blocked by a secondary mirror or by the observer's cage (if the prime focus is used). This disadvantage is far outweighed by the size advantage offered by the use of mirrors instead of lenses. A lens can be supported only by its edges, since light must be free to pass through. If a lens is large enough to be very heavy, it may sag a little under its own weight, which distorts its light-focusing properties. Even worse, the amount and direction of sag will vary as the telescope is pointed in different directions. A mirror, on the other hand, can be supported from behind and can therefore be kept much more rigid. All the largest telescopes in the world are reflectors for this reason (Fig. 7.7). Another advantage of reflectors is

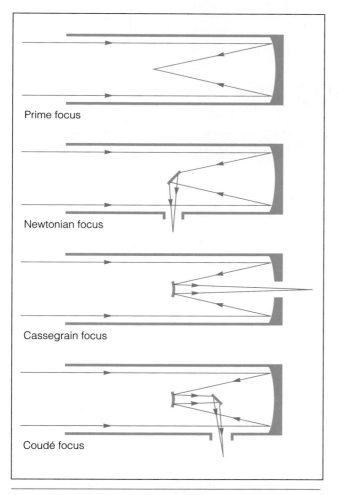

Prime focus

Newtonian focus

Cassegrain focus

Coudé focus

Figure 7.6 Various focal arrangements for reflecting telescopes.

Figure 7.7 Large reflecting telescopes. *Clockwise from upper left: The dome of the* Canada-France-Hawaii Telescope *on Mauna Kea; the 4-meter* Mayall Telescope *at Kitt Peak National Observatory; the 3.05-meter* Shane Telescope *at the Lick Observatory; and the 5-meter* Hale Telescope *on Palomar Mountain. (upper left: J. G. Timothy; upper right: National Optical Astronomy Observatories; lower right: Lick Observatory photograph; lower left: Palomar Observatory, California Institute of Technology)*

Figure 7.8 The 6-meter telescope. *This is the dome building containing the world's largest telescope, located in the Caucasus Mountains of the southern central Soviet Union. (A. G. D. Phillip)*

that only one surface has to be shaped to precision, rather than two, as in a lens (or four, in the composite lenses made of two layers of glass to prevent chromatic aberration). Also, because light reflects only from the surface and does not pass through the glass, flaws within the material of the mirror can be tolerated. The biggest telescope for many years was the 200-inch (5-meter) reflector at Mt. Palomar, in southern California (see Fig. 7.7), but a 6-meter telescope was completed in the Soviet Union (Fig. 7.8).

A very promising new kind of telescope was completed in 1979 at Mt. Hopkins, Arizona. This is the *Multiple-Mirror Telescope*, which consists of six separate 72-inch primary mirrors with secondary mirrors arranged so that each focuses light at the same point (Figs. 7.9 and 7.10) The total collecting area of the six mirrors is equivalent to a single mirror with a 176-inch (4.5-meter) diameter. This system is very complicated, and the difficulties of perfectly aligning the six mirrors are enormous, but the savings in cost and the relative ease with which the smaller mirrors can be constructed make the effort worthwhile.

Figure 7.9 The Multiple-Mirror Telescope, Mt. Hopkins, Arizona.   *This observatory is operated by the University of Arizona and the Smithsonian Institution. Six 72-inch mirrors bring light to a common focus, giving the instrument the light-collecting power of a single 4.5-meter telescope. (Multiple-Mirror Telescope Observatory, University of Arizona and the Harvard-Smithsonian Center for Astrophysics)*

Figure 7.10 A schematic of the Multiple-Mirror Telescope design.   *This drawing shows how the images from two of the six mirrors are brought to a common focus.*

Auxiliary Instruments

Once the light is brought to a focus, it must then be recorded in some way. The old-fashioned image of an astronomer peering into an eyepiece all night, perhaps taking notes, is replaced in reality by the picture of the astronomer sitting before a rather complex and impressive control panel, operating sophisticated electronic equipment.

Generally speaking, two kinds of things are done to the light after it is focused: (1) It is often spread out or sorted according to wavelength; and (2) it is then recorded by film or electronic detector.

The first of these steps may be accomplished in a variety of ways, including the use of a prism or other device to spread out the light, or the use of filters to allow only certain wavelengths to pass through. The device that spreads out, or **disperses,** the light according to wavelength and then records the resulting spectrum is called a **spectrograph.** Modern spectrographs usually employ grooved surfaces called **gratings** to disperse the light, rather than prisms. The effect of a grating may be visualized by holding a compact disc so that light glances off of it, forming a rainbow of colors (Fig. 7.11).

Besides spectrographs, other common kinds of auxiliary instruments include ordinary cameras, to photograph the field of view (sometimes this is done through a series of filters so that star brightnesses at different wavelengths can be measured); and **photometers,** which use simple photocells like those which control some kinds of automatic doors. A photocell produces an electric current proportional to the brightness of the starlight that hits it, so the brightness is determined by measuring the current.

The final stage of any instrument is the device that actually records the intensity of light. Traditionally, this job has been done with photographic film, and film is still widely used today, but several alternatives have been developed in recent years. These include a variety of electronic instruments that produce pictures like film does, but do so by storing an image electrically so that it can be analyzed easily by computer. Some of these **detectors,** as any device for measuring the intensity of light is called, are very similar to ordinary television cameras, whereas others operate on quite different principles.

The auxiliary instruments used in astronomy range from rather compact, lightweight devices that can be easily mounted directly on the telescope for use at the Cassegrain focus, to bulky, large instruments (usually spectrographs) that can only be used at the coudé focus.

Figure 7.11 Light dispersion by reflection gratings.
These photographs illustrate the spread of visible light according to wavelength that occurs when light is reflected from a surface in which parallel, closely spaced grooves have been etched. The upper panel shows light reflected from a small grating designed for use in space-based spectrographs; the lower panel shows a much larger grating (consisting of four segments, each one roughly 8 inches square) that is in use in the coude spectrograph at the Canada-France-Hawaii Telescope on Mauna Kea, in Hawaii. (Upper: Hyperfine, Inc., courtesy of J. Schwenker; lower: photo by T. P. Snow).

ASTRONOMICAL INSIGHT 7.2

The Magic of Modern Detectors

At the heart of every astonomical instrument is a device that records the light that is brought to a focus by the telescope. Virtually no research in astronomy is done by the human eye alone. The light that reaches the focus of the telescope and its instrument, whether this is a camera, a spectrograph, a photometer, or some other device, must somehow be recorded so that detailed analysis can be done.

Until recently, the only technique known for recording light was to use photographic film. Film is remarkable for its ability to react to light and record the result of this reaction, thus creating images that can be examined and analyzed away from the telescope. Film is also capable of high resolution, meaning that it can record fine details in an image, and it can cover large areas as well (some photographic plates used in astronomy as large as 17 inches square). But film also has some disadvantages: It is not very efficient, recording at best only 2 or 3 percent of the photons that strike it; its response to varying light levels is not uniform; and the range of brightness that can be recorded on a single piece of film is limited. Another important difficulty with film is that, in order to analyze images, it is usually necessary to store the information in a computer, and the process of converting photographic images into arrays of numbers suitable for computer analysis is very laborious and time-consuming.

The rise of modern electronic technology has given astronomers opportunities to develop alternatives to film. The simplest of these modern detectors are photoelectric cells, devices that produce an electrical current that is proportional to the intensity of the light that strikes them. Photocells, as they are commonly called, can record only the overall brightness of an object; they have no ability to record the variations of light from place to place, creating an image. For this it is necessary for a detector to have many picture elements, or *pixels,* each capable of recording the intensity of light at its specific location.

Some early electronic detectors consisted of cameras similar to television cameras. While these represented some improvements over film, they retained some of the disadvantages as well. They tended to be nonuniform in their response to different light levels, and they have very limited capability for simultaneously observing bright and faint objects in the same field of view.

The leading detectors in use today are *charge-coupled devices,* or CCDs. These combine virtually all of the desirable characteristics for astronomical observations. They are very efficient, recording up to 70 or 80 percent of the photons that strike them (even 100 percent is possible, for high-energy photons such as X-rays). They have a very uniform response to different light levels, so that the electrical signal generated is proportional to the intensity of light over a very large range, and they can simultaneously record the intensities of objects with very different brightnesses. Pixels of very small size can be made, so that the resolution, or ability to record fine de-

tails, rivals that of fine-grained film. Furthermore, it appears possible to use CCDs over an enormous range of wavelengths, all the way from the X-ray portion of the spectrum to the infrared. The one shortcoming of CCDs, as compared with film, is that CCDs are not very large (usually less than 1 inch square), so that large fields of view cannot be recorded in a single exposure.

A CCD consists of layers of silicon with certain impurities that cause electrons to be freed when photons of light strike them. A CCD detector, or "chip," consists of a thin wafer of silicon in which a pattern of electric fields (voltages) is embedded so that when a photon frees an electron, the electron is trapped at the location where the photon struck. Thus, as light reaches the CCD, a pattern of electrical charges builds up, corresponding to the intensity of light at each pixel. Then to record an image, it is necessary to "read out" the amount of electrical charge (i.e., the number of electrons) stored in each pixel.

The read-out procedure is accomplished by altering the voltages in adjacent rows of pixels, so that the stored electrons are transferred systematically from row to row. It is as though the image were cut into tiny strips, and then the strips were moved, one at a time, off to the edge of the picture. At the edge of the detector are electronic devices that measure the pattern of charges in each row and store this information numerically in a computer. To read out an entire image requires systematically marching all the rows of pixels off to the edge

Continued on following page

CHOOSING A SITE: MAJOR OBSERVATORIES

Most major observatories are located in remote regions, for a number of reasons. It is best to minimize absorption of light by the atmosphere, so an observatory should be situated on a high mountain, above as much of the atmosphere as possible. Of course, a site with generally clear weather is imperative, for only radio telescopes can peer through clouds. It is also desirable to build observatories well away from large cities, because the diffuse light from a densely populated area can drown out faint stars. Another consideration is the latitude of the site, which should not be too far north or south of the equator, since that would exclude a large portion of the sky.

Other factors that affect the quality of an observing site are the purity of the air, for pollution can reduce the transparency of the atmosphere; and the amount of turbulence ordinarily present in the air above the site, because turbulence creates the twinkling effect known to astronomers as "seeing." This twinkling, which tends to smear out an image, reduces the quality of photographs or spectra. On a good night, the diameter of a star image is less than 1 arcsecond, whereas when the seeing is bad, images may blow up to diameters of 2 arcseconds or more.

The largest observatories in the world (see Appendix 6) generally are located along the western portions of North and South America, in Hawaii, (Fig. 7.12) and in Australia, although the biggest single telescope is in the Caucasus Mountains of the southern central Soviet Union. The U.S. Government, through the National Science Foundation, supports large observatories in North America (Kitt Peak National Observatory, about 50 miles west of Tucson, Arizona; Fig. 7.13) and South America (the Cerro Tololo Inter-American Observatory, on the crest of the Andes, northeast of Santiago, Chile; Fig. 7.14).

TELESCOPES IN SPACE: INFRARED AND ULTRAVIOLET ASTRONOMY

As we saw in our discussion of the emitting properties of glowing objects, the radiation from a star or planet extends beyond the limits detectable by the human eye. It also extends beyond the limits set by the Earth's atmosphere, which absorbs all wavelengths shorter than about 3,100 Å, and absorbs in several wavelength regions in the infrared as well (Fig. 7.15). Some portions of the infrared can penetrate to the ground, so infrared astronomy can be done at the major observatories (two

Figure 7.12 The Mauna Kea observatory. *These are several of the large telescopes on the summit of Mauna Kea, including (from left to right on top of the ridge) the 3.6-meter Canada-France-Hawaii telescope, the University of Hawaii 2.2-meter telescope, and the 3.6-meter United Kingdom infrared telescope. In the foreground are the buildings for two new radio telescopes: a joint Dutch-United Kingdom 15-meter facility (in the large white building), and a 10-meter telescope built by Cal Tech. The planned 10-meter visible-light Keck Telescope, to be the largest in the world for several years, is being built just to the left of the 3-meter NASA Infrared Telescope Facility at the far left. (T. P. Snow)*

Figure 7.13 A panorama of telescopes at Kitt Peak. *This sunset view shows the buildings housing several of the telescopes at Kitt Peak National Observatory near Tucson, Arizona. (National Optical Astronomy Observatories)*

Figure 7.14 The Cerro Tololo Inter-American Observatory. *This mountaintop in the Andes, some 200 miles north of Santiago, Chile, is the site of the United States' Southern Hemisphere national observatory. The largest telescope here is a 4-meter instrument. (National Optical Astronomy Observatories)*

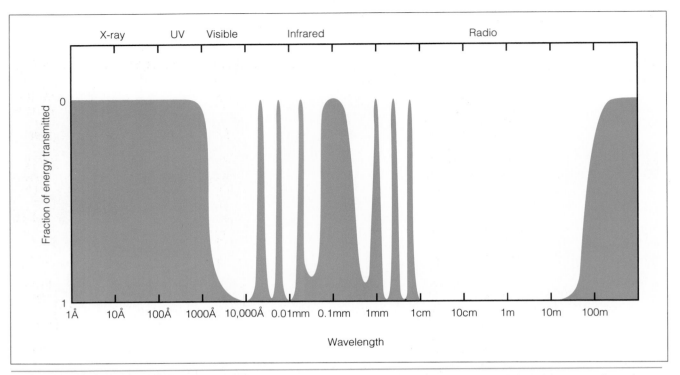

Figure 7.15 Windows through the Earth's atmosphere. *This diagram illustrates the many wavelength regions where the atmosphere blocks incoming radiation from space (shaded regions). The atmospheric "windows" allow ground-based observations only in visible and radio wavelengths, and some portions of the infrared spectrum. All other wavelengths must be observed from space.*

Figure 7.16 The Kuiper Airborne Observatory. *The aircraft is used for infrared observations at altitudes above much of the water vapor in the atmosphere. Water vapor absorbs infrared light at many wavelengths and interferes with ground-based observations at these wavelengths. (NASA)*

of the large telescopes on Mauna Kea, for example, were built specifically for infrared observations, as was the large telescope at the University of Wyoming). To cover the entire infrared spectrum or any of the ultraviolet (or X-rays), we must, however, place telescopes in orbit above the Earth's atmosphere. (Major space observatories are listed in Table 7.1). This is accomplished through the use of high-altitude planes (Fig. 7.16), balloons (which do not quite get above all the atmosphere), sounding rockets (providing only 5 minutes or so of observing time), or satellites.

Simply putting an infrared telescope above the atmosphere does not solve all the problems. As shown in the discussion of Wien's law (Chapter 6), a telescope at room temperature glows at infrared wavelengths, which can drown out the infrared light from a target object. Astronomers solve this problem, on the ground or in space, by cooling the telescope, or at least the associated instruments, so that the emission is shifted to much longer wavelengths and does not confuse the observations. The cooling is usually done by circulation of liquid nitrogen (which has a temperature of

Table 7.1 Major telescopes in space

Wavelength Region	Telescope[a]	Agency	Diameter	Dates
Gamma ray	(Gamma Ray Observatory: GRO)	NASA[b]		(1991)
X-ray	Einstein Observatory	NASA	0.56 m	1978–1981
	Exosat	ESA		1982–1986
	Roentgen Satellite: ROSAT	German-NASA		1990–
	(Advanced X-ray Astrophysics Telescope Facility: AXAF)	NASA		(1998)
	(X-Ray Multiple Mirror Mission)	ESA		(1998)
Extreme ultraviolet	(Extreme Ultraviolet Explorer: EUVE)	NASA		(1991)
	(Far-Ultraviolet Spectroscopic Explorer: FUSE)	NASA-UK-Canada	0.6	(1997)
Ultraviolet	Copernicus	NASA	0.9	1972–1980
	International Ultraviolet Explorer (IUE)	NASA-ESA-UK	0.41	1978–
	Hubble Space Telescope: HST	NASA-ESA	2.4	1990–
	(Far-Ultraviolet Spectroscopic Explorer: FUSE)	NASA-UK-Canada	0.6	(1997)
Visible	Hubble Space Telescope: HST	NASA-ESA	2.4	1990–
	Hipparcos	ESA	0.29	1989–
Infrared	Infrared Astronomical Satellite (IRAS)	Dutch-UK-NASA	0.57	1983–1984
	(Shuttle Infrared Telescope Facility? SIRTF)	NASA		(1998)
	(Infrared Space Observatory: ISO)	ESA		
	(Large Deployable Reflector: LDR)	NASA	10	(1995)
Radio	Cosmic Background Explorer: COBE	NASA		1989–
	(Orbital Very Long Baseline Interferometry: Space VLBI)	NASA		(1995)

[a]Instruments in parentheses are planned or under construction, but not yet launched.
[b]NASA is the U.S. National Aeronautics and Space Administration; ESA is the European Space Agency, a consortium of several countries.

77 K) through a series of tubes surrounding the instrument. For observations of very long infrared wavelengths, even lower temperatures are required, and are obtained by using liquid helium (4 K). The recent *Infrared Astronomical Satellite (IRAS)* used liquid helium to cool its detectors as it mapped the sky (Fig. 7.17).

Infrared and ultraviolet telescopes work on the same principles as a reflector designed for visible wavelengths, bringing the radiation to a focus by means of a similar arrangement of mirrors. Radio telescopes operate in essentially the same way, but X-rays and γ-rays require rather different designs.

Ultraviolet astronomy began in the late 1940s with sounding-rocket observations of the Sun, and has progressed steadily ever since, with the development of a series of satellite observatories for ultraviolet spectroscopic observations. Currently two such spacecraft, the *International Ultraviolet Explorer* (Fig. 7.18) and the *Hubble Space Telescope* (described briefly later in this chapter) are in operation.

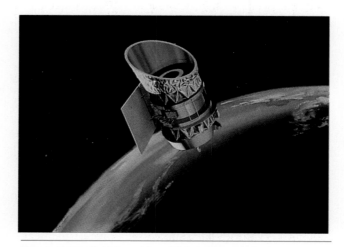

Figure 7.17 *IRAS* in orbit. *This is an artist's conception of the* Infrared Astronomical Satellite (IRAS) *as it appeared in orbit. This satellite revolutionized our knowledge of the cold matter in the galaxy. (NASA)*

Figure 7.18 The International Ultraviolet Explorer. *This NASA-operated spacecraft has been in operation since 1978, obtaining ultraviolet spectroscopic data for hundreds of astronomers. (NASA)*

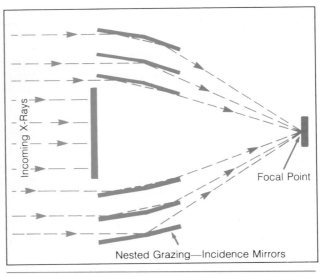

Figure 7.19 Design of an x-ray telescope. *A typical X-ray telescope consists of a set of nested rings, each able to reflect X-rays to a focal point by means of grazing-incidence reflections. Several rings are used because each one individually has little effective collecting area. (W. C. Cash)*

THE X-RAY UNIVERSE

Extremely hot gases, according to Wien's law, emit most strongly at very short wavelengths. Temperatures of a million degrees or more produce emission in the X-ray portion of the electromagnetic spectrum, between 1 and 100 Å. There are even spectral lines in this wavelength region, produced by electrons jumping between orbits with very large energy differences.

It is not simple to build an X-ray telescope, however. This is because most materials are not shiny for X-rays; a mirror used in the usual way simply absorbs X-rays rather than reflecting them. It is possible, though, to reflect X-rays if the radiation strikes the reflecting surface at a very oblique angle, and X-ray telescopes can be designed this way (Fig. 7.19). Sometimes several reflections are necessary to bring the X-rays to a focus, because only a small deflection of the radiation can be accomplished at each reflection. The process of dispersing X-rays by wavelength is also more complicated than for visible light, again because most substances absorb X-rays.

Despite the difficulties, a number of rocket and satellite observatories have successfully probed the heavens in X-ray wavelengths, and a wide range of phenomena have been discovered. Sites of violent explosions and other highly energetic regions are the main sources of X-rays, so these spaceborne observations have opened a new universe for astronomers, one whose existence was only hinted at previously.

There has been a recent lull between major X-ray observatories, but new activity has just gotten under way. The U.S.–German *ROSAT* satellite was launched in mid-1990 and is currently surveying the sky with unprecedented sensitivity. Within a few years, there will be additional large-scale X-ray observatories placed in space by the U.S. and by the European Space Agency. These spacecraft, called the *Advanced X-Ray Astrophysics Facility (AXAF)* and the *X-Ray Multiple Mirror Mission (XMM)*, will both be long-term observatories with multiple instrumental capabilities.

The shortest wavelengths of electromagnetic radiation, the γ-rays, also require special techniques for detection. Here even grazing-incidence reflections will not work, and as yet no instrument has been developed that can bring γ-rays to a precise focus. So far, only rather primitive γ-ray telescopes have been flown, but plans are being made for a major satellite mission called the *Gamma Ray Observatory*, which promises to provide important new data on the extremely violent regions in space where nuclear reactions are taking place, such as inside the stars and in supernova explosions.

RADIO TELESCOPES

We turn now to a discussion of telescopes for another invisible portion of the spectrum, the radio wavelengths. Radio astronomy has a relatively long history, dating back well before World War II. In 1931, Karl Jansky, a scientist with Bell Laboratories, noticed upon testing a radio antenna that the static he was receiving originated in the Milky Way (Fig. 7.20). This was the first indication astronomers had that celestial objects could radiate large quantities of energy outside the visible part of the spectrum, and as such was a tremendously important discovery, although its full significance was not widely recognized at first. Jansky was unable to pursue this discovery much further, but another American, Grote Reber of Wisconsin, became fascinated by Jansky's work and built his own radio antenna. Reber systematically mapped the sky and found many intense sources of radio emission; until the war was over, his work constituted the whole of the science of radio astronomy. British radar technicians during the war found that the Sun is a radio emitter. When radio techniques were further refined in the late 1940s, astronomers began to take more interest in the possibilities. One particularly provocative theoretical discovery, the prediction by the Dutch astronomer H. C. van de Hulst that hydrogen atoms in space should emit at a wavelength of 21 centimeters, spurred a great deal of activity, as several research groups worked to develop antennae capable of measuring this emission line. When the 21-centimeter line was detected in 1951, the immense promise of radio astronomy as a probe of the galaxy was realized. Now there are several major radio observatories around the world, and they provide insights into such diverse objects as the Sun, exploding galaxies, and molecules deep inside dark interstellar clouds.

The daytime sky is dark in radio wavelengths, so radio observatories need not shut down when the Sun rises, and most are on a 24-hour operating schedule.

A radio telescope works much like a reflecting telescope for visible wavelengths, except that generally everything is on a larger scale. The primary reflecting element need not be a smooth surface; instead it is usually a wire mesh. To radiation with such long wavelengths, a mesh acts like a shiny surface, as long as the wavelength observed is much longer than the spacing between wires in the mesh. At a point centered above the radio antenna, at what would be called the prime focus in a visible-wavelength reflector, is placed a

Figure 7.20 Karl Jansky with the radio antenna he used to discover the first celestial radio source. *Jansky's discovery of celestial radio emission was pivotal in inspiring the foundation of radio astronomy, although the field was not well established until after World War II. (Neils Bohr Library)*

receiver, a device that measures the intensity of the radiation. This receiver usually can be tuned so as to screen out unwanted wavelengths, serving in a sense as a filter. The intensity data can then be stored directly in a computer for later analysis.

A radio antenna must be very large in diameter to produce clear pictures of the sky. The resolution of a telescope is usually expressed in terms of how close together two objects can be and still be seen as separate objects, and is determined by the ratio of the wavelength being observed to the diameter of the telescope. Thus if long wavelengths, such as radio emission, are to be observed, the telescope must be very large in order to yield good resolution (that is, sharp images with fine detail). For a visible-wavelength telescope, a diameter of 10 centimeters (about 4 inches) provides a resolution of about 1 arcsecond, but a radio telescope would have to be several kilometers in diameter to achieve comparable resolution. Therefore even the largest radio telescopes (up to around 100 meters in diameter) have resolutions of several minutes of arc, and a radio picture of the sky tends to look fuzzy, with all the fine details smeared out.

A solution to this problem has been found and is being exploited at several radio observatories. By using

Figure 7.21 Radio interferometry. *A raw interferometric map (top) consists of patterns created by the interference of radio waves received at separate antennae. Complex computer analysis is required in order to produce the final image (bottom). This is the image of a double-lobed radio galaxy (see Chapter 30). (National Radio Astronomy Observatory)*

two or more radio antennae simultaneously, it is possible to simulate the effect of one very large antenna. The farther apart the antennae are, the better the resolution that can be achieved. Whereas a visible-wavelength telescope looking through the Earth's atmosphere typically has a resolution of 1 to 2 seconds of arc (that is, two stars about 1 or 2 arcseconds apart can be distinguished as separate objects), radio telescopes used in combination have managed resolutions of a small fraction of an arcsecond. The techniques,

called **interferometry,** is obviously very powerful, but it also is very complex, requiring precise knowledge of when particular radio waves arrive at the separate antennae and sophisticated computer processing to produce images or maps with high resolution (Fig. 7.21). Radio telescopes at various separations ranging from a few hundred feet to many thousands of miles have been employed; imagine the complexity involved in simultaneously using radio telescopes on opposite sides of the Atlantic!

Figure 7.22 Radio telescopes at Green Bank, West Virginia. *This is one of the major facilities of the U.S. National Radio Astronomy Observatory. The large dish at left is now out of service, having collapsed catastrophically in late 1988. (National Radio Astronomy Observatory)*

Figure 7.23 The Very Large Array. *This arrangement of twenty-seven radio dishes spread over a 15-mile wide section of New Mexico desert provides high resolution radio maps through the use of interferometry. The individual telescopes can move along railroad tracks, allowing the separations to be adjusted, depending on the type of data needed. (National Radio Astronomy Observatory)*

The largest radio observatory in the United States is the National Radio Astronomy Observatory (Fig. 7.22), sponsored by the National Science Foundation, with several antennae at Green Bank, West Virginia; a 30-foot dish at Kitt Peak (used mostly for observing interstellar molecules, whose emission lines occur at relatively short radio wavelengths; see Chapter 26); and the Very Large Array (VLA), near Socorro, New Mexico (Fig. 7.23). The VLA has 27 movable 82-foot antennae arranged in a Y-shaped pattern spanning some

15 miles of the New Mexico desert, and is designed for interferometry, ultimately achieving a resolution of better than 0.1 arcsecond.

Another large U.S. radio observatory, at Arecibo, Puerto Rico, consists of a 1,000-foot dish built into a natural bowl in the mountains (Fig. 7.24). This large diameter provides fairly high resolution and sensitivity, but since the antenna cannot be pointed (it is immovable), observational coverage is somewhat limited in the north–south direction (the Earth's rotation provides east–west coverage).

Major radio observatories outside the United States include the Jodrell Bank Observatory, operated by the University of Manchester in England; the Westerbork Observatory in the Netherlands; the Parkes Observatory in Australia; and the 100-meter dish at Bonn, West Germany.

THE NEXT GENERATION

What kinds of telescopes will supercede those already described in this chapter? We can speculate about this; astronomers have a flair for dreaming up new and better ways to overcome technical difficulties, so in some cases plans for new and wonderful instruments are already well developed.

Plans are now being made for a number of visible-wavelength telescopes in the 8-to-16-meter range (imagine a telescope 16 meters, or about 50 feet, in diameter!). The most advanced of these plans, being developed by Cal Tech and the University of California, calls for the construction of a 10-meter telescope on Mauna Kea. Rather than having a single 10-meter mirror, the *Keck Telescope* is being built using a **segmented mirror** design, in which the primary mirror consists of several segments (Fig. 7.25). Each segment can be separately controlled, so that all of them have a common focus. This is somewhat similar to the multiple-mirror concept, except that in this case the multiple mirrors are joined together. The *Keck Telescope* is expected to be in operation by 1991.

The U.S. National Observatories, whose largest telescopes now are 4 meters in diameter, have been developing plans for larger national facilities. The most ambitious idea was to build a multiple-mirror telescope consisting of four 8-meter mirrors, mounted on a single structure. Cost considerations have forced a change, however, and the current plan is to build two 8-meter telescopes, one in the Northern Hemisphere, and one in the Southern Hemisphere. The site selected for the Northern Hemisphere is Mauna Kea, while the Southern Hemisphere national 8-meter instrument will be located near Cerro Tololo, the site of the existing Inter-American Observatory. Funding for these telescopes is currently being reviewed by the National Science Foundation.

The *European Southern Observatory*, operated by a consortium of several European nations (and already

Figure 7.24 The giant Arecibo radio dish.   *This is the largest single radio antenna in the world. The 1,000-foot dish is built into a natural bowl in the mountains of Puerto Rico. Since the telescope cannot be pointed in arbitrary directions, it must rely on the Earth's rotation to scan the sky. (Arecibo Observatory, part of the National Astronomy and Ionosphere Center operated by Cornell University under contract with the National Science Foundation)*

Figure 7.25 The 10-meter Keck telescope. *The photo at left shows a model of the completed telescope; the photo above shows the actual telescope in August, 1990, when three of the thirty-six hexagonal mirrors had been installed (in this photo the three mirrors are covered by white, padded insulation to avoid accidental damage). It is hoped that the first astronomical observations (using only nine mirrors) will be made during the fall of 1990. (Left: Keck Observatory, University of California and California Institute of Technology; above: photo by T. P. Snow, arranged through the courtesy of the California Astronomical Research Association)*

having several telescopes in Chile), is making plans for a 16-meter telescope, to be located in the Southern Hemisphere, using a multiple-mirror design. This instrument, called the *Very Large Telescope,* will consist of four 8-m telescopes operated together so that the total light-gathering power will be equivalent to that of a single mirror with a 16-m diameter. It is hoped that the four will be operable with sufficient precision to allow for infrared and visible-wavelength interferometry, in the same manner as a radio observatory such as the *Very Large Array.* Meanwhile, several groups of U.S. universities are planning 8- to 10-meter telescopes, using a single mirror. Mirrors this large may soon be made using a novel technique being developed at the University of Arizona. The molten glass for the mirror is poured into a rotating mold (Fig. 7.26), so that the rotation forces the glass into a concave shape as it hardens. This produces a mirror that is lightweight (because its internal structure is a honeycomb, rather than being solid glass) and that is nearly in the correct shape to begin with, so that many months of grinding

to the correct shape are eliminated. Other novel techniques employing very thin glass mirrors are also being developed.

In radio astronomy, there have been recent developments, particularly in the direction of observing the shortest radio wavelengths, around 1 millimeter and below. This is a rich region of the spectrum, because many molecules in space emit at these wavelengths, but it has been a difficult one in which to work. The Earth's atmosphere provides a great deal of interference, and the short wavelengths require more precision in shaping the reflector than is normal for radio telescopes. Despite these difficulties, however, two submillimeter telescopes have recently been built on Mauna Kea, and are now in full operation. One, having a diameter of 10 meters, is operated by Cal Tech; the other, a 15-meter dish, is a joint Dutch–British venture (Fig. 7.27).

In another development, the U.S. National Radio Astronomy Observatory is developing plans to replace the large dish at Green Bank, West Virginia, which

Figure 7.26 Casting a mirror on a spinning table. *Rotation of the table containing the mirror mold while the glass hardens facilitates construction of large mirrors with concave surfaces for use as telescope primary mirrors. This technique, being developed at the University of Arizona, makes possible relatively low-cost construction of large mirrors. Mirrors as large as 8 meters are being planned using this method. (University of Arizona)*

Figure 7.27 A new radio telescope. *This is the reflecting dish for the Dutch–United Kingdom 15-meter radio telescope, now in operation on Mauna Kea. Telescopes of this size are usually designed for observing very short wavelength radio waves, with wavelengths of a few millimeters to a few centimeters, and the reflecting surface must be far more precisely shaped than in radio telescopes designed for much longer wavelengths. (T. P. Snow)*

collapsed spontaneously in 1988 (this is the large dish shown at the left in Figure 7.22).

In the more distant future, radio astronomers are hoping to have telescopes orbiting the Earth, so that interferometry can be done with telescope separations even greater than the intercontinental distances now available.

In ultraviolet and visible astronomy, the new workhorse instrument is the *Hubble Space Telescope* (Fig. 7.28), finally launched in April, 1990, after years of delays due to slow funding, technical problems and, above all, the *Challenger* disaster. At this writing, the *HST* has completed most of its initial checkout phase and is beginning to carry out the first scientific observations. Some problems with the focusing ability of the telescope are forcing some revision of the initial scientific program, but these problems can be corrected when new instruments are placed on board in about three years. The *HST* is being operated as a national observatory, accessible (through competitive proposal reviews) to astronomers from the United States. There is European participation as well; the European Space Agency (ESA) provided one of the principal cameras (the Faint Object Camera), and in return European astronomers are invited to compete for observing time on the *HST*. Recent results from the *HST* are described in the Special Report that begins on page 150.

Looking a bit farther into the future, another ultraviolet orbiting observatory is now under development, for launch in the late 1990s. This instrument, variously called *Lyman* or the *Far Ultraviolet Spectroscopic Explorer* (FUSE), will observe wavelengths that are not reached by the *HST*. There are many important spectral lines of atoms, ions, and molecules at the shortest ultraviolet wavelengths, and it is the prime goal of *FUSE* to allow these features to be studied. Lyman/*FUSE* is primarily a NASA venture, but there is some

HIGH-GAIN ANTENNA
SECONDARY MIRROR
PRIMARY MIRROR
APERTURE DOOR
LIGHT SHIELD
EQUIPMENT SECTION
FINE GUIDANCE OPTICAL CONTROL SENSORS (3)
AFT SHROUD
SCIENTIFIC INSTRUMENTS
AXIAL MODULES (4)
RADIAL MODULE WITH RADIATOR (1)
FIXED HEAD STAR TRACKERS (3)
DOUBLE ROLL-OUT ARRAY
MSFC-2/79-5

Figure 7.28 The Hubble Space Telescope   *This is a schematic drawing of the 2.4-meter telescope launched into Earth orbit in April, 1990. The open end of the telescope is pointed away from us in this view; the near end houses control systems, cameras, and other scientific instruments. The large "wings" are solar panels that provide electrical power for the spacecraft. (NASA)*

participation by the United Kingdom and by Canada.

X-ray astronomers are awaiting three major instruments. One, a joint German–U.S. operation called *ROSAT,* is making images of the sky in X-ray wavelengths, and was placed in orbit in 1990. A second major X-ray observatory, called the *X-Ray Multiple Mirror Mission* (or *XMM*), is being planned for a late-1990s launch by the European Space Agency, and will also include some U.S. participation. The third, called the *Advanced X-Ray Astrophysics Facility,* or simply *AXAF,* will be similar to the *Space Telescope* in that it will be a large telescope with several interchangeable scientific instruments. Its launch date is expected in the late 1990s. The very short wavelength ultraviolet spectrum, called the **extreme ultraviolet,** will be observed by a satellite called the *Extreme Ultraviolet Explorer,* to be launched in 1991. In the realm of higher energy photons, a satellite called *Gamma Ray Observatory* is now under development and will be in orbit by the early 1990s.

Finally, in infrared astronomy, plans are under way for two major orbiting facilities. One, an ESA project called *Infrared Space Observatory* or *ISO,* is now being

designed; the other, a major facility akin to the *Space Telescope* and called the *Space Infrared Telescope Facility* or *SIRTF,* is a NASA project now in the development stage. *SIRTF* will have a large telescope with several scientific instruments for observing infrared wavelengths that do not reach the Earth's surface. Even farther in the future is the *Large Deployable Reflector,* a 10-to-15-meter lightweight dish that will be unfolded in orbit, becoming the largest space telescope yet. This project may not get off the ground until around the turn of the century.

We have purposely discussed telescopes for all wavelengths together, in the preceding section on telescope design and in this one on future instruments, because the traditional segmentation of astronomers into one wavelength region or another is breaking down. Formerly an astronomer was usually considered an optical (visible) astronomer or an infrared astronomer, a radio astronomer, and so on. Today, however, and certainly in the future, astronomers take a broader approach to scientific problems, using data from any wavelength region or instrument that is available. If there is a revolution today in astronomy, this is it.

The Saga of the *Hubble Space Telescope.*

It is rare for any astronomical phenomenon to attract as much attention as the *Hubble Space Telescope* has. The drama has been in the public spotlight for many reasons, including the immense promise of the instrument for making breakthrough discoveries, the many long and frustrating delays, the huge expense (by normal research standards), and, finally, the technical problems that have limited the effectiveness of the *HST*. Along the way many lessons are being learned, about how to manage large-scale technological projects, about the trade-offs between large and small projects, and about conveying information to the tax-paying public. In the end, we have a functioning telescope that is capable of producing far more forefront research than most people realize, and which will be corrected within a few years so that all of the original goals can be achieved. We also have a space agency that is reeling from the shock of the problems with *HST*, the continuing problems with the Space Shuttle program, and heavy pressure from Congress and the public to get the U.S. space program into better shape.

The rationale for building the *HST* need not be dwelled upon here; there are many references to it throughout this text. Put simply, the *HST* has been envisioned as a national observatory in space, capable of taking advantage of the fine resolution available without the blurring effects of the atmosphere, and able to provide extended wavelength coverage into ultraviolet and infrared regions that cannot be reached from the ground. Much emphasis has been placed on "edge-of-the-universe" research, programs designed to answer the biggest questions about the fundamental nature of space, time, and the origins of the cosmos, and this has tended to obscure the many other areas in which the *HST* will make basic breakthroughs. Important questions in planetary science, studies of stars and their evolution, the properties of the interstellar material from which new stars form, the nature of bizarre and exotic objects such as black holes and neutron stars, and the overall content of the universe can all be addressed with the *HST*.

The *HST* was originally envisioned as long ago as 40 years, but serious planning (and Congressional funding approval) did not begin until the late 1970s. The concept was always to build a multipurpose facility having a variety of scientific instruments to capture the light gathered by a large primary mirror. Over the years of planning and development, this concept evolved into the complex orbital observatory described in this chapter; the *HST* has five primary instruments and can also perform fundamental research on stellar positions and motions with its fine pointing system. The *HST* became the flagship of a new NASA program called the "Great Observatories"; this first optical/ultraviolet facility is to be followed by major observatories in X-ray, far-infrared, and gamma-ray wavelengths. Meanwhile the cost of *HST* climbed, sometimes because of technical difficulties (usually these increases came about because of overruns by contractors responsible for major components such as the mirrors), and sometimes because of delays (the enormous infrastructure of people and organizations had to be kept going during all the lengthy slow periods). The final cost quoted by NASA of producing the *HST* and getting it launched was about $1.5 billion. This is an enormous sum by conventional research standards, but is small compared with other portions of the U.S. budget.

Launch of the *HST* finally occurred in May of 1990, from Cape Canaveral, Florida (the John F. Kennedy Space Center). The required orbit was reached, and the *HST* was carefully released from the Shuttle bay a couple of days later (see the first figure). After some delays, the solar panels were unfurled and the power systems checked out, and the Shuttle crew returned to Earth, leaving the *HST* on its own.

A lengthy in-orbit check-out phase was planned and began to be executed within a few weeks (some time had to be allowed for residual gases to escape before the high-voltage instruments were turned on). During check-out, image tests were performed to test the focus. These proved troublesome; a good focus remained elusive, and it was finally concluded that one of the mirrors did not have precisely the correct shape (instead of being parabolic, it is a

SPECIAL REPORT

Release of the *Hubble Space Telescope* by the space shuttle.
In this dramatic photograph, the Hubble Space Telescope *is floating free in space for the first time, having just been released by the Shuttle's mechanical arm. The solar panels have not yet been unfurled to their full length. (NASA)*

bit too close to being spherical, which causes light entering at off-center locations not to be focused into a single point). The error is tiny; it has been determined that the edges of the primary mirror are 2 microns (2×10^{-6} m) lower than they should be, relative to the mirror's center. But by telescopic standards, this is a huge error, sufficient to smear the images so that only about 15 percent of the light falls into a circle 0.1 arcsecond in radius (over 90 percent of the light was supposed to have been focused into an area this small). The result is serious for research programs demanding the fine resolution of which the *HST* was supposed to be capable. On the other hand, for relatively bright targets, the innermost part

of the image still contains enough light to be useful, and by using only this portion, it is still possible to achieve very high resolution images, far better than can be done with ground-based telescopes. Furthermore, the spectroscopic instruments aboard *HST* have small entrance holes to allow the light beam in, and these ensure that the full spectral detail will still be available. The only disadvantage for these instruments caused by the focus problem is that it takes longer to get the same amount of light through the hole; exposure times for spectroscopic observations will have to be increased. But if they are, all the original scientific goals can be achieved. The high-speed photometer can also still do useful work,

but it will not be able to monitor very rapid variations in the faintest sources. Thus, on balance, the *HST* is still a very useful scientific observatory and will make many important discoveries that could not be made by any other instruments.

Some interesting results have already been obtained. Stars that appear single in ground-based photos are now seen to be double (see the second figure); the core of the "normal" galaxy NGC7457 has been found to contain an unexpected concentration of stars (see the third figure) which might indicate the presence of a supermassive black hole; the compact cluster of stars in the Tarantula Nebula in the Large Magellanic Cloud has been resolved into individual stars (see the fourth figure); and, perhaps most spectacular, the remnant of Supernova 1987A has a tiny (but easily resolved by *HST*) ring of glowing gas surrounding the site of the explosion (see the fifth figure), consisting of swept-up gas from the star's earlier stellar wind phase, excited to glow by ultraviolet radiation from the blast.

The two cameras are the only instruments yet put into action (as of October, 1990), but soon the spectrographs will begin their research programs, and we can expect many new and exciting results.

The cause of the focus problem has now apparently been found, by an investigative committee appointed by NASA. During construction and polishing of the primary mirror, tests were conducted to determine that the mirror had the proper shape. To conduct these tests, optical engineers (at

SPECIAL REPORT

Perkin-Elmer Corporation, now a subsidiary of Hughes Aircraft) made special lenses called null correctors, which allowed the mirrors to be imaged in such a way that any deviation from the desired shape could be seen. The problem was that the null corrector used to test the primary mirror was itself incorrectly shaped. The *HST* primary mirror was polished so that its shape, as analyzed with the null corrector lens, was as accurately formed as possible. But because the null corrector lens was formed improperly, so was the *HST* primary mirror. Thus the *HST* mirror was perfectly made to the incorrect shape.

The cause of the improper construction of the null corrector lens was that a tiny gap went unnoticed in the framework that was used to set its focal length while it was being made; a small metal cap did not quite slip all the way into place over a glass rod, and the resulting displacement caused the focal length error, which in turn caused the incorrect shape of the *HST* mirror.

As mentioned earlier in this Special Report, the *HST* is still capable of first-rate research, despite the rather negative image being projected by the news media, and furthermore the problem can be almost completely overcome when new cameras and spectrographs are installed. The so-called "second-generation" instruments had already been under development before the *HST* was launched, because the plan had always been to replace instruments from time to time, updating the capabilities of the *HST*. The relatively minor corrections to these instruments needed to counteract

The *space telescope* resolves a binary star.
The image at left was obtained from the ground and shows only a single star. At right is the HST image of the same object, revealing that it is double. (NASA/Space Telescope Science Institute)

An unexpectedly bright galactic core.
These are two versions of the same HST image, of the galaxy NGC7457. The print at left was processed for high contrast, so that a large range of brightnesses could be shown on a single print. At right, the contrast has been adjusted to emphasize small brightness differences, revealing a tiny galactic core that is brighter than its surroundings. It is estimated that stars are crowded into the center of this galaxy at a density at least 30,000 times greater than in the Sun's neighborhood in our own galaxy. The cause of such a concentration is not known but may be related to other galaxies that have active nuclei thought to be associated with supermassive black holes (see Chapter 30). (NASA/Space Telescope Science Institute)

SPECIAL REPORT

The structure of a dense star cluster. *This tight grouping of stars lies in the Large Magellanic Cloud, about 170,000 light-years from Earth, and was for a while thought to consist of a single, gigantic star instead of many normal stars crowded together. The HST images shown here have provided the first direct proof that the object is made up of many separate stars. In this illustration, panel A (upper left) shows the HST image of the full field of view of the Wide Field and Planetary Camera. In panel B (upper right) the central portion of this image has been expanded so that the individual stars in the core of the cluster can be seen. Panel D (lower right) is the same as panel B, except that the extended fuzzy regions around the stars, caused by the HST mirror defect, have been removed through computer processing. Panel C (lower left) is a ground-based image of the same region, showing how difficult it is to resolve the individual stars in the cluster. (NASA/Space Telescope Science Institute)*

the focus problem are not technically difficult, nor are they very expensive, so it seems certain that the *HST* will regain its intended fine focus within two or three years, when the new instruments are placed on board. Thus the major costs of the mirror problem lie in lost time, because the full capabilities will have to wait an additional two or three years to be realized; and in public confidence, because of the widespread perception that NASA has difficulty managing complex programs.

What are the lessons to be learned from the escapades of

HST? To the scientific community, perhaps the most important one is that we should question whether huge, expensive, multipurpose missions are the best way to invest our limited research resources. Many smaller-scale missions, usually dedicated to a much more limited range of scientific tasks than the *HST*, have accomplished remarkable research programs at only a fraction of the cost of *HST*. An excellent example is the *International Ultraviolet Explorer* (*IUE*), now in its thirteenth year of ultraviolet spectroscopic observations, launched for under $200 million in 1978 and

costing less than $6 million per year to operate. Another good example is the *Cosmic Background Explorer*, launched in 1989 and now making fundamental discoveries about the basic nature of the universe, at a cost similar to that of the *IUE*.

One important aspect of the high cost of the *HST* is its total reliance on the Space Shuttle; had *HST* been designed for launch on an unmanned vehicle, it might well have been in orbit for five years or more by now and probably would not have cost nearly as much as it has (but this would not have avoided the mirror

SPECIAL REPORT

The remnant of Supernova 1987A.
A massive star in the Large Magellanic Cloud exploded in February of 1987, creating one of the most spectacular events for astronomers in recent memory (see Chapter 22). The star that exploded had previously lost gas through a dense outflow known as a stellar wind. Here we see a ring of leftover material from this earlier phase, excited by ultraviolet photons from the explosion so that it glows in certain emission-line wavelengths. This image was made at the wavelength of twice-ionized oxygen, in the blue-green part of the spectrum. (NASA/Space Telescope Science Institute)

problem). Many scientists have advocated for some time that research missions not be tied to the Shuttle unless human involvement is necessary; ironically, the Challenger disaster has helped accomplish this goal, by forcing NASA to explore the use of unmanned, expendable launch vehicles. One of the first benefactors of this new (actually old, but unused) option was the *ROSAT* mission, a joint German-U.S. X-ray observatory launched in late May of 1990, now well under way in its survey of the X-ray skies.

The problems with *HST*, along with other issues related to Shuttle technology and the costs of large-scale missions, appear to be forcing NASA and the Congress into a new period of evaluation of the space program and the best methods for doing fundamental space-based research. Perhaps an important legacy of the *HST*, along with the scientific triumphs it is sure to produce, will be its impact on the way in which science research will be done in the twenty-first century.

PERSPECTIVE

We have seen, in broad outline, how astronomers gather the light, in all wavelengths, that carries the information they seek. We have learned that, for many wavelengths, the same basic telescope design will work, although it may be necessary to place the instrument in orbit, above the Earth's atmosphere. To observe other portions of the electromagnetic spectrum, new technology, such as radio telescopes and X-ray and γ-ray instruments, has had to be developed. Today astronomers can look at the universe in nearly any wavelength. We are ready now to see what has been learned.

SUMMARY

1. Telescopes are better than the human eye for astronomical observations because they have larger collecting areas, have superior resolution, and can accumulate light over longer periods of time.

2. Telescopes for visible light use lenses or mirrors to bring light to a focus, but all big telescopes are reflectors, which can be made larger and are free of chromatic aberration.

3. Reflecting telescopes can bring light to a focus in a variety of ways for different purposes.

4. Auxiliary instruments are used to analyze the light once it is brought to a focus, by photographing the image, measuring its brightness, or dispersing the light for spectral analysis.

5. An observatory site must be chosen for clear weather, high atmospheric transmissivity, low turbulence in the air overhead, minimal pollution, remoteness from city lights, and the proper latitude for viewing the desired part of the sky.

6. Infrared and ultraviolet observations can be made through the use of the same optical designs as visible-light telescopes but must be carried out from above the Earth's atmosphere, which blocks these wavelengths from reaching the ground.

7. Observations in X-ray and γ-ray wavelengths require novel telescope designs and must be carried out from space.

8. Most radio telescopes are similar in design to reflecting telescopes for visible light, but must be built very large in order to provide good resolution.

9. In all wavelengths, but most commonly in the radio portion of the spectrum, several telescopes can be used together to produce high-resolution images by interferometry.

10. Planned future telescopes include several giant reflectors that will be built on the ground, and a variety of radio, X-ray, ultraviolet, infrared, and γ-ray telescopes, many of them to be launched into orbit on the Space Shuttle.

REVIEW QUESTIONS

Thought Questions

1. Summarize the benefits of using a telescope for astronomical observations, compared with using only the naked eye.

2. What are the advantages of reflecting telescopes over refractors?

3. Summarize the various focal arrangements used in reflecting telescopes, and give the advantages of each.

4. Discuss the contrasts between observational astronomy today and what it was like in the time of Tycho Brahe.

5. List the reasons why the summit of Mauna Kea, in Hawaii, is an excellent observatory site.

6. Drawing upon information from Chapter 6, explain why telescopes are needed for observations in wavelengths other than visible light. Try to think of examples of objects that might best be observed in other wavelengths such as the ultraviolet, the infrared, the radio, and the X-ray portions of the spectrum.

7. How is an ultraviolet telescope similar to an ordinary visible-light reflector, and how is it different? How does an infrared telescope compare with a normal visible-light reflector? How does a radio telescope compare?

8. Explain why the design of an X-ray telescope must be very different from that of a visible-light telescope to bring X-rays to a focus.

9. Explain why radio telescopes ordinarily cannot map the sky with as much precision (that is, with resolution as high) as visible-light telescopes can. How do you think ultraviolet telescopes compare in resolving ability with visible-light telescopes?

10. Explain why the *Space Telescope* will have advantages over ground-based telescopes, even for visible-wavelength observations.

Problems

1. The largest refracting telescope has a lens diameter of 40 inches, while the largest reflector has a diameter of 6 meters, or 236 inches. How much greater is the light-collecting power of the 6-meter reflector?

2. Show that the six 72-inch mirrors of the *Multiple Mirror Telescope* have a total light-collecting area equivalent to that of a single mirror 176 inches in diameter.

3. Suppose the cage in which an astronomer sits to make prime-focus observations with the 200-inch Hale Telescope (on Mt. Palomar) is a cylinder with a diameter of 40 inches. What fraction of the incoming light is blocked by the observer's cage?

4. Suppose you want to observe a planet whose surface temperature is 80 K. At what wavelength does this planet emit most strongly? Compare this wavelength with the wavelength of maximum emission for the telescope detector if it is cooled with liquid nitrogen (77 K) or with liquid helium (4 K). Which coolant is better to use for this observation?

5. The resolving power of a telescope, defined as the smallest separation between two objects that can be seen as separate objects, is proportional to the wavelength being observed and inversely proportional to the diameter of the telescope. For a telescope of 10 cm diameter, the resolving power for blue light (wavelength 4,000 Å) is 1 arcsecond. What is the resolving power for a telescope of 1-meter diameter, observing at the same wavelength?

6. What is the resolving power for a 1-m infrared telescope, observing at a wavelength of 40,000 Å? What is the resolving power for a police radar detector, a microwave telescope having a diameter of 10 cm, observing at a wavelength of 4 cm?

7. What is the frequency of a 21-cm photon from a hydrogen atom?

8. How large would a radio telescope have to be to obtain a resolving power of 1 arcsecond for observations at a wavelength of 21 cm?

9. The angular diameter of Pluto, when the planet is at opposition, is 0.1 arcsecond. Will the *Space Telescope* be capable of making out surface markings on Pluto?

10. If an interstellar probe could maintain an average speed of one-half the speed of light, how long would it take to reach Alpha Centauri? How long after its launch from Earth would it be until we received the first data from the star?

ADDITIONAL READINGS

Bartusiak, Marcia. 1984. Very Large Array (VLA). *Science 84* 5(6):64.

Beatty, J. K. 1990. ROSAT and the x-ray universe, *Sky and Telescope* 80:128.

Burns, J. O., N. Duric, G. J. Taylor, and S. W. Johnson. 1990. Observatories on the Moon. *Scientific American* 262(3):42.

Chaisson, E. and R. Villard. 1990. Hubble Space Telescope: the mission. *Sky and Telescope* 79:378.

Field, G. and D. Goldsmith. 1990. The space telescope: eyes above the atmosphere. *Mercury* 19:34.

Fienberg, R. T. 1990. Dazzling views from Europe's NTT. *Sky and Telescope* 79:596.

———. 1990. The Hubble Space Telescope takes wing. *Sky and Telescope* 80:31.

Fischer, D. 1989. A telescope for tomorrow. *Sky and Telescope* 78:248.

Gordon, Mark A. 1985. VLBA-A continent-size radio telescope. *Sky and Telescope* 69(6):487.

Gulkis, S., P. M. Lubin., S. S. Meyer, and R. F. Silverberg. 1990. The Cosmic Background Explorer. *Scientific American* 262(1):132.

Gustafson, J. R., and W. Sargent. 1988. The Keck Telescope: 36 mirrors are better than one. *Mercury* 17:43.

Habing, H. J., and G. Neugebauer. 1984. The infrared sky. *Scientific American* 251(5):48.

Harrington, S. 1982. Selecting your first telescope. *Mercury* 11(4):106.

Janesick, J., and M. Blouke. 1987. Sky on a chip: The fabulous CCD. *Sky and Telescope* 74(3):238.

Kellerman, K. I., and R. A. Thompson. 1988. The Very-Long-Baseline-Array. *Scientific American* 258(1):54.

King, H. C. 1979. *The history of the telescope.* New York: Dover.

Krisciunas, K. 1989. Two astronomical centers of the world: Mauna Kea and LaPalma. *Mercury* 18(2):34.

Labeyrie, A. 1982. Stellar interferometry: A widening frontier. *Sky and Telescope* 63(4):334.

Longair, Malcolm. 1985. The scientific challenge of Space Telescope. *Sky and Telescope* 69(4):306.

McAlister, H. A. 1988. Seeing stars with speckle interferometry. *American Scientist* 76:166.

McLean, I. S. 1988. Infrared astronomy's new image. *Sky and Telescope* 75:254.

Mims, S. S. 1980. Chasing rainbows: The early development of astronomical spectroscopy. *Griffith Observer* 44(8):2.

Nelson, J. 1989. The Keck Telescope. *American Scientist* 77:170.

O'Dell, C. R. 1989. Building the Hubble Space Telescope. *Sky and Telescope.* 78:31.

Physics Today. 1982. 35(11). (Several review articles on telescopes for various wavelengths).

Ridpath, I. 1990. The William Herschel Telescope. *Sky and Telescope* 80:136.

Shore, Lys Ann. 1987. IUE: Nine years of astronomy. *Astronomy* 15(4):14.

Silk, J. 1990. Probing the primeral fireball. *Sky and Telescope* 79:600.

Sinnott, R. W. 1990. The Keck Telescope's giant eye. *Sky and Telescope* 80:15.

Spradley, J. L. 1988. The first true radio telescope. *Sky and Telescope* 76:28.

Tucker, Wallace, and Riccardo Giacconi. The birth of X-Ray astronomy. Parts 1 and 2. *Mercury* 14(6):178; 15(1):13.

Tucker, Wallace, and Karen Tucker. 1986. *The cosmic inquirers: Modern telescopes and their makers.* Cambridge, Mass.: Harvard University Press.

Verschuur, Gerritt L. 1987. *The invisible universe revealed: The story of radio astronomy.* New York: Springer-Verlag.

MATHEMATICAL METHODS

Resolving Power

In this chapter, we have stressed the need for making telescopes as large as possible. There are two general reasons for doing this: (1) to collect as much light as possible, thereby enhancing the ability to observe faint objects; and (2) to make the resolution as sharp as possible, so that fine details can be seen. The light-collecting power is simple to calculate, as indicated in the text: This is proportional to the area of the primary mirror, which is usually circular, and the area of a circle is πr^2. Hence we can easily compare the light-collecting areas of different telescopes, by comparing the squares of their diameters or radii (thus a 10-m telescope has 4 times the collecting area of a 5-m telescope, for example.)

The resolving power of a telescope is limited by an effect called *diffraction,* which is caused by the fact that light rays bend a little when they pass an obstacle such as the edge of the telescope opening. Because they bend, rays that pass near the edge can overlap and interfere with other rays, causing images to be blurred. This blurring sets a limit on the angular size of the smallest feature that can be seen; the limiting angular sharpness, or resolving power, of a telescope is inversely proportional to the diameter of the telescope, and is proportional to the wavelength being observed. In equation form, this is

$$\theta = 250,000 \ (\lambda/D),$$

where θ is the resolution of the telescope (in arcseconds), λ is the observed wavelength, and D is the diameter (λ and D can be expressed in any units, as long as they are the same; i.e., both can be expressed in meters, centimeters, inches, or anything else, so long as the same units are used).

We can calculate the resolving power of the human eye, by taking 1 mm (= 0.001 m) as the value for D, corresponding to the typical diameter of the pupil of the eye, and assuming for the wavelength 5,000 Å or 5×10^{-7} m, an average wavelength for visible light. Inserting these values into the above equation yields

$$\theta = 250,000 \ (5 \times 10^{-7} \ m)/(0.001 \ m)$$
$$= 125 \ \text{arcseconds}.$$

Thus the human eye is capable of seeing details as small as 125 arcseconds, or about 2 arcminutes, in angular size. This is about the limiting accuracy of all astronomical positional measurements made before the invention of the telescope (see the discussion in Astronomical Insight 4.1).

Now let's consider a typical home telescope, of diameter 10 cm or 0.1 m (about 4 inches). Using the equation for resolving power, we find that this telescope has a resolving power of 1.25 arcseconds (this can be readily seen by noting that the replacement of 0.001 m with 0.1 m for D in the equation means that the result will be 100 times smaller). Thus, even a relatively small telescope provides an enormous advantage in angular resolution over the human eye.

In principle, a large telescope should have very good angular resolution: Using the above equation, we find that a 4-m telescope has a resolving power of 0.03 arcsecond. Unfortunately, it is virtually impossible to achieve such a high resolution for a visible-wavelength telescope, because the turbulence of the Earth's atmosphere blurs images of objects in space. Under the best of conditions, the actual angular resolution may be slightly less than one arcsecond, far from the theoretical limit for a large

Continued on following page

MATHEMATICAL METHODS

Resolving Power (continued)

telescope. Modern techniques for telescope construction, in which the curvature of the primary mirror is controlled in response to the blurring caused by the atmosphere, can improve the resolving power to about one-half of an arcsecond or slightly better, but the atmosphere still provides a fundamental limit. This explains why it is important to orbit instruments such as the *Hubble Space Telescope;* even in visible wavelengths (which reach the ground) it is important to achieve the superior resolving power that can be obtained if the blurring caused by the atmosphere is eliminated. The *Hubble Space Telescope* has a diameter of 2.4 m; inserting this into our equation for resolving power yields $\theta = 0.05$ arcsecond,

about 10 times better than the best resolution available from ground-based telescopes.

It is interesting to consider telescopes for wavelengths other than visible light. In the radio portion of the spectrum, for example, wavelengths are typically 10^8 times longer than visible light. This means that a radio telescope has intrinsically poor resolving power; the largest radio telescope in the world, the 1,000-ft (300-m) dish at Arecibo, has a resolving power at 10-cm wavelength of only 83 arcseconds or about 1.3 arcminutes. Thus the images obtained by radio telescopes are inherently fuzzy. Fortunately, at these long wavelengths, interferometry can be used to improve the resolving power of an array of radio telescopes, as mentioned in the text (currently efforts are being made to develop interferometry for infrared and visible-wavelength telescopes; if these are successful, then the quality of images obtained in these wavelength regions will eventually rival or exceed those obtained today by using multiple radio telescopes).

Exploring the Universe at All Wavelengths

BY FRANCE CÓRDOVA

Dr. Córdova has taken a somewhat unusual path into astronomy, having obtained a bachelor's degree in English from Stanford before entering graduate school in astrophysics at Cal Tech, where she received her Ph.D. in 1979. She then spent 10 years as a researcher at the Los Alamos Laboratories in New Mexico (associated with the University of California), where she became a pioneer of a modern field of astrophysics in which data are obtained from many different wavelength bands in the electromagnetic spectrum. This is a difficult endeavor, as described in her article here, because of many practical constraints, yet the multiwavelength approach is clearly the best way to develop a full understanding of any astronomical object. During her tenure at Los Alamos, Dr. Córdova developed a strong international reputation as a leader, not only in the observational development of multiwavelength astronomy, but also in research on the bizarre remnants of dying stars which we know as neutron stars and black holes (discussed in Chapter 23). Along the way, she was frequently called upon to serve on various national and international advisory committees, where she has advocated

the establishment of Earth- and space-based observatories dedicated to multiwavelength observations. In 1989 Dr. Córdova left Los Alamos to become head of the Astronomy and Astrophysics Department at the Pennsylvania State University. With colleagues there, she has successfully proposed to build a visible-wavelength telescope to be flown with the European X-ray observatory called the X-Ray Multiple Mirror Mission, thus converting this orbiting astronomy laboratory into a multiwavelength facility. In her guest editorial, Dr. Córdova explains why multiwavelength astronomy is important and discusses the difficulties and rewards of involvement in this field.

There are few celestial objects that radiate at only one frequency (HI regions, CO clouds, and masers, all discussed in later chapters, come to mind). Isolated stars emit thermal, or blackbody, radiation that, although it peaks at a specific frequency that depends on the temperature of the star, is emitted over a wide range of frequencies. The amount of blackbody radiation that a star emits at each frequency can be approximated by the Planck function (described in Appendix 5 of this book). An example follows (see next page). The filled dots show the intensity of radiation measured from Zeta Puppis, a very hot supergiant star. This star has been observed all the way from radio to X-ray frequencies. The bold line shows the Planck function that best fits the optical and ultraviolet observations. If these were all the data that we had on Zeta Puppis, we would say that it was clearly emitting thermal radiation, and not much more. But the longer and shorter wavelength observations reveal that radiation is being emitted over and above the Planck radiation. It turns out that *all* hot supergiant stars have excess radio, infrared, and X-ray radiation. Detailed spectroscopy and modeling of these stars indicate that the excess radiation is a

(continued on next page)

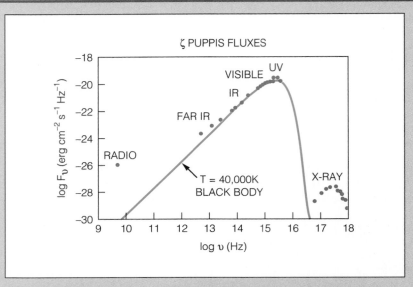

ζ PUPPIS FLUXES

catalog of sources, or the morphology of supernova remnants, radio galaxies, and planetary nebulae. Yet much of the stuff of the universe varies so quickly that the combination of multiwavelength data taken at different times would give erroneous conclusions about the behavior of the objects. In the time it takes for me to type one line of this editorial, the Crab nebula pulsar has rotated on its axis about 100 times. In the blink of an eye, a gamma-ray burster can turn on, with a hundredfold increase in luminosity. The universe is, as this volume is appropriately named, *dynamic*. And to observe a dynamic universe at many wavelengths requires simultaneous, or nearly simultaneous, measurements. It takes detailed preparation and a lot of time to coordinate a host of human and technical resources. It also takes the ability to perceive the goal so clearly and describe it so articulately that convincing proposals can be written to and accepted by a large number of telescope time-allocation committees. Multiwavelength science has to be good indeed to survive peer review committees in every wavelength band. An ad for a "dynamic" multiwavelength astrophysicist might read, *"Wanted: Leader for taxing enterprises. Must have imagination, stamina, optimistic spirit. Prefer some managerial skills, ability to organize and lead team effort. Good proposal-writing skills a must. Require background in radio, infrared, optical, ultraviolet, extreme ultraviolet, X-ray gamma-ray instrumentation, and data analysis. Expedition mountain climbers who have authored bestsellers, funded their trips with*

consequence of substantial mass loss in these stars, as evidenced in very high velocity winds that are flowing from them. These winds play an important role in the evolution of these objects.

In what has come to be called a "multiwavelength" approach to astrophysics, astronomers are boldly exploring a myriad of wavebands to try to understand the complete behavior of objects as small as neutron stars and as large as the cosmos itself. The results, as in the case of hot stars, are surprising and revealing of a universe that is comprised of much more than meets the eye.

Although it is becoming increasingly common, multiwavelength astronomy is by no means easy. The multiwavelength astronomer must be well versed in a number of very different analysis methods, as each wavelength regime has its specific ways of obtaining and analyzing data. The aperture synthesis technique of radio astronomy, for example, dif-

fers greatly from the use of charge-coupled devices at optical wavelengths, multilayer normal-incidence optics in the extreme ultraviolet, or multiple pinhole cameras at high-energy X-ray wavelengths. It helps for the astronomer to be very familiar with instrumentation in a variety of wavelength bands. It is almost axiomatic that it is difficult to analyze data well, especially to appreciate the significance of its uncertainties, without a good understanding of the instrument that took the data.

There are different kinds of multiwavelength studies, depending on the time scale for change in the object being studied. In the quasistatic or correlative approach, one can take or analyze multiwavelength data whenever one chooses, because the object of interest either changes very slowly or its changes are not of interest to the point of the study. Examples might be positional measurements, identification of a large *government or foundation monies, and hold a degree in astronomy or related field are encouraged to apply."*

In the last decade, global "campaigns" have been formed by astronomers to observe rapid, bursting neutron stars, flare stars, comets, Be stars, cataclysmic variable binaries, and active galactic nuclei. One campaign on a flare star involved 31 astronomers representing 20 institutions and using 15 different telescopes. Coordinated observations of an X-ray burster lasted 2 months. No campaign has achieved 100 percent efficiency. This is because of variables like inclement weather, equipment malfunction, observer malfunction, scheduling inconsistencies, unapproved portions of the program, lapses in organization, coordination, or communication, and lack of funding for one or more aspects of the program.

Multiwavelength astronomers, ever resourceful, are continually looking into new ways of improving their art. In the past decade, a number of proposals to establish multiwavelength platforms in space have been submitted to space agencies in this country and abroad. It is held that such a platform will get around many of the problems incurred in coordinating ground and space observations. The *Solar Maximum Mission* was such a platform. Its capability extended from gamma rays to visible light and its purpose was to study solar phenomena. Japan's *Ginga* satellite carried both X-ray and gamma-ray detectors to study gamma-ray bursts, and the multinational *International Ultraviolet Explorer* carries a small optical sensor that provides invaluable

support to the ultraviolet spectroscopic observations of this long-lived mission. In this decade, at least one multiwavelength platform has an excellent chance of flying. This is the European Space Agency's *X-Ray Multiple Mirror Mission,* which will carry not only X-ray instruments but also an imaging optical telescope that will be able to detect extremely faint stars in short exposure times.

Another approach is to try to encourage observatories and scheduling committees to be mindful of the problems in coordinating simultaneous multiwavelength observations, or simply correlative studies (as in the identification of high-energy sources), and acquire dedicated telescopes or dedicated time on telescopes for multiwavelength pursuits. A number of the new, very large ground-based telescopes under construction are being built in part to provide vital support for the space missions of every country.

Each sortie into a new wavelength band has produced unanticipated information on virtually every kind of celestial system that we knew of beforehand, and introduced a number of new objects as well. A course-resolution map of the sky in each wavelength band, for example, emphasizes very different aspects of the universe. A few sky maps, taken in different wavebands, are pictured in Astronomical Insight 24.1. In the radio portion of the spectrum (at 73 cm), we see the manifestation of synchrotron emission produced by high-speed electrons traveling through the magnetic field of the galaxy. At 21 centimeters we can see the structure of

the galaxy. In the microwave portion of the spectrum we see the remnant radiation of the big bang. A map in the wavelength of carbon monoxide emission (5 mm) reveals the birthplaces of stars and the effects of differential rotation of our galaxy. An infrared map shows the glow from dust all over the sky. The X-ray sky, similar to the microwave sky, shows a pervasive emission from every direction, superimposed upon tremendously bright point sources of high-energy radiation. The latter are principally accreting binary star systems.

These maps are striking for their differences, and they are full of information. But how are the contributions of data from various wavelengths being synthesized to produce new understandings? The difficulty of this task can be illustrated with an analogy: Given two road maps of small regions of the Earth—Albany, New York, and Toulouse, France, for example—can we then form a picture of what the Earth looks like? Telescopic photographs of these cities may reveal the existence of houses, cars, and people, but the photographs do not reveal how these things function, how hot they are, what they are made of, how they were made, and how long they will live. This is the situation we are faced with now in trying to understand the objects that comprise the cosmos—indeed, in trying to understand the cosmos itself, its birth, evolution, and future, from the part and parcel of the electromagnetic spectrum.

(continued on next page)

There is not sufficient space here to illustrate the broad spectrum of science that is coming from a multiwavelength approach. Indeed, entire conferences and books have been devoted to that subject. Here I give only one example, drawn from my knowledge of a field in which I have participated for over a decade. This is the study of accreting X-ray binary star systems. Coordinated multiwavelength observations have been essential to the development of this field.

In these close binaries, at least one of the component stars is a stellar remnant, that is, a white dwarf, neutron star, or black hole (see Chapter 23). The other component, often a fairly ordinary star, transfers matter to the stellar remnant because of the latter's extremely high gravitational field. One such binary, the bright X-ray source Cygnus X-3, has probably been detected over 20 powers of 10 in wavelength, all the way from radio waves to ultrahigh-energy gamma-rays. It is often studies in the visible band that yield the nature of the components of a binary system and the size and shape of the binary orbit. Cygnus X-3, however, is located about 35 thousand light-years from us, behind an obscuring spiral arm of our Milky Way. It therefore cannot be detected at all in the visible part of the spectrum, even though it can be at times the brightest point source of radio emission in the sky and may be a dominant contributor to the highest energy cosmic rays detected in our Galaxy. After a quarter of a century of intense observing at radio and X-ray wavelengths, we do not know for certain the nature of the stars in the Cygnus X-3 system, and we are far from understanding the source of its prodigious output of high-energy photons or the reason for its giant radio outbursts.

One of the most fruitful areas of multiwavelength research on X-ray binaries is the study of the structure of the accretion disk that surrounds the compact stellar remnant. An accretion disk is a disk of gas that is falling into a compact, massive object such as a neutron star or black hole; the rotation of the central object forces the infalling material to form a flattened, disklike structure (see Chapter 24). Accretion disks abound among numerous subfields of astrophysics: The solar system, newly forming stars, and galaxies are some examples. X-ray binary systems, including the cataclysmic variable binary stars, are excellent places to study accretion disk physics because these systems are fairly numerous and bright. There are few better ways to study accretion disk physics than by making a "light curve," or a plot of the intensity of a binary's light versus time. If this is done at a multitude of wavelengths simultaneously, one can study changes due to the turbulent accretion process in all the regions of the disk: its hot inner part where matter accretes onto the stellar remnant, and its colder outer part where matter is transferred from the companion star. Interpreting the combined data can yield a model for the geometry of the disk and the accretion flow from the companion. For example, some X-ray binaries show long-term transient low states in intensity. For these systems, only the combination of optical and X-ray data can tell us whether the low intensity results from a cessation of accretion or because the accretion disk becomes thicker and hides the X-ray source.

Multiwavelength astrophysics is a new field, one whose potential is just beginning to be realized. Truly "beautiful" multiwavelength observations, ones in which everything—the instruments, the observers, the weather, the star—performs perfectly, are rare. The knowledge gained from these observations demonstrates forcibly that the sum is better than the parts. The students who join this enterprise are privileged to learn about the tools of all of astronomy, privileged to be every kind of astronomer simultaneously. Most of all, they are lucky to be able to explore the universe from a far more encompassing point of view than ever possible before.

THE SOLAR SYSTEM

In this section of the book, we examine the properties of our localized region of the universe, the solar system. We will build a picture of what the planets are like, of the many smaller bodies that orbit the planets or pass among them, and of the vast cloud of debris that orbits in the cold, dark region beyond Pluto. Throughout, we will keep in mind the overall similarities and themes that we encounter, so that in the end we can tie them together into a coherent notion of how this complex system came to be.

The first chapter in this section provides an overview of the solar system, so that we have the context in which the more detailed discussions of individual objects can be placed. In this chapter we also introduce many of the general principles of planetary science that are to be applied to the planets in subsequent chapters. The notion of terrestrial and gas giant planets is brought into play, and serves as a basis for organizing the rest of the section.

Chapter 9 describes our home planet, including not only the basic data on its properties, but also discussion of overall processes that might apply to other planets as well, since the Earth serves as the prototype for the terrestrial planets as a class. Chapter 10 then discusses two small terrestrial bodies, our own Moon and the innermost planet, Mercury. These two barren worlds bear some superficial resemblance, but have very different histories and geologies. The next two chapters then present information on Earth's nearest neighbors, Venus and Mars. Both are found to have substantial similarities with the Earth in structure and geology, yet both also show sharp contrasts with the Earth, either internally or in surface properties.

With Chapter 13 we move into the realm of the gas giant planets, discussing the two largest, Jupiter and Saturn, together. These enormous bodies, lacking solid surfaces and composed of volatile gases, resemble the terrestrial planets only in that they also orbit the Sun. But perhaps surprisingly, many of the same physical processes that affect the terrestrial planets are also at work in the gas giants, so much of our earlier discussion carries over. In Chapter 14 we examine the three outermost planets, Uranus, Neptune, and Pluto. Only Pluto, of all the planets in the solar system, has not yet been visited by a spacecraft from Earth, so we know relatively little about this oddball of the outer solar system. New information from a unique era of eclipses by its satellite, however, provide an unprecedented wealth of data even on this remote object.

Chapter 15 summarizes what is known of the interplanetary bodies in the solar system: the asteroids (or minor planets), the comets, the fragments called meteorites that fall to the ground on Earth, and the microscopic dust grains that lie in the plane of the planetary orbits.

The big picture, that of solar system formation and evolution, is addressed in Chapter 16. Here we summarize the various phenomena that must be explained by any successful theory of solar system formation, and then we examine the current state of the leading theories. In this chapter we discuss the rings and moons of the outer planets collectively, because these bodies have been found to be a general part of the formation and evolution of the outer solar system, not isolated flukes.

Having completed this excursion through our own system of planets and associated bodies, we will be ready to move outward, into the realm of the stars.

CHAPTER 8

Overview of the Solar System and Planetary Science

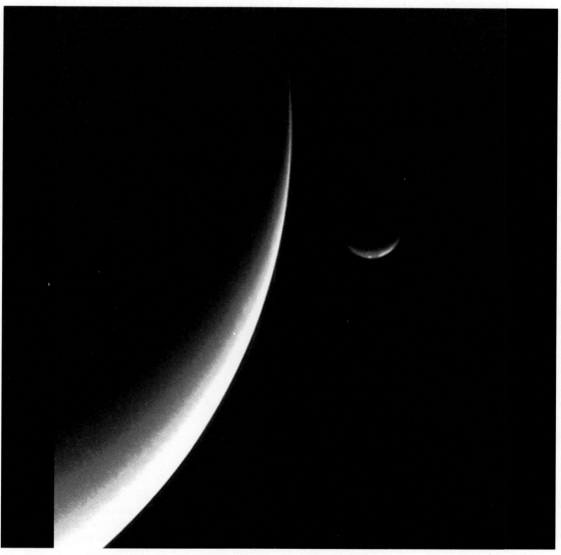

Neptune and Triton, observed from the far side of Voyager 2. (NASA/JPL)

As we prepare to explore the system of planets and interplanetary bodies that comprise the solar system, it is useful for us to begin with an overall description of the system and the general processes that shape the planets. This is particularly useful now, at a time when a major era of planetary exploration has ended, with the completion of the *Voyager* missions to the outer planets. With the exception of the currently operating *Magellan* radar mapper, which is orbiting Venus, it will be some time until the next major probe visits any of the planets and revises our current views of the Earth's neighbors. The *Mars Observer* probe to the red planet is due to carry out its mission in the mid-1990s, and the *Galileo* probe to Jupiter will not arrive at the giant planet until 1995.

In this chapter, we start with a general description of the solar system, and follow this with a summary of the methods by which we learn about the planets and interplanetary bodies. The rest of the chapter is then devoted to an outline of the basic processes that mold the planets. We refer repeatedly to these general processes in the chapters that follow, as we discuss the individual objects that make up the solar system.

OVERVIEW OF THE SOLAR SYSTEM

The solar system is dominated by the Sun, which contains more than 99 percent of the total mass, and which provides the major source of energy that heats the surfaces of most of the planets. There are nine known

planets (Table 8.1), orbiting the Sun at average distances ranging from about 58 million kilometers to almost 6 billion kilometers. In addition to these major bodies, there are innumerable smaller objects orbiting the Sun; these include the **asteroids** or **minor planets,** most of which orbit between Mars and Jupiter; the **comets,** which follow highly elongated paths through the solar system and which originate in a vast cloud far outside the orbit of Pluto; interplanetary dust, a tenuous medium that is confined to the plane of the planetary orbits; and assorted additional bodies, such as the icy asteroid-like objects occasionally found among the outer planets. In addition to this family of Sun-orbiting bodies, the solar system contains many satellites, moonlets, and rings, which orbit planets rather than the Sun directly.

Most of the major bodies in the solar system conform to a few general trends in location and motion. The system as a whole has a disklike structure, much like a record on a turntable, except that in this case the record is not rigid but instead acts like a fluid. Each planet and interplanetary object follows its own distinct orbit about the center, rather than being locked into a rigid overall rotation as a record is. The disk is very thin compared to its overall diameter (Fig. 8.1): Only the orbits of Mercury and Pluto deviate from the plane established by the Earth's orbit (i.e., the ecliptic) by more than about 3°, and even these two renegades have orbital planes lying within a few degrees of the ecliptic (the tilt of Mercury's orbit is about 7° and that of Pluto is just over 17°).

Despite the individual idiosyncrasies of the planets, there are several general ways in which they conform.

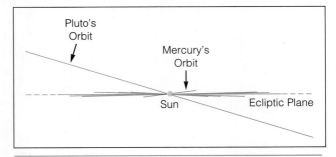

Figure 8.1 Alignment of planetary orbits. *This is an edge-on view of the planes of the planetary orbits, illustrating the fact that seven of the nine planets have orbital planes closely aligned (within about 3° or less). Two planets, Mercury (7°) and Pluto (17°) deviate significantly. The relative sizes of the orbits are not drawn to scale here.*

Table 8.1 The planets

Planet	Semimajor Axis	Sidereal Period	Mass*	Diameter*	Density	Temperature
Mercury	0.387 AU	0.241 yr	0.0558	0.381	5.50 g/cm³	100–700 K
Venus	0.723	0.615	0.815	0.951	5.3	730
Earth	1.000	1.000	1.000	1.000	5.518	200–300
Mars	1.524	1.881	0.107	0.531	3.96	145–300
Jupiter	5.203	11.86	317.9	10.86	1.33	165
Saturn	9.555	29.46	95.2	8.99	0.69	134
Uranus	19.22	84.01	14.54	3.97	1.27	76
Neptune	30.11	164.79	17.15	3.86	1.64	74
Pluto	39.44	248.5	.0022	0.18	2.03	40

*The masses and diameters of the planets are expressed relative to the mass and diameter of the Earth, which are 5.974×10^{27} grams and 12,756 km, respectively.

All of the planets orbit the Sun in the same direction, for example, and most of them spin in the same direction as well. We call this direction, which is counterclockwise as seen from above the North Pole, the **prograde** direction (Fig. 8.2). The rare exceptions of planetary of satellite motions that go against this trend are referred to as **retrograde** motions. As we shall see in Chapter 16, this general uniformity of orbital orientations and motions provides a powerful clue to the origin of the solar system.

Most of the planetary orbits are nearly circular; that is, they are ellipses with small **eccentricities,** or elongations (the formal definition of eccentricity has to do with the comparison of the long and short axes of the ellipse; see the Mathematical Methods section at the end of this chapter). Pluto and Mercury deviate significantly from circular paths about the Sun, but even for these two planets the orbits are far closer to circular than is the path of a typical comet. The comets, especially the long-period ones, usually follow very elongated elliptical paths around the Sun, indicating some significant difference in the way these interplanetary bodies formed as compared to the planets.

The spin axes of most of the planets and their moons are oriented approximately perpendicular to their orbital planes, although there is quite a large range of **obliquities,** as the tilts of the planetary rotation axes are called. Except for Uranus and Pluto, all of the planets have obliquities of less than 30° (the Earth's tilt is 23°5). The unusual tilts of Pluto and Uranus (both

Figure 8.2 Prograde motions.   *In this scale drawing of the inner six planets, arrows show the direction of orbital motion and the direction of spin. All of these motions are in the same direction, except that Venus (second planet from the Sun) spins backwards (i.e., in the retrograde direction).*

over 90°) will invoke some singular explanation, some manner in which the formations of these two planets differed from the formation of the others.

Most of the interplanetary bodies, such as asteroids, the periodic comets, and interplanetary dust grains, obey the same general rules as the planets, following prograde orbits that lie near the plane of the ecliptic. There are some exceptions, however: The long-period comets (discussed in Chapter 15) trace out paths that are randomly oriented, and some of the interplanetary dust has apparently been disturbed enough to deviate from strict alignment with the ecliptic plane.

Classifying the Planets

Of the nine planets, eight fall cleanly into two general categories, based on their general characteristics. The innermost four are called the **terrestrial planets** because they are Earth-like in their overall nature. These planets, like the Earth, are relatively small (compared to the giant planets), have high densities, and have hard, rocky surfaces (Fig. 8.3). The Moon (Fig. 8.4) is often classified as a terrestrial planet, given its large size, its similar overall characteristics, and its similar geological history.

The terrestrial planets are all **differentiated,** meaning that the denser materials of which they are made have sunk to their centers. Differentiation can occur only if a planet is fully or partially molten inside, so we conclude that each of the terrestrial planets underwent some melting at some time in its past. Some are still quite hot in their interiors today, and at least the Earth is still partially molten.

The terrestrial planets can have atmospheres, if their surface gravities are sufficiently strong to trap gas particles. (Their ability to retain atmospheres depends also on the surface temperature, as discussed later in this chapter.) Only the Moon and Mercury fail to retain any significant atmosphere.

The next four planets beyond Mars are the **gaseous giant planets** (Fig. 8.5), also called the **Jovian** planets after the largest and closest one, Jupiter. These planets are radically different from the terrestrial planets, being very much larger, having no solid surface, and having much lower densities. Where the terrestrial planets have densities ranging from around 3 grams/cm³ to over 5 grams/cm³, the densities of the gaseous giants are in the range from 0.7 to 1.6 grams/cm³ (for comparison, the density of water is 1 gram/cm³, and that of ordinary rocks is in the range of 2–3 grams/cm³).

Figure 8.3 The terrestrial planets. *All of these images were obtained from space: Mercury (upper left) and Venus (upper right) were both photographed by* Mariner 10; *Earth (lower left) by one of the* Apollo *missions; and*

All of the giant planets have atmospheres; in fact, as noted already, they do not have solid surfaces, but consist instead of a fluid structure nearly all the way to their centers (it is thought that each of the giants has a small, rocky core). At the highest levels, these atmospheres are very cold and contain complex molecules consisting primarily of the simplest elements such as hydrogen, carbon, nitrogen, and oxygen (there is also a good deal of helium, in atomic form). By contrast, the terrestrial planets contain very little hydrogen and helium in their atmospheres, because these lightweight gases can easily escape the low gravitational fields of the small, warm planets.

The ninth planet, Pluto, does not fit readily into either category, but instead more closely resembles some of the giant moons of the outer planets (Fig. 8.6). These bodies consist of ice and rock in comparable proportions (giving them overall densities ranging from 1 to 2 grams/cm³). They have hard surfaces and are generally differentiated, and some of them (Pluto included, at least some of the time) have gaseous atmospheres. It is thought that Pluto and the giant satellites have rather different geological histories than either the gaseous giants or the terrestrial planets.

Figure 8.3 Continued *Mars (lower right) by one of the* Viking *orbiters. The terrestrial planets are all relatively small and dense, with rocky surfaces. (NASA)*

Figure 8.4 The full Moon. *This is a high-quality photograph taken through a telescope on the Earth. (Lick Observatory, University of California)*

Observational Techniques

In Chapter 7, we discussed general principles of observational astronomy, making use of the properties of light described in Chapter 6. In those discussions, we assumed that our only means of studying objects in the universe is to collect light that reaches the Earth and analyze it. This passive approach is indeed our only choice for most of the universe, but for the planets we can apply other, more direct techniques.

Clearly the best and most direct method of studying a planet is to go there and examine it at close hand.

Figure 8.5 The giant planets. *The images of Jupiter (upper left), Saturn (upper right), and Neptune (lower left) and Uranus (lower right) were all obtained at close range by the Voyager probes. The giant planets are very different from the terrestrial planets, being much larger, far less dense, and having no solid surfaces. (NASA)*

This direct sampling method, with human interaction, has now been applied to just four bodies in the solar system, the Earth, the Moon, Venus, and Mars (the latter two have been sampled by robot probes) (Fig. 8.7). Naturally, we find that these are the best-studied objects, though in the case of the Moon at least, per-

haps not necessarily the best understood. Direct analysis of surface samples has been achieved for Mars through the U.S. *Viking* missions in the mid-1970s (Fig. 8.8), and for Venus through the Soviet *Venera* landers (see Chapter 11). Plans are being discussed, in both the United States and the Soviet Union, for future

Figure 8.6 Giant moons of the outer planets.   *These four satellites are all larger than the Earth's Moon and very different in nature. Where the Moon is rocky throughout, these giant satellites are partially composed of ice. While the Moon is inactive geologically, some of the giant satellites are quite active. Here we see Ganymede of Jupiter (upper left), Europa of Jupiter (upper right), Triton of Neptune (lower left), and Titan of Saturn (lower right). Ganymede and especially Europa show fault lines due to crustal shifting; Triton has active ice geysers and a dimpled crust due to slumping of the surface; and Titan has a thick atmosphere, possibly created by outgassing that is still active. (NASA)*

Figure 8.7 Human exploration of the Moon. *This is astronaut Harrison Schmidt on the last Apollo mission to the Moon (Apollo 17), collecting samples of small lunar rocks. (NASA)*

manned exploration of Mars, but this is certainly at least one or more decades away from reality at the present time (the United States has recently embarked on a course that will lead us back to the Moon before we go on to Mars).

Even if it is impractical as yet for us to personally visit the rest of the planets, we can send unmanned probes to visit them and make observations from close range. This has now been accomplished for all of the planets except Pluto; the August, 1989 *Voyager 2* flyby of Neptune marked the completion of a remarkable era of planetary exploration. Both the United States and the Soviet space programs made stunning discoveries through their robot missions to the planets. There are several advantages to be gained by making close-up observations: Vastly more detail can be seen; it is possible to observe an object from different angles (this is particularly useful in the interpretation of light-scattering properties); direct capture and measurement of charged particles trapped in planetary magnetic fields is possible; direct measurements of the planetary gravitational field can be made, and from that the internal structure of the planet may be determined; and it may be possible to measure weak emissions (such as radio noise from particle belts) that cannot be detected from the Earth.

Figure 8.8 Robots on Mars. *This is a U.S. **Viking** lander in a simulated Martian environment. Two of these robotic vehicles landed successfully on Mars in 1976, and carried out numerous experiments and observations. A mechanical arm was able to pick up soil samples and deposit them in test chambers where experiments were conducted in search of microscopic life forms. **Viking's** cameras took many panoramic photographs of the surface, and kept operating long enough to show how the dust and soil deposits vary with the seasonal winds. (NASA)*

Even though the "grand tour" of the outer planets by *Voyager 2* is over, there are several other sophisticated planetary probes in action or planned. There are two spacecraft orbiting Venus now; the *Pioneer-Venus* probe has been in operation since 1978, and the *Magellan* spacecraft has recently arrived (August 1990) and is mapping the cloud-covered planet using radar. Meanwhile the *Galileo* probe is on its lengthy and circuitous path to Jupiter, where it will arrive in late 1995. It will drop a probe into the atmosphere while the main spacecraft goes into orbit, allowing for long-term exploration of Jupiter and its major moons. The United States is preparing a new mission to Mars called *Mars Observer,* and the Soviets are planning a series of probes (some expected to land and collect surface samples) to follow their recent *Phobos* missions to the red planet.

Farther in the future, there are plans for a joint U.S.–European Space Agency mission to Saturn, called *Cassini.* And another spacecraft, called the *Comet Rendezvous and Asteroid Flyby (CRAF)* mission, is being designed to make a close encounter with an asteroid and then accompany a comet in its orbit about the Sun. *Cassini,* expected to be launched in 1996, will put a probe into the atmosphere of the giant, cloud-shrouded moon Titan, and will also place an advanced spacecraft into orbit around Saturn to explore the planet, its rings and moons, and map Titan's surface with radar. *CRAF,* to be launched in 1995, is being designed to fly a spacecraft past an asteroid and then to match orbits with a comet. An instrumented penetrator will be shot into the comet's nucleus to make direct measurements of its composition.

Despite enormous successes of the manned and unmanned probes to the Moon and the planets, there is still a lot to be learned from Earth-based observations. Of course, for a long time, telescopic data were all we had on any of the planets, so our first "explorations" of each body were of the traditional astronomical kind. From these observations, scientists were able to learn a great deal: Surface temperatures could be derived from Wien's law (Chapter 6); surface compositions could be roughly analyzed from spectroscopy of reflected light (particularly in infrared wavelengths); and atmospheric composition could be deduced, again from spectroscopy. One of the great advantages of Earth-based observations, still important even in the age of space exploration, is the ability to make observations repeatedly over long time periods. This allows us to gather detailed information on seasonal variations, for example, and also allows us to use orbiting satellites as probes of the atmospheres of the planets (Fig. 8.9). Earth-based radio observations complement the visible and infrared data by providing information on charged particle belts (and indirectly on planetary magnetic fields, which create the belts).

Even space-based observations from the vicinity of the Earth are important. Orbiting spacecraft such as the *International Ultraviolet Explorer (IUE)* have obtained extensive ultraviolet spectroscopic observations of several of the planets, and sounding rocket observations have done ultraviolet work as well. The *Hubble Space Telescope,* orbiting the Earth since May 1990, will make enormous contributions to our knowledge of the planets, providing both extended wavelength coverage (ultraviolet as well as visible) and unprecedented clarity of images, revealing surface details and allowing them to be monitored over time.

Even though the completion of the *Voyager* mission signaled the end of a particular phase of planetary exploration, the wide assortment of continuing and planned studies of the planets will continue to produce a steady stream of new information. There is no doubt that future astronomy textbooks will continue to require extensive revisions as the exploration of the solar system continues.

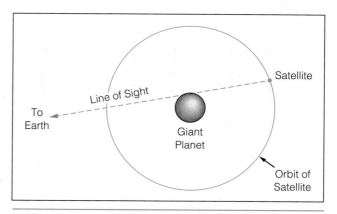

Figure 8.9 Using a satellite to probe a planetary atmosphere. *As the satellite passes behind the disk of its parent planet, the satellite is blocked by the planet. As it passes behind the edge of the planet's disk as seen from Earth, its light passes through the planetary atmosphere on its way to us. Scientists on the Earth can analyze the effects of the atmosphere on the light from the satellite to gain information on the gases in the atmosphere of the planet.*

ASTRONOMICAL INSIGHT 8.1

Voyager's Final Look Back.

After *Voyager 2* encountered Neptune in August of 1989, its mission was completed. Both *Voyagers* will continue to operate, measuring charged particles and the solar wind in the outermost reaches of the solar system, and making some observations of stars as well. But from the point of view of plenetary research, the job was done. The cameras which obtained all the magnificent images of the planets and their satellites and ring systems were not to be used further.

Mission scientists and engineers could not resist one final imaging project, however. The opportunity for an overview of the solar system, as seen from the outside, was unprecedented (and another may be a very long time in coming). Hence as *Voyager 2* raced beyond the orbit of Neptune, its scan platform was swiveled back for one last look at the Sun and planets. Final images of six of the nine planets were obtained, along with an image of the Sun (drastically enlarged due to multiple reflections of its brilliant light inside the camera optical system; even from 40 AU the Sun is about 8 million times brighter than Sirius, the brightest star in our sky other than the Sun).

The first illustration presented here shows the positions of the six planets (indicated by the first letters of their names) when the

NASA

images were obtained (Mercury, Mars, and Pluto were either in poor position relative to the Sun or were too faint). It also shows the sequence of 39 individual wide-angle images that were obtained in order to create a mosaic that encompasses the Sun and all six planets. This mosaic is shown as a jagged strip in the illustration. Next to the position of each planet is an enlarged version of that planet's color image, obtained with *Voyager*'s narrow-angle camera.

The second illustration shows, in greater detail, the six narrow-

angle color images of the planets. Jupiter and Saturn were both resolved; that is, their images are larger than the size of the individual picture elements (pixels; see discussion in Astronomical Insight 7.2) of the camera. For Saturn it is possible to tell that the image is elongated due to the ring system. The images of the Earth and Venus are not resolved; their angular diameters from such a great distance were far smaller than the size of the picture elements of the camera, so these two planets appear as single points. The elongated shapes of Uranus and Neptune are not real,

ASTRONOMICAL INSIGHT 8.1

VENUS   EARTH   JUPITER

SATURN   URANUS   NEPTUNE

NASA

but are due to spacecraft motion during these exposures, which were relatively long.

While the individual images obtained from so far away are not very detailed or informative, the perspective this set of pictures gives us is invaluable. It is comparable in a way to the impact of the first photographs of the whole Earth that were obtained in the 1960s by the Apollo missions.

Then we saw for the first time the outsider's view of our tiny planet, and for the first time many people gained a real appreciation of the limited extent and fragile resources of "spaceship Earth". Now we have a similar perspective on our entire solar system; we have a better understanding of the vastness of space and the very minor role played by planets. It is quite remarkable to think that our

technology has made this view possible only 87 years after the first powered flight off of the Earth's surface.

PLANETARY SCIENCE: THE PROCESSES THAT SHAPE THE PLANETS

The nine planets offer an amazing variety of appearances and phenomena, yet there are a few underlying physical mechanisms that explain most of the phenomena. In an important sense, the planets have more similarities than differences. As we prepare for our more detailed tour of the solar system, it is useful now to discuss the general processes that affect all the planets in varying degrees. This will not only save us from having to repeat some of the discussions from chapter to chapter, but more importantly, will equip us to understand the underlying simplicity and the many connections and similarities among the planets as we go.

Formation of the Planets

A more complete discussion of the formation of the solar system will be found in Chapter 16, but here it is useful to gain an overview of the basic processes by which the planets formed. There are general themes that apply to all of the planets, but there are some differences as well, particularly between the terrestrial and giant groups.

The solar system apparently formed from a disk of gas and dust that orbited the young Sun at the time of its formation (Fig. 8.10). This disk, composed of material gathered together by gravitational forces, was initially dominated by hydrogen, the most abundant element in the universe. The composition (by mass fraction) of the original preplanetary material was approximately 73 percent hydrogen and 25 percent helium, with all the other elements comprising the remaining 2 percent of the mass. This is the composition of the Sun and most of the stars, as well as of the interstellar matter in the galaxy today, and it is thought that 5 billion years ago, when the solar system formed, it was about the same.

The disk of atoms and small, solid particles that orbited the young Sun had a natural tendency to become clumpy, which caused the condensation of larger solid objects than the original tiny dust grains. With time, these larger objects built up so that first snowball-sized, then basketball-sized, then even larger bodies were formed. The rate of growth of these objects increased until the disk became a collection of **planetesimals,** or preplanetary bodies ranging in diameter up to several thousand kilometers. During the buildup, or accretion, process, some elements were driven out by heat, particularly in the inner solar system. The lightweight, **volatile** (i.e., easily vaporized) elements were the least likely to become trapped in the solid state. Hence, the planetesimals that formed in the inner solar system never contained large quantities of lightweight elements such as hydrogen and helium.

The planetesimals themselves collided occasionally, often with small relative speeds (because they orbited the Sun in the same direction and at nearly the same speed at a given distance from the Sun). These collisions led to the coalescence, or sticking together, of planetesimals, eventually producing the planets of today. The entire accretion process is thought to have taken place fairly rapidly by astronomical standards; it is believed that the planets formed within the first few hundred millions years after the Sun was born.

Note that the planets formed in the inner solar system started out with small abundances of the volatile elements, since these elements had been unable to condense into solid form in the first place, due to the high temperatures in the inner part of the system. In the outer solar system, however, the temperature was low enough so that even hydrogen and helium could become trapped, so the giant planets started out with a radically different composition from the terrestrial planets. The ability of the cold outer planets to trap these abundant gases gave the outer planets much larger initial masses than the inner planets. These larger masses helped the outer planets to gravitationally capture ad-

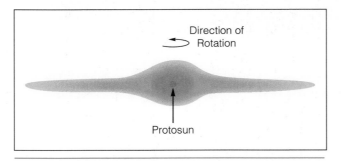

Figure 8.10 The solar nebula. *This is an edge-on view of the disk of gas and dust from which the Sun and planets formed. Rotation of a collapsing interstellar cloud caused the flattening. Gravitational compression created a hot concentration of gas at the center which was to become the Sun. The disk of surrounding material became clumpy as gas began to condense into solid form, and the clumps then merged, eventually creating the planets.*

ditional quantities of gas from their surroundings. Thus, the outer planets grew into much larger, more massive bodies than the inner planets, and had rather different compositions as well.

In the inner solar system, proximity to the Sun created strong tidal forces (discussed later in this chapter) that prevent capture of surrounding gas and dust, and the higher temperature of this material made it more difficult to trap anyway. The result was a set of inner planets that contained mostly heavy elements and that turned out to be dense and rocky, and a set of outer planets that were far more massive, were cold, and contained ample quantities of lightweight gases. The trapped material surrounding the outer planets formed into extensive systems of rings and moons, while the inner planets tended not to have significant satellites (except for the Earth, whose large moon is thought to have formed largely by accident; see Chapter 10).

The asteroids, which orbit the Sun between the paths of Mars and Jupiter, along with a few outlying bodies (including the planet Pluto) are thought to be planetesimals that survived without merging into the larger planets. These bodies, along with the comets, represent primitive material from the early days of solar system formation.

All of the planets, and many of their satellites as well as the largest of the asteroids, are round. This is because the inward force of gravity controls the shapes of these bodies. Gravity presses inward from all sides, resulting in compression in the interior. As the material inside a developing planet is compressed, pressure builds up that counteracts gravity, eventually stopping the contraction. The result is a spherical body, denser in the center than in the outer regions, in which there is a balance between gravity and pressure at every point. Only if there is another force to counteract gravity will a body end up in a nonspherical shape. For very small objects (i.e., the smaller asteroids or the tiniest moons), the strength of the material of which the body is made may exceed the inward force of gravity, allowing the body to retain a nonspherical shape. Thus some of the smallest objects in the solar system have irregular shapes, while all of the larger ones are round.

While a planet formed, gravitational forces squeezed it together tighter and tighter, causing compression and heating in its interior. Radioactive elements in the interior of a planet released additional heat as they decayed, helping to keep the interior of the planet hot to this day. For a terrestrial planet, this heating caused the interior to be partially molten over long periods of time, which in turn allowed the bulk of the heaviest elements (such as iron, nickel, cobalt, and the superheavy species such as lead, uranium, and so on) to sink to the center of the planet. In this process, called **differentiation,** each terrestrial planet gained a dense core made primarily of iron and nickel. In some of the terrestrial planets (including the Earth and probably Mercury and Venus), the core remains at least partially molten, whereas others (probably Mars and the Moon) have cooled enough to be essentially solid throughout (Fig. 8.11).

The outer planets, composed as they are of lightweight gases, remained fluid throughout most of their interiors. Thus, these planets have no outer solid surfaces, and differentiation was able to segregate out the heavy elements very thoroughly, forming small, dense solid cores deep within (Fig. 8.12). The outer planets are thought to have been largely unchanged since their earliest days, while the terrestrial planets continue to be active geologically.

GEOLOGICAL EVOLUTION

Tectonic Activity

The terrestrial planets are still alive geologically, in varying degrees. The Earth appears to be the most active, while the Moon (sometimes classified as a ter-

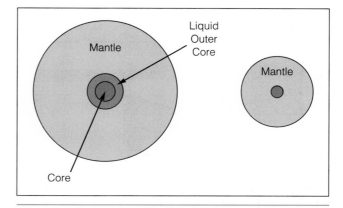

Figure 8.11 Internal structure of the terrestrial planets.
At left is a cross-sectional view of a planet like the Earth, which has a liquid zone in its outer core. Other terrestrials, particularly the small ones such as Mars and the Moon, have internal structures more like the sketch at right, with no liquid zone.

restrial body, even though it is a satellite) is probably the most inert. There are large uncertainties today regarding the current state of activity in Mercury, Venus, and Mars; perhaps these questions will be answered soon, through space probes such as *Magellan* to Venus and *Mars Observer* to Mars.

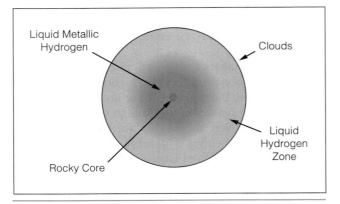

Figure 8.12 Internal structure of a gaseous giant planet.
Beneath the visible cloud tops, the density and temperature increase inward. As the density increases, hydrogen is forced into unusual forms, which are discussed in Chapter 13. Because the planet is fluid throughout, differentiation occurs readily, and virtually all of the planet's heavy elements have sunk to the center, forming a small, rocky core.

Figure 8.13 The Earth's crust. *The outermost layer of the Earth is very thin (10–50 km) relative to the radius of the planet. The crust, unlike the mantle below, is quite rigid and brittle, and it has fragmented into several segments (plates), which move about in response to motions in the underlying mantle.*

Let us consider the Earth as the prototype of a geologically active terrestrial planet. Recall that the interior is still quite hot, largely because of radioactive decay, and that a portion of the core is molten. The surface consists of a brittle crust overlying a thick layer that is in a semisolid state (Fig. 8.13). This underlying region, called the **mantle,** is the source of most geological activity such as earthquakes and volcanoes.

The crust is made of rocky material that is less dense than the mantle and therefore floats atop it. The crust has fractured into about a dozen segments, called **plates,** which are somewhat free to move about over the mantle. As they do so, the continents move (a phenomenon known as **continental drift**), and the plates rub against each other or collide, creating surface activity such as earthquakes and volcanic eruptions (Fig. 8.14). Where

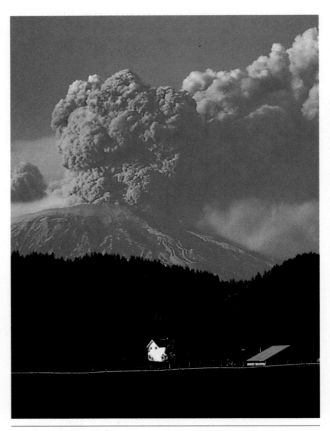

Figure 8.14 A recent volcanic eruption. *This is a photograph of the explosive eruption of Mt. St. Helens, in Washington state, during its outburst in 1980. Events such as this provide graphic evidence that the Earth is very active geologically. (John H. Meehan/Science Source, Photo Researchers)*

plates move alongside each other, such as along the famous San Andreas fault in California, they alternately stick together as stress builds up, then slip suddenly, causing earthquakes (Fig. 8.15). Volcanic eruptions also tend to occur near plate boundaries, where molten rock, or **magma** can find its way to the surface (Fig. 8.16). Where a seafloor plate, which is denser than continental material, collides with a continental plate, it tends to slip underneath it, creating deep sea trenches (Fig. 8.17) and often volcanic activity. In places where two continental plates collide, the edges are uplifted, creating the loftiest mountain ranges on the Earth (Fig. 8.18).

Figure 8.15 Crustal shifting during an earthquake. *(Right) This abrupt tear in the Earth's crust was the result of an earthquake that struck the Andes mountains of Peru. (Carl Frank, Photo Researchers)*

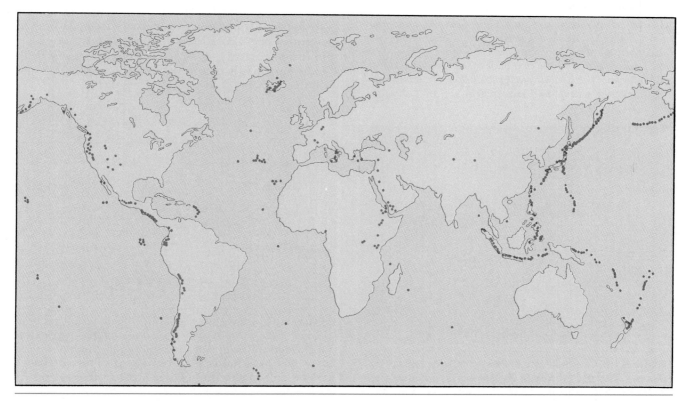

Figure 8.16 The distribution of active volcanoes. *This map shows the locations of volcanoes that have been active in recent years. Note that the volcanoes are concentrated along the boundaries of the Earth's crustal plates, particularly around the edges of the huge Pacific plate.*

The driving force behind these motions of the Earth's plates is not completely established, but is thought most likely to be slow **convection** currents in the mantle. Convection is the overturning motion that can occur naturally when a fluid is heated from the bottom (and is subjected to a gravitational field at the same time).

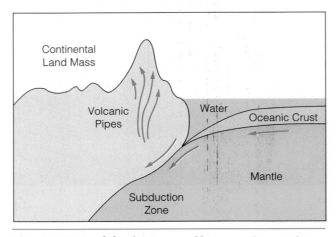

Figure 8.17 A subduction zone. *Here oceanic crust (sea-floor) is shown sinking below continental crust as two adjacent crustal plates move toward each other. Along the line of intersection of the two plates (i.e., along the subduction zone) the oceanic crust slips below the continental crust, creating a deep, undersea trench.*

Figure 8.18 The result of a collision between crustal plates. *This is a portion of the Himalaya mountains, the Earth's tallest range, including Mt. Everest (at right). These mountains are being uplifted by pressure created in a slow collision between the Indian plate and the Eurasian plate, and are growing taller each year. (G. J. James/BPS)*

Under the right conditions, bubbles of heated fluid become more buoyant than their surroundings and rise. As they rise they cool, eventually losing their buoyancy and then sinking again after moving a small distance sideways to allow room for rising bubbles from below (Fig. 8.19). In a very viscous (i.e., thick) fluid such as the Earth's semisolid mantle, the process is slow, resulting in motions of the crustal plates that are measured in units of a few centimeters per year.

The **tectonic activity,** as the action of moving crustal plates is called, takes places in the mantle and the crust. Underneath the mantle is the core, part of which is molten (see Fig. 8.11). Here, too, fluid motions occur. Some of the material in the inner Earth carries electrical charges, so that the motions in the outer core result in the creation of electrical currents. These in turn create a **magnetic field** that permeates the entire Earth and extends for some distance into space. Some of the other terrestrial planets also have magnetic fields, indicative that they, too, have partially molten cores. (A molten region is not the only requirement, however; the planet must also rotate rapidly enough to create sufficient electrical current.)

As we have already said, the other terrestrial planets are apparently not as active internally as the Earth. On the other hand, both Mars and Venus show some signs of tectonic activity, and Mercury has a magnetic field, indicating that it must have a molten zone inside. On

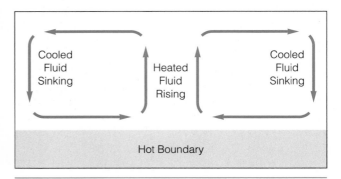

Figure 8.19 Convective overturn. *A fluid (i.e., a gas or a liquid) subjected to heating from below (in the presence of a gravitational field) may experience convective overturn, as shown here. Heated fluid, less dense than its surroundings, rises to higher levels, where it cools, becomes denser, and sinks. The result is a steady, overturning motion. In planetary atmospheres, the source of heat from below may be absorbed sunlight or internal heat from within the planet.*

Mars there are surface structures that resulted from uplift and crustal fracture, but these processes may have become inactive early in the planet's history. On the contrary, there are some indications that Venus may be active today (the *Magellan* radar maps of Venus will help to answer this question by providing detailed information on surface structures). The Moon is certainly inert geologically at present, and probably never underwent extensive tectonic activity, although there was an early era when lava flows blanketed much of the surface.

Because the giant planets have no real surfaces and no solid interiors (except for their cores), we do not speak of geological or tectonic activity in the same manner as we do for the terrestrial planets. On the other hand, there are some similar processes taking place. For example, all of the giant planets have strong magnetic fields, indicating that there are electrical currents flowing in their interiors. This is not difficult to understand, since all of these planets are fluid throughout most of their interiors, and all of them rotate very rapidly.

Rocks and Surface Processes

In addition to tectonic activity, which can modify the surface of a terrestrial planet through quakes and volcanic eruptions, there are other forces that can act to alter the hard surfaces of the terrestrial planets. The surfaces of these planets are largely composed of rock, which may be covered by water over large regions (on today's Earth, for example, and perhaps on the early Mars and the young Venus as well), and which may be decomposed into soil, dust, or sand in other regions.

On the Earth, we recognize three general forms of rock, based on origin: (1) **igneous rocks,** those formed from molten material that has hardened; (2) **sedimentary rocks,** those formed from the deposition of mud and clay deposits on the floors of oceans and lakes; and (3) **metamorphic rocks,** those that have been altered by heat and pressure resulting from large-scale crustal movements. Continual geological activity can gradually change rocks from one of these forms to another. For example, a rock may first form as an igneous rock as the result of an eruption, but later may be broken down by wind and water into sand that is deposited underwater and forms sedimentary rock, and yet later may be uplifted by geological forces and squeezed and compressed so that it becomes a metamorphic rock.

Apart from its form as a rock, we may also classify a rock according to its chemical composition. Thus we define **minerals** based on composition. By far the most common minerals on the surface of the Earth and the other terrestrial planets are the **silicates,** which contain compounds of silicon and oxygen. The common basalts and granites on the Earth are silicates of igneous origin. Other common mineral types include the **carbonates,** which are dominated by carbon-bearing compounds, and **oxides,** consisting largely of oxygen compounds.

We have seen that tectonic activity can alter the face of a terrestrial planet significantly, as pieces of the crust move about, and at the same time that local processes may alter the nature of surface rocks. Additional mechanisms are at work as well. For example, the solar system still contains a large number of small bodies orbiting the Sun between the planets, and occasionally these objects strike the surface of a planet. We call this a **meteoroid impact,** and the surfaces of all the terrestrial planets, as well as the satellites of the giant planets, are scarred by the craters that result (Fig. 8.20). On the Earth, weathering processes such as wind and water erosion, as well as tectonic activity, have obliterated most of these craters, except those that have formed most recently.

On the Earth and Mars, erosion by water has been an important force affecting the structure of the surface, and on these planets as well as Venus the atmosphere has also had an effect on the surface. The flowing water that once carved channels and flood plains into the Martian surface (Fig. 8.21) has not been present for a very long time (perhaps 3 billion years), but on Earth, water is still a very important factor in surface modification. The atmospheres of Venus, the Earth, and Mars alter the surface through winds and chemical processes.

It is believed that the planets are all about the same age, having formed at about the same time some 4.5 billion years ago. On the other hand, the surfaces of the planets have been altered to varying degrees. Some surfaces (such as that of the Earth) are relatively young, since the time scale for complete reorganization of the continents due to plate motions is around 200 million years, while other surfaces are very old (for example, parts of the lunar surface appear to have been unchanged for nearly 4 billion years). Thus, throughout this text, we will speak of the ages of planetary and satellite surfaces in contrast with the ages of the bodies themselves.

Figure 8.20 Impact craters throughout the solar system. *No planet or satellite is immune from impacts due to objects from space. Here we see a recent crater on the Earth, near Winslow, Arizona (upper left); a very ancient crater on the Moon (upper right); a crater on Venus (lower left), discovered to have an impact origin by recent* **Magellan** *radar images; and ancient craters on Saturn's icy moon Dione (lower right). Impact craters are characterized by high rim walls and central peaks, which can be seen in the examples shown here; volcanic craters may not have high walls and never have central peaks. (upper left: Meteor Crater Enterprises, Inc.; upper right: NASA; lower left: NASA/JPL; lower right: NASA/JPL)*

PLANETARY ATMOSPHERES

Most of the planets have atmospheres. An atmosphere is an envelope of gas, held in place by the planetary gravitational field. Only Mercury is without an atmosphere (and even here there are traces of gas present). The terrestrial planets, as we have seen, have solid surfaces that mark a clear boundary between the atmosphere and the interior, whereas the giant planets are essentially fluid throughout, with no clear-cut boundary. Pluto's atmosphere is apparently seasonal, being partially frozen into solid form for most of its 249-year-long "year," and existing in gaseous form only during the time of the planet's closest approach to the Sun (which has just occurred in 1989, so that at present Pluto has an atmosphere).

Figure 8.21 An ancient river channel on Mars. *This is one of the most striking examples of evidence for flowing water in the Martian past. (NASA)*

A planet gains an atmosphere through a combination of capture of gases from its surroundings and the escape, or **outgassing,** of volatile elements from its interior. For the terrestrial planets, outgassing during the early, partially molten phase was probably the more important process, whereas for the giant planets much of the atmospheric gas was captured or accreted at the time of planetary formation. Furthermore, the atmospheres of the terrestrial planets have in some cases changed radically since their initial formation, while those of the giant planets have probably remained essentially constant.

On the Earth, for example, the early atmosphere was dominated by hydrogen-bearing compounds, and only later did the present-day oxygen-nitrogen atmosphere develop. (Interestingly, the formation of life on the Earth's surface had a lot to do with this, providing some of the nitrogen and all of the oxygen.) On Mars, we find evidence of an early atmosphere that was much denser than today's thin carbon dioxide atmosphere and, as noted earlier in this chapter, there is also evidence that Mars once had a lot of liquid water on its surface. Venus has the densest atmosphere among the terrestrials, and a very unpleasant one as well; the main constituent is carbon dioxide, but there are other compounds, such as sulfuric acid, that make this planet a very inhospitable place. As we shall see (in Chapter 11), this atmosphere evolved largely because Venus is just a bit closer to the Sun than the Earth is.

The composition of the atmospheres of the giant planets is fairly uniform, except that local conditions have created some variations. In all cases, hydrogen is the most common element, comprising nearly three-fourths of the mass of the atmospheres, and helium is the next most abundant element. The hydrogen is mostly in its molecular form (H_2), but some hydrogen also appears in trace forms such as water vapor (H_2O), ammonia (NH_3), and methane (CH_4), while the helium is completely in the form of single atoms (helium does not easily combine chemically with other elements). Depending on the temperature in the outer layers of these atmospheres, more complex molecules are present in varying degrees.

We have already noted that the heavy elements in the giant planets have undergone differentiation and sunk to the planetary cores. In addition, some slow differentiation is still taking place in at least one case (Saturn), as helium atoms gradually sink through the lighter-weight hydrogen.

So far, we have described how gases enter an atmosphere, but not how they can leave it. For the terrestrial planets, the escape of certain elements has been an important factor in determining their compositions today, while very little has escaped the strong gravitational fields of the giant planets. The likelihood that a particular kind of gas will escape depends on how massive the individual particles are, how strong the planetary gravitational field is, and how hot the atmosphere is. The dependence on the planet's gravity is probably obvious; for a given planetary mass and radius, there is a specific **escape speed,** which is easily calculated (see Chapter 5). The escape speed is the upward speed needed for a particle to completely escape the planet's gravitational field and fly away into space. The escape speed itself does not depend on the mass of the escaping particle but, as we shall see, the likelihood that a gas particle will reach escape speed does depend on its mass.

The temperature of a gas is intimately related to the speed with which the individual particles move about. The hotter the gas, the faster the individual motions (the particles do not all move at precisely the same speed, but there is a well-defined average speed that is closely related to temperature). A planet that is very hot, such as Mercury (which also has a low mass and thus a low escape speed) cannot retain a substantial atmosphere because any gas particles that might be temporarily trapped can eventually exceed the escape speed. (It is not actually necessary for the average speed of the particles to exceed the escape speed; it can be shown that if the average speed exceeds one-sixth of the escape speed, eventually all the particles will escape, because there are always a few that are moving much faster than the average.)

For the other terrestrial planets, the situation is not that simple; these planets have lost some gases while retaining others. Here is where the masses of the gas particles themselves come into the picture. For a given temperature, the average speed depends on the mass of the particles as well as the temperature. The lightweight particles move fastest, and the heaviest particles move most slowly. Thus it is possible to have a situation in which some, but not all, of the particles in an atmosphere exceed the escape speed (or one-sixth of the escape speed, as noted above). This results in the selective loss of certain gases, while others remain trapped. For example, the Earth has no free hydrogen or helium in its atmosphere (but does have hydrogen in a tenuous region above the atmosphere, called the *geocorona*). However, the Earth retains the heavier compounds such as O_2 and N_2 (molecular oxygen and nitrogen), as well as a number of trace gases that are heavy enough not to escape. Similarly, it is thought that both Venus and Mars have lost some lightweight gases while maintaining their current carbon dioxide (CO_2) atmospheres.

As already mentioned, the giant planets have probably lost little or none of their original atmospheres. This is due to a combination of two factors: the high escape speeds of these massive planets, and the very low temperatures of their outer atmospheres.

Atmospheric Circulation

All planetary atmospheres undergo flowing motions, and in each case the basic causes of these motions are the same. There are two general driving forces for atmospheric circulation: heat-driven convection and planetary rotation.

Convection has already been mentioned, as the possible cause of continental drift on the Earth. In any situation in which a fluid under the influence of a gravitational field is heated from below, it is possible for

bubbles of the fluid to become buoyant and to rise, creating an overturning motion. For the atmospheres of the terrestrial planets, the source of heat is the surface, which is warmed by the Sun's radiation. For the giant planets, sunlight has some effect, but there is also an important contribution from internal heat. These planets have hot interiors because of gravitational compression, and this heat slowly makes its way outward, helping to create convection in the outer atmospheres.

As gas rises and falls due to convection, it is forced to veer to the side by the rotation of the planet. It is actually not the vertical motion that causes this veer, but the horizontal motion in the north–south direction (Fig. 8.22). Imagine, for example, an air current flowing northward from the Earth's equator. Because of the Earth's rotation, this air current is moving rapidly to the east at the same time it flows northward. The rotational speed of the Earth's surface at the equator is about 1,670 km/hr, but it decreases with distance from the equator. Hence, as the air current makes its

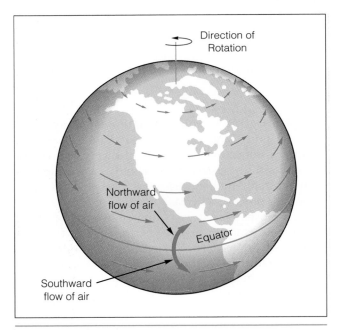

Figure 8.22 The coriolis force. *The lengths of the thin arrows indicate the relative speed of rotation of the Earth at different distances from the equator. (In a rigid sphere, points at the equator move fastest because they have farther to travel in a single rotation.) Wind flowing away from the equator is curved in the direction of rotation, because it comes from a region of more rapid rotational velocity.*

way north, it finds itself moving eastward relative to the surface, because it still has the eastward velocity given to it by the Earth's equatorial rotation. A current moving southward in the Southern Hemisphere would also veer to the east as it moves away from the equator. In both hemispheres, air currents flowing toward the equator veer toward the west. The "force" that causes these veers is called the **coriolis force,** and the result is a circular motion for atmospheric flows everywhere on the Earth.

The effect of the coriolis force depends on how rapidly a planet rotates. The Earth and Mars have fairly rapid rotations, whereas Venus rotates extremely slowly. Thus the Earth and Mars have well-defined circular flows in their atmospheres, while on Venus these patterns are weaker. The giant planets all rotate extremely rapidly, and the coriolis force has a huge effect. There the circular flow patterns are stretched out very far in the east–west direction, resulting in elongated circulation patterns that stretch all the way around these planets (Fig. 8.23).

TIDAL FORCES

There is one additional phenomenon that affects many of the planets and their satellites. As discussed in Chapter 5, the gravitational force exerted on one body by another is not equal at all locations, but is instead stronger on the near side than on the far side (Fig. 8.24). This creates a **differential gravitational force,** commonly called a **tidal force.** In Chapter 5, we saw how this force, exerted on the Earth by the Moon, creates tides in the Earth's oceans. There are many other situations in the solar system where tidal forces come into play.

Only the Earth and Pluto have satellites large enough to cause significant tidal effects on their parent planets. On the other hand, many of the satellites in the solar system are strongly affected by tidal forces due to their parent planets, and one planet, Mercury, is influenced by tidal forces due to the Sun.

Tidal forces can cause internal stress in bodies, and they can cause alterations in their motions. The stretching force that is exerted on a satellite by its parent planet can be enough to cause internal heating and, consequently, geological activity such as active volcanic eruptions. Io, the innermost large moon of Jupiter, is perhaps the most famous example; this satellite was found by the *Voyager* spacecraft to be undergoing

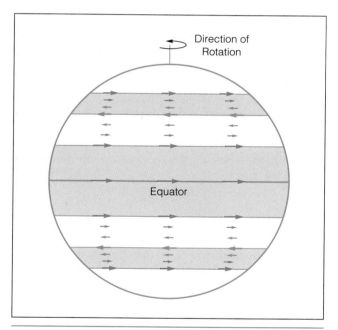

Figure 8.23 Atmospheric flow on a giant planet. *The lengths of the arrows indicate relative wind speed. The fastest winds are typically in the equatorial zone, where the flow is in the direction of rotation. North and south of this zone are circulation patterns that stretch all the way around the planet. These are driven by a combination of convection and the coriolis force created by the rapid planetary rotation.*

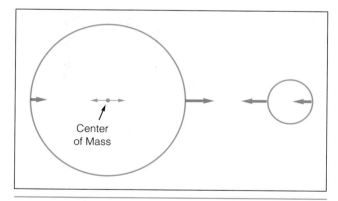

Figure 8.24 Tidal force. *This shows two spherical bodies near each other in space, such as a planet and its moon. The lengths of the dark arrows indicate the relative strength of the gravitational force at different points on each body due to the other. Note that the side nearest the other body feels a stronger force than the far side; the difference between these forces acts as a stretching force. The sketch of the larger body shows how this differential gravitational force acts as though it pulls on opposite sides of the center of mass.*

almost continual eruptions. There is evidence that at least one satellite of Saturn may be undergoing similar activity, and that the second large moon of Jupiter, Europa, is warm and liquid in part of its interior. Our own Moon is stretched out of round by tidal forces due to the Earth, but is not heated enough in its interior to cause current volcanic activity.

The tidal stretching of the Moon has had another effect, however. The Moon used to spin much more rapidly than it does now, but its elongated shape has caused its spin to slow until the same side always faces the Earth. As long as the Moon spun more rapidly, the Earth kept tugging at its elongated sides, trying to force them to point directly along the Earth–Moon line (Fig. 8.25). In time, the Earth succeeded in this, "locking in" the Moon's rotation rate to equal its orbital period. This situation is called **synchronous rotation.**

Most of the satellites in the solar system are in synchronous rotation because of the tidal effects of their parent planets. In one case—the Pluto-Charon system—the parent planet itself is also locked into synchronous rotation, so that the planet and its satellite perpetually keep the same sides facing each other. This also means that at a given point on the surface of Pluto, Charon is overhead permanently, since Pluto's day equals the orbital period of the satellite (this is similar to the situation with communications satellites placed in Earth orbit with 24-hour periods).

The result of the Sun's tidal force on Mercury is a bit more complex than simple synchronous rotation. Here the rotation has been altered by tidal forces, but not to the point where the rotational and orbital periods are equal. Instead, the rotation period is equal to exactly two-thirds of the planet's orbital period, so that the same side of Mercury faces the Sun at a particular point every other orbit. The reasons for this unusual situation are discussed in Chapter 10.

Not only can tidal forces deform bodies and alter their spins, they can also prevent them from forming in the first place. The myriad small particles that make up the ring systems of the giant planets might have merged to form satellites, except that they are too close to the parent planet to do so. Within a certain distance called the **Roche limit,** the tidal forces keeping particles apart can exceed the gravitational forces trying to pull them together. This is discussed further in Chapter 13.

PERSPECTIVE

In this chapter, we have gained an overview of the overall structure of the solar system, and of the general properties of the objects in it. We have also learned something of how these properties are observed and measured. By examining the general processes that have formed and altered the planets, we have set the stage for a more detailed look at the individual bodies in the system. In the next chapter, we examine our own home planet, the Earth.

SUMMARY

1. The principal bodies in the solar system are the Sun and nine planets, each orbiting the Sun in the same direction and nearly in the same plane. Most of the spin motions are also in the same direction, and most of the orbits are nearly circular. In addition to the planets, there are myriad minor bodies, including asteroids and comets.

2. The planets (except Pluto) fall into two general categories: (1) the terrestrial, or rocky planets; and (2) the giant planets, which are fluid throughout (except for small rocky cores) and which have much lower average densities than the terrestrial planets. Pluto is similar to certain giant planetary satellites.

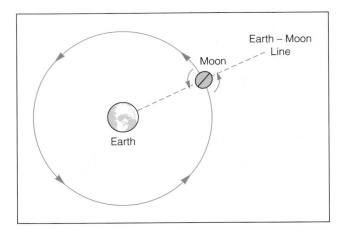

Figure 8.25 Tidal effect on the moon's spin. *Here we see the Moon, its shape elongated by the tidal force due to the Earth, at a time in the past when the Moon's spin was faster than it is today. The spin of the Moon causes its elongated axis to point away from the Earth-Moon line, as shown. The Earth's gravity acts to keep the Moon's bulge pointed directly at the Earth; thus the Earth's gravity acts to slow the Moon's spin. Eventually the slowing led to synchronous rotation, with the same side of the Moon always pointing toward the Earth. (Not to scale.)*

3. Planets can be observed through direct sampling (if robot or manned probes are sent), through close-up observations by spacecraft, and by Earth-based and Earth-orbiting telescopic observations.

4. The planets are thought to have formed by the coalescence of planetesimals, which were the first solid bodies to condense out of the disk of gas and dust that orbited the young Sun. The inner planets contain few light-weight volatile elements because they formed under high-temperature conditions, whereas the outer planets, formed under low-temperature conditions, were able to retain even the volatile elements.

5. In all the planets, heavy elements have sunk to the cores in a process called differentiation. This can occur only in a fluid, which implies that even the terrestrial planets have undergone molten or partially molten phases.

6. The terrestrial planets have similar internal structures, in each case consisting of a thin, rocky crust overlying a mantle (which may be somewhat elastic) and a core. These planets either now undergo or once underwent tectonic activity to varying degrees; this activity may involve volcanic eruptions, earthquakes, and the shifting of crustal plates. Tectonic activity may be driven by convection in the mantle.

7. Rocks, which compose the surfaces of the terrestrial planets, are classified according to origin as igneous, sedimentary, or metamorphic. Rocks are also classified according to their mineral composition.

8. The surfaces of the terrestrial planets are modified by several processes, in varying degrees: (1) water erosion; (2) meteroid impacts; (3) wind erosion; and (4) geological processes such as continental drift and uplift. Because of these processes, the age of a planet and the age of its surface can be very different.

9. Planetary atmospheres formed through a combination of outgassing and accretion. The composition of an atmosphere is determined by which gases are supplied and which can escape (which depends, in turn, on temperature and molecular mass, as well as planetary escape speed). The terrestrial planets that have atmospheres contain only relatively heavy elements, whereas the giant planets have retained even light-weight species in their atmospheres.

10. The circulatory motions of all planetary atmospheres are driven by two forces: (1) convection due to heating from below; and (2) planetary rotation, which creates circular motions through the action of the coriolis force.

11. Tidal forces play many important roles in the solar system, including the creation of synchronous rotation in many satellites; the inducement of internal heating and volcanic activity in some satellites; and the prevention of satellite formation close to the outer planets, resulting in ring systems there.

REVIEW QUESTIONS

Thought Questions

1. At the beginning of this chapter, we outlined several general trends among the planets, in terms of their motions and the locations and orientations of their orbits. As we noted, however, there are several distinct exceptions to the general rules. Using information in this chapter and in Appendix 6, make a list of the planetary and satellite motions and orbits that do not conform to the general trends.

2. Summarize the distinctions between the terrestrial and giant planets.

3. Explain differentiation in your own words. Can you think of any everyday examples of differentiation, perhaps in your kitchen?

4. Describe the various methods scientists can use for observing the planets.

5. Explain why hydrogen is not an important component of the terrestrial planets, despite the fact that it is the most common element in the universe and in the solar system.

6. List all the processes that modify the surface of the Earth. Which ones also work on Mercury? On Venus? On Mars? On the Moon?

7. Explain the difference between the age of a body and the age of its surface. On which terrestrial planets are the two ages essentially the same?

8. Compare the primitive atmosphere of the Earth with the present-day atmosphere of Jupiter.

9. Using your own words, explain how the coriolis force arises.

10. Explain why the Moon's tidal force on the Earth creates *two* bulges in the oceans and two high tides per day at any given location.

Problems

1. Add up the masses of the planets (Appendix 7), and compare the total to the mass of the Sun (Appendix 1), as a means of checking the claim made in this chapter that the Sun contains more than 99 percent of the total mass of the solar system.

2. If the core of a terrestrial planet contains half of the total mass of the planet and extends halfway out from the center, how does the density in the core compare to the density in the outer portion of the planet?

3. What is the wavelength of maximum emission from the surface of Mercury, if the temperature is 500 K? Compare this with the wavelength of maximum emission from the cloudtops of Jupiter, where the temperature is 130 K. What kinds of telescopes would be needed to directly observe this thermal emission from each of these planets?

4. Suppose that major meteoroid impacts occur at the average rate of one per square kilometer per 100 million years, and that this rate has been constant since the planets formed. If the Moon has forty-five craters per square kilometer, how old is the Moon? Note that you had to assume the Moon has retained all the craters ever formed on its surface. Now assume that the Earth has only one crater for every million square kilometers of surface area. How old is the surface of the Earth? You should get an unrealistically small value for the age of the Earth's surface; how can you explain this?

5. The average speed of gas particles (in cm/sec) is given (approximately) by the formula

$$v = \sqrt{\frac{3.5 \times 10^{-16}T}{m}}$$

where T is the gas temperature (in Kelvins; i.e., on the absolute scale) and m is the mass of an individual gas particle (atom or molecule) in grams. If the Earth's atmosphere has a temperature of 300 K and the Earth has an escape speed of 11.2 km/sec, determine which of the following molecules should have escaped: (1) hydrogen (H_2; mass 3.3×10^{-24} grams); (2) carbon monoxide (CO; mass 4.7×10^{-23} grams); (3) helium (He; mass 6.7×10^{-24} grams); and oxygen (O_2; mass 5.3×10^{-23} grams).

Note that you must convert the calculated average speeds to km/sec to compare with the escape speed, and recall the rule of thumb that if the particle speed exceeds one-sixth of the escape speed, the species will escape entirely.

ADDITIONAL READINGS

Battan, L. J. 1979. *Fundamentals of Meteorology.* Englewood Cliffs, NJ: Prentice-Hall.

Beatty, J. K., B. O'Leary, and A. Chaikin, 1990. *The new solar system.* 2d ed. Cambridge, England: Cambridge University.

Beatty, J. K. 1979. The foremost worlds of *Voyager 1. Sky and Telescope* 57:423 (Part 2); 57:516 (Part 2).

Head, J. W., C. A. Wood, and T. A. Mutch 1977. Geologic evolution of the terrestrial planets. *American Scientist* 65:21.

Kasting, J. F., B. Toon, and J. B. Pollack. 1988. How climate evolved on the terrestrial planets. *Scientific American* 258(2): 90.

Killian, A. M. 1989. Playing dice with the solar system. *Sky and Telescope* 78:136.

Montoya, E. and R. Fimmel 1988. *Pioneers* in space: the story of the *Pioneer* missions. *Mercury* 17:56.

McLaughlin, W. I. 1989. *Voyager's* decade of wonder. *Sky and Telescope* 78:16.

MATHEMATICAL METHODS

Sizes and Shapes of Orbits

In this chapter, we mention the fact that most of the planets have elliptical orbits that are nearly circular. We have stressed the one or two exceptions—Mercury and Pluto—whose orbits are somewhat elongated, and we mentioned the comets, whose orbits are very long and stretched out by comparison with those of the planets. All of these orbits are ellipses, and all obey Kepler's laws of planetary motion. The differences in size and shape are described by two very simple quantities: the *semimajor axis* (already discussed) and the *eccentricity*. The semimajor axis is simply one-half of the long axis of an ellipse, and in a sense it represents the average separation between two bodies that are orbiting a common center of mass. The eccentricity is a measure of how elongated, or *eccentric,* an ellipse is.

By definition, the eccentricity is the ratio of the distance between the two foci to the full length of the ellipse (i.e., to the major axis; see the figure).

A circle has both foci at the same place, so the distance between them is 0 and the eccentricity is 0. None of the planets has an eccentricity exactly equal to 0, but for most of them, the value is very small, less than 0.1. For these planets the orbits are nearly circular.

Appendix 7 tabulates the eccentricities for the planets. We can use those values to calculate some interesting properties of their orbits. For example, given the eccentricity and semimajor axis of an ellipse, it is very simple to calculate how close a planet

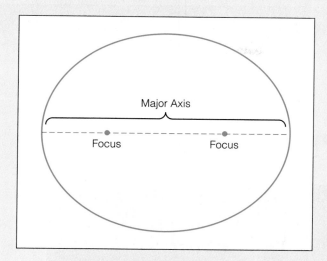

comes to the Sun at its closest approach *(perihelion)*, and how far away from the Sun it is at its most distant retreat *(aphelion)*. The perihelion distance is given by $a(1-e)$, where a is the semimajor axis and e is the eccentricity; and the aphelion distance is given by $a(1+e)$. We can apply this to the orbit of the Earth, whose semimajor axis is 1.000 AU and whose eccentricity is 0.0167: At perihelion the Earth is $1.000(1-0.0167) = 0.9833$ AU from the Sun; and at aphelion the Earth is $1.000(1+0.0167) = 1.0167$ AU from the Sun. Similarly, Mars orbits between a closest approach of $1.524(1-0.0934) = 1.382$ AU and a greatest distance from the Sun of $1.524(1+0.0934) = 1.666$ AU.

The planets with the largest eccentricities are Mercury $(e = 0.2056)$ and Pluto $(e = 0.2496)$. For Mercury, this means that at aphelion it is $(1+e)/(1-e) = 1.52$ times farther from the Sun than at perihelion; for Pluto aphelion is 1.67 times farther from the Sun than perihelion. For these planets, the wide range of distances from the Sun has a significant effect on the amount of solar heating that oc-

curs during the orbital cycle; these effects are discussed in the upcoming chapters (even Mars has a significant seasonal effect due to the variation of its distance from the Sun).

The orbits of most comets are far more elongated than those of any planet. Some comets, like Comet Halley, are only moderately stretched out compared to others; the eccentricity of Halley is 0.967, meaning that its most distant excursion from the Sun is $(1+0.967)/(1-0.967) = 60$ times farther than its closest approach! This comet, whose semimajor axis is 17.95 AU, approaches as close to the Sun as $17.95(1-0.967) = 0.59$ AU, and recedes to a distance of 35.3 AU (and because it moves very much slower in the outer portions of its orbit, Halley spends the vast majority of its time far from the Sun, where it exists as a cold, dark chunk of ice and rock). The so-called long-period comets (discussed in Chapter 15) have eccentricities of more than 0.9999 (these comets range between 1 AU or less and some 100,000 AU from the Sun). If we allow the eccentricity to equal 1, then the comet follows a shape called a *parabola*, and if e is greater than 1, the shape is called a *hyperbola*. Both are open curves instead of closed orbits, meaning that if a comet follows such an orbit, it will pass by the Sun once and then fly off into space, never to return.

All of the properties of orbits discussed here apply also to other orbital situations, such as a moon or artificial satellite orbiting a planet, or two stars orbiting each other. Keep in mind that the two bodies in each of these situations actually orbit a common center of mass, and that the semimajor axis and the eccentricity refer to the size and shape of the orbit relative to that common center. For the planets or comets orbiting the Sun, the center of mass is very near the center of the Sun, so our calculations of distances from the Sun are quite accurate, but they would have to be modified if the two orbital bodies were comparable in mass.

CHAPTER 9

The Earth As a Planet

A Space Shuttle view of the Earth's atmosphere. (NASA)

CHAPTER PREVIEW

The Atmosphere
   Heating and Atmospheric Motions
The Earth's Interior
Surface Terrain and Tectonic Activity
Rocks, Minerals, and the Age of the Surface
The Magnetic Field
   Outer Structure and the Magnetosphere
   Radiation Belts
The Evolution of the Earth

Of all the bodies in the heavens, the Earth (Table 9.1) has been the best studied, although it is equally true that many mysteries remain. Understanding the planet we live on has practical, as well as philosophical, importance. Furthermore, working at such close quarters with the object of study provides us with numerous advantages, including the ability to observe in great detail and over long periods of time, and the possibility of making direct experiments by probing and sampling the surface of the Earth.

It would be beyond the scope of this text to attempt to summarize all that is known about the Earth. Our discussion will emphasize those aspects that can be compared with what we know of the other planets. We should keep in mind, however, that not only is the

Table 9.1 Earth

Orbital semimajor axis:
1.000 AU (149,600,000 km)
   Perihelion distance: 0.983 AU
   Aphelion distance: 1.017 AU
Orbital period: 365.256 days (1.000 years)
Orbital inclination: 0°0′0″
Rotation period: 23 hr, 56 min, 4.1 sec
Tilt of axis: 23°27′
Diameter: 12,756 km (1.000 $D_\oplus$)
Mass: 5.974 × 10^{27} g (1.000 $M_\oplus$)
Density: 5.518 g/cm³
Surface gravity: 980 cm/sec² (1.000 Earth gravity)
Escape speed: 11.2 km/sec
Surface temperature: 200–300 K
Albedo: 0.37 (average)
Satellites: 1

Figure 9.1 The Earth as seen from space   *(NASA)*

Earth much more complex than it is portrayed to be here, but so, undoubtedly, are the other planets. Each is a unique and varied world, probably as diverse as the Earth, except for the influences of life.

From afar, the Earth is a brilliant planet, with an overall bluish color created by the scattering of light in its atmosphere (Fig. 9.1). Much of it is usually cloud covered, and where it is clear, brownish land masses and large regions of water appear. The Earth appears so bright because it has a high **albedo,** the fraction of light that is reflected from it. The Earth's albedo is 0.31, meaning that about 31 percent of the light that strikes it is reflected.

THE ATMOSPHERE

We begin by examining the thin layer of gas that surrounds the solid Earth. The Earth's atmosphere is composed of a variety of gases (Table 9.2), as well as a distribution of suspended particles called **aerosols.** The gases, which are evenly mixed together up to an altitude of about 80 kilometers (km), are primarily nitrogen and oxygen, in the form of the molecules N_2 and O_2. Nearly 80 percent of the gas is nitrogen, about 20

Table 9.2 Composition of Earth's atmosphere

Gas	Symbol	Fraction[a]
Nitrogen	N_2	0.77
Oxygen	O_2	0.21
Water vapor	H_2O	0.001–0.028
Argon	Ar	0.0093
Carbon dioxide	CO_2	3.3×10^{-4}
Neon	Ne	1.8×10^{-5}
Helium	He	5.2×10^{-6}
Methane	CH_4	1.5×10^{-6}
Krypton	Kr	1.1×10^{-6}
Hydrogen	H_2	5×10^{-7}
Ozone	O_3	4×10^{-7}
Nitrous oxide	N_2O	3×10^{-7}
Carbon monoxide	CO	1.2×10^{-7}
Ammonia	NH_3	1×10^{-8}

[a]The listed values are fractions by number.

percent is oxygen, and all the other species exist only as traces, although in many cases they are important traces. In addition to this standard mixture of gases, a number of others appear in the atmosphere. Among these, water vapor (H_2O) is important, because of its direct role in supporting life. Others, such as ozone and carbon dioxide, are also essential, but for less direct reasons. Ozone (designated O_3) is a form of oxygen in which three oxygen atoms are bound together instead of two. Unlike ordinary oxygen (O_2), ozone in the upper atmosphere acts as a shield against ultraviolet light from the Sun, which could be harmful to life forms if it penetrated to the ground.

The importance of carbon dioxide (CO_2), which is produced by respiration in animals, by fires and decomposition of organic matter, and by many industrial processes, lies in its ability to trap the Sun's radiant energy within the atmosphere, creating overall heating by a process called the **greenhouse effect.** This heating process has progressed to a much greater extent on the planet Venus than on the Earth (and will be discussed more fully in Chapter 11).

The principal atmospheric constituents, nitrogen and oxygen, are both released into the air by life forms on the surface. This process has been important in the evolution of the Earth's atmosphere and leads to the expectation that other planets in the solar system are unlikely to have the same composition (although nitrogen is produced also by processes that do not involve

life forms, so we might find it elsewhere). Nitrogen is released into the Earth's atmosphere by the decay of biological material and by emission from volcanic eruptions. Oxygen comes almost exclusively from the photosynthesis process in plants, a process in which carbon dioxide is converted into oxygen with the assistance of radiant energy from the Sun.

Heating and Atmospheric Motions

Although there are obvious fluctuations in the temperature of the atmosphere, particularly near the surface, there is a definite, stable temperature structure as a function of height (Fig. 9.2). This structure includes several distinct layers in the atmosphere: the **troposphere,** from the surface to about 10 km in altitude, where the temperature decreases slowly with height and where the phenomena that we call **weather** occur; the **stratosphere,** between 10 and 50 km, where the temperature increases with height; the **mesosphere,** ex-

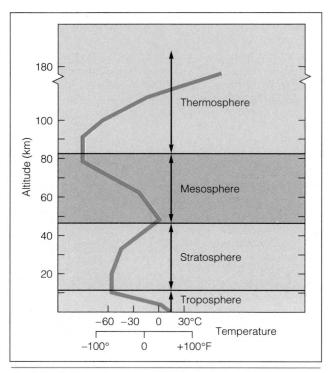

Figure 9.2 The vertical structure of the Earth's atmosphere. *The distinct zones are defined according to the temperature variation with height, which influences atmospheric motions.*

tending from 50 to 80 km, where the temperature again decreases; and the **thermosphere,** above 80 km, where the temperature gradually rises with height to a constant value above 200 km. In layers where the temperature decreases with height, there can be substantial vertical motion of the air, whereas in layers where it gets warmer with height, the air is stable, with only lateral motions.

The principal source of heating throughout the atmosphere is the Sun's radiation, but the energy is deposited in different ways at different levels. Much of the heat (that is, infrared radiation) is absorbed by the ground. This heating, which is concentrated at low and middle latitudes on the Earth, is responsible for driving the large-scale motions of the atmosphere, which produce the global wind patterns. Heat is also deposited in the stratosphere where ozone (O_3) absorbs ultraviolet light from the Sun. This absorption is responsible for the increase of temperature with height in the stratosphere.

There are many scales of motion in the atmosphere, ranging from small gusts a few centimeters in size to the continent-spanning flows thousands of kilometers in scale. Here we will discuss only the large-scale motions, since generally we are able to study only motions of similar scale in the atmospheres of other planets.

The primary influences on the global motions in the Earth's atmosphere are heating from the Sun, as already noted, and the rotation of the Earth. The heating creates regions where the air rises. The air must later cool and fall somewhere else, and a pattern of overturning motions called **convection** is created (Fig. 9.3). Generally, air rises in the tropics and descends at more northerly latitudes, although the situation is more complicated than that. For example, the continents tend to be warmer than the oceans in the summer, so that high-pressure regions (characterized by descending air) preferentially lie over water, whereas low-pressure regions (where air rises and cools) tend to be over land. In winter this is reversed.

These tendencies, combined with the rotation of the Earth, create horizontal flows at several levels in the atmosphere. Air that is rising or falling is forced into a rotary pattern by the Earth's rotation, as described in Chapter 8. Where air rises in a low-pressure zone, the resultant swirling motion is called a **cyclone** (Fig. 9.4). In the Northern Hemisphere, the motion of a cyclone is counterclockwise; south of the equator, it is clockwise. Descending air flows in high-pressure regions are forced into oppositely directed circular patterns, and

Figure 9.3 Convection in the Earth's atmosphere. *This is a simplified illustration of the principle of convection, in which warm air rises, cools, and descends.*

are called **anticyclones.** Cyclonic flows sometimes intensify into storms of great strength; hurricanes and typhoons are examples.

The net result of the vertical motions caused by the Sun's heating and the spiral motions caused by the Earth's rotation is the creation of a complex pattern of flows (see Fig. 9.5) with vertical and horizontal components.

Figure 9.4 A storm system on Earth. *This is in the Northern Hemisphere, so the circulation about this low-pressure region is counterclockwise. (NASA)*

ASTRONOMICAL INSIGHT 9.1

Technology and the Atmosphere

Life forms of the Earth have had a profound impact on the past evolution of the atmosphere, as we have discussed in the text. Much of the nitrogen that dominates the composition of the atmosphere comes from the decay of life forms, and virtually all of the oxygen is the by-product of photosynthesis in plants. Since life first appeared on the Earth, the interplay between biological activity and atmospheric processes has been rich and diverse. Today the interaction continues, with the added impact of our technological society.

Human activities result in many kinds of emissions into the atmosphere that would not occur naturally. These emissions include both gases and tiny particles, or aerosols. The atmosphere is a complex, dynamic entity, and scientists have difficulty predicting how it will react to these emissions. Furthermore, it takes a long time for significant global changes to occur, so it is not easy to see right away what the effects of a change in conditions might be. Despite these difficulties, two gases in particular have been identified as potentially capable of changing the atmosphere enough to alter the Earth's climate, and many others are known or suspected to have other kinds of effects.

One of the gases having known effects is carbon dioxide (CO_2), a by-product of burning fossil fuels (i.e. oil, gas, coal, and wood). Carbon dioxide is an effective absorber of infrared radiation, which means that a high abundance of CO_2 in the atmosphere can trap heat near the Earth's surface, increasing the temperature of the lower atmosphere. In this process, generally known as the **greenhouse effect,** infrared radiation emitted from the surface is not allowed to escape as freely into space as it normally would, because the atmospheric CO_2 absorbs it. The resulting temperature increase in the lower atmosphere could have far-reaching effects. For example, the abilities of different regions to support farm crops might be vastly altered. In addition, even a relatively small increase could, over time, result in substantial melting of the ice in arctic regions, in turn raising sea levels worldwide enough to inundate coastal areas.

As an illustration of the complexity of atmospheric processes, it is noteworthy that not all scientists agree on what the effects of increased CO_2 would be. Some argue that there would be an increase in cloud cover which would block incoming sunlight sufficiently to decrease the heating at the Earth's surface.

An even more controversial technological addition to the atmosphere is created by the so-called **chlorofluorocarbons,** complex molecules containing chlorine and other elements. These are commonly used as refrigerants (such as freon) and as propellants in making foam products. In this case, scientists are concerned about complex chemical reactions that may occur in the upper atmosphere involving by-products of the chlorofluorocarbons. One of these by-products is chlorine oxide (ClO), which can react with ozone (O_3), destroying the ozone. For the past decade or so, there has been a growing worry that the release of chlorofluorocarbons into the atmosphere has already decreased the abundance of ozone in the stratosphere (at an altitude of about 16 to 18 km). If so, this could have serious consequences, because the ozone layer is our primary shield against ultraviolet radiation from the Sun, and an increase in the amount of ultraviolet reaching the ground could be quite harmful to life on the Earth.

Recently it has been discovered that indeed there is a deficiency in the quantity of ozone in the atmosphere, at a particular time and at a particular place. During the spring (October) in the Southern Hemisphere, the quantity of ozone over the south polar region diminishes. To learn more about this south polar ozone deficiency, teams of scientists spent several weeks in Antarctica in late 1986, and will continue to do so for some years.

The first of these intensive studies produced detailed measurements of the quantity of ozone and of various molecular species known to react with ozone. Balloons and other devices made measurements at various altitudes in the atmosphere. As a result, we know that the quantity of ozone over the pole has dropped by about 50 percent since the late 1960s, confirming that the ozone hole is either a new phenomenon, or one that varies, perhaps with the Sun's cycle of activity (see Chapter 13).

The conclusion of some of the scientists involved in the study is that the ozone hole is caused by chlorofluorocarbons. This conclusion is based on measurements of other molecular species, on inconsistencies in other hypotheses, and on direct measurements of freon

ASTRONOMICAL INSIGHT 9.1

in the atmosphere, showing that its abundance has increased by 50 percent over the same time period that the south polar springtime ozone abundance has decreased. All of these arguments are circumstantial and not completely conclusive, but they are consistent with the idea that human technology has begun to erode the Earth's protective ozone shield. There remain many important

questions about whether this conclusion is valid, and if so, whether the ozone deficiency will be confined to the south polar region or will spread in time.

Very recently, new studies of ozone over the north polar region were begun, and it was found that there may also be a deficiency in that region.

The changes in the Earth's atmosphere caused by natural pro-

cesses related to life forms took millenia to occur, and they took place in concert with the development and evolution of those life forms. Now it seems we must take measures to prevent much more rapid changes due to unnatural processes from altering our atmosphere in ways that would make the Earth inhospitable to life.

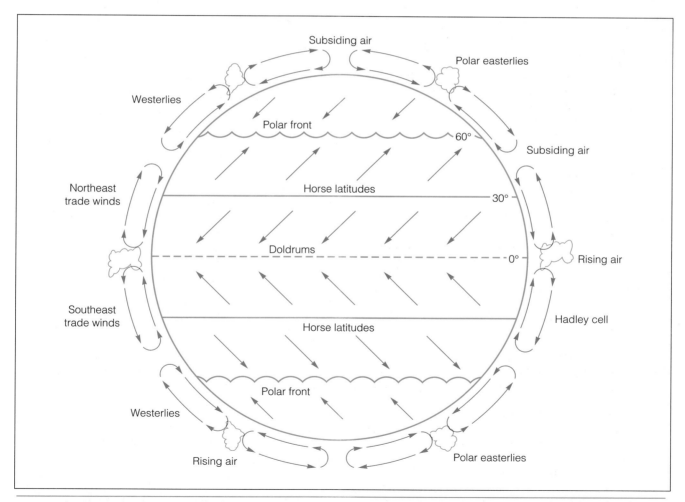

Figure 9.5 The general circulation of the Earth's atmosphere. *The circulation is made even more complex than shown here by temperature contrasts between seas and land, and by surface features such as mountains. (Based on Fig. 7.3, p. 150, F. K. Lutgens and E. J. Tarbuck 1979,* The Atmosphere: An Introduction to Meteorology *[Englewood Cliffs: Prentice Hall])*

Seasonal shifts in the distribution of the Sun's energy input cause changes in the flow patterns, creating our well-known seasonal weather variations. Apparently other factors can influence the location of the high- and low-pressure regions, because there are longer-term fluctuations in the climate whose cause is not well understood. Some of these may be related to cyclic behavior in the Sun (see Chapter 17).

In later chapters, we will examine the flow patterns in the atmospheres of the other planets, and we will see that in many ways they behave like the Earth, except for differences in the amount of solar-energy input and the rate of rotation.

THE EARTH'S INTERIOR

At first it may seem surprising that much is known about the deep interior of the Earth. Substantial quantities of data have been collected, however, mostly by indirect means. It is interesting that the same indirect techniques could be applied (through the use of unmanned space probes) to other planets, and in principle it is possible for us to know as much about their interiors as that of our own planet (this would not be true of the outer, gaseous planets, but it does apply to the other terrestrial planets and the Moon).

The primary probes of the Earth's interior are **seismic waves** created in the Earth as a result of major shocks, most commonly earthquakes. These waves take three possible forms, called P, S, and L waves. Studies of wave transmission in solids and liquids have shown that P waves, which are the first to arrive at a site remote from the earthquake location, are **compressional** waves, which means that the oscillating motions occur parallel to the direction of motion of the wave, creating alternating regions of high and low density without any sideways motions (Fig. 9.6). Sound waves are examples of compressional waves. The S waves are **transverse** or **shear**, waves, the vibrations occurring at right angles to the direction of motion (see Fig. 9.6). These waves require that the material they pass through have some rigidity and, unlike the P waves, they cannot be transmitted through a liquid. The L waves travel only along the surface of the Earth and thus do not provide much information on the deep interior.

By measuring both the timing and the intensity of these seismic waves at various locations away from the site of an earthquake, scientists can determine what the Earth's interior is like. The speed of the P waves

depends on the density of the material they pass through, and the distribution of the P and S waves reaching remote sites provides data on the location of liquid zones in the interior (Fig. 9.7).

The general picture that has developed from studies is that of a layered Earth (Fig. 9.8 and Table 9.3). At the surface is a crust whose thickness varies from a few kilometers beneath the oceans to perhaps 60 km under the continents. There is a sharp break between the crust and the underlying material that is called the **mantle**. The mantle transmits S waves, so it must be solid; but on the other hand, it undergoes slow, steady, flowing motions in its uppermost regions. Perhaps it is best viewed as a plastic material, one that has some rigidity, but which can be deformed, given sufficient time (this is discussed more fully in the next section).

The uppermost part of the mantle and the crust together form a rigid zone called the **lithosphere**. The

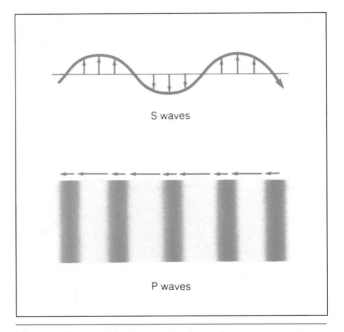

Figure 9.6 S and P waves. *An S, or shear, wave (top) consists of alternating motions transverse (perpendicular) to the direction of propagation. Waves in a tight string or on the surface of water are S waves. Compressional, or P, waves (bottom) have no transverse motion but consist of alternating dense and rarefied regions created by motions along the direction of propagation. The arrows at the top represent the relative speeds in and between the dense zones. Sound waves are P waves.*

part of the mantle itself where the fluid motions occur, just below the lithosphere, is called the **asthenosphere.** Below the asthenosphere, there is a more rigid portion of the mantle that extends nearly halfway to the center of the Earth. The lower mantle is called the **mesosphere** (not to be confused with the level in the Earth's atmosphere bearing the same name).

Beneath the mantle is the **core,** consisting of an **outer core** (between 2,900 and 5,100 km in depth) and an **inner core.** The S waves are not transmitted through the outer core, which is therefore thought to be liquid. As a result, the inner core can be probed only with the P waves. Since these waves travel more rapidly through the inner core, its density is thought to be greater than that of the outer core, and the innermost region is thought to be solid.

The overall density of the Earth is 5.5 times that of water (that is, 5.5 grams per cubic centimeter, or 5.5 g/cm^3). The rocks that make up the crust have typical densities of less than 3 g/cm^3, and the mantle material is thought to have relatively low density also, perhaps 3.5 g/cm^3. From these facts, it is deduced that the core must have a density of roughly 13 g/cm^3. The most likely substance that could give the core this high density is iron, probably combined with some nickel and perhaps cobalt.

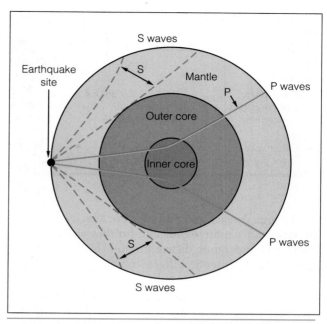

Figure 9.7 Seismic waves in the Earth. *This simplified sketch shows that P waves can pass through the core regions (although their paths may be bent), while S waves cannot. This led to the deduction that the outer core is liquid, because S waves, which require an elastic or solid medium, cannot penetrate liquids.*

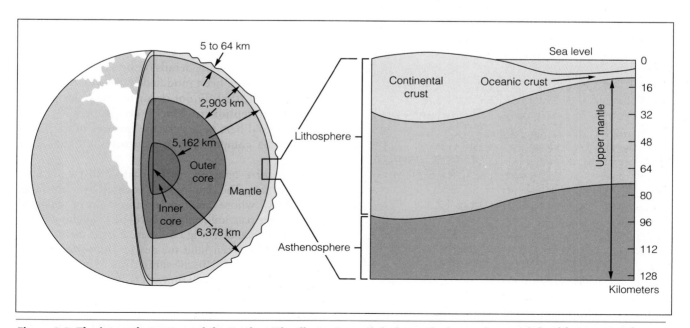

Figure 9.8 The internal structure of the Earth. *The illustration at left shows the layers that are defined by contrasts in density or physical state, as deduced from observations of seismic waves. The expanded view illustrates the relative thickness of the crust in continental areas as compared to the seafloor.*

Table 9.3 Earth's interior

Layer	Depth (km)	Density (g/cm³)	Temperature (K)	Composition
Crust	0–30	2.6–2.9	300–700	Silicates and oxides
Mantle				
Lithosphere	30–70	2.9–3.3	700–1,200	Basalt, silicates, oxides
Asthenosphere	70–1,000	3.9–4.6	1,200–3,000	Basalt, silicates, oxides
Mesosphere	1,000–2,900	4.6–9.7	3,000–4,500	Basalt, silicates, oxides
Core				
Outer core	2,900–5,100	9.7–12.7	4,500–6,000	Molten iron, nickel, cobalt
Inner core	5,100–6,378	12.7–13.0	6,000–6,400	Solid iron, nickel, cobalt

The high density and probable metallic composition of the core are quite significant, for they show that the Earth has undergone differentiation (Fig. 9.9). As explained in Chapter 8, differentiation can occur only when the planet is in a molten state, so it shows that the Earth was once largely liquid, probably early in its history, soon after it formed. The principal cause of the Earth's molten state most likely is radioactive heating in the interior. Several naturally occurring elements, such as uranium, thorium, and potassium, are radioactive, which means that their nuclei spontaneously emit subatomic particles and, over a long period of time, produce substantial quantities of heat. Another source of heat early in the Earth's history was created by the energy of impact as the smaller bodies that merged to form the Earth came together. Additional heating was caused by the process of differentiation itself, as the heavy elements sank to the center of the Earth, releasing gravitational potential energy. We shall see that differentiation has not occurred in all the bodies in the solar system, and this tells us something significant about their histories.

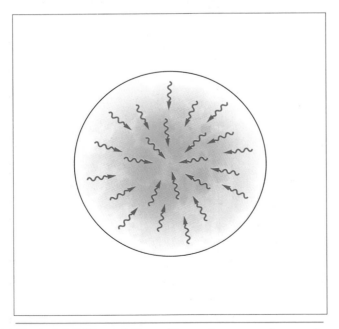

Figure 9.9 Differentiation. *Heavy elements sank toward the Earth's center, creating the dense core. This process required that the Earth be partially or wholly liquid, and took place very early in the Earth's history.*

SURFACE TERRAIN AND TECTONIC ACTIVITY

As in previous sections, here we will deal primarily with large-scale features. One reason for this is that many of the small-scale details of the surface structures of the Earth have been modified by processes, such as flowing water, that do not occur on the other planets (with the exception of Mars; see Chapter 12). Our principal goal here is to describe features of the Earth that can be compared with structures on the other planets.

About three-quarters of the Earth's surface is covered by water. The remainder is covered by a few large continents, whose relative positions have changed steadily over past millennia and are still changing today.

The evidence favoring continental drift includes the obvious fit between land masses on opposite sides of the Atlantic (Fig. 9.10), the similarities of mineral types and fossils, and the alignment of vestigial magnetic fields in rocks, now separated by the ocean, that once were together in the same place. In modern times, more direct evidence, such as the detection of seafloor spreading away from undersea ridges and the actual measurement (through the use of sophisticated laser-

Figure 9.10 A relief map of the Earth. *This map shows relative elevations over the entire crust of the Earth, including the seafloors. (M. Kobrick, JPL)*

ranging techniques) of continental motions, has removed any trace of doubt that pieces of the Earth's crust are in motion, and that the continents were once arranged differently (Fig. 9.11).

From all this evidence has arisen a theory of plate tectonics, which postulates that the Earth's crust (that is, the lithosphere) is made of a few large, thin, pieces that float on top of the asthenosphere (Fig. 9.12). The flowing motions in the asthenosphere cause these plates to move about constantly and occasionally to crash into each other. The rate of motion is only a few centimeters per year at the most, and the major rearrangements of the continents have taken many millions of years to occur (both the Americas and the European-African system became separated from each other 150 to 200 million years ago)

The driving force in the shifting of the plates is not well understood. The most widely suspected cause is convection currents in the asthenosphere (Fig. 9.13). This is the same process that occurs in the Earth's atmosphere: Temperature differences between levels cause an overturning motion. If a fluid in a gravitational field is hot at the bottom and much cooler at the top, warm material will rise, cool, and descend again, creating a constant churning. The speed of the overturning motions depends largely on the **viscosity** of the fluid, that is, the degree to which it resists flowing freely. The Earth's mantle, as we have already seen, is sufficiently rigid to transmit S waves, and must therefore have a high viscosity as well. Hence, if convection is occurring in the mantle, it is reasonable that the motions should be very slow.

Whether convection is the cause or not, the effects of tectonic activity are becoming well known. When plates collide, one may submerge below the other in a process called **subduction,** which creates an undersea

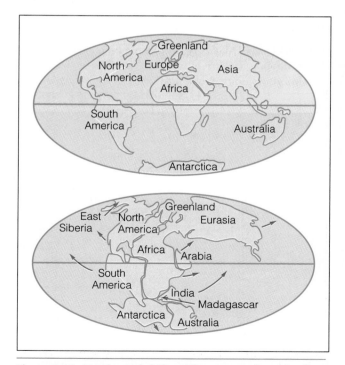

Figure 9.11 Continental drift. *These maps show the distribution of the continents today and as it was some 200 million years ago.*

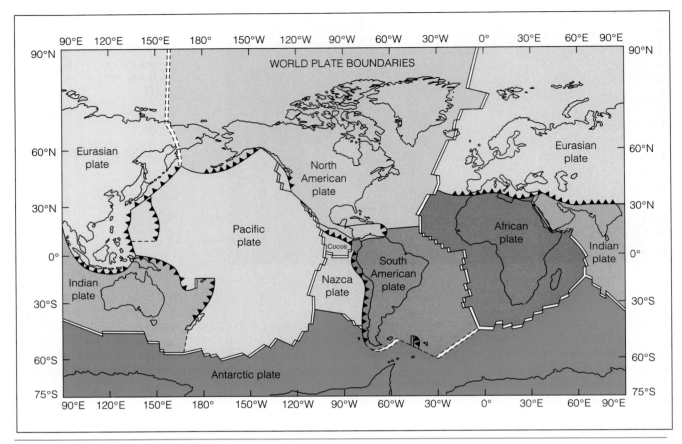

Figure 9.12 The Earth's crustal plates. *This map shows the plate boundaries and the direction in which the plates are moving.*

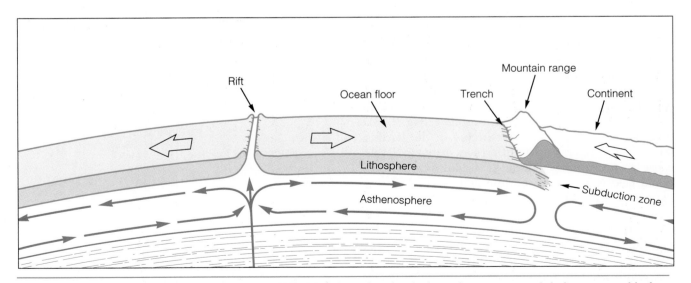

Figure 9.13 Schematic of the mechanisms of continental drift. *This sketch shows how continental drift is responsible for uplifted mountain ranges and parallel undersea trenches, where one crustal plate sinks below another (subduction zone), and how a mid-ocean ridge is built up where two plates move away from each other.*

trench, or, particularly if both plates carry continents, the collision may force the uplifting of mountain ranges. The lofty Himalayas are thought to have been created when the Indian plate collided with the Eurasian plate some 50 million years ago. The boundaries where plates either collide or separate are marked by a wide variety of geological activity. Earthquakes are attributed to the sporadic shifting of adjacent plates along fault lines, and volcanic activity is common where material from the mantle can reach the surface, most often along plate boundaries. It has long been recognized, for example, that a great deal of earthquake and volcanic activity is concentrated around the shores of the Pacific, defining the so-called "Ring of Fire". Now it is understood that this region represents the boundaries of the Pacific plate.

Chains of volcanoes, such as the Hawaiian islands, are also attributed to the action of plate tectonics. In several locations around the world, volcanic hot spots that lie deep and are fixed in place bring molten rock to the surface. As the crust passes over one of these hot spots, volcanoes are formed and then carried away, and a chain of mountains is created as new material keeps coming to the surface over the hot spot. This process is still active, and the Hawaiian islands, for example, are still growing. There is volcanic activity on the southeast part of the largest (and youngest) island, Hawaii (Fig. 9.14). Furthermore, a new island, already given the name Loihi, is rising from the seafloor about 20 km south of Hawaii. Its peak has already

risen 80 percent of the way to the surface and has only about 1 km to go, but it will take an estimated 50,000 years to break through.

Plate tectonics, then, accounts for many of the most prominent features of the Earth's surface. Of course, other processes, such as running water, wind erosion, and glaciation, are also important in modifying the face of our planet.

ROCKS, MINERALS, AND THE AGE OF THE SURFACE

When we examine the surface of the Earth, we see little outward evidence of the large-scale processes. We do not see the continents moving, and we cannot directly perceive the interior. What we do see are rocks. The basic substance of the Earth's crust is rock, although in many areas, soil covers the bedrock. Even in those places, however, gravel, boulders, and rocky stream and lake beds are evident.

As we learned in Chapter 8, rocks fall into three categories according to their origins: igneous (volcanic), sedimentary (from seafloor deposits), and metamorphic (altered by heat and pressure). These three forms are interchangeable through a continual recycling process (Fig. 9.15), and all three can be eroded by wind and water into soil, sand, or gravel.

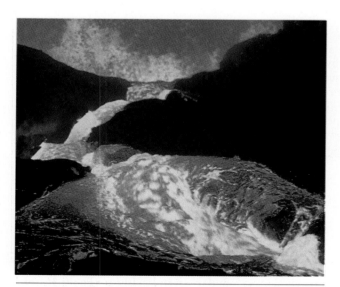

Figure 9.14 Volcanic eruption in Hawaii *(Rick Golt, Photo Researchers)*

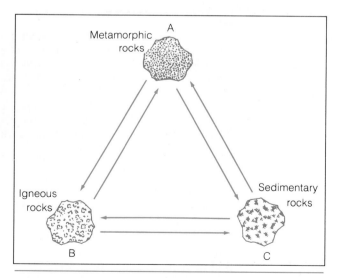

Figure 9.15 The rock cycle. *This shows schematically how the three basic types of rocks can be converted from one to another.*

ASTRONOMICAL INSIGHT 9.2

The Ages of Rock

 We have spoken of the ages of rocks and of the Earth itself, but we have said little about how these ages are measured. The most direct technique, **radioactive dating,** involves the measurement of relative abundances in rocks of closely related atomic elements.

Recall from our earlier discussions of atomic structure (Chapter 6) that the nucleus of an atom consists of protons and neutrons, and that the number of protons (that is, the **atomic number**) determines the identity of the element. Different **isotopes** of an element have the same number of protons but differing numbers of neutrons. In most elements, the number of protons is not very different from the number of neutrons, but there are exceptions. If the imbalance between protons and neutrons is large enough, the element is unstable, meaning that it has a natural tendency to correct the imbalance by undergoing a spontaneous nuclear reaction. The reactions that occur in this way usually involve the emission of a subatomic particle, and the element is said to be **radioactive.** The emitted particle may be an al-pha particle (that is, a helium nucleus consisting of two protons and two neutrons) or, as is more often the case, an electron or a **positron,** a tiny particle with the mass of an electron but with a positive electrical charge. When a positron or an electron is emitted, the reaction is called **beta decay,** and at the same time a proton in the nucleus is converted into a neutron (if a positron is emitted) or a neutron is changed into a proton (if an electron is released). If the number of protons in the nucleus is altered, then this changes the identity of the element. A typical example is the conversion of potassium 40 (^{40}K, with 19 protons and 21 neutrons in its nucleus) into argon 40 (^{40}Ar, with 18 protons and 22 neutrons).

Several elements are known to be radioactive, naturally changing their identity by emitting particles. The rate of change, expressed in terms of **half-life,** is known in most cases. The half-life is the time it takes for half of the original element to be converted. It may be as short as fractions of a second or as long as billions of years. The very slow reactions are the ones that are useful in measuring the ages of rocks.

If we know the relative abundances of the elements that were present in a rock when it formed, then measurement of the ratio of the elements in that rock at the present time can tell us how long the radioactive decay has been at work; that is, the age of the rock. In a very old rock, for example, there might be almost no ^{40}K, but a lot of ^{40}Ar. The decay of ^{40}K to ^{40}Ar has a half-life of 1.3 billion years, so from the exact ratio of these two species in a rock, we can infer how many periods of 1.3 billion years have passed since the rock formed.

Other decay processes that are useful in dating rocks include the decay of rubidium 87 (^{87}Rb) into strontium 87 (^{87}Sr), which has a half-life of 47 billion years; and the decays of two different isotopes of uranium to isotopes of lead, ^{235}U to ^{207}Pb and ^{238}U to ^{206}Pb, with half-lives of 700 million years and 4.5 billion years, respectively. (Note that these decays each involve several intermediate steps and that the half-lives given represent the total time for all of those steps.)

Use of these dating techniques has shown that the oldest rocks on the Earth's surface are about 3.5 billion years old and that ages of a few hundred million years are more common. The age of the solar system, and therefore of the Earth itself, is estimated, from isotope ratios in meteorites, to be about 4.5 billion years.

As a result of these processes that alter rocks, along with the rearranging and overturning caused by tectonic activity, the surface of the Earth is continually being renewed. Whereas the age of the Earth is thought to be some 4.5 billion years, the ages of most surface rocks can be measured in the millions or hundreds of millions of years. As noted in Chapter 8, the distinction between the age of a planet and the age of its surface is an important one.

We also discussed in Chapter 8 the fact that rocks and soil are composed of minerals, which are identified by their chemical composition and internal structure. The various kinds of atoms in a mineral are usually arranged in a regular crystal pattern (Fig. 9.16). In most cases, these crystal patterns are confined to very small, microscopic regions that are randomly stuck

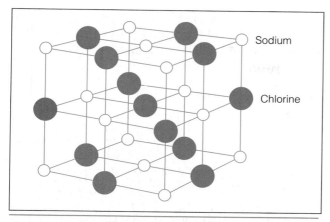

Figure 9.16 A typical crystal. *This shows the regular arrangement of atoms in the crystalline structure of sodium chloride, common table salt.*

together in a rock, giving it an overall amorphous appearance. Rarely we find large specimens of crystals, some of which are hard and brilliant enough to be used in jewelry.

Some 2,000 minerals have been identified in rocks, but the bulk of the Earth's surface is composed of a very limited number, about a dozen. The most common, accounting for about 90 percent of the Earth's rocks, are the **silicates,** minerals containing silicon, oxygen, and some metallic elements. Other classes of common minerals include the **oxides,** simple combinations of oxygen and other elements; the **sulfides,** similar combinations of sulfur and other elements; and the **carbonate** and **sulfate** minerals, which contain special combinations of carbon and oxygen or sulfur and oxygen, along with metallic elements. Finally, there are the **halides,** minerals formed by combinations of the haline elements chlorine, iodine, bromine, and fluorine with other elements. Common table salt, sodium chloride, is a halide. Any rock can be classified according to both its type of origin and its mineral content. Two very common examples are basalt and granite, which are both igneous rocks with silicate compositions.

THE MAGNETIC FIELD

The magnetic field of the Earth has guided travelers since antiquity. Certain metallic substances, usually containing iron, are sensitive to the presence of a magnetic field and are used as pointers in compasses; thus,

we take advantage of the near alignment of the Earth's magnetic field with the poles (the angle between the magnetic axis and the rotation axis of the Earth is about 11.5°). To see how a compass works, it is convenient to imagine magnetic lines along which iron filings lie when placed near a small bar magnet.

Outer Structure and the Magnetosphere

In cross-sectional view, the Earth's magnetic field is reminiscent of a cut apple, but one that is lopsided, because of the flow of charged particles from the Sun that constantly sweeps past the Earth (see the discussion of the solar wind in Chapter 17). The region enclosed by the field lines is called the **magnetosphere** (Fig. 9.17), and it acts as a shield, preventing the charged particles from reaching the surface.

The past alignment of the magnetic field can be ascertained from studies of certain rocks that contain iron-bearing minerals whose crystalline structure is aligned with the direction of the magnetic field at the time the rocks solidify from a molten state. Traces of the Earth's ancient magnetic activity, called **paleomagnetism,** thus can be deduced from the analysis of magnetic alignments of rocks. Such studies reveal not only that the continents have moved, but also that the magnetic poles of the Earth have moved. Furthermore, the north and south magnetic poles have completely and rather suddenly reversed from time to time, so that the North Pole has moved to the south and vice versa. This flip-flop of the magnetic poles seems to have happened at irregular intervals, typically thousands or hundreds of thousands of years apart.

Figure 9.17 The Earth's magnetic field structure. *This cross-section shows the Earth's field and how its shape is affected by the stream of charged particles from the Sun known as the solar wind.*

While a reversal of the poles is taking place, there is, for a short period of time, a much-weakened magnetic field. The magnetosphere is greatly diminished, and charged particles from the solar wind and from **cosmic rays,** highly energetic particles from interstellar space (see Chapter 25), are more likely to penetrate to the ground. These particles have important effects on life forms, including genetic mutations. The sporadic reversals of the Earth's magnetic field may have played a major role in shaping the evolution of life on the surface of our planet.

The source of the Earth's magnetic field is not fully understood, although it is nearly certain that it arises in the nickel-iron core. A magnetic field is produced by flowing electrical charges, as in the current flowing through a wire wound around a metal rod, which is the basis of an electromagnet. Convection and the Earth's rotation may combine to create systematic flows in the liquid outer core, giving rise to the magnetic field, if the core material carries an electrical charge. This general type of mechanism is called a **magnetic dynamo.**

The cause of the reversals of the poles is not well understood but presumably could be related to occasional changes in the direction of the flow of material in the core.

Radiation Belts

We have referred to the magnetosphere and its ability to control the motions of charged particles. When the first U.S. satellite was launched in 1958, zones high above the Earth's surface were discovered to contain intense concentrations of charged particles, primarily protons and electrons. There are several distinct zones containing these particles, and they are now called the **Van Allen belts** (Fig. 9.18), after the physicist who, in the late 1950s, first recognized their existence and deduced their properties. The charged particles, or ions, in these radiation belts are captured from space (primarily from the solar wind), and spiral around the lines of force of the Earth's magnetic field.

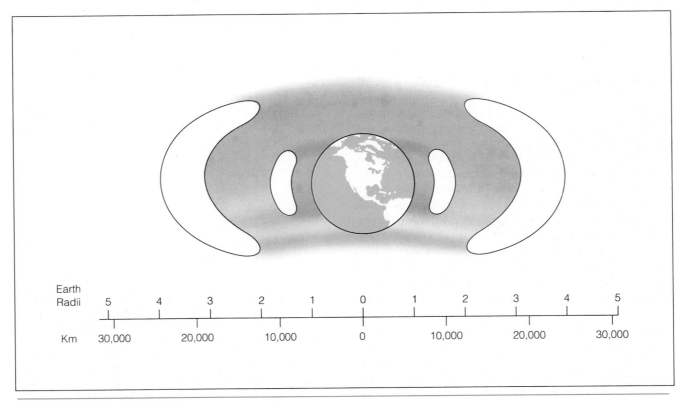

| Earth Radii | 5 | 4 | 3 | 2 | 1 | 0 | 1 | 2 | 3 | 4 | 5 |
| Km | 30,000 | | 20,000 | | 10,000 | 0 | | 10,000 | | 20,000 | | 30,000 |

Figure 9.18 The Van Allen belts. *This cross-sectional view illustrates the sizes and shapes of the two principal charged-particle belts that girdle the Earth.*

Far below the Van Allen belts, the uppermost portion of the atmosphere also contains charged particles, and is called the **ionosphere.** The ionosphere, which extends upward from an altitude of about 60 km, has important effects for us on the surface, particularly in terms of our radio communications. Short-wave radio signals are able to travel around the world because they are reflected back to the ground from the underside of the ionosphere. When the ionosphere is disrupted, as it can be after a particularly violent solar flare, its reflective properties may be destroyed temporarily, interrupting communications. (This problem has been largely overcome by communications satellites, which do not rely on the ionosphere at all.) At the same time, emissions caused by the charged particles may be enhanced, creating displays of the **aurora borealis** (Fig. 9.19), or northern lights (in the Southern Hemisphere they are called the **aurora australis,** or southern lights). The aurorae are always most prominent close to the magnetic poles, because that is where the Earth's magnetic field lines, and the charged particles that are forced to travel along them, come closest to the surface. In middle latitudes, these particles are kept well outside the atmosphere, and aurorae are rarely seen.

THE EVOLUTION OF THE EARTH

A great deal has been learned about the general evolution of the Earth. We know it was once largely a molten body with an atmosphere different from that of the present day. Along the way, the atmosphere has undergone major changes, some of them triggered by the development of life on the surface.

The age of the Earth, estimated from geological evidence, is about 4.5 billion years. As discussed in Chapter 8, the Earth probably formed from the coalescence of a number of **planetesimals,** small bodies that were the first to condense in the earliest days of the solar system. As a result of the heat of impact as these bodies merged, and in time, of the heating caused by radioactivity, most or all of the Earth was molten at some point during the first billion years. During this early period, differentiation occurred, with the densest materials (such as iron) sinking to the core. At the same time, **volatile** gases were emitted by the newly formed rocks at the surface. Volatile gases are those that vaporize most easily when heated; they formed the earliest atmosphere of the Earth. The most common of

these gases were hydrogen (H_2), ammonia (NH_3), methane (CH_4), water vapor (H_2O) and carbon dioxide (CO_2). It is noteworthy that the present-day atmospheres of the cold, outer planets are very similar to that of the primitive Earth.

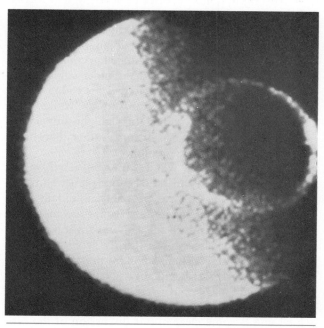

Figure 9.19 The aurora borealis. *The upper photograph shows the aurora borealis, or northern lights, as seen from the ground at middle latitudes in the Northern Hemisphere. The lower image shows the circle of aurorae around the North Pole as seen from space in ultraviolet wavelengths. The Earth's magnetic field prevents the ions that create the aurorae from reaching the surface except near the poles, where the magnetic field reaches low altitudes. (Upper: © Ned Haines, Photo Researcher; lower: NASA)*

The dominant element in the Earth's early atmosphere was hydrogen, but nearly all of it escaped into space by the time the Earth was about 1 billion years old. (This is soon after the first simple life forms apparently arose.) Recall from Chapter 8 that individual molecules in a gas whiz around randomly at speeds that depend on both the temperature of the gas and the mass of the particles. At a given temperature, the lightweight molecules move fastest and are therefore most likely to escape the planet's gravitational attraction. This is what happened to the hydrogen in the early atmosphere of the Earth (whereas the heavier gases that now dominate the atmosphere are too massive to escape). Another gas that escaped early in the Earth's history is helium. Helium is the second most abundant element in the universe, but is now so rare on Earth that it was first discovered in the Sun, where spectroscopic observations revealed its presence (hence the name, derived from *helios,* the Greek word for "sun").

According to some theories of solar system formation, there was not much water vapor contained in the solid material from which the Earth condensed, so the Earth's oceans require another explanation. It is possible that most of the water originated in comets, which were much more numerous in the inner solar system during the first billion years than they are today. Hence in the view of some scientists, most of the water in the Earth's oceans was accreted through the impacts of comets early in the Earth's history, rather than being present from the very beginning.

Whatever its origin, water vapor collected on the surface, and from its earliest days our planet had seas. Carbon dioxide (CO_2) was probably abundant for a time but was eventually removed from the atmosphere by being absorbed into rocks. This absorption process, of critical importance to the further evolution of the Earth's atmosphere, depended on the presence of liquid water; if the early Earth had not had oceans, the carbon dioxide might never have left the atmosphere. This would have had fateful consequences for the Earth, which will become clear in our discussion of the evolution of Venus (Chapter 11).

The critical reactions that led to the development of life must have taken place before all the hydrogen escaped, because the types of reactions that were probably responsible involve this element. The earliest fossil evidence for primitive life dates back nearly 4 billion years, when some hydrogen was still left.

Essentially, no free oxygen was present in the atmosphere until life forms that release this element as a by-product of their metabolic activities had developed. Most plants release oxygen into the atmosphere, and as a supply of this element built up, the opportunity arose for complex animal forms to evolve. Besides being a substance that was necessary for the metabolism of living animals, the oxygen buildup created a reservoir of ozone (O_3) in the upper atmosphere, which in turn began to screen out the harmful ultraviolet rays from the Sun. Once life forms gained a toehold on the continental land masses, the process of converting the atmosphere to its present state began to accelerate. Soon nitrogen from the decomposition of organic matter began to be released into the air in large quantities, and by the time the Earth was perhaps 2 billion years old, the atmosphere had reached approximately the composition it has today.

By this time also the mantle had solidified and the crust had hardened (the oldest known surface rocks are nearly 4 billion years old). The interior has remained warmer than the surface, because heat escapes slowly through the crust, and because radioactive heating of the interior took place over a long period of time and is still occurring today.

PERSPECTIVE

We have itemized the Earth's overall properties, from its outermost atmosphere to its core, always keeping in mind the facets of its character that will provide a basis for comparison with what we know about the other planets. We have seen that the Earth is an active world, seemingly placid only because of the long time-scales over which significant changes occur. We are ready now to examine other bodies in the solar system, to see whether they are similarly dynamic. We will begin with the Moon, our nearest neighbor, and Mercury, the smallest terrestrial planet (and one that is very similar to the Moon in many ways).

SUMMARY

1. The Earth, seen from space, is brilliant because of its high albedo.
2. The atmosphere consists of 80 percent nitrogen and about 20 percent oxygen, traces of H_2O and CO_2, and fine suspended particles.

3. The atmosphere is often divided into four vertical zones, based on the variation of temperature with height: the troposphere, stratosphere, mesosphere, and thermosphere.

4. The Sun's heating and the Earth's rotation create the global wind patterns.

5. The Earth's interior, explored with seismic waves, consists of a solid inner core, a liquid outer core, a mantle, and a crust.

6. The crust is broken into tectonic plates that shift around, which accounts for continental drift and most of the major surface features of the Earth.

7. Rocks can be igneous, metamorphic, or sedimentary in origin, and most are silicate minerals by composition.

8. The Earth has a magnetic field, probably created by currents in the molten core, which traps charged particles in zones, called radiation belts, above the atmosphere.

9. The Earth's evolution from a largely molten planet with a hydrogen-dominated atmosphere to its present state was caused by the loss of lightweight gases into space and by the development of life forms on the surface of the Earth, which helped convert the atmospheric composition to nitrogen and oxygen.

REVIEW QUESTIONS

Thought Questions

1. Summarize the Earth's atmosphere, in terms of chemical composition and physical conditions such as temperature, density, and pressure.

2. Explain how heat from the Sun and rotation of the Earth create the global circulation pattern of the atmosphere. How do the oceans influence the circulation?

3. How do we know anything about the Earth's interior? Do you think the same methods can be applied to other planets? Explain.

4. Explain the differences between transverse and compressional waves. Can you think of everyday examples of each kind?

5. If the density of typical surface rocks is 3.5 grams/cm³, and the Earth's average (overall) density is 5.5 grams/cm³, what does this tell us about the density in the deep interior? How did this situation arise?

6. Summarize the evidence that tectonic activity, including continental drift, occurs on the Earth.

7. How are the motions in the Earth's atmosphere and those in its interior similar?

8. Explain the distinction between the age of the Earth's surface and the age of the Earth itself.

9. How do you think magnetic field measurements for other planets might be used to give us clues to their internal structure?

10. Summarize the effects of life forms on the evolution of the Earth's atmosphere.

Problems

1. Verify that the Earth's average density is about 5.5 grams/cm³ by using the figures given in Table 9.1 for the mass and diameter of the Earth. The volume of a sphere is $4\pi r^3/3$, and the density is the mass divided by the volume.

2. From the figures in Table 9.3, calculate the volume of the core (including both inner and outer cores) of the Earth. What fraction of the Earth's total volume is occupied by the core?

3. Assume that all of the Earth outside of the core has a density of 3.5 grams/cm³, and that the core (inner and outer) has a constant density as well. What must the density of the core be in order to give the Earth its overall density of 5.5 grams/cm³? (You will need to use your answer to the previous problem.)

4. If North America is approaching Japan at a rate of 3 cm per year, and the present distance between North America and Japan is 3,000 miles, how long will it take for the two to collide?

5. If the half-life of a radioactive element is 5 million years, and a rock sample contains one-half of its original quantity of that element, how old is the rock? How old is it if only one-fourth of the original quantity of the element is left? How old is it if one sixty-fourth is left?

6. We refer to hydrogen and helium as "light" gases. Using data from Appendix 3, compare the masses of hydrogen molecules (H_2) and helium atoms with the masses of the common molecules in today's atmosphere, oxygen (O_2) and nitrogen (N_2). (Hint: Use the atomic weight figures in Appendix 3, along with the number of each type of atom in the molecule in question.)

ADDITIONAL READINGS

Badash, L. 1989. The age-of-the-Earth debate. *Scientific American* 261(2):90.

Battan, L. J. 1979. *Fundamentals of meteorology.* Englewood Cliffs, N.J.: Prentice-Hall.

Ben-Avraham, Z. 1981. The movement of continents. *American Scientist* 69:291.

Bloxham, J. and D. Gubbins. 1989. The evolution of the Earth's magnetic field. *Scientific American* 261(6):68.

Bonatt: E. 1987. The rifting of continents. *Scientific American* 256(3):96.

Burchfiel, B. C. 1983. The continental crust. *Scientific American* 249(3):86.

Carrigan, C. R., and D. Gubbins. 1979. The source of the Earth's magnetic field. *Scientific American* 240(1):118.

Clark, W. C. 1989. Managing the Earth. *Scientific American* 261(3):46 (also pp. 58, 70, 80, 108, 118, 128, 136, 144, 154, and 166).

Friedman, Herbert. 1986. *Sun and earth*. New York: W. H. Freeman.

Goldreich, P. 1972. Tides and the Earth-Moon system. *Scientific American* 226(4):42.

Hoffman, K. A. 1988. Ancient magnetic reversals: clues to the geodynamo. *Scientific American* 258(5):76.

Ingersoll, A. P. 1983. The atmosphere. *Scientific American* 249(3):114.

Jeanloz, R. 1983. The Earth's core. *Scientific American* 249(3):40.

Johnston, A. C. and L. R. Kanter. 1990. Earthquakes in a stable continental crust. *Scientific American* 262(3):68.

Jones, P. D. and T. M. L. Winglee. 1990. Global warming trends. *Scientific American* 263(2):84.

Jordan, T. H., and J. B. Minster. 1988. Measuring crustal deformation in the American west. *Scientific American* 259(2):48.

Kasting, J. F., B. Toon, and J. B. Pollack. 1988. How climate evolved on the terrestrial planets. *Scientific American* 258(2):90.

Macdonald, K. C. and P. C. Fox. 1990. The mid-ocean ridge. *Scientific American* 262(6):72.

McKenzie, D. P. 1983. The Earth's mantle. *Scientific American* 249(3):50.

Nance, R. D., T. R. Worsley, and J. B. Moody. 1988. The supercontinent cycle. *Scientific American* 259(1):72.

Nervell, R. E., H. G. Reidule, Jr., and W. Feiler. 1989. Carbon monoxide and the burning Earth. *Scientific American* 261(4):82.

Repetto, R. 1990. Deforestation in the tropics. *Scientific American* 262(4):36.

Siever, R. 1983. The dynamic Earth. *Scientific American* 249(3):30.

Toon, O. B., and S. Olson. 1985. The warm Earth. *Science 85* 6(8):50.

White, R. M. 1990. The great climate debate. *Scientific American* 262(1):132.

White, R. S. and D. P. McKenzie. 1989. Volcanism at rifts. *Scientific American* 261(1):67.

MATHEMATICAL METHODS

Half-Lives and Radioactive Dating

In Chapter 8, we made reference to the fact that the age of a planetary surface can be quite distinct from the age of the planet itself. The best method available for measuring the age of the Earth's surface is *radioactive dating*. This technique has also been applied to rocks from the Moon's surface and, we may hope, will in due course be used in the analysis of rocks from Mars or even Venus and Mercury. Here we describe how the radioactive dating method is used, and how ages can be calculated from a measurement of isotope ratios.

If a rock is formed initially with a quantity N_o of an isotope, then after one half-life has passed, the number of this isotope is reduced to $\frac{N_o}{2}$. If N is the number after one half-life, then we say that $N = \frac{N_o}{2}$, which can be rewritten as $\frac{N}{N_o} = \frac{1}{2}$. After two half-lives have passed, we have $N = \frac{N_o}{4}$ or $\frac{N}{N_o} = \frac{1}{4}$, and so on. We see that for every half-life that passes, the quantity N is reduced by another factor of 2. Mathematically this statement can be written as

$$\frac{N}{N_o} = \left(\frac{1}{2}\right)^n,$$

where n is the number of half-lives that have passed.

In all our examples so far, we have always assumed that n is a whole number, such as 1, 2, 3, and so on. But of course in nature this is rarely so. The equation above works just as well for any value of n, but the calculation is more difficult when n is not a whole number, and usually a calculator must be used.

Normally when we want to find the age of a rock sample the quantity that we measure is N/N_o, and what we want to find is n, the number of half-lives that have passed since the rock was formed. The age of the rock would then be n times the half-life, or $nT_{1/2}$. Therefore, we have to solve the equation above for n; to do this most simply, we take the logarithm of both sides, finding

$$\log\left(\frac{N}{N_o}\right) = n\log\left(\frac{1}{2}\right) = -0.301n,$$

or

$$n = -3.32\log\left(\frac{N}{N_o}\right).$$

The age of the sample is then

$$\text{age} = nT_{1/2} = -3.32T_{1/2}\log\left(\frac{N}{N_o}\right).$$

Now let us apply this to a hypothetical example, using the decay of ^{40}K to ^{40}Ar. We can assume that all of the ^{40}Ar we find in any rock is the result of decay from ^{40}K, because argon does not easily bond with other elements and therefore would not be incorporated into the rock as it formed. In that case, the quantity of ^{40}Ar measured in our sample corresponds to N in the equations above, while N_o corresponds to the sum of ^{40}Ar plus ^{40}K found in the sample.

Suppose a rock is found in which there are 32 parts per billion of ^{40}K and 7 parts per billion of ^{40}Ar. This leads to

$$\frac{N}{N_o} = 7/(32 + 7) = 0.22.$$

The age of our sample is then

$$\begin{aligned}
\text{age} &= -3.32T_{1/2}\log(0.22) \\
&= -3.32(1.3 \text{ billion yrs})(-0.66) \\
&= 2.85 \text{ billion years.}
\end{aligned}$$

For isotopes other than ^{40}Ar and ^{40}K, the problem is a bit more complex, because the product of the decay can be present when a rock forms. For example, when we find a certain quantity of ^{206}Pb (lead) in a rock, it is not immediately obvious how much of this lead came from the decay of ^{238}U (uranium) and how much was in the rock to begin with. To determine the age of this sample, it is necessary to estimate how much ^{206}Pb was initially present in the rock. This can sometimes be done indirectly, by assuming that the initial rock had a certain ratio of lead to some other, nonradioactive element whose abundance has not been altered by decay.

CHAPTER 10

The Small Terrestrials: The Moon and Mercury

A false-color image of the Moon, enhancing the contrast between the highlands and the maria (F. Espenak)

It may seem odd to group the Moon and Mercury together in our discussions, but there are good reasons for doing so. Both can be viewed as terrestrial bodies, although the Moon cannot properly be called a planet, since it is not a Sun-orbiting body. Yet, as we shall see, the Moon might have become a planet had it not been for an unlikely occurrence early in the history of the solar system. (Furthermore, Mercury may have been strongly affected by a very similar, equally unlikely, occurrence.) In addition, the Moon is nearly as large as Mercury, and it bears an almost uncanny resemblance to it as well, at least superficially.

In this chapter we will discuss the Moon first, and then use what we know of the Moon as a basis for pointing out the similarities and contrasts between it and Mercury. We will find that, despite the outward similarities, there are profound differences between the interior structures of the two bodies. These differences reflect the contrasts in the environments of the two as they formed, but may also indicate some fundamental differences in their geological histories.

The Moon is, of course, the best-studied object in the solar system other than the Earth: the Moon is close enough to allow very detailed Earth-based observations; many unmanned probes have been sent to the Moon to make close-up photographs; and, most significantly, a dozen people have walked on the Moon and brought back samples of its surface material. By contrast, Mercury is very difficult to observe from the Earth, and has been visited by only a single unmanned spacecraft. Thus it is very helpful if we can draw on our knowledge of the Moon when we try to determine the properties of Mercury from our meager collection of data on that planet.

EXPLORING THE MOON

Seen from afar, the Moon (Table 10.1) is a very impressive sight, especially when full (Fig. 10.1). Its ½° diameter is pockmarked with a variety of surface features, the most prominent of which are the light and dark areas. The latter, thought by Renaissance astronomers to be bodies of water, are called **maria** (singular; **mare**), after the Latin word for "seas." Closer examination of the Moon with even primitive telescopes also revealed numerous circular features called **craters,** after similar structures found on volcanoes. The lunar craters are not volcanic, however, but were formed by the impacts of bodies that crashed onto the surface from space.

The pattern of surface markings on the face of the Moon is quite distinctive, and because it never changes, it was recognized long ago that the Moon always keeps one face toward the Earth. The far side cannot be observed from the Earth, but has now been thoroughly mapped by spacecraft (Fig. 10.2).

The lunar surface in general can be divided into lowlands, primarily the maria, and highlands, which are vast mountainous regions, not as well organized into chains or ranges as mountains on the Earth. With

Table 10.1 Moon

Mean distance from Earth: 384,401 km (60.4 $R_\oplus$)
   Closest approach: 363,297 km
   Greatest distance: 405,505 km
Orbital sidereal period: $27^d\ 7^h\ 43^m\ 12^s$
Synodic period (lunar month): $29^d\ 12^h\ 44^m\ 3^s$
Orbital inclination: 5°8′43″

Rotation period: 27.32 days
Tilt of axis: 6°41′ (with respect to orbital plane)

Diameter: 3,476 km (0.273 $D_\oplus$)
Mass: 7.35×10^{25} grams (0.0123 $M_\oplus$)
Density: 3.34 grams/cm³
Surface gravity: 0.165 Earth gravity
Escape speed: 2.4 km/sec
Surface temperature: 400 K (day side); 100 K (dark side)
Albedo: 0.07

Figure 10.1 The full Moon. *This high-quality photograph was taken through a telescope on the Earth. (Lick Observatory, © Regents University of California)*

Figure 10.2 The Moon as seen from space. *Here is a view obtained by one of the* **Apollo** *missions, showing portions of the near (left) and far (right) sides. On the left horizon is Mare Crisium, and in left center are Mare Marginis (upper) and Mare Smythii (lower). No maria are seen in the right-hand half of this image; there are almost none on the entire lunar far side. (NASA)*

only a very tenuous atmosphere (because the escape velocity is so low that all the volatile gases were able to escape long ago), erosion due to weathering does not occur on the Moon, so large features such as craters and mountain ranges retain a jagged appearance when seen from afar. The surface consists everywhere of loosely piled rocks ejected and redistributed by impacts. Other types of features seen on the Moon include **rays,** light-colored streaks emanating from some of the large craters, and **rilles,** winding valleys that resemble earthly canyons.

The very thin atmosphere of the Moon is known to be present through Earth-based spectroscopic measurements. The principle gases detected are hydrogen, helium, and argon. None can remain; the Moon's low escape speed and high daytime surface temperature ensure that these gases escape readily. Thus their presence indicates that they are constantly replenished from some source. It is thought most likely that these gases are gradually seeping out from the Moon's interior.

The Moon's mass is only about 1.2 percent of the mass of the Earth, and its radius is a little more than one-fourth the Earth's radius. The Moon's density, 3.34 grams per cubic centimeter, is below that of the Earth as a whole, more closely resembling the density of ordinary surface rock. This indicates that the Moon probably does not have a large, dense core, and that it therefore has a low iron content. Infrared measurements showed, even before the space program allowed direct measurements, that the lunar surface is subject to hostile temperatures, ranging from 100 K ($-279°$F) during the two-week night to 400 K ($261°$F) during the lunar day.

Of course, all of the long-range studies of the Moon were almost instantly antiquated by the success of the space program, which has featured extensive exploration of the Moon, first by unmanned robots, and then by manned landings (Figs. 10.3 and 10.4). The historic first manned landing occurred on July 20, 1969. This was the *Apollo 11* mission; it was followed by

ASTRONOMICAL INSIGHT 10.1

Lunar Cartography*

All the major topographic features on the Moon, and many minor ones, have names. These names are used in maps of the Moon and are recognized by the international community of astronomers. It is interesting to examine the history of how they were assigned.

Galileo was the first to name lunar features, because he was the first to look at the Moon through a telescope that enabled him to see them. He is responsible for the terms *maria* (the dark areas that he thought were seas) and *terrae* (the highlands). Following Galileo's lead, as telescope technology improved and more and more people examined the Moon, a number of early lunar cartographers chose names for prominent craters, mountain ranges, and other regions. Some of the names that were assigned, such as those devised by the court astronomer to the king of Spain (who named features after Spanish nobility), were doomed to be forgotten. Others, like those based on the suggestion of a German astronomer that lunar mountain ranges be named after terrestrial features, have persisted to the present time.

The most influential of the lunar cartographers of Medieval times was the Italian priest

*Based on information from El-Baz, F. 1979. Naming Moon's features created oceans of storms. *Smithsonian* 9(10):96.

Joannes Riccioli. With his pupil, Francesco Grimaldi, Riccioli named maria after human moods or experiences, and craters after famous scientists. Thus, we have Mare Imbrium (Sea of Showers), Mare Tranquilitatis (Sea of Tranquility, where the first manned Moon landing occurred), and Mare Serenitatis (Sea of Serenity); and craters named Tycho, Hipparchus, and Archimedes. There were political overtones to the nomenclature of Riccioli and Grimaldi, who published their map in 1651, when the heliocentric hypothesis was not something that people publicly embraced. With this in mind, Riccioli and Grimaldi gave geocentrists like Ptolemy and Tycho large, prominent craters but assigned Galileo only a small one, and placed the name of Copernicus on a crater in Mare Nubium (Sea of Storms or Sea of Clouds).

Other names were added during the eighteenth and nineteenth centuries, but until 1921 there was no formal international agreement on a specific, uniform naming system. In that year, the newly formed International Astronomical Union designated a committee to oversee and standardize lunar nomenclature. Things were uneventful thereafter, until 1959, when the Soviets obtained the first photographs of the far side of the Moon and promptly named numerous features after prominent Soviets. Many of these names (but not all) were later approved by the International Astronomical Union. Perhaps the most controversial was the Sea of Moscow, which reportedly was finally approved (amid laughter) when the

Soviet representative declared—in response to criticism that it was inconsistent with tradition to name a mare after a city—that Moscow is a state of mind just as tranquility and serenity are.

Numerous additional features have been mapped and named as the sophistication of lunar probes has improved and as the manned exploration of the Moon has proceeded. The *Apollo* astronauts assigned rather colloquial names to features of each landing site, and because of all the attendant publicity these names became well known before there was time for the International Astronomical Union even to consider them for approval. Most of them were based on characteristics of the features themselves, such as a terraced crater called Bench and one with a bright rim dubbed Halo. Groups of craters were named for their pattern, such as a pair called Doublet and a group named Snowman, which included individual craters with names such as Head. Some of the features of the terrain mapped by the *Apollo* astronauts received names of prominent scientists, in keeping with the tradition established by Riccioli and Grimaldi more than 300 years earlier. Many of the names invented by the *Apollo* astronauts were eventually approved by the International Astronomical Union, as were those of a dozen craters named for American and Soviet astronauts and cosmonauts. Seven small craters have recently been named for the seven American astronauts who lost their lives in the

(continued on next page)

five more manned landings, the last being *Apollo 17,* which took place in late 1972. Each mission incorporated a number of scientific experiments, some involving observations of the Sun and other celestial bodies from the airless Moon, but most devoted to the study of the Moon itself.

A Battle-Worn Surface and a Dormant Interior

Viewed on any size scale, from the largest to the smallest, the Moon's surface is irregular, marked throughout by a variety of features. We have already mentioned

Figure 10.3 Man on the Moon. *The* Apollo *missions, six of which included successful manned landings on the Moon, represent humankind's only attempt so far to visit another world. (NASA)*

Figure 10.4 The Lunar Rover. *The later* Apollo *missions used these vehicles to travel over the Moon's surface, allowing the astronauts to explore widely in the vicinity of the landing sites.*

the maria, the large, relatively smooth, dark areas (Fig. 10.5). The maria appear darker than their surroundings because they have a relatively low albedo (that is, they reflect less sunlight). Despite their smooth appearance relative to the more chaotic terrain seen elsewhere on the Moon, the maria are marked here and there by craters.

Outside the maria, much of the lunar surface is covered by rough, mountainous terrain. Even though the maria dominate the near side of the Moon, there are almost none on the far side, and the highland regions actually cover most of the lunar surface.

Craters are seen everywhere (Figs. 10.6 and 10.7). They range in diameter from hundreds of kilometers to microscopic pits that can be seen only under intense magnification. In some regions, the craters are so densely packed together that they overlap. The lack of erosion

Figure 10.6 Craters on the Moon.   *This fine Earth-based photograph reveals the details of a heavily cratered region on the lunar surface. (Palomar Observatory, California Institute of Technology)*

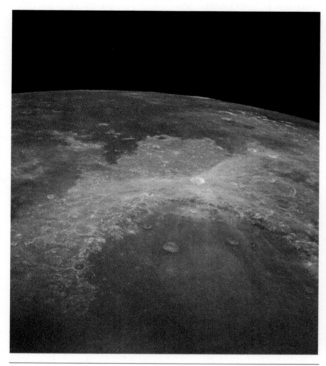

Figure 10.5 The lunar "seas."   *Here is a broad vista encompassing portions of three maria. Mare Crisium (foreground); Mare Tranquilitatis (beyond Mare Crisium); and Mare Serenitatis (on the horizon at upper right). These relatively smooth areas are younger than most of the lunar surface, having been formed by lava flows after much of the cratering had already occurred. (NASA)*

Figure 10.7 *Apollo 12* descending near the crater Herschel   *(NASA)*

on the Moon allows craters to survive for billions of years, providing plenty of time for younger craters to form within the older ones. The fact that relatively few craters are seen in the maria indicates that the surface in these regions was transformed after most of the cratering had already occurred.

All of the craters on the Moon are **impact craters,** formed by collisions of interplanetary rocks and debris with the lunar surface, rather than by volcanic eruptions. This conclusion is based on the shapes of the craters, the central peaks in some of them, and the trails left by **ejecta,** or cast-off material created by the impacts (Fig. 10.8). The rays seen stretched away from craters are stings of smaller craters formed by the ejecta from the large, central crater.

Some of the lunar mountain ranges reach heights greater than any on Earth. Their contours are rounded by a form of erosion, due to continual impacts from space throughout their long history.

The **rilles** are rather interesting features, resembling dry riverbeds (Fig. 10.9). Apparently they were formed by flowing lava rather than water. In some cases, there are even lava tubes that have partially collapsed, leaving trails of sinkholes. The rilles and the maria indicate that the Moon has undergone stages when large portions of its surface were molten.

The *Apollo* astronauts found a surface strewn in many areas with loose rock, ranging in size from pebbles to boulders as big as a house (Figs. 10.10 and 10.11). The rocks generally show sharp edges, due to the lack of erosion, and occasional cracks and fractures. In most cases the large boulders appear to have been ejected from nearby craters, and are therefore thought to represent material originally from beneath the surface.

Almost 800 pounds of small lunar rocks were returned to Earth by the *Apollo* astronauts, giving scientists an opportunity to study the lunar surface characteristics in detail comparable to that possible for Earth rocks and soils (Fig. 10.12). Thus, extensive chemical analysis and close-up observation of rocks in place on the Moon were both possible. The rock samples from the Moon are presently housed in numerous scientific laboratories around the world, where analysis continues. Specimens have also found their way into museums, and at the Smithsonian's Air and Space Museum in Washington, D.C., a lunar rock can be touched by visitors.

The lunar soil, called the **regolith,** consists of loosely packed rock fragments and small glassy mineral deposits probably created by the heat of meteor impacts. In addition to the loose soil, a few distinct types of surface specimens were recognized on the basis of morphology. The most common of these are the **breccias** (Fig. 10.12), which consist of small rock fragments ce-

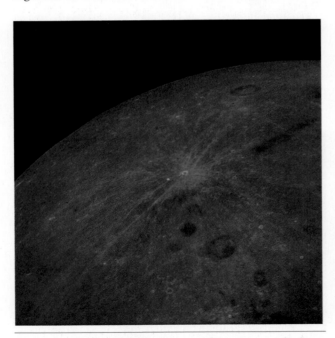

Figure 10.8 A crater with ejecta. *This crater on the lunar far side is a good example of a case in which material ejected by the impact has created rays of light-colored ejecta. Close examination of such features often reveals secondary craters, formed by the impacts of debris blasted out of the lunar surface by the primary impact. (NASA)*

Figure 10.9 Rilles. *Here is shown the intersection of two rilles, riverlike valleys on the Moon that are thought to have formed as a result of lava flows early in the Moon's history. (NASA)*

Figure 10.10 A field of boulders.   *Rocks in a wide variety of sizes are strewn over much of the Moon's surface. Most have been blasted out of the surface by impacts. (NASA)*

Figure 10.11 A large boulder.   *Rocks on the lunar surface range in size from tiny pebbles to massive objects like this. (NASA)*

mented together and resembling chunks of concrete. There are similar kinds of rock on Earth, except that those on Earth are formed in streambeds where water plays a role in shaping them. The lunar breccias contain jagged, sharp rock fragments, and they are probably

fused together by pressure created in meteor impacts.

All lunar rocks are pitted, on the side that is exposed to space, with tiny craters called **micrometeorite craters.** These are formed by the impact of tiny bits of interplanetary material no bigger than grains of dust.

Figure 10.12 Moon rocks.   *One of the thousands of lunar samples brought back to Earth by the **Apollo** astronauts, the rock at left is an example of a breccia. At right is a thin slice of a Moon rock, illuminated by polarized light, which causes different crystal structures to appear as different colors. (NASA)*

Radioactive dating techniques showed the lunar rocks to be very old by earthly standards: as old as 3.5 to 4.5 billion years. Rocks from the maria are not quite so old, but still date back to 3 billion years ago or earlier.

Some of the experiments done on the Moon by the *Apollo* astronauts were aimed at revealing the interior conditions by monitoring seismic waves caused by earthquakes on the Moon. Therefore, the astronauts carried with them devices for sensing vibrations in the lunar crust, and, because it was not known whether natural moonquakes occurred frequently, they also brought along devices for thumping the surface to make it vibrate. It turned out that natural moonquakes do occur, although not with great energy. The seismic measurements continued after the *Apollo* landings, with data radioed to Earth by instruments left in place on the lunar surface.

The measurements showed that the regolith is typically about 10 meters thick and is supported by a thicker layer of loose rubble. The crust is 50 to 100 kilometers thick at the *Apollo* sites, and may be somewhat thicker than this on the far side where there are no maria (Fig. 10.13). Beneath the crust is a mantle, consisting of a well-defined lithosphere, which is rigid, and beneath that an asthenosphere, which is semiliquid. The innermost 500 kilometers consist of a relatively dense core, but not as dense as that of the Earth.

The seismic measurements indicate no truly molten zones in the Moon at the present time, although temperature sensors on the surface discovered a substantial heat flow from the interior. This is probably caused by radioactive minerals below the surface.

There is no detectable overall magnetic field, further evidence against a molten core. The chief distinction among the internal zones in the lunar interior is density, with the densest material closest to the center. Thus moderate differentiation has occurred in the Moon, implying that it was once at least partially molten.

There is no trace of present-day lunar tectonic activity, perhaps the greatest single departure from the geology of the Earth. The force that has had the greatest influence in shaping the face of the Earth has no role today on the Moon. Instead, the lunar surface is entirely the result of the way the Moon was formed (which may have involved some tectonic activity in the early stages) and the manner in which it has been altered by lava flows and by the incessant bombardment of debris from space.

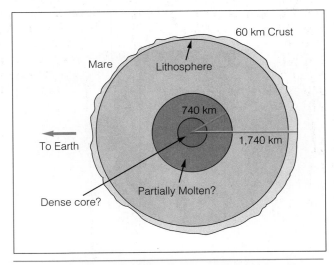

Figure 10.13 The internal structure of the Moon. *This cross-section illustrates the interior zones inferred from seismological studies. The existence of a dense core is not certain. Note that the maria lie almost exclusively on the side facing the Earth, where the lunar crust is relatively thin.*

ORIGIN OF THE MOON

The Moon's beginnings are much more obscure than the Earth's, despite the close-up examination of the Moon afforded by the *Apollo* missions. The reasons for the difficulty in explaining how the Moon formed lie in the very unusual chemical composition of the Moon, and the large amount of angular momentum of the Earth-Moon system. Recall that angular momentum is related to the mass, orbital speed, and separation of two bodies in mutual orbit. In comparison to other planet-satellite pairs in the solar system, the Moon is very large relative to its parent planet, and the Earth-Moon system has a high angular momentum. This has been difficult to explain in any theory calling for the formation of the Moon by splitting off from the Earth, for example, or in the hypothesis that the Earth and Moon simply formed together as a double planet. In other words, it has been difficult to explain why the material that formed the Moon remained separate from the Earth, instead of coalescing with it.

The chemical composition of the Moon is characterized by a low overall abundance of heavy elements

ASTRONOMICAL INSIGHT 10.2

Colonizing the Moon

 The history of human exploration of the Moon has taken a course that few would have predicted some 30 years ago, when the space age was in its infancy. Responding to the Soviet challenges created first by the launch of *Sputnik* in 1957, and then the first views of the lunar farside (by unmanned Soviet spacecraft, in 1959), the U.S. space effort went into high gear during the 1960s. Throughout that decade, the primary goal, as set by President John F. Kennedy, was to send people to the Moon. This was a highly visible and clearly identified goal, and it was also a technological challenge that was difficult but within the reach of the U.S. aerospace industry. Sending people to the Moon combined political and technological advantages that stimulated an all-out effort and won the support of the American public.

Unfortunately, no plans were made for the post-Apollo era. The entire goal was to get people to the Moon and back. No follow-up exploration or long-term colonization of the Moon was planned, even though these might have seemed natural steps to take after the initial successes. The goal of landing humans on the Moon and bringing them back safely was achieved with the success of Apollo 11 in 1969, and no one has been on the Moon since the Apollo 17 mission in 1973. Budget constraints caused the early cancellation of the Apollo program, and the U.S. space effort went into unmanned exploration of the planets and remote observatories in Earth orbit.

Now, nearly 30 years after President Kennedy's announcement of the lunar program, and 20 years after the landing of Apollo 11 on the Moon, the United States has once again unveiled plans for sending people to the Moon. The hope among scientists is that this time the goal will be long-term and systematic exploration and discovery, rather than a single-purpose, partially political mission.

No definite plans have been made, and the program faces serious challenges in Congress. NASA's reputation has been hurt by a series of problems (which somehow always seem to overshadow the successes), and the new lunar exploration is linked to the very expensive Space Station and Mars Exploration initiatives, both of which are receiving careful scrutiny because of their immense cost. Nevertheless, it is exciting to think that humans may soon walk on the Moon again.

From the scientific point of view, the Moon has much to offer. In addition to furthering studies of the Moon itself and its many mysteries, establishment of a base there, as has been proposed, would benefit other scientific areas as well. Perhaps astronomers stand to gain more than scientists in most other fields. For example, it would be highly desirable to build large telescopes on the Moon, perhaps on the far side, to avoid contamination by Earth light and human radio noise. Enormous instruments could be built in the weak lunar gravity, which would make all operations easier. Of course, the lack of a lunar atmosphere would also be beneficial, allowing astronomers to achieve very sharp images and to observe wavelengths that do not penetrate the Earth's atmosphere. Combining these capabilities with the flexibility of having humans present who can maintain and repair and update the instruments would create an enormous advance for our science.

But perhaps the greatest benefit of establishing a lunar base would be more philosophical. We are an Earth-bound species, tied intimately to our small planet and its limited resources. Many have said and written that it is human destiny to expand from our planet, to colonize the solar system and perhaps the reaches beyond. To those who believe this, a lunar base is an essential first step. We cannot talk about permanent habitats on other planets unless we can show that human beings can live independent of Earth over long periods of time. The new lunar exploration program has a chance to demonstrate this.

such as iron, and a higher relative abundance of the **refractory** elements, those that do not vaporize easily. The readily vaporized **volatile** elements are under-abundant in comparison with the Earth, and water is almost completely absent. These contrasts might suggest that the Moon formed somewhere else in the solar system rather than near the Earth, yet there are some strong similarities, such as the relative proportions of different isotopes of oxygen, that argue for formation in the same general vicinity. It seems that most of the chemical properties of the Moon could be explained if the Earth and Moon formed out of very similar material originally, but then somehow the Moon lost much of its iron and was subjected to substantial heating early in its history.

Traditionally there have been three classes of models for lunar formation: (1) the fission hypothesis, in which the Moon somehow split off of the Earth; (2) the capture theory, in which the Moon formed elsewhere and was then captured by the Earth's gravitational field; and (3) the coeval formation, or double planet model, in which the two bodies formed together. Each of these has always had serious difficulties. The fission theory lacks a mechanism for causing the split of the young Earth, lacks an explanation for the high angular momentum of the Earth-Moon system, and cannot easily explain the low iron content of the Moon or the other chemical dissimilarities. The capture hypothesis can easily account for the high angular momentum, since the Moon would have approached the Earth with a substantial velocity that would have been transformed into orbital angular momentum upon capture, but that same velocity is the downfall of the theory, for it is difficult or impossible for an initially free Moon to be slowed enough to be caught by the Earth's gravity. This model also has difficulty explaining the chemical similarities between the Earth and the Moon, as well as the low iron content of the Moon. The coeval formation hypothesis has grave problems with the chemical contrasts between Earth and Moon, since in this view the two are thought to have formed together from the same material, and this theory also founders on the angular momentum problem.

A recent suggestion appears to resolve all of the problems related to angular momentum and chemical composition, and is therefore becoming the favorite among lunar scientists. This idea has some elements of the capture model, in that it invokes a large body approaching the Earth, but there the similarity ends. In this model, the body that approaches is a large planetesimal, at least 10 percent the mass of the Earth (hence at least the size of Mars). It is proposed that this large body, already differentiated and having an iron core, collided with the young Earth at a grazing angle (Fig. 10.14) and was partially vaporized by the energy of the impact. Material from the body's iron-poor mantle formed a hot disk around the Earth, which subsequently coalesced to form the Moon. The result is a Moon having a large quantity of angular momentum, derived from the speed (about 10 km/sec) of impact; a low overall iron abundance, due to the fact the incoming planetesimal had already differentiated and only its mantle material formed the Moon; and a very low content of volatile elements, since these were vaporized by the heating that occurred on impact, and lost to space.

It may seem contrived to invoke a singular event such as a collision with a Mars-size planetesimal, but studies of the process by which early solar system material condensed into solid form and then built up planetesimals shows that this hypothesis is quite reasonable. There should have been a period of time when most of the preplanetary material was concentrated in a few rather large planetesimals, all orbiting the Sun in the same direction. The final formation of the planets came about as these large planetesimals merged, so the early history of each planet would have involved collisions between large bodies. Evidence for this is found in some other anomalous planetary characteristics: Venus has a backwards, slow rotation, for example, and Uranus is tipped over on its side.

HISTORY OF THE MOON

The Moon's evolution after its formation has not been as difficult to deduce. Detailed study of the surface structure of the Moon, along with analyses of surface rocks and information on the interior structure gleaned from seismic data, have revealed a comprehensive picture.

The Moon underwent a period when it was molten to a depth of about 400 km (this was another factor that was more easily explained by the planetesimal-collision formation model than by the others). As the crust cooled and solidified, impacts began to sprinkle it with craters. Within the first 1 or 2 billion years, extensive lava flows occurred on the surface, welling up in regions where the crust was relatively thin. Much of one side of the Moon was covered by the flows, which we know today as maria. The rate of impacts from space began to taper off at about the time the maria formed; hence, they have fewer craters than the highlands, where the high concentration of craters records the high frequency of collisions suffered by the young Moon. Impacts have continued throughout the history of the Moon, but following that early, intense period, the rate has been much slower and quite steady.

The Moon was much closer to the Earth when it formed than it is today, but tidal stresses exerted on it by the Earth have gradually slowed the Moon's spin,

causing it to retreat from the Earth as angular momentum due to its spin has been converted into orbital angular momentum. The Moon's rotation slowed to the point where it keeps one face toward the Earth as it orbits; this is the face containing the maria, whose high density causes the Moon to be slightly asymmetric in its mass distribution. The relatively higher density of the side with the maria caused this side to "lock in" facing the Earth; thus it is no coincidence that the near side is the one with the maria.

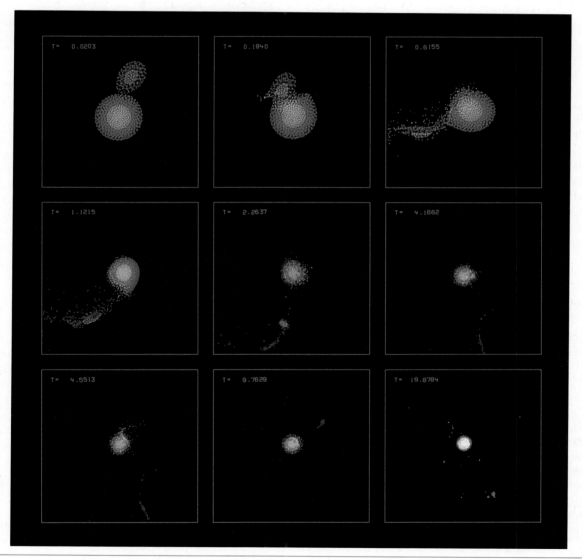

Figure 10.14 A computer simulation of the Moon's formation.   *In the newest theory of lunar formation, it is envisioned that a planetesimal as massive as Mars or larger collided with the young Earth and was partially vaporized on impact. This sequence of computerized illustrations shows how the resulting debris could then have coalesced to form the Moon. (W. Benz and W. Slattery, Los Alamos National Laboratories)*

Geologically, the Moon now is quite inert, having no molten core and insufficient internal heat to drive such processes as tectonic activity. The only changes that will occur over the next few billion years will be the addition of new craters.

MERCURY

The Moon is very large relative to its parent planet, and as we have just discussed, it is perhaps only an accident of fate that the Moon did not become a separate, Sun-orbiting planet rather than a trapped satellite of the Earth. If it had done so, then it probably would have very closely resembled the innermost planet, Mercury. The many similarities between the Moon and Mercury justify grouping the two bodies together in this chapter.

Mercury (Table 10.2) has been known since ancient times, even though it is rather difficult to observe it from the Earth (Fig. 10.15). Its greatest elongation is only 28° (Fig. 10.16), so the planet can only be seen shortly before sunrise or just after sunset. For some time, the ancient observers thought Mercury was two different planets, because of the fact that it alternately appeared as a morning object and as an evening object.

THE EARTH-BASED VIEW

Few people ever see Mercury, despite the fact that it is rather bright (a little brighter than Sirius, the bright-

Figure 10.15 Mercury as seen from the Earth. *The planet is never far from the Sun and is therefore difficult to observe. (NASA)*

est star in the sky). The difficulty of observing it has made Mercury the most enigmatic of the terrestrial planets, with little of its nature observable by Earth-based telescopes. The best photographs are taken during daylight so that Mercury can be viewed well above the distortion caused by Earth's atmosphere near the horizon, but are degraded by the high degree of atmospheric turbulence or poor "seeing" that prevails during the day due to the Sun's heating of the atmosphere. These photographs show little except a fuzzy disk with a hint of surface markings possibly resembling the lunar maria.

The orbit of Mercury is unusually eccentric; that is, its elliptical shape is more highly elongated than those of all the other planets except distant Pluto, and the Sun is quite far off center in Mercury's orbit. Mercury is over 50 percent father away from the Sun at its greatest distance than when it is closest. As we will see, this variation in distance from the Sun has had some important consequences. The orbit is also tilted by the unusually large angle of 7° with respect to the ecliptic. At an average distance from the Sun of 0.387 AU, Mercury has an orbital period of 88 days, the shortest of all the planets.

Mercury is small, as planets go, with a diameter only slightly larger than that of the Moon. At least three satellites in the solar system, Ganymede and Callisto of Jupiter and Titan of Saturn, are bigger. The mass of Mercury has been determined from its gravitational influence on nearby objects such as comets, and especially the asteroid Icarus, which passed nearby in

Table 10.2 Mercury

Orbital semimajor axis: 0.387 AU (57,900,000 km)
  Perihelion distance: 0.308 AU
  Aphelion distance: 0.467 AU
Orbital period: 87.97 days (0.241 years)
Orbital inclination: 7°0'15"

Rotation period: 58.65 days
Tilt of axis: 28°

Diameter: 4,864 km (0.382 $D_{\oplus}$)
Mass: 3.33 × 10^{26} g (0.0558 $M_{\oplus}$)
Density: 5.50 g/cm³
Surface gravity: 0.38 Earth gravity
Escape speed: 4.3 km/sec

Surface temperature: 700 K (day side); 100 K (dark side)
Albedo: 0.106

Satellites: None

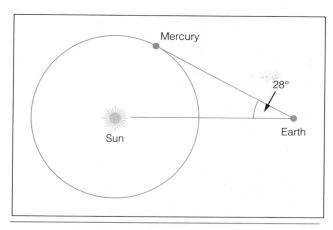

Figure 10.16 Mercury's greatest elongation. *The farthest the planet ever gets from the Sun is 28°, which means that it always rises and sets within 2 hours of the Sun.*

1968. More accurate measurements were later possible, when the U.S. spacecraft *Mariner 10* flew by Mercury. Mercury's mass is about 5.6 percent of the Earth's mass, and its density is 5.4 grams/cm³, very similar to that of the Earth, implying that Mercury probably has a large, dense core.

Orbital Eccentricity and Rotational Resonance

As we have seen, observations of Mercury are difficult to make, and no surface features on the planet can be seen distinctly. Careful observations of the fuzzy markings that do appear led some observers in the past to the conclusion that Mercury was in synchronous rotation, always keeping the same side facing the Sun. As close to the Sun as Mercury is, this was not a surprising idea, for the tidal forces acting on it are immense.

Radio observations of Mercury done in the early 1960s allowed its surface temperature to be calculated, through the use of Wien's law (see Chapter 6). The daylit side was hot, as expected, with a temperature of about 700 K. Because observers thought that the other side never saw the light of day, they expected to find very low temperatures there. It was a surprise, therefore, when a temperature of about 100 K was deduced for the dark side. This is certainly cold, but not as cold as it would have been if this side of the planet never faced the Sun.

Radar observations soon provided the solution to this mystery. The Doppler shift of radar signals re-

flected off of Mercury showed that the planet's rotation period is shorter than its 88-day orbital period. Mercury is not in synchronous rotation, but instead rotates once every 59 days, so that all portions of its surface are exposed to sunlight some of the time.

But why 59 days? This was a bit of a puzzle in itself, because most planets spin much more rapidly. It was noticed that 59, or, more precisely, 58.65 days, is precisely two-thirds of the orbital period of 88 (actually 87.97) days. Thus, Mercury rotates exactly three times for every two trips around the Sun (Fig. 10.17).

This is no mere coincidence; clearly the gravitational tidal forces exerted by the Sun are at work. The key is what happens at the point where Mercury is closest to the Sun, called **perihelion,** for it is here that the tidal forces are the strongest. If Mercury has one side denser than the other, as the Moon does, then the tidal forces exerted by the Sun will act to ensure that this side is aligned with the Sun's direction when Mercury is at perihelion. One way to accomplish this, of course, could be synchronous rotation, so that this side would always point toward the Sun, but because Mercury's orbit is

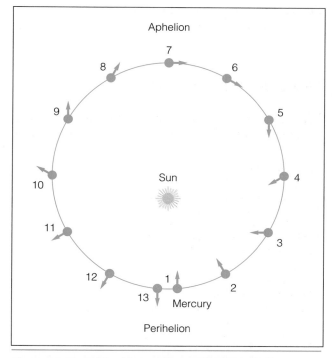

Figure 10.17 Spin-orbit coupling. *This sketch shows how Mercury spins one-and-a-half times while completing one orbit around the sun. At perihelion it always has the heavy side facing directly toward or away from the Sun.*

so elongated, the urge to keep its heavy side facing the Sun is most significant at perihelion.

The planet spins one-and-a-half times during each orbit; thus the heavy side either faces the Sun or exactly the opposite direction at each perihelion passage. In either case, the tidal force on Mercury is balanced, so that there is no tendency to alter the rotation further. Apparently the planet once rotated much faster, but whenever it passed through perihelion with the heavy side pointed in some random direction, tidal forces exerted a tug that tended to change its spin. This went on until the rotation slowed to the present rate, so that the tidal forces were always balanced when Mercury was at perihelion. If the orbit had not been so eccentric, or if Mercury had been symmetric instead of having a relatively dense side, the tidal forces would have been more uniform throughout, and Mercury might today be in synchronous rotation.

The 3:2 relationship between Mercury's orbital and spin periods is an example of **spin-orbit coupling,** a general term applied to any situation where the spin of a body has been modified by gravitational forces so that a special relationship between the orbital and the spin periods is maintained. Synchronous rotation is the most common example, and we will find several cases of it as we explore the rest of the solar system.

One remaining question regarding Mercury and its spin is why the planet should be asymmetric, with one side denser than the other. We will find a hint of a possible answer in a later section.

The combination of orbital and rotation speeds on Mercury produces a very unusual occurrence for anyone who might visit this planet. At closest approach to the Sun, when Mercury is moving most rapidly in its orbit, the orbital speed is actually greater than the rotation speed at its surface. For a short time (lasting a few hours), the Sun would appear to turn around and move backward across the sky, from west to east. Imagine the difficult time our ancestors would have had explaining this retrograde motion of the Sun if a similar phenomenon had occurred on the Earth!

RENDEZVOUS WITH MARINER

Mercury has not drawn the concentrated attention of the U.S. space program that Venus and Mars have, but nevertheless it was observed at close range by one probe, *Mariner 10*. This spacecraft, launched in late

1973, flew by Venus in February of 1974 and then set its course for Mercury. The gravitational pull of Venus was used to modify *Mariner 10*'s path so that this could be done. The spacecraft arrived at Mercury in late March, 1974.

An extraordinary opportunity presented itself when *Mariner 10* reached Mercury and continued into solar orbit. It proved possible to adjust its orbit so that the period was 176 days, exactly twice that of Mercury. This meant that the spacecraft made one orbit about the Sun for every two of Mercury, and the two therefore encountered each other repeatedly (Fig. 10.18). There was sufficient maneuvering gas to operate the instruments aboard *Mariner 10* on each of its next two meetings with Mercury, so there were actually three close encounters during which observations were made. The approaches to the planet were very close on the first and third encounters, when *Mariner 10* passed only a few hundred kilometers above the surface, and much further away on the second flyby, when the spacecraft passed by some 50,000 kilometers away.

There was one drawback to the *Mariner 10* orbit. Although it did allow ample opportunity for close-up observation of Mercury, only one side of the planet

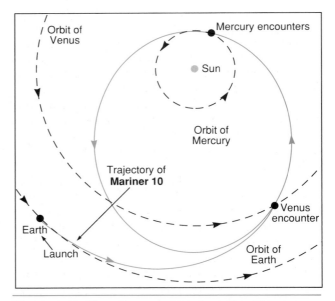

Figure 10.18 The trajectory of *Mariner 10*. *The spacecraft flew by Venus and then encountered Mercury repeatedly because its orbital period was adjusted to equal with twice the orbital period of Mercury.*

could be viewed. Because the spacecraft visited Mercury every other orbit, the same side of the planet was facing the Sun each time, since Mercury rotates three times in every two orbits around the Sun. If *Mariner 10* could have been placed in an orbit with an 88- or 264-day period, it could have viewed opposite sides of Mercury each time the two met. As it was, though, only one side of the planet was charted photographically; close-up examination of the other side will have to wait for some future mission.

MAGNETIC FIELD AND INTERNAL CONDITIONS

The relatively high density of Mercury, known long before the *Mariner 10* encounters, indicated the likelihood of a dense core inside this planet, similar to that of the Earth and Venus. This might have led us to believe that Mercury had a magnetic field, except for the fact that the slow rotation was thought insufficient to produce the internal electrical currents needed to activate a magnetic dynamo. Furthermore, the small mass of Mercury, and the resulting low internal pressure due to gravity, should have allowed the core to cool and solidify, leaving no liquid zone to create a magnetic dynamo.

Thus it was a bit of a surprise when *Mariner 10* detected a magnetic field around Mercury. Its strength is not great, about 1 percent as strong as that of the Earth, but nevertheless it is there. Like the Earth, Mercury has a magnetosphere whose shape is modified by the charged particles flowing past from the Sun. It has been suggested that the core has a liquid zone despite its low temperature because the iron there is combined with another element, such as sulfur, that lowers the melting point.

Scientists concluded that Mercury's core must be relatively large (Fig. 10.19), in order to produce a magnetic field despite the slow rotation of the planet. The core apparently extends about three-fourths of the way out from the center to the surface, whereas the Earth's core is contained within the inner half of its radius.

The implication of such a large core is not only that Mercury is differentiated, but that it contains a relatively high overall abundance of heavy elements, especially iron, to begin with. This says something about its formation, namely, that a greater portion of its volatile, lightweight gases escaped than was the case

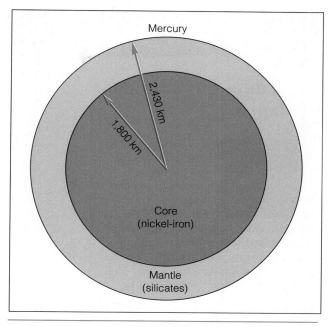

Figure 10.19 Mercury's internal structure.   *The presence of a magnetic field, along with the planet's relatively high density, implies the presence of a large core.*

for the other terrestrial planets. The temperature in the vicinity of Mercury must have been relatively high early in the history of the solar system, higher than at the places where Venus, the Earth, and Mars formed, so some of the volatile elements would have been prevented from condensing simply because it was too hot. It appears that another mechanism for dispersing these elements is needed, however, and one has recently been suggested. A collision with a large planetesimal early in Mercury's lifetime could have vaporized much of the planet's mantle, leaving a smaller planet with a disproportionately large iron core, as is the case today. As in the other cases where well-developed cores exist, Mercury must have been fully molten, staying in that state long enough for differentiation to occur.

Not much is known about the remaining mantle of Mercury, which occupies the outer 25 percent or so of its radius. The density there is lower than in the core, but its other properties are unknown. The photographs taken by *Mariner 10* show no obvious evidence of tectonic or volcanic activity, although we must keep in mind that half of the surface has yet to be examined.

ASTRONOMICAL INSIGHT 10.3

Visiting the Inferior Planets

In Chapter 5 there was some discussion of the mechanics of placing a satellite in the Earth's orbit, but nothing on how to aim a spacecraft that is designed to visit another planet. The principles involved are particularly simple when an inferior planet is to be the target.

The concept of energy was discussed in Chapter 5, where we pointed out that in a system of two orbiting bodies, there is kinetic energy due to the motions of the bodies, and potential energy due to their gravitational attraction for each other. For a planet orbiting the Sun, the greater the total energy (the sum of kinetic plus potential), the larger the orbit. Thus for an object to go from the orbit of the Earth to that of an inner planet, it must lose energy.

A spacecraft sitting on the launchpad is moving with the Earth in its orbit, at a speed of 29 km/sec. To make the rocket fall into a path that intercepts the orbit of an inner planet, we must diminish this speed. Therefore, we launch the rocket backward with respect to the Earth's orbital motion, thereby lowering its velocity with respect to the Sun, and decreasing its orbital energy. If the launch speed is properly chosen, the spacecraft will fall into an elliptical orbit that just meets the orbit of the target planet (see figure). In order to reach Mercury from the Earth, the rocket must

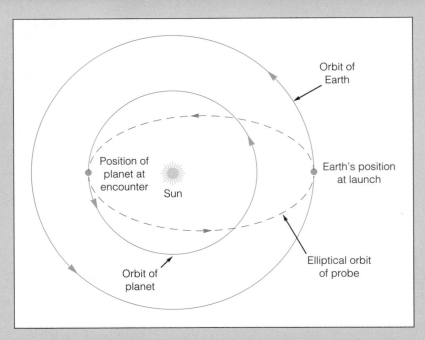

be launched backward with a speed relative to Earth of 7.3 km/sec.

There are, of course, a few additional considerations. For one thing, the rocket has to be launched with enough speed to escape the Earth's gravity. The escape speed for the Earth is 11.2 km/sec, so in fact we have to launch our Mercury probe with a speed in excess of this value, which is more than the speed needed to attain our trajectory to Mercury. The launch speed must therefore be calculated to take the Earth's gravity into account, so that the rocket escapes the Earth, but in the process is slowed just the right amount to give it the proper course. The gravitational pull of the target planet must also be taken into account, for this speeds the spacecraft up as it approaches.

Another important considera-

tion is timing; the target planet must be in the right spot in its orbit at just the moment when the spacecraft arrives. It is for this reason that the term **launch window** is used. The Earth and the target planet must be in specific relative positions at the time of the launch. The interval between launch windows for a given planet is simply its synodic period. For a probe to Mercury, the travel time is about 211 days, so Mercury actually makes a little more than one complete orbit while the probe is on its way.

The gravitational pull of the target planet can be used to good advantage in modifying the orbit of a spacecraft that flies by. If the probe is aimed properly, its trajectory may be altered as it goes by in just the right way to send it on to some other target. This technique was used with *Mariner 10*, first to send it from Venus to

ASTRONOMICAL INSIGHT 10.3

Mercury, and then to modify its orbit again so that it returned to Mercury repeatedly thereafter.

Sending spacecraft to the outer planets is a bit more difficult, because the spacecraft has to have more energy than it gets from the Earth's motion. The launch is therefore made in the forward direction, so that the rocket has the speed of the Earth in its orbit plus its own launch speed with respect to the Earth.

Ironically, the seemingly simple tasks of launching a probe to the Sun is one of the most complex. The most straightforward solution would be to launch the spacecraft backward from the Earth with a speed of 29 km/sec, entirely canceling out the Earth's orbital speed, so that the probe would then fall straight in toward the Sun. This is a prohibitively high launch speed, however, so in practice we will probe the Sun by

first sending the spacecraft out around Jupiter. The spacecraft will fly by the massive planet in such a way (backward with respect to Jupiter's orbital motion) that its own orbital energy is reduced; the craft can then fall into the center of the solar system. A mission called *Ulysses* to explore the outer layers of the Sun in this manner is planned for a late 1990 launch.

SURFACE FEATURES AND GEOLOGICAL HISTORY

The surface of Mercury, at first glance, strongly resembles the lunar landscape (Fig. 10.20). It even looks quite similar on second glance, the differences being revealed only under rather detailed scrutiny.

Perhaps the most obvious distinction is found in the prominent and extensive cliffs, referred to as **scarps** (Fig. 10.21), which exist in many locations and form a network that girdles the planet. These are quite unlike anything seen on the other terrestrial planets or the Moon. Their appearance suggests that Mercury shrank a little after its crust hardened, causing it to shrivel and crack. Another clear distinction is that Mercury has extensive, rolling plains, quite unlike either the maria or the highlands on the Moon.

A less obvious difference between Mercury and the Moon is that on Mercury the craters tend to be more widely separated, with more smooth space between them (Fig. 10.22). Apparently, when impacts occur on Mercury and form craters, the ejecta do not travel as far as on the Moon, so there are less extensive rays or other ejecta debris around the large craters (Fig. 10.23). The reason for this is clear: Mercury has a higher surface gravity than the Moon, by about a factor of 3, so that the rubble blasted out of the surface by an impact does not travel as far before falling back down. The craters on Mercury also tend to have a flatter appearance than those on the Moon, for the same reason.

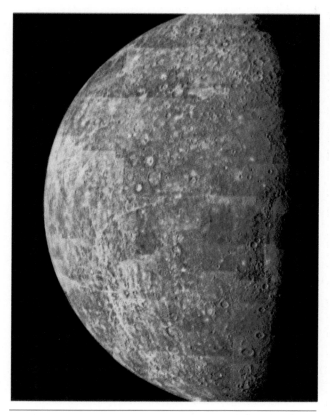

Figure 10.20 A photomosaic of Mercury. *This is a composite image of a large portion of the side of Mercury that was observed by* Mariner 10. *The brownish color has been artificially introduced to simulate the actual color that would be seen by the human eye. (NASA)*

There are extensive smooth areas on Mercury which were apparently formed by lava flows, much like the lunar maria.

One very large impact crater was seen by *Mariner 10,* just at the dividing line between daylight and darkness (Fig. 10.24). The basin, formed by the tremendous crash of a massive body hitting the surface, resembles a giant bull's-eye, with concentric rings around it, and it extends over a diameter of some 1,400 kilometers. The gigantic crater, called Caloris Planitia, happens to lie on the side of Mercury that is facing the Sun at perihilion every other orbit (Fig. 10.25). (The name Caloris was given to the crater for this reason, because this position is the hottest place on Mercury every other time it passes close to the Sun.) We noted earlier that Mercury is apparently lopsided, with an asymmetric distribution of mass; the position of this huge impact basin suggests that perhaps the object that crashed into Mercury was responsible for making the planet denser on one side.

Directly opposite Caloris Planitia, on the far side of Mercury, is a curiously wavy region, so strange and unprecedented in appearance that is called the **weird terrain.** The rippled appearance there may have been created by seismic waves that raced around the planet when the impact occurred that formed Caloris Planitia. When the waves met on the far side of the planet, they distorted the surface there, resulting in the odd terrain seen today. Similar, but less prominent, features are seen on the Moon, at positions opposite some of the largest impact craters. On alternate orbits of Mercury about the Sun, when Caloris Planitia is on the dark side, the weird terrain is at the point directly beneath the Sun, and temporarily holds the dubious distinction of being the hottest place on the planet.

Infrared measurements made by *Mariner 10* indicate that there is a fine layer of dust on Mercury, much like that on the Moon. The composition of the surface rocks is unknown, but probably not identical to lunar surface rocks, since those on Mercury show a higher albedo.

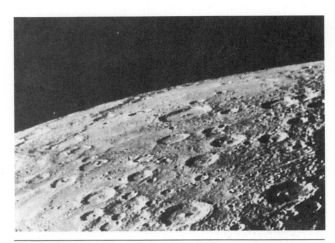

Figure 10.21 A Scarp System on Mercury. *The lengthy series of cliffs extending along the upper left is one of many on the planet's surface. (NASA/JPL)*

Figure 10.22 Craters on Mercury. *Because Mercury has a greater surface gravity than the Moon, impact craters have lower rims and are shallower, and ejecta do not travel as far. (NASA/JPL)*

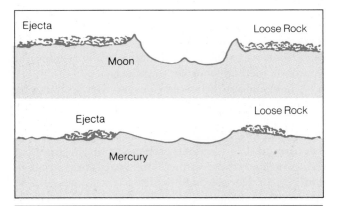

Figure 10.23 Crater formation on the Moon and on Mercury. *The Moon's lower surface gravity allows higher rim walls to remain, and allows ejecta to travel farther from the point of impact than on Mercury.*

Despite the combination of relatively low surface gravity and high temperature, which allowed nearly all the gases present on Mercury to escape long ago, a trace of an atmosphere does exist. *Mariner 10* detected very small quantities of a number of gases, and recently (in 1988) sodium and potassium were detected through Earth-based measurements. The amounts are small (less than 20,000 atoms per cubic centimeter, or about 10^{15} times less than the atmospheric density at sea level on the Earth). These atoms in the atmosphere of Mercury are swept away by the solar wind, and therefore must be replenished in order to be present in the observed quantities. It is though that these gases most likely originate from crustal rocks, which may simply be hot enough that the relatively volatile gases such as sodium and potassium are literally baked out. The presence of a very thin atmosphere on Mercury provides another point of similarity between this smallest terrestrial planet and the Moon.

All these data indicate that the evolution of Mercury's surface must have been quite similar to that of the Moon. Mercury and the Moon went through comparable stages of accretion and melting at the beginning, followed by hardening of the crust along with major lava flows (perhaps triggered by large meteorite impacts), and succeeded finally by a long period of cooling and decreased cratering.

The interiors of the two bodies are very different, however, as a result of contrasts in their formation processes. Whereas the Moon has a low overall iron content and only a modest core, Mercury has a very high iron content and a huge core. Ironically, it is suggested that both of these characteristics are the results of collisions with major planetesimals. In the

Figure 10.24 Caloris Planitia. *This photo shows half of the immense impact basin known as Caloris Planitia. This region is directly facing the Sun at perihelion on every other orbit. (NASA/JPL)*

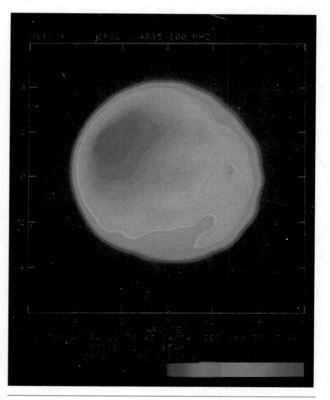

Figure 10.25 Hot spot on Mercury. *As explained in the text, the two points on Mercury's surface which face the Sun at alternating perihelion passages receive much more solar radiation than the rest of the planet's surface, and are therefore much hotter. This radio image of Mercury, in which false colors have been used to indicate relative temperatures, clearly shows one of the two hot spots (the other is on the far side of the planet). (J. O. Burns, University of New Mexico).*

Moon's case, the remaining body is thought to consist of mantle material from an object that had already differentiated; in Mecury's case, the planet may be the core that was left after much of its mantle was vaporized in a collision.

We are not sure just when the tremendous impact that created Caloris Planitia took place, nor how long ago tidal forces succeeded in locking the planet's rotation into its 3:2 coupling with the orbital period. We can surmise that there must have been substantial internal heating before then, as tidal forces exerted by the Sun created internal friction. Perhaps there is still some excess internal heating caused by tidal forces, which helps to maintain the large molten core needed to create the planet's magnetic field.

PERSPECTIVE

Among the terrestrial planets, the Moon and Mercury are superficially very close to being twins, although there are substantial differences in overall composition and internal structure. In this chapter, we have discussed the similarities and the differences between these two airless worlds, finding them to be very different from the Earth. Whereas the Earth has an atmosphere and is very active geologically, the Moon and Mercury have tenuous atmospheres and are relatively quiescent internally.

We now turn our attention to the intermediate terrestrial planets, Venus and Mars, which have significant atmospheres and fall between the Earth and the small terrestrials in geological activity. We will begin with Venus, which in some ways very closely resembles the Earth, but which in others is remarkably different.

SUMMARY

1. The Moon is an airless, rocky body having a mass of just over 1% of the Earth's mass and a diameter of about one-quarter that of the Earth.

2. The lunar surface consists of relatively smooth areas called maria and mountainous regions and is marked everywhere by impact craters.

3. The lunar soil is called the regolith, and rocks on the surface are all igneous, mostly silicates, with low abundances of volatile gases.

4. Seismic data show that the Moon has a crust 50 to 100 kilometers thick, a mantle, and a core extending about 500 kilometers from the center.

5. The Moon has no present-day tectonic activity and no magnetic field, which indicate that there is probably no liquid core.

6. The Moon has significant chemical contrasts with the Earth, which argues against a common origin for the two. The true mechanism for the Moon's formation is unknown, but the best current explanation appears to be the hypothesis that the Moon formed as the result of a collision between the young Earth and a large planetesimal.

7. The Moon's evolution consisted of a molten state, followed by hardening of the crust and subsequent large-scale lava flows that created the maria. Since that time (about 1 billion years after the Moon's formation), the Moon has been geologically quiet.

8. Mercury is so close to the Sun that it is difficult to observe from Earth, and only very general properties have been derived from telescopic observations.

9. Radar measurements showed that Mercury rotates exactly one-and-a-half times for each orbit around the Sun, in a form of spin-orbit coupling that was created by a combination of the Sun's tidal force on Mercury and the highly elliptical orbit of the planet.

10. Mercury has a magnetic field and a high density, indicating that it has a large, dense core, probably partially molten. The origin of the high iron content and large core is uncertain, but may be the result of a collision which vaporized much of the original mantle.

11. The surface of Mercury is nearly identical in general appearance to that of the Moon, except for a planetwide system of scarps, minor differences in the heights of impact craters and the distances traveled by ejecta, and extensive rolling plains on Mercury that do not exist on the Moon.

12. The great impact basin Caloris Planitia is located at the subsolar point at perihelion on alternate orbits of the Sun, and may be the site of the mass imbalance that caused Mercury's spin to fall into resonance with its orbital period.

REVIEW QUESTIONS

Thought Questions

1. What is the evidence that the craters on the Moon are formed by impacts, rather than by volcanic eruptions?

2. Why does the Moon have much more extreme fluctuations in surface temperature than does the Earth?

3. Why does the Earth not have as many craters as the Moon?

4. Why are there relatively few craters in the maria?

5. Compare the Moon's interior structure with that of the Earth.

6. Even though the craters on the Moon are thought to have been formed by impacts, there is evidence for volcanic activity in the past. What is the evidence?

7. Explain how the shape of Mercury's orbit has led to the unusual spin-orbit coupling of the planet's rotation with its orbital period.

8. What do you think the rotation period of Mercury would be if the planet had a circular orbit?

9. Explain why *Mariner 10* was able to photograph only one side of Mercury, despite the fact that it visited the planet three separate times.

10. What could you say about the internal structure of Mercury, based only on the knowledge that the planet's overall density is relatively high?

11. Discuss the similarities and contrasts between the surface of Mercury and that of the Moon.

12. Why does Mercury not have a significant atmosphere?

Problems

1. If the Moon and the Earth had equal diameters, so that they would appear the same size to an outside observer, the Earth would appear brighter because of its higher albedo. How much brighter would the Earth appear to be? Now take into account the fact that the Earth's diameter is approximately 4 times greater than that of the Moon, and estimate how much brighter the Earth appears to an outside observer.

2. Using information given in Chapter 5, calculate the surface gravity on the Moon, compared to that on the Earth. If you dropped a rock from the edge of a high cliff on the Moon, how fast would it be falling after 3 seconds?

3. If impacts by objects from space occurred at a uniform rate over the entire history of the Moon, and the maria have only one-fourth as many craters (per square kilometer) as the rest of the Moon, how much younger than the rest of the lunar surface would the maria be? Is this consistent with what the text says about the ages of the maria and of the Moon itself? What does this imply about the rate of impacts over the Moon's history?

4. If Mercury is 50 percent farther from the Sun at aphelion than at perihelion, how much more intense is sunlight on its surface at perihelion?

5. *Mariner 10* was put into a solar orbit having twice the orbital period of Mercury. What was the semi-major axis of the orbit of *Mariner 10*?

6. Using information from Chapter 5, calculate the surface gravity on Mercury and compare it to the surface gravity on the Moon.

7. Calculate and compare the escape velocities of Mercury and the Moon.

ADDITIONAL READINGS

Anderson, D. L. 1974. The interior of the Moon. *Physics Today* 27(3):44.

Beatty, J. Kelly. 1986. The making of a better moon. *Sky and Telescope* 72(6):558.

Beatty, J. K., O'Leary, and A. Chaikin, eds. 1990. *The new solar system*. 2d ed. Cambridge, England: Cambridge University Press.

Brownlee, Shannon. 1985. The whacky theory of the Moon's birth. *Discover* 6(3):65. (the Moon's origin attributed to a giant impact)

Burgess, E., and J. E. Oberg. 1976. Science on the Moon. *Astronomy* 4:684.

Burt, D. M. 1989. Mining the Moon. *American Scientist* 77:574.

Cadogan, P. 1983. The Moon's origin. *Mercury* 12(2):34.

French, B. M. 1977. What's new on the Moon? *Sky and Telescope* 53:164.

Goldreich, P. 1972. Tides and the Earth-Moon system. *Scientific American* 226(4): 42.

Goles G. G. 1971. A review of the Apollo project. *American Scientist,* 59:326.

Hammond, A. L. 1973. Lunar science: analyzing the Apollo legacy, *Science* 179:1313.

Hartmann, W. K. 1976. The significance of the planet Mercury. *Sky and Telescope* 51(5):307.

Head, J. W., C. A. Wood, and T. A. Mutch. 1977. Geologic evolution of the terrestrial planets. *American Scientist* 65:21.

Morrison, D., and Owen, T. 1988. Our ancient neighbor, the Moon. *Mercury* 17:66 (Part I); *Mercury* 17:98 (Part II).

Murray, B. C. 1975. Mercury. *Scientific American* 233(3):58.

Register, Bridget Mintz. 1985. The fate of the moon rocks. *Astronomy* 13(12):14.

Runcorn, S. K. 1987. The Moon's ancient magnetism. *Scientific American* 257(6):60.

Strom, R. G. 1990. Mercury: The forgotten planet. *Sky and Telescope* 80:256.

Taylor, S. R. 1987. The origin of the Moon. *American Scientist* 75:468.

Weaver, K. F. 1975. Mariner unveils Venus and Mercury. *National Geographic* 147:848.

MATHEMATICAL METHODS

Particle Motions and Atmospheric Escape

In Chapter 8, we discussed the relationship between the temperature of an atmosphere, the masses of the gas particles in the atmosphere, and the likelihood that a given type of gas will escape. The essential point made was that if the average speed for a certain type of particle exceeds one-sixth of the planet's escape speed, that type of particle will eventually escape completely.

In this chapter, we have discussed two terrestrial bodies, the Moon and Mercury, which have virtually no atmosphere (except for very rarefied gaseous envelopes, which we shall ignore for the present discussion). It is of interest to see how the average particle speed is calculated, and then to see whether the lack of an atmosphere for both bodies is consistent with expectations.

The great Scottish physicist James Clerk Maxwell made many important contributions to our understanding of electromagnetic radiation and of the behavior of particles in gases. He developed a fundamental relation between gas temperature and particle speed, known as the **Maxwell distribution,** which describes the range of particle speeds for a given temperature and which can be used to calculate the average speed. The equation representing the Maxwell distribution is complex, and beyond the scope of the present discussion. It is worth stressing here, though, that there is indeed a range of particle speeds for a given gas temperature, which is shown graphically below.

The particles do not all move at the same speed; instead, there is a most probable particle speed, with the probability decreasing for either higher or lower speeds. The most probable speed, which is not quite the same thing as the average speed (see graph), is given by a simple equation involving the gas temperature T and the particle mass m:

$$v_p = \sqrt{\frac{2kT}{m}},$$

where v_p is the most probable speed and k is called the *Boltzmann constant* and has the value $k = 1.38 \times 10^{-16}$ in the system of units where masses are measured in grams and distances in centimeters. The average speed is very similar in value, being given by

$$v_a = \sqrt{\frac{8kT}{\pi m}}.$$ We see that there are always some particles traveling much faster than the average or most probable speed; these particles are represented by the long "tail" to the right in the graph above. Because of the high-speed tail of the Maxwell distribution, some particles will exceed the escape speed even in a gas where the average speed is much lower. As we have already mentioned, detailed calculations show that if the average speed is at least one-sixth of the escape speed, eventually all of the particles will escape. Now we see that this is because of the small fraction (which remains constant) of particles that exceed the escape speed; the gas escapes by "leaking" out through the tail of the Maxwell distribution.

Technically, it is the most probable speed, rather than the average speed, that must satisfy the one-sixth criterion, so the criterion for a given gas of particle mass m to escape an atmosphere of temperature T is

$$\sqrt{\frac{2kT}{m}} > 0.17v_e,$$

where v_e is the escape speed, given by $v_e = \sqrt{\frac{2GM}{R}}$

for the planet in question (see discussion in Chapter 5).

Now let us see how this can be applied to the Moon and to Mercury. Each of these bodies has large temperature differences between night and day, but we take the daytime temperatures (400 K for the Moon and 700 K for Mercury) in order to set the most stringent limitations for retention of gases. The escape speed for the Moon is 2.4×10^5 cm/sec, and for Mercury it is 4.3×10^5 cm/sec. Let us first consider a lightweight gas such as hydrogen ($m = 1.67 \times 10^{-24}$ grams), and then a heavier one such as carbon dioxide (7.35×10^{-23} grams). For hydrogen, we find from the equation above for most probable speed a value of 2.6×10^5 cm/sec for the Moon's temperature of 400 K, and 3.4×10^5 cm/sec for Mercury's 700 K. The value for the Moon is actually in excess of the escape speed, and for Mercury it is about 80 percent of the escape speed, so it is easy to see why neither body can retain an atmosphere of hydrogen atoms.

For carbon dioxide the same calculations yield most probable speeds of 3.9×10^4 cm/sec for the Moon and 5.1×10^4 cm/sec for Mercury. These values correspond to 0.16 and 0.12 of the escape speeds for two bodies, respectively. Thus we see that carbon dioxide is only marginally unable to remain at the Moon's surface, and that it should be retained by Mercury. We conclude that if Mercury had a source of carbon dioxide, it would have a carbon dioxide atmosphere today. (But, as mentioned in the text, it is thought that the solar wind sweeping past Mercury also helps to remove gases in its vicinity.)

For the Earth, hydrogen has a most probable speed of 0.20 times the escape speed, while carbon dioxide has a most probable speed of only 0.03 times the escape speed; these figures explain why the Earth does not have free hydrogen in its atmosphere, but can retain carbon dioxide and other heavy gases.

CHAPTER 11

Venus: A Cloud-Covered Inferno

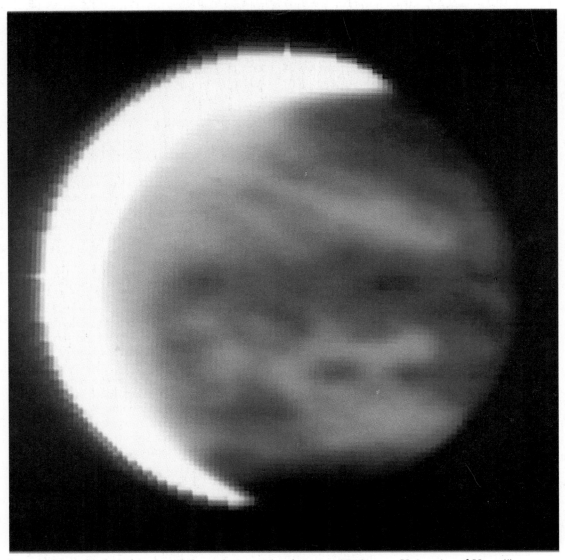

A ground-based infrared image of the nighttime side of Venus. (W. Sinton, University of Hawaii)

CHAPTER PREVIEW

Observations from Near and Far
The Atmosphere of Venus
The Structure of the Atmosphere
Atmospheric Motions
The Surface and Interior
The Earth and Venus: So Near and Yet So
  Different

Of all the planets, Venus (Table 11.1 and Fig. 11.1) is the most prominent in our nighttime sky. It has played a role in the astronomical lore since antiquity. Because Venus is an inferior planet, it never appears far from the Sun, as seen from the Earth (its greatest elongation is 47°). Thus it is visible to us either early in the evening, shortly after sunset, or in the wee hours just before dawn. In ancient times, the morning and evening stars were thought to be two separate planets, named Phosphorous and Hesperus, but by the sixth century B.C. the two had been identified as a single planet alternately appearing on either side of the Sun.

Bright as it is, Venus has become a natural subject for careful scrutiny by astronomers throughout history, and it has played an important role in some major developments. For example, the fact that it never wanders far from the Sun prompted early speculation that perhaps it orbits the Sun rather than the Earth, and Galileo's observation of the phases of Venus was an

Figure 11.1 The crescent Venus.   *This photograph, taken from Earth, shows Venus when it is near inferior conjunction; thus we see only a sliver of its sunlit side. (NASA/JPL)*

important bit of evidence favoring the heliocentric theory.

In modern times, a great deal of scientific effort has gone into studies of Venus, because its proximity to Earth, both in location and in general characteristics, means that Venus may teach us valuable lessons about our own planet.

Table 11.1 Venus

Orbital semimajor axis: 0.723 AU (108,200,000 km)
  Perihelion distance: 0.718 AU
  Aphelion distance: 0.728 AU
Orbital period: 224.7 days (0.615 years)
Orbital inclination: 3°23'40"
Rotation period: 243 days (retrograde)
Tilt of axis: 3°
Diameter: 12,104 km (0.951 $D_\oplus$)
Mass: 4.87 × 10²⁷ grams (0.815 $M_\oplus$)
Density: 5.3 g/cm³
Surface gravity: 0.90 Earth gravity
Escape speed: 10.3 km/sec
Surface temperature: 750 K
Albedo: 0.65
Satellites: none

OBSERVATIONS FROM NEAR AND FAR

A substantial body of knowledge about Venus was developed, particularly within the last two decades, from remote observation. Its size and orbital parameters, as well as some data on the composition of its upper atmosphere and an estimate of its mass, all were derived from Earth-based observations, as were some surprising discoveries about its rotation and its surface temperature.

Venus is brilliant because it is shrouded in clouds that efficiently reflect sunlight. Its albedo is 0.76. The same clouds that create the brilliance also hide the surface from telescopic view, and for this reason special techniques have been required to penetrate the clouds.

Although it was once popular to imagine that Venus is a pleasantly tropical planet, with lots of water and, no doubt, plant and animal life, what was discovered from early radio observations was vastly different. Radio waves penetrating the cloud cover brought the message to Earth that the surface of Venus is very hot, about 750 K (477°C, or about 900°F). This discovery made it obvious that, despite outward appearances of similarity, Venus and the Earth had followed different paths. No longer could Venus be viewed as a likely abode for life.

The reason for the high temperature was quickly ascertained (Fig. 11.2). Spectroscopic measurements dating back to the 1930s showed that carbon dioxide (CO_2) is a major constituent of the atmosphere. This molecule has many spectral lines in the infrared part of the spectrum, so that a CO_2 atmosphere is effectively opaque to infrared radiation.

Visible light from the Sun can penetrate the clouds of Venus from above and is absorbed at the surface,

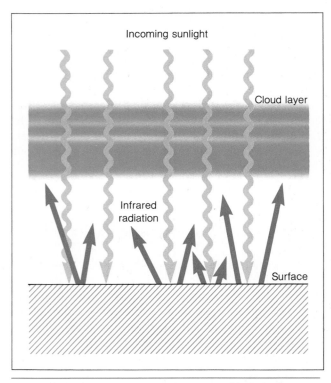

Figure 11.2 The greenhouse effect. *Visible light from the Sun reaches the surface of Venus and heats it, which causes the ground to emit infrared radiation. The carbon dioxide in the atmosphere efficiently absorbs infrared radiation, so heat is trapped near the surface.*

heating it. When the ground responds by emitting infrared radiation, as it must according to Wien's law (Chapter 6), this radiant energy cannot rapidly escape because of absorption by the CO_2, and heat is trapped near the surface. This heating mechanism is called the **greenhouse effect,** because it is similar to what happens in a greenhouse when light enters through the glass and heats the interior.

A second surprising discovery was made from Earth, again through the use of radio (specifically radar) wavelengths. The first radar measurements of the surface were made in 1961, and the Doppler effect was used to determine the rate of rotation. Contrary to expectations, the planet was found to rotate very slowly, and even more astonishingly, to do so in the backward direction, the direction opposite that of the other orbital and rotational motions in the solar system. This is called **retrograde rotation.**

The spin of Venus takes 243 Earth days, so slow that its day is actually longer than its year (the orbital period is 225 days). Here we must distinguish between the sidereal day, the 243 days just mentioned, and the solar day, the apparent time it takes for the Sun to go through one daily cycle, as seen from a point on the surface of Venus. The length of the solar day on Venus is 116.8 days.

The reason for the unusual rotation of Venus is not understood. Because it is so strongly at odds with the spins of the other planets, it seems likely that something unusual happened either during the formation of Venus or at some later time.

Both the Soviet Union and the United States focused on Venus as a logical and interesting target for exploration by spacecraft, and a number of missions have been flown there by both nations (Table 11.2). Spacecraft can be launched toward Venus most easily when the relative positions of the Earth and Venus are favorable. The United States launched *Mariner 2* toward Venus on such an occasion in 1962, and this spacecraft flew within 35,000 kilometers of the cloud tops, making various measurements. The next opportunity came in 1965–1966, when both the United States and the Soviet Union took advantage by launching Venus probes. The U.S. *Mariner 5* mission was another flyby, but the Soviets chose instead to have their *Venera 4* probe descend into the atmosphere of Venus, and it relayed back to Earth important information on conditions there, before it failed on its way down to the surface. *Veneras 5* and *6*, launched in 1969, also entered the atmosphere of Venus, as did *Veneras 7* and *8* in 1970 and 1972.

ASTRONOMICAL INSIGHT 11.1

Velikovsky and Venus

The planet Venus played a major role in the scientific controversy that began in the 1950's. The controversy was not so much over the proper interpretation of scientific data, although that played a role, but was centered instead on questions of the proper role of science in responding to pseudoscientific theories, which occasionally catch the public imagination. Science must be open to new ideas, but at the same time it must be on watch against those which are not supported by evidence or rational analysis. In the case of Immanuel Velikovsky and the origin of Venus, scientists perhaps stepped beyond the boundaries of proper response.

In 1950, following years of study of ancient records, Velikovsky, trained as a medical doctor and psychologist, published a book called *Worlds in Collision*. In this book, extrapolating from records of catastrophic upheavals that apparently affected many different ancient cultures at about the same time (between the fifteenth and eighth centuries B.C.), Velikovsky suggested that these catastrophes had extraterrestrial origins. He described a sequence of events involving a "comet," released from the atmosphere of Jupiter, that twice nearly collided with the Earth, causing its rotation to be stopped temporarily, and which then lost its tail in a collision with Mars before entering a circular orbit about the Sun

and becoming the planet Venus. Velikovsky hypothesized that the interactions between the Sun and planets were dominated by electromagnetic forces, that the motions of both Mars and the Earth were altered to their present state by the near-collisions with the comet, and that these near-collisions (there was also supposed to have been a collision between the Earth and Mars in 687 B.C.) accounted for the upheavals in human society whose occurrence had originally inspired Velikovsky.

Despite its absolute and total disregard for astronomical observation and for the laws of physics, *Worlds in Collision* attracted a great deal of public attention and interest. Fearing that this attention would be followed by general acceptance, the scientific community reacted quickly. Numerous well-known astronomers, historians, and physicists made public statements criticizing the book, its publisher was effectively boycotted, and in general the scientists gave the appearance of feeling strongly threatened. Unfortunately, critics of Velikovsky often used questionable tactics. Also, in many cases they were not as well-versed in his writings as they should have been, and they frequently left themselves open for countercriticisms. Matters became more heated when new evidence, in the form of data from planetary space probes launched in the early 1960's, supported some of Velikovsky's predictions. These included the fact that Venus has a high surface temperature, that the Earth's magnetic field extends very far into space, and that Jupiter is an emitter of radio waves.

The Velikovsky affair has by

now faded from public view, but it has not been forgotten. There are probably people who still believe the ideas expressed in *Worlds in Collision* and in Velikovsky's second book, *Earth in Upheaval*. The continued development of the space program and the successful probes of other planets (including, of course, Venus) have helped a great deal, since other predictions of Velikovsky, such as his notion that electromagnetic forces control planetary motions, have now been refuted by direct measurement. There is much less room now for Velikovsky's supporters to avoid conflict with observation.

Responding to works such as *Worlds in Collision* and similar unfounded pseudoscientific treatises (popular books on ancient astronauts and astrology, for example) is a difficult problem for scientists. There is clearly an obligation to keep the public as well-informed as possible, so that people can see for themselves the folly of unsupportable and unprovable ideas but this must be done in a dignified manner, without creating the impression of having something to hide. Thus, Velikovsky could not be ignored, but he and his ideas probably should have been treated with less emotion and more logic.

Table 11.2 Probes to Venus

Spacecraft	Launch	Encounter	Nature of Encounter
Mariner 2	Aug. 26, 1962	Dec. 14, 1962	Flyby, 34,833 km
Venera 4	June 12, 1967	Oct. 18, 1967	Hard landing, dark side
Mariner 5	June 14, 1967	Oct. 19, 1967	Flyby, 4,100 km
Venera 5	Jan. 5, 1969	May 16, 1969	Hard landing, dark side
Venera 6	Jan. 10, 1969	May 17, 1969	Hard landing, dark side
Venera 7	Aug. 17, 1970	Dec. 15, 1970	Soft landing, dark side
Venera 8	March 27, 1972	July 22, 1972	Soft landing, day side
Mariner 10	Nov. 3, 1973	Feb. 5, 1974	Flyby, 5,700 km
Venera 9	June 8, 1975	Oct. 22, 1975	Orbiter plus soft lander
Venera 10	June 14, 1975	Oct. 25, 1975	Orbiter plus soft lander
Pioneer Venus 1	May 20, 1978	Dec. 4, 1978	Orbiter
Pioneer Venus 2	Aug. 8, 1978	Dec. 9, 1978	Four hard landers, both sides
Venera 11	Sept. 9, 1978	Dec. 21, 1978	Flyby (25,000 km), soft lander
Venera 12	Sept. 14, 1978	Dec. 25, 1978	Flyby (25,000 km), soft lander
Venera 13	Oct. 30, 1981	March 1, 1982	Flyby, soft lander
Venera 14	Nov. 4, 1981	March 5, 1982	Flyby, soft lander
Venera 15	June 2, 1983	October 10, 1983	Orbiter, radar mapper
Venera 16	June 7, 1983	October 14, 1983	Orbiter, radar mapper
Vega 1	Dec. 15, 1984	June 11, 1985	Lander, atmos. probe
Vega 2	Dec. 21, 1984	June 15, 1985	Lander, atmos. probe
Magellan	May 4, 1989	Aug. 1990	Orbiter, radar mapper
Galileo	Oct. 18, 1989	Feb. 10, 1990	Flyby (10,000 km)

Figure 11.3 A close-up portrait of Venus. *This is a full-disk ultraviolet image of Venus obtained by the* Mariner 10 *spacecraft; it shows structure in the clouds.* (NASA)

The U.S. *Mariner 10* mission in 1973–1974 flew by Venus (and later Mercury), obtaining detailed data on the structure of the clouds and their motions. This and earlier *Mariner* spacecraft showed that Venus has no magnetic field.

The ultraviolet images obtained by the early *Mariner* flyby missions showed structure in the clouds (Fig. 11.3). Continued observations of this structure revealed that the cloud tops are moving rapidly, circling the planet every 4 days at speeds of about 100 meters per second (360 kilometers per hour). This atmospheric circulation occurs in the retrograde direction, the same direction as the rotation of the planet. The temperature of the cloud tops was derived from infrared measurements, which led to an estimate of 240 K (−27°F).

Recent probes sent to Venus have provided spectacular information on conditions beneath the clouds. Whereas the earlier Venera instruments had failed (as a result of the intense heat and pressure) before reaching the surface, the *Venera 9* and *10* missions (in the mid-1970s), *Veneras 11* and *12* (1978), and most recently, *Veneras 13* and *14* (1982), all succeeded in landing and operating on the surface. *Veneras 9, 10, 13,* and *14* carried cameras and sent back photographs of the ground surrounding the landing sites. These images showed that the surface is brightly lit, with enough sunlight penetrating the clouds to actually cast shadows, and that the ground is strewn with rocks, many of them with surprisingly sharp edges. This indicates that, despite the hostile conditions, there is very little erosion acting on the surface of Venus, or that some geological process renews the surface from time to time.

Before the current *Magellan* mission, the most recent U.S. probe to Venus was called *Pioneer Venus*, and consisted of two separate spacecraft, an orbiter and a package of probes designed to penetrate the atmosphere. As of late 1990, the orbiter was still in operation, gathering data on the cloud structure, composition, and motions, as well as on conditions in the upper atmosphere, above the clouds. The five small probes that descended into the atmosphere when the spacecraft arrived at Venus in late 1978 gathered data on physical conditions such as temperature and pressure, and data on chemical composition of the atmosphere as a function of altitude. The *Pioneer Venus* probes reached the ground safely and continued to operate there for awhile (up to 1 hour).

The *Magellan* spacecraft was launched toward Venus in April, 1989, and arrived at the cloudy planet in August, 1990. The primary goal of *Magellan* is to obtain a detailed radar map of the entire surface of Venus, with much finer detail than any previous maps, even those obtained of parts of the Northrn Hemisphere by the Soviet *Venera* probes.

THE ATMOSPHERE OF VENUS

We can summarize what has been learned about the atmosphere of Venus by discussing three distinct aspects: its composition, vertical structure, and motions.

The most important ingredient in the atmosphere of Venus, already identified from Earth-based measurements, is carbon dioxide (Table 11.3). The *Venera* and *Pioneer Venus* probes showed that CO_2 dominates all through the atmosphere to the ground and makes up about 96 percent of the atmosphere. It is interesting that the overall quantity of CO_2 on the Earth is about the same as that on Venus, except that on Earth the CO_2 is mostly in carbonate rocks or dissolved in seawater, and not in the atmosphere.

The next most abundant gas in the atmosphere of Venus is nitrogen (in the form of N_2), which accounts for most of the remaining 4 percent or so. In addition, traces of water vapor and other species such as oxygen,

Table 11.3
Composition of the atmosphere of Venus

Gas	Symbol	Fraction[a]
Carbon dioxide	CO_2	0.96
Nitrogen	N_2	0.035
Sulfur dioxide	SO_2	1.5×10^{-4}
Water vapor	H_2O	1×10^{-4}
Argon	Ar	7×10^{-5}
Carbon monoxide	CO	4×10^{-5}
Neon	Ne	5×10^{-6}
Hydrogen chloride	HCl	4×10^{-7}
Hydrogen fluoride	HF	1×10^{-8}

[a]Values are fractions by number.

argon, neon, and sulfur (in the form of compounds such as sulfur dioxide and sulfuric acid) exist there. The sulfuric acid (H_2SO_4) apparently condenses out of the lower clouds, forming droplets like rain. These circumstances, along with the high temperature and pressure at the surface, make it difficult for spacecraft to operate on Venus.

THE STRUCTURE OF THE ATMOSPHERE

We have already seen that the atmospheric conditions on Venus vary quite a lot from the top of the clouds to the surface. The temperature increases from 240 K at cloud tops to 750 K at the ground, and the pressure at the bottom of the atmosphere reaches the rather incredible value of 90 Earth atmospheres, or more than 1,300 pounds per square inch, equivalent to the pressure about 3,000 feet below the surface of the ocean. The temperature decreases smoothly from the surface to the top of the clouds.

The clouds in the atmosphere of Venus are composed of small droplets of sulfuric acid and some other, as yet unidentified, kinds of particles. The H_2SO_4 droplets are about 2 millionths of a meter (about 0.0002 cm) in diameter.

There are three distinct zones of clouds at separate altitudes (Fig. 11.4). The uppermost layer, the one visible from afar, lies some 60 km above the surface of Venus and averages about 10 km in thickness. The middle cloud layer, about 6 km thick, is suspended at an altitude of roughly 53 km, whereas the lowest layer lies at 48 km and is relatively thin, extending only 2 km

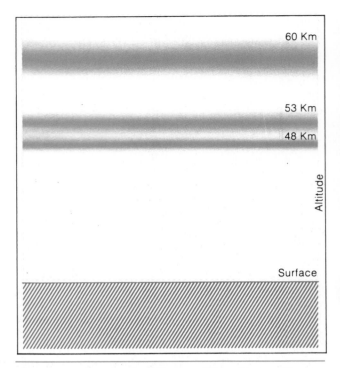

Figure 11.4 The clouds of Venus. *The sulfuric acid clouds are separated into three distinct layers. These layers occur at altitudes where the combination of temperature and pressure causes sulfuric acid to condense.*

Figure 11.5 An ultraviolet image from *Pioneer Venus*. *The dark regions seen here are places where atmospheric sulfur dioxide, which absorbs ultraviolet light, is exposed. (NASA)*

or so vertically. Below the lower cloud is clear atmosphere, all the way to the ground. Above the uppermost clouds is a haze layer, consisting of much smaller particles than the droplets that form the clouds, but of unknown composition.

Clouds form in the atmosphere wherever the combination of temperature, pressure, and relative abundance of H_2SO_4 corresponds to the conditions required for this gas to condense. This is exactly analogous to the formation of clouds in the Earth's atmosphere, except that on the Earth, the gas that condenses is water vapor rather than sulfuric acid. It happens that the proper conditions for condensation occur at three different levels in the atmosphere of Venus.

The clouds also contain quantities of the gas sulfur dioxide (SO_2), which is an efficient absorber of ultraviolet light. We see from ultraviolet photographs (Figs. 11.3 and 11.5) that there is considerable structure in the clouds, because the SO_2 abundance varies with depth: therefore, in places where the winds stir up regions with quantities of SO_2, or where the cloud structure allows us to see to greater depths, the clouds

look dark (see Fig. 11.3). Thus, ultraviolet photographs show some contrast in the clouds, and are useful for mapping the motions of the upper regions (Fig. 11.6).

The atmosphere seems to be less changeable than that of the Earth, where cloud structures come and go with changes in the weather. The relative stability of Venus is probably a result of a combination of factors, including the greater density and pressure of the atmosphere, and the slow rotation of the planet, which minimizes circulation patterns driven by rotation.

Another important distinction between Earth and Venus is that there is no magnetic field around Venus. There is no magnetosphere and no protective shield of magnetic field lines to prevent charged particles from entering the atmosphere. This creates a complex electrical environment at the top of the atmosphere, perhaps including lightning discharges, which probably affects the chemical processes there as well.

Atmospheric Motions

We have seen that the upper atmosphere of Venus has a general circulation pattern at the cloud tops, moving in the same direction as the planet's rotation (Fig. 11.6).

Figure 11.6 Circulation of the atmosphere of Venus.
These four ultraviolet views show two rotations of the planetary cloud cover. The upper two are separated by one day; the lower two were obtained about a week after the first pair and are also separated by a day. The atmospheric motion is from right to left in these images. (NASA)

The velocity is high at the cloud tops but diminishes to nearly zero at the surface.

There are complex vertical motions in addition to these horizontal ones. Because of solar heating near the equator, there is a rising current there that spreads out in all directions, descending when it reaches cooler regions near the poles and on the dark side of the planet. This pattern is similar to the circulation cells in the Earth's atmosphere, except that on Venus there is insufficient rotation of the planet to break up the flow into numerous small cells. On Venus it is more of a global circulation pattern, with warm gas rising in the location called the **subsolar point,** where the Sun is directly overhead, and descending in a very broad region away from that point (Fig. 11.7). Besides minimizing rotational forces that might otherwise break up this simple flow pattern into smaller cells, the slow spin of Venus also means that the subsolar point moves very slowly as the planet rotates. Thus, the heating from the Sun lingers, enhancing the flow pattern.

Some of the gas that rises at the subsolar point flows all the way around to the dark side of Venus before cooling and descending. As it does so, free atoms combine to re-form molecules, which then emit light as they cool further. This creates aurorae on the dark side of Venus, a phenomenon that was suspected from Earth-based observations, but which was not confirmed until the flight of *Pioneer Venus* (Fig. 11.8). These aurorae are quite distinct from those in the Earth's upper atmosphere, which are created by charged particles from space.

In and between clouds, additional vertical circulation patterns exist, depending on where the Sun's heat energy is deposited. This, in turn, depends on the cloud composition; we have already seen that SO_2 is particularly effective as an absorber of solar energy, so there is extra heating in levels where SO_2 is abundant. There appear to be at least four flow patterns at different levels involving circulation between the equatorial and polar regions.

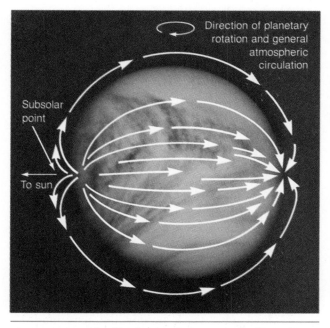

Figure 11.7 High-altitude circulation. *Well above the clouds of Venus, which are rotating from right to left in this sketch, gases heated at the subsolar point circulate around the planet toward the dark side, where they cool and descend. (Photo from NASA)*

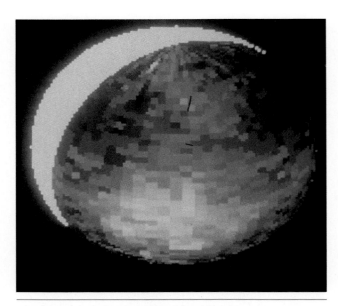

Figure 11.8 Airglow on the night side of Venus. *Atoms produced by the disruption of molecules on the sunlit side circulate around to the dark side, where they cool and combine once again into molecules. The newly formed molecules have excess energy, which is released in the form of light, creating a glow in the upper atmosphere on the night side. (LASP, University of Colorado, sponsored by NASA)*

THE SURFACE AND INTERIOR

Radar experiments have provided most of the available information on the surface of Venus. The radar instrument on the *Pioneer Venus* orbiter is capable of making fairly detailed measurements and can detect features as small as 30 kilometers. Such measurements have made possible the construction of global relief maps showing many interesting features of Venus (Fig. 11.9). Some of the Soviet *Venera* spacecraft were equipped with radar devices and were able to obtain much more detailed maps of portions of the surface. In addition, radar measurements made from Earth have provided very detailed maps of small regions on Venus.

The surface has three general types of terrain: (1) rolling plains, which cover about 65 percent of the planet; (2) highlands, which cover about 8 percent; and (3) lowlands, which occupy the remaining 27 percent.

The rolling plains are characterized by numerous craters and circular basins (Fig. 11.10), many of them apparently filled with lava and resembling small-scale lunar maria. Many of the craters were probably caused by volcanic activity rather than meteor impacts, but some are clearly impact craters (see Fig. 11.10).

The highlands are concentrated in three major regions, two of which—Ishtar Terra and Aphrodite Terra—are comparable in size to the continents of Africa and Australia; the third, Beta Regio, is much smaller. The maximum elevation above the mean surface level, the summit of a mountain called Maxwell Montes (Fig. 11.11), is greater than that of Mt. Everest above sea level, but the numerous canyons are much shallower than the deepest undersea trenches on Earth. There are many adjacent parallel valley and trench systems in the highlands of Venus, forming systems up to 9,000 kilometers long.

Given the similarities in size and average density of Venus and the Earth, it is interesting to consider whether the internal structure and processes are also similar. One point of comparison is tectonic activity. Initial data have suggested that Venus may not have such active continental drift as the Earth, but more recent observations, particularly by the Soviet radar mapping instruments on *Venera 15* and *16*, reveal structures that closely resemble the Earth's mid-ocean ridges. It is not yet known whether there is a planetwide system of moving crustal plates; it is hoped that the *Magellan* mission, which will map the entire surface, will answer this question. Early results from *Magellan* (Fig. 11.12) show a surface having many rifts and fractures, suggesting that tectonic activity has been both energetic and widespread.

There is some evidence, in the form of the valley and trench systems in the highlands, that tectonic activity has played a role in the formation of these regions. The highlands consist in part of large volcanic mountains that built up as a result of prolonged lava flow episodes. The smallest of the three highland areas, Beta Regio, appears to consist entirely of a few large volcanoes. The combination of the valley and ridge systems and the large volcanoes indicates that there are (or were) convection currents in the upper mantle, and that there are (or were) volcanic hot spots where lava flowed upward for long periods. Thus, the internal activity and structure may be much like the Earth's, but the crust on Venus may not have broken up into plates that move about (further results from *Magellan* should clarify this).

We noted earlier that variations in the sulfur dioxide level in the atmosphere may imply that there are active

Figure 11.9 A relief map of Venus.   *This image, constructed on the basis of* Pioneer Venus *radar maps of the surface, shows the three major highland areas: Ishtar Terra at the top, Aphrodite Terra at right center, and Beta Regio at left center. (NASA/JPL)*

Figure 11.10 Impact craters on Venus.   *These high-resolution rader images, obtained by a recent Soviet* Venera *mission, (left) and by the currently-operating U.S.* Magellan *mission (right) show craters having the distinctive rim walls and central peaks characteristic of impact craters. (left: U.S. Geological Survey; right: NASA/JPL)*

Figure 11.11 A portion of Ishtar Terra. *This computer reconstruction illustrates (on an exaggerated vertical scale) the vertical relief of this highland region, and includes the largest mountain on Venus, Maxwell Montes, more than 11 kilometers above the mean surface level. (U.S. Geological Survey)*

Figure 11.12 Fractured surface terrain. *The ridges and valleys seen here are formed by intersecting faults. The irregularly-shaped dark region near the right is probably an ancient lava flow. The strip shown here, derived from Magellan radar data, is about 20 km wide and 75 km long; the smallest details visible are about 120 m (400 feet) across, which represents an improvement over previous surface maps of a factor of ten. (NASA/JPL)*

volcanoes on Venus. Some of the early radar mapping data have been interpreted as showing crater shapes characteristic of recent volcanic eruptions; new images obtained by *Magellan* clearly confirm this, showing numerous craters that are definitely volcanic in origin (Fig. 11.13). The portions of Venus that were mapped by the Soviet *Venera* missions are peppered with an enormous number of low mounds whose shape is very similar to that of shield volcanoes on the Earth. These findings lend support to the suggestion that Venus has been volcanically active, but as yet do not tell us whether it is now. We hope that the question of current volcanic activity on Venus will be settled by the high-resolution radar maps being obtained by *Magellan;* the orbiter will map much of the surface twice, so that changes in structures caused by lava flows taking place between the times of the two maps will be evident, if they occur.

If continental drift is not occurring on Venus, as some believe, the reason is not known, although there are some interesting speculations. One suggestion is that, because Venus has such a high surface temperature, its crust is warmer and therefore more buoyant than that of the Earth. On both planets, the highlands (or continents) consist of old material that essentially floats in equilibrium with the denser material of the lowlands (or seafloors). On Earth, the seafloor material is as dense as the underlying asthenosphere and therefore is easily forced below continental margins in the subduction zones that produce the great undersea trenches. On Venus the lowland material is apparently less dense than the underlying mantle and therefore cannot easily be forced downward. Thus, it appears that convection currents inside Venus have not succeeded in creating crustal plates that move about by slipping over one another, because subduction cannot take place.

Some information about surface rocks was derived from the photographs (Fig 11.14) and other measurements obtained by the *Venera* landers. Somewhat to the surprise of scientists studying the photographs, the rocks have sharp edges, which indicates a lack of erosion. This was understood when it was realized that the winds at the surface of Venus are almost nonexistent.

Measurements of the composition of surface rocks are consistent with a basaltic composition, similar to the predominant crustal rocks on the Earth and the Moon. Basalt is an igneous silicate rock with a moderate density. Combined with the fact that the average density of Venus is comparable to that of the Earth (5.3 grams per cubic centimeter compared to 5.5 for

Figure 11.13 A volcano on Venus. *This image obtained in late 1990 by the* **Magellan** *probe shows a mountain that is distinctly shaped like many volcanoes on Earth, and certainly is of volcanic origin. The crater seen here at left center is about 1 km in diameter. Debris ejected by the volcano is seen as a light-colored region stretching away from it, toward the top of the picture. The pattern of light lines in the lower portion is created by numerous faults and fractures in the plains adjacent to the volcano. The entire region shown here is about 40 km wide and 112 km long; the finest details visible are about 100 m in size. (NASA/JPL)*

the Earth), this implies that Venus is differentiated, with a large, dense core.

Other than this very indirect evidence, little is known about the interior of Venus. The probable similarity to the internal structure of the Earth might lead us to suspect that Venus has a magnetic field, but we have seen that it does not. Perhaps the slow rotation of the planet is insufficient to create a magnetic dynamo in its core.

EARTH AND VENUS: SO NEAR AND YET SO DIFFERENT

We are now equipped with enough information to discuss the history of Venus and develop some under-

ASTRONOMICAL INSIGHT 11.2

Volcanoes on Venus

Until the *Voyager* spacecraft flew past Jupiter in the late 1970s, the Earth was the only body known to have current volcanic activity. Gigantic volcanoes had been found on Mars in the early 1970s, but they are apparently extinct. *Voyager* found volcanoes in the act of erupting on Io, one of the moons of Jupiter, making that satellite the first nonterrestrial body known to have current volcanic activity. Now Venus has become the second, if a recent suggestion proves to be correct. (There is now little doubt that Venus has been volcanically active in the past, as indicated by recent radar maps from the *Magellan* and *Venera* missions.)

No space probe or other device has actually observed an eruption in progress on Venus, but a variety of evidence leads to the conclusion that eruptions have occurred recently. As we see in this chapter, there are surface features on Venus that appear to be volcanoes. It has been argued that the number of these now present implies that they are still forming from time to time. Other data, including observations of lightning storms and determinations of atmospheric chemical abundances, lend support to the idea that vol-

canoes may erupt occasionally on Venus.

The strongest argument, however, comes from recently reported variations in the quantity of sulfur dioxide (SO_2) in the atmosphere of Venus. This molecule was detected in abundance by the *Pioneer Venus* orbiter when it arrived at Venus in 1978, but its abundance has declined since then as the orbiter has continued to make measurements. Furthermore, ultraviolet measurements made by other instruments before 1978 failed to reveal any SO_2, even though it should have shown up if it had been as abundant as it was later found to be by *Pioneer Venus*. Apparently there was a sudden increase in the quantity of atmospheric SO_2 in 1978 or shortly before. This molecule is one of the major products of volcanic venting, so it has been suggested that a large-scale eruption took place on Venus in the late 1970s.

Other evidence supporting this idea comes from observations, also by *Pioneer Venus*, of a haze layer above the clouds of Venus. This haze has been attributed to tiny particles suspended in the atmosphere, and such particles are another common product of volcanic eruptions. It is known that no such haze existed in the 1960s, when observations were made from Earth that would have revealed its presence if it had been there.

Based on the quantity of SO_2

and haze apparently injected into the atmosphere, estimates have been made of the magnitude of the proposed eruption. These estimates suggest that the energy released was comparable to that of the famous Krakatoa explosion that occurred on Earth (near Java) in 1883. This was a very violent event, devastating an entire island, producing widespread tidal waves, creating sound waves heard thousands of miles away, and putting enough dust into the atmosphere to affect the passage of sunlight through it for several years. It may prove possible within the next few years to test the hypothesis that Venus is volcanically active. Of course, one possible form of verification will have been found if a new influx of SO_2 is discovered by the still-operating *Pioneer Venus* or by some other ultraviolet instrument. The typical time between eruptions major enough to be detected in this indirect way is not known, so there is no way of guessing whether a new rise in the SO_2 abundance will occur soon.

Another test of the theory will be provided by the *Magellan* high-resolution radar data, which may find evidence of ongoing changes in the shapes of surface features, if such changes occur during *Magellan's* observational period.

standing of what might be responsible for the extreme differences between conditions there and on the Earth. This is important, for we need to know how precarious our situation is and how to avoid turning the Earth onto the path that Venus has followed.

Like the Earth, Venus is thought to have formed from rocky debris orbiting the infant Sun in a great disk. The early evolution must also have been similar with Venus undergoing a molten period during which its dense elements sank to the center, leaving a lighter

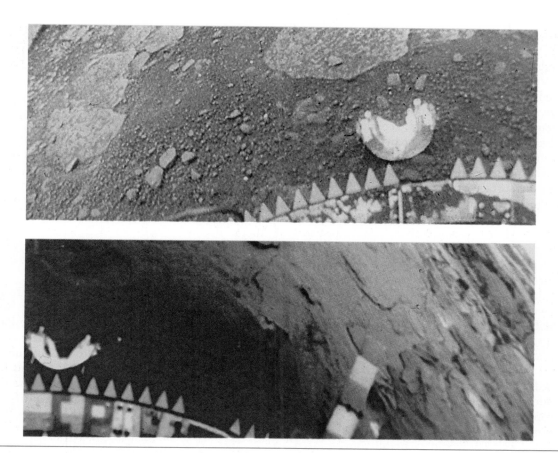

Figure 11.14 Color photos of surface rocks. *These photographs were obtained by the Soviet* Venera 13 *and* 14 *landers. The yellowish color seen here is an artifact of the camera response; reprocessing of these images has shown that the surface rocks are gray, like similar rocks on Earth. (TASS from SOVFOTO)*

crust composed largely of carbonates and silicates. Before the crust cooled, volatile gases escaped from the surface, forming a primitive atmosphere of hydrogen compounds and carbon dioxide much like the earliest atmosphere on Earth.

At this point a major contrast developed between the Earth and Venus. As oxygen escaped from the rocks on Earth and combined with hydrogen to form water, much of it could persist in the liquid state, and the Earth had oceans from that time onward. Like the Earth, Venus may have had large quantities of water early in its history, either due to outgassing from the interior or due to the impact of many waterbearing comets. As we discussed in the previous chapter, Earth had liquid oceans from an early time. On Venus, which is closer to the Sun, it is commonly thought that liquid water could not exist but stayed in the atmosphere as

water vapor. Venus is 0.72 AU from the Sun, so the intensity of sunlight at its surface is $1/.72^2 = 1/.52 = 1.92$ times greater than at the surface of the Earth. (Recall that the intensity of light from a source like the Sun drops off as the square of the distance.) This is not a great difference, but it was extremely significant.

Once the atmosphere of Venus became contaminated with water vapor, the greenhouse effect was enhanced, further heating the surface. Furthermore, the lack of liquid water meant that there was no mechanism for the atmosphere to lose its carbon dioxide, which was continuing to build up due to outgassing. On the Earth, the carbon dioxide in the primitive atmosphere was dissolved in the oceans and deposited in rocks, where it still resides. The development of life on Earth also had a major effect that was absent on Venus, gradually altering the Earth's atmospheric composition to one dominated by oxygen and nitrogen, which do not create a strong greenhouse effect.

The large quantity of carbon dioxide in the atmosphere of Venus caused increased greenhouse heating. This helped make the atmosphere hot during a time when there was a lot of water vapor present. The water vapor began to dissociate, that is, separate into free atoms of hydrogen and oxygen. The hydrogen escaped from the atmosphere, while the oxygen became trapped in surface rocks through a process called **oxidation.** Meanwhile the surface of Venus started to follow an evolution much like that of the Earth. Convection apparently developed soon after the interior differentiated and caused the tectonic activity that shaped the highland masses. On Venus, however, these continental areas did not begin to move about as they do on Earth. Volcanic activity may have been a continuing feature of the environment, however, as it has been on Earth. In a strictly geological sense, Venus may be a twin of the Earth, except for the lack of water erosion.

Contrary to the picture just described, some scientists have argued that Venus might once have had liquid oceans, rather hot ones. The high surface temperature caused by the carbon-dioxide induced greenhouse effect would have boiled these oceans, except that the high atmospheric pressure could have raised the boiling point enough so that liquid water could survive. Such oceans, if they existed, would eventually have caused their own demise, however. Carbon dioxide dissolves in water so the oceans would have removed this gas from the atmosphere, lowering the atmospheric pressure to the point where eventually the oceans would boil away and the water vapor would be lost as it dissociated and the hydrogen escaped. If this idea is correct, and Venus once had oceans that eventually boiled away due to greenhouse heating, it adds urgency to the need to understand and prevent a similar sequence of events from occurring on the Earth.

PERSPECTIVE

The study of Venus has been very instructive, not only for what it tells us of the planet itself, but also for the lessons learned about the Earth. We must not forget that, except for a difference in the intensity of sunlight at an early stage in its history, the Earth might have ended up with a crushing, hot CO_2 atmosphere like that on Venus.

We turn our attention next to another of the terrestrial planets, the next closest to Earth. In examining Mars we will be able to draw other parallels and contrasts with Earth.

SUMMARY

1. Observations of Venus from Earth, particularly radio data, revealed that it is extremely hot at the surface, and rotates slowly in the retrograde direction.
2. Soviet and U.S. spacecraft have flown by Venus and probed its atmosphere. Several Soviet probes have landed on the surface and sent back information on conditions there.
3. The atmosphere of Venus consists mostly of carbon dioxide, with a bit of nitrogen and traces of sulfur dioxide and sulfuric acid.
4. The atmospheric temperature increases from 240 K at the cloud tops to 750 K at the surface, where the pressure is 90 times that at sea level on the Earth. The heating is caused by the greenhouse effect.
5. The clouds, composed of droplets of sulfuric acid, form three distinct layers between 48 and 60 km in altitude. The atmosphere is clear below the clouds.
6. At the cloud tops, high winds circle the planet in four days, whereas it is calm at the surface. Above the clouds there are global convection currents.
7. The surface of Venus, consisting of rolling plains, highlands, and lowlands, shows evidence of limited tectonic activity but no continental drift. There is possible evidence of current volcanic activity.
8. Venus evolved in a manner very different from the Earth, primarily because it is so close to the Sun that liquid water could not exist on its surface. The lack of water allowed CO_2 to stay in the atmosphere and create a strong greenhouse effect, as well as discouraging the development of life.

REVIEW QUESTIONS

Thought Questions

1. Summarize the composition of the atmosphere of Venus, and compare it with the composition of Earth's atmosphere.
2. How does the role of sulfur dioxide (SO_2) in the atmosphere of Venus compare with the role of ozone in the Earth's atmosphere?
3. Why does Venus have so much carbon dioxide (CO_2) in its atmosphere, while the Earth has so little?

4. Why could the mass of Venus not be deduced from Earth-based observations in the same manner as the masses of most of the other planets?

5. How are the clouds in the atmosphere of Venus like clouds in the Earth's atmosphere, and how are they different?

6. Would you expect Venus to have charged-particle belts surrounding it, comparable to the Earth's Van Allen belts? Explain.

7. Summarize the evidence that Venus has undergone tectonic activity, and the extent of that activity compared to that on the Earth.

8. What is the evidence that Venus may have current volcanic activity?

9. Explain why Venus has never had liquid oceans, and the effect this has had on the evolution of the planet's atmosphere.

10. Why are scientists concerned about increasing levels of carbon dioxide in the Earth's atmosphere?

Problems

1. What is the wavelength of maximum emission from the surface of Venus, whose temperature is 750 K?

2. If Venus and the Earth are the same size, how much brighter than the Earth is Venus, given that Venus has a higher albedo (0.76 compared to 0.31) and receives more intense sunlight (almost twice as much; the figure is given in the text)?

3. Sea level atmospheric pressure on the Earth is 14.7 lb/in², and on Venus it is about 1,300 lb/in². How many times higher is the pressure on Venus?

ADDITIONAL READINGS

Allen, D. A. 1987. Laying bare Venus' dark secrets. *Sky and Telescope* 74:350.

Bazílevsky, A. J. 1989. The planet next door. *Sky and Telescope* 77:360.

Beatty, J. Kelly. 1985. A radar tour of Venus. *Sky and Telescope* 69(2):507.

————. 1985. A Soviet space odyssey. (Vega at Venus). *Sky and Telescope* 70(4):310.

Beatty, J. Kelly, B. O'Leary, and A. Chaikin, eds. 1990. *The new solar system.* 2d ed. Cambridge, England: Cambridge University Press.

Head, J. W., C. A. Wood, and T. A. Mutch. 1977. Geologic evolution of the terrestrial planets. *American Scientist* 65:21.

Kasting, J. F., B. Toon, and J. B. Pollack. 1988. How climate evolved on the terrestrial planets. *Scientific American* 258(2):90.

Montoya, E., and R. Fimmel. 1988. *Pioneers* in space: the story of the *Pioneer* missions. *Mercury* 17:56.

Pettingill, G. H., D. B. Campbell, and H. Masursky. 1980. The surface of Venus. *Scientific American* 243(2):54.

Schubert, G., and C. Covey. 1981. The Atmosphere of Venus. *Scientific American* 245(1):66.

Weaver, K. F. 1975. Mariner unveils Venus and Mercury. *National Geographic* 147:848.

Young, A., and L. Young. 1975. Venus. *Scientific American* 233(3):70.

MATHEMATICAL METHODS

Radar Measurements of Planetary Rotation

It is mentioned in the text that radar measurements from the Earth have been used to determine the rotation period of Venus. Here we discuss how this is done in a bit more detail.

The sketch below shows the basic situation. Radar beams are sent to a planet such as Venus, with reflected signals coming back from different parts of the planetary surface (in reality, it is not usually possible to send distinct, sharply-pointed

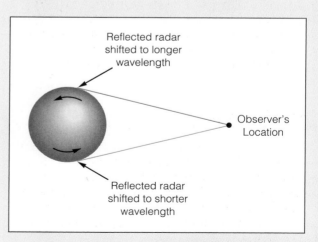

Reflected radar shifted to longer wavelength

Observer's Location

Reflected radar shifted to shorter wavelength

(continued on next page)

MATHEMATICAL METHODS

Radar Measurements of Planetary Rotation (continued)

beams to different sections of the planet; instead the beam covers the entire planet, and the reflected portions from different parts have to be sorted out by complex computerized methods). The reflected radar waves are Doppler-shifted if they bounce off of a portion of the planet that is moving away from us or toward us. From the amount of the Doppler shift, it is possible to determine the speed of approach or recession at the point where the beam is reflected, and from that the rotation period can be found, if the planetary diameter is known.

Let us use Venus as an example. Suppose radar having a rest wavelength of precisely 10.000 cm is beamed toward Venus. The return beam from one edge of the planet has a wavelength that is shorter than 10 centimeters by 6×10^{-8} cm, while the beam from the opposite edge has a longer wave-

length by an equal amount (in reality, frequency shifts are measured rather than wavelength shifts, because the measurement of such small shifts in frequency is a bit easier). For a shift $\Delta\lambda/\lambda = (6 \times 10^{-8}$ cm$)/(10.00$ cm$) = 6 \times 10^{-9}$, the speed of approach or recession is

$$v = \frac{\Delta\lambda}{\lambda}c = (6 \times 10^{-9})c,$$

where c is the speed of light, equal to 300,000 km/sec. Multiplying yields a speed of the surface of Venus of 0.0018 km/sec. One edge of Venus is approaching us at this speed, while the other edge recedes at this speed.

The radius of Venus is 6,055 km, so the circumference is 2π times this, or 38,040 km. The rotational period of Venus is the time it takes for a single point on the surface to travel the circumference of the planet. Thus the period of Venus is

$$P = (38{,}040 \text{ km})/(0.0018 \text{ km/sec})$$
$$= 2.1 \times 10^7 \text{ sec} = 243 \text{ days.}$$

This is the value listed in Table 11.1.

CHAPTER 12

Mars and the Search for Life

A portion of the giant Martian canyon Valles Marineris. (NASA; image processing by the U. S. Geological Survey)

The planet Mars, while not as brilliant as Venus, is nevertheless a prominent object in the heavens (Fig. 12.1). The fourth planet from the Sun, and our next nearest neighbor after Venus, Mars has played an important role in the development of astronomy, first lending itself to mythological interpretations, and later providing the basis of Kepler's discoveries about the motions of the planets. In more recent times Mars again became a center of considerable human speculation as people mused over the possibility that life

might exist there. The most recent chapter in the story, the exploration of Mars by unmanned spacecraft, has dispelled such notions but has not diminished our fascination with the red planet. Mars is likely to be the first planet to be visited by human beings.

OBSERVATIONS AND GENERAL PROPERTIES

From antiquity, Mars has attracted special attention because of its distinct and unusual reddish color, apparent even to the unaided eye. The Greeks named this planet Ares after their god of war; the Roman name Mars has the same meaning.

Telescopic observations began in the seventeenth century with Galileo and Huygens, who was able to discern surface markings. By 1666 the rotation period was known to be slightly more than 24 hours, and in the late 1700s the British astronomer William Herschel had deduced the tilt of the Martian rotation axis (very similar to the Earth's tilt). In 1877, the two Martian moons were discovered, which allowed accurate measurement of the planet's mass through the use of Kepler's third law (Table 12.1).

There was a period, peaking about 1900, when speculation over the possibility of a planetwide system of canals was popular. Even then, logical consideration showed that the linear surface features identified by some as canals would have to be hundreds of miles

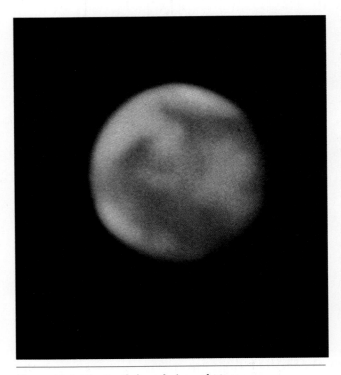

Figure 12.1 The Earth-based view of Mars.

Table 12.1 Mars

Orbital semimajor axis: 1.524 AU (227,900,000 km)
  Perihelion distance: 1.381 AU
  Aphelion distance: 1.667 AU
Orbital period: 1.881 years (687.0 days)
Orbital inclination: 1°51′0″
Rotation period: $24^h37^m22.6^s$
Tilt of axis: 23°59′
Diameter: 6,768 km (0.531 $D_\oplus$)
Mass: 6.42×10^{26} grams (0.107 $M_\oplus$)
Density: 3.96 grams/cm³
Surface gravity: 0.38 Earth gravity
Escape speed: 5.0 km/sec
Surface temperature: 145–300K
Albedo: 0.15 (average value)
Satellites: 2

Figure 12.2 Variations in the Martian surface markings. *This series of photos shows the changes in the polar ice caps and surface markings that occur with the seasons on Mars. (NASA)*

wide to be visible from Earth and that the notion of a Martian civilization had no basis in observable evidence.

Spectral analysis of the light reflected to Earth from Mars started in the first decades of this century. Photographs taken through red and blue filters revealed a haze in the Martian atmosphere that scatters blue light. The first spectroscopic measurements in the 1930s revealed an absence of water vapor and oxygen, of particular interest in view of the prevailing idea that life might exist on Mars. More recent observations identified absorption lines of carbon dioxide in the spectrum of Mars.

Seasonal changes in the coloration of certain regions on Mars (Fig. 12.2), well away from the polar caps, were once interpreted as due to foliation of green plants, which were thought to undergo the same sort of annual cycle as earthly plants. By the 1960s, however, good photographs had revealed evidence of large-scale dust storms in the Martian atmosphere, which inspired the American astronomer Carl Sagan to explain the seasonal color variations in terms of seasonal shifts in the Martian wind patterns, covering and uncovering some

areas of the surface. This explanation satisfactorily accounts for most of the observed variations, although some remain unexplained.

Infrared observations of Mars yielded data on its surface temperature, which was found to vary between 145 K (about −198°F) and 300 K (81°F). Even when the surface is warm, the Martian air remains cold, rarely rising above freezing.

Given our profound interest in Mars and the history of speculation over the possibility that life exists there, it is not surprising that the red planet became an early target for unmanned interplanetary probes (Table 12.2). The first successful mission to Mars was by *Mariner 4,* launched in 1964. This was followed by other *Mariner* missions, most notably *Mariner 9,* which orbited the planet and surveyed it photographically. This task was stymied for several months by a global dust storm, which completely obscured the surface of Mars (Fig. 12.3). The storm finally abated, however, allowing *Mariner 9* to make many profound discoveries, including several gigantic volcanoes and a canyon that dwarfs anything on the Earth.

ASTRONOMICAL INSIGHT 12.1

Percival Lowell and the Martian Civilization

 With its reddish color and brilliance when near the Earth in its orbit, Mars has always been a focus for speculation and whimsy. Ancient philosophers saw Mars as the representative of bloodshed and war; modern astronomers think Mars may once have housed life forms, though perhaps only on the microscopic scale. But probably the height of public interest in Mars occurred around the turn of the century, as the result of the writings and studies of a wealthy Boston aristocrat named Percival Lowell.

Lowell was an amateur astronomer whose fascination with the skies (and especially with Mars) would eventually lead him to establish a major observatory (in Flagstaff, Arizona), which today is still a leading research center. In the late 1800s, he learned of reports of dark, linear features on the surface of Mars that had been especially well observed during the opposition of 1877, when Mars came unusually close to the Earth (this was also the time when the two tiny moons of Mars were discovered). The features on the Martian surface had been described by one astronomer, the Italian Giovanni Schiaparelli, as *canali,* an Italian word simply

meaning "channels," usually applied to natural bodies of water. Some people misunderstood the word to mean *canals*—that is, artificial channels. Soon the notion of canals on Mars caught the public fancy and inspired Lowell in particular to speculate that an advanced civilization flourished on Mars. In 1895, Lowell published a book on the Martian civilization in which he theorized that the canals had been built in response to a global drought, as Martian farmers and ranchers desperately sought to preserve their crops and livestock by bringing in water from the melting polar ice caps. Lowell was not bothered in the least by the difficulties others had in seeing the canals, nor by the many rational arguments that the markings could not be canals in any event. (One difficulty is the enormous width— more than 50 miles—that the canals would have to reach in order to be seen from Earth with the telescopes of the day).

Lowell's writings increased public interest in Mars and Martians, and eventually fueled much of today's fascination with the possibilities of alien life forms. (How many books and movies have you encountered that involved Martians?) In 1898 the noted philosopher and science-fiction pioneer H. G. Wells published his novel about an invasion of Earth by Martians entitled *The War of the Worlds,* which has served as inspiration for generations of science-fiction fans and writers ever since. Perhaps the most famous spin-off of Lowell's ideas

was the 1938 radio broadcast depicting a Martian invasion of New Jersey, produced by Orson Welles. This broadcast, which was done in the manner of a news report (actually, a series of bulletins with ever-increasing urgency), thoroughly frightened many people who had not heard the disclaimers that were aired at the beginning.

Today's science-fiction stories and movies involving alien life forms can trace their roots to the early yarns of H. G. Wells, Edgar Rice Burroughs, Jules Verne, and others, some of whom in turn were inspired by Percival Lowell. Lowell himself pursued his interests in astronomy and in Mars, devoting much of the rest of his life to the field. His observatory, which housed some fine telescopes, was initially dedicated to furthering studies of Mars, but eventually it also became involved in other areas such as early studies of galaxies outside the Milky Way. The observatory's most notable moment came in 1930, when staff astronomer Clyde Tombaugh discovered the ninth planet, Pluto. The search for a ninth planet, inspired by Lowell himself and overseen by him until his death in 1916, is described in Chapter 14.

Table 12.2 U.S. Probes to Mars

Spacecraft	Launch	Encounter	Type of Encounter
Mariner 4	Nov. 28, 1964	July 14, 1965	Flyby, 10,000 km
Mariner 6	Feb. 24, 1969	July 31, 1969	Flyby, 3,200 km
Mariner 7	Mar. 27, 1969	Aug. 5, 1969	Flyby, 3,200 km
Mariner 9	Mar. 30, 1971	Nov. 13, 1971	Orbiter
Viking 1	Aug. 20, 1975	June 16, 1976	Orbiter plus lander
Viking 2	Sept. 9, 1975	Aug. 7, 1976	Orbiter plus lander
Phobos 2	July, 1988	June, 1989	Orbiter

A spectacular achievement of the Martian exploration program came some four years later, with the launch of the two *Viking* spacecraft in the late summer of 1975. Each consisted of an orbiter destined to circle Mars as *Mariner 9* had done, and a lander designed to descend to surface and land intact (Fig 12.4). The orbiters and the landers each carried a complex assortment of instruments designed for a variety of measurements. The orbiters improved on the quality of the *Mariner 9* photographic survey of the entire planet (Fig. 12.5), and the landers were able to carry out many sophisticated experiments, including panoramic photographs of the Martian landscape and sensitive tests for the presence of microscopic life forms in the soil. The results will be discussed later in this chapter.

The latest unmanned exploration of Mars took place

Figure 12.3 A global dust storm. *This is the view of Mars that confronted the* Mariner 9 *spacecraft for the first several weeks after it went into orbit around the planet.*

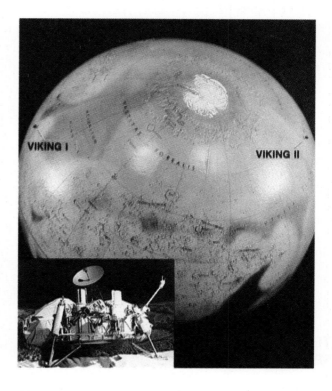

Figure 12.4 The Viking landing sites. *This globe of Mars indicates the locations of the landing sites for the two Viking spacecraft. One of the landers is depicted in the inset. (NASA)*

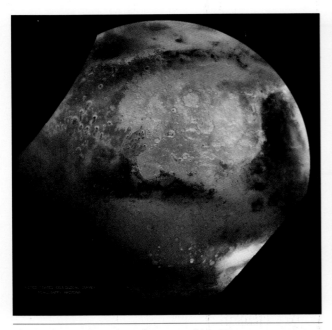

Figure 12.5 A Martian photomosaic. *Here numerous impact craters are clearly visible, as are color variations due to varying dust cover from place to place. The white area at lower right is a frost-covered lowland region. (U.S. Geological Survey; data from NASA)*

Figure 12.6 Phobos: Mars surface (Tass/Sovfoto). *This is one of the images of Mars obtained by the short-lived Phobos mission.*

very recently. Launched in the summer of 1988, two Soviet *Phobos* spacecraft were sent to Mars to survey the planet over an extended period and also to examine Phobos, one of the two tiny satellites of Mars. One of the spacecraft lost its communications with Earth on the way, and was unable to complete its mission, but the other one was successful in reaching Mars. Unfortunately, however, it also failed before its mission of close-up exploration of Phobos could be completed, although it did succeed in obtaining many fine images (Fig. 12.6).

THE MARTIAN ATMOSPHERE AND SEASONAL VARIATIONS

Mars has a very thin atmosphere compared to that of the Earth, with a surface pressure of 0.006 times the sea-level pressure on Earth. The Martian air is about 95 percent CO_2 and is similar to Venus in this regard (Table 12.3). Besides CO_2, other species include nitro-

gen, which accounts for 2.7 percent; argon, which constitutes another 1.6 percent; and other trace gases, which together total less than 1 percent. Water vapor is present in variable quantities, reaching levels of nearly one-tenth of 1 percent in the Martian summer. This is occasionally sufficient to form clouds of ice particles.

The Martian polar caps (Fig. 12.7) have been somewhat controversial because of the difficulty of determining whether they are made of water ice or frozen

Table 12.3 Composition of the Martian Atmosphere

Gas	Symbol	Fraction[a]
Carbon dioxide	CO_2	0.95
Nitrogen	N_2	0.027
Argon	Ar	0.016
Oxygen	O_2	0.0013
Carbon monoxide	CO	7×10^{-4}
Water vapor	H_2O	3×10^{-4}
Neon	Ne	2.5×10^{-6}
Krypton	Kr	3×10^{-7}
Ozone	O_3	1×10^{-7}
Xenon	Xe	8×10^{-8}

[a]Values are fractions by number

Figure 12.7 The south polar cap on Mars. *This is a composite of several images from the* Viking *missions, showing cratered terrain and the southern polar cap. (U.S. Geological Survey).*

CO$_2$ (dry ice). Based on temperature measurements, the conclusion was finally reached that the surface cap material is frozen CO$_2$, but that some water ice is also present, lying beneath the CO$_2$, particularly in the northern cap. The seasonal variations in the polar caps are due to the accumulation of CO$_2$ ice during the Martian winter, followed by evaporation in the summer. The northern and southern caps experience different seasonal variations, due in part to differences in the intensity of the seasonal changes in the weather, and in part to differences in the concentration of dust impurities, which affect the albedo of the caps, hence their ability to absorb sunlight and to heat up during the summer. Whereas the northern cap annually loses all of its CO$_2$, exposing the water ice cap beneath, in

most southern summers the south polar cap retains enough CO$_2$ to keep any underlying water ice protected.

The general question of water on Mars is complex and for a long time has been the basis of a serious scientific mystery, quite apart from the older speculations about canals. Enough water should have been released from surface rocks early in the history of the planet to cover its entire surface to a depth of tens to thousands of meters, so the lack of water on the planet is difficult to explain. There was probably no energy source on Mars sufficient to dissociate all of the H$_2$O molecules, as happened on Venus, but on the other hand, the atmospheric pressure is too low to allow liquid water to exist on the surface. The atmosphere, which gets rather cold at night, cannot support much water in the form of vapor, and the amounts in the ice caps is much too small to explain the discrepancy. The answer must be sought elsewhere. We will discuss current ideas about the fate of the Martian water supply in the next section.

The existence of clouds in the atmosphere of Mars was recognized long ago, from Earth-based observations. While some are due to blowing dust, many are white clouds formed of ice crystals, primarily H$_2$O, although some CO$_2$ clouds also exist. The clouds are found in lowlands in the early Martian morning and in regions where winds push the air up mountain slopes, forcing the H$_2$O vapor to cool and condense. In late summer in each hemisphere, extensive cloud systems form over the polar caps.

Winds on Mars are always present, often with substantial velocities (although the thinness of the atmosphere reduces the effect of the winds on surface features). The global flow patterns are created by the daily variations in solar heating at the surface and are strongly modified by terrain such as the large mountains that exist on Mars. The prevailing winds tend to move from west to east, but with a veer either toward or away from the poles, depending on the season. Although typical wind velocities are 35 to 50 km/hr, they can reach speeds of over 200 km/hr.

The fact that Mars has seasons was recognized long ago, when variations in the sizes of the polar ice caps were noticed. The cause of the seasons is a bit more complex than on Earth, however, because of the elliptical shape of the orbit of Mars. Earth's orbit is so nearly circular that the intensity of sunlight hitting its surface is almost constant year-round (in fact, the Earth is closest to the Sun in early January, and the slight extra solar input surely does not moderate the winter weather in the northern hemisphere). The orbit of Mars is more elongated, as we have already learned, with the result that at its closest approach to the Sun, Mars is about 17 percent nearer than when it is farthest away, on the other side of its orbit (Fig. 12.8). Therefore, the Sun's intensity on the Martian surface is $1/.83^2 = 1.45$ times greater at closest approach. This is enough to have a pronounced effect on the climate. Mars approaches the Sun most closely during winter in the northern hemisphere and is farthest away during the northern summer; hence the shape of the orbit tends to moderate both seasons in the north. By the same token, the southern seasons are enhanced by the orbital shape, so that this hemisphere has much more extreme variations from summer to winter. The southern polar cap grows larger than the northern one in its winter, but it diminishes to a smaller remnant in summer. As noted, however, only the southern cap retains sufficient CO_2 ice during the summer to keep the underlying water ice covered.

The most striking seasonal variations on Mars are the annual global dust storms, which have been observed for several decades to occur at the time when Mars is closest to the Sun, during the southern summer. The extra solar heating in the south creates winds that can begin to lift fine particles into the air, where they can stay suspended for lengthy periods. Some of the dust is raised by small-scale circular winds known on Earth as **dust devils,** which have been detected in photographs from the *Viking* orbiters. Once the seasonal winds begin, the dust storm grows to cover at least the southern hemisphere, and often the entire planet, and

it lasts for weeks. As we noted earlier, one result of these annual storms is seasonal variations in the dust cover on the ground in certain regions, creating the well-known seasonal color variations seen from Earth. Another result is substantial dust contamination of the northern polar cap, which is building up for the winter at the time of the dust storms. This makes the northern cap "dirty," allowing it to absorb more sunlight than if it were clean, and the result is enough heating to cause complete evaporation of the CO_2 ice during the summer. In contrast, the southern cap remains cold enough all summer to retain more CO_2 ice.

WATER, TECTONICS, AND THE MARTIAN SURFACE

The surface of Mars (Fig. 12.9) can be classified into two types of areas: plains, which dominate much of the northern hemisphere; and rough, cratered terrain, which covers much of the south. The plains are low-elevation regions covered with lava flows, and altogether they spread over some 40 percent of the planet. The remainder is occupied by the rougher cratered land, much of which consists of highlands that are older than the lowlands. One huge uplifted region, called Tharsis, covers nearly one-quarter of the entire Martian surface and averages some 6 km elevation above the mean surface level. The formation of this gigantic bulge on Mars is not well understood, but most suggestions invoke tectonic activity driven by convection in the mantle that occurred early in the planet's history.

A major valley system (Fig. 12.10), extending some 4,500 km in length, was named Valles Marineris in honor of *Mariner 9,* whose photographs revealed its gargantuan proportions. This valley would dwarf the Grand Canyon, which is 5 to 10 times smaller in every dimension. Valles Marineris is 700 km across at its widest and 7 km deep at its deepest. It probably was formed when the Tharsis highland was uplifted, creating extensive cracks in the planetary crust.

Another major feature of the surface is created by the astounding volcanoes that exist on Mars, primarily in three locations. These are the most conspicuous features on the planet, except for the polar caps. The grand champion is Olympus Mons (Fig. 12.11), a monster with a height of 27 km (nearly 90,000 feet) above the mean surface level, and a base some 700 km across. (Mt. Everest, on Earth, rises 8.8 km or 29,000

Figure 12.8 The Martian seasons. *This exaggerated view shows that in the northern hemisphere both summer and winter are moderated by the varying distance of the planet from the Sun, whereas in the southern hemisphere both seasons are enhanced. The extreme temperature fluctuations in the southern hemisphere give rise to the winds that cause the seasonal dust storms.*

Figure 12.9 A relief map of Mars. *The highest elevations are shown in white in this view; note the massive Tharsis plateau (left), and the huge volcanoes. (M. Kobrick, JPL)*

Figure 12.10 Valles Marineris. *A tremendous valley as long as the breadth of North America is put into perspective in the left-hand global mosaic, which shows the valley at lower center. To the left are three of the giant volcanoes of Mars. The image at right shows detail in a portion of Valles Marineris. (left: U.S. Geological Survey, data from NASA; right: NASA)*

Figure 12.11 Olympus Mons. *This is the largest mountain known to exist in the entire solar system. Its base is comparable in size to the state of Colorado, and its height is about 3 times that of Mount Everest. (NASA)*

feet). Olympus Mons is in the northern hemisphere, not far from the Tharsis plateau. The other great volcanic structures are formed by chains of peaks stretching across vast areas of the Martian plains.

The largest of the giant volcanoes apparently formed because of a lack of continental drift at the time and location of their formation. If they had formed on moving plates in the Martian crust, they should have moved away from the subsurface hot spots that created them before they got so big.

The entire question of Martian tectonic activity is unsettled. There is no well-developed system of crustal plates, and no evidence of current continental drift. It has generally been believed that the volcanoes are quite old—perhaps 3 billion years—but there is evidence that some, including Olympus Mons, are less than half a billion years old. This implies that volcanic activity took place as recently as that. Some scientists have even suggested that volcanic activity on Mars is not yet completely finished, and that future eruptions may be possible. The Tharsis plateau and the nearby Valles Marineris formed about 4 billion years ago, indicating that Mars was geologically active at that early stage in its history. All of the age estimates are based on crater counts and are therefore quite uncertain. It would be helpful, indeed, to be able to perform radioactive dating measurements on rocks from various parts of the Martian surface.

The present-day lack of tectonic activity is consistent with the absence of any detected magnetic field, since

most theories of magnetism invoke a fluid core, something else that Mars may not have.

Close examination of the surface of Mars reveals some interesting details. In most regions, there are craters created by impacts, as on the Moon. Some regions appear completely disorganized, resembling loosely piled rocks and rubble, as though landslides had occurred there. In some locations, often adjacent to this so-called **chaotic terrain,** there are winding valleys and flood plains that appear to have been formed by flowing water (Figs. 12.12, 12.13, and 12.14). These features have created a great deal of excitement because, as we learned earlier, the lack of water on Mars has been a major puzzle. All of the channels are quite old, having formed around the time when the Tharsis plateau was uplifted.

The fact that some of the channels are adjacent to chaotic regions suggests that water has been present on Mars in the form of ice underneath the surface, probably in the form of **permafrost.** One theory of the formation of the channels is that, early in the planet's history, underground heating caused the occasional sudden melting of a pocket of permafrost, creating a large reservoir of water. This water quickly ran off, carving a channel in the surface and leaving behind empty caverns or porous soil in the subsurface volume that it evacuated. The soil there collapsed, creating the kind of jumbled, disorganized appearance associated with the chaotic terrain. This hypothesis seems satisfactory for the situations in which a channel emanates from the chaotic region, but there are channel systems not associated with chaotic terrain, so at least in some cases another explanation may be needed. One possibility is that Mars had substantial quantities of surface ice at times in the past, and that melting of this ice created the channels. Another suggestion is that Mars once had a sufficiently thick and warm atmosphere that liquid water could have persisted on its surface; that is, Mars may once have had seas or extensive lakes. In this view, long-term changes in the atmosphere reduced its pressure, so that eventually the water evaporated and escaped or froze out, forming permafrost.

It is not clear why the climate of Mars would have changed so dramatically that lakes or oceans of liquid water would have been lost. One possibility is that a catastrophic event such as a collision with a large planetesimal could have stripped away the early, thick atmosphere, creating the low pressure that today prevents liquid water from existing on the surface. Another possibility is that water created its own downfall

Figure 12.12 Chaotic terrain.   *(Above) This* Viking *orbiter photo shows a region that apparently subsided when permafrost below the surface suddenly melted and flowed away (toward the left). NASA*

Figure 12.13 An ancient river channel on Mars.   *(Left) This is one of the most striking examples of evidence for flowing water in the Martian past. (NASA)*

Figure 12.14 A flood plain.   *Here is clear evidence that water once inundated this region of the Martian surface. The surface patterns indicate that the prominent craters in this photo already existed when the water flowed. The two mosaics show parts of Kasei Valles. (NASA)*

ASTRONOMICAL INSIGHT 12.2

Meteorites from Mars?

Meteorites (discussed in Chapter 15) are pieces of rock or metal which occasionally fall to the Earth's surface from space. Detailed analysis of the chemical and isotopic composition of meteorites has led scientists to the realization that some groups of meteorites may have common origins. There are "families" of meteorites classified according to their composition, that are thought to have come from the same parent body. In most cases, the parent body is believed to have been an asteroid that was destroyed in a collision (see discussion in Chapter 15), but there is one family of meteorites whose parent body may be the planet Mars.

This grouping of meteorites actually is a combination of three originally distinct groups: (1) the shergottites; (2) the nakhalites; and (3) the unique object that fell in Chassigny, in France, in 1815, called simply Chassigny. (The other two groups are named after particular minerals that occur in their compositions.) These three groups have many features in common and are now lumped together as the **SNC meteorites.**

The SNC meteorites are characterized by crystalline structures whose ages are estimated to be no greater than 1.3 billion years. They also lack chondrules, small spherical inclusions found in many meteorites, which form at low temperatures. These two facts together suggest that the parent body was subjected to high temperatures at fairly recent times (i.e., 1.3 billion years ago or more recently) in the history of the solar system. Scientists believe that asteroids, the source of most meteorites, should not have been hot enough to undergo crystallization so recently, whereas a planet of sufficient size could have been hot enough for long times after the formation of the solar system. A massive planet has large quantities of internal heat left over from its formation, has internal sources of radioactive heating, and may also have ongoing geological processes such as volcanic activity that create rocks through heating and melting. Therefore it is thought that the SNC meteorites must be fragments of a planet-sized body.

Another argument in favor of this point of view is based on the detailed chemical composition of this family of meteorites. Certain gases that are highly volatile and also lightweight are present in the SNC meteorites but are absent in families of meteorites thought to have originated in small bodies such as asteroids. The best explanation for the presence of these elements in the SNC meteorites is that these meteorites are fragments of a parent body that had a large gravitational field, strong enough to retain the lightweight gases. This suggests that the parent body was a planet rather than an asteroid.

The clue that Mars may be the parent body comes from the analysis of tiny volumes of gas found trapped within the rocky structure of the SNC meteorites. The composition of this gas is very similar to that of the Martian atmosphere, suggesting that the gas may have been trapped there when the rocks originally solidified on Mars.

The weight of current evidence regarding the origin of the SNC meteorites is that they were formed as the result of a major impact or impacts on Mars sometime in the past 1.3 billion years or less. It is possible that the SNC meteorites originated from more than one impact. The crystalline structure of the fragments retains information on the energy of impact through distinctive features due to shock heating. Variations in the degree of shocking suggest that the SNC meteorites may have arisen from two or more distinct impact events on Mars.

Through indirect arguments, it is possible to determine where on Mars the impact or impacts occurred. The fact that the crystalline structure of the SNC meteorites must have formed less than 1.3 billion years ago suggests that the region where the fragments originated must have undergone volcanic activity within the past 1.3 billion years. The only region on Mars that appears to be geologically young enough to allow this is the Tharsis upland area, a huge, elevated region containing the immense volcanoes discussed in this chapter.

If the SNC meteorites are indeed fragments of violent impacts on the surface of Mars, then they provide scientists with an unparalleled opportunity to study the surface of Mars, far in advance of the day when Earth-based probes travel to the red planet and bring back samples of surface material. Thus, we may expect to learn a lot about the geological history of Mars even before we go there.

by dissolving CO_2 from the atmosphere and depositing it in carbonate rock, thereby reducing the heating influence of the greenhouse effect. As the atmosphere cooled, it became less able to support water vapor, further reducing the greenhouse heating. This combined loss of atmospheric CO_2 and H_2O and the consequent reduction of the greenhouse heating effect would have led, in time, to the dry, thin atmosphere of Mars today. One test of this theory will be a search for carbonate rocks on the surface, which should have formed if CO_2 was removed from the atmosphere by the action of water.

Another entire class of surface features on Mars are those created by winds. We have already learned about the seasonal dust storms and the variations in the surface coloration that they cause, but there are other features created by the winds, including light and dark streaks and the **laminated terrain** (Fig. 12.15), curious areas where the surface has apparently been deposited in layers and then eroded at the edges. The stair-step structures that result are especially prominent around the edges of the polar caps and may be formed of thin layers of dust and ice that are precipitated by the annual storms.

There are also sand-covered regions where the surface has been shaped into dunes, just like those seen on earthly deserts and beaches. (Fig. 12.16). The sand in the dune regions is distinct from the much finer dust

Figure 12.16 Martian sand dunes.   *These dunes are testimonials to the incessant action of winds on Mars. (NASA)*

that is suspended in the air during the major storms. A vast field of dunes girdles the planet just outside the north polar cap, creating both a dark belt visible from afar and a mystery for scientists, who can offer no plausible explanation of its origin.

A great deal more will be learned about the surface of Mars and the role of water on the red planet, when the U.S. *Mars Observer* spacecraft reaches Mars (in 1993, according to the current schedule). This orbiter will be equipped with a variety of instruments designed to probe the composition of Mars' atmosphere and its changing water vapor content, and to perform sophisticated analyses of the nature and composition of the surface. It is planned to operate long enough to monitor the seasonal variations and should clear up many of the uncertainties pointed out in this text.

SURFACE ROCKS AND THE INTERIOR OF MARS

The *Viking* landers were able to closely analyze Martian soil and surface rocks. Both landing sites were in the plains regions of the northern hemisphere, and at both places the landscape consisted of dusty soil, with low ridges, a dense scattering of rocks and boulders, and craters here and there (Figs. 12.17 and 12.18). The rocks lying around were probably ejected by the impacts that formed nearby craters.

Figure 12.15 Laminated terrain.   *This photo shows some of the curious layered surface features seen near the edges of the polar ice caps. (NASA)*

Figure 12.17 A ground-level panorama. *This is the first photograph made by the* Viking 1 *lander. It shows a rock-strewn plain extending in all directions. (NASA)*

Figure 12.18 Drifting dust on Mars. *This sequence of Viking lander photos reveals subtle changes in the quantity and color of fine dust particles, which vary with shifting winds. (U.S. Geological Survey; data from NASA)*

The chemical composition of the surface minerals was measured (Fig. 12.19), and striking differences were found between Mars and the Earth. The soil of Mars appears to have formed from the breakdown of igneous rocks such as basalt, but it contains unusually high concentrations of silicon and iron. The iron, in the form of iron oxide (ie., rust) is responsible for the reddish color. Until scientists have a chance to search for carbonate rocks on Mars, the theory that early lakes or oceans depleted atmospheric CO_2 by forming carbonate rocks will remain unverified.

The high abundance of iron is quite significant, for it indicates that Mars is not as highly differentiated as the Earth. Apparently the crust was not sufficiently molten for long enough to allow the heavy elements to sink to the center of the planet, and Mars does not have a large iron-nickel core. There probably is a dense core, but it must be considerably smaller than the cores of the other terrestrial planets (measurements of the gravitational field show evidence for this, as does the absence of a planetary magnetic field). Ironically, even though Mars has a high surface content of iron, its *total* iron content is smaller than that of the other terrestrial planets. The explanation for this lies in the history of element condensation during the formation of the solar system, something we will discuss in Chapter 16.

These data lead to a picture of an early Mars that, due to its low mass or its distance from the Sun, was not fully molten for long and which formed a very

thick crust early in its history. Some tectonic activity has occurred, but there have not been extensive recent motions of the crust. Because of this, massive volcanoes formed, some of them building up to tremendous heights because of the lack of continental drift to carry them away from the hot spots that created them.

PROSPECTING FOR LIFE: THE *VIKING* EXPERIMENTS

The most widely publicized aspect of the dual *Viking* missions was the attempt made by the robot landers to detect evidence of life forms on Mars. This was done in several ways, the first and most straightforward being simply to look with the television cameras for any large plants or animals. None were seen, and much more sophisticated tests were tried.

The tests for life were done by three different experiments on board the landers, each of which used soil scooped up by a mechanical arm (Fig. 12.20) and deposited in containers for analysis. The basic idea of all three was similar: to look for signs of metabolic activity in the sample. All living organisms on Earth,

Figure 12.19 A geochemical map of Mars.   *This map, reconstructed from* Viking *orbiter data, uses colors to indicate predominant minerals in an equatorial strip between 30°N and 60°S latitude on Mars. The red colors indicate iron oxides, and the blues are less-oxidized basalts. Turquoise indicates fogs and frosts, and oranges, browns, and yellows show where sand or dust covers the surface. (NASA/U.S. Geological Survey)*

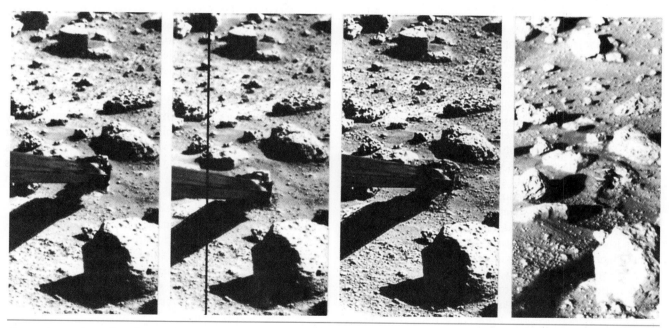

Figure 12.20 Sampling the Martian soil.   *Here the scoop on* Viking 2 *is shown as it digs up a sample of Martian soil for one of its life-detection experiments. (NASA)*

even microscopic ones, alter their environment in some way just by existing. Usually the effects involve chemical changes as the organism derives sustenance from its surroundings and ejects waste material.

Two of the three experiments, the *pyrolytic release experiment* and the *gas-exchange experiment,* showed no evidence for possible life forms. The third, however,

the *labeled release experiment,* created much excitement among scientists because its initial results duplicated the expected effects of active life forms. In this experiment, a sample of Martian soil was placed in a container, and a nutrient solution was added, in the expectation that organisms in the soil would metabolize the nutrient and release waste gases in the chamber. To make such released gases detectable, a small quantity of radioactive carbon (^{14}C, each atom consisting of the usual six protons but eight neutrons instead of six) was added to the nutrient. If metabolic activity took place in the soil sample, then some ^{14}C should appear in gases emitted by the sample, and that is just what happened. As a check, the same experiment was performed on other samples that had been heated so that any life forms present would have been killed, and indeed no tracer gases were emitted from those samples. The combination of activity in the normal samples and the lack of activity in the sterilized examples was consistent with the presence of life forms in the Martian soil.

The only flaws in the picture were that the other experiments failed to show positive results, and that the speed with which the reactions occurred in the labeled release experiment was much higher than expected for metabolic activity. Gases containing ^{14}C were released more quickly and in greater quantity than would have been possible for any earthly microorganisms. Furthermore, and perhaps most telling, a given sample would only react positively *once*, even though nutrient was added to some samples several times. Real life forms should be capable of eating more than one meal. Eventually it was concluded that the observed activity must have been due to an unexpected type of rapid chemical reaction between the soil and the nutrient. The activity probably involved oxides in the soil whose chemical properties were altered by heating, so that the reaction did not occur in the sterilized samples.

A further test for evidence of life was carried out by a device called a **mass spectrometer,** which is capable of analyzing a sample to determine the types of molecules it contains. The mass spectrometers aboard the *Viking* landers found no evidence in Martian soil samples of **organic molecules,** those containing certain combinations of carbon atoms that are always found in plant or animal matter on Earth. Hence, no doubt with some reluctance, the *Viking* scientists concluded

that no evidence for life on Mars had been found.

The lack of evidence for life on Mars is not the same as evidence for lack of life. After all, these experiments could test samples at only two localities on a whole planet, and furthermore, the tests were predicated on the assumption that Martian life forms, if they exist, would be in some way similar to those on Earth. Now that a great deal is known about the chemical properties of the Martian soil, new experiments probably could be devised that would be relatively free of the confusion created by the nonbiological activity detected in the Viking experiments. Perhaps one day soon, further attempts to find living organisms on Mars will be made.

THE MARTIAN MOONS

The two satellites of Mars, discovered telescopically by Asaph Hall in 1877, are rather insignificant compared to the Earth's moon, or to the major satellites of the outer planets (Table 12.4). Both are very near to the parent planet and therefore have relatively short periods: Phobos (Fig. 12.21), the inner moon, circles Mars in just 7 hours, 39 minutes; while Deimos (Fig. 12.22) has a period of 30 hours, 18 minutes. Because Phobos orbits the planet in a time less than a Martian day, to an observer on the planet's surface this satellite would cross the sky twice daily, in the west-to-east direction.

Close-up photos of Phobos and Deimos reveal that each is a small, irregularly shaped chunk of rock, pitted with craters, and, in the case of Phobos, covered by linear grooves. Both satellites are somewhat elongated, Phobos having a length of some 28 km and a width of 20 km, while Deimos measures roughly 16 by 10 km. Both are in synchronous rotation, keeping one end permanently pointed toward Mars.

A striking and unusual characteristic of the surface

Table 12.4 Satellites of Mars

No.	Name	Distance (R_M*)	Period (days)	Diameter (km)	Mass (g)	Albedo
1	Phobos	2.77	0.3189	27 × 21.6 × 18.8	9.6×10^{18}	0.06
2	Deimos	6.94	1.2624	15 × 12.2 × 11	1.9×10^{18}	0.07

*The distances given are orbital semimajor axes, in units of the radius of Mars, which is $R_M = 3,384$ km.

ASTRONOMICAL INSIGHT 12.3

Primordial Life on Mars?

The *Viking* landers failed to find any conclusive evidence for the existence of living forms in the Martian soil. Some scientists have argued that the experimental techniques employed were not sensitive enough, and that microscopic life could still be present on Mars. Many others are more pessimistic, pointing out that all life forms we know of need liquid water, and that water does not exist in this form on Mars.

Even if Mars has no life today, some scientists have suggested that life may once have thrived on the red planet. We believe that life on Earth began through complex chemical reactions that took place in the early oceans (this is discussed in more detail in Chapter 32). The oldest fossil evidence for life on Earth is found in rocks in western Australia that are about 3.5 billion years old. Thus, life began within the first billion years after the formation of the Earth.

As discussed in the text, we believe that Mars originally contained as much water and carbon dioxide (in proportion to its mass) as Venus and the Earth. Today we do not know where all of this water and carbon dioxide is, but we believe some of the water is stored below the surface in the form of ice. Additional quantities of water and carbon dioxide may have escaped into space, as discussed in the text. The ancient river channels on Mars testify to the fact that liquid water has existed there, although there is some question as to whether this was a temporary phenomenon or a long-term condition.

If Mars really did have an early period when the atmosphere was dense, then the planet would very likely have been much warmer than it is today, because of greenhouse heating due to the high carbon dioxide abundance. The elevated atmospheric pressure, combined with the warmer temperatures, would have allowed lakes or oceans to exist. In short, conditions much like those on the early Earth might have prevailed. If so, there is no reason to rule out the possibility that primitive life forms may have developed. The search for traces of such life forms is one of the important reasons that scientists are making plans for new missions to Mars.

To seek evidence of fossil life forms on Mars, soil samples must be scrutinized in various ways, so the task demands a landing on the surface of the planet. The most comprehensive testing could be done by using sophisticated equipment, so the idea of returning samples of the soil to the Earth has been proposed. Even more daring is the proposal that a manned mission go to Mars and conduct searches there. People in vehicles equipped for roving the surface could explore areas where standing water once might have been, and could probe depths where rocks with fossil traces might have been undisturbed.

Both the U.S. and the Soviet Union are making plans for future missions to Mars. At the moment, the only definite U.S. mission is the *Mars Observer* (scheduled for a 1992 launch from Earth), an orbiter which will tell us little about possible evidence for primordial life. The Soviets are planning three types of missions, two of which involve making probes of the Martian surface. One will insert shafts with scientific instruments deep into the soil, to measure conditions below the surface; the other will be a sample return mission. Scientists in both nations have begun to discuss the possibility of a joint manned mission, so perhaps in our lifetimes we will see a definitive search for ancient life on Mars. If evidence for such life is found, this will tell us that the Earth is not the only place where life started, as profound and far-reaching a discovery as any ever made.

of Phobos is that it is covered, over most of its area, with parallel grooves some 100 to 200 meters wide and 10 to 20 meters deep. These features are most prominent in the vicinity of a very large crater on Phobos and are least evident on the opposite side. Several ideas for their origin have been suggested, but the one that appears to fit the data best is the hypothesis that they were created by the same impact that carved out the major crater. It has been suggested that this impact fractured Phobos along planes where its material was weakest, and that heat from the impact caused some melting of subsurface rock, which then partially filled

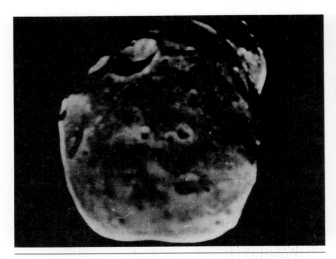

Figure 12.21 A new view of Phobos. *This is one of the images obtained by the Soviet mission Phobos. Here we see greater detail than was afforded by the U.S. Viking photographs, including several impact craters. Not seen here is a larger crater (described in the text) that resulted from an impact that almost destroyed the satellite. (Soviet Academy of Sciences)*

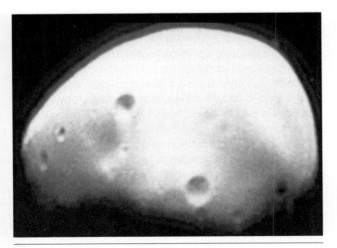

Figure 12.22 Deimos. *The more distant of the two moons of Mars, Deimos has a relatively featureless surface. (NASA)*

the fractures. The complete lack of similar grooves on Deimos, in this view, is due to the fact that this satellite has never suffered such a major impact, as shown by its lack of large craters.

Clearly Phobos and Deimos were not formed in the same way as the Earth's moon. The Moon has a much larger mass relative to its parent planet and evidently went through some geological evolution, whereas Phobos and Deimos are tiny and show no signs of having been acted on by the kinds of forces, such as vulcanism, that have forged the surface of the Moon. The origin of the two Martian moons is not well understood, but many have pointed out the general similarities between them and typical asteroids.

PERSPECTIVE

The mysterious red planet is still mysterious, but in new ways. Most of the age-old questions related to its changing appearance, the possibility of civilizations thriving on its surface, and its distinctive color have been answered through close-up examination. What we have found is a planet like the Earth, but one not as fully developed, in the geological sense. Tectonic

activity was arrested in the planet's youth, and the low temperatures and low escape velocity combined to leave Mars with only a thin atmosphere of CO_2, its water either escaped or hidden away underground.

We have now completed our tour of the terrestrial planets, and we are ready to consider the outer planets of the solar system. These are vastly different from the terrestrial planets, as we shall see.

SUMMARY

1. Telescopic observations of Mars show a thin carbon dioxide atmosphere, polar caps that vary in size with the Martian seasons, and a very low abundance of water vapor.
2. The Martian atmosphere has almost the same composition as that of Venus, but has only 0.6 percent (0.006) of Earth's sea-level pressure.
3. Seasonal variations on Mars are strongly influenced by the relatively elongated shape of its orbit.
4. The Martian surface consists of plains and rough highlands, with evidence for ancient tectonic activity, but no continental drift.
5. The Martian water is stored in the permanent ice caps, and possibly beneath the surface, in the form of permafrost. Much of the original water may have escaped Mars long ago.
6. Surface rocks have a high iron content, and Mars

has only a moderate average density, both facts indicating that the planet has undergone little differentiation.
7. The Viking experiments found no evidence for life forms in the Martian soil.
8. Mars has two tiny satellites, resembling asteroids in size and shape.

REVIEW QUESTIONS

Thought Questions

1. Before close-up observations of Mars were made by the *Mariner* and *Viking* probes, what were the best arguments against the idea that the dark markings on Mars were canal systems?
2. Compare the atmospheric composition of Mars with that of Venus. How do the physical conditions (pressure and temperature) compare?
3. The composition of the atmosphere of Mars (and other planets) can be determined by measuring the spectrum of sunlight reflected from the planet. Gases in the atmosphere of the planet create absorption lines that allow scientists on Earth to identify the gases. How do you think the absorption lines created by the planet's atmosphere are distinguished from absorption lines in the Sun's spectrum? How do you think they are distinguished from absorption lines created by the Earth's atmosphere?
4. Summarize the role of water on Mars; that is, where is it and how does its distribution vary with the seasons?
5. Explain why the seasonal variations on Mars are so different in the northern and southern hemispheres of the planet.
6. Compare the atmospheric circulation and winds on Mars with those on the Earth.
7. Discuss the evidence for tectonic activity on Mars, and whether it is thought to be occurring today.
8. Why is Mars red?
9. In the *Viking* searches for life on Mars, what assumptions were made about the nature of life forms that might be present there?
10. Mars has little or no magnetic field. Discuss the effect of this on possible Martian life forms, and on humans who might someday visit there.

Problems

1. From Figure 12.2, estimate how wide some of the dark features are, relative to the diameter of Mars. Knowing that the planet's diameter is 6,768 km, calculate how wide these dark features must be. Is your answer reasonable if the features are canals?

2. To illustrate why Earth's seasonal variations are not affected by the varying distance of Earth from the Sun, calculate the relative intensity of sunlight on the Earth when it is closest to the Sun, compared to when it is farthest. At closest approach, the Earth is 0.983 AU from the Sun, and at its farthest, it is 1.017 AU away. Compare this variation with that on Mars (figures are given in the text).
3. Calculate the mass of Mars from data on the period and semimajor axis of one of its satellites, using Kepler's third law. How does your answer compare with the value given in Table 12.1?

ADDITIONAL READINGS

Arvidson, R. E., A. B. Binder, and K. L. Jones. 1978. The surface of Mars. *Scientific American* 238(3):76.

Beatty, J. K., B. O'Leary, and A. Chaikin, eds. 1990. *The new solar system.* 2d ed. Cambridge, England: Cambridge University Press.

Carr, M. S. 1983. The surface of Mars: A post-*Viking* view. *Mercury* 12(1):2.

————. 1976. The volcanoes of Mars. *Scientific American* 234(1):32.

Cordell, B. M. 1986. Mars, Earth and ice. *Sky and Telescope* 71(1):17.

Dick, S. J. 1988. Discovering the moons of Mars. *Sky and Telescope* 76:242.

Goldman, S. J. 1987. In search of Martian Seas, *Sky and Telescope* 73:6.

————. 1990. The legacy of *Phobos. Sky and Telescope* 79:156.

Hartmann, W. K. 1989. What's new on Mars? *Sky and Telescope* 77:471.

Horowitz, N. H. 1977. The search for life on Mars. *Scientific American* 237(5):52.

————. 1986. *To utopia and back: The search for life in the solar system.* New York: W. H. Freeman.

Kasting, J. F., Toon, B., and Pollack, J. B. 1988. How climate evolved on the terrestrial planets. *Scientific American* 258(2):90.

Murray, B. C. 1973. Mars from Mariner 9. *Scientific American* 228(1):48.

Leovy, C. B. 1977. The atmosphere of Mars. *Scientific American* 227(1):34.

Pollack, J. B. 1975. Mars. *Scientific American* 233(3):106.

Veverka,d J. 1977. Phobos and Deimos. *Scientific American* 236(2):30.

Young, R. S. 1976. Viking on Mars: A preliminary survey. *American Scientist* 64:620.

MATHEMATICAL METHODS

Planetary Energy Budgets

 We learned in Chapter 6 that every body radiates according to a few simple laws that depend on surface temperature alone. The radiation is emitted over a range of wavelengths, with a maximum at a specific wavelength that is inversely proportional to the temperature (this is Wien's law), and with an amount of energy per square centimeter of surface area that is proportional to the fourth power of the temperature (Stefan's law). The total rate of energy emission (in units of ergs/sec), known as the luminosity of an object, is equal to its emission per square centimeter times the surface area in square centimeters. The emission per square centimeter is σT^4, where σ is the Stefan-Boltzmann constant, having a value of 5.67×10^{-5} in the centimeter-gram-second system of units. The surface area of a spherical body is $4\pi R^2$, where R is the radius. Putting these together, we see that the luminosity of any spherical object of temperature T is given by the Stefan-Boltzmann law:

$$L = 4\pi R^2 \sigma T^4$$

There are many useful applications of this relationship, as we can demonstrate by examining the observed properties of Mars. For example, we can predict what the wavelength and intensity of emission from Mars should be, if we were to observe the radiation that it is emitting. Here we have to distinguish between *emitted* radiation and *reflected* sunlight; fortunately, this is easy to do observationally, because Wien's law tells us that the emitted radiation for a body of average temperature 200 K (approximately correct for Mars) occurs in the infrared part of the spectrum (λ_{max} for $T = 200$ K is 145,000 A, well into the infrared). The Sun's major energy output occurs in the visible portion of the spectrum, near 5,000 A.

Let us first calculate the luminosity of Mars, using the above equation and a value of 200 K for the temperature and 3.38×10^8 cm for the radius; we find $L = 1.3 \times 10^{23}$ ergs/sec.

If this quantity of energy is emitted by Mars every second, then Mars must be receiving an equal amount from absorbed sunlight (neglecting any internal source of energy Mars may have, such as radioactive heating). The Sun's luminosity is approximately 4×10^{33} ergs/sec; if we use the inverse square law to calculate how much energy reaches Mars, we find that the surface of Mars receives 5.9×10^5 ergs/sec/cm². The cross-sectional area of a sphere is πR^2, so the total energy Mars receives from the Sun is 2.1×10^{23} ergs/sec (we get this by multiplying the cross-sectional area times the amount of energy per square centimeter that is reaching Mars from the Sun.)

We note, with some satisfaction, that the amount of energy received from the Sun by Mars, 2.1×10^{23} ergs/sec, is comparable to the amount being emitted, 1.3×10^{23} ergs/sec. The planetary energy budget is almost balanced, even in these crude calculations. However, we know that the two values must be precisely equal (again, if we can safely neglect internal energy sources). The difference is largely due to two factors: (1) Not all of the incoming sunlight is absorbed by Mars; and (2) the precise temperature to use in the calculation of luminosity is difficult to determine for a planet whose temperature varies over its surface and also changes with time.

The albedo, or fraction of sunlight that is reflected, is 0.15 for Mars, meaning that 0.85 of the incoming energy is absorbed. This reduces the discrepancy, since we now revise our calculation of the absorbed energy to 0.85 of the value we got above; this gives us an absorbed energy of 1.8×10^{23} ergs/sec. To remove the remaining discrepancy, we must adjust the value of the surface temperature of Mars that we use in calculating its luminosity; we find that a value of 217 K (instead of 200 K) would provide a luminosity to match the calculated absorbed energy. We refer to a temperature derived in this way as an *effective temperature* because it represents the temperature required to produce the observed luminosity of an object. The concepts discussed here are applied to stars as well as to planets, with the exception that stars have powerful internal energy sources, and absorbed energy from outside sources is usually negligible.

It should be noted that the energy budgets for some of the planets are more complex than described here; some planets do have internal energy sources and some have surface temperatures that are elevated because of the greenhouse effect.

CHAPTER 13

The Gaseous Giants: Jupiter and Saturn

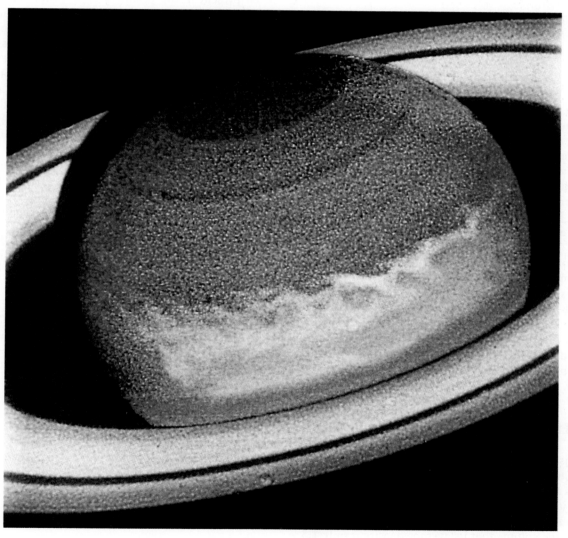

A Hubble Space Telescope image of Saturn, showing the great equatorial storm system that developed during the fall of 1990 (NASA/Space Telescope Science Institute)

Jupiter and Saturn are the largest planets in the solar system, and they offer many striking contrasts with the terrestrial planets. Besides their enormous sizes, Jupiter and Saturn have radically different compositions and physical characteristics, compared to the small, rocky terrestrials.

Both Jupiter and Saturn have complex systems of moons and rings, again much unlike the terrestrial planets. The rings of Saturn are very prominent, and images of the planet with its rings have long symbolized astronomy and outer space for many people.

In this chapter, we discuss the two giants but defer most of our treatment of the rings and moons until Chapter 16. Here we begin with Jupiter, the fifth planet, and then we discuss Saturn, often using Jupiter as a standard for comparison. We will find many similarities, but also some significant contrasts, between the two.

JUPITER

Jupiter is truly the king of the planets, containing more mass than all the other eight put together (Table 13.1). It orbits at an average distance of 5.2 AU from the Sun, so that it is always at least 4.2 AU from the Earth. Its enormous mass (318 Earth masses) is spread out over such a large volume that its average density is only 1.3 grams/cm³, not much greater than the density of water (and only about one-fourth the density of the Earth).

Observation and Exploration

Even a small telescope reveals colorful structure in the Jovian atmosphere, with reddish brown **belts** and light-colored **zones** encircling the planet (Fig. 13.1). The Great Red Spot, a giant reddish oval in the southern hemisphere that dwarfs the Earth in size, has apparently been a persistent feature for at least 300 years. The thick atmosphere prevents any hope of our seeing a solid surface underneath; as we will see, modern theories hold that there is no well-defined solid surface, anyway. Spectroscopy of Jupiter's atmosphere reveals compounds of the general type that are thought to have been present in the primitive atmospheres of the terrestrial planets—i.e., hydrogen-bearing molecules such as methane (CH_4), ammonia (NH_3), and hydrogen itself (H_2), as well as helium. This composition is rather significant, for it means that the evolution of the Jovian atmosphere was frozen at a very early stage, since the composition today is still essentially identical to that of the material from which the solar system formed. Jupiter is certainly a cold planet by our earthly standards, with a temperature at the cloud tops of about 130 K.

Jupiter's rotation is very rapid, despite the large size of the planet. The rotation period is difficult to pinpoint, because the outer regions of the planet spin differentially, meaning that the surface goes around more quickly at the equator than near the poles. This can happen only in a nonrigid object. The Jovian day is just under 10 hours long, which means that the cloud

Table 13.1 Jupiter

Orbital semimajor axis: 5.203 AU (778,300,000 km)
  Perihelion distance: 4.951 AU
  Aphelion distance: 5.455 AU
Orbital period: 11.86 years (4,333 days)
Orbital inclination: 1°18′17″
Rotation period: 9^h 55^m 30^s
Tilt of axis: 3°5′
Mean diameter: 138,346 km (10.86 D$_\oplus$)
  Polar diameter: 133,708 km
  Equatorial diameter: 142,984 km
Mass: 1.901 × 10³⁰ grams (318.1 M$_\oplus$)
Density: 1.327 g/cm³
Surface gravity: 2.64 Earth gravities
Escape speed: 60 km/sec
Surface temperature: 165 K (cloud tops)
Albedo: 0.52
Satellites: 16

Figure 13.1 The view of Jupiter from Earth. *This is a good photograph from an Earth-based telescope, revealing the characteristic color and cloud patterns. (NASA)*

tops near the equator must travel at a speed close to 45,000 km/hr. This rapid spin has flattened the planet a bit, giving it an oblate shape, so that the diameter through the equator is noticeably larger than the diameter through the poles (Figs. 13.1 and 13.2).

Jupiter was accidentally discovered to be a source of radio emission, and further observations have shown a variety of activity in the radio portion of the spectrum, some of it understood, and some not. A general radio glow from the planet, at a wavelength of a few millimeters, is simply the thermal radiation expected according to Wien's law (although the peak emission occurs at shorter wavelengths, in the infrared).

In addition to the thermal radiation, however, Jupiter emits bursts of radio static at wavelengths of several meters, much like the emission associated with lightning in the Earth's atmosphere, and it is now known that there is lightning on Jupiter. A form of radio emission at wavelengths of a few centimeters is caused by charged particles moving through a magnetic field. This **synchrotron radiation** showed that Jupiter has both a strong magnetic field and intense radiation belts surrounding the planet, analogous to the Van Allen belts on Earth.

The intensity and spectrum of the infrared radiation from Jupiter imply that the giant planet is producing some energy in its interior. The total amount of energy being emitted by the planet is about twice the amount

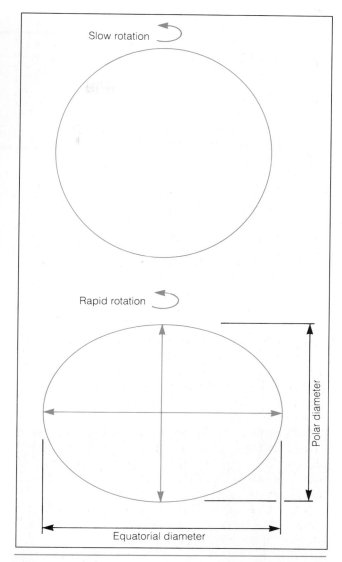

Figure 13.2 The effect of rapid rotation. *An elastic or fluid body is distorted out of a spherical shape by rotation. Jupiter is "flattened" by its rotation so much that its equatorial diameter is 6 percent larger than its polar diameter.*

it receives from the Sun. The probable explanation for this is discussed later.

Observations from Earth also revealed that Io, the innermost of the four Galilean satellites, has strange properties, including the fact that it modulates the Jovian radio emission as it passes in front of the planet, and the fact that Io is accompanied by a cloud of atomic and ionized gas that streams along behind it in its orbital path around Jupiter.

Table 13.2 Probes to Jupiter

Spacecraft	Launch	Encounter	Type of Encounter
Pioneer 10	March 2, 1972	Dec. 3, 1973	Flyby, 131,000 km
Pioneer 11	April 6, 1973	Dec. 2, 1974	Flyby, 46,400 km
Voyager 1	Sept. 5, 1977	March 5, 1979	Flyby, 280,000 km
Voyager 2	Aug. 20, 1977	July 9, 1979	Flyby, 960,000 km
Galileo	Oct. 18, 1989	(1995)	Probe, orbiter

Like the terrestrial planets, Jupiter has been a target of the U.S. space program (Table 13.2). Sending probes to the outer planets requires some modifications to the technology that has been so successful in reaching the inner planets. The spacecraft must endure a passage through the asteroid belt, they must have their own internal power source (small nuclear energy sources have been used) because they travel too far from the Sun to use solar panels, and they must be able to communicate with the Earth over great distances.

These challenges were successfully met by the *Pioneer 10* and *11* missions (an image of Jupiter from *Pioneer 11* is shown in Fig. 13.3), and more recently by the *Voyager 1* and *2* spacecraft (Fig. 13.4). *Pioneers 10* and *11* reached Jupiter in late 1973 and late 1974, and carried out an assortment of studies of the planet's atmosphere, its magnetic field and radiation belts, and its satellites. The *Voyager* spacecraft flew past Jupiter in March and July of 1979, and performed similar types of research, as well as sending back some 30,000 sensational images of Jupiter and its moons, obtained with their fine high-resolution cameras. Perhaps more than any other facet of the planetary exploration program, these pictures of Jupiter (and later of Saturn, Uranus, and Neptune) have caught the public imagination, and today copies are seen everywhere.

Pioneer 11, as well as both *Voyager* spacecraft, went on to visit Saturn, using the gravitational pull of Jupiter to correct their courses and to provide a boost along

Figure 13.3 A close-up view of Jupiter. *This image of Jupiter was obtained by the* Pioneer *spacecraft from a distance of just over a million kilometers. While far superior to any Earth-based photos, this image was soon to be surpassed by those from the* Voyager *missions. (NASA)*

Figure 13.4 A Voyager image of Jupiter. *This photo was obtained by* Voyager *when it was still millions of kilometers from Jupiter. Already a vast amount of detail is evident in the atmospheric structure. (NASA)*

ASTRONOMICAL INSIGHT 13.1

The *Galileo* Mission to Jupiter

Perhaps more than any other planetary probe, the Galileo mission represents both the frustrations and the high hopes of the U.S. planetary exploration program. This unmanned robot probe to the giant planet was planned for launch in 1986, but the disastrous failure of the Space Shuttle *Challenger* in January of that year changed everything. Not only did the *Challenger* loss delay the launch of *Galileo,* but it also caused major revisions in the manner of its launch and the means by which it will travel to Jupiter. Launched in October, 1989, Galileo will take some 6 years to reach Jupiter, instead of the 2 years or less that was originally expected.

The role of the Space Shuttle was to lift *Galileo* into Earth orbit. From there *Galileo's* own propulsion system accelerated it toward the outer solar system. Originally this propulsion system was to consist of a liquid-fuel rocket having sufficient power to launch *Galileo* into a direct trajectory toward Jupiter, but new safety regulations following the *Challenger* disaster have ruled out the use of such rockets as cargo in the Shuttle bay. Instead, *Galileo* was launched from Earth orbit toward Jupiter with a much less powerful solid-fuel rocket. This requires that a very circuitous and roundabout path will be followed by *Galileo,* taking advantage of

The Galileo probe entering Jupiter's atmosphere.   *This is an artist's rendition of what the entry probe from the Galileo mission might look like as it enters the dense atmosphere of the giant planet.*

gravitational boosts as it passes close by two of the inner planets.

When *Galileo* was released from the Space Shuttle, at first it orbited the Earth, along with the Shuttle. When *Galileo's* rocket engine fired, it was aimed so that *Galileo* escaped the Earth's gravitational field, but in a direction opposite to the Earth's orbital motion about the Sun. This caused *Galileo* to slow down with respect to the Sun, so that the probe dropped in toward the inner solar system and flew by Venus at close range. This encounter, which occurred in February of 1990, gave *Galileo* a boost, altering its solar orbit, so that in December of 1990 *Galileo* flew close to the Earth, giving it another

boost in speed. From there *Galileo* is following an enlarged solar orbit, bringing it past the Earth once more, about 2 years later (ie., in December of 1992). This final encounter with the Earth will give *Galileo* enough speed to follow a trajectory to the outer solar system, where it is expected to reach Jupiter in the late 1995 or very early 1996. Thus *Galileo* will reach Jupiter with the assist of no less than three close encounters with planets in the inner solar system.

When *Galileo* arrives at Jupiter, it will not yet be finished with its progression of multiple gravitational boosts. The probe will slow

(continued on next page)

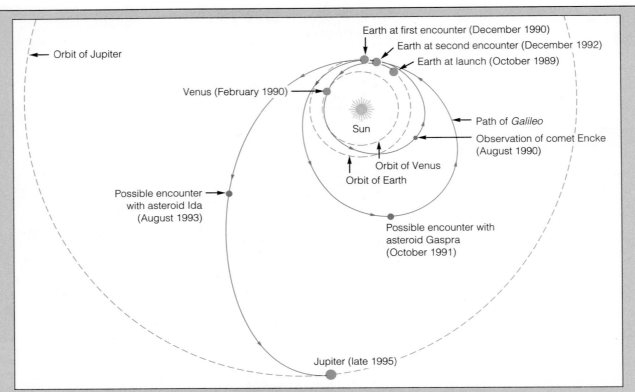

Earth at first encounter (December 1990)
Earth at second encounter (December 1992)
Earth at launch (October 1989)
Orbit of Jupiter
Venus (February 1990)
Sun
Path of *Galileo*
Observation of comet Encke (August 1990)
Orbit of Venus
Orbit of Earth
Possible encounter with asteroid Ida (August 1993)
Possible encounter with asteroid Gaspra (October 1991)
Jupiter (late 1995)

Galileo's path from Earth to Jupiter. *The Galileo spacecraft will follow a complex trajectory, using gravitational assists by Venus (once) and the Earth (twice) to gain sufficient speed to reach Jupiter. Along the way, Galileo will encounter one or two asteroids and will observe at least one comet.*

ASTRONOMICAL INSIGHT 13.1

The *Galileo* Mission to Jupiter (continued)

its speed so that it drops into orbit about Jupiter, but it will then make use of the gravitational fields of the major Jovian satellites to follow a complex series of orbits around the giant planet. Thus *Galileo* will be able to visit each major moon several times, at close range as well as at greater distances. The first ten orbits around Jupiter (with typical periods of 3 months) will be arranged so that each major satellite is visited at close range. The elev-

enth orbit will be dedicated to probing the magnetic field of Jupiter in the direction away from the Sun (that is, in the downstream direction of the solar wind). Following that time, the probe will be low on propellant for its small guiding motors, and it is not known what kind of orbit will be established for the following months. The scientific instruments will continue operating for an extended period, and we may expect to receive continuous information on the giant planet for several years as *Galileo* continues to orbit and send back data.

Galileo actually consists of two distinct instrument packages. One is the orbiter, containing devices for obtaining images and for measuring the spectra of Jupiter and

its satellites in ultraviolet, visible, and infrared wavelengths, as well as instruments to detect and analyze charged particles and the planet's magnetic field. The other instrument package will separate from the orbiter some 5 months before arrival at Jupiter, and then will plunge into the planet's dense atmosphere, on the way in. The instruments on this probe will directly measure the composition and physical state of the atmosphere. Because the pressure is so high, this probe will be subjected to extreme conditions that will eventually stop it from operating. The deceleration upon entry into the atmosphere will be about 250 times the gravity of Earth, and within an hour the spacecraft will have encountered pressures as

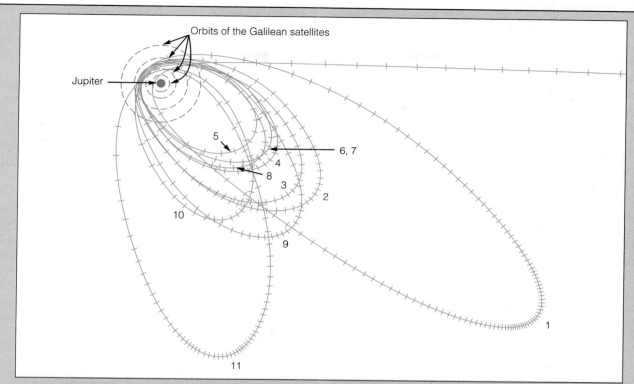

Galileo's orbital path around Jupiter. *As Galileo orbits Jupiter, it will repeatedly encounter the four major moons (orbits shown here by dashed circles), each time gaining speed and altering its orbit. This sketch illustrates the first 11 orbits about Jupiter, showing how the path of the spacecraft will be varied to allow close encounters with several satellites and to provide extensive observations of the planet's magnetic field and charged particle belts.*

high as 20 Earth atmospheres, which will ultimately disable it. Before this happens, however, the spacecraft will have penetrated several hundred kilometers into the clouds and will have sent back (via the orbiter) invaluable infor-mation on the outer regions of Jupiter.

Galileo is the first U.S. probe to the outer solar system since the *Voyager* spacecraft were launched in 1977, and the arrival at Jupiter will occur more than 16 years after the two *Voyager* probes reached the giant planet in 1979. The long delays have been frus-trating, so the excitement and an-ticipation as *Galileo* approaches Jupiter will be enhanced by all the waiting.

the way. Through much of the 1980s, all of the outer planets are on the same side of the Sun, and this led to the idea of a "grand tour." It was realized that a space probe could visit all the outer planets, using each along the way to provide a gravitational boost toward the next. While the full tour is not being tried, *Voyager 2* has traveled most of it. This spacecraft, having flown by Saturn in August of 1981, encountered Uranus in January of 1986 and visited Neptune in August of 1989.

The next probe to Jupiter will be the U.S. *Galileo* spacecraft, which will reach the giant planet in 1995.

Galileo, already on its way, carries a probe designed to send back data as it drops into the atmosphere, and an orbiter that will circle Jupiter for an extended pe-riod, obtaining detailed data on the planet and its sat-ellites. *Galileo* is traveling to Jupiter by a very circui-tous route, so that it can fly by Venus once and the Earth twice in order to boost the speed of the spacecraft as it is accelerated by the gravitational fields of the planets. (Another U.S. probe, the solar mission *Ulysses,* will fly past Jupiter in 1993 and make close-up mea-surements of the planet's magnetic field.)

Atmosphere in Motion

Let us now return to Jupiter itself, and consider the structure and especially the complex motions in its atmosphere. The upper portion of the atmosphere is made of hydrogen and hydrogen compounds, as already mentioned. Helium is also present in large quantities. The presence of these gases represents a major contrast with the terrestrial planets, which long ago lost to space most of their original hydrogen and helium. The overall chemical composition of Jupiter (Table 13.3) is essentially identical to that of the Sun, except that on Jupiter it is cold enough for most of the material to be in molecular form, whereas the Sun is so hot that few molecules can exist, and most atoms are ionized. It is noteworthy that the principal species in the Jovian atmosphere, H_2, CH_4, NH_3, and helium, are the ones thought to be present in the Earth's atmosphere at the time life forms first appeared.

Heavy elements, such as metals, are not observed in the Jovian atmosphere but are probably present deep inside, having sunk to the core of the planet. Jupiter, being fluid nearly all the way to the center, is highly differentiated.

The bright and dark colors in the upper atmosphere (Fig. 13.5) are probably caused by slight differences in

Table 13.3 Composition of Jupiter's atmosphere

Gas	Symbol	Fraction[a]
Hydrogen	H_2	0.861
Helium	He	0.138
Water vapor	H_2O	1×10^{-6}
Methane	CH_4	9×10^{-4}
Ammonia	NH_3	2×10^{-4}

[a]The values listed are the fraction of the total number of molecules in the atmosphere. Note that hydrogen and helium themselves total 100 percent; this is not precisely true but reflects the dominance of these two species.

molecular composition, which in turn are created by subtle differences in temperature and pressure. The dark **belts** are regions where the atmospheric gas is descending, while the light-colored **zones** are regions where the gas is rising (Fig. 13.6). The banded appearance of the planet is created by the rapid rotation, which stretches what would otherwise be circular flow patterns into elongated strips that circle the planet. The direction of the flow alternates from each belt or zone to the adjacent one (Fig. 13.7) and the wind speeds typically reach 350 km/hr or more for the strong zones at the equator and at intermediate latitudes.

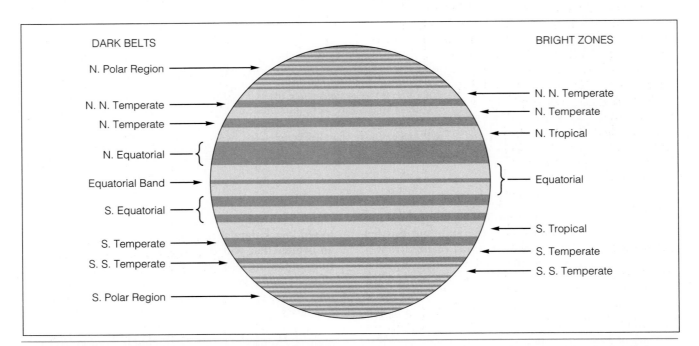

Figure 13.5 Belts and zones on Jupiter. *This schematic drawing names the major features that give Jupiter its banded appearance. (NASA)*

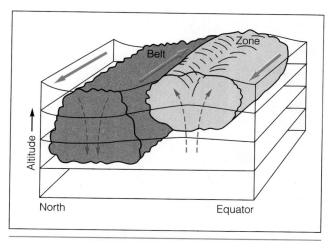

Figure 13.6 Vertical motions in the Jovian atmosphere. *This cross-section indicates how the horizontal circulation pattern is related to vertical convective motions in the outer layers of the atmosphere. This is analogous to circulation in the Earth's atmosphere, except that the rapid rotation of Jupiter stretches rotary patterns into belts and zones. (NASA)*

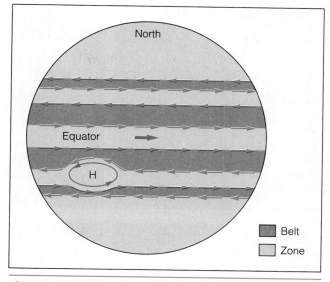

Figure 13.7 Circulation of the Jovian atmosphere. *The global circulation pattern shown here indicates the location and names of the belts and zones in Jupiter's cloud layer.*

Besides the regular pattern of belts and zones, Jupiter's atmosphere displays a wide variety of spots and other features, the Great Red Spot (Fig. 13.8) being the largest and most prominent. The Great Red Spot is a circulatory pattern that rises above the level of the surrounding clouds. It may be analogous to a storm system on Earth (Fig. 13.9), but one that has lasted at least 300 years. Other spots on Jupiter are known to have persisted for at least several decades.

Very detailed photographs of Jupiter's cloud tops reveal a fantastic picture of turbulent, chaotic flows where the belts, zones, and spots interact (Fig. 13.10). Movie sequences reconstructed from images obtained from the *Voyager* spacecraft as they approached Jupiter show an incredible complexity of motions all happening at once, as in some gigantic machinery of gears and wheels.

Internal Structure and Excess Radiation

No direct probes of the interior of Jupiter have been made, so what we know has been surmised largely from theoretical calculations. This may seem like an exceedingly uncertain basis for any detailed ideas about interior conditions, and of course there is room for error, but in fact the laws of physics dictate rather explicitly what the internal structure must be like. The

Figure 13.8 The Great Red Spot. *This is a* Voyager *image of the rotating column of rising gas that has been present on Jupiter for at least 300 years. (NASA)*

general procedure for calculating a theoretical model of Jupiter is very much like that described in Chapter 20 for determining the internal properties of a star. A set

Figure 13.9 Formation of the Great Red Spot.   *This computer simulation, showing Jupiter from above the south pole, successfully recreates the Great Red Spot. A pair of vortices (upper left), rotating in opposite directions are modified by the strong global winds so that a clockwise blue vortex is destroyed while a counterclockwise (red) vortex grows and persists. (P. Marcus and N. Socci, University of California, Berkeley and Harvard-Smithsonian Center for Astrophysics; computations done at the National Center for Atmospheric Research.)*

of equations describing the relationships among various properties such as pressure and temperature, as well as processes such as heat flow, is solved, starting with the known size, mass, composition, and surface characteristics of the planet.

While different assumptions made by people working in the field lead to somewhat different results, the major features come out the same (Fig. 13.11). Below a relatively thin layer of clouds (perhaps 1,000 km thick) is a zone consisting of hydrogen in liquid form, extending more than one-third of the way towards the center. The pressure in this region is as great as 3 million times sea-level pressure on Earth, and the temperature is as high as 10,000 K or more. An even thicker zone below the liquid hydrogen has such extreme pres-

sure that hydrogen is forced into a state called **liquid metallic hydrogen,** which has a somewhat rigid structure like certain metals, and which is an electrical conductor because its electrons flow freely. Liquid metallic hydrogen can be produced on Earth only in very tiny quantities, because it is very difficult to create the necessary pressure, but the simplicity of hydrogen creates confidence in the theoretical calculations that predict its behavior under these extreme conditions. The central core of Jupiter, below the liquid metallic hydrogen, is probably solid rocky material of very high density, containing all the heavy elements of the entire planet, which sank to the center as the planet differentiated. In the core, the temperature is expected to be as great as 30,000 K. The mass of the core is probably in the

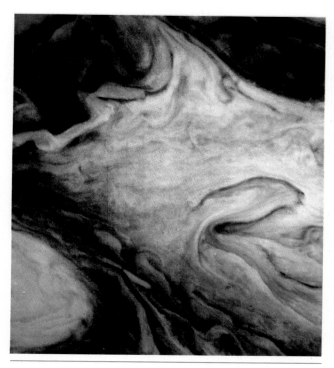

Figure 13.10 Turbulent motions in the Jovian atmosphere. *In this* Voyager *image, colors are altered to enhance contrast and reveal complex details. (NASA)*

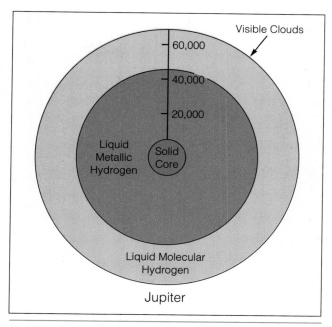

Figure 13.11 Internal structure of Jupiter. *This cutaway illustrates the rough model of the Jovian interior described in the text.*

range of 10 to 20 times the mass of the Earth, indicating that as much as 6 percent of the planet's mass is contained in only 0.5 percent of its volume.

Except possibly for the boundary of this rocky central zone, Jupiter has no solid surface, instead merging gradually from gas to liquid to metallic liquid as we go inward from the outside.

As noted earlier, the planetary radio emission shows the presence of an internal heat source, and this was included in the theoretical calculations just described. The source of this internal heat is difficult to understand, however, because Jupiter is not sufficiently massive to have nuclear reactions taking place in its core, as in a star. For a while, it was thought that Jupiter might be shrinking gradually, converting gravitational energy into heat in the process, but this now appears to be ruled out. The most likely explanation, based on calculations of the heat capacity and compressibility of the interior, is that the interior is still hot from the time of the planet's formation some 4.5 billion years ago. Such a prolonged retention of heat is made possible by the characteristics of the outer layers, which do not efficiently transmit heat from the interior to be

lost to space. The atmospheric circulation, already described, is caused in part by heating of the lower atmosphere from below.

The Magnetic Field and Radiation Belts

It has been known for some time that Jupiter has a magnetic field, as mentioned earlier. This was originally deduced from observations of radio emission and was later verified by the instruments on board the *Pioneer* and *Voyager* probes.

The Jovian field at the planet's cloud tops is about 10 times the strength of the Earth's magnetic field, and is oriented in the opposite direction, with the north magnetic pole pointing south, and vice versa. This should not be considered too surprising, since we know that the Earth's magnetic poles have reversed themselves from time to time. Perhaps the Jovian field does the same, and has its poles aligned with those of the Earth some of the time.

The magnetic axis of Jupiter does not exactly parallel the rotation axis, but is instead tilted by 11° (Fig. 13.12). This is also similar to the Earth, but on Jupiter it has some important effects on the distribution of charged particles in the radiation belts, as we shall discuss shortly.

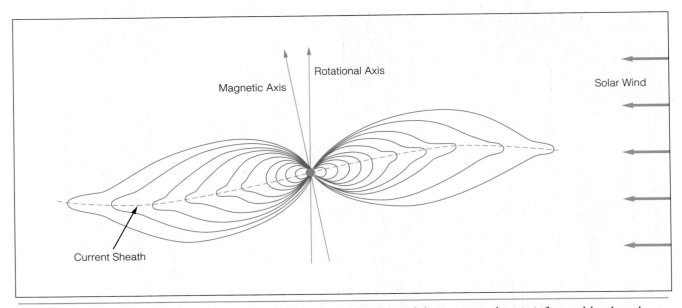

Figure 13.12 The Jovian magnetosphere and radiation belts. *The shape of the magnetosphere is influenced by the solar wind and by the rapid rotation of Jupiter. The result is a sheet of ionized gas that is closely confined to the equatorial plane but wobbles as the planet's off-axis magnetic field rotates. (NASA/JPL)*

The origin of the magnetic field is thought to be roughly similar to that of the Earth's field; that is, electrical currents created by flows inside the planet create a magnetic dynamo. The structure of the field indicates that the dynamo operates not far below the surface of the planet, probably in the liquid hydrogen or liquid metallic hydrogen zone.

Within the magnetosphere of Jupiter are very intense radiation belts (Fig. 13.12), with much higher densities of charged particles than in the Van Allen belts of Earth. The passages of the *Pioneer* and *Voyager* spacecraft through these zones were rather perilous, since electronic circuits can be damaged by an environment full of speeding electrons and ions. Some data transmissions were garbled or lost, but on the whole the craft survived intact and were able to continue functioning normally.

One mystery concerning the Jovian radiation belts was where the great quantities of charged particles come from. A number of possibilities have been considered, but most were not feasible, and gradually it was realized that the answer had to do with the inner Jovian satellites. The five moons closest to the planet all orbit within the magnetosphere, where they sweep up charged particles as they move along their paths.

The discovery, from Earth-based observations, that Io is accompanied by a cloud of glowing gas led to the speculation that perhaps some of the satellites emit particles that enrich the radiation belts. As we shall see in the next section, the *Voyager* missions were soon to provide spectacular confirmation that just such a process is occurring.

The Satellites of Jupiter

Jupiter has 16 known satellites. (Table 13.4), ranging from Metis, closest to Jupiter, to faraway Sinope, the outermost. Most of the 16 are small, icy objects, with diameters of less than 100 km. The outermost four have retrograde orbits that are quite eccentric and inclined by several degrees to the orbital plane of the more regular, inner satellites. The fact that these four vary from the norm may suggest that their origin is different from the others, which probably formed from interplanetary debris swirling around the infant Jupiter at the time the solar system coalesced. (One theory is that the four retrograde moons, along with four of the prograde satellites, are fragments of two larger bodies which collided.)

Four of the Jovian satellites, the fifth through the

Table 13.4 Satellites of Jupiter

No.	Name	Distance (R$_J$)[a]	Period (days)	Diameter (km)[b]	Mass (g)	Density (g/cm³)
16	Metis	1.790	0.295	20:	9.5×10^{19}	
15	Adrastea	1.804	0.298	20 × 20 × 15	1.9×10^{19}	
5	Amalthea	2.532	0.498	270 × 170 × 150	7.2×10^{21}	
14	Thebe	3.105	0.675	110 × ? × 90	7.6×10^{20}	
1	Io	5.903	1.769	3,630	8.92×10^{25}	3.53
2	Europa	9.386	3.551	3,138	4.87×10^{25}	3.03
3	Ganymede	14.967	7.155	5,262	1.49×10^{26}	1.93
4	Callisto	26.30	16.689	4,800	1.08×10^{26}	1.70
13	Leda	155.26	238.7	16:	5.7×10^{18}	
6	Himalia	160.85	250.6	186:	9.5×10^{21}	
10	Lysithea	163.65	259.2	36:	7.6×10^{19}	
7	Elara	165.05	259.7	76:	7.6×10^{20}	
12	Ananke	296.54	631	30:	3.8×10^{19}	
11	Carme	316.12	692	40:	9.5×10^{19}	
8	Pasiphae	328.71	735	50:	1.9×10^{20}	
9.	Sinope	331.51	758	36:	7.6×10^{19}	

[a]The symbol R$_J$ stands for the equatorial radius of Jupiter, which is 71,492 km.
[b]A colon(:) following a value for diameter indicates an uncertain value, and no density has been calculated for these cases (or for satellites having irregular shapes).

eighth in order from Jupiter, are much larger than the rest, all being comparable to or larger than the Earth's Moon (Fig. 13.13). These are the four spotted by Galileo with his primitive telescope, and they are easily bright enough to be seen with the unaided eye, except that the brilliance of nearby Jupiter drowns them out.

The *Pioneer* and *Voyager* spacecraft were sent along trajectories aimed at providing close-up views of these Galilean satellites. Of particular interest was Io, the innermost of the four, because of its mysterious influence on radio emissions from Jupiter and because of the glowing cloud of gas that surrounds it. The color photographs taken by the *Voyager* probes (Fig. 13.14) showed tremendous complexity, revealing the Galilean satellites to be wondrous worlds, each with its own complex history.

The sizes of the Galilean satellites, as well as their masses, have been most accurately determined by the *Pioneer* and *Voyager* probes. While each of the four has its own distinctive character, the Galilean satellites fit a pattern, with systematic trends from one to the next. Their densities decrease steadily with distance from Jupiter, Io having the greatest density and Callisto, the outermost of the four, the least. Their albedos also decrease in the same manner. These two facts,

along with measurements of the surface temperatures, indicate that the outer satellites contain large quantities of water ice, which can be very dark when seen from afar, especially if it contains some gravel and grit. The inner Galilean satellites contain less ice and greater fractions of rock. The surfaces of these bodies also show marked contrasts with one another.

Unique though each of the Galilean satellites is, each also fits in with a general class of large satellites that accompany the outer planets. The mixture of ice and rock that characterizes the Galilean satellites is also typical of the larger moons of Saturn, Uranus, and Neptune and, as we shall see, appears also be found in the ninth planet, Pluto. The overall characteristics of these large moons are discussed in Chapter 16.

In the Jovian system, the trend of decreasing ice content and increasing density going inward from Callisto to Io has been explained at least partially in terms of tidal effects exerted on the satellites by Jupiter. (It may also be true that Io and Europa had less water to begin with, having formed close to Jupiter, which was probably very hot initially.) These effects are maximized for Io, and decrease with increasing distance from the giant planet. Io has the additional influence of being in **orbital resonance** with its neighbor, Europa.

Figure 13.13 The major Jovian satellites.   *These profiles indicate the relative distances of the major satellites from Jupiter and their relative sizes. The Earth's Moon and the planet Mercury are included for comparison. (NASA)*

Figure 13.14 The Galilean satellites of Jupiter.   *These are the four moons discovered by Galileo, shown in correct relative size. They are Io (upper left), Europa (upper right), Ganymede (lower left), and Callisto (lower right). Ganymede is the largest satellite in the solar system. From Callisto, the farthest from Jupiter, to Io, the innermost of the four, there is a wide range of geological and physical properties, which is discussed in the text. (NASA)*

Orbital resonance refers to a situation where two satellites have orbital periods that are simple multiples of one another, so that the two are aligned with the parent planet at frequent, regular intervals. In this case, Europa's period is exactly twice that of Io, so that every second time Io makes a trip around Jupiter, it finds Europa lined up in the direction opposite Jupiter. This adds to the tidal stresses exerted on Io (and to those exerted on Europa as well), causing intense internal friction and heating. As a result, Io is largely molten inside, and is undergoing very active volcanism. Europa is thought to be warm enough inside to have vast liquid water seas under its outer layer of ice.

Ganymede and Callisto, the outermost two of the Galilean satellites, are less affected than the inner two by tidal stresses, and have been able to retain most of their ice. Hence these two have lower average densities than Europa and Io, and their surfaces have not been altered as much by geological forces. Whereas Io has essentially no impact craters, and Europa has only a few, Ganymede and especially Callisto are pockmarked over their entire surfaces with ancient craters. This tells us that Ganymede and Callisto have old surfaces, whereas the surfaces of Europa and Io are much younger. Io's surface is continually being recoated with deposits from the volcanic eruptions, and Europa appears to be undergoing some sort of crustal plate shifting, a bit like the continental drift that occurs on the Earth.

The discovery of Io's volcanic eruptions by *Voyager 1* was one of the most spectacular events of the *Voyager* explorations. The initial discovery (Fig. 13.15) was made on one of the last images obtained before *Voyager 1* headed off toward Saturn, but further examination of other *Voyager 1* images as well as a large number obtained later by *Voyager 2* revealed many volcanic vents scattered over the surface of Io (Fig. 13.16). It has been estimated that as many as eight to ten volcanoes are erupting at a time on this satellite.

The gases being released by these eruptions, consisting largely of sulfur and related compounds, become ionized by charged particles as they escape Io. Once ionized, they become subject to forces exerted by Jupiter's intense magnetic field. The magnetic field rotates with Jupiter, which has a spin period of just under 10 hours. Io's orbital period is much longer than this (about 42.5 hours), so the gas from the eruptions, once ionized, is pulled *ahead* of Io by the rotating magnetic field. This gas fills Io's entire orbital path around Jupiter. Because the magnetic field axis of Jupiter is tipped by about 11° with respect to the planet's orbital plane, the magnetic field wobbles or precesses, causing the charged particles to be spread out into a doughnut-shaped distribution (Fig. 13.17). This structure is called the **Io torus.** The existence of the torus was known before the *Voyager* missions, but its cause was not. One of the observational effects of the torus is the periodic variation in Jupiter's radio emission,

Figure 13.15 An eruption on Io.   *This image of an eruption from a volcanic vent with a plume of ejected gas dramatically illustrates Io's present state of dynamic activity. (NASA)*

Figure 13.16 Volcanic vents on Io. *The darkest material seen here indicates the locations of recently active volcanoes. (NASA)*

which was detected long before the *Voyager* encounters.

Besides finding the second known active volcanic body in the solar system, *Voyagers 1* and *2* discovered the third known ring system. Saturn's brilliant rings had been known for centuries, and only two years before *Voyager 1* reached Jupiter, several thin, dark rings had been discovered encircling Uranus (see Chapter 14). The earlier *Pioneer* encounters with Jupiter had led some scientists to suspect the presence of a ring there also, but this was not confirmed until the *Voyager* probes. A single thin, dark ring was found (Figs. 13.18 and 13.19).

SATURN

We now turn our attention to the sixth planet, Saturn. This brilliant object is the outermost of the planets

Figure 13.17 The Io Torus. *This remarkable series of photographs, taken through a ground-based telescope on Earth, shows the cloud of sodium atoms surrounding Io as it orbits Jupiter. The yellow color of the cloud is due to the fact that sodium atoms emit most strongly in a pair of emission lines in the yellow part of the spectrum (this is why certain types of street lights appear yellow). Once this gas, which also includes other atoms such as sulfur, escapes from Io, it becomes ionized and then spread all the way around Io's orbit by magnetic forces. This ionized gas is not visible here, so we do not see the full extent of the torus. The sizes of Jupiter, Io (dot inside the cross-hair), and Io's orbit are all to correct relative scale. (B. A. Goldberg, G. W. Garneau, and S. K. LaVoie, JPL)*

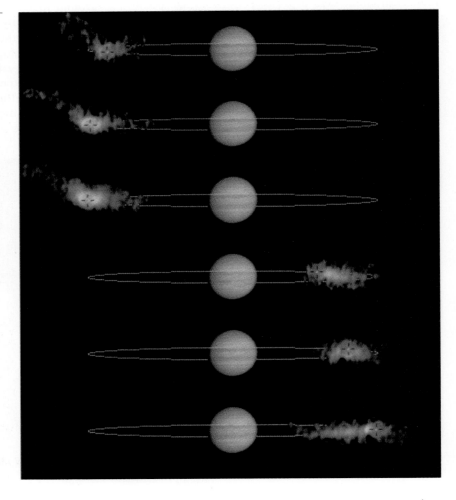

easily seen by the unaided eye (it happens that Uranus is barely visible under good conditions, but it was not discovered until telescopic observations were possible; see Chapter 14). With its banded atmosphere and delicate ring system, Saturn (Fig. 13.20), perhaps more than any other object, has come to represent the realm of outer space in our imagination.

Like Jupiter, Saturn (Table 13.5) is truly gigantic, containing nearly 100 times the mass of the Earth (but spread out so far that its average density is actually less than that of water). It orbits the Sun at a distance of 9.5 AU, taking nearly 30 Earth years for each trip around.

General Properties

As in many previous cases, it was Galileo who made the first important discoveries about Saturn, using his telescope to observe it in 1610. He saw a brilliant yellowish object that appeared strongly elongated in shape. He later referred to Saturn as the planet with "ears."

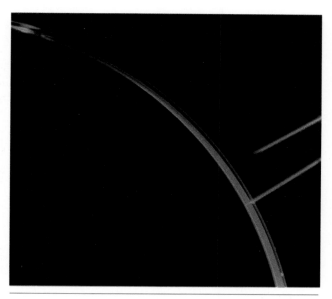

Figure 13.18 The Jovian ring. *This* Voyager 1 *image revealed the first confirmation of a suspected ring girdling Jupiter. Here a segment of the ring is seen at the right. (NASA)*

Figure 13.19 The geometry of the Jovian ring. *Here the ring has been drawn in, showing its size relative to Jupiter. The ring radius is about 1.8 times the radius of the planet; it is estimated to be no more than 30 kilometers thick. (NASA)*

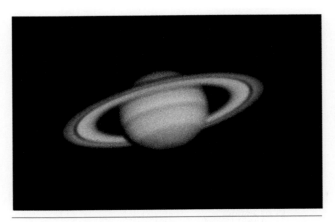

Figure 13.20 Saturn as seen from Earth.   *This photo was taken under good conditions and shows some structure in the ring system, including the prominent Cassini division (Lick Observatory photograph)*

Table 13.5 Saturn

Orbital semimajor axis: 9.555 AU (1,427,000,000 km)
   Perihelion distance: 9.023 AU
   Aphelion distance: 10.086 AU
Orbital period: 29.46 years (10,759 days)
Orbital inclination: 2°29′33″

Rotation period: 10^h 39^m .4
Tilt of axis: 26° 44′

Mean diameter: 114,632 km (9.00$D_\oplus$)
   Polar diameter: 108,728 km
   Equatorial diameter: 120,536 km
Mass: 5.69×10^{29} grams (95.12 $M_\oplus$)
Density: 0.69 g/cm
Surface gravity: 1.13 Earth gravities
Escape speed: 36 km/sec

Surface temperature: 134 K
Albedo: 0.47

Satellites: 18

Some 50 years after Galileo's observations, Huygens, with more complete data, was able to deduce the ringlike structure of the extensions on the planet, and thus explained the shape noted by Galileo. Extended observations showed the rings to be so thin that when viewed edge-on, they are nearly impossible to see. Saturn's rotation axis is tilted with respect to its orbital plane by 27°, and the rings, which lie in its equatorial plane, are therefore tilted through a range 27° above and 27° below the orbital plane as Saturn goes around the Sun.

Huygens also discovered Titan, the largest of Saturn's satellites, and shortly thereafter the French astronomer G. D. Cassini found four more satellites and a prominent gap in the rings (called the Cassini division to this day).

The mass of Saturn was deduced from the satellite orbits, using Kepler's third law, and this led to the realization that the planet's density is very low, only about 0.7 grams/cm³. This is about half the density of the Sun, and little more than half that of Jupiter. The density of Saturn is actually less than that of water.

Careful study of photographs showed that the rings, as seen from the Earth, appear to be three in number, with two inner ones (one of which is brighter than the other), and a third that lies outside the Cassini division. From observations made when the rings are viewed edge-on, it was determined that they can be no more than about 10 km thick. It was also discovered that when they are seen nearly face-on, background stars

and the disk of Saturn itself are visible through them, as though they were made of sheer fabric. This supported a suggestion made in the late 1800s that the rings actually consist of innumerable small chunks of debris, each orbiting Saturn in accordance with Kepler's third law. Spectroscopic measurements, and later radar observations, used the Doppler effect to determine the speeds of the particles, and confirmed this idea by directly showing that they travel at orbital speeds consistent with Kepler's laws. The radar data also revealed that at least some of the particles are chunks of material some centimeters or meters in size.

The planet itself rotates very rapidly, with a period of about 10 hours, similar to that of Jupiter, and resulting in a similarly oblate shape. The rotation is also differential, with a shorter period at the equator than at the poles.

When spectroscopic measurements of the atmosphere were first made, familiar chemical compounds such as molecular hydrogen (H_2), methane (CH_4), and some ammonia (NH_3), were found. The lower temperature of Saturn compared to Jupiter caused some differences, however, in that much of the ammonia on Saturn has apparently crystallized and precipitated out of the atmosphere in the form of snow, and a few more complex molecular species are present on Saturn. No doubt the overall composition of Saturn is the same as that of Jupiter, since both formed out of the same material originally, and neither has lost any significant fraction of its gases to space. (However, Saturn has a lower helium content in its clouds, probably because some of the helium has sunk to lower levels in the planet's interior; this is discussed later.) With their large masses and low temperatures (about 130 K at the Jovian cloud tops, and 95 K in the upper atmosphere of Saturn), the combination of high escape velocity and low average particle speed has prevented any large quantity of molecules from escaping.

Like Jupiter, Saturn is a source of radio emission, with a combination of different origins. Synchrotron emission signifies the presence of radiation belts, where the planetary magnetic field has captured an extensive cloud of high speed electrons and ions, and thermal radiation shows that Saturn also emits excess energy from its interior.

Another similarity between Saturn and Jupiter is the attention each has received from the U.S. space program. Three of the four probes sent to Jupiter so far have also visited Saturn, using the gravitational pull of Jupiter to assist and guide them on their way. Only *Pioneer 10* did not do this, having taken a path through the Jovian system that was unsuited for such a maneuver.

Pioneer 11, with its charged particle monitors, radio antennae, and camera, uncovered several new details about Saturn and its attendant rings and satellites. At least one new moon was discovered, and an additional ring was found lying outside those previously known or suspected. The magnetic field of Saturn was directly measured for the first time and was found to be somewhat stronger than that of the Earth, with the north and south magnetic poles upside down compared to Earth (as in the case of Jupiter), and the magnetic axis well aligned with the rotation axis, unlike both Jupiter and the Earth. The radiation belts are largely confined to the equatorial plane by the rapid rotation of Saturn, again mimicking Jupiter.

Hardly more than a year after *Pioneer 11* encountered Saturn in September, 1979, *Voyager 1* arrived, and a great wealth of additional information quickly followed *Pioneer's* discoveries. The high-resolution color cameras on *Voyager* provided new revelations as sweeping in their impact as their previous portraits of Jupiter. *Voyager 2* reached Saturn in August of 1981, following a different trajectory and viewing it from a different perspective.

As in the case of Jupiter, the *Voyager* and *Space Telescope* missions have added breadth, color, and a certain amount of fantasy to our picture of a newly explored planetary system (Fig. 13.21). Now we can discuss in some detail the weather patterns on Saturn,

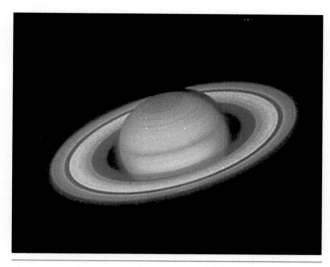

Figure 13.21 A Space Telescope Portrait. *This image of Saturn was obtained by the Hubble Space Telescope in Earth orbit, and it shows much more detail than can be seen in the best ground-based images. This image, together with Fig. 13.20, graphically illustrates the blurring effect of the Earth's atmosphere, since both photos were taken from approximately the same distance. (NASA/Space Telescope Science Institute)*

the composition and structure of its atmosphere and interior, and the nature of its dazzling array of rings and satellites.

Atmosphere and Interior: Jupiter's Little Brother?

To the faraway observer, the atmospheres of Jupiter and Saturn look similar in overall color and banded appearance. Although these aspects are generally similar, there are also many contrasts (Table 13.6).

As noted earlier, the chemical compositions of the two atmospheres are much alike, except that the colder temperature on Saturn has caused much of the ammonia to precipitate out and has allowed the formation of somewhat more complex molecules. An example that was detected from Earth-based spectroscopic data is ethane (C_2H_6), and other derivatives of methane are probably present.

Whereas Jupiter has a colorful and distinct contrast between belts and zones, and turbulent, swirling winds that show up with great clarity, Saturn has a muted overall appearance. The belts and zones are not so distinct, and the dark and light spots that can be seen do not contrast strongly with their surroundings.

Table 13.6 Comparison of Saturn and Jupiter

Phenomenon	Comparison
Mass	Jupiter is 3.34 times more massive
Diameter	Jupiter is 1.2 times larger
Density	Jupiter is almost twice as dense
Atmosphere	
Structure	Saturn has deeper clouds, muting the contrast of the belts and zones
Composition	Nearly the same, except that ammonia has precipitated out of Saturn's atmosphere, and Saturn has more complex molecules, in both cases because of Saturn's lower temperature. Saturn also has less helium in its cloud layers.
Circulation	Wind speeds are greater on Saturn; belts and zones extend closer to poles on Saturn
Internal structure	Very similar, but Jupiter has higher internal density and more extended liquid metallic hydrogen zone
Internal heat	Both have excess internal heat, but apparently for different reasons: Jupiter's is thought to be simply left over from its formation; Saturn's is thought to be due to continuing differentiation as helium separates and sinks
Seasons	Jupiter has none; Saturn has significant seasons because of its tilt of nearly 27°
Magnetic field	Both strong and of similar origin, although Saturn's is less extensive and may form deeper in the interior.
Rings	Saturn has extensive, complex system of rings; Jupiter has just one thin, dim ring

Figure 13.22 The contrasting cloud thickness on Jupiter and Saturn. *The deeper cloud zone on Saturn is largely responsible for the relative lack of contrast in its visible surface features. (Data from Beatty, J. K., B. O'Leary, and A. Chaikin, eds. 1981. The new solar system. Cambridge, England: Cambridge University Press.)*

Observations show that Saturn's cloud layer is much thicker than that of Jupiter (Fig. 13.22), although it is lacking the high haze layer that characterizes the larger planet. The lower surface gravity of Saturn allows the cloud layer to persist to much greater heights, with the result that the structure of the lower atmosphere is much more heavily obscured than on Jupiter. Hence, Saturn does not have the striking banded appearance of Jupiter, and the atmospheric flow patterns are correspondingly more difficult to discern (Fig. 13.23).

Enough can be seen through the haze on Saturn to show that the atmospheric flow patterns, while similar in some ways to those on Jupiter, also display significant differences.

The speed of the flow patterns on Saturn is much greater than on Jupiter. At the equator, for example, where Jupiter has wind velocities of about 100 meters/sec (360 km/hr), Saturn has winds moving as rapidly as 400 to 500 meters/sec (up to 1,800 km/hr). The belt and zone structure on Saturn is different in overall pattern, having a much wider equatorial stream, but also having narrow belts and zones extending closer to the poles (Fig. 13.24).

There are also contrasts between the northern and southern hemispheres. The belts and zones are broader in the north, and at a latitude of 40°N the *Voyager* images reveal a large number of spots and eddies. These are caused by an abrupt change in flow speed at the

Figure 13.23 Circulation in the atmosphere of Saturn. *Even though the contrast is not as vivid on Saturn as on Jupiter, the ringed planet does have complex atmospheric motions, driven by a combination of convection and rapid planetary rotation. (NASA)*

Figure 13.24 Bands in the atmosphere of Saturn. *These are* Voyager 1 *(upper) and* Voyager 2 *(lower) images, showing atmospheric bands. Computer-enhanced colors have been used to bring out subtle contrasts between adjacent regions. (NASA)*

northern edge of the equatorial stream. These differences between the north and south could be seasonal effects, since Saturn's axis is tilted by about 27° (Jupiter's axis is nearly perpendicular to its orbital plane, so seasonal effects on Jupiter are minimal).

There are contrasts in the internal structures of the two planets as well. First and foremost, the density inside Saturn is much lower, because its much smaller mass is spread out over nearly the same volume as Jupiter. The solid core inside Saturn is probably not as dense as the core of Jupiter and may contain some ice (Fig. 13.25). Outside this core, the internal pressure, while not so great as inside Jupiter, is thought to be sufficient to form liquid metallic hydrogen, which may inhabit a zone extending about halfway out to the surface. Above this is thought to be a thick layer of liquid hydrogen and helium, topped off by the layer of clouds that is visible from afar.

Like Jupiter, Saturn also radiates substantially more energy than it receives from the Sun, by a factor of 1.78 in this case. The explanation offered for Jupiter does not work as well here; Saturn, with its lower mass, would not have been as hot when it formed and could not have retained primordial heat as efficiently. Some additional source of heat must be present. This source

is thought to be ongoing differentiation in the metallic and liquid hydrogen zones, as the heavier helium atoms gradually sink towards the center of the planet, releasing gravitational energy as they do so. The likelihood that this process is really at work inside Saturn is enhanced by the observation that Saturn has relatively less helium in its outer layers than does Jupiter, consistent with the idea that Saturn's helium has sunk, while Jupiter's has not. The probable reason that this has not occurred inside Jupiter is that the interior of the larger planet is warmer and more turbulent,

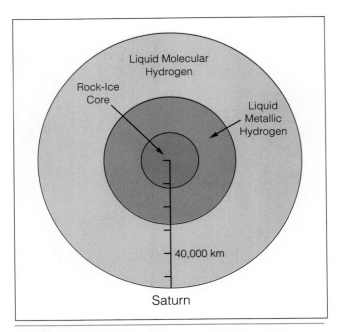

Figure 13.25 The internal structure of Saturn. *This cross-sectional view should be compared with the structure of Jupiter shown in Figure 13.11.*

preventing the helium from quietly sinking towards the center.

The magnetic field of Saturn is, no doubt, formed by a dynamo in the interior. The structure of Saturn's magnetic field suggests that the depth at which the field forms is greater than in the case of Jupiter. The zone of trapped ions and electrons around Saturn is still quite large, and most of the satellites and all of the rings are within this zone, sweeping up charged particles as they move along their orbits. Because of variations in the flow of ions from the Sun, the magnetosphere of Saturn fluctuates in size, with the result that Titan, the largest satellite, is sometimes within the magnetosphere and sometimes outside of it.

Filling the orbit of Titan and extending quite a bit farther inward and outward is a cloud of hydrogen atoms resembling in shape the Io torus that girdles Jupiter. In this case, however, the gas is not ionized, so its structure is not controlled by the planetary magnetic field.

The Satellites of Saturn

Before the *Pioneer* and *Voyager* encounters, Saturn was thought to have at least 10 satellites, some of them

small and rather difficult to see from Earth. (An eleventh satellite was suspected, having been reported in 1978, but was not considered confirmed until *Pioneer 11* also detected it.) The first of Saturn's moons to be discovered was Titan, noted by Huygens in 1655. Four more were found by Cassini in the latter part of the seventeenth century, and in the next 300 years an additional five were seen.

The *Pioneer* and *Voyager* encounters, along with new Earth-based observing techniques, have added to the total, which is currently 18 (Table 13.7). There have been reports of additional satellites, but these have not yet been confirmed. Thus Saturn may have as many as 19 or 20 moons (or possibly more), but for now we accept the 18 that are officially recognized (the eighteenth was confirmed very recently, in late 1990, based on *Voyager* images from 1981 that were re-examined). There are one giant moon and seven intermediate ones, and the rest are tiny and usually irregularly shaped (Fig. 13.26). Some of these were found in unexpected configurations; some share orbits with larger moons, and others are found in closely spaced orbits with narrow rings in between. The moons that share the same orbit are called **co-orbital satellites** and are held in place 60° ahead of or behind the large satellite in the same orbit. The combined gravitational forces of Saturn and the larger satellite (such as Tethys, which has two co-orbital satellites, Telesto and Calypso), make these positions unusually stable. The closely spaced moons such as Prometheus and Pandora can trap ring particles between them, maintaining a thin ring at that position. This is discussed more fully later in Chapter 16.

The major satellites, those easily observed from Earth, turned out to be as varied in their individual qualities as the Galilean satellites of Jupiter. The biggest of them all, Titan (Fig. 13.27), was especially important for *Voyager 1* to examine closely, since it was already known to have an atmosphere, making it nearly unique in that respect among all the satellites of the solar system. Recall that, until *Voyager 1* revealed the great depth of Titan's atmosphere, this satellite was thought to be larger than Ganymede, whereas in fact Ganymede has the honor of being the largest satellite in the solar system.

It was hoped that the vertical structure of Titan's atmosphere (Fig. 13.28) could be probed and the surface below examined, but it turned out that the atmosphere is so thick that it is impossible to see to any great depth. In addition, there is an obscuring haze layer at the top of the atmosphere that makes direct

Table 13.7 Satellites of Saturn

No.	Name	Distance (R_s)[a]	Period	Diameter (km)	Mass (g)	Density (g/cm³)[b]	Albedo
18	1981S13 (Pan)	2.207	0^d.573	20: km			—
15	Atlas	2.285	0.602	40 × 20			0.9
16	Prometheus	2.313	0.613	140 × 100 × 80			0.6
17	Pandora	2.351	0.629	110 × 90 × 80			0.9
11	Epimetheus	2.512	0.694	140 × 120 × 100			0.8
10	Janus	2.512	0.695	220 × 200 × 160			0.8
1	Mimas	3.078	0.942	392	4.5×10^{22}	1.43	0.5
2	Enceladus	3.951	1.370	500	7.4×10^{22}	1.13	1.0
13	Telesto	4.890	1.888	34 × 28 × 26			0.5
14	Calypso	4.890	1.888	34 × 22 × 22			0.6
3	Tethys	4.890	1.888	1,060	7.4×10^{23}	1.19	0.9
4	Dione	6.262	2.737	1,120	1.05×10^{24}	1.43	0.7
12	Helene	6.274	2.737	36 × 32 × 30			0.7
5	Rhea	8.746	4.518	1,530	2.50×10^{24}	1.33	0.7
6	Titan	20.28	15.945	5,150	1.35×10^{26}	1.89	0.21
7	Hyperion	24.57	21.277	410 × 260 × 220	1.71×10^{22}		0.3
8	Iapetus	59.09	79.330	1,460	1.88×10^{24}	1.15	0.5/0.05
9	Phoebe	213.2	550.48	220			0.06

[a]The symbol R_s stands for the equatorial radius of Saturn, taken here to be equal to 60,268 km. A colon (:) following a value for distance indicates an uncertain value.

[b]No density value is given when the diameter is uncertain or the satellite shape is irregular.

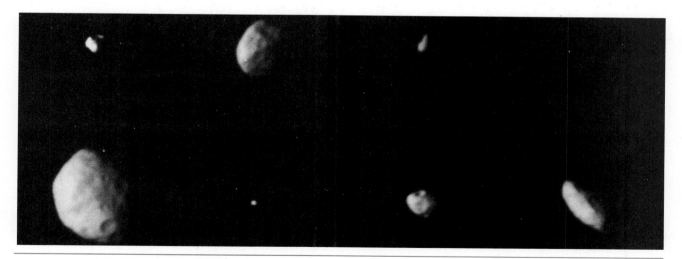

Figure 13.26 Some of the small satellites of Saturn.   *Taken at various ranges, these images do not reflect the true relative sizes. The upper middle and lower left moons are the co-orbital satellites (Epimetheus and Janus), and the two at lower right (Prometheus and Pandora) are the pair that "shepherd" the F ring (discussed later in this chapter). The upper left satellite (Helene) is the one that shares the orbit of Dione, and at upper right and lower left center are the pair (Telesto and Calypso) that share the orbit of Tethys. (NASA)*

Figure 13.27 Titan. *The largest of Saturn's moons, Titan is almost unique among all the satellites in the solar system in having an atmosphere. (NASA)*

observations more difficult. Nevertheless, radio and infrared sensors could penetrate deeply, providing data which, along with spectroscopy of the upper layers, were sufficient to allow some rather detailed estimates of the surface conditions to be made.

First, regarding the solid body of Titan itself, it was ascertained that its density is just under 2 grams/cm³, corresponding to an internal composition of about half rock and half ice. This is very similar to Ganymede, the largest Jovian satellite.

We can only speculate as to what the surface of Titan might be like. There is methane in the atmosphere, and radio data showed that the surface temperature is about 90 K, very near the value where methane can exist in any of the three forms: liquid, solid, or vapor. Thus, there may be liquid methane on the surface, along with methane ice formations such as glaciers, as well as gaseous methane in the atmosphere. In short, methane on Titan may play very much the same role as does water on the Earth (although any liquid methane on Titan would probably be mixed with other compounds

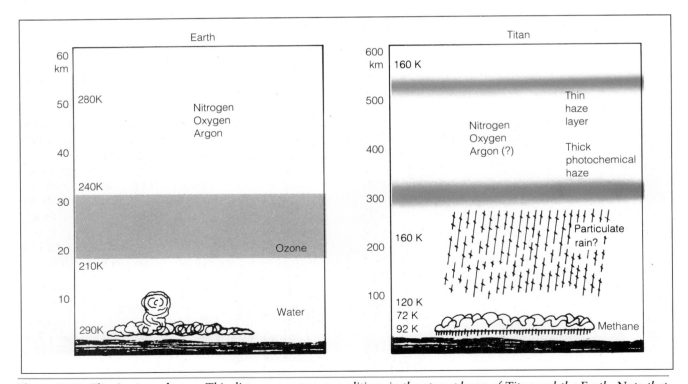

Figure 13.28 Titan's atmosphere. *This diagram compares conditions in the atmospheres of Titan and the Earth. Note that the vertical scales are not the same; Titan's atmosphere is much deeper. (Data from Beatty, J. K., B. O'Leary and A. Chaikin, eds. 1990. The new solar system. 2d ed. Cambridge, England: Cambridge University Press.)*

as well, such as nitrogen and more complex hydrocarbons, which should also be in liquid form). If surface liquids exist, there is probably substantial erosion, so that old impact craters and tectonic formations, if they are present, are continually being smoothed over. Very recent radar observations have shown some variation in the ability of the surface to reflect radar beams, indicating that the giant moon is not entirely covered with liquid, but must instead have surface features that vary from place to place. Further radar observations are planned, and might answer the question of lakes or seas on Titan.

Besides Earth, only Mars, Venus, and now Titan, are known to have any significant abundance of nitrogen in their atmospheres (recently Neptune's satellite Triton has been added to this list). The origin of the nitrogen on Titan may be volcanic, since it is a volatile gas that can be emitted from certain kinds of rocks when they are heated. Therefore, there may also be volcanic features on the surface of this satellite.

The atmospheric pressure is about 60 percent greater at the surface of Titan than at sea level on Earth, and Titan's atmosphere extends to 5 times the height of Earth's atmosphere above the surface. It is not well understood why Titan should have been able to retain such a thick atmosphere, when no other satellite (except Neptune's moon Triton) has any atmosphere at all. Certainly the low temperature and large mass have helped, but in these regards Titan does not appear to have any great advantage over Ganymede, for example. Perhaps the atmosphere of Titan is being continually replenished by some ongoing process, such as volcanic activity.

The seven intermediate satellites, named Mimas, Enceladus, Tethys, Dione, Rhea, Iapetus, and Hyperion, are actually quite varied in some respects (Fig. 13.29). They have diameters ranging from about 400 to 1,500 km (whereas Titan's diameter is 5,150 km). Masses for most of them were determined from their gravitational influences on the trajectory of the *Pioneer* and *Voyager* spacecraft, and from these measurements the densities were estimated. In all cases the densities are rather low, indicating a high ice content, and a few of these moons may be nearly pure ice, with densities near 1 gram/cm³.

Among the intermediate-sized satellites, a variety of surface features were seen in the *Voyager* photographs, including heavy cratering in many cases, and also including some peculiar structures. On Dione were found sinuous, branching valley systems and white, wispy patterns that look deceptively like the high-altitude cirrus clouds of Earth. These may be frozen gases that crystallized immediately upon being emitted from surface fissures or volcanoes, falling to the ground in cloudlike arrangements. Linear trenches were found on a few of the satellites, including Dione, Tethys, and Rhea, possibly the result of fractures caused by massive impacts. The innermost of this group of satellites, Mimas is marred by a tremendous impact crater that is fully one-fourth as large as the diameter of the satellite itself.

Mysteries abound. Iapetus, one of the larger members of this group, has a bright and a dark side, the one some 10 times more reflective than the other. *Voyager 2* images indicate that the dark side of Iapetus may be covered by some sort of deposit, perhaps resembling soot. Enceladus is the shiniest object in the solar system, with an albedo near 1, meaning that it reflects just about all the light that hits it. Such a bright surface indicates that it must be very fresh. The orbit of this satellite is coupled to that of Dione, with Enceladus orbiting Saturn once for every two orbits of Dione, so that the two are frequently lined up with Saturn. This may create sufficient tidal friction inside Enceladus to make it volcanically active, much like Io. In this case, the eruptions may consist of ejections of water vapor, so that the surface of Enceladus may be continually recoated by ice. Another hypothesis is that Enceladus is being continuously coated by ice particles from Saturn's rings, within which its orbit lies. The satellite Hyperion was found to be strangely asymmetric in shape despite its moderately large size, perhaps indicating that it was never molten.

Perhaps the strangest of all the satellites of Saturn is Phoebe, nearly four times farther out than all the rest, and orbiting in the retrograde direction, with the plane of the orbit tilted nearly perpendicular to the others. This oddball among oddballs, which has a dark surface, may be a stray asteroid that was captured by Saturn's gravitational field and forced into orbit.

The Rings

We turn our attention now to the stunning ring system of Saturn, probably the most spectacular sight of all (Fig. 13.30). The *Voyager* data proved these structures to be even more complex and fascinating than had been imagined.

The rings consist of countless individual particles, ranging in diameter from a fraction of a centimeter in

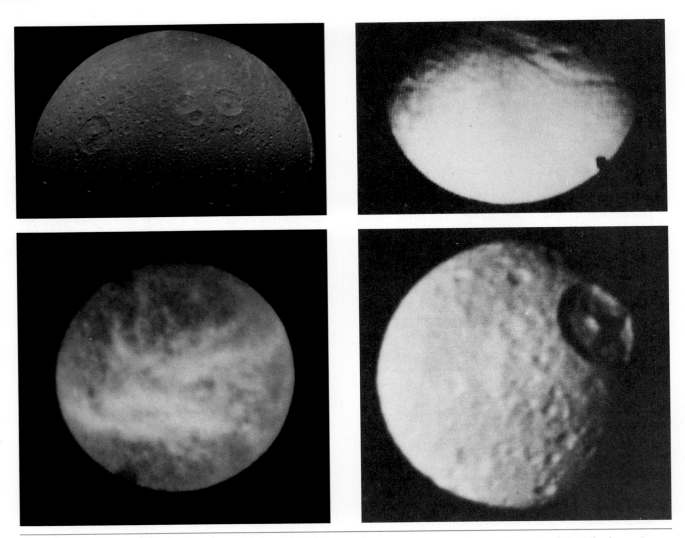

Figure 13.29 The seven intermediate moons of Saturn.   *Clockwise from top left: Dione, Tethys, Mimas, and Rhea. (Continued on next page.)*

some portions of the system to a few meters in others. Each particle has its own orbit, with its period determined by its distance from Saturn according to Kepler's third law. Thus, the rings do not rotate like rigid hoops around the planet, in which case all the particles would have the same period (meaning that the outer ones would have to travel fastest). Instead, they are fluid structures, moving slowest in the outermost portions.

The three distinct rings visible from the Earth had been labeled the A, B, and C rings, in order from the outer one inward (Table 13.8). The possibility of a faint ring inside the C ring was raised some time ago,

and this D ring was actually found, as were new E, F, and G rings outside of the A ring (the letters are assigned in the order of discovery).

Even before the full complexity of the ring system was known, some idea of how the rings formed had been developed. The rings are all very close to Saturn, and this led to the suspicion that tidal forces may have had something to do with their formation. There is a certain distance from any massive body, called the **Roche limit,** inside which the tidal forces pulling an object of a given size apart are greater than the gravitational force holding it together. The exact location of the

Roche limit depends on the mass and size of the orbiting object. We have discussed the fact that the Moon has a tidal bulge on it because of the Earth's gravitational pull; to envision what the Roche limit means, imagine that the bulge becomes so severe that the Moon is actually pulled apart and broken into pieces.

The rings of Saturn are all within the Roche limit for a satellite of any significant size. Therefore, it may be that there once was a satellite that wandered too close to Saturn and was broken apart, leaving behind the debris that now forms the rings. More likely, the rings are made of particles that have been orbiting Saturn since its formation but that were so close to it that tidal forces prevented their ever merging together to form a satellite.

Another possibility, recently proposed but gaining some acceptance, is that the rings consist of debris from former moons that were destroyed by major impacts with large meteoroids (see Chapter 15). Such impacts are more probable near a massive planet than elsewhere, because the gravitational field of the planet attracts passing objects. Furthermore, there is evidence that the rings of Saturn (and of Uranus) are not permanent features, but may be rather young.

Regardless of the origin of the rings, the many mechanisms that govern their structure are the same. Most of these mechanisms involve gravitational effects of Saturn's many satellites. These effects are discussed in Chapter 16.

Figure 13.29 (Continued).   *Clockwise from top left: Iapetus, Enceladus, and Hyperion. (NASA)*

Gravitational forces are responsible for the major gap in the ring system of Saturn, although the details are not entirely clear. The Cassini division, between the A and B rings, lies at just the right distance from Saturn that any particle orbiting there would be in orbital resonance with Mimas, the innermost of the larger satellites. By this we mean that the orbital period of such a particle would be precisely half the period of Mimas, so that every two trips around, the particle

ASTRONOMICAL INSIGHT 13.2

The Elusive Rings of Saturn*

In the text we said little about the discovery of the ring system encircling Saturn, except that Galileo was perplexed by the appearance of the planet, and that, some years after his death, the problem was solved by Christian Huygens. The story is actually much more complex and interesting than is implied by that simple description. The mystery of Saturn's appearance, and the extensive efforts devoted to solving it, was the leading problem in astronomy from the 1630s to the end of the seventeenth century.

Galileo first observed Saturn through a telescope in 1610, and immediately saw that its appearance was strange. He believed that he saw a spherical planet attended on either side by smaller spherical bodies, and he concluded that these were two satellites of Saturn. He made repeated observations thereafter, but to his surprise, he saw no changes in the positions of the suspected satellites, at least not for a while. Frustrated by the lack of change, Galileo stopped observing Saturn

*Based on information from Van Helden, A. 1974, Saturn and his anses, *Journal Hist. Astronomy*, 5(2):105; and Annulo Cingitur: the solution to the problem of Saturn, *Journal Hist. Astronomy*, 5(3):155.

regularly and was therefore shocked when, in 1612, he found it to be a simple sphere, with no trace of the peculiar extensions or satellites seen earlier (the rings were edge-on to the Earth at that time). Galileo resumed frequent observations, and saw the shape of Saturn change steadily. In 1616, he saw an elliptical overall shape, with dark spots near each end, and this is what gave him the impression of ears (the dark spots were the small gaps between the rings and the planet, where the blackness of space can be seen in the background).

Galileo exchanged correspondence with other observers, and soon Saturn was being watched by a number of astronomers, all of whom were puzzled by its unorthodox behavior. When the planet regained a simple spherical appearance in 1642, interest was further heightened, and by this time Saturn was the central problem in astronomical research and discussion. After 1642, many sketches of the planet were published by a number of observers, and soon a variety of suggestions were made as to the cause of its variable and strange appearance.

It is interesting that the resolution of most of the telescopes used was actually good enough to provide some detail, and the explanation of the changing appearance in terms of a ring circling Saturn quite possibly could have been developed years before it actually was put forth by Huygens. The problem was that the idea of a ring was completely alien to the

usual astronomical phenomena, and nobody made the mental leap to this idea. Meanwhile, the data on the question continued to accumulate.

In early 1655, soon before the ring was to disappear again as it presented itself edge-on to the Earth-based observer, Huygens had his first look at Saturn. His telescope, contrary to popular belief, was really little better than those of many of his contemporaries, and he was able to see no more detail than they. Furthermore, he made his first observation at a time when the rings were so nearly edge-on that he had no chance of seeing the true nature of Saturn's peculiar shape. Nevertheless, on the basis of his own limited data and especially his knowledge of the appearance seen by others, Huygens deduced that all the observations could be explained by a flat ring that circled the planet in its equatorial plane (the inclination of the equator of Saturn had been deduced from the orbital motions of the satellite, Titan, which had been discovered by Huygens during his observations of 1655).

A number of alternative views were proposed at about the same time, and it took years for Huygens' solution to become generally accepted. It is interesting that it was essentially a theoretical solution, rather than something that he saw simply by looking through a good telescope, as is commonly assumed.

Figure 13.30 The rings of Saturn. *This* Voyager *image reveals some of the fantastically complex structure of the ring system. (NASA)*

Table 13.8 The rings of Saturn

Feature	Distance (R,)[a]
D ring inner edge	1.11
C ring inner edge	1.233
B ring inner edge	1.524
B ring outer edge	1.946
A ring inner edge	2.021
A ring gap center	2.212
A ring outer edge	2.265
F ring center	2.326
G ring center	2.8
E ring inner edge	3:
E ring outer edge	8:

[a]R, stands for the equatorial radius of Saturn, taken here to be equal to 60,330 km. A colon (:) following a value for distance indicates an uncertain value.

and Mimas would line up on the same side of Saturn. These frequent alignments periodically subject particles in this region to extra gravitational forces. The result is that particles are cleared out of this region by the small gravitational tugs exerted on them at each alignment with Mimas. Calculations of this process are not entirely satisfactory, however, because it appears to take too long to clear out the observed gaps. It is possible that collisions between particles are enhanced by the alignments, so that the particles are deflected more efficiently than would be the case without collisions. There is little doubt that orbital resonance with Mimas is responsible for the gap, despite the present uncertainty in some of the details. Other satellites, or other fractions of the orbital period of Mimas (such as the location where an orbiting particle would have exactly one-third the period of Mimas, for example) could create other gaps in the rings, and indeed such gaps have been found. Orbital resonances with satellites of Saturn led to the expectation that the rings might have complex structure, even before *Pioneer* and *Voyager* data were available.

Despite this expectation, when the first *Voyager* photos of the rings were obtained, scientists were amazed by what they saw. None of the rings had a smooth appearance when viewed from nearby, but instead the prominent ones visible to us from Earth are each composed of a number of thin rings that came to be called "ringlets" (see Fig. 13.30). The photopolarimeter experiment on *Voyager 2* revealed much more detail in the rings, showing structures as small as 100 meters in size. There are thousands of ringlets in the entire system, some appearing dark and others light. Even the Cassini division is not completely empty, but instead is inhabited by several dark ringlets. There apparently is a narrow gap as well that truly is empty, probably created by Mimas in the manner just described.

Whether a ring will appear dark or light depends on how the particles in it reflect sunlight and the angle from which it is viewed. Some of the rings tend to reflect light straight back, while others allow it to go through in a forward direction. A ring that reflects sunlight back will appear bright when viewed from the Earth, but will look dark when viewed from the far side, looking toward the Sun. Conversely, a ring that scatters light in the forward direction will appear dark from the Earth (as do the narrow ringlets in Cassini's division) and bright if viewed from behind. Whether a given ring will scatter light forward or backward depends on the size of particles it is made of and also on their shape and composition.

From consideration of their light-scattering properties, it has been possible to deduce the average sizes of the particles in the various rings of Saturn. The outermost ones tend to be made of very tiny particles, a few ten-thousandths of a centimeter in diameter, while the inner rings, in contrast, seem to consist mostly of chunks of material a few meters across. The composition of all the particles is probably mostly ice, and their shapes are irregular. The total mass of all the ring material together is very small, only about a millionth of the mass of our Moon, so if the ring particles could have formed into a satellite of Saturn, it would have been a rather small one.

The complex pattern of rings and ringlets seen by the *Voyager* cameras required a more sophisticated explanation than the orbital resonance hypothesis described above, although that process is undoubtedly at work. Apparently at least three other mechanisms exist for shaping the rings of Saturn.

Two of these involve gravitational forces, while the third may be due to electromagnetic forces acting on ring particles that have acquired electrical charges. As we will see in Chapter 16, small satellites orbiting within the domain of the rings can trap thin rings between them. We will also find that gravitational tugs from satellites can create standing, spiral-shaped wave patterns. We will defer a more detailed discussion of these processes until after we have examined the properties of rings and satellites around the remaining outer planets, which we do in the next chapter.

PERSPECTIVE

In learning of the many complexities and beauties of Jupiter and Saturn, we have encountered two planets so rich and varied as to rival miniature solar systems themselves. The assortment of phenomena ranging from the giant planets themselves, with their swirling, colorful atmospheres, to their immense and energetic charged particle belts, to their fantastic families of satellites, and to their intricate, wispy ring systems, have presented us with many new insights and many new challenges to our understanding.

While the two giants are similar in many ways, we have also uncovered many significant contrasts. Next we turn our attention to the remaining three outer planets, where once again we will find some similarities and many differences.

SUMMARY

1. Jupiter is gigantic, with 318 times the Earth's mass, but it is so large that its average density is only one-fourth that of the Earth.

2. Jupiter is totally unlike the terrestrial planets, having no solid surface, a thick atmosphere of hydrogen compounds, and rapid rotation.

3. Jupiter's atmosphere has a pattern of belts and zones, which are planet-girdling wind systems created by a combination of convection and the rapid rotation of the planet.

4. Internally, Jupiter has a small, probably solid core formed by differentiation, a thick zone of liquid metallic hydrogen outside the core, an even thicker layer of liquid hydrogen above that, and a thin layer of clouds at the top.

5. Jupiter's strong magnetic field and its rapid rotation combine to create and maintain intense radiation belts.

6. Jupiter has sixteen known satellites, of which the four Galilean moons are by far the largest.

7. The Galilean satellites show a range of properties indicating a decreasing degree of tectonic and geological activity from the innermost (Io) to the outermost (Callisto).

8. Io, the innermost Galilean moon, has volcanic activity and is accompanied in its orbit by a cloud of gas that forms a torus around Jupiter, whose size and orientation is controlled by Jupiter's magnetic field. This volcanic activity is caused by internal heating due to gravitational stress imposed on Io by Jupiter and Europa.

9. Jupiter has a thin, dim ring.

10. Saturn's atmosphere and internal structure are different from those of Jupiter in the lack of contrast in the global wind patterns, the fact that Saturn has seasonal variations, the greater velocities of its winds, and its different source of excess internal heat.

11. Saturn has eighteen known satellites, including the giant Titan, seven intermediate moons, and a large number of tiny, irregular satellites.

12. Titan has a thick atmosphere dominated by nitrogen, but it also includes methane, which may exist in gaseous, liquid, and frozen states on the surface.

14. The intermediate moons all have very low densities, indicating that they have icy compositions, and some show evidence of resurfacing, by frozen internal gases that were released, or by debris from space.

15. The complex structure of Saturn's ring system is created by a combination of effects: orbital resonances with the satellites; spiral density waves; shepherding by pairs of satellites; and electromagnetic forces acting on charged particles. The rings may not be permanent features, but instead may be relatively young.

REVIEW QUESTIONS

Thought Questions

1. Compare the composition of the atmospheres of Jupiter and Saturn with that of primitive Earth's atmosphere.
2. What is the significance of the fact that Jupiter is a source of radio emission?
3. Discuss the similarities and contrasts between the atmospheric circulation in Jupiter's atmosphere and that in the Earth's atmosphere.
4. How are the atmospheric circulations of Jupiter and Saturn similar, and how do they differ?
5. How is anything at all known about the interiors of Jupiter and Saturn?
6. How is the internal structure of Saturn thought to differ from that of Jupiter? Are the sources of excess internal heat thought to be the same? Explain.
7. Compare the extent to which Jupiter and Saturn are differentiated with the degree of differentiation in the terrestrial planets.
8. How is methane on Titan comparable to water on Earth?
9. Explain why Io has volcanic activity. What is the role of Europa in this explanation?
10. Describe the structure and the origin of the Io torus.

Problems

1. Sum up the masses of the other eight planets, and compare the total with the mass of Jupiter alone.
2. If the radius of Jupiter's solid core is 12,000 km and the mass of the core is fifteen Earth masses, what is the average density of the core? Compare your answer with the typical densities of the terrestrial planets and their cores.
3. Show that Kepler's third law is valid for Jupiter's satellites by computing P^2 and a^3 for several of them, and showing that P^2 is proportional to a^3 (that is, that the ratio of P^2 to a^3 is constant). Can you think of a system of units in which P^2 will equal a^3 for the satellites of Jupiter?
4. Io has the highest density among the Galilean satellites, largely because it has little or no ice in its interior, while the others have retained ice in varying proportions. What would the density of Io be if its volume were doubled by adding ice to it? (Ice has a density of 1 gm/cm³.) Compare with the densities of the other Galilean satellites.
5. Under the best observing conditions, astronomers on Earth can make out details as small as one-twentieth of the angular diameter of Saturn. How many kilometers thick can the rings be without being visible when viewed edge-on?

6. Given the rotation period and diameter of Saturn as listed in Table 13.5, calculate how rapidly a point on the equator is moving as the planet spins. Compare your answer with the wind speed given in the text for atmospheric circulation near the equator.
7. Calculate and compare the wavelengths of maximum emission for Jupiter and Saturn, from their cloudtop temperatures (130 and 95 K, respectively).
8. Calculate the surface gravity and the escape velocity on Titan, and compare with that of the Earth. Do your values explain how Titan can have a thicker and denser atmosphere than Earth?

ADDITIONAL READINGS

Beatty, J. K., B. O'Leary, and A. Chaikin, eds. 1990. *The new solar system*. 2d ed. Cambridge, England: Cambridge University Press.

Carroll, M. W. 1987. Project *Galileo:* The phoenix rises. *Sky and Telescope*. 73:359.

Elliot, J., and R. Kerr. 1985. How Jupiter's ring was discovered. *Mercury* 14(6):162.

Esposito, Larry. 1987. The changing shape of planetary rings. *Astronomy* 15(9):6.

Ingersoll, A. P. 1981. The meterology of Jupiter. *Scientific American* 245(6):90.

Ingersoll, A. P. 1981. Jupiter and Saturn. *Scientific American* 145(6):90.

Johnson, T. V., and L. A. Soderblom. 1983. Io. *Scientific American* 249(6):60.

Lanzerotti, L. J., and C. Uberoi. 1989. The planets' magnetic environments. *Sky and Telescope* 77:149.

Miner, E. D. 1990. *Voyager 2's* encounter with the gas giants. *Physics Today* 43(7):40

Morrison, David. 1981. The new Saturn system. *Mercury* 10(6):162.

————. 1985. The enigma called Io. *Sky and Telescope* 69(3):198.

Pollack, J. B., and J. N. Cuzzi. 1981. Rings in the solar system. *Scientific American* 145(5):104.

Sanchez-Larega, A. 1989. Saturn's great white spot. *Sky and Telescope* 78:141.

Soderblom, L. A. 1980. The Galilean moons of Jupiter. *Scientific American* 242(1):68.

Soderblom, L. A., and T. V. Johnson. 1982. the moons of Saturn. *Scientific American* 246(1):100.

Squyres, S. W. 1983. Ganymede and Callisto. *American Scientist* 71(1):56.

MATHEMATICAL METHODS

Tidal Forces

The concept of a tidal force, or differential gravitational force, was introduced in Chapters 5 and 8, and has been referred to in other chapters since then. We have learned that tidal forces can modify the spin period of a satellite or a planet, as in the cases of the Moon or Mercury (Chapter 10), and in this chapter we have discussed the strong influence of tidal forces on the geological evolution and activity of the satellites of the giant planets.

To help see why the tidal forces are so much more important for satellites such as Io or Europa compared to the Moon or Mercury, it is useful to consider the mathematical expression for the tidal force exerted on one body by another. The tidal force is given by

$$\Delta F = \frac{4GmMR}{r^3},$$

where m is the mass of the affected body, M is the mass of the body that is creating the tidal force, r is the separation of the two bodies, and R is the radius of the affected body (see figure). As usual, G is the gravitational constant. In this expression, the force F is the *difference* between the gravitational forces exerted on the near and far sides of the body in question.

If we substitute values into this expression, using the centimeter-gram-second system of units, we will calculate a force that is in units of dynes. This may not be very meaningful, since few of us can visualize how strong a force of one dyne is, so it is more meaningful to carry out comparative calculations. We can do this by writing expressions for the tidal forces to be compared, and then dividing. When we do this, we have the advantage that we can visualize the comparison that results, and that the constant terms cancel out. For example, we can compare the tidal force exerted on the Earth by the Sun with that exerted on the Earth by the Moon:

$$\frac{F_{SE}}{F_{ME}} = \left(\frac{r_{ME}}{r_{SE}}\right)^3 \left(\frac{M_S}{M_M}\right) = 0.43$$

Here F_{ES} is the tidal force exerted on the Earth by the Sun, and F_{EM} is the tidal force exerted on the Earth by the Moon; r_{EM} is the Moon–Earth distance, and r_{ES} is the Sun–Earth distance. Otherwise, subscript M refers to the Moon, S to the Sun, and E to the Earth. The mass and radius of the Earth have canceled out. This result tells us that the Sun's tidal force on the Earth is less than half as strong as the tidal force due to the Moon, even though the Sun's *total* gravitational force on the Earth is stronger by a factor of about 170 (you can check this by using the law of universal gravitation, again dividing in order to compare the two forces).

We can also compare the tidal force exerted on the Moon by the Earth with that exerted on the Earth by the Moon:

$$\frac{F_{ME}}{F_{EM}} = \frac{R_E}{R_M} = 3.7$$

In this calculation, the $1/r^3$ factors canceled out, since the distance is the same in the expressions for both tidal forces; and all the mass factors canceled out, because the same two masses appear in the expressions for both forces. Hence the Earth exerts a tidal force on the Moon that is almost 4 times stronger than that exerted on the Earth by the Moon. This helps us to understand why the Moon has slowed its spin enough to be in synchronous rotation, while the Earth has not yet slowed as much.

Now let us consider a satellite of one of the giant planets. For example, the tidal force exerted on Io by Jupiter may be compared to that exerted on our Moon by the Earth. If we write expressions for the two tidal forces in question and then divide, we find

$$\frac{F_{IJ}}{F_{ME}} = \left(\frac{r_{ME}}{r_{IJ}}\right)^3 \left(\frac{m_I}{m_M}\right) \left(\frac{M_J}{M_E}\right) \left(\frac{R_I}{R_M}\right) = 298$$

Thus, Io is subjected to a tidal force almost 300 times greater than that exerted on the Moon by the Earth (and in this calculation, we left out the effect of Europa, which adds to the tidal force on Io when the two are aligned).

As we have noted, tidal forces play many important roles in the solar system. With the mathematical expressions discussed here, you can compute and compare the tidal forces exerted on various bodies, helping you to understand why the effects are so different in different cases.

CHAPTER 14

The Outer Planets

Neptune and Triton as seen by Voyager 2 from a distance of 76 million kilometers. (NASA/JPL)

The three planets that we have not yet discussed, orbiting in the distant reaches of the solar system far from the warmth of the Sun, are the least understood. These three—Uranus, Neptune, and Pluto—all lie well beyond the orbit of Saturn and were not known to the ancient astronomers. Until very recently, they were also little understood by modern astronomers. Now two of them, Uranus and Neptune, have been visited by the *Voyager 2* spacecraft, and we have detailed information on these planets, much as we do for Jupiter and Saturn.

URANUS

The seventh planet from the Sun, Uranus (Table 14.1) is marginally bright enough to be seen with the naked eye and was actually included in a number of star charts compiled from telescopic observations in the seventeenth and eighteenth centuries. At its distance of nearly 20 AU, its period is so long (about 84 years), and its motion so slow, that for quite a while it was not noticed that its position constantly changes with respect to the background stars.

Discovery and Observations from Earth

The planet was finally discovered in 1781, when the English astronomer William Herschel found an object that seemed unusual for a star during a routine sky survey. It had a disklike appearance, rather than being a single, twinkling point of light, and it appeared blue-green. Herschel at first suspected that he had found a comet, but after observing it for an extended time and then calculating its orbit, he found it to be moving about the Sun in a nearly circular path, about twice as far out as Saturn. This established that Herschel had found a planet, for no comet is visible at such a great distance from the Sun, nor would a comet move in a circular orbit. Herschel initially named the new planet after King George III of England, but the name was never widely accepted, and the recognized name became Uranus, after the Greek god representing the heavens.

Some information was derived from Earth-based telescope observations of Uranus. For example, its mass was estimated using the orbital periods of its satellites and Kepler's third law (Uranus had five known satellites until the recent *Voyager* encounter, which revealed ten more), and its radius was estimated from its angular diameter measured from photographs made by high-altitude balloon experiments (Fig. 14.1) and from knowledge of its distance from Earth. Spectroscopic observations revealed some information on the composition of the atmosphere, and infrared data provided estimates of its surface temperature and indications of surface markings (Fig. 14.2), which are very difficult to see from Earth. A system of dark, relatively thin rings was discovered in 1977, when observations of a star that was about to pass behind the planet showed a series of occultations due to the rings. Five major

Table 14.1 Uranus

Orbital semimajor axis: 19.22 AU (2,875,000,000 km)
  Perihelion distance: 18.31 AU
  Aphelion distance: 20.13 AU
Orbital period: 84.01 years (30.685 days)
Orbital inclination: 0°46′23″
Rotation period: 17^h 14^m
Tilt of axis: 97°52′
Mean diameter: 50,532 km (3.97 $D_\oplus$)
  Polar diameter: 49,946 km
  Equatorial diameter: 51,118 km
Mass: 8.68×10^{28} g (14.54 $M_\oplus$)
Density: 1.27 g/cm³
Surface gravity: 0.89 Earth gravities
Escape speed: 21.2 km/sec
Surface temperature: 76 K (average)
Albedo: 0.50
Satellites: 15

Figure 14.1 Uranus. *This exceptionally fine photograph was obtained from a high-altitude balloon, which was able to avoid much of the blurring effect of the Earth's atmosphere. No surface markings are evident, except for some darkening at the edges of the disk. (Project Stratoscope, Princeton University, supported by the National Science Foundation)*

Figure 14.2 Infrared views of Uranus. *The image at left, obtained through a filter matching a wavelength at which methane absorbs light, shows zonal structure in the planet's atmosphere. At right is an image in which colors have been used to indicate relative temperatures. The blue disk of the planet is surrounded by red for the rings, which are colder. (NASA, left photo courtesy of H. Reitsema)*

satellites, including two found by Herschel in 1787, were discovered from Earth-based observations, although more were suspected because of the structure in the ring system. (Recall the role of Saturn's moons in governing the structure of its rings.)

One of the most striking peculiarities about Uranus, known from spectroscopic observations, is that its rotation axis is tipped over very far. The pole that we normally think of as the north pole (where rotation is counterclockwise as seen from above) actually points slightly (8°) *below* the plane of the planet's orbit. We say, therefore, that the inclination of the planet is 98°, which technically means that the rotation of the planet is retrograde, since when viewed from above the ecliptic plane, the rotation is backward. Seasons on Uranus, of course, are quite extreme (Fig. 14.3). Presently, the "south" pole of the planet is pointed almost directly toward the Sun, and this was the orientation of the planet when *Voyager 2* visited in January of 1986.

The extreme tilt of Uranus stands as an anomaly in the solar system and is not easily explained by theories of planetary formation (see Chapter 16). The fact that all five of the known satellites, as well as the recently discovered rings of Uranus, orbit in the equatorial plane

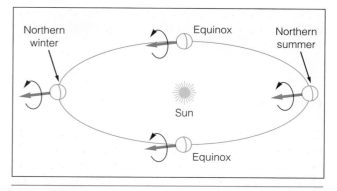

Figure 14.3 Seasons on Uranus. *The unusual tilt of Uranus creates bizarre seasonal effects. The solid arrow in each case indicates the direction of the planet's north pole, and the curved arrow, the direction of its rotation.*

of the planet (that is, the plane of the satellite and ring orbits is nearly perpendicular to the ecliptic) suggests that whatever disturbed the planet away from the usual orientation (possibly a collision with a large planetesimal) must have happened at some time before the satellites had formed, perhaps when they were still part of a disk of orbiting debris.

Atmosphere and Interior

The atmosphere of Uranus has a composition similar to that of Jupiter and Saturn: Hydrogen molecules dominate, helium is the next most abundant species, and methane (CH_4) is common. Ammonia is not seen in large quantities, presumably because it has condensed into solid form and precipitated out. The blue-green color of the planet is caused by methane, which absorbs red light, so that light reflected from the upper layers of atmosphere is deficient in red. Jupiter and Saturn contain methane as well, but both have cloud layers of other gases that overlie the methane. By contrast, methane dominates the upper atmosphere of Uranus, so the reflected light from the top of the clouds is blue.

Uranus has no source of excess internal heat, yet it has atmospheric circulation patterns much like those of Jupiter and Saturn (although far less vivid). This indicates that, even at the remote distance of Uranus from the Sun, sunlight can provide enough heat to cause convection, which, along with the planet's rapid rotation (its period is 17.24 hours, as derived from *Voyager* measurements of radio emission), creates belts and zones that girdle the planet. One major surprise was the discovery that the direction of flow in the atmosphere is just backward from what had been expected on the basis of theoretical calculations. Theory indicates that when the poles of a rotating planet are warmer than the equator, which should be the case on Uranus, the winds near the equator should be slower than the rotation, meaning that they should flow backward with respect to the body of the planet. Instead, the equatorial winds on Uranus flow faster than the planet's rotation, so they are in the forward direction with respect to Uranus. This unexpected wind pattern is probably caused by an equally unexpected distribution of temperatures over the planet: Rather than being much warmer at the poles than at lower latitudes, the temperature was found to be quite uniform over most of the planet, with the coldest zones at intermediate latitudes (about 30° north and south of the equator).

The *Voyager* images of Uranus (Fig. 14.4) showed almost no surface markings, except when computer-processing techniques were used to enhance contrast (Fig. 14.5). Then the zonal flow patterns were revealed. Also seen were occasional white clouds near the equa-

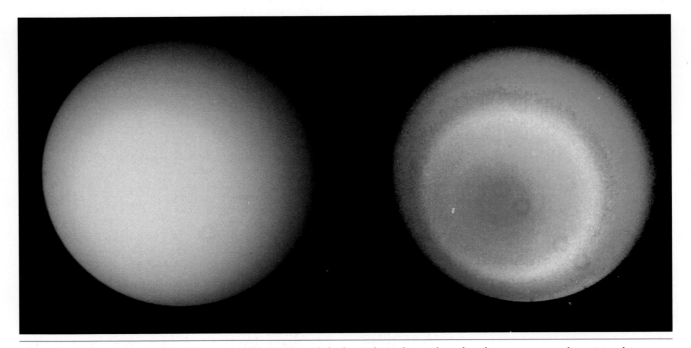

Figure 14.4 Uranus as seen by Voyager. *The image at left shows how featureless the planet appears when viewed in ordinary visible light, even when seen close up. At right, color enhancement has been used to increase the contrast and show bands. The fuzzy, circular features in these and other images are caused by imperfections in the Voyager camera and are not real features on the planet. (NASA/JPL)*

Figure 14.5 Zonal flows on Uranus. *The computer-enhanced image shows the belts and zones on Uranus, which are the equivalent of those on Jupiter and Saturn except that the contrast is very low. Despite the lack of internal heating on Uranus, the flow patterns in its atmosphere are very similar to those of the larger giant planets. (NASA/JPL)*

tor, apparently due to rising plumes of methane ice crystals which are quickly spread out by the planetary rotation. (Fig. 14.6).

A tenuous cloud of hydrogen atoms surrounds Uranus, produced by the breakup (dissociation) of hydrogen molecules (H_2) due to collisions with rapidly moving electrons in the charged particle belts that surround the planet.

The magnetic field of Uranus was not detected until *Voyager's* approach, although the presence of a field had been suspected and was no surprise. What was a surprise, however, was its orientation: The magnetic axis is tipped some 60° with respect to the rotation axis. Thus, the magnetic axis is almost perpendicular to the plane of the planet's orbit, so that the magnetosphere, the volume surrounding the planet where charged particles are trapped by the magnetic field, has much the same appearance as for any other planet, with an extended tail pointing away from the Sun because of the solar wind (Fig. 14.7). As the planet rotates on its highly tipped axis, this tail rotates about.

Figure 14.6 Clouds in the atmospheres of Uranus. *The white streak seen here is a cloud of methane ice crystals. Such clouds occasionally rise through the atmosphere near the equator and are then carried around the planet by the winds, as seen in this sequence. (NASA/JPL)*

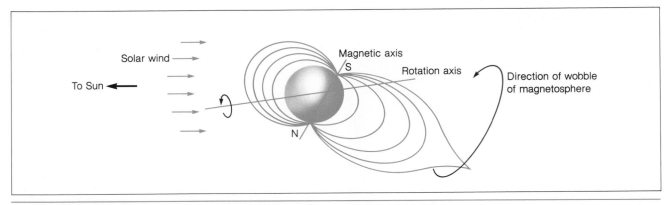

Figure 14.7 The magnetosphere of Uranus. *Due to the combination of the high tilt of the rotation axis of Uranus and the misalignment between its rotation and magnetic axes, the magnetosphere of Uranus resembles that of one of the planets whose rotation axis is more normal. The tail of the magnetosphere rotates as the planet spins around its highly tilted axis.*

The magnetic field of Uranus is thought to arise in an interior level where water may exist, and may be sufficiently compressed to conduct electricity. Data from *Voyager 2* have been used to construct models of the internal structure of Uranus (Fig. 14.8). It is thought that there is a core of rock and ice, probably about the size and mass of the Earth, outside of which is a liquid hydrogen zone, perhaps overlain by a water zone, and, in the lower atmosphere, layers of methane ice crystals (along with other ices such as ammonia and water). There is probably no liquid metallic hydrogen zone, because the pressure inside Uranus is not sufficient.

Satellites and Rings

Uranus has fifteen known satellites (Table 14.2), five of which were detected from Earth (Figs. 14.9 and 14.10). All fifteen are in synchronous rotation because

Figure 14.8 The interior of Uranus. *This shows a model based on Voyager 2 results. The outer zone is gaseous, and is underlain by a thick ice and liquid zone extending from about 7,800 km out to 18,000 km from the center. This zone contains about 65 percent of the planet's mass, while the rocky core contains roughly 24 percent of the mass.*

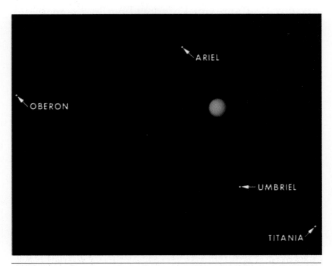

Figure 14.9 The major moons as seen by *Voyager*. *This is a Voyager 2 image, showing the planet and the positions of the major satellites during approach. (NASA/JPL)*

Table 14.2 Satellites of Uranus

Name	Distance (R_U)[a]	Period (Days)	Diameter (km)	Mass (g)	Density (g/cm³)	Albedo
Cordelia	1.945	0.336	40			0.1
Ophelia	2.105	0.377	50			0.1
Bianca	2.316	0.435	50			0.1
Cressidia	2.418	0.465	60			0.1
Nesdimona	2.453	0.476	60			0.1
Juliet	2.528	0.494	80			0.1
Portia	2.586	0.515	80			0.1
Rosalind	2.735	0.560	60			0.1
Belinda	2.946	0.624	60			0.1
Puck	3.365	0.764	170			0.07
Miranda	5.078	1.413	484	7.5×10^{22}	1.26	0.34
Ariel	7.481	2.520	1,160	1.4×10^{24}	1.65	0.40
Umbriel	10.407	4.144	1,190	1.3×10^{24}	1.44	0.19
Titania	17.051	8.706	1,610	3.5×10^{24}	1.59	0.28
Oberon	22.794	13.463	1,550	2.9×10^{24}	1.50	0.24

[a]The distances are orbital semimajor axis, given in units of the equatorial radius of Uranus, which has the value $R_U = 25{,}559$ km.

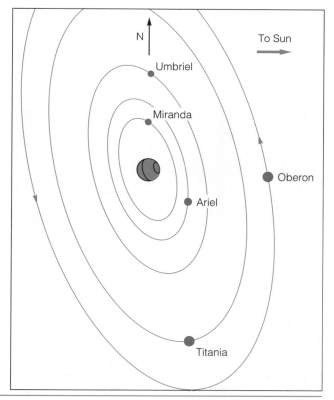

Figure 14.10 Uranus and its satellites.   *This is an Earth-based photo, showing all five of the large moons that are identified in the sketch. (NASA)*

of tidal forces from Uranus. The ten new satellites discovered by *Voyager* are all small objects, some of which are within the system of rings. These satellites, as well as the five large ones, govern the structure of the ring system, just as the many moons of Saturn influence its rings. There are shepherd satellites, for example (Fig. 14.11), and cases of orbital resonances which create gaps in the rings.

Each of the five major satellites was observed in some detail by *Voyager*. The outermost of the five, Oberon (Fig. 14.12), has a heavily cratered surface and shows little sign of recent geological activity, yet it has huge mountain and scarp systems indicative of past episodes of activity. Next in is Titania (Fig. 14.13), which shows evidence of having a younger surface. It appears that somehow Titania has undergone a more recent episode of cratering from impacts than Oberon, and Titania has an extensive series of cliffs that may be the result of expansion due to the freezing of water ice in the interior. The middle one of the five major moons, Umbriel (Fig. 14.14), shows a surface equally cratered and apparently as old as that of Oberon. Umbriel is the darkest of the major satellites, with an albedo of only 0.19, which indicates that its surface may have been coated with dark dust or debris at some time in the past. There is a very bright region on Umbriel, however (see Fig. 14.14), which may be a freshly exposed ice deposit.

The next satellite, Ariel (Fig. 14.15), is the brightest of the Uranian moons, evidence of a young surface coated by ices that have probably escaped from the interior. Ariel also has extensive faults, and there is evidence of icy flows on the surface.

The strangest of the large satellites is Miranda, the smallest and innermost (Fig. 14.16). Miranda has several remarkable features on its surface that indicate recent or current geological activity, more than can be easily explained by tidal stresses due to Uranus and the other moons. However, at the same time, other portions of the surface are heavily cratered and apparently are very old. Possible explorations of Miranda's unusual geology will be discussed in Chapter 16.

The rings of Uranus (Fig. 14.17 and Table 14.3) are thinner and darker than those of Saturn, perhaps resembling the ring of Jupiter more closely. The thinness of the rings of Uranus probably indicates that small satellites orbiting within the ring system are affecting the structure. In Chapter 13 we mentioned that pairs of small moons can trap ring particles between them; this will be discussed in more detail in Chapter 16.

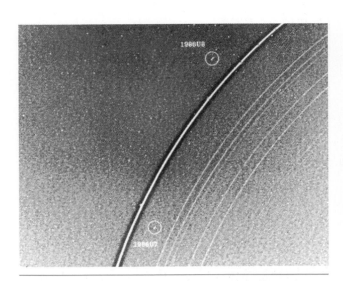

Figure 14.11 Shepherd satellites. *Here we see two small moons on either side of a thin ring around Uranus. The gravitational effects of these moons keep the ring confined. (NASA/JPL)*

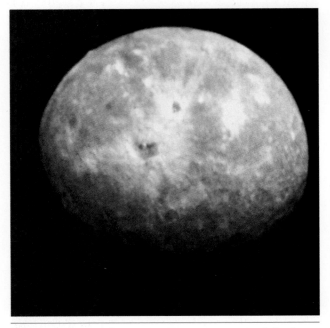

Figure 14.12 Oberon. *This is the outermost of the five major satellites of Uranus. (NASA/JPL)*

Figure 14.13 Titania. *This is the next major satellite inward from Oberon. Titania shows signs, such as more recent cratering, of having a younger surface than Oberon and Umbriel, the next moon inward. (NASA/JPL)*

Figure 14.15 Ariel. *This satellite shows signs of geological activity. Its bright surface and evidence of recent ice flows indicate probable internal motions and outgassing. (NASA/JPL)*

Figure 14.14 Umbriel. *The third of the five major moons of Uranus, Umbriel is also the darkest, its surface apparently having been coated by some sort of low-reflectivity deposit. The many craters show that the surface is old and has not undergone major geological activity in recent times. (NASA/JPL)*

Figure 14.16 Miranda. *The innermost of the five large moons of Uranus, Miranda shows a surprising degree of recent geological activity. Tidal stresses are an inadequate explanation for the youth of the surface of this satellite, and more exotic explanations are needed. (NASA/JPL)*

Figure 14.17 The rings of Uranus. *Both views show the epsilon ring at the top, and include all the other major rings inward toward the planet. The color-enhanced image brings out greater detail. (NASA/JPL)*

Table 14.3 The rings of Uranus

Ring	Distance (R_U)[a]	Width (km)
1986U2R	1.41–1.51	2,500
6 ring	1.60	1–3
5 ring	1.62	2–3
4 ring	1.63	2
Alpha ring	1.71	8–11
Beta ring	1.75	7–11
Eta ring	1.81	2
Gamma ring	1.82	1–4
Delta ring	1.85	3–9
1986U1R	1.91	1–2
Epsilon ring	1.96	22–93

[a]Distances are in units of the equatorial radius of Uranus, which is $R_U = 26,145$ km.

rings that are not perfectly circular in shape. Both kinds of structures are unstable and should not be able to persist; thus it has been suggested that the rings are young, rather than being a permanent fixture that has not changed for long periods. Similar arguments may be made about the rings of Saturn and Neptune. The new picture of evolving planetary rings that has been developed is discussed in Chapter 16.

NEPTUNE

Voyager 2 completed its planetary exploration mission in spectacular fashion in late August, 1989, when it passed close to Neptune and its satellites and ring system. Thus, of the nine known planets, eight have now revealed their intimate secrets to us. In a very real sense, the past two decades have been a new era of exploration comparable to any in humankind's past.

Neptune (Table 14.4) orbits at an average distance of about 30 AU from the Sun, and takes 165 years to make one circuit. In size and appearance through an Earth-based telescope, Neptune resembles Uranus very closely, but we now know that this resemblance is superficial. The two planets, long thought to be nearly identical twins, are now seen to be very different from one another in significant ways.

The Discovery: A Victory for Newton

Neptune is a rather dim object, too faint to be seen with the unaided eye, and even too inconspicuous to

The rings of Uranus appear dark because the particles do not reflect light efficiently. Particles that scatter light preferentially in the forward direction tend to be relatively large; it is estimated that typical particles in the rings are about a meter in diameter, whereas the rings of Saturn contain a mixture of sizes, including some very small ones, perhaps only a millionth of a meter in diameter. It is possible that small particles have been swept out of the rings of Uranus by an extended cloud of hydrogen atoms that surrounds the planet.

Several features of the ring system appear to be temporary, unable to last for a long time. These include a broad, dark dust ring lying close to the planet, and

Table 14.4 Neptune

Orbital semimajor axis: 30.11 AU (4,504,300,000 km)
  Perihelion distance: 29.85 AU
  Aphelion distance: 30.37 AU
Orbital period: 164.79 years (60,188 days)
Orbital inclination: 1°46′22″

Rotation period: 16ʰ 3ᵐ
Tilt of axis: 29°34′

Mean diameter: 49,104 km (3.86 D⊕)
  Polar diameter: 48,680 km
  Equatorial diameter: 49,528 km
Mass: 1.03×10^{29} grams (17.15 M⊕)
Density: 1.64 grams/cm³
Surface gravity: 1.13 Earth gravities
Escape speed: 23.5 km/sec

Surface temperature: 74 K
Albedo: 0.5:

Satellites: 8

have been included on star charts made before the mid-1800s. Rather than being found by accident, as Uranus was, the existence of Neptune was first deduced indirectly.

When Uranus was found by Herschel in 1781 and its orbit subsequently computed, astronomers were naturally eager to see whether its path conformed perfectly to the predictions of Newton's laws of motion and gravitation. Since Uranus had been included on star charts dating back some 90 years before Herschel's discovery, it was possible to compare its observed and computed positions over a long period. Much to the dismay of the scientists who did this, discrepancies were found. The mismatch between the expected and observed positions of the planet were not very large, amounting to 2 minutes of arc by 1840. This was nevertheless considered serious, for the error was far too large to be consistent with the laws of motion. Something was amiss, and people began to question the validity of Newton's mechanics. Eventually it was suggested that the discrepancies were caused by the gravitational influence of an eighth planet. Soon Neptune was discovered by the German astronomer Johann Galle, within 1° of the predicted location. The name *Neptune* was taken from that of the Roman god of the seas.

The *Voyager* Encounter and Properties of Neptune

Earth-based studies of Neptune reveal very little detail; the best photographs show only an indistinct object with few noticeable surface features (although some evidence for atmospheric structure was found through the use of special filters that isolate the wavelengths at which methane absorbs light; Fig. 14.18). Even the rotation period was poorly known, and there was no information on whether the planet had a magnetic field.

Voyager 2 has ended our state of near-total ignorance about Neptune. Besides the many images of the planet (Fig. 14.19) and its moons obtained by

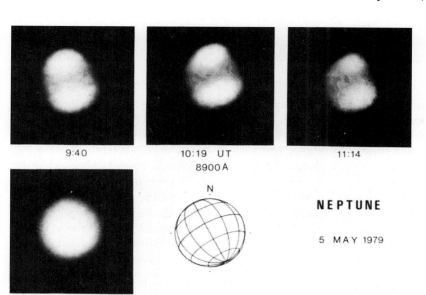

9:40   10:19 UT   11:14
8900 A

10:07
7550 A

N

NEPTUNE

5 MAY 1979

Figure 14.18 Earth-based views of Neptune. *These images, obtained with the use of special filters to isolate the wavelengths at which methane molecules absorb light, represent the greatest amount of detail observed from the Earth. Despite the great distance to Neptune and the tiny size of its image, data such as shown here demonstrated the presence of atmospheric structure, even before the* Voyager 2 *encounter. (NASA)*

ASTRONOMICAL INSIGHT 14.1

Science, Politics, and the Discovery of Neptune

Many times, throughout the history of science, a great discovery was just waiting to be made, as soon as the necessary technology or preliminary information had been developed. If Galileo had not turned his telescope to the heavens, surely someone else would have done so very soon, and would have found the wonders that he discovered. If Newton had not developed his laws of mechanics when he did, they would have become apparent to some other student of the motions of the Moon and planets.

The discovery of Neptune was such a case. The irregularities in the orbit of Uranus were well known by the mid-1800s, and the mathematical techniques needed to calculate from them the posi-tion of the eighth planet were also standard. It was only a matter of someone recognizing the possibil-ity and carrying out the computa-tion. As it happened, two people did so, unbeknownst to each other, and the result was one of the more fascinating disputes in the history of science.

In England, a young college graduate named John C. Adams made the prediction in 1845 that an eighth planet should be at a certain location in the constella-tion Aquarius. Being young and unknown, Adams had difficulty convincing the established astron-omers of England that an effort should be made to find the pre-dicted planet. A contributing fac-tor was the development of a mi-nor tiff with the Astronomer Royal, with the result that the lat-ter simply ignored the affair.

Within a few months, a French scientist named Urbain Leverrier carried out calculations that led him to the same conclusion, al-though he was totally unaware of Adams' work, which had not been published. Leverrier's work did appear in scientific journals, however, and this prompted the Astronomer Royal in England to take the idea of an eighth planet a bit more seriously, and a search was begun.

The English astronomers were hindered by the lack of good star charts for the appropriate region of the sky, and substantial prelim-inary work was necessary before an earnest search could begin. Meanwhile, Leverrier contacted an acquaintance at the observa-tory in Berlin, who had excellent charts of Aquarius, and discov-ered Neptune on the very first night that he looked for it. This was in 1846.

England and France were inter-national rivals, and years of con-tentious arguments followed over who should be credited with the prediction of Neptune's existence. There never was an amicable set-tlement of this question; it just gradually faded from the forefront of French-English affairs. Today, the heat of the debate having cooled with the passage of time, Adams and Leverrier are generally credited equally with the discovery.

Voyager's cameras, the other instruments on board the spacecraft provided information on the atmospheric structure, the magnetic field, and the planetary rotation period, as well as detailed data about the structure of the planet's rings.

In overall appearance, Neptune bears little resem-blance to Uranus. Whereas Uranus is nearly featureless, Neptune shows light and dark spots, clouds, and a more distinctly banded appearance (Fig. 14.20). The color is an overwhelming, deep blue, caused by the presence of methane, which absorbs and scatters red light (Fig. 14.21). The light blue color of Uranus is also caused by methane absorption, but it is not as intense because we cannot see to as great a depth in the at-mosphere. It is evident that the atmosphere of Neptune is dynamic, with more weather than is exhibited by Uranus. This is probably caused by the greater amount of internal heat being produced inside Neptune; this planet emits more excess energy relative to its mass than either Jupiter or Saturn, whereas Uranus shows little or no excess internal heating.

The outermost portion of the atmosphere of Nep-tune is transparent, which allowed the *Voyager* images to reveal structures lying deep in the atmosphere. There are high-level, white wispy clouds (Fig. 14.22), prob-ably composed of methane crystals, which resemble earthly cirrus clouds. Shadows of these clouds could be seen projected onto the lower levels of the atmo-sphere (Fig. 14.23). There is an enormous dark feature in the southern hemisphere (Fig. 14.24, see also

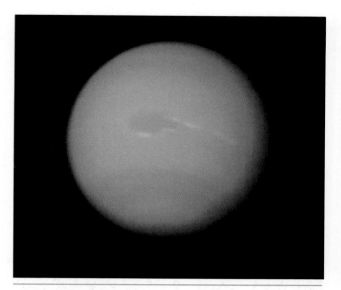

Figure 14.19 A *Voyager* view of Neptune. *This is one of the images of Neptune obtained as Voyager 2 approached. The color is natural and is due to the absorption of red light by methane in the atmosphere of the planet. Even at this distance (over 21 million miles), surface features were readily visible. (JPL/NASA)*

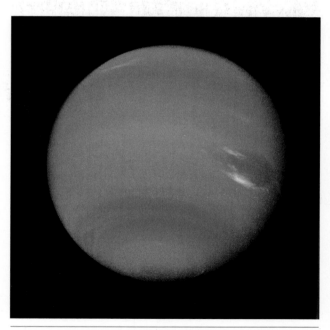

Figure 14.20 A detailed image of Neptune. *This Voyager image, obtained from a range of about 9 million miles, has been contrast-enhanced to bring our surface features. We see the Great Dark Spot, bright clouds in the upper atmosphere, and a banded appearance similar to those of Jupiter and Saturn. (JPL/NASA)*

Figure 14.21 The effect of methane on the appearance of Neptune. *This is a false-color image of Neptune, designed to illustrate the effect of methane in the atmosphere. The blue color of most of the disk is natural, but the red around the edges has been enhanced to show light scattered by methane molecules. This gas absorbs red photons, causing the remaining color to be blue in regions where we see light that has passed through a region of methane absorption, such as the central portions of the disk. In regions where we see the reemitted (scattered) light from the methane, we see red. (JPL/NASA)*

Figure 14.22 Cirrus clouds on Neptune. *The white, wispy clouds seen here are similar to cirrus clouds in the Earth's atmosphere. They lie at high altitudes and are probably composed of methane ice crystals. (JPL/NASA)*

Figure 14.23 Vertical structure in the atmosphere of Neptune. *This close-up image of Neptune shows that the white "cirrus" clouds lie above other levels of the atmosphere that are observed. Here we see that the sides of the clouds that face the Sun are brighter, and that the clouds cast shadows on the lower atmosphere in the direction away from the Sun. (JPL/NASA)*

Figure 14.24 The Great Dark Spot. *This* Voyager *image shows the large, dark oval feature that is very reminiscent of Jupiter's Great Red Spot, as well as a couple of smaller features. South of (below) the Great Dark Spot are a white feature known as the "scooter" because of its rapid motion around the planet, and the "dark spot 2," a second counterclockwise storm system. Each of these features moves around the planet in the eastward direction (to the right in this view), but at different speeds, so that they are not often grouped together as in this image. (JPL/NASA)*

Fig. 14.20), dubbed the Great Dark Spot, which bears a strong resemblance to the Great Red Spot of Jupiter. Like the Great Red Spot, Neptune's Great Dark Spot appears to be a longstanding storm system, rotating counterclockwise. There are also smaller dark and bright spots that move around the planet at different speeds, depending on the velocity of the atmospheric flow at the latitude where they lie. Wind speeds up to 2,160 km/hr were measured.

One goal of the *Voyager* exploration was to determine whether Neptune has a magnetic field. It was expected that it would, but Earth-based observations had revealed no sign of the charged particle belts that would be present. It was not until near its closest approach to Neptune that *Voyager 2* finally detected charged particle belts and hence was able to confirm that there is a magnetic field. The field is weak and, to the surprise of the *Voyager* scientists, is tilted by some 55° with respect to the rotation axis of the planet. Furthermore, the magnetic axis is offset from the planet's center by about one-fourth of the radius of Neptune, again a surprise. Recall that *Voyager* made very

similar discoveries about the magnetic field of Uranus. When there was only one planet known to have such an unexpected magnetic field structure, it was possible to explain it by invoking some low-probability event, but now that there are two such cases, some more natural explanation seems to be needed. One of the low-probability occurrences suggested to explain the field of Uranus was that *Voyager* had just happened to fly by at a time when the planetary field was undergoing a reversal of direction, as we suspect many planets do from time to time. It seems very unlikely indeed that Neptune was similarly caught in the process of reversing its field, however, so some other mechanism for the strange tilt and offset must be found.

This is one of the outstanding questions left by the *Voyager* encounters with the outer planets.

The magnetic field observations did settle another longstanding question, that of the rotation period of Neptune. The planet is too distant for Earth-based observations to track surface features in order to measure the time it takes for the planet to spin once. Furthermore, the differential rotation of the atmosphere would make it difficult to decide which surface features to follow, since the rotation period varies with distance north or south of the equator. Therefore the convention that has been adopted for the gaseous giant planets is to measure the rotation period of the magnetic field, which reflects the spin rate at some interior level in the planet. For Neptune, the *Voyager* data show that the rotation period is 16 hours and 3 minutes, with an uncertainty of 4 minutes. This is very rapid, but quite consistent with the rotation periods of the other outer planets.

Neptune's Moons and Rings

Before the *Voyager 2* encounter, Neptune had only two known satellites and sketchy evidence for a ring system. The satellites have very unusual properties: Triton, a giant known to have an atmosphere, orbits the planet in the retrograde direction, while Nereid, much smaller, follows a very elongated elliptical orbit, varying its distance from Neptune by a factor of 5 during its year-long trip around the parent planet. Triton in particular was a high-priority target of the *Voyager* exploration because of its atmosphere and its strange orbit.

Stellar occultation measurements made from Earth had revealed some evidence for rings around Neptune, but these appeared not to be full circles, consisting instead of ring segments or arcs. Thus there was an expectation that these rings arcs were transient features, probably associated with small moons that perturb their structure. As we will see, this expectation was partly correct.

Voyager 2 discovered six new satellites (Table 14.5), one of which, temporarily named 1989N1, is actually larger than Nereid (it was not discovered from Earth because it is much closer to Neptune). All of the new satellites orbit the parent planet in the normal direction, and all lie fairly close in, within a distance of five times the radius of Neptune (by contrast, Triton's orbit lies 14 radii away from Neptune and Nereid's semi-major axis is 218 planetary radii). The newly discovered moons are all small (N1 has a diameter of just over 400 km) and irregularly shaped. It is thought therefore that these objects never underwent heating and melting, which would have allowed them to assume spherical shapes. There has been some geological evolution, however, in the form of impacts from meteoroids; many craters were seen in places where sufficient image quality was obtained (Figs. 14.25 and 14.26). In one case, a crater nearly half as large as the diameter of the satellite was seen; clearly this was caused by an impact nearly sufficient to destroy the satellite.

Triton, the large moon of Neptune, provided a wealth of surprises. The surface, which could be seen clearly through the atmosphere (Fig. 14.27), displays signs of geological activity consisting of both tectonic and volcanic processes. It was immediately suspected that Triton's surface was young, because the albedo is quite high (about 0.7), and most aging processes such as long-term exposure to radiation cause icy surfaces to grow dark. The southern polar region is especially bright; this area is covered by a polar ice cap that reflects light

Table 14.5 Satellites of Neptune

No.	Name	Distance (R_N)[a]	Period	Diameter	Mass
8	1989N6	1.938	0.^{d}296	54 km	
7	1989N5	2.019	0.313	80	
5	1989N3	2.120	0.333	150	
6	1989N4	2.504	0.429	180	
4	1989N2	2.972	0.554	190	
3	1989N1	4.749	1.121	400	
1	Triton	14.327	5.875	2705	2.21×10^{25} gm
2	Nereid	222.64	360.13	340	2.1×10^{22}

[a]Distances are semimajor axis in units of the equatorial radius of Neptune, which is $R_N = 24{,}764$ km.

Figure 14.27 The surface of Triton.   *This close-up view shows a variety of surface features which indicate that some reprocessing has taken place recently. The surface is bright, unlike old surfaces, and there are relatively few impact craters. There are also linear, grooved features suggestive of crustal plate motion. Most intriguing of all are the dark streaks near the south pole (bottom), which resemble volcanic plumes or ejecta. It has been suggested that these are either current or recently active geysers, which emit frozen nitrogen. If so, Triton becomes the third body in the solar system known to be volcanically active. (The Earth and Io are the other two, but recall that Venus may also have active volcanoes; see Chapter 11.) (JPL/NASA)*

Figure 14.25 Neptune's satellite N1.   *The first new satellite to be discovered by Voyager 2 was this one, which is actually larger than Nereid, which had been discovered many years earlier from the Earth. The diameter of N1 is just over 400 km. Like the other new satellites, N1 is very dark and is irregular in shape, showing evidence for cratering on its surface. (JPL/NASA)*

Figure 14.26 Satellite N2.
This is the second of the six new moons discovered by Voyager 2. Its diameter is about 200 km, but its shape is irregular, indicating that this moon, like the other new ones, probably never was molten. (JPL/NASA)

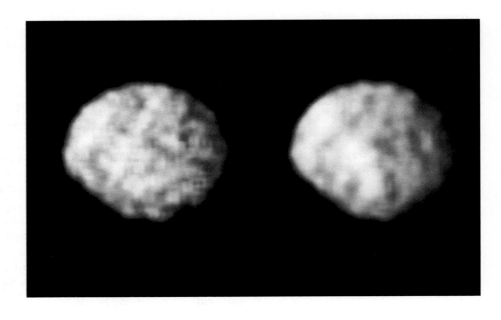

efficiently (the presence of a polar cap had actually been suggested earlier, based on images obtained from Earth). The *Voyager* images revealed few impact craters on Triton, adding support to the notion that the surface is young. The most intriguing evidence for this came in the form of dark streaks near the south pole (see Fig. 14.27) which bear a close resemblance to debris ejected by volcanic vents. Detailed analysis showed that at least two vents were actively erupting at the time of the *Voyager* encounter, demonstrating that Triton, like the Earth and Io among solar system bodies, is in an active volcanic phase. It is speculated that nitrogen ice beneath the surface can turn to gas and erupt if minor additional heating is supplied. This is more analogous to terrestrial geysers than volcanoes, since no crustal melting appears to be involved. Because the ejected debris consists of frozen gases, scientists are now speaking of "ice volcanoes" on Triton. Further detailed examination of the *Voyager* images will help to establish whether other vents were active when *Voyager* passed by.

Other features on Triton's surface indicate that crustal plates have been fractured and shifted by internal forces. There are also some curious circular depressions (Fig. 14.28), clearly not impact craters, whose origins are unknown. The only explanation offered so far is that these may be regions where the crust has subsided, perhaps as the result of subsurface melting and collapse.

The atmosphere of Triton is very thin, having a pressure at the surface of only about 10^{-5} times the atmospheric pressure on the Earth. Earth-based observations had led to the detection of both methane and nitrogen in Triton's atmosphere. This was confirmed by *Voyager 2,* which found that nitrogen is the more abundant of the two gases. Thus, like Titan and the Earth, Triton has a nitrogen-dominated atmosphere. Triton also has aurorae in its upper atmosphere, caused not by its own magnetic field (it has none), but instead due to charged particles from Neptune's radiation belts.

The mass of Triton was determined accurately from *Voyager's* observations, allowing the first reliable estimate to be made of its density. The resulting value, 2 grams/cm³, suggests that, internally, Triton is about half rock and half ice. This is very similar to Pluto (next section) which, along with the strange backwards orbit of Triton, suggests that Triton is a captured body. Recent theories and observational evidence argue that the outer solar system once contained a number of planetary objects that were left over from the time of the solar system's origin, which were not initially incorporated into the major planets. Pluto and Triton

Figure 14.28 Detailed structures on Triton's surface. *Features as small as one-half mile across are visible in this close-up image from* Voyager 2. *Most of the area shown here is covered with circular depressions that are separated by ridges. No surface like this has been seen elsewhere in the solar system, and the cause is not clear. One suggestion is that these are regions where subsurface melting has allowed the surface to collapse. The grooves and ridges cutting across the landscape indicate that there has been crustal fracturing in the past. (JPL/NASA)*

may both be remnants of this era in the early history of the solar system.

As expected, the *Voyager* data provided definitive information on the nature of Neptune's ring system. To the surprise of the scientific team, the rings were found to be complete circles around the planet, rather than the partial rings or arcs that had been suggested by Earth-based observations. The rings are very dim, making it difficult to identify distinct features or to make a definite count of their number. There appear to be two or three distinct, thin rings, and also a broad particle zone or "plateau" as well as a wide inner belt of particles extending inward very close to Neptune itself. At least one of the distinct, thin rings has bright regions (Fig. 14.29), which may indicate that it is clumpy (and which may account for the impression of ring arcs deduced from Earth-based observations). Some of the newly discovered moons lie within the ring system, and no doubt their gravitational influence is helping to create the structure that is seen. As in the *Voyager* encounters with Saturn and Uranus, the photopolarimeter experiment was used to make a stellar occultation observation of Neptune's rings, revealing complex fine detail (Fig. 14.30).

The apparent lumpiness of the ring system, combined with the fact that several small satellites were found orbiting within the rings, fits into the picture developed through the study of the ring systems of Saturn and Uranus. In this scenario, it is suggested that ring systems are ephemeral features, changing continually because of the gravitational interactions with small satellites and even being replenished occasionally by the destruction of small moons through meteoroid impacts. We saw that some of the ring features of Uranus are too short-lived to be permanent; perhaps detailed analysis of the new data on Neptune's rings will lead to a similar conclusion. One of the lasting contributions of the *Voyager* encounters will be the realization that rings are not forever.

Images obtained as *Voyager* was leaving the area of Neptune, in which the Sun was seen shining through the rings, showed that the rings tend to scatter sunlight in the forward direction. This means that the rings look dark when the Sun's rays are coming from behind the viewer, and bright when the observer looks through

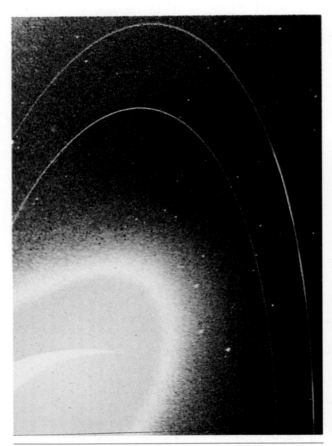

Figure 14.29 The rings of Neptune. *This* Voyager *image shows the thin, dim rings of Neptune in some detail. Here we see two distinct rings, one of which has three bright segments, which may have led to earlier indications that the rings are not complete circles. Other views show that there is an indistinct, broad region encompassing some of the thin rings and known as the "plateau," and that there is an inner belt of ring particles that extends close to the planet. (JPL/NASA)*

Figure 14.30 Detailed structure in the ring. *This is a reconstruction of a portion of Neptune's ring system, based on a stellar occultation observation made using the photopolarimeter experiment aboard Voyager 2. A background star was observed as it was viewed through the rings, so that fluctuations in the measured brightness of the star could be used to determine how the density of the ring system varies from place to place. (JPL/NASA)*

the rings toward the Sun's direction. Forward scattering of light is characteristic of very small particles, suggesting that the rings of Neptune are composed of smaller particles than the majority of the rings girdling the outer planets.

As *Voyager 2* departed the vicinity of Neptune, a brilliant chapter in the history of space exploration came to a close. A few more images were made as the spacecraft pointed its cameras back toward the inner solar system (see Astronomical Insight 8.1), but now the major task of this remarkable robot is over. *Voyager 2* will continue into interstellar space and will have sufficient power to communicate with Earth until at least the year 2010, so that we will continue to gather information on places none of us can hope to explore in person.

PLUTO

The ninth and last of the known planets is Pluto (Table 14.6), orbiting at an average distance of nearly 40 AU from the Sun. From that perspective, the Sun would appear as a very bright star but surely would not produce much warmth. This planet is the most obscure of them all, and we will raise many questions about its nature.

In 1930 Clyde Tombaugh (Fig. 14.31) found a dim spot of light that displayed the motion expected for a planet (Fig. 14.32). The ninth planet was found by chance near the position predicted years earlier by Percival Lowell (Tombaugh probably would have found

it eventually in any event, because he was undertaking a systematic search of the entire ecliptic). The new member of the Sun's family was given the name *Pluto*, after the Roman god of the underworld (the name also honors Percival Lowell, since its first two letters are his initials).

Planetary Misfit

The little we know about Pluto shows it to be a nonconformist. Throughout our discussions of the other planets, certain systematic regularities have shown up, and Pluto violates many of them.

First, its orbit is highly irregular, compared to those of the other planets. It is both noncircular and tilted with respect to the ecliptic. At its greatest distance from the Sun, Pluto is nearly 70 percent farther away than at its closest, when it actually moves closer to the Sun than Neptune for a time. (In fact, Pluto is temporarily the eighth planet from the Sun right now, having moved inside Neptune's orbit in 1979, not to reemerge until 1999.) Pluto's orbit is tilted by 17° to the ecliptic, whereas the tilts of all the other planetary orbits are 8° or less.

Figure 14.31 Clyde Tombaugh at the blink comparator. *This photo shows the discoverer of Pluto peering into the eyepiece of the device he used to find the planet on photographic plates. The blink comparator allows two images of the same field of stars to be examined alternately (with the use of a mirror that flips between two positions) so that moving objects stand out. (Lowell Observatory photograph)*

Table 14.6 Pluto

Orbital semimajor axis: 39.44 AU (5,900,000,000 km)
  Perihelion distance: 29.58 AU
  Aphelion distance: 49.30 AU
Orbital period: 248.5 years (90,700 days)
Orbital inclination: 17°10'12"
Rotation period: 6.39 days
Tilt of axis: 122°
Diameter: 2302 km (0.18 $D_\oplus$)
Mass: 1.3×10^{25} gm (0.0022 $M_\oplus$)
Density: 2.03 gm/cm³
Surface gravity: 0.06 Earth gravity
Escape speed: 1.1 km/sec
Surface temperature: 40 K
Albedo: 0.6:
Satellites: 1

In physical characteristics, Pluto also seems not to fit in with the other planets. The other outer members of the solar system are giant planets, with thick atmospheres. While it has been difficult to pinpoint the mass of Pluto, it has long been clear that this planet does not resemble the other outer planets. For a long time, the only information on the mass of Pluto came from the lack of any significant effect on the motion of Neptune, which implied that Pluto's mass had to be less than about 10 percent of that of the Earth. The diameter has been difficult to determine, with results from interferometric measurements (see Chapter 6), indicating a value near 3,000 km. Very recently, new techniques, discussed later, have revised this value.

Spectroscopy has shown that Pluto has methane either on its surface or in its atmosphere, or both. The question of whether there is an atmosphere or not was settled by a stellar occultation which occurred in 1988 when Pluto passed in front of a background star. The manner in which the star's light dimmed (gradually instead of abruptly) demonstrated conclusively that Pluto has an atmosphere, one that is quite extended. Methane is certainly present, and there is evidence of some other, heavier gas, probably carbon monoxide or nitrogen.

Pluto has just passed its point of closest approach to the Sun, which means that currently the solar energy input is much higher than it is when Pluto is at its greatest distance from the Sun (the ratio of solar intensity at perihelion to that at aphelion is 2.8—almost a factor of 3!). Some scientists speculate that Pluto's atmosphere may largely freeze out while the planet is farther from the Sun, so that the atmosphere it now has may be a transient phenomenon that exists only briefly every 249 years.

Brightness variations appear to indicate that Pluto has some surface markings and is rotating with a period of 6.39 days. The orientation of the rotation axis has been difficult to determine, but new information indicates a large tilt, probably about 122° (with respect to the perpendicular to the planet's orbital plane, which is itself tilted 17° with respect to the ecliptic).

The small size and apparent low mass of Pluto, combined with its unusual orbit, led to the speculation that perhaps it was once a satellite of one of the giant planets, and that it somehow escaped to follow its own path around the Sun. More likely is the suggestion that Pluto is a leftover planetesimal, in some ways comparable to an asteroid. As we shall see, some asteroids are thought to be double, as is Pluto (this will be discussed shortly). Furthermore, the composition of Pluto is similar to that of certain classes of asteroids (asteroids are discussed in the next chapter).

In 1978 Pluto was discovered to have a satellite (Fig. 14.33), which was given the name *Charon*, after the mythical boatman who ferried the dead into the realm of the underworld. Charon is unusually large compared to its parent planet, and both Pluto and

Figure 14.32 The discovery of Pluto. *These are portions of the original photographs on which the ninth planet was discovered in 1930. The position of the planet in each is indicated by an arrow. (Lowell Observatory photograph)*

ASTRONOMICAL INSIGHT 14.2

The Search for Planet X

The discovery of Pluto was by no means the end of the search for new planets. Quite to the contrary, this event sharpened the interest in continuing the search, and it was carried on for an extended period.

The technique by which Pluto was discovered was tedious but efficient, in that it was thorough. Tombaugh accomplished the task by comparing photographs taken at different times (usually a few days apart), to see whether any of the objects in the field of view moved. This would ordinarily be very difficult to do, particularly in areas of the sky that are crowded with stars, but was made easier with the aid of a special instrument called a **blink comparator**. This is a microscope for viewing two photographic plates simultaneously, with a small mirror that flips back and forth, providing, in rapid succession, alternating views of the two plates. If the plates are mounted so that the images of fixed stars are perfectly aligned, then an object that moved between the times of the two exposures will appear to blink on and off at two separate locations, while all the fixed objects are steady. Examining a region of the sky using this technique, therefore, consists of mounting in the blink comparator two plates taken at different times, then sys-

tematically scanning them with the movable eyepiece so that each object could be checked, to see if it is blinking. The entire job could take days or weeks for each pair of plates.

The blink comparator technique, while time-consuming, has the capability of allowing thorough searches of large regions of the sky. The search for a tenth planet, often referred to as Planet X, continued at Lowell Observatory for 13 years after the discovery of Pluto. Clyde Tombaugh, who had discovered Pluto, was the principal worker in the extended search. Nothing was found.

During and since that time, occasional reports of evidence for a tenth planet have surfaced, but always the evidence has been indirect. Reanalysis of the motion of Neptune has convinced some astronomers that there really may be discrepancies attributable to gravitational perturbations caused by another planet. Other researchers have pointed to the motions of certain comets as indicative of gravitational influence exerted by Planet X. Most of these predictions have suggested a distance from the Sun of 50 to 100 AU. The motion of Halley's comet, a bright and very famous object that visits the inner solar system every 76 years, has been analyzed by several scientists, who find that its arrival near the Sun is usually several days later than it should be, according to the laws of motion. This has been attributed by some to the gravitational influence of a tenth planet, but no such ob-

ject could be found at the predicted position. One of the predictions based on the motion of Halley's comet was rather spectacular: The inferred Planet X had a mass 3 times that of Saturn, and an orbit inclined by 120° to the ecliptic!

Modern information on comets has explained the discrepancy in the motion of Halley's comet another way. As it nears the Sun and heats up, volatile gases vaporize and stream outwards from the heated side, acting like a rocket exhaust to slow the comet's motion (this process is described in more detail in Chapter 15).

The routine search for new objects continues today, with occasional success, but no new planets have been found. There was a brief flurry of excitement in 1977, when a Sun-orbiting body called Chiron was discovered whose elliptical path carries it almost as far from the Sun as Uranus, and as close in as Saturn. It is clearly not a planet, but the exact nature of Chiron is uncertain. It may be an asteroid in a highly unusual orbit, or an old cometary nucleus that never approaches the Sun closely enough to develop a luminous tail (see Chapter 15). In 1988, Chiron was observed to flare up in brightness. It is now generally accepted that this object is a very large comet. Perhaps eventually Planet X will be found, but if so, it will undoubtedly be a dim, cold, distant object, much too faint to be seen without a large telescope.

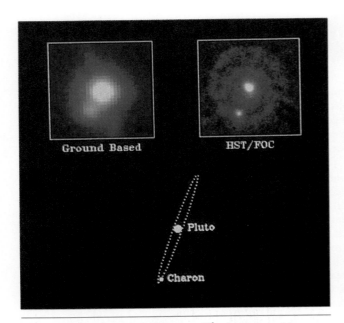

Figure 14.33 Pluto and Charon. *The image at upper right, obtained by the* Hubble Space Telescope, *shows Pluto and Charon clearly separated, whereas the best ground-based photo (upper left) barely resolves the two bodies. (NASA/European Space Agency: Space Telescope Science Institute)*

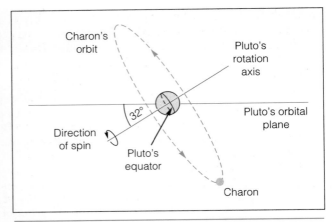

Figure 14.34 The orientation of Pluto's spin axis and Charon's orbit. *Like Venus and Uranus, Pluto has its rotation axis tipped over so far that it points below the plane of the planet's orbit (which is itself tipped by an unusually large angle, 17 degrees, relative to the ecliptic). Thus the spin of the planet is technically retrograde (i.e. backwards). Charon orbits in the equatorial plane of the planet. During the late 1980s the plane of Charon's orbit was aligned with the Pluto-Earth direction so that Pluto and Charon alternately passed in front of each other. Observations of these repeated transit events provided a wealth of information on the nature of both bodies and the transient atmosphere of Pluto.*

Charon are locked into synchronous rotation, each keeping the same side facing the other. The orbital and rotational periods of both are 6.39 days. To an observer on the surface of Pluto, Charon would appear fixed in position overhead, while the background stars streamed past because of the planet's rotation.

The orbital plane of Charon is tilted about 58° with respect to the plane of Pluto's orbit about the Sun, and Charon orbits in the retrograde direction (Fig. 14.34). The conclusion that Pluto's rotation axis is tipped 122° with respect to the perpendicular to its orbital plane is based on the assumption that Charon orbits in the plane of Pluto's equator.

The discovery of Charon gave astronomers a chance to determine the mass of the planet, using Kepler's third law. It was found that Pluto's mass is only about 0.0021 of the Earth's mass. The diameter of Pluto has recently been accurately measured as well, again with help from Charon. It happens that the orbital plane of Charon has recently been aligned with the direction toward Earth, so that for a time Charon eclipsed Pluto each orbit. The duration of the eclipses allowed astronomers to determine the diameter of Pluto, which was found to be 2,300 km (Charon's diameter is

1186 km). This, combined with the mass of Pluto, yields a density for the outermost planet of about 2.0 grams/cm³. Evidently, Pluto is composed partially of ice and partially of rock, much like the satellites of Jupiter and Saturn, and much like the expected composition of the planetesimals that formed the cores of the giant planets.

Measurements of the spectrum of sunlight reflected from Pluto and Charon reveal that Pluto has a methane frost surface (probably covering a water ice layer below), and that Charon's surface is coated with water ice. Pluto also has gaseous methane in its atmosphere (as already noted), but the presence of any atmosphere at all may be temporary—it appears likely that the atmosphere freezes out when Pluto is farther from the Sun.

PERSPECTIVE

We have discussed all of the available information on the planets. We certainly have many unanswered questions about them, but on the other hand, a number of systematic trends have revealed themselves to us. Two of the outer three planets fit the pattern established by the others, being gaseous giants with many apparent similarities to Jupiter and Saturn. The last planet, Pluto, is mysterious, and remains an anomaly to be reckoned with.

Nearly all the pieces of the puzzle are now in place. We need only examine the nonplanetary bodies roving through the solar system to see what they can tell us of its origins, and then we will address the formation of the system. We turn our attention now to the minor bodies, the asteroids, comets, and meteoroids.

SUMMARY

1. Uranus has been examined by the *Voyager 2* space probe and has been found similar to the other giant planets in that it has a banded atmosphere and an internal structure that is highly differentiated, with a solid core and layers of liquid and gaseous light elements above the core.

2. The spin axis of Uranus, as well as the orbital plane of its fifteen satellites, is tilted 98°, creating strange seasonal effects and an unusual magnetosphere.

3. Uranus has fifteen known satellites and at least nine thin, dim rings, whose structure is governed by the satellites in the same ways as the rings of Saturn.

4. The major satellites of Uranus show varying degrees of geological activity; the innermost of these moons, Miranda, is surprisingly active geologically, with many very young surface features.

5. Neptune was discovered because of its gravitational effects on Uranus, a discovery that vindicated the laws of motion and gravitation developed by Newton.

6. *Voyager 2* data on Neptune show that the atmosphere has more distinct bands than Uranus, has several spots (including the Great Dark Spot), and has high-lying white methane ice clouds.

7. The weak magnetic field of Neptune, like that of Uranus, is tilted very far away from the rotation axis, and is offset from the center of the planet.

8. *Voyager* discovered six new satellites of Neptune, bringing the total to eight. Triton, the largest, shows evidence of recently active geysers in the form of ice eruptions, and appears to be geologically active in other ways as well. All of the new satellites have normal prograde orbits lying close to Neptune, unlike Triton and Nereid, which have very peculiar orbits.

9. The previously detected ring arcs were found by *Voyager* to be dim, complete circular rings, made of small particles. There is some evidence that this ring system, like those of Saturn and Uranus, is young.

10. Pluto is unusual in that its orbit is highly eccentric and tilted, and is much smaller and probably different in structure than the other outer planets.

11. Pluto has a tenuous, possibly transient atmosphere consisting of methane and perhaps other gases.

12. Pluto has a relatively large satellite that provides data on the planet's mass and density, showing that Pluto is probably composed of roughly half rock and half ice.

REVIEW QUESTIONS

Thought Questions

1. Summarize what has been learned about Uranus and Neptune from Earth-based observations, and discuss the advances made when a probe such a *Voyager 2* is able to make close-up observations.

2. How long is a day at the north pole of Uranus? What is the length of day at the equator when the north pole is pointed 90° away from the Sun?

3. Summarize the major differences between Uranus and Neptune, and between Jupiter and Saturn.

4. The innermost major satellite of Jupiter, Io, is undergoing current geological activity. The innermost major moon of Uranus, Miranda, apparently has undergone geological activity recently. Why is such activity readily explained for Io, but quite unexpected and unexplained for Miranda?

5. How are the rings of Uranus similar to those of Saturn, and in what ways are they different?

6. Contrast the means by which Uranus, Neptune, and Pluto were discovered.

7. Discuss the kinds of information *Voyager 2* might have gained on Pluto if the spacecraft could have been re-routed to encounter the ninth planet.

8. How are the satellites and rings of Neptune different from those of the other major planets?

9. Summarize the many ways in which Pluto is unlike any other planet in the solar system.

10. Explain why Pluto has been forced into synchronous rotation with Charon, while no other planet is in synchronous rotation with its satellite.

Problems

1. Suppose that when Uranus occulted the light from a background star, it took 1 hour, 40 minutes to pass in front of the star. The apparent motion of Uranus as seen from the Earth is 0.0005 arcseconds per second of time. Calculate the angular diameter of Uranus.

2. When Uranus is situated directly between the Sun and Neptune, the gravitational disturbance on its orbit caused by Neptune is maximized. At this time, Neptune is one-half as far from Uranus as the Sun is. The mass of the Sun is about 20,000 times greater than the mass of Neptune. Calculate the fraction by which the Sun's attraction on Uranus is reduced because of Neptune's gravitational force on Uranus.

3. Use Kepler's third law to calculate the semimajor axis of Nereid's orbit around Neptune. (Note: This will be easiest if you use the form of the law derived by Newton, including the sum of the masses.) How does your answer compare with the typical sizes of the orbits of other satellites of the major planets?

4. By how much does the intensity of sunlight on Pluto vary during the planet's orbit around the Sun (its distance from the Sun varies from about 30 AU to about 50 AU)? How does the Sun's intensity on Pluto at perihelion compare with the intensity of sunlight on the Earth?

ADDITIONAL READINGS

Allen, D. A. 1983. Infrared views of the giant planets. *Sky and Telescope* 65(2):110.

Beatty, J. Kelley. 1986. A place called Uranus. *Sky and Telescope* 71(4):333.

_____. 1990. Getting to know Neptune. *Sky and Telescope* 79:146.

_____. 1985. Pluto and Charon: The dance begins. *Sky and Telescope* 69(6):501.

_____. 1987. Pluto and Charon: The dance goes on. *Sky and Telescope* 74(3):248.

_____. 1986. Voyager's triumph. *Sky and Telescope* 72(4):336.

Beatty, J. Kelley, B. O'Leary, and A. Chaikin, eds. 1990. *The new solar system.* 2d ed. Cambridge, England: Cambridge University Press.

Beatty, J. Kelley, and A. Killian. 1988. Discovery Pluto's atmosphere. *Sky and Telescope* 76:624.

Binzel, R. 1990. Pluto. *Scientific American* 262(6):50.

Chaikin, Andrew, 1986. Voyager among the ice worlds. *Sky and Telescope* 71(4):338.

Croswell, Ken. 1986. Pluto: Enigma on the edge of the solar system. *Astronomy* 14(7):6.

Cuzzi, J. N., and L. W. Esposito. 1987. The rings of Uranus. *Scientific American.* 257(1):52.

Elliott. J. L., E. Dunham, and R. L. Mills. 1977. The discovery of the rings of Uranus. *Sky and Telescope* 53(6):412.

Harrington, R. S., and B. J. Harrington. 1980. Pluto: Still an enigma after 50 years. *Sky and Telescope* 59(6):452.

Hunten, D. M. 1975. The outer planets. *Scientific American* 233(3):130.

Ingersoll, Andrew P. 1987. Uranus. *Scientific American* 256(1):38.

Johnson, T. V., R. H. Brown, and L. A. Soderblom. 1987. The moons of Uranus. *Scientific American* 256(4):48.

Kinashita, J. 1989. Neptune. *Scientific American* 261(5):82.

Lanzerotti, L. J., and C. Uberoi. 1989. The planets' magnetic environments. *Sky and Telescope* 77:149.

Littman, M. 1989. Where is Planet X? *Sky and Telescope* 78:596.

Miner, E. D. 1990. *Voyager 2's* encounter with the gas giants, *Physics Today,* 43(7):40.

Moore, P. 1989. The discovery of Neptune, *Mercury* 18:98.

Mullholland, Derrall. 1982. The ice planet: Pluto. *Science 82* 3(10):64.

Overbye, Dennis, 1986. Voyager was on target again: At Uranus. *Discover* 7(4):70.

Tombaugh, Clyde W. 1986. The discovery of Pluto. *Mercury* 15(3):66.

_____. 1979. The search for the ninth planet. *Mercury* 8(1):4.

MATHEMATICAL METHODS

The Roche Limit

In this chapter, as well as in Chapter 13, we have cited the Roche limit in our discussions of the formation of planetary ring systems. There the Roche limit was described as the closest distance of approach to a planet where a satellite could survive without being torn apart by tidal forces. We have said that the rings may be either the remnants of satellites that strayed too close to their parent planets and were destroyed by tidal forces (unlikely), or that the rings consist of particles too close to the parent planet to coalesce into satellites (the more likely scenario). In either case, it is useful to compute the position of the Roche limit, and see how that compares with the locations of the planetary ring systems.

In its simplest form, the Roche limit is defined as the distance from a massive body where the self-gravitation of a smaller body is just equaled by the tidal force due to the major body. That is, the Roche limit corresponds to the closest distance where a collection of particles held together only by their own mutual gravity can stay together despite the tidal force that is acting to pull them apart. Equating the tidal force (formula given in the Mathematical Methods section at the end of Chapter 13) to the force of self-gravitation for a spherical object yields an equation for the Roche limit r_R:

$$r_R = \left(\frac{2M}{m}\right)^{1/3} R \quad ,$$

where M is the mass of the parent planet, m the mass of the satellite, and R is the radius of the satellite.

If the satellite and the parent planet have similar average densities, it can be shown that this equation reduced to

$$r_R = 1.26 R_P,$$

where R_P is now the radius of the parent planet.

We now compare this with the locations of the ring systems of the outer planets. For all except Jupiter, we find that the rings extend farther out than 1.26 times the radius of the planet. Thus, according to the equation above, satellites should be able to survive in the region where the rings are seen.

A more complete and correct calculation of the Roche limit, made originally by Edouard Roche himself, includes a more realistic treatment of the forces that hold a satellite together, and yields the result

$$r_R = 2.44 R_P.$$

This is outside the radii of most of the planetary rings.

Note that a satellite that is denser than its parent planet can survive closer in than can one that has the same density. It is also worth keeping in mind that even the more complex treatment leading to the limit of $2.44 R_P$ does not take into account all the complex forces that hold solid bodies together. Nevertheless, calculations using the equations given here are very informative in explaining the origins of planetary ring systems.

CHAPTER 15

Space Debris

A gigantic meteor crater in Canada, called the Manicougan crater. (U.S. Geological Survey; NASA photo)

The planets are the dominant objects among the inhabitants of the solar system (except, of course, for the Sun), but they are not entirely alone as they follow their regular paths through space. The abundant craters on planetary and satellite surfaces have shown us that there must have been a time when interplanetary rocks and gravel were more abundant than they are today, covering any exposed surface with impact craters. Today the rate of cratering is much lower than it once was, but some vestiges of the space debris that caused it still remain, orbiting the Sun and occasionally becoming obvious to us as they pass near the Earth or enter its atmosphere.

There are at least four distinct forms of interplanetary matter, some of which are closely related. In this chapter we will discuss the **asteroids** (more commonly, the **minor planets**), myriad rocky chunks up to several hundred kilometers in diameter that orbit the Sun between Mars and Jupiter; the **comets**, whose ephemeral and striking appearances have caused us to pause and marvel since ancient times; **meteors,** the bright streaks often visible in our nighttime skies, along with the objects that create them; and the **interplanetary dust,** a collection of very fine particles that occupy the void between the planets.

THE MINOR PLANETS

As we have already seen, some of the early students of planetary motions were concerned with the dis-tances of the planets from the Sun. Kepler spent enormous amounts of time and energy seeking a mathematical relationship that would describe the distances of the planets in terms of geometrical solids. What he finally did discover was something different, a relationship between the distances and orbital periods. Kepler's third law can be used to predict the period of a planet, given its distance from the Sun; or its distance, if its period is known. But it did not provide any underlying basis for explaining why there are planets only at certain distances from the Sun.

In 1766, a German astronomer named J. D. Titius found a simple mathematical relationship that seemed to accomplish what Kepler had set out to do. Titius discovered that if we start with the sequence of numbers 0, 3, 6, 12, 24, 48, and 96 (obtained by doubling each one in order), then add 4 to each and divide by 10, we end up with the numbers 0.4, 0.7, 1.0, 1.6, 2.8, 5.2, and 10.0, corresponding closely to the observed planetary distances in astronomical units. A few years after Titius found this numerological device, it was popularized by another German astronomer, Johann Bode, and eventually became known as Bode's law, or the Titius-Bode law.

The sequence of numbers dictated by Bode's law includes one, 2.8 AU, where no planet was known to exist. After the discovery of Uranus in 1781, and the recognition that its distance fits the sequence (the next number is $(192 + 4)/10 = 19.6$, and the semimajor axis of Uranus's orbit is 19.2 AU), a great deal of interest arose in the idea that there ought to be a planet at 2.8 AU, since Bode's law was doing so well in predicting the positions of the others.

A deliberate search for such a planet was begun in 1800, but the discovery of the sought-after object came accidentally, when an Italian astronomer named G. Piazzi noted a new object on the night of January 1, 1801, and within weeks found from its motion that it was probably a solar system body.

When the orbit of the new object was calculated, its semimajor axis turned out to be 2.77 AU, placing the object in solar orbit between Mars and Jupiter, where a prediction had been made that a new planet might be found. The new fifth planet was named Ceres.

A little over a year after the discovery of Ceres, a second object was found orbiting the Sun at approximately the same distance, and was named Pallas. Because of their faintness, both Ceres and Pallas were obviously very small bodies, and were not respectable planets. By 1807, two more of these **asteroids,** as they were then being called, had been found and were designated Juno and Vesta. A fifth, Astrea, was discovered in 1845, and in the next decades, vast numbers of these objects began to turn up. The efficiency of finding them was improved greatly when photographic techniques began to be used. A minor planet or asteroid will leave a trail on a long-exposure photograph, because of its orbital motion (Fig. 15.1).

Today the number of known asteroids is in the thousands, with almost 5,000 of them sufficiently well observed to have had their orbits calculated and logged in catalogs. The total number is probably much more, perhaps 100,000. The vast majority have orbits lying between the orbits of Mars and Jupiter (Fig. 15.2).

It is possible to deduce some of the properties of the asteroids from a variety of observational evidence. Measurements of their brightness can lead to size estimates; spectroscopy of light reflected from their surfaces provides information on their chemical makeup; and direct analysis of meteorites that may be remnants of asteroids adds data on their internal properties.

The largest asteroids were the first to have their diameters estimated, simply by calculating what size they had to be to reflect the amount of light observed. Much more recently, infrared brightness measurements of both large and small asteroids have provided information on their sizes, since at temperatures of a few hundred degrees they glow at infrared wavelengths. The Stefan-Boltzmann law (Chapter 6) then can be used to determine the total surface area from the intensity of the emission.

Direct size measurements are possible when an asteroid occults a background star, or when one passes close to the Earth so that its angular size can be directly measured. Interferometry has also yielded angular sizes of some asteroids.

The result of using these assorted techniques is that asteroids are known to have a wide range of diameters (Table 15.1), a few as large as several hundred kilometers. Ceres, the largest, is nearly 1,000 km in diameter, and the majority are rather small, with diameters of 100 km or less. The largest are apparently spherical, whereas the smaller ones often are jagged,

Figure 15.1 Asteroid motion. *The elongated image in this photo is the trail made by an asteroid that moved relative to the fixed stars during the twenty-minute exposure. Most new asteroids are discovered on photos of this type, although the first few were spotted visually. (Eleanor F. Helin, Palomar Observatory)*

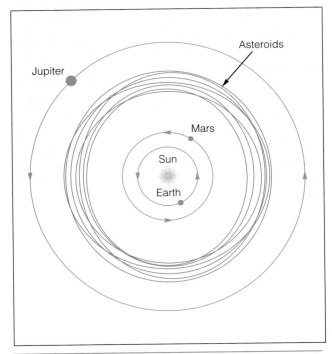

Figure 15.2 Asteroid orbits. *This shows typical orbital paths for the majority of asteroids, which orbit the Sun between Mars and Jupiter.*

Table 15.1 Selected asteroids

No.	Name	Year of Discovery	Diameter (km)	Mass (g)	Period (years)	Distance (AU)[a]
1	Ceres	1801	933	1×10^{24}	4.60	2.766
2	Pallas	1802	538	3×10^{23}	4.61	2.768
3	Juno	1804	200	2×10^{22}	4.36	2.668
4	Vesta	1807	561	2×10^{23}	3.63	2.362
6	Hebe	1847	220	2×10^{22}	3.78	2.426
7	Iris	1847	200	2×10^{22}	3.68	2.386
10	Hygiea	1849	320	6×10^{22}	5.59	3.151
15	Eunomia	1851	280	4×10^{22}	4.30	2.643
16	Psyche	1852	280	4×10^{22}	5.00	2.923
51	Nemausa	1858	80	9×10^{20}	3.64	2.366
511	Davida	1903	260	3×10^{22}	5.67	3.190

[a]Periods are orbital periods; distances are orbital semimajor axes.

irregular chunks of rock (Fig. 15.3), varying in brightness as they tumble through space, reflecting light with different efficiencies on different sides. A few asteroids are thought to be binary, consisting of two chunks orbiting each other as they circle the Sun. The total mass of all the asteroids together is small by planetary standards, amounting to only an estimated 0.04 percent of the mass of the Earth.

The composition derived from spectroscopic analyses are highly varied. The principal ingredients range from metallic compounds to nearly metal-free ones, from carbon-dominated to silicon-bearing minerals, and there are several classes whose composition is not known. Among the biggest asteroids in the main belt between Mars and Jupiter, the majority (about three-quarters) are carbonaceous, meaning that they contain carbon in complex molecular forms. Most of the rest are composed chiefly of silicon-bearing compounds, and about 5 percent are metal-rich, the primary metals being nickel-iron mixtures.

Figure 15.3 A portrait of an asteroid? *Many asteroids probably bear a general resemblance to this irregularly shaped object, which is Deimos, one of the tiny Martian moons. (NASA)*

KIRKWOOD'S GAPS: ORBITAL RESONANCES REVISITED

As increasing numbers of asteroids were discovered and cataloged throughout the Nineteenth century, calculations of their orbits showed remarkable gaps at certain distances from the Sun. One such gap is at 3.28 AU, and another is at 2.50 AU.

The explanation for these was offered by Daniel Kirkwood in 1866, when he realized that these distances correspond to orbital periods that are simple fractions of the period of Jupiter (Fig. 15.4). The giant planet, at its distance of 5.2 AU from the Sun, takes 11.86 years to make a trip around it, whereas an asteroid at 3.28 AU, if one existed there, would have a period exactly half as long, 5.93 years. Thus this asteroid and Jupiter would be lined up in the same way frequently, every time Jupiter made one orbit. The asteroid would therefore be subjected to regular tugs by Jupiter's gravity. Apparently this has prevented

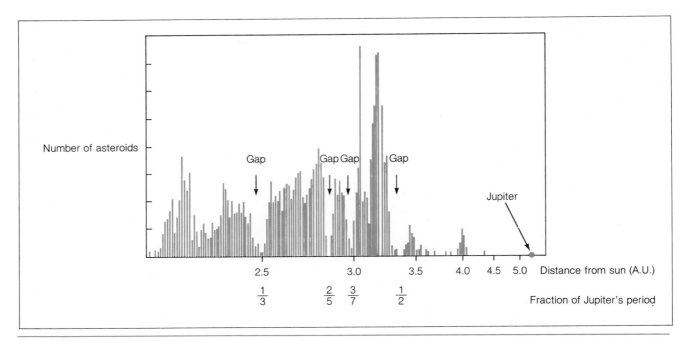

Figure 15.4 Kirkwood's gaps. *This graph shows the distribution of asteroid orbits. Gaps appear at the distances from the Sun where asteroids would have exactly one-half, one-third, or other simple fractions of the orbital period of Jupiter.*

asteroids from staying in such orbits, at this and other distances where the orbital periods would result in regular alignments. The gap at 2.50 AU corresponds to orbits with a period one-third that of Jupiter. These orbital resonances are exactly analogous to the ones created by the moons of Saturn, which are responsible for some of the gaps in the ring system of that planet (see Chapter 13). Since Jupiter is the most massive and the nearest of the outer planets to the asteroid belt, it has by far the greatest effect in producing gaps, but in principle the other outer planets could do the same thing.

THE ORIGIN OF THE ASTEROIDS

The most natural explanation of the asteroids seemed for a long time to be the breakup of a former planet, creating a swarm of fragments that continued to orbit the Sun. This argument was weakened when the total mass of the asteroids was estimated and found to be much less than that of any ordinary planet. An addi-

tional, rather strong argument against the planetary remnant hypothesis can be cited: There is no known reasonable way to break a planet apart once it has formed, whereas it is easy to understand why material orbiting the Sun at the position of the asteroid belt could never have combined into a planet in the first place. It became simpler to accept the idea that the debris never was part of a planet, rather than to find a way to have it first form one and then break apart.

The invisible hand of Jupiter's gravity is invoked again. If, as it is suspected, the planets formed from a swarm of debris orbiting the young Sun in a disk, no planet could have formed in a location where the pieces of debris could not stick together. Calculations show that once Jupiter had formed, its immense gravitational force would have stirred up the material near its orbit, so that collisions between particles would occur at speeds too great to allow them to stick together. It is as though Jupiter wielded a giant spatula, stirring up the debris near it and keeping it spread out as a loose collection of rocky fragments. Even today, Jupiter is still at work, keeping the asteroids mixed up, occasionally causing collisions between them that can sometimes break them up.

The study of meteorites, some of which probably originated as pieces of asteroids that broke apart in collisions, gives us a chance to examine material from the early solar system. It is interesting to note that some of the asteroids apparently have undergone differentiation, developing nickel-iron cores, because some meteorites are almost pure chunks of this material. There must have been sufficient heat inside some of the asteroids to create a molten state, allowing the heavy materials to segregate from the rest.

In contrast with the nickel-iron asteroids, some of the other types, particularly the carbonaceous ones, apparently have undergone almost no heating, since they contain high quantities of volatile elements that would have been easily cooked out.

COMETS: FATEFUL MESSENGERS

Among the most spectacular of all the celestial sights are the comets. With their brightly glowing heads and long, streaming tails, along with their infrequent and often unpredictable appearances, these objects have sparked the imagination (and often the fears) of people through the ages.

In antiquity, when astrological omens were taken very seriously, great import was attached to the occasion of a cometary appearance (Fig. 15.5). Ancient descriptions of comets are numerous, and in many cases these objects were thought to be associated with catastrophe and suffering.

Included among the teachings of Aristotle was the notion that comets were phenomena in the Earth's atmosphere. There was no good evidence for this idea; apparently Aristotle adopted it because he believed that any phenomena that were changeable were associated with the Earth, whereas the heavens were perfect and immutable. In any case, this idea was accepted for centuries to come. Tycho Brahe in 1577 was able to prove that comets were too distant to be associated with the Earth's atmosphere, because they do not exhibit any parallax when viewed from different positions on the Earth. If a comet were really located only a few kilometers or even a few hundred kilometers above the surface of the Earth, its position as seen from the Earth would change from one location to another. Tycho was able to show that this was not the case, and that therefore comets belonged to the realm of space.

Figure 15.5 Calamity on Earth associated with the passage of comets. *This drawing is from a seventeenth-century book describing the universe. (The Granger Collection)*

Halley, Oort, and Cometary Orbits

A major advance in the understanding of comets was made by a contemporary and friend of Newton, Edmond Halley. Aware of the power of Newton's laws of motion and gravitation, Halley reviewed the records of cometary appearances and noted one outstanding regularity. Particularly bright comets seen in 1531, 1607, and 1682 seemed to have similar properties, and Halley suggested that in fact all three were appearances of the same comet, orbiting the Sun with a 76-year period.

Calculations showed, using Kepler's third law, that for a period of 76 years, this object must have a semimajor axis of nearly 18 AU. Halley realized, therefore,

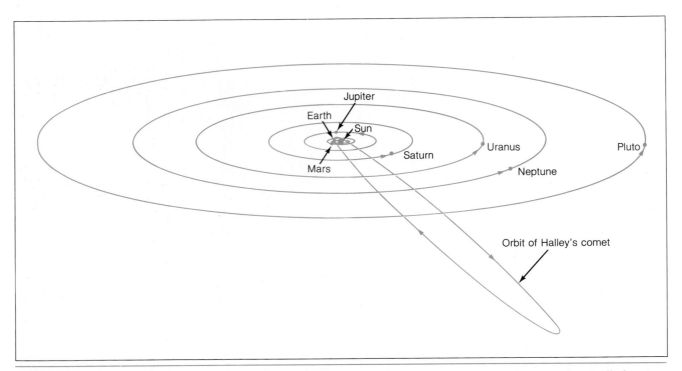

Figure 15.6 A cometary orbit. *This is a rough scale drawing of the orbit of Halley's comet. Most comets actually have much more highly elongated orbits than this, and correspondingly longer periods.*

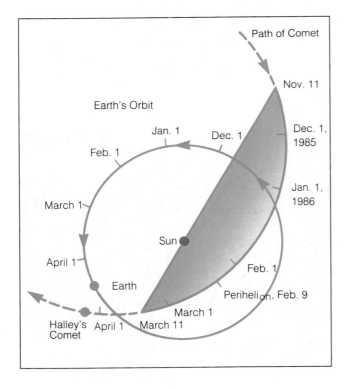

Figure 15.7 The path of Halley's comet in 1986. *This graph shows the path of Halley's comet and the relative position of the Earth during the recent appearance. Note that the comet was on the far side of the Sun when it passed closest to the Earth.*

that in order to appear as dominant in our skies as the comet does, it must have a highly elongated orbit (Fig. 15.6), so that it comes close to the Sun at times, even though its average distance is well beyond the orbit of Saturn, nearly as far out as Uranus. Such an eccentric orbit, as a very elongated ellipse is called, had not previously been observed, even though Newton's laws clearly allowed the possibility.

Since Halley's time, searches of ancient reports of comets have revealed that Halley's comet has been making regular appearances for many centuries. The earliest records are provided by the ancient Chinese astronomers, who apparently observed its every appearance for well over 1,000 years, possibly beginning as early as the fifth century B.C.

Figure 15.8 Halley's comet in 1986. *Here are a few of the fine photographs made from Earth during the apparition of Halley's comet in 1985 and 1986. The two negative images on top, obtained only days apart, show changes in the detailed structure of the tail, including a disconnection event, as part of the ion tail broke free of the comet (top right). The two images at the bottom were both made from the same photograph, the color-coded version being made to enhance slight brightness variations. (top: © 1986 Royal Observatory, Edinburgh; bottom: U. Fink, University of Arizona)*

The most recent visit of Halley's comet was in 1985 and 1986, when it did not pass very close to the Earth (Fig. 15.7), and the best views were from the Southern Hemisphere (Fig. 15.8). A much more spectacular appearance occurred on the comet's previous visit in 1910 (Fig. 15.9), when the Earth actually passed through the rarefied gases of its tail. This will happen again during the 2061 appearance of Halley.

Other spectacular comets have been seen (Fig. 15.10), and a number have rivaled Halley's comet in brightness. Traditionally, a comet is named after its discoverer, and there are astronomers around the world who spend long hours peering at the nighttime sky through telescopes, looking for a piece of immortality.

When a new comet is discovered, a few observations of its position are sufficient to allow computation of

Figure 15.9 Halley's comet as it appeared in 1910. *(Palomar Observatory, California Institute of Technology)*

Figure 15.10 Another bright comet. *This is comet IRAS-Araki-Alcock, which appeared in 1983, and was first discovered in infrared images obtained with the* IRAS *satellite. The colored streaks are star trails created by the telescope's motion as it tracked the comet, using color filters to make multiple exposures. (National Optical Astronomy Observatories)*

its orbit. The results of many years of comet-watching have shown that there are many comets whose orbits are so incredibly stretched out that their periods are measured in thousands or even millions of years. These comets, for all practical purposes, are seen only once. They return thereafter to the void of space well beyond the orbit of Pluto, there to spend millenia before visiting the inner solar system again.

The orbits of comets, particularly these so-called long-period ones, are randomly oriented. These comets do not show any preference for orbits lying in the plane of the ecliptic, in strong contrast with the planets, and about half go around the Sun in the retrograde direction.

Consideration of these orbital characteristics, especially the large orbital sizes, led Dutch astronomer Jan Oort to suggest that all comets originate in a cloud of objects that surrounds the solar system. The **Oort cloud,** as it is now called, was envisioned to be a spherical shell with a radius of about 100,000 AU, extending, therefore, a significant fraction of the distance to the nearest star, which is almost 300,000 AU from the Sun.

Modern calculations show the Oort cloud to lie primarily between 1,000 and 30,000 AU from the Sun, although its tenuous outer portions may extend as far out as Oort thought.

Occasionally, a piece of debris from the Oort cloud is disturbed from its normal path, either by a collision with another object or perhaps by the gravitational tug of a nearby star, and it begins to fall inward towards the Sun. Left to its own devices, a comet falling in from the Oort cloud would follow a highly elongated orbit with a period of millions of years, appearing to us as one of the long-period comets when it made its brief, incandescent passage near the Sun. In many cases, a comet is not left to its own devices, however. Instead it runs afoul of the gravitational pull of one of the giant planets—most often Jupiter. When this happens, the comet may be speeded up, so that it will escape the solar system entirely after it loops around the Sun. Or it may be slowed down, dropping into a smaller orbit with a shorter period, so that it becomes one of the numerous comets that are seen to reappear frequently (Table 15.2).

Detailed simulations of the process by which comets fall into the inner solar system from the Oort cloud show that most of the short-period comets, whose orbits tend to lie close to the plane of the planetary orbits, cannot originate in a spherical Oort cloud. Instead these comets must arise in a region that is disk-shaped, so that they start out with orbits that are aligned with the planetary orbits. This has led astronomers to deduce the existence of an inner cloud of comets, called the **Kuiper belt** after the American astronomer Gerard Kuiper, lying just outside the orbits of the outer planets.

Thus, in the modern picture of cometary origins, we distinguish between the long-period comets, those with very elongated, randomly oriented orbits and periods of millions of years, and short-period comets, which have smaller orbits lying close to the plane of the planets. The long-period comets arise in the inner Oort cloud, a spherical shell concentrated between 1,000 and 30,000 AU from the Sun, while the short-period comets originate in the Kuiper belt, a disk some 30 to 50 AU in radius. The total mass in the Kuiper belt is estimated to be about equal to the mass of the Earth, while the Oort cloud is thought to contain about 10^{11} objects totaling 100 or more Earth masses.

Both the Oort cloud and the Kuiper belt consist of bodies left over from the time of solar system formation, as will be discussed in the next chapter. The Kuiper belt objects are thought to have formed at roughly

Table 15.2 Selected periodic comets

Comet[a]	Period (years)	Semimajor Axis (AU)	Year of Next Appearance
Encke	3.30	2.21	1991.0
Temple (2)	5.26	3.0	1994.2
Schwassmann-Wachmann (2)	6.52	3.50	1994.3
Wirtanen	6.65	3.55	1994.6
Reinmuth (2)	6.72	3.6	1994.5
Finlay	6.88	3.6	1995.1
Borrelly	7.00	3.67	1995.5
Whipple	7.44	3.80	1993.1
Oterma	7.89	3.96	1990.0
Schaumasse	8.18	4.05	1993.0
Wolf	8.42	4.15	1992.9
Comas Sola	8.58	4.19	1995.6
Vaisala	10.5	4.79	1991.9
Schwassmann-Wachmann (1)	16.1	6.4	1998.6
Neujmin (1)	17.9	6.8	2002.7
Crommelin	27.9	9.2	2012.6
Olbers	69	16.8	2025
Pons-Brooks	71	17.2	2025
Halley	76.1	17.2	2062.5

[a]A number in parentheses following the name of a comet indicates cases where a single observer has discovered more than one comet.

their present positions, while the Oort cloud bodies were ejected from the region of the outer planets by gravitational encounters and by tidal forces due to the Milky Way galaxy.

THE ANATOMY OF A COMET

A remarkably accurate picture of the nature of a comet was developed many years ago by American astronomer Fred Whipple. Whipple's model, sometimes called the "dirty snowball," envisions that for most of its life, a comet is just a frozen chunk of icy material, probably consisting of small particles like gravel or larger boulders, embedded in frozen gases. This picture was developed on the basis of observations of many comets over the years, and its basic features were verified by the close-up examination of Halley's comet in early 1986, when several space probes were able to make observations from close range (Fig. 15.11). Of course, a great many new details were discovered as well.

As a comet passes through the outer reaches of its orbit, far from the Sun, it does not glow, has no tail, and is not visible from Earth. As it approaches the Sun, however, it begins to warm up as it absorbs sunlight, and the added heat causes volatile gases to escape. A spherical cloud of glowing gas called the **coma** develops around the solid **nucleus** (Fig. 15.12). The principal gases found in the coma of Halley's comet (Table 15.3) are H_2O (about 80 percent by number), carbon dioxide (CO_2; about 3.5 percent), and carbon monoxide (CO; most of the remaining 16.5 percent). Other slightly more complex species, such as ammonia (NH_3) and methane (CH_4) are probably present also, along with hydrogen molecules (H_2). The observed species glow by a process called **fluorescence**. The molecules absorb ultraviolet light from the Sun, causing them to be excited to high energy levels, and then they emit visible light as they return to low energy states. A cloud of hydrogen atoms, resulting from the breakup of the molecules in the coma, extends out to great distances from the nucleus. The visible coma may be as large as 100,000 km in diameter, while the halo of hydrogen atoms may extend as much as 10 times farther from the nucleus, as demonstrated by ultraviolet observations of Halley's comet in 1986 (Fig. 15.13). The solid nucleus is relatively tiny, having a diameter of a few

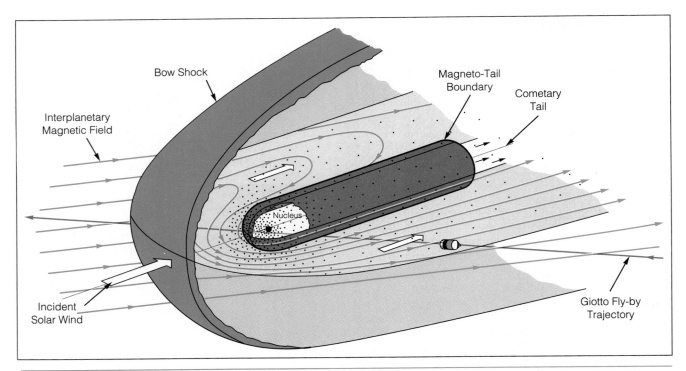

Figure 15.11 Probes to comet Halley. *This shows the path taken by the European Space Agency's* Giotto *spacecraft, which made the closest approach to the nucleus of the several probes that encountered the comet. The other probes flew along similar paths, crossing ahead of the nucleus, but not as close to it. (European Space Agency)*

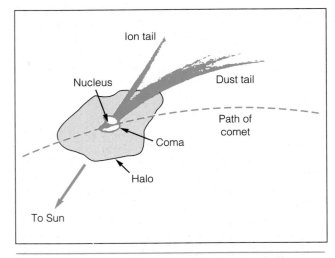

Figure 15.12 Anatomy of a comet. *This sketch illustrates the principal features of a comet, though not necessarily to correct scale.*

Table 15.3 Detected constituents of comets

Coma		Ion Tail	
Atoms	**Molecules**	**Ions**	**Molecular Ions**
Hydrogen (H)	H_2O	C^+	CO^+
Oxygen (O)	CO_2	Ca^+	CO_2^+
Sulfur (S)	C_2	O^+	H_2O^+
Carbon (C)	C_3	S^+	OH^+
Sodium (Na)	CH	Fe^+	CH^+
Iron (Fe)	CN	Na^+	CN^+
Potassium (K)	CO		N_2^+
Calcium (Ca)	CS		H_3O^+
Vanadium (V)	NH		CS_2^+
Chromium (Cr)	OH		S_2^+
Manganese (Mn)	NH_2		CS^+
Cobalt (Co)	NCN		
Nickel (Ni)	CH_3CN		
Copper (Cu)	S_2		
	HCO		
	NH_3		

Figure 15.13 The halo of Halley's comet. *This is an ultraviolet image, made by the* Pioneer Venus *spacecraft from Venus orbit, showing (in yellow, red, and orange) the extent of the hydrogen cloud surrounding the comet's nucleus. The hydrogen is a by-product of water vapor that is ejected from the nucleus and then dissociated by the Sun's ultraviolet radiation. The halo was about 500,000 kilometers, or roughly ⅓ AU in diameter, when this image was obtained. (LASP, University of Colorado, sponsored by NASA)*

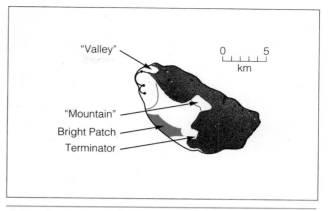

Figure 15.14 The nucleus of Halley's comet. *This sketch, based on images obtained by the* Giotto *spacecraft, illustrates the size and shape, as well as some surface features, on the nucleus. (Based on data from the European Space Agency).*

kilometers. The nucleus of Halley was found to be irregularly shaped, with a long axis of about 15 km and a width ranging up to about 10 km (Fig. 15.14). The mass of the nucleus is estimated to be about 10^{17} grams, and its density is extremely low—0.5 to 1.0 grams/cm³—much less than the density of ordinary ice. The nucleus was found to have a coating of very dark material (albedo only 0.04), probably a matrix of carbonaceous dust left behind by the evacuation of ices that evaporated previously. Hence, the dirty snowball idea has been modified to incorporate a complete outer coating of dirt, along with the traditional picture of mixed dirt and ice in the interior of the nucleus.

Another modification of the picture was that the surface of the nucleus is quite hot, instead of being cold and icy. The high temperature had been suspected from Earth-based infrared measurements and was confirmed by the spacecraft observations made at close range. The high temperature is the result of the black-

ness of the crust; low reflectivity means that sunlight is efficiently absorbed, causing heating.

Apparently the gas escaping from the nucleus of a comet is released in jets that can be so powerful that they alter the orbital motion of the comet. It has been known for years that comets can be slowed in this way as they approach the Sun, but the close encounters with Halley's comet provided dramatic confirmation, the images revealing spectacular fountains of gas and dust erupting from beneath the crust of the nucleus (Fig. 15.15). Apparently these gas jets are created by heat that is absorbed by the dark crust and conducted into the icy interior, where the heat causes ices to evaporate in a process called **sublimation,** in which solid ice is converted directly into gas. The released gases expand and create outward pressure, which is relieved as the gases break through the crust, forming the jets that are observed. As the nucleus rotates, jets are active only on the sunlit side, so each individual jet turns on and off as it rotates toward and then away from the Sun. Surprisingly, examination of photos of Halley's comet in 1910 indicate that the jets occurred in the same locations then as they did in 1986. Apparently, the jets occur at weak points in the crust of the comet, creating craters that persist over the years between encounters with the Sun's warming radiation.

As a comet nears the Sun, the solar radiation and the solar wind force some of the gas from the coma to

ASTRONOMICAL INSIGHT 15.1

Comets and Water

In this chapter we have learned that comets consist largely of water ice. Halley's comet released enormous quantities of water vapor as the icy nucleus of the comet was vaporized by solar heating during the weeks when the comet was close to the Sun. Spectroscopy of the comae and tails of other comets reveals the hydrogen that is released when water molecules are dissociated by the Sun's heat. If comets contain so much water, some scientists argue, then comets could be a major source of water for the planets.

It is known from studies of cratering patterns on the Moon, and from theoretical models of solar system formation, that there was an early period in the history of our planetary system when comets were far more common than they are today. The rate of impacts on the Moon during the first several hundred million years was as high as 10,000 times the current rate. Studies of major craters on the Moon suggest that up to 50 percent of them were caused by cometary bodies, rather than remnants of asteroids. If even 10 percent of the impacts during the earliest days of the solar system were due to comets, then the quantity of water that fell on the Earth was large enough to account for virtually all the water in the seas of today.

Interestingly, if the Earth's oceans formed as the result of cometary bombardment, then Venus must also have received large quantities of water, yet that planet today is extremely dry. Recall that when we discussed Venus (Chapter 11), we assumed that water never existed there in liquid form, but that water vapor released from the surface was soon dissociated by solar heating, allowing the hydrogen atoms to escape into space. Now some scientists are suggesting that there may have been liquid seas on Venus during its early history. If so, the Sun's heating was still sufficient to gradually vaporize the water, allowing it to dissociate and the hydrogen to escape. Under this hypothesis, Venus and the Earth were even more similar at first than was previously thought, and only gradually have they become so different.

The rate of impacts by comets and asteroidal bodies was greatly reduced by the time the solar system was a billion years old, and it is normally assumed that today water is not being added to the Earth (or Venus) at a significant rate. A recent controversial suggestion to the contrary has been made, however, based on ultraviolet observations from the *Voyager* spacecraft.

The *Voyager* probes have made many observations of targets other than the outer planets; recall that months have gone by between planetary encounters. During these "cruise" phases, measurements have been made of interplanetary gas in the inner solar system, and some of these measurements indicated an excess quantity of hydrogen atoms in space. The quantity discovered was too great to be maintained, and it appeared that there must be a source of hydrogen to replenish the supply. It was suggested that a steady infall of comets from the outer solar system could supply the hydrogen by releasing water vapor that was then dissociated. These would have to be small comets, or else they would be seen, and they were dubbed "minicomets."

Subsequently the observations of excess hydrogen have been called into question, and it is unclear now whether or not there really is a need to invoke a supply of new comets. The idea is intriguing, though, and new measurements are being planned. Among these will be ultraviolet searches for interstellar hydrogen using instruments aboard the *Galileo* probe as it travels its long journey to Jupiter. *Galileo* is spending many months traveling through the inner solar system, as it passes close to Venus and the Earth to gain speed for its outward journey to Jupiter, and during this time it measured hydrogen in interplanetary space. The results should settle the issue of whether the inner solar system is being bombarded with minicomets.

Whether the hypothetical minicomets are verified or not, there is little doubt that comets in the early solar system played an important role in establishing the water content of the planets. This is interesting to contemplate while swimming or taking a drink of water.

Figure 15.15 The nucleus of comet Halley.   *These are two versions of the same composite of some 60 images obtained by the Giotto spacecraft as it flew past the nucleus of the comet. Detailed surface markings in the crust of the nucleus can be seen, as well as the bright jets of warm dust that were being ejected toward the Sun (to the left). (Max Planck Institute for Aeronomie, courtesy of A. Delamere, Bell Aerospace Corporation)*

flow away from the Sun, forming the tail, which in some cases is as long as 1 AU. Often there are two distinct tails (Fig. 15.16), one formed of gas from the coma, usually containing molecules that have been ionized, such as CO^+, N_2^+, CO_2^+, and CH^+: and another formed of tiny solid particles released from the ice of the nucleus. The **ion tail,** the one formed of ionized gases, is shaped by the solar wind, and therefore points almost exactly straight away from the Sun at all times. It has been observed many times, for Halley and other comets, that all or part of the ion tail can detach itself and dissipate, then to be replaced by a new tail (see Fig. 15.8). These "disconnection events" may be related to variations in the Sun's magnetic field, which exert forces on charged particles in the ion tail.

The other tail, called the **dust tail,** usually takes on a curving shape, as the dust particles are pushed away from the Sun by the force of the light they absorb. This **radiation pressure** is not strong enough to force the dust particles into perfectly straight paths away from the Sun, so they follow curved trajectories that are a combination of their orbital motion and the outward push caused by sunlight. The dust particles ejected from

the nucleus are apparently quite fragile, being broken apart readily in collisions or when heated. Streams of gas resulting from the evaporation of icy dust particles were observed to extend as far as 50,000 km from the nucleus. Particles captured by the *Giotto* spacecraft appeared to be low-density, fluffy grains, and the farther from the nucleus they were captured, the smaller they tended to be.

One of the major discoveries made by the probes that encountered Halley's comet and Comet Giaccobini-Zinner was that the ionized gases surrounding a comet create a trapped magnetic field. A "bow shock," analogous to a boundary of a planet's magnetosphere, builds up on the sunward side of a comet, where the solar wind encounters the trapped magnetic field and flows around it. In the direction away from the Sun, the magnetic field that is wrapped around the comet forms a sheet of trapped charged particles called a **current sheet.**

The gases that escape from the nucleus of a comet as it approaches the Sun are highly volatile and would not be present in the nucleus if it has ever undergone any significant heating. This tells us that comets must

Figure 15.16 Two tails. *These photos of Comet Mrkos (1957) show distinctly the two types of tails that characterize many comets. The ion tail points straight up in these views, while the dust tail curves to the right. (Palomar Observatory, California Institute of Technology)*

have formed and lived their entire lives in a very cold environment, probably never getting as warm as even 100 K before falling into orbits that bring them close to the Sun. If a comet is so easily vaporized, then once it has begun to follow a path that regularly brings it close to the Sun, its days are numbered. It may take many round trips, but eventually it will dissipate all of its volatile gases, leaving behind nothing but rocky debris. It has been estimated that Halley's comet was losing some 50 tons per second of water ice when it passed nearest the Sun in early 1986! (Despite this high mass-loss rate, the comet should last at least another 100,000 years before the nucleus is entirely dissipated. There is some speculation that a comet as large as Halley may eventually build up a crust thick enough to halt further sublimation of its interior ices. If this

is so, Halley's comet may be immortal, but it will eventually stop developing a coma and tail on its visits to the inner solar system.) Several times, a comet has failed to reappear on schedule, but has been replaced by a few pieces or perhaps a swarm of fragments (Fig. 15.17). In time, the remains of a dead comet will be dispersed all along the orbital path, so that each time the Earth passes through this region, it encounters a vast number of tiny bits of gravel and dust, and we experience a meteor shower.

METEORS AND METEORITES

Occasionally one of the countless pieces of debris float-

Figure 15.17 The breakup of Comet West. *This dramatic sequence shows the nucleus of Comet West fragmenting into four pieces. (New Mexico State University)*

Figure 15.18 Two bright meteors. *The streaks of light in this photo are created by tiny particles entering the Earth's atmosphere from space. (Yerkes Observatory)*

Figure 15.19 A meteorite. *This is a stony meteorite; the black coloring is due to heating as the object passed through the atmosphere. The light-colored spots are breaks in the fusion crust where interior material is exposed. (NASA)*

ing through the solar system enters the Earth's atmosphere, creating a momentary light display as it evaporates in a flash of heat created by the friction of its passage through the air. The streak that is seen in the sky is called a **meteor** (Fig. 15.18). Most of us are familiar with this phenomenon, commonly called a shooting star, since it is often possible to see one in just a few minutes of sky-gazing on a clear night. On rare occasions an especially brilliant meteor is seen, possibly persisting for several seconds, and these spectacular events are called **fireballs** or **bolides.**

The piece of solid material that causes a meteor is called a **meteoroid.** Most are very small, amounting to nothing more than tiny grains of dust or perhaps fine gravel. A few, however, are larger, solid chunks, which are responsible for the bright fireballs.

Occasionally one of the larger meteoroids survives the arduous trip through the atmosphere and reaches the ground intacts. Such an object is called a **meteorite** (Fig. 15.19), and examples can be found in museums around the world. Meteorites have been the subject of intense scrutiny, for until the past 20 years or so, they were the only samples of extraterrestrial material scientists could get their hands on.

Historically, the notion that rocks could fall from the sky was scoffed at by scientists, until a meteorite that was seen to fall near a French village in 1803 was found and examined just after it fell. Such falls are rare, but they are observed occasionally, and have even been known to cause damage (but so far, few injuries).

Primordial Leftovers

Meteorites are old, much older than most surface rocks on the Earth. This fact enhances the interest of scientists in studying them, for in doing so they get a glimpse into the history of the solar system.

Meteorites generally can be grouped into three classes: the stony meteorites, which comprise about 93 percent of all meteorite falls; the iron meteorites, accounting for about 6 percent; and the stony-iron meteorites, which are the rarest. These relative abundances were determined indirectly, because the different types of meteorites are not equally easy to find on the ground. The majority of all meteorites found are the iron ones, which, as we have just seen, comprise only a small fraction of those that fall. The stony meteorites look so much like ordinary rocks that they are usually difficult to pick out, and some are burned up on their way through the atmosphere. A particularly productive place to search for meteorites is Antarctica, where a thick layer of ice conceals the native rock. Meteorites that fall there are relatively easy to find, and there is little chance to confuse them with Earth rocks.

The stony meteorites are mostly of a type called **chondrites,** so named because they contain small spherical inclusions called **chondrules** (Fig. 15.20). These are mineral deposits formed by rapid cooling, most likely

Figure 15.20 Cross-section of a chondrite. *This shows the many chondrules (light patches) embedded within the structure of this type of stony meteorite. (Griffith Observatory, Ronald A. Oriti Collection)*

Figure 15.21 A carbonaceous chondrite. *This example, not of the type in which amino acids have been found, shows a large chondrule (the light-colored spot, upper center), which is about 5 millimeters in diameter. (Griffith Observatory, Ronald A. Oriti Collection)*

at an early time in the history of the solar system, when the first solid material was condensing. A few of the stony meteorites are **carbonaceous chondrites** (Fig. 15.21), thought to be almost completely unprocessed since the solar system was formed, and therefore representative of the original stuff of which the planets were made. The primordial nature of the carbonaceous chondrites, like the carbonaceous asteroids mentioned in an earlier section, is deduced from their highly volatile contents, which indicates that they were never exposed to much heat. One particularly fascinating aspect of these meteorites is that in at least one case, complex organic molecules called **amino acids** were found inside a carbonaceous chondrite meteorite, showing that some of the ingredients for the development of life were apparently available even before the Earth formed.

The iron meteorites have varying nickel contents and sometimes show an internal crystalline structure (Fig. 15.22) that indicates a rather slow cooling process in their early histories. This has important implications for their origin, as we shall see.

Figure 15.22 Cross-section of a nickel-iron meteorite. *This example shows the characteristic crystalline structure indicative of a slow cooling process from a previous molten state. Such meteorites are thought once to have been parts of larger bodies that differentiated. (Griffith Observatory, Ronald A. Oriti Collection)*

Dead Comets and Fractured Asteroids

The origins of the meteoroids that enter the Earth's atmosphere can be inferred from what we know of the properties of meteorites and of the asteroids and comets.

As we have seen, most meteors are caused by relatively tiny particles that do not survive their flaming entry into the Earth's atmosphere. During **meteor showers**, when meteors can be seen as frequently as once per second, all seem to be of this type. As noted

ASTRONOMICAL INSIGHT 15.2

The Impact of Impacts

 It is well established that impacts by objects from space have occurred throughout the solar system, because every exposed surface has craters. Recently scientists have found evidence that impacts may have played a far more important role in the shaping of solar system bodies than just to create craters. Impacts may have modified biological and geological processes on Earth, might have caused the formation of our Moon, may explain various anomalies in the motions of other planets, and might provide evidence for a companion star orbiting the Sun. All of these conclusions are unproven, but the amount of attention they are receiving is testimony to their appeal among scientists in many fields.

Mathematical analyses of the process by which planetesimals formed and built up by coalescence during the early history of the solar system shows that there was a time when most of the preplanetary material was in the form of very large planetesimals several thousand kilometers in diameter. These orbited the Sun in a disk, so that planetesimals near each other traveled in the same direction with only relatively small velocity differences between near neighbors. Planets formed when these largest planetesimals collided and merged. Even with low relative speeds, vast amounts of heat were released upon im-

pact, making much of the material molten and allowing the two bodies to merge into a single spherical one. In most cases, the planet that resulted had its rotation axis roughly perpendicular to the plane of the solar system disk, and its rotation in the prograde direction. It is easy to see, however, that the tilt of the axis and the direction of spin could have been modified by the direction of impact of the major planetesimals that merged to form the planet. We see that the slow retrograde spin of Venus, and the large tilt (98°) of Uranus could have resulted from the off-center impacts of large planetesimals. It was also suggested (as discussed in Chapter 10) that the Earth's Moon might have formed as the result of the impact of a Mars-size planetesimal during the latter stages of formation of the Earth.

While the very massive impacts of planetesimals occurred only during the very early stages of solar system formation, we know that impacts by much smaller bodies still take place in modern times. The Barringer crater in Arizona is only about 25,000 years old, a major impact occurred in Siberia in 1908, and we find traces of craters scattered over the Earth (these are described in the text). In the past few years, it has been suggested that impacts of large meteorites or cometary nuclei on the Earth have been responsible for major modifications in the climate, which in turn may have been linked to mass extinctions of life forms on the Earth. Geological evidence for impacts (consisting of thin layers of debris deposited over large areas of the Earth) coincide in age with times of major extinctions. The most

widely discussed event seems to have occurred at just about the time (65 million years ago) when the dinosaurs became extinct. This suggestion is highly controversial, because it appears that the extinctions were not as sudden as the impact scenario supposes, but there is strong evidence that such an impact occurred. A location for the impact (in the Indian Ocean) has even been suggested.

Very recently it was hypothesized that impacts have been responsible for the sporadic reversals in the Earth's magnetic field (described in Chapter 9). There is a pattern of very striking coincidences between the times of major impacts in the past and the times when reversals of the Earth's magnetic field occurred. The impacts are dated by the deposition of thin layers of debris, and the magnetic field reversals are dated by measurements of the magnetic orientation of rocks which show what the field direction was when the rocks formed. If indeed impacts somehow triggered magnetic field reversals, the actual mechanism is not well understood. It has been suggested that changes in climate caused by dust in the atmosphere could have trapped more of the world's water in the form of polar ice, thus changing the mass distribution of the Earth enough to affect its rotation. The change in rotation of the Earth, it is argued, could disrupt the internal currents that give rise to the magnetic field, causing the field to vanish temporarily. When the field reestablished itself later, its orientation would have an even chance of being the opposite of what it had been before the impact.

(continued on next page)

earlier, these showers are associated with the remains of comets that have disintegrated (Table 15.4), leaving behind a scattering of gravel and dust. Therefore, it is thought that the most common meteors, those created by small, fragile meteoroids, are a result of cometary debris.

The larger chunks that reach the ground as meteorites may have a different origin. It is likely that there are occasional collisions among the asteroids, some-times sufficiently violent to destroy them, dispersing the rubble that is left over throughout the solar system. Most meteorites are probably fragments of asteroids. The iron meteorites apparently originated in asteroids that had undergone differentiation, while the stony ones either came from the outer portions of differentiated asteroids, or from smaller bodies that never underwent differentiation at all. The chondrites probably fall into the latter category, since the chondrules reflect a rapid cooling that would be characterized by very small bodies. This is why the chondrites are thought to be the most primitive of the meteorites, having undergone no processing in the interiors of large bodies.

From our studies of the other planets and satellites, we know that there was a time long ago when frequent impacts occurred, forming most of the craters seen today. Certainly the Earth was not immune, and no doubt was also subjected to heavy bombardment. The difference, of course, is that the Earth has an atmosphere, along with flowing water and glaciation, all of which combine to erase old craters in time. A few traces are still seen, however; there is a very large basin under the Antarctic ice that is probably an ancient impact crater, and a portion of Hudson's Bay in Canada shows a circular shape thought to have a similar origin. A number of other suspected ancient impact craters have been found throughout the world (Fig. 15.23 and Table 15.5).

Table 15.4 Major meteor showers

Shower	Approximate Date	Associated Comet
Quandrantid	January 3	—
Lyrid	April 21	Comet, 1861 I
Eta Aquarid	May 4	Halley's Comet
Delta Aquarid	July 30	—
Perseid	August 11	Comet 1862 III
Draconid	October 9	Comet Giacobini-Zinner
Orionid	October 20	Halley's Comet
Taurid	October 31	Comet Encke
Andromedid	November 14	Comet Biela
Leonid	November 16	Comet 1866 I
Geminid	December 13	—

Figure 15.23 Impact craters on the Earth. *This map shows the locations of major craters thought to have been created by impacts of massive objects. (Griffith Observatory)*

Although the frequency of impacts has decreased, major impacts still occur on rare occasions. The Barringer crater (Fig. 15.24), near Winslow, Arizona, was formed only about 25,000 years ago, for example, and the possibility exists that other large bodies could still hit the Earth. Given a long enough time, it is almost inevitable.

MICROSCOPIC PARTICLES: INTERPLANETARY DUST AND THE INTERSTELLAR WIND

The space between the planets plays host to some very tiny particles, in addition to the larger ones we have

Table 15.5 Some known or suspected impact craters on the Earth

Location	Diameter (km)	Location	Diameter (km)
Amirante Basin, Indian Ocean	300	Deep Bay, Saskatchewan	13.7
Sudbury, Ontario	140	Deep Bay, Saskatchewan	12
Vredefort, Orange Free State, South Africa	100	Lake Bosumtwi, Ashanti, Ghana	10.5
Manicouagan	100	Chassenon Structure, Haut-Vienne, France	10
Sierra Madera, Texas	100	Wolf Creek, Western Australia	8.5
Charlevoix Structure, Quebec	46	Brent, Ontario	3.8
Clearwater Lake West, Quebec	32	Chubb (New Quebec), Quebec	3.2
Mistastin Lake, Labrador	28	Steinem, Swabia, Germany	2.5
Gosses Bluff, Northern Territory, Australia	25	Henbury, Northern Territory, Australia	2.2
Clearwater Lake East, Quebec	22	Boxhole, Central Australia	1.8
Haughton, Northwest Territories	20	Barringer Crater, Winslow, Arizona	1.2
Wells Creek, Tennessee	14		

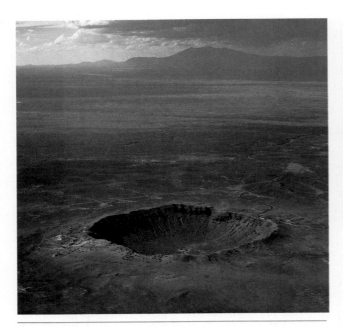

Figure 15.24 Meteor crater near Winslow, Arizona. *The impact that created this crater occurred about 25,000 years ago. (Meteor Crater Enterprises)*

Figure 15.25 The zodiacal light. *The diffuse glow of reflected sunlight from dust grains in the ecliptic plane is clearly visible in this photograph made from the summit of Mauna Kea. (W. Golisch)*

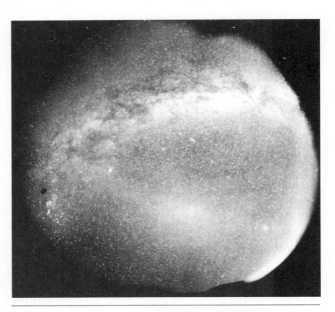

Figure 15.26 The gegenschein. *This photo shows the Milky Way stretching across the upper portion and a diffuse concentration of light at lower center, which is the gegenschein. It is created by light reflected directly back to Earth from interplanetary dust in the direction opposite the Sun. (Photo by S. Suyama, courtesy of J. Weinberg)*

just described. There is a general population of small solid particles, perhaps a millionth of a meter in diameter, called interplanetary dust grains. There is also a very tenuous stream of gas particles flowing through the solar system from interstellar space.

The presence of the dust has been known for some time from two celestial phenomena, both of which can be observed with the unaided eye, though only with difficulty. The dust particles scatter sunlight, so that under the proper conditions a diffuse glow can be seen where the light from the Sun hits the dust. This is analogous to seeing the beam of a searchlight stretching skyward; you see only the beam where there are small particles (either dust or water vapor) that scatter its light, so that some of it reaches your eye.

One of the phenomena created by the interplanetary dust is the **zodiacal light** (Fig. 15.25), a faintly illuminated belt of hazy light that can be seen stretching across the sky (along the ecliptic) on clear, dark nights, just after sunset or before sunrise. The second observable phenomenon created by the dust is a small bright spot seen on the ecliptic in the direction opposite the Sun. This diffuse spot, called the **gegenschein** (Fig. 15.26), is created by sunlight that is reflected straight back by the interplanetary dust, which is concentrated in the plane of the ecliptic. This is analogous to seeing a bright spot on a cloud bank or low-lying mist when you look at it with the Sun directly behind you; the bright spot is just the reflected image of the Sun and is the counterpart of the gegenschein.

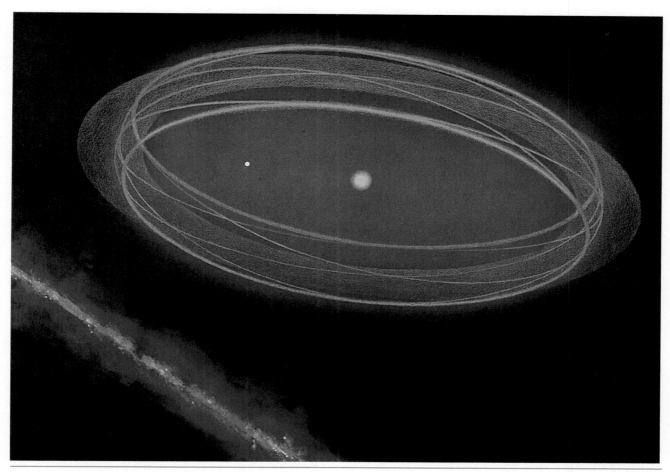

Figure 15.27 Dust bands in the outer solar system.   *The* IRAS *satellite's infrared sensors detected emission from this spiral-shaped tail of dust, which may have been created by a collision between an asteroid and a comet. At lower left is the Milky Way, as seen in the infrared by* IRAS. *(NASA)*

The *IRAS* satellite (see Chapter 7), which mapped the sky in infrared wavelengths, observed interplanetary dust. The grains of dust are cold and emit light at infrared wavelengths, so these observations reveal much more information about the dust than observations at other wavelengths. The *IRAS* sky maps show that the zodiac glows at infrared wavelengths due to the concentration of dust in the ecliptic plane. In addition, the infrared maps revealed streaks of dust following cometary orbits, and a huge spiraling ring of dust between the orbits of Mars and Jupiter (Fig. 15.27). This band of dust, which extends above and below the plane on the ecliptic, is thought to have been created by a collision between asteroids or between a comet and an asteroid.

It has been possible to collect interplanetary dust particles for direct examination (Fig. 15.28). This used to be done most commonly by the use of high-altitude balloons, but new techniques have been developed. Interplanetary particles that fell into the oceans can be scooped off of the sea floor, and very recently it has been found that rich deposits of the particles lie in arctic lakes, particularly in Greenland, where they can be dredged up in vast numbers. The Earth is constantly being pelted with dust particles (which add about 8 tons per day to its mass!), and those that fall into the oceans or on arctic ice sheets can lie undisturbed for long times. Studies of the grains show that they are probably of cometary origin, having been dispersed throughout space from the dead nuclei of old comets.

Figure 15.28 An interplanetary dust grain. *This is a microscopic view of a tiny particle from interplanetary space. The amorphous structure is highly variable from one grain to another. (NASA photograph, courtesy of D. E. Brownlee)*

These grains have a complex structure, consisting of many segments stuck together (see Fig. 15.28). The grains are low-density and fragile, having the general characteristics attributed to grains ejected from Comet Halley (as discussed earlier in this chapter).

As we will learn in Chapter 25, the space between stars in our galaxy is permeated by a rarefied gas medium. In the Sun's vicinity, the average density of this gas is far below that of any artificially created vacuum; it amounts to only about 0.1 particle per cubic centimeter (i.e., there is one atom, on average, in every volume of 10 cm³, corresponding to a cube about 1 inch on each side). Because of the motion of the Sun, the interstellar gas streams through the solar system with a velocity of about 20 km/sec. The presence of this ghostly breeze, consisting mostly of hydrogen and helium atoms and ions, was discovered in the early 1970s, when observations made from satellites revealed very faint ultraviolet emission from the hydrogen and helium atoms in the gas. The interstellar wind, tenuous as it is, has very little effect on the other components of the solar system but is nevertheless studied with some interest for what it may tell us about the interstellar medium.

PERSPECTIVE

The interplanetary wanderers discussed in this chapter

have given us insight into the history of the solar system and have told us much about its present state as well. We have found two primary origins of the various objects: comets and asteroids. The former account for most of the meteors and for the interplanetary dust, and the latter are responsible for the meteorites, including the massive bodies that formed the major impact craters in the solar system.

In the next chapter we consider the overall properties of the solar system as we discuss theories of its formation and evolution.

SUMMARY

1. Thousands of minor planets have been discovered and cataloged, displaying a variety of sizes (up to 1,000 km in diameter) and compositions (ranging from metallic ones to rocky minerals).

2. Gaps in the asteroid belt are created by orbital resonances with Jupiter, whose gravitational influence was probably also responsible for preventing the asteroids from coalescing into a planet in the first place.

3. Comets are small, icy objects that develop their characteristic comae and tails only when in the inner part of the solar system.

4. Comets apparently originate in a cloud of debris very far from the Sun, occasionally falling in, either to bypass the Sun and return to the distant reaches of the solar system for millenia, or to be perturbed by the gravitational influence of one of the planets, becoming periodic comets.

5. When near the Sun, a comet ejects gases that glow by fluorescence. Periodic comets eventually lose all of their icy substance in this process and disintegrate into swarms of rocky debris.

6. A comet may have two tails, one created by ionized gas, and the other made of fine dust particles.

7. A meteor is a flash of light created by a meteoroid entering the Earth's atmosphere from space, and a meteorite is the solid remnant that reaches the ground in some cases.

8. Meteorites are either stony, stony-iron, or iron in composition, and are very old, providing information on the early solar system.

9. Most meteors are created by fine debris from comets, but most meteorites are fragments of asteroids.

10. Interplanetary space is permeated by fine dust particles and by an interstellar wind of hydrogen and helium atoms from the space between the stars.

REVIEW QUESTIONS

Thought Questions

1. How is the asteroid belt similar to the rings of Saturn?
2. Why is the study of asteroids, comets, and meteorites of particular interest to scientists concerned with the history of the solar system?
3. Discuss how Bode's law stands in comparison with scientific theories; that is, how is it like a theory and how is it not?
4. Summarize the role of Jupiter in influencing asteroids, meteorites, and comets.
5. How do the motions of comets differ from those of the planets?
6. Discuss how the "dirty snowball" model of comets compared with what was found from the detailed close-up observations of Comet Halley in 1986.
7. Summarize the life story of a periodic comet.
8. Why are carbonaceous chondrites especially important clues to the early history of the solar system?
9. Would an observer on the surface of Mercury see meteors in the nighttime sky? Might he or she find meteorites on the ground there?
10. Summarize the evidence for the existence of interplanetary dust in the solar system.

Problems

1. What is the orbital period of an asteroid whose semimajor axis is 2.8 AU?
2. At what distance from the Sun would an asteroid have one-tenth the orbital period of Jupiter? Is there a gap in the asteroid belt at this distance?
3. Use Wien's law and the Stefan-Boltzman law to find the surface temperature and diameter of an asteroid whose wavelength of maximum emission is 0.003 cm and whose luminosity is 2×10^{20} erg/sec.
4. Suppose a comet in the Oort cloud has a semimajor axis of 100,000 AU. The nearest star like the Sun is Alpha Centauri, 4 light-years (about 300,000 AU from the Sun. Compare the gravitational force on the comet due to the Sun with that due to Alpha Centauri, when the comet lies in the same direction from the Sun as Alpha Centauri.

ADDITIONAL READINGS

Anatomy of a comet. 1987. *Sky and Telescope* 73(3).

Beatty, J. Kelly. 1986. An inside look at Halley's Comet. *Sky and Telescope* 71(5):438.

Cassidy, W. A., and L. A. Rancitelli, 1982. Antarctic meteorites. *American Scientist* 70(2):156.

Chapman, C. R. 1975. The nature of asteroids. *Scientific American* 23(91):24.

Chapman, R. P., and J. C. Brandt. 1985. An introduction to comets and their origins. *Mercury* 14(1):2.

Gehrels, Tom. 1985. Asteroids and comets. *Physics Today* 38(2):32.

Gingerich, Owen. 1986. Newton, Halley and the comet. *Sky and Telescope* 71(3):230.

Greenstein, George. 1985. Heavenly fire: Tunguska. *Science 85* 6(6):70.

Grieve, R. A. F. 1990. Impact cratering on the Earth. *Scientific American* 262(4):66.

Hartmann, W. K. 1975. The smaller bodies of the solar system. *Scientific American* 233(3):142.

Larson, Stephan, and David H. Levy. 1987. Observing Comet Halley's near nucleus features. *Astronomy* 15(5):90.

McSween, H. Y., Jr. 1989. Chondritic meteorites and the formation of planets. *American Scientist* 77:146.

Morrison, D. 1976. Asteroids. *Astronomy*, June 1976, p. 6.

Morrison, D. and C. R. Chapman. 1990. Target Earth: it will happen. *Sky and Telescope* 79:261.

Ronan, C., and M. Mohs. 1986. Scientist of the year: Edmund Halley. *Discover* 7(1):52.

Sagan, Carl, and Ann Druyan. 1985. *Comet*. New York: Random House.

Tatum, J. B. 1982. Halley's comet in 1986. *Mercury* 11(4):126.

Van Allen, J. A. 1975. Interplanetary particles and fields. *Scientific American* 233(3):160.

Wagner, Jeffrey K. 1984. The sources of meteorites. *Astronomy* 12(2):6.

Whipple, Fred L. 1986. Flying sandbanks or dirty snowballs: Discovering the nature of comets. *Mercury* 15(1):2.

————. 1974. The nature of comets. *Scientific American* 230(2):49.

MATHEMATICAL METHODS

Orbital Resonance and Kirkwood's Gaps

 We have referred to a phenomenon known as orbital resonance in this chapter as well as in the chapters in which the ring systems of the giant planets are discussed. As explained in the text, orbital resonance refers to a situation in which two orbiting bodies have periods that are simple multiples of one another. When this happens, the two bodies are frequently lined up on the same side of the parent body, and this creates extra gravitational forces between them that can nudge the smaller body out of its orbit. This phenomenon has been cited in explaining some of the gaps in the ring systems of the giant planets, and in explaining the voids in the asteroid belt known as Kirkwood's gaps.

To calculate the positions of gaps created by orbital resonance is really quite simple; we just use Kepler's third law. In most cases we can use the law in its simplified form

$$P^2 = a^3,$$

where P is the orbital period in years and a is the semimajor axis in AU (we would have to use the more complex form involving the sum of the masses in cases where either of the orbiting bodies had a mass comparable to that of the parent body).

Suppose we want to calculate the distance from the Sun where there should be a gap in the asteroid belt corresponding to asteroids which would have a period equal to half the period of Jupiter. Jupiter's orbital period is 11.86 years, so the problem is to find the semimajor axis of an orbit having half this period, or 5.93 years. We solve Kepler's third law for a, finding

$$a = P^{2/3},$$

where the exponent ⅔ means "P raised to the two-thirds power" or "the cube root of P squared." To compute this, you first square P and then find the cube root (you will need a calculator equipped to compute exponential values such as this, or else you will have to use logarithms). If we substitute for P the value 5.93 years, we find $a = 3.28$ AU. This is the position of one of Kirkwood's gaps in the asteroid belt; it is about halfway between Mars and Jupiter.

Other gap locations are calculated in similar fashion. For example, if we seek the gap corresponding to periods of one-third that of Jupiter, we set $P = 11.86/3 = 3.95$ years and solve for a, finding $a = 2.50$ AU. In the same manner, we find that there are gaps at distances of $a = 2.06$ AU (corresponding to a period equal to one-fourth that of Jupiter), 1.78 AU (for periods equal to one-fifth that of Jupiter), and so on.

Other fractions of the period of Jupiter also create orbital resonances. For example, an asteroid with two-thirds the period of Jupiter would also find itself frequently aligned between Jupiter and Sun, and would also be tugged out of its orbit eventually. Two-thirds of 11.86 years is 7.91 years; substituting this into Kepler's third law yields $a = 3.97$ AU. A period equal to three-fourths the period of Jupiter, or 8.90 years, corresponds to a semimajor axis of 4.29 AU. All of these computed distances, as well as several others we have not calculated here (see Fig. 15.4), coincide with the positions of observed gaps in the asteroid belt (although those beyond 3.5 AU are difficult to notice, since there are not many asteroids this far out).

Very similar calculations can be made for any other system; for example, we can find the positions of orbital resonances in the rings of a giant planet such as Saturn. We note that the fractional values are always the same; that is, when the period is half that of the major body, the semimajor axis for the resonance will always be $(½)^{2/3} = 0.63$ times the semimajor axis of the major body. Similarly, the $P = ⅓$ resonance will always occur at a distance of $(⅓)^{2/3} = 0.48$ that of the major body; and so on. If we consider Mimas, the innermost major moon of Saturn, we calculate that the $P = ½$ resonance should lie at a distance from Saturn of 0.63 times the semimajor axis of Mimas, which is 1.855×10^5 km. Hence we expect to find a gap in Saturn's rings at a distance from the center of the planet of 0.63 $(1.855 \times 10^5$ km), which equals 1.94 times the equatorial radius of Saturn. This is just the location of the Cassini division (the outer edge of the B ring), as listed in Table 13.8.

CHAPTER 16

Formation of the Solar System: Disks, Rings, and Moons

An artist's view of the dust around Vega. (NASA)

We have collected a considerable quantity of information about the solar system and its contents. Along the way, we have uncovered various systematic trends, such as the similarities in internal structure and composition of the terrestrial planets and the comparisons among the outer planets, and in Chapter 8 we briefly outlined the modern picture of solar system formation. In discussing the histories of the individual planets, we have referred to some of the important processes thought to have been involved, such as the condensation of the first solid material from a cloud of gas surrounding the young Sun, but it remains for us to put together a coherent story of the formation and evolution of the entire system.

We have so far said little about the Sun, the central object in the solar system. For the purposes of this chapter, we need to know something about the Sun, although most of the discussion of it is deferred until the next chapter, which opens the section of this text on stars and their properties. For our present discussion, we need to keep in mind that the Sun is by far the most massive body in the solar system, having a mass over 300,000 times that of the Earth. It is composed primarily of hydrogen (about 70 percent, by mass) and helium about 27 percent), with the other elements in trace amounts, carbon, nitrogen, and oxygen being the most abundant. (Table 13.2 shows the composition of the Sun in detail). The Sun is gaseous throughout, with a surface temperature of just under 6,000 K and a central temperature of roughly 10 million degrees. It has hotter layers above the surface; of particular interest for this chapter is the **corona**, an extended zone as hot as 1 to 2 million degrees, and the **solar wind**, a stream of charged particles (electrons and protons) flowing outward from the Sun throughout the solar system.

Before we describe the modern theory of solar system formation, we will review the overall properties of the system that must be accounted for, and then we will have a look at the historical developments and alternative origin theories that scientists have considered over the years.

A SUMMARY OF THE EVIDENCE

Several facts must be explained by any successful theory regarding the formation of the solar system. In a way, this is much like a mystery story, in which the detective (the scientist seeking the correct explanation) has certain clues that reveal isolated parts of the story, from which past events must be reconstructed.

One category of clues in our mystery has to do with the orbital and spin motions of the planets and satellites in the solar system (Figs. 16.1 and 16.2, Table 16.1). First and most obvious, the planetary orbits lie in a common plane, all are nearly circular, and all go around the Sun in the same direction. Pluto violates these rules to an extent, because its orbit is both elongated and tilted (by 17°) with respect to the ecliptic, but even this misfit goes around the Sun in the same direction as the other planets. The spins of nearly all the planets are in this same direction, as are the orbital and spin motions of the vast majority of the satellites in the solar system. The direction of these motions, counterclockwise when viewed from above the North Pole, is said to be **direct,** or **prograde,** motion.

Another thing we must understand in a successful theory is why the planets formed where they did, and not elsewhere. A final clue having to do with the motions of the planets is that nearly all of them have small **obliquities,** meaning that their equatorial planes are nearly aligned with their orbital planes. Even the Sun has its axis of rotation nearly perpendicular to the plane of the planetary orbits. Uranus, with its 98° tilt, does not fit the pattern, nor do Venus and Pluto.

Table 16.1 Properties of planetary orbits

Progression of planetary distances from the Sun.
Planetary and satellite orbits lie in a common plane.
Nearly all orbital and spin motions are in the same direction.
Spin axes of most planets and satellites are nearly perpendicular to the ecliptic.
Planetary and satellite orbits are nearly circular.

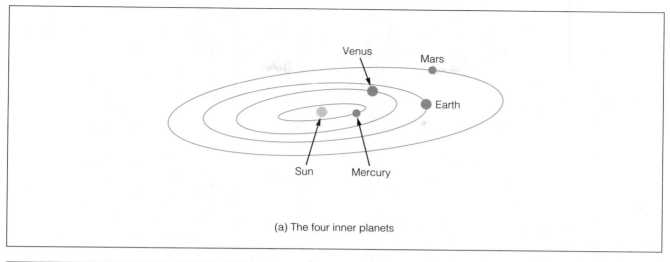

(a) The four inner planets

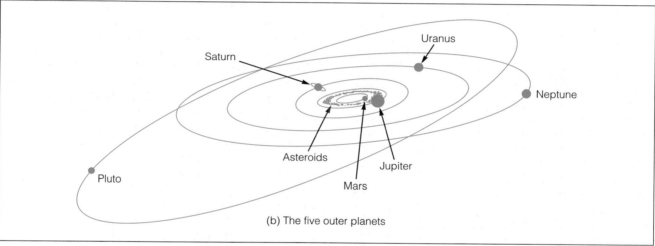

(b) The five outer planets

Figure 16.1 The coalignment of planetary orbits. *All but one of the nine planets have nearly circular orbits in a common parallel plane. The exception is Pluto, whose orbit is tilted 17° with respect to the ecliptic, and is sufficiently eccentric (elongated) that Pluto is actually closer to the Sun than Neptune at times.*

A second general type of clue in the mystery is concerned with the natures of the individual planets, and the systematic trends from planet to planet that we have discussed (Table 16.2). First, we must consider the planetary compositions, seeking an explanation for why the inner planets—the terrestrial ones—have high densities, indicating that they are made mostly of rock and metallic elements; whereas the outer planets have low densities and consist primarily of lightweight gases such as hydrogen and helium, as well as ices composed of these and similarly volatile gases. Within each of the two groups of planets, there are relatively minor differences in chemical composition, such as the var-

ious isotope ratios we have discussed, which must also be accounted for.

The Sun rotates in a time of about 25 days, which posed a major mystery for a long time. According to many early theories, the Sun should be spinning much faster. The difficulty arises in theories that suggest that the solar system formed from the collapse of a cloud of gas and dust; these theories, as we shall soon see, are the most successful ones. The laws of physics tell us that the angular momentum of an object (see Chapter 5) must remain constant. This means that, if a spinning object shrinks in size, it must spin faster to compensate, thereby maintaining constant angular mo-

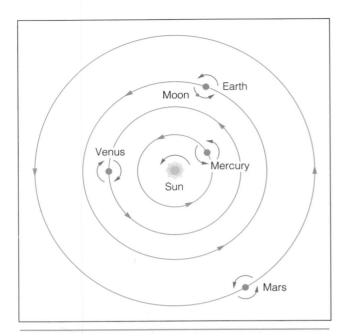

Figure 16.2 Prograde motions in the inner Solar System. *The rotations and orbital motions of the planets are generally in the same direction. This sketch of the orbits of the inner four planets shows that Venus has retrograde rotation but that the other three planets and their satellites obey the rule. The outer planets (except for Pluto) and nearly all their satellites do also.*

mentum. Thus the Sun, thought to have formed at the center of a collapsing cloud, should have a very rapid rotation rate, rather than the leisurely 28-day period it has. As we will see, the explanation for this was a long time in coming, and for a while astronomers were sidetracked in their effort to understand the formation of the solar system.

Table 16.2 Physical properties of the planets

There are two major types of planets: terrestrial and gaseous giant.

Giant planets are much larger and more massive than the terrestrial planets but have lower densities.

Terrestrial planets have low abundances of volatile elements; giant planets have plentiful volatiles.

Giant planets have more satellites than terrestrial planets.

Giant planets tend to have ring systems; terrestrial planets do not.

Giant planets rotate more rapidly than do terrestrial planets.

A final category of information that bears on the origin of the solar system is the distribution and nature of the various kinds of interplanetary objects that were described in Chapter 15. We learned there that asteroids, comets, meteoids, and interplanetary dust inhabit the space between the planets, and we found that the motions and compositions of these objects reflect conditions very early in the history of the solar system. Somehow this information must fit into our overall picture.

Let us now take a look at some of the ideas that have been developed historically, in the face of all this evidence.

CATASTROPHE OR EVOLUTION?

Since the time of Copernicus, many scientists and philosophers have tackled the problem of explaining the origin of the solar system. While there have been many different hypotheses, most of them have fallen into one or the other of two general types: those invoking some singular event, which are called **catastrophic theories;** and those calling for the formation of the solar system through gradual, natural processes, in what are called **evolutionary theories.**

The first speculations and hypotheses involved evolutionary theories in which it was suggested that the Sun and the planets formed from a gigantic whirlpool or vortex that flattened into a disk from which the planets condensed. First the French philosopher Rene Descartes (in 1644) and later the German Immanuel Kant (in 1755) developed this picture (Fig. 16.3).

Ironically, even though the evolutionary model was then pushed aside for some time by catastrophic theories, it was the evolutionary hypothesis that eventually won out. The chief reason it fell into difficulty for some time was the realization that if the Sun formed at the center of a collapsing, spinning cloud of material, conservation of angular momentum (discussed in Chapter 5) should have given it a much faster spin rate than it has. The catastrophic scenarios that were suggested (mostly during the nineteenth and early twentieth centuries) all invoked a near-collision between the Sun and some other body, with the result that gaseous material was dragged out of the Sun and then left to condense into planets that continued to orbit it.

Soon several serious objections were raised. For example, gas from the Sun's interior should have too much internal energy to condense into solid material, and furthermore such material should be completely

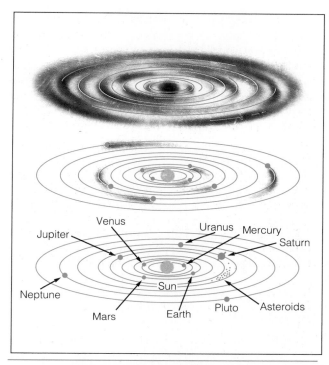

Figure 16.3 The hypothesis of Kant and Laplace.
Descartes's simple vision of a vortex was refined by Kant, who realized that rotation should cause a collapsing cloud to take a disklike shape; and by Laplace, who hypothesized that a rotating disk would form detached rings that could then condense into planets.

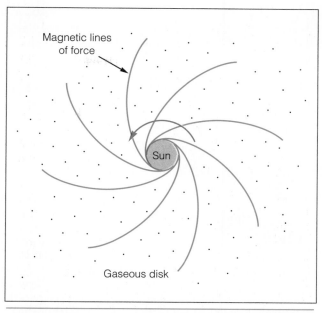

Figure 16.4 Magnetic braking. *The young, rapidly spinning Sun has the magnetic field structure shown, the field lines rotating with the Sun. The early solar system was permeated with large quantities of debris and gas, some of which was ionized and therefore subject to electromagnetic forces. The Sun's magnetic field exerted a force on the surrounding gas, which in turn created drag that slowed the Sun's rotation.*

devoid of the hydrogen isotope **deuterium,** yet deuterium is found in the planets. Deuterium is a form of hydrogen having a neutron and a proton in the nucleus; it is much less abundant than normal hydrogen (with no neutrons and only a single proton), but deuterium is present on the Earth and in meteoritic material. Deuterium undergoes nuclear reactions very easily, and because of this it cannot survive at temperatures typical of the Sun's interior. Therefore, any gas pulled out of the Sun should have no deuterium in it. Because the planets contain deuterium, it has been argued that planetary material could not have formed from gas released from the solar interior.

The slow spin of the Sun remained a mystery until the 1960s, when the solar wind was discovered. This stream of charged particles from the Sun consists of mostly electrons and protons (see discussion in Chapter 17). The Sun's magnetic field exerts forces on the charged particles in its vicinity, creating a form of

drag that acts to slow the Sun's rotation (Fig. 16.4). As the Sun spins, its magnetic field spins with it, but the field is pulled behind by its interaction with charged particles. The result can be likened to trying to spin a pinwheel under water; drag caused the pinwheel to turn more and more slowly. The drag caused by the action of the solar magnetic field on solar wind particles is called **magnetic braking,** and the process can easily explain why the Sun today spins with a leisurely 28-day period.

Given the difficulties with the catastrophic theories for the formation of the solar system and the success of the evolutionary models in accounting for the observed nature of the Sun and planets, only the evolutionary theory is seriously considered today. It is interesting to note, however, that the modern picture of solar system formation and evolution includes several very important events that could only be called catastrophes.

ASTRONOMICAL INSIGHT 16.1

The Explosive History of Solar System Elements

 Even according to the evolutionary theory that is now widely accepted, the development of the solar system required explosive events. The evolution of the universe itself may be viewed as the result of a series of explosions, including the initial one from which the universe has been expanding ever since. A variety of astronomical evidence (discussed thoroughly in Chapter 31) shows that the universe began some 10 to 15 billion years ago in a single point, and, at some time in the past, began to expand outward. The initial temperature and density were so extreme that the birth of the universe must have been an explosive event, commonly called the big bang.

For a brief time during the early stages of the universal expansion, conditions were suitable for nuclear fusion reactions to occur, and the primordial substance was converted from a soup of subatomic particles to one of recognizable elements, primarily hydrogen and helium, with almost no trace of heavier species. Eventually, as the universe continued to expand, galaxies and individual stars formed. Just as nuclear reactions in the Sun form a heavier element (helium) from a light one

(hydrogen), similar reactions take place inside all stars, gradually enriching the universal content of heavy elements. Stars more massive than the Sun undergo more reaction stages, going from the production of helium to that of carbon, and from carbon to other elements. When these stars die, often explosively, their newly formed heavy elements are returned to interstellar space and are available for inclusion in new stars that form later. Thus, during the 5 to 10 billion years that passed between the time of the big bang to the time of the solar system formation, the chemical makeup of our galaxy gradually changed, so that roughly 2 percent of the total mass was in the form of elements heavier than helium.

Since this roughly represents the composition of the Sun and the primordial material in the solar system, it might seem that there is no more to the story of the origin of the elements. There are compelling reasons to believe, however, that the formation of the solar system was influenced by at least two local explosive events. Certain atomic isotopes that are present in solar system material are the result of explosive nuclear reactions that must have occurred relatively recently, compared to the time that had passed since the big bang. These isotopes, one a form of aluminum (with 26 instead of the normal 27 neutrons and protons) and the other a form of plutonium (with 244 protons and neutrons instead of 242) are radioactive, with half-lives short

enough that they could not have been present when the solar system formed unless they had been created relatively recently in explosive events. The plutonium isotope in question probably formed in a stellar explosion called a supernova that occurred a few hundred million years before the solar system formed, whereas the aluminum must have been created in a nearby supernova only a million years or so before the solar system was born.

The second of these inferred explosions may have played a crucial role in triggering the formation of our Sun and planetary system. We have said that the solar system formed from an interstellar cloud that collapsed. What we have not discussed, however, is what caused this collapse to occur. Normally, an interstellar cloud is quite stable as an interstellar cloud, and it will not fall in on itself unless something happens to compress it. Various mechanisms can bring this about, and one of them is a shock wave from an explosive event such as a supernova. Thus, in a singular event sometimes called the "bing bang," our solar system may have been born as the direct consequence of the death of a nearby star. We derive from this event not only some minor anomalies in the abundance of nuclear isotopes, but also the very existence of the solar system. It is interesting to note that, even in an evolutionary origin as discussed in this chapter, the solar system has been strongly influenced by what can only be described as catastrophic events.

A MODERN SCENARIO

Table 16.3 summarizes the current theory of solar system formation. The first step was the collapse of an interstellar cloud, one that must have been rotating before it began to fall in on itself. Our galaxy has a large quantity of interstellar material in it that contains both gas and small dust grains (about the size of the interplanetary grains described in Chapter 15). The interstellar medium is not uniformly distributed throughout space, but instead is concentrated here and there in large amorphous regions called interstellar clouds, some of them sufficiently dense to block out starlight, appearing as dark regions in photographs of the sky (Fig. 16.5). Even in these areas of relatively high concentration, the material is so rarefied that a quantity of mass equal to that of the Sun is spread over a volume up to 10 or more light-years across.

The composition of the interstellar material is apparently quite uniform, consisting of about the same mixture of elements as the Sun. Hence the Sun and other stars are born with a standard composition of about 73 percent hydrogen (by mass), about 25 percent helium, and just traces of all the other elements. The planets must also have formed out of material with this composition, yet, as we have seen, their present makeup is quite different from this, especially in the case of the terrestrial planets.

Somehow the extended, tenuous gas in interstellar space had to become concentrated in a very small volume on its way to becoming a star. Gravitational forces were responsible for pulling the cloud together, but there must have been an initial push or a chance condensation to get the process started. (One possibility is that a shock wave from a violent event such as a supernova explosion caused the presolar cloud to begin its collapse; see Chapter 25 for details). Once the cloud was falling in on itself, the rest of the process leading to the formation of the Sun was inevitable, dictated by physical laws.

A combination of theory and observation of the formation of other stars tells us how the collapse proceeded. The innermost portion of the cloud fell in on itself very quickly, leaving much of the outer material still suspended about the center. The rotation of the cloud sped up as its size diminished, and if the cloud had magnetic field to begin with (as most do), the field was intensified in the central part as a result of the condensation.

The core of the cloud began to heat up from the energy of impact as the material fell in. Eventually it began to glow, first at infrared wavelengths, and finally, after a prolonged period of gradual shrinking, at visible wavelengths. Nuclear reactions began in the center when the temperature and pressure were sufficiently high, and the Sun began its long lifetime as a star, powered by these reactions.

The steps in the formation of the Sun described up to this point are fairly well understood and have been observed to be taking place today in many cloudy regions of the galaxy. In a number of stellar nurseries, infrared sources are found embedded inside dark interstellar clouds, indicating that newly formed stars are hidden there, still heating up. The details of planet formation are sketchier, however, because it is difficult to observe the process as it occurs. All that we know about it has had to be inferred from observations of the end product: the present-day solar system.

As the central part of the interstellar cloud collapsed to form the Sun, the outer portions were forced into a disk shape by rotation, just as Kant had argued. At this stage there was an embryonic Sun surrounded by

Table 16.3 Timetable for solar system formation

Approximate Time	Event
0	Collapse of interstellar cloud; disk formation
400,000 years	Central condensation becomes hot enough for pressure to balance gravity; matter still falling in
1 million years	Second collapse stage is followed by new nearly stable phase, with heating by slow gravitational contraction; central object is now called the protosun; infalling gas and dust shrouds protosun from exterior view
1–10 million years	Planetesimals form and accrete to form planets
1–100 million years	T Tauri wind
100 million years	Start of nuclear reactions in core
0.1–1 billion years	Magnetic braking slows Sun's rotation; major cratering occurs throughout solar system

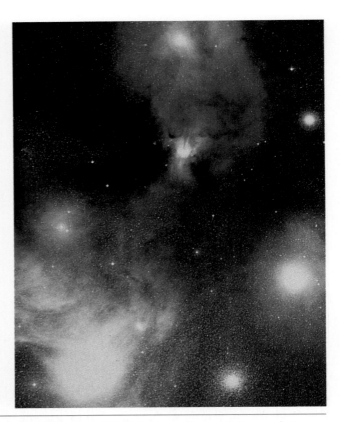

Figure 16.5 Regions of star formation.   *At left is the Orion nebula, in the "sword" of the constellation Orion, and at right is the region near the M supergiant Antares (bright object at right center). In both regions, newly formed stars are known from infrared observations to lie within the dark, dusty clouds seen here. (left: © 1979 ROE/AAT Board)*

a flattened, rotating cloud called the **solar nebula** (Fig. 16.6). The inner portions of the nebula were hot, but the outer regions were quite cold.

Throughout the solar nebula, the first solid particles began to form, probably by the growth of the interstellar grains that were mixed in with the gas. For every element or compound, there is a combination of temperature and pressure at which it "freezes out" of the gaseous form, in direct analogy with the formation of frost on a cold night on Earth. The **volatile elements** require a very low temperature in order to condense, so these materials tended to stay in gaseous form in the inner portions of the solar nebula but condensed to form ices in the outer portions. The elements that condense easily, even at high temperatures, are called **refractory elements,** and these were the ones that formed the first solid material in the warm inner portions of the solar nebula. This material therefore consisted of rocky debris containing only low abundances of the volatile species.

In due course, rather substantial objects built up that resembled asteroids in size and composition, and these are referred to as **planetesimals.** Mathematical analysis of the process by which particles orbiting the young Sun collided and built up their sizes shows that the growth of large planetesimals happened rather quickly (taking place in only a few thousand years), and that most of the material in the disk would have gone into large planetesimals. Thus a stage was reached in which collisions between large bodies were likely. Because the planetesimals were all orbiting the Sun in a disk, their relative speeds would have been rather small, so that when collisions occurred, the colliding bodies could stick together. Even at low speeds (perhaps 10 km/sec, typically), enough energy would be released in a collision between large planetesimals to partially vaporize or melt both objects. This would help them merge into a single spherical body and would also aid the process of differentiation, which requires that a body be molten. In most cases, the planet that was formed in this

Figure 16.6 Steps in the Sun's formation.   *An interstellar cloud, initially very extended and rotating very slowly, collapses under its own gravitation. This happens most quickly at the center. As the collapse occurs, the internal temperature rises and the rotation rate increases. Eventually the central condensation becomes hot and dense enough to be a star, with nuclear reactions in its core.*

way rotated in the same direction as the overall rotation of the disk, but it is possible that an off-center collision could alter the direction of spin. This may have happened during the formation of Venus and Uranus, accounting for the slow retrograde rotation of Venus and the large tilt of Uranus. As we learned in Chapter 10, many scientists now think that the Earth's Moon formed as the result of a grazing collision between the young Earth and a Mars-sized planetesimal. There have even been recent suggestions that Mercury was involved in such a collision during its formation (see Astronomical Insight 16.2).

The scenario just described apparently applies to the terrestrial planets, which therefore seem to have formed in two stages: (1) the condensation of refractory elements, leading to the development of planetesimals; and (2) the accretion of the planetesimals to form planets. The low quantity of volatile elements that characterizes the terrestrial planets was already established when the planets formed, and then was exaggerated by the release of volatile gases that occurred during their early histories, when they underwent molten periods.

There is more uncertainty about the sequence of

events that led to the formation of the outer planets. These planets contain a much higher proportion of volatile gases, which would naturally be expected because of the lower temperatures in the outer portions of the solar nebula. Thus, planetesimals that formed there would have contained higher relative abundances of volatiles and therefore would have been partly rock and partly ice, much like the large satellites of the giant planets, and similar to Pluto.

The most recent theory of the formation of the giant planets suggests that their cores formed from the coalescence of planetesimals, just as the terrestrial planets themselves formed. In the outer solar system, however, where the temperatures were much lower than in the inner portions of the system, the solid objects that formed from the coalescence of planetesimals were able to gravitationally trap gas from large volumes in their vicinities, and these planets therefore accreted extended, gaseous atmospheres, including the very lightweight and volatile gases hydrogen and helium. Tidal forces from the Sun played an important role in this scenario; for the inner planets, these forces helped prevent the accretion of gases from the surrounding

ASTRONOMICAL INSIGHT 16.2

Catastrophe Revisited: Violent Origins of the Planets

In the text we reviewed some early theories in which it was hypothesized that the planets formed from material pulled out of the Sun by some catastrophic event such as a near-collision with another star. These theories are now considered untenable, largely because it has been demonstrated convincingly that solar matter could not and did not condense to form the planets. Furthermore, the need to invoke such theories was reduced when the solar wind was discovered and it was realized that the Sun's slow spin could be explained by magnetic braking.

With this discovery, the so-called evolutionary theory of solar system formation became widely accepted, because it appeared to explain most of the properties of the system quite well. There were some nagging problems, however, and today scientists are modifying the evolutionary theory by adding some catastrophes back in.

The most noticeable departures from the expectations of the evolutionary theory are the unusual spins of Venus, Uranus, and Pluto. Venus spins very slowly in the backward, or retrograde, direction, while Uranus is tilted so far over that its axis lies almost in the plane of its orbit around the Sun. Both planets are quite anomalous because in the evolutionary theory it is expected that each planet's axis would be nearly perpendicular to its orbital plane, and that the spin of each planet would be in the prograde, or forward, direction, the same direction as its orbital motion about the Sun. Thus the strange proper-

ties of both Venus and Uranus have long been attributed to unusual events that might have occurred during their formation, such as a collision with a large planetesimal.

Now similar collisions are being invoked to explain much more subtle anomalies in other planets. The Moon is quite unusual among inner solar system bodies, being the only major satellite. Furthermore, it has many properties that have proven difficult to explain: It carries far more angular momentum that can be understood on the basis of standard models for the formation of the Earth, and it has chemical and isotopic anomalies that distinguish its material from that of the Earth's outer layers. Hence, theories hypothesizing that the Moon formed by splitting off of the Earth, or that the two bodies formed together, are faced with fatal difficulties. The large angular momentum of the Moon could be explained if the Moon were some-

---

nebula, but in the outer solar system, farther from the Sun, the tidal forces were weaker and did not prevent the trapping of gases by the giant planets.

It is thought that the outer planets were able to gain mass very rapidly due to a runaway effect created by their own gravitation. Once a massive body began to grow by accretion of planetesimals, its gravitational field would tend to pull in additional planetesimals, which would merge, adding mass to the original planetary body, further enhancing its gravitational attraction for new material. Thus, once a major planet began to form, it would have grown very rapidly, quickly providing it enough mass to disrupt the remaining solar nebula. In other words, Jupiter and Saturn quickly reached sufficient mass to disrupt the asteroid belt and to eject cometary bodies to the Oort cloud. Current models suggest that Uranus and Neptune formed some-

what later (perhaps 100 million years) than Jupiter and Saturn. Because the two giant planets had already trapped much of the volatile gas in their vicinity, less was available for Uranus and Neptune, which consequently ended up with smaller masses and higher proportions of heavy elements.

This viewpoint is prompted in part by the extensive ring and satellite systems of the outer planets, which resemble miniature solar nebulae. The suggestion is that these planets, in accreting gas from their surroundings, formed disks much like the solar nebula itself. Gravitational and rotational effects would have caused the material falling in to form a disk in the equatorial plane of each planet, and then instabilities in the disk would have caused planetesimals to form, leading to the coalescence of satellites (Fig. 16.7). Ring systems formed wherever the disks extended closer to the par-

ASTRONOMICAL INSIGHT 16.2

how captured by the Earth, but this theory has other problems in that such a capture is either difficult or impossible. (Models for the formation of the Moon are reviewed in Chapter 10).

The most recent suggestion for the formation of the Moon calls for a catastrophic origin. It is now thought that the Moon may be the remnant of a very large planetesimal, perhaps the size of Mars or larger, that underwent a grazing collision with the young Earth, vaporizing its own mantle and part of the Earth's outer layers. The debris from the collision would have formed a disk around the Earth, which subsequently would have condensed to form the Moon. This theory is still quite new, and therefore subject to possible revision or rejection, but so far it appears to explain the Moon's formation better than any of the others that have been proposed.

The planet Mercury also presents anomalies that have been difficult to explain. Foremost among them is the high density of Mercury and the existence of a detectable magnetic field, which have led to the conclusion that Mercury's nickel-iron core may be much larger (relative to its overall diameter) than for any other terrestrial planet (see Fig. 10.20, where the model of Mercury's interior is shown). While it is true that Mercury formed closer to the center of the solar system than the other planets, so that it should have had a lower abundance of the lightweight volatile elements to begin with, it has proven difficult to understand how Mercury could be so deficient in these elements. Now a new theory suggests that a major impact that occurred just as Mercury was forming could explain the anomaly. This impact, involving the young Mercury and a large planetesimal, would have vaporized much of the material of Mercury's outer layers, leaving behind a very dense planet with an anomalously large core and thin outer layers.

There is little dispute over the fact that the early solar system underwent a period when impacts were important. The scarred surface of the Moon and other satellites alone is sufficient to establish that there must have been a time when bombardment from space was far more common than it is today. The evolutionary theory of solar system formation has always included an early period when planetesimals formed in the disk of solid material that orbited the young Sun; what is new today is the realization that during the age of planetesimals there must have been some very catastrophic collisions between large bodies. These catastrophes have molded the planets into the form they have today.

---

ent planet than the Roche limit, where tidal forces prevented the formation of large satellites. The rapid rotation rates of the giant planets are explained in this view by the fact that angular momentum conservation would force them to spin more rapidly as the disks contracted (and by the fact that no magnetic braking process would have acted, in contrast with the early Sun).

It is interesting that Uranus, despite its large (98°) tilt, has its satellites and rings lying in its equatorial plane. This indicates that this planet accumulated its disk of icy debris after the event occurred that caused its highly tilted orientation, and this is consistent with the picture of outer planet formation just described. In this view, material trapped by Uranus would be forced into a disk in the equatorial plane of the planet, regardless of the orientation of that plane. Hence the tilt of Uranus must have been established before it trapped

the material that was to form its satellites and rings.

The asteroids probably formed as planetesimals, similar to those that eventually created the terrestrial planets, but they were prevented from coalescing into a planet by the gravitational effects of Jupiter, as we noted in Chapter 15. Jupiter's gravity also would have altered the orbits of many of these planetesimals, causing them to be ejected from the solar system. Hence the total mass of the asteroids is small. The comets probably formed farther out, though not at the distance of the Oort cloud, where they now reside. At such a great distance (up to 30,000 AU) from the center of the solar nebula, it is doubtful that any condensation could have occurred, because the density would have been too low. Therefore the comets are believed to have condensed at intermediate distances, probably near the orbits of Uranus and Neptune, and then were forced

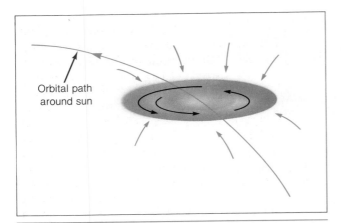

Figure 16.7 Formation of a giant outer planet. *Here a rotating disk has formed around a condensation in the outer solar system. Lumps in the disk grow to become satellites, except in the innermost portion where tidal forces prevent this, and a ring system forms instead. The entire system of planet, rings, and moons orbits the Sun.*

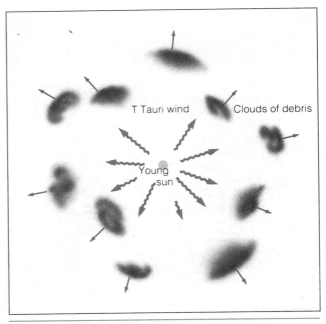

Figure 16.8 The T Tauri phase. *A star like the Sun is thought to go through a phase very early in its lifetime during which it violently ejects material in a high-velocity wind. This wind sweeps away matter left over from the formation process.*

out to their present great distances by the gravitational effects of the major planets. Both Jupiter and Saturn, lying closer to the Sun, would have forced the cometary bodies farther out by speeding them up a little each time they passed by. Eventually, most of the comets retreated to a great distance, forming the Oort cloud. The Kuiper belt, a disk of cometary bodies lying between 30 and 50 AU, probably represents leftover planetesimals that were not ejected to greater distances, but which have remained in the region where they formed. The Kuiper belt is thought to be the source of periodic comets, which have orbits closely aligned with the ecliptic plane.

Once the planets and satellites were formed, the solar system was nearly in its present state, except that the solar nebula had not totally dissipated. A lot of gas and dust were still swirling around the Sun, along with numerous planetesimals that had not yet accreted onto planets. The Sun's magnetic field, pulling at ionized gas in the nebula, was very effective during this time in slowing the solar rotation through the magnetic braking process that we have already described. Meanwhile, the remaining planetesimals were orbiting about, now and then crashing into the planets and their satellites. Most of the cratering in the solar system occurred during this period, in the first billion years.

The leftover gas in the solar nebula was dispersed rather early, when the infant Sun underwent a period of violent activity, developing a strong wind that swept the gas and tiny dust particles out into space (Fig. 16.8 and the Guest Editorial following Chapter 23). This left behind only the larger solid bodies, the newly formed planets and the planetesimals. This wind phase has been observed in other newly formed objects called **T Tauri** stars and is apparently a natural stage in the development of a star like the Sun. Because of rotational forces, the wind escaped primarily in directions perpendicular to the equatorial plane of the young Sun creating a bipolar flow similar to those seen today coming from newly-formed stars.

Once the majority of the remaining interplanetary debris was eliminated, either by the T Tauri wind or by accretion onto planetary bodies, the solar system looked much like it does today. Each planet went its own way, either continuing to evolve geologically, as most of the terrestrial planets have, or remaining perpetually frozen in nearly its original condition, as the outer planets apparently have.

MINIATURE SOLAR SYSTEMS: RINGS AND MOONS

We have seen that the formation of the giant planets began in a fashion similar to formation of the terrestrial bodies, but that conditions in the outer solar system caused some very significant differences as well. Lightweight elements were able to condense in the outer solar system, and gravitational trapping allowed the giant planets to gain enormous quantities of mass, becoming the gaseous, cold worlds we see today. Along the way, the giant planets acquired disks of material that produced rings and satellites. In this section, we discuss the formation and evolution of these disks and the rings and moons in further detail.

Satellites of the Giant Planets

The satellites of Jupiter, Saturn, Uranus, and Neptune seem to fall into two general classes: those that follow nearly circular orbits lying in the equatorial plane of the parent planet; and those that have orbital peculiarities such as high tilts to the equatorial plane or retrograde directions. All are composed largely of ice, although many have densities greater than 1 gram/cm³, indicating that rocky material is mixed in to varying degrees. Pluto may be more like one of the giant satellites than a proper planet; note, for example, the many similarities between Pluto and Triton.

We have seen that each giant planet is thought to have been surrounded by a disk of residual material when it formed. This disk consisted of small bodies that could coalesce to form satellites. Thus the formation of satellites is not difficult to understand, given the prior formation of the disks. The satellites formed this way would have circular orbits lying in the equatorial plane of the parent planet, placing them in the first of the two categories discussed above.

The satellites having irregular orbits may have a different origin. These may be planetesimals that have been captured gravitationally; such captures would lead to orbits that did not necessarily conform to the normal pattern. It was pointed out earlier (Chapter 10) that capture of a free body is not easy and in fact cannot happen unless there is some way for the incoming body to lose some of its kinetic energy. Energy can be lost through a near-collision with a third body, or through friction if the incoming body encounters an atmosphere or a medium of fine particles, such as an equatorial disk. Most of the satellites that have irregular orbits are small moons such as Phoebe of Saturn or Nereid

of Neptune, but on the other hand, at least one giant satellite also falls into this category. Triton, a very large satellite having a thin atmosphere and an assortment of geological scars on its surface, follows a retrograde orbit around Neptune and may have been captured.

If Triton is indeed a captured satellite, it may resemble the planetesimals that had built up in the outer solar system at late stages in the formation of the planets. Hence this giant moon may represent a sample of the intermediate stages in the creation of planets such as Jupiter, Saturn, Uranus, and Neptune. There are some other examples of this stage. In 1977, a Sun-orbiting body in the outer solar system was discovered, and for a while there was speculation that it might be a new planet. Further analysis showed that it was not large enough to be a planet, and this body, named Chiron, came to be thought of as an asteroidlike or cometary body much farther from the Sun than the main asteroid belt. There may be other such bodies in the outer solar system, yet to be discovered. Further, as already noted, Pluto is far more similar to one of the giant satellites (such as Triton in particular) than to any of the planets, and may also represent a planetesimal, a leftover from the era of planet formation.

The larger satellites are spherical, indicating that gravitational compression, rather than internal rigidity, dominates their structure. Their inner conditions are difficult to analyze, but probably most have a central core region where most of the rocky material resides, and outer mantles and crusts consisting of mixtures of ice and rock (Fig. 16.9).

Geological activity has proceeded in varying degrees, for reasons that are not always well understood. Tidal stresses can account for many of the contrasts from one satellite to another, as we have seen in the case of the four Galilean moons of Jupiter. In that case, we saw that the amount of heating and hence of geological activity increases from the outermost, Callisto, to the innermost, Io. Triton is active geologically, with evidence of slushy outflows on its surface as well as having active geyser eruptions in which plumes of nitrogen or methane gas are released. Enceladus, the very bright satellite of Saturn, may also be undergoing volcanic activity in which water is released, recoating the surface with ice.

Water, particularly in the form of ice, plays a major role in the geology of the outer satellites. The degree to which a moon undergoes internal convection, hence the movement of crustal plates at its surface, depends on whether or not the ice in the interior is heated enough to melt and flow. Among the Galilean satellites

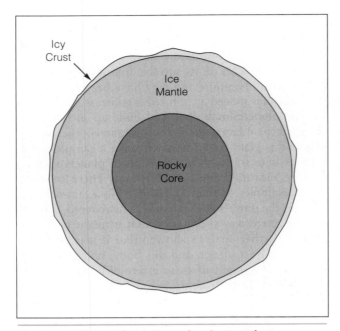

Figure 16.9 Internal structure of a giant satellite. *Substantial differentiation has occurred, so that the heavy elements have congregated to form a rocky core. The lightweight elements, in the form of ice, constitute the mantle and crust (if there is significant internal heating, because of tidal stress, for example, the mantle may be partially liquid water). In a typical case, the mass of the satellite is about half rock and half ice.*

Figure 16.10 Features on Ganymede. *This close-up of the surface shows a complex pattern of grooves and ridges, which were probably created by deformations of an icy crust. Some impact craters are also visible. (NASA)*

of Jupiter, this apparently happened inside Ganymede, which has old surface fractures due to plate motion (Fig. 16.10), but did not occur in Callisto, where it appears that no crustal shifting has ever taken place. The inner moons, Europa and Io, probably never had as much water in the first place (although Europa has a thin water or ice mantle), since the inner part of the disk that surrounded Jupiter when it formed is thought to have been too hot for water to condense. These two satellites are heated enough due to tidal stresses that their interiors are sufficiently elastic to undergo convective flows, even in the absence of large quantities of ice.

Other satellites undergoing ice-related geological activity include Enceladus of Saturn, as already mentioned, Triton of Neptune, possibly Titan of Saturn, and perhaps Pluto (which, as we have noted, closely resembles the giant, icy satellites of the outer planets). The source of internal heating for these objects is not well understood; Triton may undergo sufficient tidal stress due to Neptune, and may have some residual heating due to radioactive elements in its interior. For Enceladus, it appears unlikely that sufficient tidal stress is taking place, yet it is clear that water is occasionally being ejected from the interior.

Ice may play another important role related to the formation of atmospheres on some of the large satellites, as well as on Pluto. Ice that is very cold (below 135 K) can trap large quantities of other gases within its crystal structure. If the ice is then heated to a temperature above 135 K, the trapped gas escapes. It has been suggested that this mechanism helped to form the atmospheres of Titan, Triton, and Pluto. If these bodies were cold enough at the time of their formation to trap gases in their icy material, then subsequent heating (by tidal stresses or internal radioactivity) could have released the gases that had been trapped, forming the atmospheres of these bodies.

There is one additional mechanism that may be affecting the geological evolution of the moons of the giant planets, one that has only recently been suggested in the light of some of the bizarre findings of the *Voyager* probes. Some of the moons show perplexing varieties and degrees of geological activity, which is difficult to explain by the conventional processes such as tidal stress. Miranda, the innermost of the five large moons of Uranus, is the best example of this; there are a variety of unusual and dramatic features on its surface which must represent recent and energetic activity (Fig. 16.11). At the same time these features were being discovered, a growing body of evidence was mounting to indicate that planetary ring systems are transient, changing with time and somehow being replenished occasionally. Putting these observations together has led to the suggestion that the satellites of the giant planets, especially the inner satellites, are occasionally destroyed by major impacts. If the fractured moon is outside the Roche limit, it can soon re-form, as appears to have happened in the case of Miranda; if the debris left over from the collision is too close to the parent planet, then tidal forces prevent reaccumulation of the material, and a new ring results.

The innermost moons are most likely to undergo such major collisions because the immense gravity of the parent planet attracts incoming meteoroids, pulling them close in. Each giant planet can be viewed as a gravitational trap, altering the paths of bypassing objects so that they speed up and pass by at very close approaches (Fig. 16.12). This **gravitational focusing** process may account for the many gigantic craters found on inner satellites such as Mimas of Saturn (Fig. 16.13), for the strange geology of Miranda, and for the very existence of the ring systems.

Figure 16.11 The "Circus Maximus." *This unusual feature (right) on Miranda consists of an elevated inner portion, and sharply sloping terraces around it, resembling a Roman racetrack. The cause of this feature is unknown, but it would require some source of uplift in the interior. At lower left is a similar, chevron-shaped feature. (NASA/JPL)*

Rings and Moonlets

We now know that all four of the giant planets have rings. The detailed characteristics of these rings vary from planet to planet, ranging from the broad, bright rings of Saturn to the single thin, dim ring of Jupiter. The differences in appearance are in part due to the varying sizes of the ring particles; the very small particles in the rings of Saturn are much more effective at reflecting light in the backward direction than are the forward-scattering large particles that dominate the rings of the other planets.

In Chapter 13, we suggested that rings originate in the disks of debris that surrounded each giant planet when it formed. Close to the parent planet (inside the Roche limit), tidal forces prevent free particles from merging together, remaining forever as a thin sheet or series of belts confined by gravity and rotational forces to the equatorial plane of the parent planet. In this picture, the rings are primordial features, left over from the time of planetary formation, some 4.5 billion years ago.

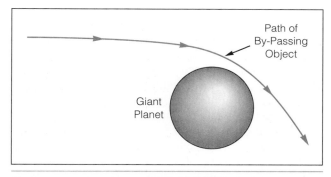

Figure 16.12 Gravitational focusing. *The immense gravitational field of a giant planet causes objects such as asteroids, comets, or large meteoroids (see Chapter 15) to veer in close to the planet. This focusing effect increases the likelihood of collisions between such bypassing bodies and satellites orbiting the giant planet.*

Recent evidence suggests, however, that this picture needs to be modified. It is likely that rings started in the way just described, but now it is apparent that they

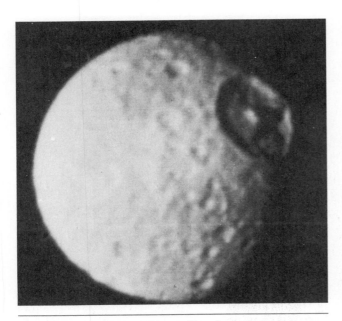

Figure 16.13 Mimas. *This satellite shows many impact craters, including one very large one that has about one-fourth the diameter of the satellite itself. This moon is probably composed chiefly of ice, as are several of the other satellites of Saturn. (NASA)*

are dynamic, evolving entities, occasionally replenished with new injections of particles. The evidence for this first became apparent after the *Voyager* encounters with Saturn, when it was discovered that some of the structure in the ring system there is unstable, unable to persist for very long times. The noncircular rings (Fig. 16.14) and the irregularly shaped F ring (Fig. 16.15) cannot be maintained indefinitely; these features must be varying with time. At Uranus, *Voyager 2* found a broad inner belt or ring that is losing its material to friction and infall towards Uranus, and must either be relatively new or receiving new material from some source. The rings of Neptune include some that are noncircular and some containing clumpy portions, again suggesting that variations with time are occurring. The very fine structure found in the rings of Saturn (Fig. 16.16) is by itself convincing evidence that the rings must change with time, because these structures are probably caused by variable phenomena, as discussed shortly.

The idea of fragmented satellites again comes into our discussion. If, as we have already mentioned, inner satellites of the giant planets are susceptible to destruction from impacts of large meteoroids, then such collisions can be a source of new ring particles. A satellite orbiting close to or within the Roche limit, if destroyed, would not easily be able to re-form because of tidal forces, and would remain a cloud of debris orbiting the planet. In this way, a ring system can be replenished from time to time.

The rings of Saturn revealed an astounding array of detailed structure (see Fig. 16.16). Major gaps had been seen from Earth, but the vast and intricate detail of the many tiny ringlets and small gaps came as a surprise. There are many proposed mechanisms to explain the fine structure.

First, as already discussed, orbital resonance is probably responsible for some of the major features. Most notable among these is the Cassini division, the prominent dark gap between the A and B rings (this "gap" is not truly empty, however; *Voyager* images reveal several dim, forward-scattering rings there). The Cassini division lies at the distance from Saturn where ring particles would have orbital periods exactly equal to half the period of Mimas. Hence, even though the mechanism is not fully understood, it appears that orbital resonance with Mimas is responsible for maintaining this gap.

In addition to orbital resonance, at least three other mechanisms have been invoked to explain the fine structure in the rings of Saturn.

One of these is also caused by gravitational effects of Saturn's moons. The clue to this was presented by the orbit of the two small satellites, discovered by *Voyager 1,* which orbit just inside and just outside the F ring (Fig. 16.17). The gravitational interaction between the ring particles and these satellites acts to keep the particles confined to orbits between them (Fig. 16.18). The moon just outside the ring, according to Kepler's laws, moves a little more slowly than the ring particles. Whenever a ring particle overtakes this satellite at a close distance, the particle loses energy due to the gravitational tug of the satellite, and drops to a slightly lower orbit. On the other hand, the satellite on the inside of the ring slowly overtakes the ring particles, and when it passes close by one, the particle gains energy and moves out to a higher orbit. In this way the two satellites ensure that the ring particles stay in orbit between them, because any particle that strays too close to either one will have its orbit altered in just such a way that it moves back into the ring. For this reason these two satellites are called "shepherd" satellites, because they behave like sheep dogs, keeping their flock in line by nipping at their heels.

Voyager 2 data showed that another mechanism,

Figure 16.14 Asymmetries in the rings.   *This composite shows that the rings are not perfectly circular; here the thin ring within the dark gap is thinner on one side of the planet than on the other and slightly displaced. (NASA)*

Figure 16.15 The F ring.   *Here we see the unusual braided appearance of this thin, outlying ring. The cause of the asymmetric shape is not well understood. (NASA)*

Figure 16.16 The rings of Saturn.   *This Voyager image reveals some of the fantastically complex structure of the ring system. (NASA)*

suspected before *Voyager* missions, was also influencing the structure of the rings. Mathematical calculations show that, in any flattened disk system such as Saturn's rings, small gravitational forces can create a spiral wave pattern, something like the grooves on a

Figure 16.17 The shepherd satellites and the F ring. *The two small satellites (Pandora and Prometheus) that maintain the F ring through gravitational interaction are both visible in this image. (NASA)*

Figure 16.18 The action of shepherd satellites. *This illustrates how a pair of small moons can keep particles trapped in orbit between them. The inner shepherd satellite overtakes the particles, accelerating them if they fall inward, thus moving them out. Similarly, the outer shepherd slows particles that move too far out, so that they fall back in. (NASA/JPL)*

phonograph record. The waves take the form of alternating dense and rarefied regions, and the entire spiral pattern rotates about the center of the system. Similar spiral density waves have long been thought responsible for the spiral arm structure of galaxies like our own, and so the *Voyager* data have shown that the rings of Saturn can be considered a small-scale model of a spiral galaxy.

It was predicted from calculations that the spacing between spiral density waves should increase steadily with distance from the center. The photopolarimeter data from *Voyager 2* (Fig. 16.19) revealed that the expected increase in spacing between ringlets occurs in several portions of the ring system, confirming that at least some of the ring structure is created by spiral density waves. The ringlets that are formed by this mechanism are not separate, circular structures, but instead are part of a continuous, tightly wound spiral pattern that slowly rotates. Individual ring particles, by contrast, follow elliptical orbits, and alternately pass through the ringlets and the spaces between. (It is important here to distinguish between the motion of the wave and the motions of the particles, which do not move with it; a reasonable analogy would be a cork bobbing on water waves, but not moving along with them). The reason there are more particles in the ringlets than between them at any given moment is that they move move slowly as they pass through the density waves, and tend to congregate there, just like cars in a traffic jam that build up at a point where the flow is slowed down.

The gravitational disturbances that create the density waves in portions of Saturn's ring system are caused by satellites. The strongest disturbances occur at resonance points where the particles have orbital periods that are a simple fraction of the periods of satellites orbiting at greater distances. These are also the locations of gaps in the rings in some cases, as explained earlier, so we sometimes find that a gap lies just at the inner edge of a series of ringlets that are created and maintained by a spiral density wave.

Although the ring system of Saturn is the most extensive and complex, those of the other giant planets show some of the same features, and are probably governed by many of the same processes. For example, the rings of Uranus tend to be thin, which indicates that shepherd satellites are responsible for confining them. Indeed some of the shepherd satellites were found in the *Voyager 2* images (Fig. 16.20), and the presence of others is suspected. Asymmetric ring shapes were found for both Uranus and Neptune, an indication that

Figure 16.20 Shepherd satellites. *Here we see two small moons on either side of a thin ring around Uranus. The gravitational effects of these moons keep the ring confined. (NASA/JPL)*

Figure 16.19 Very fine structure in the rings. *This is a synthetic image of a small section of the ring system, reconstructed from stellar occultation data obtained by the photopolarimeter experiment on Voyager 2. The finest details visible here are only about 100 meters in size, whereas the best Voyager camera images have a resolution of about 10 kilometers. Data like these have established that spiral density waves play a role in shaping the rings. (NASA)*

perturbations due to moons are acting on the rings, as well as contributing to the hypothesis that the rings are changing with time.

In all the ring systems, structures were found that cannot be permanent features, so all the systems share the common link that they are evolving. If a new *Voyager*-like mission is launched in the distant future, the images sent back might look very different from those we have today. It may even be found that some minor differences in the structure of Saturn's rings will be discovered by the *Cassini* probe, planned to orbit and observe the ringed planet for several years.

There was one additional new phenomenon found by the *Voyager* probes, associated only with the rings of Saturn. Several dark, tenuous features were seen in the rings, pointing away from Saturn (Fig. 16.21). It appears that these so-called "dark spokes" lie above and below the plane of the rings, rather than in the plane. This suggests that the forces holding them there

are not gravitational, since gravity (combined with the rotation of the system) acts to keep the rings tightly confined to the equatorial plane of Saturn. The spokes are highly variable; one may appear very quickly (in a matter of minutes), then gradually dissipate over a time of several hours. It is suspected that the spokes consist of very tiny particles that are electrically charged, so that the magnetic field of Saturn can exert forces on them to suspend them above the ring plane. The major question is where these tiny particles come from; apparently, they are injected into the ring system every now and then. One suggestion is that each spoke's appearance is triggered by the impact of a small meteoroid (a few centimeters in diameter) on the ring particles, pulverizing some of them into a fine dust. The dust particles accumulate electrical charge from free electrons in the vicinity and are then spread out into the observed spokelike pattern as the planetary magnetic field acts on them.

The many discoveries made by the *Voyager* probes about the moons and rings of the outer planets represent a real revolution in our understanding of the outer solar system, because in revealing their complexities and their continuing evolution, the *Voyagers* taught us that our solar system is still changing. Next we consider the possibility that there may be other solar systems, just as complex, elsewhere in the galaxy and the universe.

ASTRONOMICAL INSIGHT 16.3

Resolving the Rings

The *Voyager* missions to Saturn produced a wealth of new information on the rings, primarily because of the high resolution of the images that they sent back to Earth. Recall, from our discussions in Chapter 7, that resolution is defined as the ability to make out details. In our past discussion, we emphasized that resolution can be enhanced by the use of large telescopes, by placing observatories above the Earth's turbulent atmosphere, or by the use of special interferometry techniques. There is another way to achieve high resolution, however: getting close to the subject of the observations. This is what the *Voyager* spacecraft did.

The cameras aboard *Voyagers 1* and *2* yielded images of Saturn's rings with a maximum resolution of about 10 km; that is, details as small as 10 km in size could be seen. This was quite an improvement over the best photographs taken from Earth, which have a resolution about 4,000 km. The *Space Telescope* will resolve details in the ring system about 10 times smaller than this, roughly 400 km, still far inferior to *Voyager* images.

There was an instrument on the *Voyager* spacecraft that provided much better resolution than even the cameras. This was the photopolarimeter, a device for measuring the intensity of light, as well as its polarization. The photopolarimeter on *Voyager 1* failed before that spacecraft reached Saturn, but the one on *Voyager 2* was able to carry out an observation that has provided by far the most detailed information yet on the ring system.

The observation was rather simple: As *Voyager 2* approached Saturn, the photopolarimeter measured the brightness of a star (Delta Scorpii) that happened to lie behind the rings, from the point of view of the spacecraft. As *Voyager 2* swept by the planet with the photopolarimeter trained on the star, the spacecraft motion caused the ring system to move across in front of the star. As it did so, the ringlets and gaps caused the brightness of the star to fluctuate. When a dense ringlet passed in front of the star, its brightness temporarily dimmed; when a gap passed in front, the star appeared brighter. This kind of observation is known as a **stellar occultation.**

The photopolarimeter was able to record the changes in the star's brightness very rapidly. This, combined with the rate at which the rings appeared to move across in front of the star, determined how fine the details were that could be detected in the ring structure. As it turned out, the resolution of this experiment was about 100 meters, a factor of 100 better than the best images obtained by the *Voyager* cameras! This meant that ringlets or gaps as small as the length of a football field could be detected.

Of course, the photopolarimeter could only measure the ring structure along a single cross section; it could not provide a full picture of the entire system. Its image of the rings was one-dimensional, but it nevertheless led to profound new understanding, particularly of the spiral density waves described in the text.

The data produced by the photopolarimeter were originally in the form of plots showing the variations in the star's intensity as a function of time (see Fig. 14.21). Dips in the intensity corresponded to the passage of ringlets in front of the star. This information was then translated into a plot of ring density as a function of distance from the parent planet. The data were processed a step further so that things were easier to visualize; synthetic images of the rings, simulating those taken by a camera, were produced (it was assumed that the rings were circular, and different colors were used to represent varying degrees of ring density; see Fig. 16.18). In this way, false-color images of the rings were produced, somewhat similar to the actual photographs obtained by the cameras, but showing details a factor of 100 smaller.

The *Voyager 2* photopolarimeter was able to use the same stellar occulation technique to analyze fine structure in the rings of Uranus and of Neptune, during the encounters in January, 1986 and August 1989.

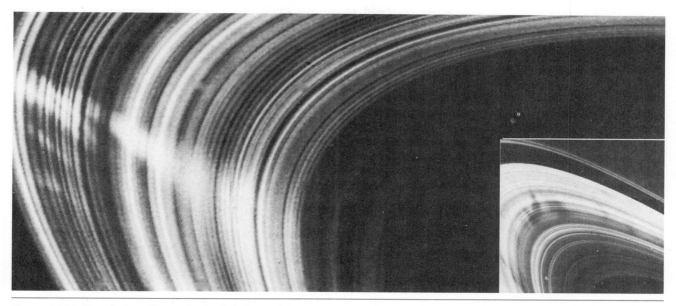

Figure 16.21 The dark spokes. *At the left the spokes appear bright because they scatter light in the forward direction, and they were photographed looking toward the Sun. At right they are viewed looking away from the Sun, so they appear dark. (NASA)*

ARE THERE OTHER SOLAR SYSTEMS?

As far as we know, the entire process by which our planetary system formed was a by-product of the Sun's formation, requiring no singular, catastrophic event. We might expect, therefore, that the same processes have occurred countless times as other stars were born, and therefore there may be many planetary systems in the galaxy. Even if we rule out double- and multiple-star systems, in which perhaps all the material was either ejected or included in the stars, a vast number of candidates remain.

Detecting planets orbiting distant stars is not easy. For a long time astronomers suspected one particular star of having a planet. The gravitational pull of a very massive planet would cause its parent star to wobble a little in position as the planet orbited it. A nearby star called Barnard's star was thought for a long time to be wobbling in such a way, but a recent reexamination of the data have shown that there is no clear-cut indication of such a motion, and at the moment the case for the existence of a massive planet circling Barnard's star is not well established.

Another technique is to search carefully for variations in the velocities of stars suspected of having planets. Here the Doppler shift is used to measure the speed of the star relative to Earth (see Chapter 6). Small, regular changes in this velocity could indicate that the star is orbiting a small companion (Fig. 16.22). Very recently, a study using this technique has revealed several nearby stars with dim companions, some of which may be massive planets. Planets as small as the Earth are much more difficult to detect in this way, because they would cause only minute variations in the velocity of their parent star.

Infrared observation shows some promise as a technique for detecting other planetary systems. This was illustrated by the *Infrared Astronomical Satellite* (IRAS), which mapped most of the sky in far-infrared wavelengths and found several stars that are surrounded by disks containing tiny solid particles of the sort thought to be the precursors of planetesimals (Fig. 16.23). Other kinds of new technology may also provide new information in this area. The *Space Telescope*, for example will be (when new cameras are installed) free of the distortion and fuzziness that plagues ground-based telescopes, and may be able to detect bright planets that are widely separated from their parent stars. The *Space Telescope* is expected to allow astronomers to see objects comparable to Jupiter, if they orbit nearby stars. There is little hope of directly detecting smaller objects such as the terrestrial planets, however.

The fine resolution of the *Space Telescope* images

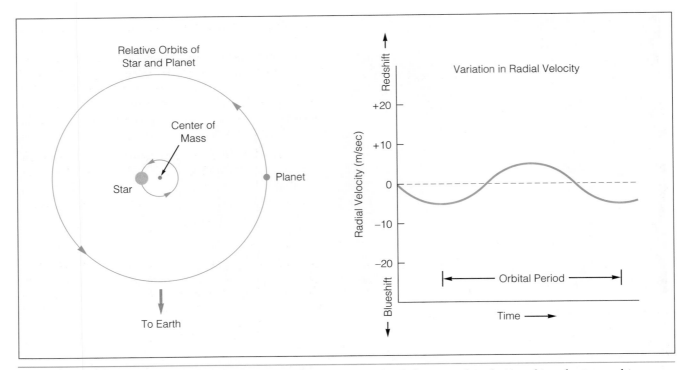

Figure 16.22 Using radial velocity measurements to detect planets. *At left we see the relative orbits of a star and its planet about their common center of mass (here the relative size of the star's orbit has been greatly exaggerated). As the star orbits the center of mass, its spectrum undergoes small Doppler shifts (alternately blueshifts and redshifts) as the star approaches us and then recedes. At right is a graph showing how the Doppler shift (radial velocity; see Chapter 6) changes with time. Note that the size of the velocity shifts is very small, only a few meters per second in this hypothetical case. Such small shifts are very difficult to detect and measure.*

will also allow much more precise measurements of stellar motions than have been achieved from the ground, so that tiny oscillations in the positions of stars will be more easily detected. Again, this will probably not allow the detection of small planets like the Earth, but could locate objects as large as Jupiter. The use of interferometry at ground-based observatories may also provide new hope of detecting small oscillations in star positions indicative of the presence of planets. Astronomers are now waiting eagerly for the chance to use these new instruments and techniques in a search for planets in other solar systems. It will be fascinating to learn the results in the coming years.

istics of the planets, their compositions, the Sun's slow rotation, and the nature and motions of the interplanetary material have all been explained. Many details are yet to be worked out, but the overall scenario is probably correct in essence.

We have tied together the many facets of the satellites and rings of the outer planets, finding several important underlying mechanisms to explain their complexity. One of the most important themes in this has been the growing realization that the ring systems are evolving, dynamic entities, changing with time.

With this, we have completed our examination of the solar system, and we are ready to move on. The next logical step is to explore the realm of the stars.

PERSPECTIVE

In this chapter, we have outlined a natural, evolutionary process that seems to account for all of the observed properties of the solar system. The orbital character-

SUMMARY

1. The facts that must be explained by a successful theory of the formation of the solar system include the

Figure 16.23 A star with a dust disk. *This image of the star β Pictoris, obtained with an infrared telescope on the ground, shows a disk of solid particles edge-on, much like the solar nebula. Infrared data have revealed a number of stars with surrounding material, supporting the suggestion that planetary systems may form around many stars. (NASA/JPL, courtesy of B. A. Smith)*

systematics of the orbital and spin motions of the planets and satellites, the contrasts between the terrestrial and giant planets, the slow rotation of the Sun, and the existence and properties of the interplanetary bodies.

2. There are two general classes of theories: catastrophic theories and evolutionary theories.

3. Catastrophic theories require some singular event, such as a near-collision between stars, to distort the Sun by tidal forces, pulling out matter that can then condense to form the planets.

4. Evolutionary theories postulate that the planets formed as a natural by-product of the formation of the Sun. These theories, although they are simpler, were not accepted for a while because of their failure to explain the slow spin of the Sun.

5. When the solar wind and its creation of magnetic braking was discovered in the 1960s, the evolutionary theory became most widely accepted.

6. In the modern theory, the planets formed from condensations in the solar nebula, the flattened disk of gas and dust that formed around the young Sun.

7. The terrestrial planets formed from the coalescence of planetesimals, solid, asteroidlike objects that condensed from the hot inner portions of the solar nebula.

8. The giant planets are thought to have formed when solid cores coalesced from planetesimals, then trapped surrounding gas, which formed their atmospheres. The extensive ring and satellite systems of these planets formed from disks of gas and solid particles in their equatorial planes.

9. Most of the moons of the giant planets probably formed by accretion in disks that were left around the planets when they formed; these moons have circular orbits lying in the equatorial plane of their parent planets. Other moons, having irregular orbits, may be captured objects left over from the era of planetesimals during planet formation.

10. Some of the large satellites are or have been geologically active, due to internal heating from tidal stresses or from radioactivity. The geological activity consists of crustal plate motions or volcanic eruptions in which ice or gas is released.

11. Some satellites show evidence that occasional large impacts shatter inner moons of the giant planets; this would account for the strange surface geology in some cases (especially Miranda of Uranus), and can also serve to replenish the rings.

12. Many kinds of evidence indicate that the ring systems of the giant planets are not stable over long periods of time, but instead must be evolving and changing. The evidence includes the existence of very fine structure on the rings, the presence of asymmetric rings, and the inference that particles are falling inward from some rings toward the parent planet.

13. It is thought likely that other planetary systems exist, orbiting other stars. Careful radial velocity measurements have begun to reveal evidence confirming this, in the form of possible detections of Jupiter-class planets circling a few nearby stars.

REVIEW QUESTIONS

Thought Questions

1. Summarize the overall pattern of orbital and spin motions in the solar system that must be explained by any successful theory of solar system formation. Describe any exceptions to the normal pattern.

2. Describe the general contrasts between the terrestrial planets and the giant outer planets.

3. The acceptance of evolutionary models of solar system formation was delayed considerably because of the

difficulty in explaining why the Sun's rotation is so slow. Explain why this was difficult. Discuss the parallels between this, the difficulty early astronomers had in accepting the heliocentric model, and the long delay in the general acceptance that continental drift occurs.

4. What are the chief arguments against the catastrophic theories of solar system formation?

5. If the solar system formed from an interstellar cloud of uniform chemical composition, explain how the terrestrial planets developed compositions that are so different from that of the Sun.

6. Based on the information in this chapter regarding the formation of the Sun, how do you think astronomers can observe other stars in the process of formation?

7. Summarize the contrasts between the formation of the terrestrial planets and the formation of the giant gaseous planets.

8. Discuss the many ways in which the moons and rings of the outer planets interact with each other.

9. Explain why it is sometimes argued that Pluto may have more in common with the large satellites of the giant planets than it has in common with other planets, either terrestrial or giant.

10. Discuss the methods that can be used to search for planets orbiting other stars. Which method do you think is most sensitive, that is, most likely to succeed?

Problems

1. We already know (from Chapter 8) that Jupiter contains more mass than all the other planets combined. Now compare the orbital angular momentum of Jupiter with the total orbital angular momentum of the other planets. Assume that each planetary orbit is a circle having a radius equal to the semimajor axis given in Appendix 7, and recall that the orbital angular momentum for a circular orbit is mvr, where m is the body's mass, v is its orbital speed, and r is the radius of the orbit. Discuss your result.

2. To illustrate how much of the volatile material in the inner solar system was not captured by the terrestrial planets, calculate what the mass of the Earth would be if it had the other elements in the same proportion to iron as the Sun has. Assume that the mass of the Earth's iron core is 7×10^{26} g, then calculate the mass of other elements needed so that they have the same relative abundance as in the Sun. (Hints: You can ignore all elements but hydrogen and helium in this calculation, and you should use the last column of Table 18.3 for the relative abundances by mass.) Compare your answer with the masses of the giant planets.

3. Suppose the Oort cloud defines the outer boundary of the interstellar cloud that was to collapse and form the solar system. Calculate what the density of this cloud would have been; that is, what is the density of a sphere having the mass of the Sun and a radius of 100,000 AU?

If the average atom in this cloud had a mass of 2×10^{-24} g, to how many atoms per cubic centimeter does your answer correspond?

4. Suppose the early Sun had a temperature of 1,000 K at some time during its collapse and heating phase. At what wavelength would it have emitted most of its energy?

5. The total mass of the rings of Saturn is estimated to be about 3×10^{22} grams. If all the ring particles came from the destruction of a single moon made of pure ice (thus having a density of 1 gram/cm^3), how large would this satellite have been (assume it was spherical)? Does this size correspond to that of any of Saturn's moons? Which one?

6. How much dimmer would Jupiter appear if it orbited the nearest star, 4.3 light-years, or 270,000 AU away? (Compare its brightness as seen from 270,000 AU with its brightness when it is closest to Earth, 4.2 AU away.) In view of your answer, discuss the feasibility of directly observing a planet like Jupiter orbiting another star.

ADDITIONAL READINGS

Beatty, J. K., B. O'Leary, and A. Chaikin, eds. 1990. *The new solar system.* 2d ed. Cambridge, England: Cambridge University Press.

Cameron, A. G. W. 1975. The origin and evolution of the solar system. *Scientific American* 233(3):32.

Falk, S. W., and D. N. Schramm. 1979. Did the solar system start with a bang? *Sky and Telescope* 58(1):18.

Head, J. W., C. A. Wood, and T. A. Mutch. 1977. Geologic evolution of the terrestrial planets. *American Scientist* 65:21.

Killian, A. M. 1989. Playing dice with the solar system. *Sky and Telescope* 78:136.

Knacke, R. 1987. Sampling the stuff of a comet. *Sky and Telescope* 73:246.

McSween, H. Y. Jr. 1989. Chondritic meteorites and the formation of the planets. *American Scientist* 77:146.

Reeves, H. 1977. The origin of the solar system. *Mercury* 7(2):7.

Sagan, C. 1975. The solar system. *Scientific American* 233(3):23.

Schramm, D. N., and R. N. Clayton. 1978. Did a supernova trigger the formation of the solar system? *Scientific American* 237(4):98.

Waldrop, M. 1985. First sightings: Solar systems elsewhere. *Science 85* 6(5):26.

Wetherill, G. W. 1981. The formation of the earth from planetesimals. *Scientific American* 244(6):162.

MATHEMATICAL METHODS

Angular Momentum

In Chapter 5, we said that for a body in circular orbit, the angular momentum is the product of the mass m of the body times the speed v of orbital motion times the radius r of the orbit: that is,

$$a.m. = mvr.$$

We can express this in terms of the orbital period P which, for a circular orbit, is $P = 2\pi r/v$; doing so gives us

$$a.m. = \frac{2\pi m r^2}{P}.$$

The angular momentum for a spinning sphere such as the Sun or a planet is a bit more complicated, since it involves contributions from all the parts of the body, from the center to its outer boundary (the methods of calculus are used to perform the summation of the angular momentum from all points within a sphere). If we assume that the internal density is uniform (which is not strictly true of any planet or of the Sun), then there is a simple formula for the angular momentum of a spinning sphere:

$$a.m. = \frac{4\pi M R^2}{5P},$$

where M is the mass and R the radius of the sphere.

Having these formulae available, we can now make some interesting comparisons. For example, in Chapter 10, we made the point that the Moon contains a lot of the total angular momentum in the Earth-Moon system, and that this has been a problem in explaining how the Moon formed. Now, by using our equations for the Moon's orbital angular momentum and for the Earth's spin angular momentum, we can see what the problem is: We find that the Moon's orbital angular momentum is more than 3 times the Earth's spin angular momentum. As discussed in Chapter 10, it is very difficult to see how the Moon could have been formed with so much of the total angular momentum of the system, unless the Moon originated from a body that approached the Earth at a high speed.

In similar fashion, we can assess the angular momenta of other solar system bodies, and make comparisons. One of the more interesting ones is to compare the orbital angular momentum of Jupiter with the spin angular momentum of the Sun; use of the preceding formulae shows us that Jupiter contains about 17 times more angular momentum than the Sun. This is due largely to the fact that Jupiter is quite far from the Sun, and that it managed to accrete much of the available material in its vicinity when the solar system was young, thereby giving Jupiter a large mass. It is also true that the Sun's spin is slower than would have been predicted from considerations of angular momentum.

We can now see why the slow spin of the Sun has been a problem for scientists. To do this, we start by assuming that the angular momentum of the Sun has remained constant since the time when the Sun's material was spread out over a much larger volume than it is now. For example, we can assume that at one time the solar nebula, containing the mass of the Sun, was a sphere having a radius about equal to the radius of the orbit of Pluto (actually the solar nebula was shaped more like a disk at this point, but this makes little difference for the comparison we are about to make). If we rearrange the equation for the angular momentum of a uniform sphere, solving for the period P, we find

$$P = \frac{4\pi M R^2}{a.m.}$$

We can simplify our calculation if we divide the expression for the rotation period of the Sun (P_s) by the expression for the rotation period of the solar nebula (P_N):

$$P_S/P_N = (R_S/R_N)^2.$$

Here we have assumed that the mass M remained constant as the nebula compressed, so that M's canceled out along with the other constant terms. Substituting values for the radius of the Sun $(R_s = 7 \times 10^{10}$ cm) and the radius of the solar nebula (assumed equal to the semimajor axis of Pluto's orbit, or approximately $R_N = 40$ AU $= 6.0 \times 10^{14}$ cm), we find the ratio of the periods:

$$P_S/P_N = 1.4 \times 10^{-8}$$

The rotation period of the solar nebula when it was equal in size to Pluto's orbit can be taken as equal to Pluto's orbital period, which is about 250 years. Hence we predict, based on the assumption of constant angular momentum, that the Sun's spin period today should be

$$P_S = 1.1 \times 10^{-8} P_N = 110 \text{ seconds}$$

Thus, if the Sun had not lost angular momentum during its collapse and formation, simple arguments tell us that it should be spinning once every 110 seconds!

GUEST EDITORIAL

The Grand Tour of the Outer Solar System . . . and Coming Attractions

BY JONATHAN I. LUNINE

Dr. Lunine is Associate Professor of Planetary Sciences and Theoretical Astrophysics at the University of Arizona. After receiving his doctorate from Caltech in 1985, he was a Research Associate at Arizona and a visiting Assistant Professor at UCLA. He then joined the Arizona faculty in 1986. His research interests include modeling of chemical and physical processes in the outer solar system, the origin of the solar system, and the nature of planet-like bodies ("brown dwarfs") outside the solar system. Additionally, some of his work has concerned terrestrial processes such as ozone chemistry and stratospheric aerosols. Dr. Lunine was a guest participant in the Voyager *Neptune encounter, working with the Ultraviolet Spectrometer team, and is a coinvestigator on the penetrator for NASA's Comet Rendezvous Asteroid Flyby mission. He has served on several national advisory committees, including the Committee on Planetary and Lunar Exploration of the Space Studies Board, a part of the National Academy of Sciences advisory structure. Dr. Lunine was awarded the Harold C. Urey Prize in 1988, given by the Division of Planetary Sciences, American Astronomical Society, for outstanding achievement in*

planetary research by a young scientist. In 1990, the international space research organization, COSPAR, selected him as one of seven recipients of the Soviet Union's Zel'dovich award.

The exploration of the solar system by unmanned spacecraft has been an important element of the United States space program for 30 years. Robot probes sent to fly by, orbit, or land on the planets have made observations that are impossible or too expensive for manned missions, and much more detailed than can be achieved from groundbased or Earth-orbital observatories. These missions also serve as sources of public excitement and national pride, pushing the frontiers of human endeavor well beyond the range

of current manned flight. No planetary program has better symbolized these twin purposes of scientific research and high adventure than the *Voyager* project, which sent two spacecraft on a decade-long tour of the outer solar system. From 1979 through 1989, *Voyagers 1* and 2 visited the solar system's four major planets—Jupiter, Saturn, Uranus and Neptune—and revealed an unexpected richness and diversity in the planets' atmospheres, moons, rings and magnetic fields. Now on their way out of the solar system, the two spacecraft continue to make observations of the Sun's magnetic field and solar wind, which scientists hope will reveal the boundary between the Sun's influence and the winds of interstellar space.

The *Voyager* project arose out of a NASA proposal to take advantage of an unusual alignment of the outer planets during the 1970s and 1980s, by which a spacecraft could use the gravitational pull of one planet to provide it sufficient momentum to reach the next. The last time the four major planets were so aligned was at the beginning of the nineteenth century, prior to the discovery of Neptune; the next such alignment, in the mid-twenty-second century, will perhaps see human colonies across the solar system. NASA's request to build two specialized "Grand

Tour" spacecraft was rejected by Congress. Engineers at Caltech's Jet Propulsion Laboratory, which designs, builds, and operates most U.S. planetary probes, developed a plan to salvage most of the Grand Tour scheme with *Mariner*-class vehicles, renamed *Voyagers 1* and *2*. The strategy was that, should *Voyager 1* complete its scientific objectives at Jupiter and Saturn, *Voyager 2* would be targeted at Saturn to achieve a flyby of both Uranus and Neptune. (Because one of the key science goals for *Voyager 1* was a very close pass by Saturn's giant moon, Titan, it was not physically possible to complete this objective and position *Voyager 1* to be flung to Uranus or Neptune by Saturn's gravity).

Launched in late summer, 1977, *Voyager 1* achieved all of its objectives at Jupiter and Saturn in 1979 and 1980. *Voyager 2*, launched two weeks earlier but sent on a longer trajectory, was routed through the Saturn system in 1981 so as to achieve a close flyby of Uranus in 1986 and Neptune in 1989. Overcoming several equipment malfunctions and re-programming the spacecraft to improve performance at each successive encounter, JPL engineers succeeded in completing the bulk of the Grand Tour with two spacecraft originally designed to return data from only Jupiter and Saturn.

Because of the great distance from Earth to the outer solar system, very little was known about the giant planets before the space age. Although Jupiter and Saturn had been visited by Pioneer probes in 1973, 1974, and 1979, these much less sophisticated machines could not provide the detailed information that Voyager's instruments were capable of obtaining. The turbulent nature of Jupiter's atmosphere was understood before Voyager, as well as the fact that both Jupiter and Saturn were mostly hydrogen. Jupiter's magnetic field had been studied from Earth, but information on the nature or even existence of magnetic fields around the other outer planets was rudimentary or nonexistent. Very little was known about the natural satellites of the giant planets, although evidence of a methane atmosphere on Titan had been detected from Earth.

The triumph of the *Voyager* program lay in revealing these four distant planets as complex systems with a richness and variety of physical processes not anticipated by scientists. The lengthy list of discoveries must include active volcanism on Io, nitrogen geysers and frost sheets on Triton, exquisitely intricate structure in the rings of the giant planets, extensive and varied tectonic features on the icy satellites, tilted and shifted magnetic fields in Uranus and Neptune, and a surprising amount of weather activity in the atmosphere of Neptune. Each of these discoveries, along with others not mentioned here, has stimulated in-depth analyses to explain the physical and chemical processes responsible for the phenomena. Most of the time, such investigations have forced scientists to sharpen their understanding of particular sets of physical laws in order to fully explain the data. The new sophistication gained in such efforts has, in many cases, revised or deepened our understanding of natural processes on Earth. In effect, *Voyager* made the outer solar system a giant laboratory for investigating physical processes that cannot be simulated on Earth.

Beyond the list of individual discoveries made by *Voyager* is the impact on profound questions such as how the planets and satellites actually formed. One example of the powerful techniques that were brought to bear on this issue is the detailed investigation of Titan from *Voyager 1*. Although the camera system was unable to penetrate the thick haze shrouding Titan, a suite of instruments provided extensive information on Titan's atmosphere and surface. A radio occultation experiment was performed in which signals were beamed through the Titan atmosphere and detected on Earth. The distortion of the beam gave scientists a detailed density profile down to Titan's surface. Combined with *Voyager's* infrared and ultraviolet measurements of the atmosphere, scientists could deduce that the atmosphere is mostly molecular nitrogen (as on Earth), that the surface pressure is 1.5 atmospheres, and the temperature 95 K. Within this dense, cold atmosphere, *Voyager* data provided a picture of a complex set of chemical cycles in which sunlight in the upper atmosphere converts methane into a mix of heavier hydrocarbons that rain down onto the surface over time. The atmospheric measurements provided by the spacecraft suggest that, if methane is being resupplied from a surface reservoir, the composition cannot be pure methane, but must be a mixture of methane and ethane, along with other hydrocarbons, and nitrogen. This "ocean," if it exists, would be a bizarre soup of nonbiologically produced organic compounds,

(continued on next page)

controlling the near-surface climate and erosively modifying the surface over geologic time. The *Voyager* data painted a tantalizing picture of a world every bit as complex and changing as that of the Earth, but in which life almost assuredly plays no role.

The preceeding discussion is just one example of the kind of detailed studies scientists are able to undertake with the *Voyager* data. In almost every discipline of planetary science, the *Voyager* Grand Tour profoundly extended our base of knowledge.

The future of planetary exploration centers on missions to directly sample bodies in the outer solar system and to use powerful remote-sensing tools mounted on orbiters to study planets of the inner solar system. As this article is being written, *Magellan* is preparing to map the surface of Venus with high resolution using a sophisticated radar system. It is now thought that plate tectonics play a significant role in cycling water and atmospheric gases through the Earth's crust and interior, which early in Earth's history may have played a key role in the evolution of the biosphere. Whether Venus ever had plate tectonics can be addressed with *Megellan* radar data. *Mars Observer* will go into orbit around the red planet in 1993 and will begin the first comprehensive study of its global surface composition and climate. Many questions about the early climate of Mars were raised by the *Viking* orbiters and landers, and the powerful new remote-sensing instruments aboard the *Observer* are designed to investigate these questions, including whether standing water existed in the past and where the water is locked up today.

The sampling of the outer solar system has begun with *Galileo*. The study will put a probe directly into the atmosphere of Jupiter in 1995, to measure abundances of gases hard to detect remotely. Such measurements attack very specific questions on the nature and origin of the giant planets. The sophisticated orbiter portion of the mission will observe the moons, atmosphere, and magnetosphere in great detail, and undoubtedly will discover new phenomena unseen by *Voyager*. The spacecraft will change its orbit repeatedly over three years by using the gravitational fields of the Galilean satellites; these "pump and crank" maneuvers, as JPL engineers call them, will allow the main spacecraft to perform a tour of the Jovian system with 60 times less rocket fuel than would be required without the gravity assists!

Beyond these missions, NASA has a twin foray to the outer solar system under development. The Comet Rendezvous Asteroid Flyby (CRAF) mission will be launched in 1995, fly past an asteroid, and then match paths with Comet Kopff in the year 2000. The spacecraft will slowly orbit about the icy nucleus of the comet, enabling the first in-depth study of the nature of the heart of the comet. Because comets are thought to be among the best-preserved samples of material left over from the origin of the solar system, it is crucial to analyze the chemistry of the material in the nucleus. A penetrator will be released from the main spacecraft and injected into the upper meter or so of the nucleus, collecting ice samples as it pierces the nucleus. A series of instruments will then analyze the structure and compo-

sition of the ice, dust, and gas, which can be compared with models of the early solar system and with the composition of interstellar clouds observed by astronomers from Earth. Additionally, experiments on the main spacecraft will analyze dust released by the nucleus as it becomes active. The goal of these chemical studies will be to better understand the history of cosmic materials as they evolve from giant molecular clouds into planetary systems.

The second of the two spacecraft is called *Cassini*. Launched from Earth in 1996, *Cassini* will go into orbit around Saturn in 2002 and study intensively the moons, atmosphere, and rings of this system. The highlight of this mission will come when the main spacecraft releases *Huygens,* the European Space Agency probe, for a descent through the atmosphere of Titan. *Huygens* will make compositional measurements of the atmosphere and take images of the clouds and surface features as it nears the ground. Should the probe survive impact, it will determine whether the impact site is liquid or solid and may yield data on the composition of any liquids present. After the 3-hour period in which the probe performs its mission, the orbiter will map the surface of Titan with a powerful radar as it makes repeated passes above the cloud-shrouded moon over several years.

Beyond these missions are others, many of which are designed to map resources on the Moon and Mars, which might one day be used for human colonies. If U.S. resolve is firm, the age of robotic planetary exploration will pave the way for a human civilization that calls the entire solar system its home.

THE STARS

This section is devoted to the stars and their properties. In its seven chapters we learn about how stars are observed, how their fundamental properties are deduced from observations, and how stars work. We see what the processes are that control the structure of stars, and we learn how stars live, evolve, and die. We discover that stars are dynamic, changing entities, and only seem fixed and immutable because the time scales on which they evolve are vastly longer than the human lifetime.

The first of the seven chapters describes our own Sun, its properties and structure, and how it compares to other stars in the universe. The next chapter describes the three basic types of stellar observations; measurements of positions, brightnesses, and spectra. The emphasis is on raw information that is gathered from observation, whereas discussion of how this information is used to derive the more basic stellar parameters is reserved for the next chapter. There we systematically consider several stellar properties, discuss how each is determined, and summarize the typical values that are found.

Chapter 20 then initiates the discussion of how stars function and evolve by describing the physical processes that occur in stars, and how these processes govern the stellar parameters previously discussed. Here we draw upon knowledge of our own Sun, whose well-observed properties provide vital information about other stars as well.

Chapter 21 illustrates how astronomers can learn about stellar evolution from observations, particularly of star clusters. We will find that despite the immensity of the time scales involved, it is possible to deduce the life stories of stars.

The final two chapters describe the evolution of stars from their births to their deaths. In Chapter 22 we have to rely on the results of theoretical calculations, but we are guided by observations discussed earlier. In Chapter 23 we discuss the remnants of stars that have finished their lives, and here we discover some of the most bizarre and exciting objects in the universe.

Having studied the stars, we will then be ready to consider large groupings such as our own galaxy, which contains many billions of stars, and whose own evolution is governed in part by that of the individual stars within it.

CHAPTER 17

The Sun

A solar image obtained through a special filter designed to reveal active regions on the surface. (National Optical Astronomy Observatories)

CHAPTER PREVIEW

Basic Properties and Internal Structure
Nuclear Reactions
Modeling the Sun
Structure of the Solar Atmosphere
The Solar Wind
Sunspots, Solar Activity, and the Magnetic
    Field

Our star, the Sun, is rather ordinary by galactic standards. It has modest mass and size, and there are stars as much as a few hundred times larger in diameter and a million times more luminous. Its temperature is also moderate, as stars go. In many respects the Sun is entirely a run-of-the mill entity.

Because the Sun is an ordinary star, and because we have far more detailed knowledge of it than of other stars, it is useful to begin our discussion of stars with a chapter on the Sun. In subsequent chapters, we will see how the information on the Sun helps us to understand the nature of other stars. Conversely, we will find also that studies of other stars help us to learn more about the nature of the Sun.

BASIC PROPERTIES AND INTERNAL STRUCTURE

Perhaps the most obvious attribute of the Sun is its brightness; it emits vast quantities of light. Its **luminosity,** the amount of energy emitted per second, is about 4×10^{33} ergs per second. (For comparison, a 100-watt light bulb emits 10^9 erg/sec.) The intensity of sunlight reaching the Earth (above the atmosphere) is about 1.4×10^6 erg/cm^2/sec. This quantity is known as the **solar constant,** although it varies slightly according to activity in the Sun's outer layers (discussed later in this chapter).

The Sun is a ball of hot gas (Fig. 17.1). It dwarfs any of the planets in mass and radius (Table 17.1). Its density, on average, is 1.41 g/cm^3, not much more than that of water, but its center is so highly compressed that the density there is about 10 times greater than that of lead. The interior is gaseous rather than solid, because the temperature is very high, around 15 million degrees at the center, diminishing to just under 6,000 K at the surface. At these temperatures, the gas is partially ionized in the outer layers of the Sun, and is completely ionized in the core, all electrons having been stripped free of their parent atoms.

The Sun is held together by gravity. All of its constituent atoms and ions attract each other, and the net effect is for the solar substance to be held in a spherical shape. Gas that is hot exerts pressure on its surroundings, and this pressure, pushing outward, balances the

Figure 17.1 The Sun. *This is a visible-light photograph of the Sun at a time when only a few sunspots were present. The overall yellow color reflects the surface temperature of almost 6,000 K. The spots and the darkening of the disk near the edges are discussed in the text. (National Optical Astronomy Observatories)*

Table 17.1 The Sun

Diameter: 1,391,980 km (109.3 D$_\oplus$)
Mass: 1.99×10^{33}g (332,943 M$_\oplus$)
Density: 1.41 g/cm^3
Surface gravity: 27.9 Earth gravities
Escape velocity: 618 km/sec
Luminosity: 3.83×10^{33} erg/sec
Surface temperature: 5780 K
Rotation period: 25.04 days (at equator)

force of gravity, which pulls the matter inwards. This balance is called **hydrostatic equilibrium,** with gravity and pressure equaling each other everywhere. The deeper we go into the Sun, the greater the weight of the overlying layers, and the more the gas is compressed. The higher the pressure, the greater the temperature required to maintain the pressure, so we find that the pressure and temperature both increase as we approach the center.

The composition of the Sun (Table 17.2) is the same as that of the primordial solar system and of most other stars: About 73 percent of its mass is hydrogen, 25 percent is helium, and the rest is made up of traces of other elements. As we have already seen, this is similar to the chemical makeup of the outer planets and would represent the terrestrial planets as well, except that they have lost most of their volatile elements such as hydrogen and helium. It is apparent that all the components of the solar system formed together, from the same material.

The ultimate source of all the Sun's energy is located in its core, within the innermost 25 percent or so of its radius (Fig. 17.2). Here nuclear reactions create heat and photons at γ-ray wavelengths. It is this radiation that eventually reaches the surface and escapes into space, but it is a laborious journey. Each photon is absorbed and reemitted many times along the way (Fig. 17.3), gradually losing energy. In the process, most γ-ray photons become visible light photons, and the energy they lose heats the surroundings. Because a photon travels only a short distance before it is absorbed, and because when it is reemitted, it is in a random direction, progress towards the surface is very slow.

Throughout most of the solar interior, the gas is quiescent, without any major large-scale flows or currents. The energy from the core is transported by the radiation wending its slow pace outwards, except in the layers near the surface, where convection occurs, and heat is transported by the overturning motions of the gas. As we shall see, the bubbling, boiling action in the outer portions of the Sun creates a wide variety of dynamic phenomena on the surface.

The Sun rotates, and it does so differentially. This is apparent from observations of its surface features,

Table 17.2 The composition of the Sun's photosphere

Element	Symbol	Number of Atoms[a]	Fraction of Mass
Hydrogen	H	1.0000	7.35×10^{-1}
Helium	He	8.51×10^{-2}	2.48×10^{-1}
Lithium	Li	1.55×10^{-9}	7.85×10^{-9}
Beryllium	Be	1.41×10^{-11}	9.27×10^{-9}
Boron	B	2.00×10^{-10}	1.58×10^{-9}
Carbon	C	3.72×10^{-4}	3.26×10^{-3}
Nitrogen	N	1.15×10^{-4}	1.18×10^{-3}
Oxygen	O	6.76×10^{-4}	7.88×10^{-3}
Fluorine	F	3.63×10^{-8}	5.03×10^{-7}
Neon	Ne	3.72×10^{-5}	5.47×10^{-4}
Sodium	Na	1.74×10^{-6}	2.92×10^{-5}
Magnesium	Mg	3.47×10^{-5}	6.15×10^{-4}
Aluminum	Al	2.51×10^{-6}	4.94×10^{-5}
Silicon	Si	3.55×10^{-5}	7.27×10^{-4}
Phosphorus	P	3.16×10^{-7}	7.14×10^{-6}
Sulfur	S	1.62×10^{-5}	3.79×10^{-4}
Chlorine	Cl	2.00×10^{-7}	5.17×10^{-6}
Argon	Ar	4.47×10^{-6}	1.30×10^{-4}
Potassium	K	1.12×10^{-7}	3.19×10^{-6}
Calcium	Ca	2.14×10^{-6}	6.26×10^{-5}
Scandium	Sc	1.17×10^{-9}	3.84×10^{-8}
Titanium	Ti	5.50×10^{-8}	1.92×10^{-6}
Vanadium	V	1.26×10^{-8}	4.68×10^{-7}
Chromium	Cr	5.01×10^{-7}	1.90×10^{-5}
Manganese	Mn	2.51×10^{-7}	1.01×10^{-5}
Iron	Fe	3.98×10^{-5}	1.62×10^{-3}
Cobalt	Co	3.16×10^{-8}	1.36×10^{-6}
Nickel	Ni	1.91×10^{-6}	8.18×10^{-5}
Copper	Cu	2.82×10^{-8}	1.31×10^{-6}
Zinc	Zn	2.63×10^{-8}	1.25×10^{-6}
(all others combined)		(less than 10^{-8} of total)	

[a]The numbers given are relative to the number of hydrogen atoms.

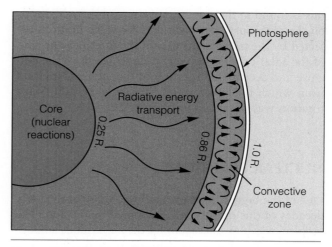

Figure 17.2 The internal structure of the Sun. *This drawing shows the relative extent of the major zones within the Sun, except that the depth of the photosphere is greatly exaggerated.*

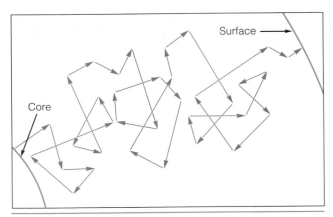

Figure 17.3 Random walk. *A photon is continually being absorbed and reemitted as it travels through the Sun's interior. Each time it is reemitted, it is in a random direction. Thus its progress from the core, where it is created, to the surface is very slow.*

which reveal that, like Jupiter and Saturn, the Sun goes around faster at the equator than near the poles. The rotation period is about 25 days at the equator, 28 days at middle latitudes, and even longer near the poles. As we will see, the differential rotation probably plays an important role in governing variations in the solar magnetic field, which in turn have a lot to do with behavior of the most prominent surface features, the sunspots.

There is evidence that the core of the Sun rotates a little more rapidly than the surface. The only clue for this is found in the presence of very subtle oscillations on the solar surface, which may be wave motions affected by the more rapid spin of the interior. The spin of the solar core is probably a direct result of the collapse and accelerated rotation of the interstellar cloud from which the Sun formed. (The origin of the Sun's rotation was discussed in Chapter 16.)

NUCLEAR REACTIONS

One of the biggest mysteries in astronomy in the early decades of this century was presented by the Sun. The problem was how to account for the tremendous amount of energy it radiates, particularly perplexing in view of geological evidence that the Sun has been able to produce this energy for at least 4 or 5 billion years. Two early ideas, that the Sun is simply still hot from

its formation, or that it is gradually contracting, releasing stored gravitational energy, were both ruled out, since neither mechanism could possibly supply the energy needed to run the Sun for a long enough time.

The first hint at the solution came in the first decade of the 1900s, when Albert Einstein developed his theory of special relativity. He showed that matter and energy are equivalent, according to the famous formula $E = mc^2$, where E is the energy released in the conversion, m stands for the mass that is converted, and c for the speed of light. If mass could be converted into energy, enormous amounts of energy could be produced. Physicists began to contemplate the possibility that somehow this energy was being released inside the Sun and stars.

By the 1920s, following the pioneering work on atomic structure by Max Planck, Niels Bohr, and others, the concept of nuclear reactions began to emerge. These are transformations, like chemical reactions, except that in this case it is the subatomic particles in the nuclei of atoms that react with each other. There can be **fusion** reactions, in which nuclei merge to create a larger nucleus, respresenting a new chemical element; and there are **fission** reactions, in which a single nucleus, usually of a heavy element with a large number of protons and neutrons, splits into two or more smaller nuclei. In either type of reaction, energy is released as some of the matter is converted according to Einstein's formula. In the 1920s, many physicists explored these possibilities, and by the 1930s Hans Bethe had suggested a specific reaction sequence that might operate in the Sun's core.

Bethe envisioned a fusion reaction in which four hydrogen nuclei (each consisting of only a single proton) combine to form a helium nucleus, made up of two protons and two neutrons. The reaction actually occurs in several steps (see Appendix 11): (1) Two protons combine to form **deuterium,** a type of hydrogen that has a proton and a neutron in its nucleus (one of the two protons undergoing the reaction converts itself into a neutron by emitting a positively charged particle called a **positron**) and another particle called a **neutrino,** which has very unusual properties described in Astronomical Insight 17.1; (2) the deuterium combines with another proton to create an isotope of helium (^{3}He) consisting of two protons and one neutron; and (3) two of these ^{3}He nuclei combine, forming an ordinary helium nucleus (^{4}He, with two protons and two neutrons in the nucleus) and releasing two protons. At each step in this sequence, heat energy is imparted to the surroundings in the form of kinetic energy of the

ASTRONOMICAL INSIGHT 17.1

The Solar Neutrino Mystery

 Nearly all that we know about the reactions in the Sun's core is based on theoretical calculations, and of course astronomers and physicists have been interested in verifying the results of these calculations. We think that the theory is basically correct, because it neatly accounts for the Sun's observed energy output, but nevertheless a more direct confirmation has been sought.

One way to do this was suggested by the early work of Enrico Fermi, who had postulated the existence of a tiny subatomic particle called the **neutrino,** which, his calculations showed, ought to be released at certain stages in nuclear reactions. This is a strange little particle, having no mass and no electrical charge, but which can carry away small amounts of energy, thus balancing the equations describing the reactions. Because of its ephemeral properties, a neutrino hardly interacts with anything at all (it can interact with other matter only through the *weak nuclear force,* which is mentioned in Astronomical Insight 5.2). Thus, neutrinos produced in the Sun's core ought to escape directly into space at the speed of light, in sharp contrast with the photons from the core, which, as we have seen, take a very long time to get out.

A possible test of the reactions going on inside the Sun, therefore, would be to measure the rate at which neutrinos are coming out.

Unfortunately, the same properties that allow them to escape the Sun so easily also make them very difficult to catch and count. There is an indirect technique, however, based on the fact that certain nuclear reactions can be triggered by the impact of a neutrino.

A chlorine atom, for example, can be converted into a radioactive form of argon when it encounters a neutrino. Accordingly, an experiment was set up a few years ago and is still in operation, in which a huge underground tank containing a chlorine compound is monitored to see how many of its atoms are being converted into radioactive argon atoms, which in turn indicates how many neutrinos are passing through the tank.

Much to everyone's surprise, neutrinos have not been detected coming from the Sun in the expected numbers, and at present no satisfactory explanation has been found. Clearly there is something not yet understood about nuclear reactions in the Sun, but it is not clear whether the problem is in our understanding of nuclear reactions or in our understanding of the internal properties of the Sun. Unfortunately, the neutrinos that can react with chlorine are produced by a minor offshoot of the main energy-producing reactions in the Sun, so it is possible that the main reaction sequence occurs as expected, but that the rate at which this other reaction occurs is slower than thought.

Ideally, we would like to detect the neutrinos produced directly in the proton-proton chain. These neutrinos are even harder to catch, however, and the first attempts to detect them have only begun recently (in 1990). The neutrinos produced by the principal reaction in the Sun can trigger reactions in a rare element called gallium, and it has taken time (and substantial funds) to accumulate the needed gallium. Very preliminary results indicate that the neutrino shortage is real; *none* have been detected so far. This is, to say the least, a major surprise, and physicists are carefully reexamining the experiment, looking for any possible flaws. It seems unlikely that instrumental problems are the explanation, however, because most such problems would be far more likely to produce extra events that could be mistakenly counted as neutrinos, rather than somehow allowing real neutrinos to go undetected.

Meanwhile, experimentalists and theorists are pursuing possible solutions to the neutrino mystery. One recent suggestion invokes another kind of elementary particle whose existence is not as well established as that of the neutrino. A family of so-called Weakly Interacting Massive Particles (WIMPs) has been postulated to exist. These particles, relatively massive by subatomic standards but, like neutrinos, capable of interacting with other matter only through the weak nuclear force, could exist in the Sun in large quantities. If so, they are capable of balancing the Sun's energy budget with the low number of neutrinos that have been observed. Thus the low numbers of one kind of exotic particle, the neutrino, could be explained by invoking the presence of another, the WIMP. WIMPs also have the potential to solve far-reaching questions about the structure of our galaxy and the distribution of

(continued on next page)

ASTRONOMICAL INSIGHT 17.1
The Solar Neutrino Mystery (continued)

mass in the universe, as we discuss in later chapters.

Another recent line of pursuit shows perhaps even more promise for solving the solar neutrino problem, however. Over the past decade or so, several scientists have suggested that neutrinos may have a property that would allow a significant fraction of them to escape detection by the methods involving chlorine or gallium detectors. There are three types of neutrinos (particle physicists often speak of them as three "flavors" of neutrinos), and the current detectors can detect only one of these types. It has been suggested that neutrinos traveling through space can change types, so that only a fraction of them would have the right "flavor" when they reached the Earth from the Sun. The others would escape detection, thus accounting for the low number being seen in the experiment. These theoretical ideas have been revitalized by the recent failure of the gallium detectors to find any solar neutrinos. The interplay between solar physics and elementary particle physics that has been stimulated by the neutrino problem has been fascinating and will almost certainly lead to new physical insights.

The WIMP hypothesis may be tested within the next several years, if the U.S. builds the **Superconducting Supercollider,** a very powerful particle accelerator planned to be capable of producing WIMPs if indeed they exist. The federal government has taken preliminary steps toward construction of the supercollider, but it is very expensive, and completion of the project, to be located in Texas, will take several years.

---

particles that are produced, and in step (2) a γ-ray photon is emitted as well.

The net result of this reaction, which is called the **proton-proton chain,** is that four hydrogen nuclei (protons) combine to create one helium nucleus. The end product has slightly less mass than the ingredients, 0.007 of the original amount having been converted into energy. We will learn in Chapters 20 and 22 that other kinds of nuclear reactions take place in some stars, but the vast majority produce their energy by the proton-proton chain.

It is easy to see that the proton-proton chain can produce enough energy to keep the Sun shining for billions of years. The total mass of the Sun is about 2×10^{33} grams. Only the innermost portion of this, perhaps 10 percent, undergoes nuclear reactions, so the Sun started out with about 2×10^{32} grams of mass available for nuclear reactions. If 0.007 of this, or 1.4×10^{30} grams, is converted into energy, then to find the total energy the Sun can produce in its lifetime, we multiply this mass times c^2, finding a total energy of $E = (1.4 \times 10^{30} \text{ g}) (3 \times 10^{10} \text{ cm/s}^2)^2 = 1.3 \times 10^{51}$ erg. The rate at which the Sun is losing energy is its luminosity, which is about 4×10^{33} ergs per second. At this rate, the Sun can last for

$$\frac{1.3 \times 10^{51} \text{ ergs}}{4 \times 10^{33} \text{ ergs/sec}} = 3.2 \times 10^{17} \text{ seconds},$$

or just about 10 billion years. Geological evidence shows that the solar system is now about 4.5 billion years old, so we can expect the Sun to keep shining for another 5 billion years or more. (In Chapter 22, we will see what happens to stars like the Sun when the nuclear fuel, hydrogen, runs out).

Nuclear fusion reactions can take place only under conditions of extreme pressure and temperature, because of the electrical forces that normally would keep atomic nuclei from ever getting close enough together to react. Nuclei, which have positive charges, must collide at extremely high speeds to overcome the repulsion caused by their like electrical charges. The speed of particles in a gas is governed by the temperature, and only in the very center of the Sun and other stars is it hot enough (around 15 million degrees) to allow the nuclei to collide fast enough to fuse. This is why only the innermost portion of the Sun can ever undergo reactions. The high pressure in the Sun's core causes nuclei to be crowded together very densely, and this means that collisions will occur very frequently, another requirement if a high reaction rate is to occur.

MODELING THE SUN

Astronomers have learned a great deal about the interior of the Sun by combining observations of its surface with theoretical calculations of the internal structure. The theoretical models are constructed by writing equations to represent the known physical processes that take place in the interior, and then solving these equations for many different positions inside the Sun. As an example of one such equation, there is a relationship stating that the inward force of gravity is equal to the outward force of pressure. Another gives the rate of energy flow outward (this depends on the density and temperature at each point), and another provides the rate of energy production by nuclear reactions (which depends on temperature, density, and composition).

It can be shown that there is only one possible solution to the set of equations. The major uncertainty in the process lies in the fact that some of the physical parameters are either too complex to be represented accurately or are not well enough known, so that approximations have to be used. Furthermore, it is possible that some physical processes, such as internal rotation or the role of the magnetic field inside the Sun, are more important than allowed for in the equations at present.

Despite these uncertainties, astronomers believe that the results of model calculations for the Sun are at least approximately correct, because they reproduce the observed surface conditions quite well. Of course, one exception is the failure to detect the predicted number of neutrinos, and this has caused some consternation about the models (see Astronomical Insight 17.1).

The results of one such model calculation are summarized in Table 17.3, which lists the temperature and density at several points inside the Sun (and extending throughout its atmosphere, which is discussed in the next section). The fraction of the Sun's mass contained inside each distance from the center is listed also.

STRUCTURE OF THE SOLAR ATMOSPHERE

Observations of the Sun's appearance when viewed in different wavelengths of light make it clear that the outer layers are divided into several distinct zones (Fig. 17.4). The "surface" of the Sun that we see in

Table 17.3 Structure of the Sun

Zone	Radius[a] (R)	Radius[a] (km)	Temperature (K)	Density (g/cm³)	Mass Fraction (M)[a]
Interior	0.00	0	1.6×10^7	160	0.000
	0.04	28,000	1.5×10^7	141	0.008
	0.1	70,000	1.3×10^7	89	0.07
	0.2	139,000	9.5×10^6	41	0.35
	0.3	209,000	6.7×10^6	13	0.64
	0.4	278,000	4.8×10^6	3.6	0.85
	0.5	348,000	3.4×10^6	1.0	0.94
	0.6	418,000	2.2×10^6	0.35	0.982
	0.7	487,000	1.2×10^6	0.08	0.994
	0.8	557,000	7.0×10^5	0.018	0.999
	0.9	627,000	3.1×10^5	0.0020	1.000
	0.95	661,000	1.6×10^5	0.0004	1.000
	0.99	689,000	5.2×10^4	0.00005	1.000
	0.995	692,000	3.1×10^4	0.00002	1.000
	0.999	695,300	1.4×10^4	0.0000001	1.000
Photosphere	1.000	695,990	6.4×10^3	3.5×10^{-7}	1.000
	1.000	+ 280	4.6×10^3	4.5×10^{-8}	1.000
Chromosphere	1.000	+ 320	4.6×10^3	3.1×10^{-8}	1.000
	1.001	+ 560	4.1×10^3	3.6×10^{-9}	1.000
(Transition)	1.002	+ 1,900	8.0×10^3	3.4×10^{-13}	1.000
	1.003	+ 2,400	4.7×10^5	4.8×10^{-15}	1.000
Corona	1.003	+ 2,400	5.0×10^5	1.7×10^{-15}	1.000
	1.2	+140,000	1.2×10^6	8.5×10^{-17}	1.000
	1.5	+348,000	1.7×10^6	1.4×10^{-17}	1.000
	2.0	+696,000	1.8×10^6	3.4×10^{-18}	1.000

[a]The radii are expressed in fractions of the Sun's radius $R_\odot$ as well as in kilometers. Above 695,990 km, the kilometer values are heights above the Sun's "surface," or lower boundary of the photosphere. The mass fractions are expressed in units of the Sun's mass $M_\odot$, which has the value 1.989×10^{33}g. (Data from C. W. Allen, 1973, *Astrophysical Quantities*, London: Athlone Press.)

ASTRONOMICAL INSIGHT 17.2

Measuring the Sun's Pulse

In the early 1960s, measurements of spectral lines formed in the solar photosphere revealed small Doppler shifts that alternated between blue- and redshifts. This seemed to indicate that the Sun's surface was rising and falling with a period of about 5 minutes. The reported Doppler shifts were quite small (less than one one-hundredth of an Ångstrom), near the limit of detectability, and the measurements were correspondingly difficult. Today the existence of this and other periodic oscillations of the Sun's surface is the basis for an entire research field, sometimes called solar seismology.

Oscillations of the solar surface can be likened to a musical instrument with a resonant cavity. When waves of any kind are created inside a cavity with reflecting walls, certain wavelengths or frequencies persist, while all others die out. The reason is that waves reflected from the walls interfere with incoming waves, the valleys and crests either adding together or canceling each other out. In an enclosed space, the wavelengths that add together constructively are determined by the dimensions of the enclosure; such wavelengths are characteristic of the particular enclosure and correspond to its so-called resonant frequency. In a musical wind in-

(National Optical Astronomy Observatories)

strument, this frequency determines the pitch of the tone that is emitted. The musician can control the pitch by altering the dimensions of the internal cavity (by opening and closing valves in most cases).

Inside the Sun, there are no walls, but physical conditions that vary with depth can create barriers as real as walls for certain types of waves. It has been deduced that sound waves (i.e., compressional waves) can persist in the Sun's convective zone, and are confined above and below by the manner in which the Sun's temperature varies with depth.

Sound waves in a certain range of wavelength are reflected back as they travel farther in or farther out, and this creates a resonant cavity in the outermost portion of the Sun for sound waves whose period of pulsation is about 5 minutes. These waves create a standing pattern of crests and valleys on the solar surface, and at any particular point on the surface, the gas rises and falls with this period.

This phenomenon is quite interesting for its own sake but is worthy of careful study for another reason as well: The properties of the waves or oscillations of the

ASTRONOMICAL INSIGHT 17.2

solar surface can reveal information on conditions in the solar interior, where the waves arise. For example, study of the frequency range of the waves can yield information on exactly how the temperature varies with depth inside the Sun, since it is this variation that determines the location of the lower boundary of the resonant cavity. Perhaps even more interestingly, analysis of the surface oscillations can provide information on the internal rotation of the Sun. This is possible because waves traveling over the solar surface in opposite directions are either spread out or squeezed closer together, depending on whether they move in the direction of the Sun's rotation or in the opposite direction. This means that waves going in opposite directions have different frequencies, and the difference in frequency depends on the speed of rotation of the Sun in the layer where the waves arise. In effect, it is possible to infer the Sun's internal rotational velocity at the lower boundary of the resonant cavity in which the waves oscillate.

The determination of internal rotational speeds using observations of surface oscillations is very complex and requires very sophisticated treatment of extensive quantities of data. At present, it has not been possible to carry this effort very far, but it has been determined from the 5-minute oscillations that the Sun is rotating more rapidly near the bottom of the convective zone than at the surface, leading to speculations that the rotation near the core

may be even more rapid. If so, this could have important implications for the Sun's overall structure and evolution, and a great deal of importance is being attached to further study of this phenomenon.

While the 5-minute oscillation of the Sun is now well established and understood, other reported oscillations are not. There is a 3-minute oscillation that may be caused by a similar resonant cavity at the level of the chromosphere, and apparently there are oscillations lasting much longer that are driven by waves in very deep resonant cavities, near the solar core. These controversial oscillations, having periods of 40 and 160 minutes, have the potential of revealing conditions such as the rotation rate very deep inside the Sun, and their study is therefore of great importance.

For a variety of reasons, the detection of long-period oscillations is very difficult. It is important, for example, to observe the Sun continuously over many cycles of the oscillations, which means it is necessary to observe without interruption for several days. This may sound impossible, because of the Sun's daily rising and setting, but in fact it can be done from one of the Earth's poles during local summer. Therefore, solar studies have been done from the scientific research station at the South Pole. During one of these studies, involving six days of continuous observations of the Sun, the 160-minute oscillation was convincingly detected.

Another approach to making continuous observations of the os-

cillations is to establish a worldwide network of observers, so that the Sun can be observed, with similar instruments, throughout the Earth's 24-hour rotation period. One such network, called the Global Oscillation Network Group (GONG), has been established by the U.S. National Solar Observatory. GONG, which will consist of several identical observing stations located at sites around the world, is expected to provide data on solar oscillations for years to come.

Beyond simply detecting the existence of the long-period oscillations, further refinements are needed if they are to be exploited as tools for probing the Sun's deep interior. To observe them in enough detail to do this requires entirely new technology, and at present there are several research groups at work developing the necessary instruments. The problem is to find a way to observe very small Doppler shifts in spectral lines at many positions on the Sun's surface, and to make these measurements repeatedly at very closely spaced intervals of time. The velocities involved are very small, only a meter per second or less, so the Doppler shift in a typical visible-wave length line is only a few millionths of an Ångstrom. To accurately measure such small shifts is a formidable task, but one that probably will be successfully completed within a few years. When this has been accomplished, a wealth of new information on the internal properties of our local star will be mined.

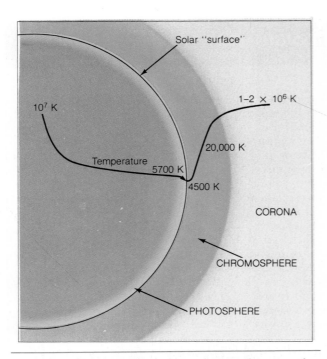

Figure 17.4 The structure of the Sun's outer layers. *This diagram shows the relative temperatures of the convective zone, the photosphere, the chromosphere, and the corona.*

visible wavelengths is the **photosphere,** with a temperature ranging between 4,000 and 6,500 K. When we view the Sun at wavelength of the strong line of hydrogen at 6,563 Å, we see the **chromosphere,** a layer above the photosphere whose temperature is 6,000 to 10,000 K. Outside of that is the very hot, rarefied **corona,** best observed at X-ray wavelengths, whose temperature is 1 to 2 million degrees. Between the chromosphere and the corona is a thin region, called the **transition zone,** where the temperature rapidly rises. Overall, the tenuous gas within and above the photosphere is referred to as the solar atmosphere. If we consider the temperature throughout this region, we find that it decreases outward through the photosphere, reaching a minimum value of about 4,000 K. From there the trend reverses itself, and the temperature begins to rise as we go farther out. The chromosphere, immediately above the temperature minimum, is perhaps 2,000 km thick. Above there the temperature rises very steeply within a few hundred kilometers, to the coronal value of over a million degrees. Clearly, something is adding extra heat at these levels; shortly we will consider where this heat comes from.

First let us discuss the photosphere, the "surface" of the Sun as we look at it in visible light. This is the level

Figure 17.5 Fraunhofer lines. *This is a portion of the Sun's spectrum, showing some of the dark lines cataloged by Fraunhofer. The two lines near the far left in the upper segment are due to ionized calcium. The center segment shows several lines of the Balmer series of hydrogen, and the strongest Balmer line, H-alpha, is seen in the red portion of the lower segment. (K. Gleason, Sommers Bausch Observatory, University of Colorado)*

we see, because it is here that the density becomes great enough for the gas to be opaque, so that we can see no farther into the interior. It is in the photosphere that the Sun's absorption lines (Fig. 17.5) are formed, as the atoms in this relatively cool layer absorb continuous radiation coming from the hot interior.

A photograph of the photosphere reveals a cellular appearance called **granulation** (Fig. 17.6). Bright regions, thought to represent areas where convection in

Figure 17.6 Solar granulation. *The upper photograph, obtained by a high-altitude balloon above much of the atmosphere's blurring effect, distinctly shows the granulation of the photosphere, which is due to convective motions in the Sun's outer layers. The six images in the lower figure, obtained from space, show how granules change from minute to minute (top: Project Stratoscope, Princeton University, sponsored by the National Science Foundation; bottom: Courtesy of A. Title, Lockheed Aerospace Corporation)*

the Sun's outer layers causes hot gas to rise, are bordered by dark zones where cooler gas is apparently descending back into the interior. Recent detailed studies of granules and their changes with time have shown their behavior to be more complex than previously thought: Granules can change appearance rapidly and can disintegrate quickly and disappear.

The temperature of the photosphere, roughly 6,000 K, is measured from the use of Wien's law and from the degree of ionization in the gas there. The density, roughly 10^{17} particles per cubic centimeter in the lower photosphere, was found from the degree of excitation, as described in Chapter 6. This is lower than the density of the Earth's atmosphere, which is about 10^{19} particles per cubic centimeter at sea level.

Most of what we know about the Sun's composition is based on the analysis of the solar absorption lines, so strictly speaking, the derived abundances represent only the photosphere. We have no reason to expect strong differences in composition at other levels, however, except for the core, where a significant amount of the original hydrogen has been converted into helium by the proton-proton reaction.

Figure 17.7 The solar chromosphere. *This is an image of the Sun, taken through a filter that allows only the wavelength of the red emission line of hydrogen to pass through. The Sun's chromosphere emits strongly at this wavelength whereas the photosphere does not; hence this image shows primarily the chromosphere and its structure. (National Optical Astronomy Observatories)*

The photosphere near the edge of the Sun's disk looks darker than in the central portions. This effect, called **limb darkening** (see Fig. 17.1), is caused by the fact that we are looking obliquely at the photosphere when we look near the edge of the disk. We therefore do not see as deeply into the Sun there as we do when looking near the center of the disk. Therefore, at the limb, the gas we are seeing is cooler than at disk center, and radiates less.

The chromosphere lies immediately above the temperature minimum (Fig. 17.7). The fact that this region forms emission lines tells us, according to Kirchhoff's laws, that the chromosphere is made of hot, rarefied gas, hotter than the photosphere behind it. When viewed through a special filter that allows only light at the wavelength of the hydrogen emission line at 6,563 Å to pass through (see Fig. 17.7), the chromosphere has a distinctive cellular appearance referred to as **supergranulation,** similar to the photospheric granulation, but with cells some 30,000 km across instead of about 1,000 km. There is also fine-scale structure in the chromosphere, in the form of spikes of glowing gas called **spicules** (Fig. 17.8). These come and go, probably at the whim of the magnetic forces that seem to control their motions.

The outermost layer of the Sun's atmosphere is the corona, which extends a considerable distance above the photosphere and chromosphere. The corona is irregular in form, patchy near the Sun's surface, but with radial streaks at great heights suggestive of outflow from the Sun (Fig. 17.9). The density of the coronal gas is very low, only about 10^9 particles per cubic centimeter.

As we have already seen, the corona is very hot, containing highly ionized gas. The source of the energy that heats the corona to such extreme temperatures is not well understood, although a general picture has emerged, as we will see.

X-ray observations reveal that the corona is not uniform, but instead has a patchy structure (Fig. 17.10). There are large regions that appear dark in an X-ray photograph of the Sun, where the gas density is even lower than in the rest of the corona. These regions are called **coronal holes** and, as we shall see, are probably created and maintained by the Sun's magnetic field. The coronal holes, as well as the overall shape of the corona, vary with time (Fig. 17.11), showing that the corona in general is a dynamic region. Another impressive sign of this is presented by the **prominences** (Figs. 17.12 and 17.13), great streamers of hot gas

Figure 17.8 Spicules. *This photograph shows the transient features in the chromosphere known as spicules. These spikes of glowing gas, which are apparently shaped by the Sun's magnetic field, come and go irregularly. (Sacramento Peak Observatory, National Optical Astronomy Observatories)*

Figure 17.9 The corona. *The upper photograph, obtained during a total solar eclipse, shows the type of structure commonly seen in the corona. There are giant loop-like features and an overall appearance of outward streaming. Bits of more intense light from the chromosphere are seen around the edges of the Moon's occulting disk. Visible at left is the planet Venus. The lower image, obtained from space using a disk to block out the Sun itself, illustrates brightness gradations with false colors. (top: High Altitude Observatory, National Center for Atmospheric Research, sponsored by the National Science Foundation; bottom: NASA)*

stretching upward from the surface of the Sun, taking on an arc-shaped appearance. These are usually associated with sunspots, and both phenomena are linked to the solar activity cycle (to be discussed shortly).

The hot outer layers of the Sun have provided astronomers with a second major mystery regarding the solar energy budget. In contrast with the mystery of the Sun's internal energy source, the mechanism for heating the chromosphere and corona has not yet been entirely deduced. There is a great deal of energy available in the form of gas motions in the convection zone just beneath the solar surface, and it is generally accepted that the heating of the corona somehow comes from this energy. It is not clear how the energy is transported to such high levels, however, although there are several ideas, each invoking some form of waves. Waves of any kind carry energy, and it has been sug-

gested that either sound waves or magnetic waves of some kind are the agents that transfer energy from the convection zone to the corona. There is certainly enough energy in the convective motions in and below the photosphere; the problem is one of understanding how this energy is transported into the higher levels.

Figure 17.10 An X-ray portrait of the corona. *This image, obtained by Skylab astronauts, shows the Sun in X-ray light, which reveals only very hot, dense regions. The bright regions are places where the corona is especially dense, and the dark regions, known as coronal holes, are places where it is much more rarefied. The structure seen here changes with time. (NASA)*

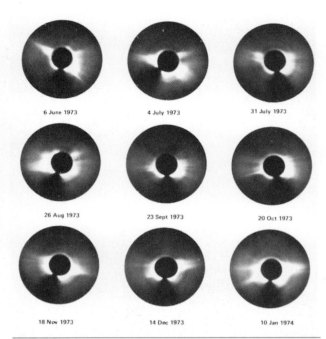

Figure 17.11 Changes in the corona. *This series of photographs was obtained using a special shutter device that blocks out the solar disk and allows the corona to be observed without waiting for a total solar eclipse. Here we see that the structure of the corona changes quite markedly over a period of a few months. (High Altitude Observatory, National Center for Atmospheric Research, sponsored by the National Science Foundation)*

Recent satellite observations in ultraviolet and X-ray wavelengths have shown that stars similar to the Sun in general type also have chromospheric and coronal zones, so if we can understand how the Sun operates, we will also gain a deeper understanding of how other stars work. Similarly, observations of other stars, particularly of the relationship of chromospheric and coronal activity to stellar properties such as age and rotation, can help us learn more about the Sun.

THE SOLAR WIND

The long, streaming tail of a comet always points away from the Sun, regardless of the direction of motion of the comet. The significance of this was fully realized in the late 1950s, when the first U.S. satellites revealed the presence of the Earth's radiation belts and the fact that they are shaped in part by a steady flow of charged particles from the Sun. The solar wind reaches a speed near the Earth of 300 to 400 km/sec. It is this flow of charged particles that forces cometary tails always to point away from the Sun.

The existence of the solar wind is evidently a natural by-product of the same heating mechanisms that produce the hot corona of the Sun. It was originally thought that particles in this high-temperature region move about with such great velocities that a steady trickle escapes the Sun's gravity, flowing outward into space. Subsequently, however, X-ray observations of the Sun have shown that the situation is not that simple. It is the solar magnetic field that governs the outward flow of charged particles. The coronal holes, mentioned earlier (see Fig. 17.10), are regions where the magnetic field lines open out into space. Charged particles such as electrons and protons, constrained by electromagnetic forces to follow the magnetic field lines, therefore escape into space only from the coronal holes. The speed of the solar wind is relatively low close to the Sun, but accelerates outward, quickly reaching the speed of 300 to 400 km/sec already mentioned, after which it is

Figure 17.12 A prominence. *Here the looplike structures thought to be governed by the Sun's magnetic field can be readily seen. This photograph was obtained by Skylab astronauts using an ultraviolet filter. (NASA)*

Figure 17.13 An eruptive prominence. *The striking sequence at top shows an outburst of ionized gas from a prominence seen on the Sun's limb. The photographs were obtained by using a special device to block out the Sun's disk. At bottom is an ultraviolet view of another eruptive prominence, showing the extent reached in less than two hours. (top: High Altitude Observatory, National Center for Atmospheric Research, sponsored by the National Science Foundation; bottom: NASA)*

Figure 17.14 The solar wind. *This far-ultraviolet image was obtained by the Solar Maximum Mission in Earth orbit. A special color separation technique was used to show the outward flow of gas. (Data from the Solar Maximum Mission, NASA and the High Altitude Observatory National Center for Atmospheric Research, sponsored by the National Science Foundation)*

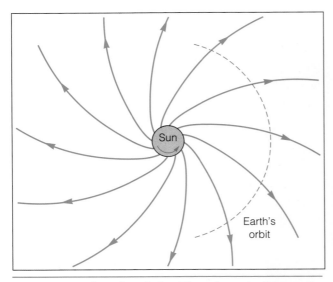

Figure 17.15 The solar wind. *This schematic diagram illustrates how ionized gas from the Sun spirals outward through the solar system in a steady stream. The solar magnetic field creates sectors of variable density in the wind.*

nearly constant. The wind nearly reaches its maximum speed by the time it passes the Earth's orbit, and beyond there it flows steadily outward, apparently persisting at least to a distance beyond the orbit of Neptune. (The *Pioneer 10* spacecraft, on its way out of the solar system, still detected solar wind particles as it crossed the orbit of Neptune in mid-1983, and it has not yet found a limit to the solar wind.) At some point in the outer solar system, the wind is thought to come to an abrupt halt where it runs into an invisible and tenuous wall of matter swept up from the interstellar medium that surrounds the Sun.

Most of the direct information we have on the solar wind comes from satellite and space probe observations (Fig. 17.14), because the Earth's magnetosphere shields us from the wind particles. Solar wind monitors are placed on board the majority of spacecraft sent to the planets. One striking discovery has been the fact that the wind is not uniform in density, but instead seems to flow outward from the Sun in sectors, as though it originates only from certain areas on the Sun's surface. This is explained by the X-ray data already mentioned, which indicate that the wind emanates only from the coronal holes. Because the base of the wind is rotating with the Sun, the wind sweeps out through space in a great curve, similar to the trajectory of water from a rotating lawn sprinkler (Fig. 17.15).

Occasional explosive activity occurring on the Sun's surface releases unusual quantities of charged particles, and some three or four days later, when this burst of ions reaches the Earth's orbit, we experience disturbances in the ionosphere that can interrupt short wave radio communications and cause auroral displays. These **magnetic storms,** as they are often called, are outward manifestations of a much more complex overall interaction between the Sun and the Earth.

SUNSPOTS, SOLAR ACTIVITY, AND THE MAGNETIC FIELD

Dark spots on the Sun's disk were observed centuries ago, by ancient Chinese astronomers. The spots were cited by Galileo as evidence that the Sun is not a perfect, unchanging celestial object, but rather has occasional flaws. Observations of the spots, which individually may last for months, have, over the centuries since, revealed some very systematic behavior. The number of spots varies, reaching a peak every 11 years, and during the interval between peak numbers of spots, their locations on the Sun change steadily from middle latitudes to a concentration near the equator. When the sunspot number is high, most of them appear in activity bands about 30° north or south of the solar equator. During the next 11 years, the spots tend to lie ever closer to the equator, and by the end of the cycle they are nearly on it (Fig. 17.16). By this time, the first spots of the next cycle may already be forming at middle latitudes. A plot of sunspot locations from cycle to cycle clearly shows this and is called a "butterfly diagram" because of the shape the pattern of spots makes (Fig. 17.17).

A hint at the origin of the spots was found when their magnetic properties were first measured. This is accomplished by applying spectroscopic analysis to the light from the spots, taking advantage of the fact that the energy levels in certain atoms are distorted by the presence of a magnetic field. This in turn causes the spectral lines formed by those levels to be split into two or more distinct, closely spaced lines. The degree

of line-splitting, which is referred to as the **Zeeman effect,** depends on the strength of the magnetic field, so the field can be measured from afar simply by analyzing the spectral lines to see how widely split they are. This technique works for distant stars as well as for the Sun (except that for distant stars we cannot measure different portions of the disk separately).

When the first measurements of the Sun's field were made in the first decade of this century, the field was found to be especially intense in the sunspots, about 1,000 times stronger than in the surrounding gas. It is now thought that the intense magnetic field associated with a sunspot inhibits convection, so that heated gas from below cannot rise up to the surface. Hence the spot is cooler than its surroundings and therefore is not as bright (Fig. 17.18). The typical temperature in a spot is about 4,000 K, compared to the roughly 6,000 K temperature of the photosphere. Using Stefan's law, we see that the intensity of light emitted in a spot compared to the surroundings is $(4,000/6,000)^4$

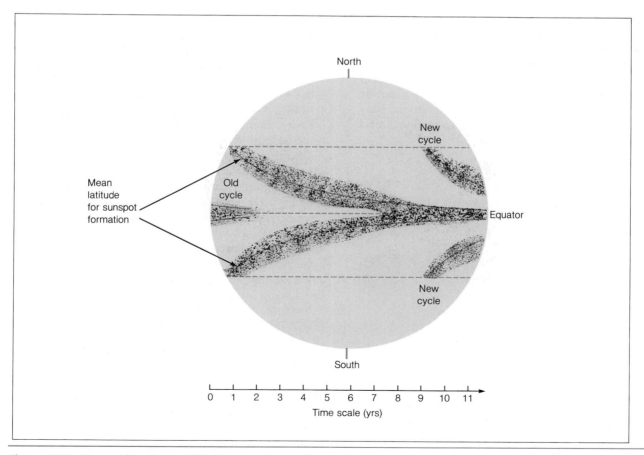

Figure 17.16 Sunspot locations.   *This drawing shows where sunspots appear at different times in the sunspot cycle.*

Figure 17.17 The butterfly diagram. *This is a plot of the latitudes of observed sunspots through several cycles of solar activity. At the beginning of each cycle, the spots tend to appear at mid-latitudes, and then as the cycle progresses, they arise at positions closer and closer to the equator. (Prepared by the Royal Greenwich Observatory and reproduced with the permission of the Science and Engineering Council, courtesy of J. A. Eddy)*

= 0.2; i.e. the brightness of the solar surface within a sunspot is only about one-fifth of the brightness in the photosphere outside.

When the sunspot magnetic fields were measured (Fig. 17.19), they were found to act like either north or south magnetic poles; that is, each spot has a specific magnetic direction associated with it. Furthermore, pairs of spots often appear together, the two members of a

Figure 17.18 Sunspots. *This is a telescopic view of a group of spots, showing their detailed structure. They appear dark only in comparison with their much hotter surroundings. (Mt. Wilson and Las Campanas Observatories, Carnegie Institution of Washington)*

pair usually having opposite magnetic polarities. During a given 11-year cycle, in every sunspot pair in the same hemisphere the magnetic polarities always have the same orientation. For example, during one 11-year cycle, in one hemisphere the spot to the east in each pair will generally be the one with north magnetic polarity, and the one to the west has south magnetic polarity. During the next cycle, all the pairs will be reversed, with the south magnetic spot to the east, and the north polar spot to the west. In the other hemisphere, these relative orientations are exactly the opposite. Between cycles, when this arrangement is reversing itself, the Sun's overall magnetic field also reverses, with the solar magnetic poles exchanging places. It is actually 22 years before the Sun's magnetic field and sunspot patterns repeat themselves, so the solar cycle is truly 22 sunspot years long (but note recent evidence that things are even more complicated, as discussed a little later in this chapter).

Sunspot groups, often called **solar active regions** (Fig. 17.20), are the scenes of the most violent forms of solar activity, the **solar flares.** These are gigantic outbursts of charged particles, as well as visible, ultraviolet, and X-ray emission, created when extremely hot gas spouts upwards from the surface of the Sun (Fig. 17.21). Flares are most common during sunspot maximum, when the greatest number of spots are seen on the solar surface. Close examination of flare events shows that the trajectory of the ejected gas is shaped by the magnetic lines of force emanating from the spot

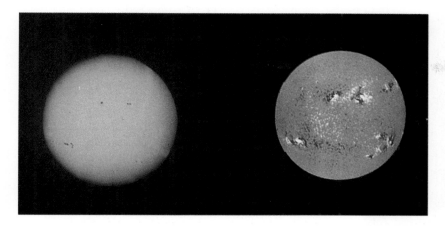

Figure 17.19 Magnetic fields on the Sun. *The magnetogram, or map of magnetic field strength (right), was made through the measurement of the splitting of spectral lines by magnetic fields (the Zeeman effect). Here colors (dark blue and yellow) are used to indicate regions of opposing magnetic polarity. Note that the polarities of east-west pairs in the southern hemisphere are reversed with respect to those in the north. At left is a visible-light photograph of the Sun taken at the same time, so that the correspondence between the surface magnetic field and the visible sunspots can easily be seen. (National Optical Astronomy Observatories)*

where the flare occurred. Charged particles flow outwards from a flare, some of them escaping into the solar wind. If the solar wind flowing from the site of the flare reaches the Earth, then the Earth is bathed by an extra dosage of solar wind particles some three days later, creating effects on radio communications that have already been described. The extra quantity of charged particles entering the Earth's upper atmosphere can also be responsible for unusually widespread and brilliant displays of aurorae. Apparently flares occur when twisted magnetic field lines suddenly reorganize themselves, releasing heat energy and allowing huge bursts of charged particles to escape into space.

Because of the connection between the solar activity cycle and reversals in the Sun's magnetic field, a model for the activity cycle was developed some years ago in which sunspots are envisioned as locations where tubes

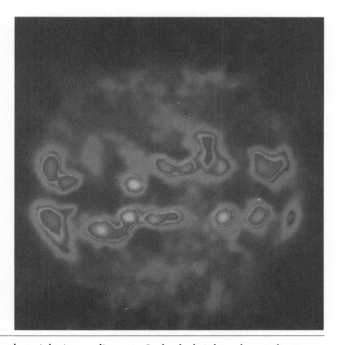

Figure 17.20 Solar active regions.   *At left is an X-ray image and at right is a radio map. In both, bright colors indicate the solar active regions, which emit strongly in X-ray and radio wavelengths. (left: NASA; right: National Radio Astronomy Observatories)*

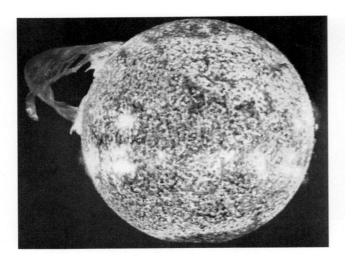

Figure 17.21 A major flare. *(Left) The gigantic looplike structure in this ultraviolet photograph obtained from Skylab is one of the most energetic flares ever observed. Supergranulation in the chromosphere is easily seen over most of the disk. (NASA)*

Figure 17.22 Flux tube model for solar activity. *(Below) This depicts the model in which sunspots are thought to be locations where magnetic flux tubes break through the surface. In this scenario, magnetic lines connecting the Sun's poles are bent by differential rotation (a) then stretched around the Sun's interior (b and c), becoming tangled and twisted so that they break through, forming pairs of oppositely polarized sunspots (d).*

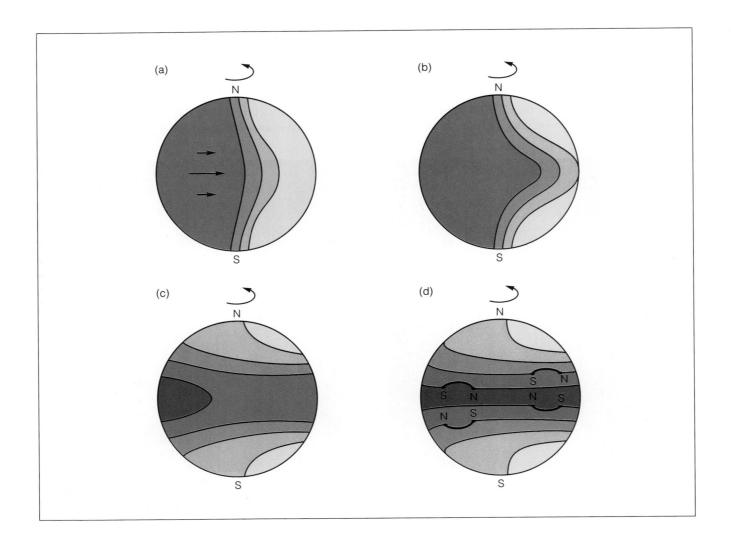

of magnetic field lines break through the solar surface. If it is imagined that these magnetic flux tubes connect the Sun's north and south poles, then we see that when the poles switch places, the tubes must become twisted and kinked. In the process they break through the surface in more and more places, ever closer to the equator, until the overall reversal of the field is completed and the flux tubes can become relatively smooth and straight again (Fig. 17.22). In this way, the flux tube model accounts for the main features of the solar activity cycle, particularly the changes in number, latitude, and magnetic field orientation of the spots during each cycle.

Unfortunately for this theory, however, recent observations have suggested that the solar activity cycle is more complicated than previously thought, and that it must be driven by some other mechanism. It has been found that, 11 years before the appearance of the

first spots of a new cycle, regions of intense magnetic fields occur at very high solar latitudes, close to the poles. The polarities of the magnetic fields are arranged in east-west pairs, with the same orientation of north and south magnetic polarities that will appear when the next spot cycle starts, 11 years later. Thus it appears that these regions of magnetic enhancement are early precursors of the next sunspot cycle. If so, then a typical cycle consists of an initial period where no spots are seen, but pairs of magnetic-enhanced regions form near the poles. After 11 years, these regions have drifted down to middle latitudes, where they begin to form spots, which then increase in number as they continue moving toward the equator over the next 11 years. In this picture, it takes 22 years for the magnetic regions in one hemisphere to form and migrate toward the equator as they increase in number, and the full cycle time is 44 years (Fig. 17.23). Furthermore, two cycles

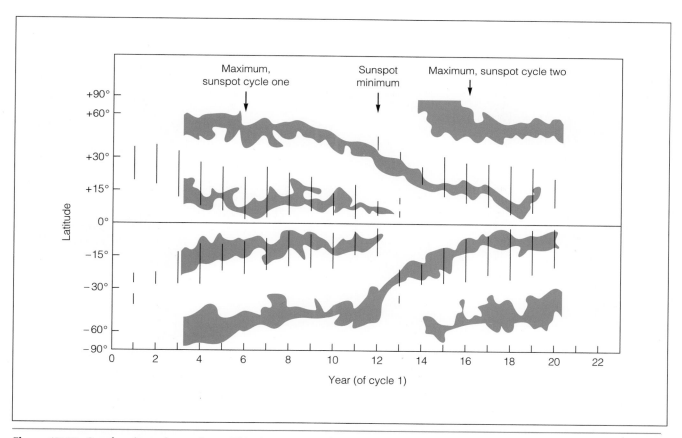

Figure 17.23 Overlapping solar cycles. *This diagram, in which the shaded areas represent the summed data from several different types of observations of solar activity, shows that there may be two 22-year cycles occurring at all times. The two are separated in time by one-half cycle, or 11 years, so that their patterns of active regions overlap as shown. (Adapted from a diagram in* Sky and Telescope, 73:590, 1987).

are present on the Sun at the same time, because as the new magnetically enhanced regions form near the poles to begin one cycle, the similar regions from the previous cycle are just beginning to appear as spots at middle latitudes.

A possible mechanism for this large-scale solar activity cycle has been suggested, based on the discovery of surface flows on the Sun, near the poles. These flows may indicate that convection in the Sun's outer regions begins near the poles with very large cells of hot gas from the interior that roll over as they spread, like gigantic doughnuts (Fig. 17.24). If these rolls form every 22 years and have alternating flow directions, then the surface magnetic field will become compressed between them in zones that start at high latitudes and then drift toward the equator, forming first the magnetically enhanced regions and then sunspots. If at the same time the Sun's overall magnetic field is reversing with a 22-year cycle time, then the behavior of the magnetic polarities of the spots is also explained.

Whether this new model of the solar activity cycle is correct or not, there remains the fundamental question of what causes the 22-year reversals of the solar magnetic field. Whatever it is, it probably originates very deep in the solar interior, and so we must look there to find the underlying cause of the activity cycle. Perhaps new information to be derived from studies of solar oscillations will eventually yield to the missing link.

There is some evidence for long-term changes in solar behavior, at intervals longer than 22 (or 44) years, and this also will require an explanation from deep inside the Sun, where the magnetic field is formed. The most striking known variation from the regular cycle occurred in the late 1600s, when sunspots seemed to stop altogether for over 50 years (Fig. 17.25). This period, called the **Maunder Minimum** after Edward W. Maunder, its discoverer (who also invented the butterfly diagram), was accompanied by extreme behavior of the Earth's climate, including a very cold period known in Europe as the "Little Ice Age." Only time will tell whether the Maunder Minimum was part

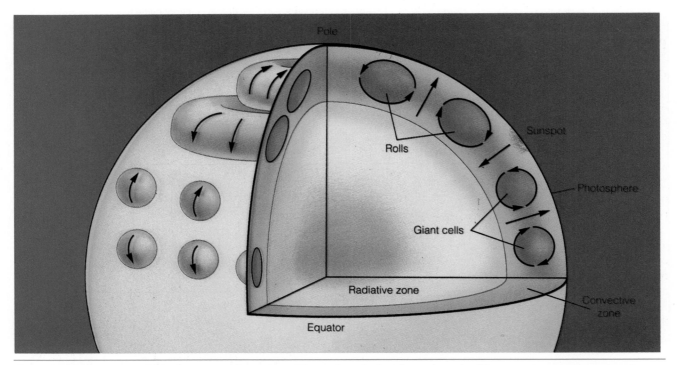

Figure 17.24 Structure within the convective zone. *These gigantic doughnut-shaped eddies, rotating in alternating directions, may be responsible for the sunspot patterns that are observed. These eddies form near the poles and then migrate toward the equator, breaking up into smaller cells in the process. The surface flows of gas that are created can compress the magnetic field in the regions where the rolls converge, causing concentrations of sunspots. (Adapted from a figure in* Sky and Telescope *73:591, 1987).*

Figure 17.25 The Sun's long-term activity.   *This diagram illustrates the relative level of activity (in terms of sunspot numbers) over three centuries. It is clear that the level varies, including a span of about 50 years (1650–1700), known as the Maunder Minimum, when there was little activity. There is evidence that the Sun has long-term cycles that modulate the well-known 22-year period. (J. A. Eddy)*

of some very long-term cyclic behavior in the Sun. The implied link between sunspot activity and weather patterns on Earth is also quite uncertain; recent studies have suggested that there may be a correlation.

PERSPECTIVE

Our Sun, the source of nearly all our energy, is a very complex body. As stars go, it is apparently normal in all respects, so we imagine that other stars are just as complex, even though we cannot observe them in such detail.

We have explored the Sun, both in its deep interior and in the outer layers that can be observed directly. We know that nuclear fusion is the source of all the energy, and that the size and shape of the Sun are controlled by the balance between gravity and pressure throughout the interior. Somehow heat is transported above the surface, keeping the chromosphere and the corona hotter than the photosphere. Perhaps most intriguing of all is the solar activity cycle and its relationship to the Sun's complex magnetic field.

We are now prepared to study stars in general, using the Sun as an example to help us understand some of what we learn about other stars.

SUMMARY

1. The Sun is an ordinary star, one of billions in the galaxy.

2. The Sun is a spherical, gaseous object whose temperature and density increase toward the center. The Sun emits light from its surface, whose ultimate source is nuclear reactions in its core.

3. Energy is produced in the core by nuclear fusion and is slowly transported outward by radiation, except near the surface, where energy is transported by convection.

4. The nuclear reaction that powers the Sun is the proton-proton chain, in which hydrogen is fused into helium.

5. The outer layers of the Sun govern the nature of the light it emits. These layers consist of the photosphere, with a temperature of about 6,000 K; the chromosphere, where the temperature ranges between 6,000 and 10,000 K; and the corona, where the temperature is over 1,000,000 K.

6. The photosphere, which is the visible surface, creates absorption lines, while the hotter chromosphere and corona create emission lines.

7. The excess heat in the outer layers is somehow transported there from the convective zone just below the surface, but the mechanism for transporting the heat is not well known.

8. The Sun emits a steady outward flow of ionized gas called the solar wind. The wind originates from coronal holes and is controlled by the solar magnetic field.

9. The sunspots occur in 11-year cycles that are part of the 22-year cycle of the Sun's magnetic field reversals. The spots are regions of intense magnetic fields where magnetic field lines from the solar interior break

through the surface. There is some preliminary evidence that the 22-year cycle of solar activity is part of a longer-term cycle (possibly a 44-year one).

10. Studies of other stars show that those similar to the Sun in temperature also have chromospheres, coronae, and activity cycles.

REVIEW QUESTIONS

Thought Questions

1. Explain in your own words why hydrostatic equilibrium causes the Sun to have a spherical shape. Do you think the rotation of the Sun might affect its shape?

2. Why do nuclear reactions occur only in the central core of the Sun?

3. Explain why the use of special filters to isolate the wavelengths of certain spectral lines allows astronomers to separately examine different layers of the Sun.

4. Why are the granules in the Sun's photosphere bright, and the descending gas that surrounds them relatively dark?

5. Explain, in the context of Kirchhoff's rules (Chapter 6), why the chromosphere forms emission lines while the photosphere forms absorption lines.

6. Describe the evidence for the existence of the solar wind.

7. Discuss the relationship among the Sun's magnetic field, coronal holes, and the solar wind.

8. Why do we say that the solar activity cycle is 22 years long, even though the sunspot maxima occur every 11 years? (For now, ignore the recent suggestion concerning a possible 44-year cycle.)

9. How would the appearance of sunspots be altered if they had weaker magnetic fields than their surroundings? Would the explanation of them given in the text have to be altered if this were the case?

10. Describe a research program that you could undertake to determine the effects of solar variations on the Earth's climate.

Problems

1. Suppose that an X-ray photon released at the Sun's center has a frequency of $f = 1 \times 10^{20}$ Hz. If half of its energy is converted into heat and the other half into visible photons with frequency $f = 5 \times 10^{14}$ Hz as the energy propagates outward to the Sun's surface, how many visible photons are produced, to be radiated away into space?

2. How would the lifetime of the Sun be altered if its luminosity were 10 times greater than it is and its mass were twice as large?

3. Calculate the wavelengths of maximum emission for the solar photosphere (temperature 6,500 K), the chromosphere (20,000 K), and the corona (2×10^6 K). How would each of these layers be best observed?

4. If the average speed of the solar wind between the Sun and the Earth is 250 km/sec, how long does it take for a particle to reach the Earth once it is emitted by the Sun? How is your answer related to the time it takes for a solar flare to begin to affect radio communications on the Earth?

5. Suppose the temperature in a sunspot were one-half the temperature in the surrounding photosphere. How much less intense would the radiation be from the sunspot than from the photosphere?

ADDITIONAL READINGS

Bahcall, J. N. 1990. Neutrinos from the Sun: an astronomical puzzle. *Mercury* 19:53.

———. 1990. The solar neutrino problem. *Scientific American* 262(15):54.

Foukal, P. V. 1990. The variable Sun. *Scientific American* 262(2):34.

Friedman, H. 1986. *Sun and Earth.* (New York: W. H. Freeman)

Giampapa, M. S. 1987. The solar-stellar connection. *Sky and Telescope* 74:142.

Gough, D. 1976. The shivering sun opens its heart. *New Scientist* 70:590.

Harvey, J. W., J. R. Kennedy, and J. W. Leibacher. 1987. GONG: To see inside our Sun. *Sky and Telescope* 74:470.

Newkirk, G., and K. Frazier. 1982. The solar cycle. *Physics Today* 35(4):25.

Parker, E. N. 1975. The sun. *Scientific American* 233(3):42.

Walker, A. B. C., Jr. 1982. A golden age for solar physics. *Physics Today* 35(11):61.

Wallenhorst, S. G. 1982. Sunspot numbers and solar cycles. *Sky and Telescope* 64(3):234.

Williams, G. E. 1986. The solar cycle in precambrian times. *Scientific American* 255(2):88.

Wilson. O. C., A. H. Vaughan, and D. Mihalas. 1981. The activity cycle of stars. *Scientific American* 244(2):104.

Wolfson, R. 1983. The active solar corona. *Scientific American* 148(2):104.

Zirin, H. 1988. Heading toward solar maximum. *Sky and Telescope* 76:335.

MATHEMATICAL METHODS

Surface Temperature, Solar Luminosity, and the Solar Constant

Our study of the Sun in this chapter affords us a useful opportunity to review and use some of the information introduced in Chapter 6 on the properties of light. In particular, we can apply to the Sun our knowledge of Stefan's law and the inverse square law, as we discuss some of the Sun's fundamental properties.

Stefan's law, as stated in Chapter 6, says that the energy emitted per second per square centimeter of surface area is proportional to the fourth power of the temperature. Here we quantify this statement by adding a constant term (whose value is determined by laboratory measurements) called the Stefan-Boltzmann constant and usually denoted by the Greek letter sigma (σ). Now Stefan's law can be written in equation form:

$$E = \sigma T^4,$$

where E is the energy emitted per second per square centimeter; σ is the Stefan-Boltzmann constant, whose value is $\sigma = 5.67 \times 10^{-5}$ erg/cm^{2}deg^{4}sec; and T is the surface temperature in absolute units (Kelvins).

We can now calculate the energy emitted by various surfaces. If a human body has a surface temperature of 305 K (about 90°F), then it emits

$$E = (5.67 \times 10^{-5})(305)^4 = 4.9 \times 10^5 \text{ erg/cm}^2\text{sec};$$

that is, each square centimeter of skin emits almost 500,000 erg/sec of energy. For comparison, a 100-watt light bulb emits 10^9 ergs/sec (over its entire surface), so a number as seemingly large as 500,000 erg/sec should not be alarming.

The solar surface, with a representative temperature of about 6,000 K, emits

$$E = (5.67 \times 10^{-5})(6,000)^4$$
$$= 7.4 \times 10^{10} \text{ ergs/cm}^2 \text{ sec.}$$

If we now multiply this by the Sun's surface area, we can calculate the total energy emitted by the Sun per second, which is its luminosity. The surface area of a sphere of radius R is $4\pi R^2$, so we write

$$L = 4\pi R^2 \sigma T^4,$$

where L is the luminosity (in erg/sec). This equation is commonly known as the Stefan-Boltzmann law. Substituting $R = 7 \times 10^{10}$ and $T = 6,000$ K yields

$$L = 5.4 \times 10^{33} \text{ erg/sec.}$$

This is very close to the measured value for the solar luminosity (the most accurate value today is 3.83×10^{33} erg/sec). If we adjust the surface temperature so that the luminosity calculated from this equation is equal to the observed value, then the adjusted temperature is called the *effective temperature*. For the Sun, the effective temperature is 5,780 K (you may want to calculate the solar luminosity as done above, using $T = 5,780$ K, to show that this yields approximately the correct value for L).

The intensity of sunlight that reaches the Earth is diminished according to the inverse square law. The solar luminosity is spread out over an ever-increasing area as the distance from the Sun is increased; at the Earth's distance of 1.5×10^{13} cm, the area covered is $4\pi R^2 = 4\pi (1.5 \times 10^{13} \text{ cm})^2 = 2.83 \times 10^{27}$ cm^2. If we divide the solar luminosity of 3.83×10^{33} erg/sec by this area, we find that at the Earth's distance the Sun's intensity is 1.35×10^6 erg/cm^{2}sec. This is the quantity of energy per square centimeter that reaches the Earth above the atmosphere and is known as the *solar constant*. At the Earth's surface, the intensity is diminished by more than half, because of the absorbing effects of the atmosphere. As a useful exercise, you might want to calculate the solar constant for some of the other planets, such as Mars or Mercury. (To do this, you will need distance data provided in Appendix 7.)

CHAPTER 18

Stellar Observations: Positions, Magnitudes, and Spectra

Spectra of the bright star Vega. (National Optical Astronomy Observatories)

CHAPTER PREVIEW

Nearly all observations of stars fall into one of three categories, involving measurements of position, brightness, or spectra. The first two of these methods have rather long histories. Positional measurements date back to the first human observations of the skies, and brightness measurements (although rather crude ones) were made by Hipparchus in the second century B.C. In the following sections, each method is discussed separately.

POSITIONAL ASTRONOMY

For many centuries, positional measurements were practically the only kind of observation made, as we learned in Chapters 2 and 3. Although some heed was paid to stellar brightnesses and especially to changes that occurred in the heavens, it was the positions of the stars and planets that were systematically observed and recorded. This was an inexact science until the first sextants (Fig. 18.1) and quadrants were available to provide accurate measurements of angular distances. In modern times, highly precise positions can be measured.

One of the principal techniques used today in **astrometry,** the science of measuring star positions, is to photograph the sky and to carefully measure the positions of the star images on the photographic plates. If a number of photographs of a given portion of the sky are taken and measured separately, the results can be averaged to produce a more accurate determination than is possible by measuring a single photograph. Today astronomers can measure relative stellar positions to precision of about 0.01 second of arc (recall that a second of arc is one-sixtieth of a minute of arc, and that an arcminute is one-sixtieth of a degree; there are 1,296,000 arcseconds in a full circle). New space-

Figure 18.1 A sextant. *Instruments of this type were used for centuries for the measure of star positions relative to each other. In modern times, sextants are used as tools for navigation. (The Granger Collection)*

based instruments, such as the European *Hipparcos* satellite (now in orbit, mapping the sky) and the *Hubble Space Telescope,* will improve the accuracy to 0.005 arcseconds or better.

All stars in the galaxy have individual motions (Fig. 18.2), consisting in part of the systematic rotation of the galaxy as a whole, and in part of random motions of individual stars. The situation could be likened to a traveling swarm of bees: The individual bees move randomly within the swarm, and the group as a whole has an overall motion as well. The apparent speed of a star (measured in units of changing angle with time) depends on how fast the star is actually moving and on how far away it is. If a star is too far from the Sun, even the most careful measurements will fail to reveal any motion. The apparent gradual changes in a star's position are called **proper motions** and are expressed in terms of arcseconds per year.

Astrometric measurements reveal another kind of motion, one referred to often in Chapters 3 and 4. From ancient times through the Renaissance, astronomers were perplexed by the fact that no stellar parallax could be measured, even with the measurement

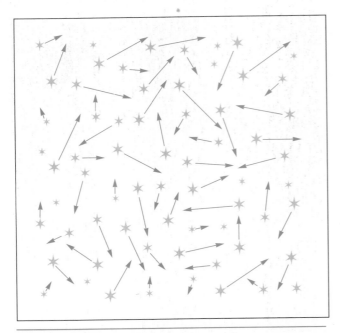

Figure 18.2 Proper motions. *Each star has its own motion on the sky. We call the angular movement of a star its proper motion. Proper motions are small (at most a few arcseconds per year), and require separate photographs taken years apart in order to be measured.*

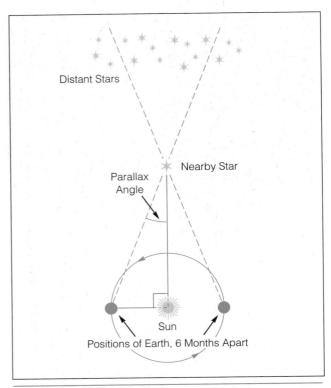

Figure 18.3 Stellar parallax. *As the Earth orbits the Sun, our line of sight to a nearby star varies in direction, so that the nearby star appears to move back and forth with respect to the more distant background stars. The parallax angle p is defined as one-half of the total angular motion. If p is one arcsecond, the distance to the star is 206,265 AU, or one parsec. This drawing is not to scale.*

precision available to Tycho Brahe. This failure to detect a parallax led many, including Brahe, to reject the notion that the Earth was orbiting the Sun, since it seemed obvious that if it were, the stars would appear to shift back and forth as a reflection of our motion. Eventually, the heliocentric theory won out, but stellar parallax was still undetected.

Finally, in 1838, parallaxes were detected independently by F. W. Bessel, F. G. W. Struve, and T. Henderson, each observing a different nearby star. This was clearly a case of a great discovery awaiting the development of suitable techniques for making the measurement. The reason stellar parallaxes had defied earlier observers became obvious: Even for the closest stars, the maximum shift in position was just over one second of arc! The stars are simply much farther away than the ancient astronomers had dreamed possible. The annual parallax motion of the star Alpha Centauri, the nearest to the Sun, is comparable to the angular diameter of a dime as seen from a distance of about 2 miles, a very small angle indeed.

The successful detection of stellar parallax led to a direct means of determining distances to stars (Fig. 18.3).

The amount of shift in a star's apparent position resulting from our own motion about the Sun depends on how distant the star is: the closer it is, the bigger the shift. The parallax p is defined as one-half the total angular shift of a star during the course of a year. This angle, once it is measured, tells us the distance to a star, in a unit of measure called the **parsec**. A star whose parallax is 1 arcsecond is, by definition, 1 parsec away (the word parsec, in fact, is a contraction of *parallax-second*, meaning a star whose parallax is 1 arcsecond). In mathematical terms, the distance to a star whose parallax is p is

$$d = \frac{1}{p}.$$

Thus, if a star has a parallax of 0.4 arcseconds, it is $1/0.4 = 2.5$ parsecs away. The closest star has a parallax of less than 1 arcsecond; we conclude, therefore, that even this star is more than 1 parsec away.

ASTRONOMICAL INSIGHT 18.1

Star Names and Catalogs

From the earliest times when astronomers systematically measured star positions, they began to compile lists of these positions in catalogs. Hipparchus developed an extensive catalog, as did early astronomers in China and other parts of the world, more than 2,000 years ago (see Chapter 3).

Listing stars in a catalog of positional measurements requires some kind of system for naming or numbering the stars, and a variety of such systems have been employed. The ancient Greeks designated stars by the constellation and the brightness rank within the constellation; for example, the brightest star in Orion is α Orionis, the next brightest is β Orionis, the next is γ Orionis, and so on. (The constellation names are spelled in their Latin versions, since this system was

perpetuated by the Romans and the Catholic Church after the time of the Greeks). Today many stars are still referred to by their constellation rankings, using the Greek alphabet to designate the rank.

After the time of Ptolemy, when Western astronomy went into decline for some 1,300 years, Arab astronomers, occupying northern Africa and southern Europe, carried on astronomical traditions. These people assigned proper names to many of the brightest stars, and these names, such as Betelgeuse (α Orionis), Rigel (β Orionis), Vega (α Lyrae), and Spica (α Virginis), are also still in use.

With the advent of the telescope, when many new stars were discovered that were too faint to be seen with the naked eye, catalogs rapidly outgrew the old naming systems. Generally, each catalog now assigned numbers to the stars, usually in a sequence related to their positions. Thus, modern catalogs, such as the *Henry Draper Catalogue*, the *Boss General Catalogue*, the *Yale Bright*

Star Catalogue, and the *Smithsonian Astrophysical Observatory Catalog*, list stars by increasing right ascension (i.e., from west to east in the sky, in the order in which the stars pass overhead at night). In some cases, separate listings are made for different zones of declination (i.e., for different east-west strips of sky, separated in the north-south direction). Because each of these catalogs includes a different particular set of stars, each has its own numbering system, and often no cross-reference to other catalogs is provided. Hence a given star may have a constellation-ranked name, an Arabic name, and numbers in a variety of catalogs, but only by matching up the coordinates is it possible to be sure you are referring to the same star in different catalogs. Actually, things are not quite so bad as that, since some modern catalogs do provide cross-references to others. Nevertheless, one of the principal tasks of a would-be astronomer is to become familiar with catalogs and their uses.

In more familiar terms, a parsec is equal to 3.26 light-years, or 3.08×10^{18} centimeters, or 206,265 astronomical units (AU). The use of parallax measurements to determine distances is a very powerful technique and is the only direct means astronomers have for measuring how far away stars are. Unfortunately, parallaxes are large enough to be measured only for stars rather near us in the galaxy. The smallest parallax that can be measured is barely less than 0.01 arcsecond, and accurate measurements are possible only for parallaxes larger than this. Thus, reliable distance measurements can be made only out to $d = 1/0.01 = 100 \ pc$ or so. The galaxy, on the other hand, is more than 30,000 parsecs (100,000 light-years) in diameter! Clearly, other distance-determination methods are

needed if we are to probe the entire galaxy. One very powerful method will be described in the next chapter.

STELLAR BRIGHTNESSES

The second general type of stellar observation is the measurement of the brightnesses of stars. This was first attempted in a systematic way by Hipparchus, who, more than 2,000 years ago, established a system of brightness rankings that is still with us today. Hipparchus ranked the stars in categories called **magnitudes,** from first magnitude (the brightest stars) to sixth (the faintest visible to the unaided eye). In his catalogue

Figure 18.4 Stellar magnitudes.
Magnitude differences are indicated above the line; below the line are shown the corresponding brightness ratios.

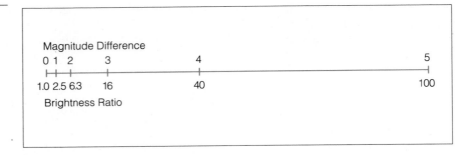

of stars, and in all since then, these magnitudes are listed along with the star positions.

By the mid-1800s, astronomers were beginning to make precise measurements of stellar brightnesses, rather than estimates made by the naked eye. It was discovered that what the eye perceives as a fixed *difference* in brightness from one magnitude to the next actually corresponds to a fixed brightness *ratio*. Measurements showed that a first-magnitude star is about 2.5 times brighter than a second-magnitude star, a second-magnitude star is 2.5 times brighter than a third-magnitude star, and so on (Fig. 18.4). The ratio between a first-magnitude star and a sixth-magnitude star was found to be nearly 100. In 1850, the system was formalized by the adoption of this ratio as exactly 100; thus, the ratio corresponding to a 1-magnitude difference is the fifth root of 100, or $100^{1/5} = 2.512$.

Thus a precise, quantitative magnitude system was developed, and with the worldwide adoption of a few standard stars used for calibration, all astronomers could record magnitudes in a comparable way. To do so no longer depended on rough estimates made by the naked eye. Today photoelectric devices are used to make the measurements (Fig. 18.5). These are devices that produce an electric current when light strikes them; they are used in many familiar applications, such as door openers in modern buildings. The amount of electrical current produced is determined by the intensity of light, so an astronomer need only measure the current to determine the brightness of a star.

Let us work out a few examples to gain familiarity with the magnitude system. If a first-magnitude star is 2.512 times brighter than a second-magnitude star, and a second-magnitude star is 2.512 times brighter than a third-magnitude star, then the first-magnitude star is $2.512 \times 2.512 = 6.3$ times brighter than the third-magnitude star. Similarly, a second-magnitude star is $2.512 \times 2.512 \times 2.512 = 15.9$ times brighter than

a fifth-magnitude star. All we need to remember is that a difference of one magnitude corresponds to a brightness ratio of 2.512. It is sometimes confusing that the brighter a star is, the smaller the magnitude is.

Once magnitude could be measured precisely, it was found that stars have a continuous range of brightnesses, and do not fall neatly into the various magnitude rankings. Therefore, fractional magnitudes must be used. Deneb, for example, has a magnitude of 1.26 in the modern system, although it was classified simply as a first-magnitude star in the old days. Each of the former categories is found to include a range of stellar brightnesses. This is especially true for the first-magnitude stars, some of which turned out to be as much as 2 magnitudes brighter than others. To measure these especially bright stars in the modern system requires the adoption of magnitudes smaller than 1. Sirius, for example, which is the brightest star in the sky, has a magnitude of -1.42. By using negative mag-

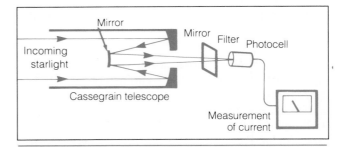

Figure 18.5 Measuring stellar magnitudes. *This is a schematic illustration of photometry, the measurement of stellar brightnesses. Light from a star is focused (often through a filter that screens out all but a specific range of wavelengths) onto a photocell, a device that produces an electric current in proportion to the intensity of light. The current is measured and, by comparison with standard stars, converted into a magnitude.*

nitudes for very bright objects, astronomers can extend the system to include such objects as the Moon, whose magnitude when full is about −12, and the Sun, whose magnitude is about −26 (thus the Sun is 25 magnitudes, or a factor of $2.512^{25} = 10^{10}$, brighter than Sirius). A simple mathematical method for dealing with any magnitudes, even fractional or negative ones, is described in Appendix 10. Table 18.1 lists the magnitudes of a variety of objects.

Of course, most stars are too faint to be seen by the human eye, so the magnitude scale must extend beyond sixth magnitude. With moderately large telescopes, astronomers can measure stars as faint as fifteenth magnitude, and with long-exposure photographs taken with the largest telescopes, they can detect objects as faint as thirtieth magnitude. Such a star is twenty-four magnitudes, or a factor of $2.512^{24} = 4 \times 10^{9}$, fainter than the faintest star visible to the unaided eye. The *Hubble Space Telescope* can detect stars as faint as twenty-eighth magnitude (ironically, ground-based instruments using modern electron detectors such as the CCDs described in Chapter 7, have exceeded the sensitivity of the Space Telescope for detecting faint objects).

The stellar magnitudes we have discussed so far all refer to visible light. As we learned in Chapter 6, however, stars emit light over a much broader wavelength band than the eye can see. We also learned that the wavelength at which a star emits most strongly depends on its temperature. Therefore, by measuring a star's brightness at two different wavelengths, astronomers can learn something about its temperature. Filters that allow only certain wavelengths of light to pass through are used for this purpose. Typically, a star's brightness is measured through two such filters, one that passes yellow light, and one that allows blue light to pass, resulting in the measurement of V (for visual, or yellow) and B (for blue) magnitudes. Since a hot star emits more light in blue wavelengths than in yellow, its B magnitude is smaller than its V magnitude. For a cool star, the situation is reversed, and the B magnitude is larger than V (remember that the magnitude scale is backward in the sense that a smaller magnitude corresponds to greater brightness).

The difference between the B and V magnitudes is called the **color index** (Fig. 18.6). The exact value of this index is closely related to the surface temperature of a star, so stellar temperatures can be estimated simply by measuring the V and B magnitudes. A very hot star might have a color index $B - V = -0.3$, whereas a cool one might typically have $B - V = +1.2$.

One additional type of magnitude should be mentioned, although it is a very difficult one to measure directly. This is the **bolometric magnitude,** which includes all the light emitted by a star at all wavelengths (Fig. 18.7). To determine a bolometric magnitude requires ultraviolet and infrared observations, as well as visible, and can best be done with the use of telescopes in space. This is particularly true for hot stars, which emit a large fraction of their light in ultraviolet wavelengths. For cool stars, which emit little ultraviolet light but much infrared radiation, bolometric magnitudes can be measured from the ground with fair accuracy. The bolometric magnitude of a star is always smaller than the visual magnitude, because more light is included when all wavelengths are considered than when only the visual wavelengths are measured. The difference between the bolometric magnitude and the visual magnitude of a star, called the **bolometric correction,** is therefore always negative. We will see in Chapter 19 how bolometric magnitudes are related to other properties of stars, particularly the luminosity.

Table 18.1 Apparent magnitudes of familiar objects

Object	Apparent Magnitude
Sun	−26.74
100-W bulb at 100 ft	−13.70
Moon (full)	−12.73
Venus (greatest elongation)	−4.22
Jupiter (opposition)	−2.6
Mars (opposition)	−2.02
Sirius (brightest star)	−1.45
Mercury (greatest elongation)	−0.2
Alpha Centauri (nearest star)	−0.1
Large Magellanic Cloud	+0.1
Saturn (opposition)	+0.7
Small Magellanic Cloud	+2.4
Andromeda galaxy (farthest naked eye object)	+3.5
Brightest globular cluster (47 Tucanae)	+4.0
Orion nebula	+4
Uranus (opposition)	+5.5
Faintest object visible to naked eye	+6.0
Neptune (opposition)	+7.9
Crab nebula	+8.6
3C273 (brightest quasar)	+12.80
Pluto (opposition)	+14.9

Figure 18.6 The color index. *Here are continuous spectra of two stars, one much hotter than the other. The wavelength ranges over which the blue* (B) *and visual* (V) *magnitudes are measured are indicated. We see that the hot star is brighter in the B region than in the V region of the spectrum; therefore, its B magnitude is* smaller *than its V magnitude, and it has a negative B − V color index. The opposite is true for the cool star.*

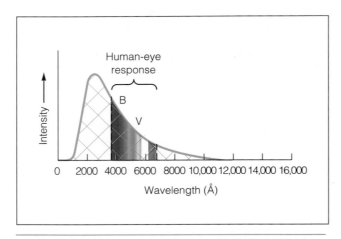

Figure 18.7 Bolometric magnitude. *The bolometric magnitude includes all the energy emitted by a star at all wavelengths, not just the visible spectrum.*

THE APPEARANCE OF STELLAR SPECTRA

Observations of stellar spectra began in the mid-1800s, well before scientists understood the physics of the atom and how it forms spectral lines. A great deal was learned about the appearance of the spectra of stars, even though the full significance was not appreciated until much later. For most of the nineteenth century,

stellar spectra were examined through the use of a **spectroscope,** which is simply an eyepiece focused on the light emerging from a prism, which in turn is placed at the focus of a telescope. The first stellar spectra to be recorded, in the 1870s, were exposed on photographic plates, a technique that is still widely used. Spectra of many stars can be photographed at the same time when a thin prism is placed in front of a telescope (Fig. 18.8). As we learned in Chapter 7, modern astronomers are turning more and more to new electronic detectors to record the spectra of stars.

Regardless of how it is done, the basic principle is the same: The light is dispersed according to wavelength, and its intensity is recorded at each wavelength. The result is called a **spectrogram** if recorded photographically (Fig. 18.9), or simply a spectrum if recorded by other means.

Normal Stars and Spectral Classification

Among the first astronomers to systematically examine spectra of a large number of stars was the Jesuit priest Angelo Secchi, who in the 1860s cataloged hundreds of spectra, using a spectroscope. He found the appearance of the spectra to vary considerably from star to star, although they were consistent in one respect: Nearly all showed continuous spectra with absorption lines. The work of Kirchhoff soon explained this to be the result of the relatively cool outer layers of a star, which absorb light from the hotter interior.

The primary differences between spectra of different stars lay in the differing strengths and patterns of the absorption lines. One star might have a particular series of strong absorption lines, whereas another might not have these lines at all, or they might be much weaker. Since Bunsen and Kirchhoff were able to identify many of the lines with known chemical elements, it was naturally assumed that stars with different patterns of lines were made of different elements. As we will see, this was not quite correct.

By the end of the nineteenth century, others had taken up and extended the work of Secchi; they used photographic plates to record the spectra, which allowed more detailed analyses (see Fig. 18.9). Most notable were a group of astronomers at Harvard University, led by Edward C. Pickering. A sequence of spectral categories or classes was established, based on the different types of spectra that were being found, and Pickering's group set out to classify the spectra of all the stars in the sky brighter than ninth magnitude, some 200,000 in all. This immense task was eventually carried out by Annie J. Cannon (Fig. 18.10), who today is regarded as the founder of modern spectral classification (and as a pioneer of woman's leadership in astronomy).

Stars were placed in categories according to which lines were prominent in their spectra, and the categories were simply labeled with letters of the alphabet. The sequence began with A stars, which showed the

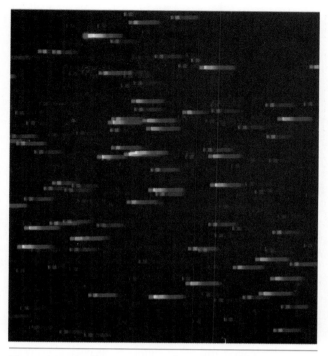

Figure 18.8 Stellar spectra. *This photograph was made through a telescope with a thin prism in the light beam, so that each stellar image was dispersed into a spectrum. It is possible to measure and classify the spectra of large numbers of stars using this technique. (University of Michigan Observatories)*

Figure 18.9 A stellar spectrum. *This photograph of a stellar spectrum shows the manner in which they are usually displayed, as negative prints. Hence light features are absorption lines, wavelengths at which little or no light is emitted by the star. (From Abt, H., W. W. Mogan, and R. Tapscott. 1968. An Atlas of Low-Dispersion Grating Stellar Spectra (Tucson-Kitt Peak National Observatory)*

Figure 18.10 Annie J. Cannon. *A member of the Harvard College Observatory for almost 50 years, Cannon classified the spectra of several hundred thousand stars. At right is shown a page from one of her notebooks. Today she is recognized as the founder of modern spectral classification. (Harvard College Observatory)*

strongest lines of hydrogen, the simplest element. Other classes followed, based on the relative strengths of the lines of hydrogen and other elements, as each class was defined by reason of similar appearance and then as-

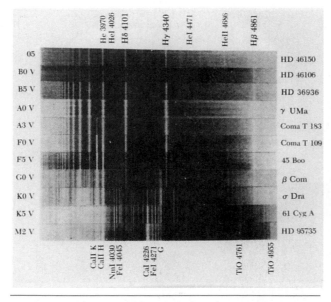

Figure 18.11 A comparison of spectra. *Here we see several spectra representing different spectral classes. A few major absorption lines are identified. These are arranged in order of decreasing stellar temperature (top to bottom). (Mt. Wilson and Las Campanas Observatories, Carnegie Institution of Washington)*

signed a letter of the alphabet. The spectral types were found, however, to represent a smooth sequence if arranged in the order O, B, A, F, G, K, M (Fig. 18.11), although at that time the reason for this sequence was unknown. The Harvard classifications were collected in a massive compilation called the *Henry Draper Catalogue,* and today stars are commonly referred to by their HD numbers. (The collection was named after an astronomer whose estate sponsored its publication. Draper had been the first, in 1872, to photograph a stellar spectrum.)

Through the work of Bohr, who showed how spectral lines are formed by atoms, and the work of later researchers such as Sir Arthur Eddington, Henry Norris Russell, and Sir James Jeans, it was later deduced that stars of different spectral classes do not differ significantly in chemical composition; they are all made of basically the same ingredients. What does differ from one spectral type to the next is temperature.

We learned in Chapter 6 that the state of ionization of a gas determines which spectral lines it forms. When an atom has lost one of its electrons, the ion, as it is now called, has an entirely different set of energy levels through which the remaining electrons can jump, so it has an entirely different set of spectral lines. The number of electrons lost by the atoms in a gas depends on its temperature; thus, the hottest stars have spectra of highly ionized gases, whereas the coolest stars show little or no ionization (Fig. 18.12). The difference in ionization from one temperature to another is responsible for the differences in the appearance of the spectra from one class to the next. Hence, the sequence of

ASTRONOMICAL INSIGHT 18.2

Stellar Spectroscopy and the Harvard Women

There is some irony in the fact that Harvard University, long a stronghold of the all-male tradition in American colleges, was the institution that nurtured some of the nation's first leading women astronomers. Even today there are few women among professional astronomers, but they were even more underrepresented at the turn of the century when the foundations of modern stellar spectroscopy were developed at the Harvard College Observatory.

When Edward C. Pickering became director of the observatory in 1877, Secchi's work (described in the text), based on visual inspection of stellar spectra through a spectroscope, had been the only attempt to classify stars according to the appearance of their spectra. Henry Draper, an American amateur astronomer, had in 1872 become the first to photograph the

spectrum of a star. Upon Draper's death in 1882, his widow endowed a new department of stellar spectroscopy at Harvard. As director, Pickering hired among his assistants a number of women, several of whom went to work on the problem of classifying spectra of stars.

An innovative technique was used to photograph spectra of large numbers of stars. A thin prism was placed in front of the telescope, so that each star image on the photographic plate at the focus was stretched, in one direction, into a spectrum. If color film had been used (it wasn't), each stellar image would have looked like a tiny rainbow. Pickering and his group refined this **objective prism** technique to the point where all stars in the field of view as faint as ninth or tenth magnitude would appear as spectra sufficiently exposed for classification. Thus it became possible to amass stellar spectra in vast quantities.

The problem of sorting out and classifying the spectra fell to Pickering's associates. One of them, Williamina Fleming, published the first *Draper Catalog of Stellar Spectra* in 1890. This catalog as-

signed some 10,351 stars in the Northern Hemisphere to spectral classes A through N, in a simple elaboration of a rudimentary classification scheme adopted earlier by Secchi. However, a number of Fleming's classes were later dropped.

While the first catalog was being prepared, a niece of Henry Draper, Antonia Maury, joined the staff and set to work on the analysis of spectra of bright stars. Spectra for these objects could be photographed using thicker prisms (actually, a series of two or three thin prisms) so that the spectra were more widely spread out according to wavelength, which allowed greater detail to be seen. Antonia Maury concluded, on the basis of these high-quality spectra, that stars should be grouped into three distinct sequences rather than just one. She had found that some stars had unusually narrow spectral lines, others were rather broad, and a third group was in between. It was later found that her sequence *c*, the thin-lined stars, are giant stars. (The lower atmospheric pressure in these stars causes the

(continued on next page)

A large group of Harvard astronomy assistants, about 1917. Sixth from the left is Henrietta Leavitt, who discovered the period-luminosity relationship for variable stars (see Chapter 25); and fifth from the right is Annie J. Cannon, whose contributions to spectral classification are described in this chapter.

ASTRONOMICAL INSIGHT 18.2
Stellar Spectroscopy and the Harvard Women (continued)

line to be less broadened than in main-sequence star.) Maury's discovery helped the Swedish astronomer Ejnar Hertzsprung (discussed in the next chapter) confirm the distinction he had recently found between giant and normal stars, and he thought her work to be of fundamental importance. Unfortunately, Maury's separate sequences were not adopted in the further work on classification at Harvard, which was thereafter based on a single sequence.

The most important of the Harvard workers in spectral classification arrived on the scene in 1896. Her name was Annie J. Cannon, and she gradually modified the classification system to the present sequence, finding that the arrangement O, B, A, F, G, K, M was a logical ordering, with smooth transitions from one type to the next. (At this time, there was still no knowledge of the fact that this was a temperature sequence.) Cannon was able to distinguish the gradations so finely that she established the ten subclasses for each major division that are in use today. In 1901 she published a catalog of classifications for 1,122 stars and then embarked on her major task: the classification

of over 200,000 stars whose spectra appeared on survey plates covering both hemispheres. By this time she was so skilled that she could reliably classify a star in a few minutes, and most of the job was done in a four-year period between 1911 and 1914. The resulting *Henry Draper Catalog* appeared in nine volumes of the *Annals of the Harvard College Observatory*, the final one being published in 1924. Pickering died in 1919, before the catalog was complete, and was succeeded as director by Harlow Shapley.

Annie Cannon continued her work, later publishing a major *Extension* of the catalog, along with a number of other specialized catalogs. She died in 1941, and the American Astronomical Society subsequently established the Annie J. Cannon prize for outstanding research by women in astronomy.

The successes of the Harvard women were not confined to stellar spectroscopy. Another major area of interest to Pickering and later to Shapley was the study of variable stars, and Henrietta Leavitt, who joined the group as a volunteer in 1894, played a leading role in this area. Following some early work on the establishment of standard stars for magnitude determinations, by 1905 Leavitt was at work identifying variables from comparisons of photographic plates. (She was eventually to discover over 2,000

of them.) In the process she examined the Magellanic Clouds for variables and noticed that their periods of pulsation correlated with their average brightnesses. This was a discovery of profound importance; the period-luminosity relationship for variable stars was to become an essential tool for establishing both the galactic distance scale and the intergalactic scale (see discussions in this chapter and in Chapters 19 and 24).

Another major figure, who came to Harvard later (1920s) was Cecilia Payne, who was a pioneer in the development of stellar astrophysics. Payne was the first to demonstrate, through detailed and complex analyses of stellar spectra, combined with theoretical considerations of ionization and radiative processes, that the chemical compositions of most stars are dominated by hydrogen, and that the make-up of most stars is the same.

The Harvard women, including those discussed here and several whose names are now obscure, were a remarkable group. They were responsible for a number of major advances in the science of astronomy at a time when its basis in physics was just becoming clear. In this day of awakening recognition of the proper role of women in all areas, it is fitting to consider and appreciate the pioneering work done by this group.

spectral types is really a sequence of stellar temperatures. Refinements in the classification guidelines, along with improvements in the quality of spectrograms, have led to the creation of subclasses, so that we have B0, B1, B2 stars, and so on. Each major division has up to ten subdivisions. The Sun, in this modern system, is a G2 star.

Peculiar Spectra

Not all stars fall into one of the standard spectral types, and astronomers generally refer to any such misfits as "peculiar" stars. The fact is that these oddballs may be rare, but in many cases they probably represent normal stages in the evolution of stars.

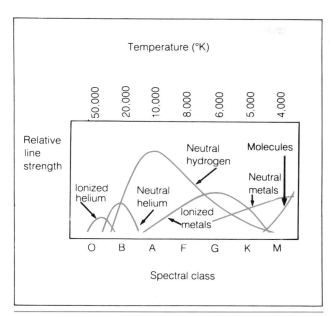

Figure 18.12 Ionization for stars of various spectral types. *This diagram shows which ions appear prominently in the spectra of stars of different classes. Note that the degree of ionization evident for the hot stars (left) is much greater than for the cool ones (right).*

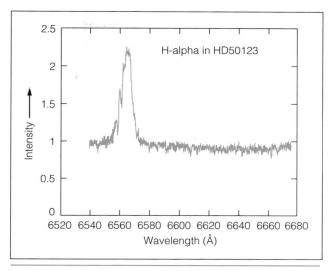

Figure 18.13 A stellar emission line. *This graph shows an emission line due to hydrogen in a star belonging to a peculiar class of stars known to have hot gas surrounding them. (T. P. Snow; data obtained at the Canada-France-Hawaii Telescope)*

Most of the peculiar stars are so designated because of unusual absorption lines in their spectra, unusual relative strengths of the lines, or emission lines. These peculiarities in many instances do represent abnormal chemical abundances in the outer layers of the star, contrary to our statement that most stars have about the same composition. Actually, the peculiar stars may have essentially normal element abundances, but for complex reasons, have undergone a kind of differentiation or sifting process that resulted in overabundances of certain elements at the surface. In other cases, elements produced by nuclear reactions inside the star (see Chapter 20) are brought to the surface by internal currents, thereby altering the surface composition.

Some stars are referred to as peculiar because they have emission lines in their spectra (Fig. 18.13). As we have seen, in a normal star the cool outer layers form absorption lines, so how do we understand these unusual cases? According to Kirchhoff's second law, a hot, rarefied gas forms emission lines, and this is how the emission lines arise in the spectra of some stars. In these stars, hot gas extends above the surface, and it is in this region that the emission lines are produced (Fig. 18.14). In most cases, astronomers do not yet

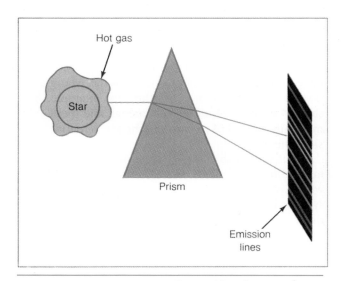

Figure 18.14 Stellar emission lines. *Very few stars have emission lines in their visible-wavelength spectra (but many do in ultraviolet wavelengths; see Chapter 20). When such lines are present, it usually signifies the existence of hot gas above the surface. According to Kirchhoff's second law, a rarefied, hot gas produces emission lines.*

Table 18.2 Types of binary systems

Type of System	Observational Characteristic
Visual binary	Both stars visible through telescope (usually requires photograph); change of relative position confirms orbital motion
Astrometric binary	Positional measurements reveal orbital motion about center of mass
Spectrum binary	Spectrum shows lines from two distinct spectral classes, revealing presence of two stars
Spectroscopic binary	Spectral lines undergo periodic Doppler shifts, revealing orbital motion. If two sets of spectral lines are seen shifting back and forth (which occurs only if the two stars are comparable in magnitude), it is a double-lined spectroscopic binary; if only one set of lines is seen (in the more common situation where one star is much brighter than the other), it is a single-lined spectroscopic binary.
Eclipsing binary	Apparent magnitude varies periodically because of eclipses, as stars alternately pass in front of each other

Figure 18.15 An optical double. *At left we see the line of sight from the Earth passing close to two stars (A and B) that are far apart but in nearly the same direction from Earth. At right we see that these two stars appear very close together on the sky. This is not a true binary system.*

understand why this extra gas is present, although they have found some indications of its source. Many of the hottest and most luminous stars, for example, have steady "winds" of outflowing material, and these winds maintain a cloud of gas around the star, creating emission lines.

Other stars with gas swirling around them are found in double-star systems (see "Binary Stars" section), where one star may lose matter to the other. Such systems have received much attention recently because they often involve exotic remnants of stellar deaths (these are discussed in Chapter 23).

BINARY STARS

About half of the stars in the sky are members of double or **binary**, systems, in which the two stars orbit each other regularly. All types of stars can be found in binaries, and their orbits come in many sizes and shapes. In some systems the two stars are so close together that they literally are touching, and in others they are so far apart that it takes hundreds or thousands of years for them to complete one revolution.

Each of the three types of observations can be used to detect binary systems (Table 18.2). Positional measurements, of course, tell us when two stars are very near each other in the sky. Sometimes this occurs by chance, when a nearby star happens to lie nearly in front of one that is in the background; this is called an **optical double** and is not a true binary system, since the two stars do not orbit each other (Fig. 18.15). When a pair of stars is seen close together, and measurements show that they are in motion about one another, this system is called a **visual binary**. Accurate positional measurements are needed to reveal the orbital motion, because the two stars are very close together in the sky, and because they appear to move very slowly about each other. The distance between the two stars must be many AUs in order for both stars to appear separately (as seen from the Earth), and therefore the orbital period is many years.

It is possible for binary systems to be recognized

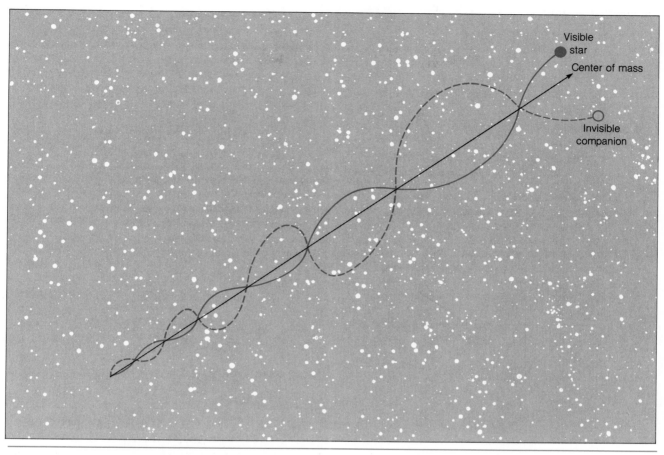

Figure 18.16 An astrometric binary. *Careful observation of the motion of a star across the sky in some cases reveals a curved path such as the one shown here. Such a motion is caused by the presence of an unseen companion star, so that the visible star orbits the center of mass of the binary system as it moves across the sky.*

even when only one of the two stars can be seen. If positional measurements over a lengthy period of time reveal a wobbling motion of a star as it moves through space (Fig. 18.16), it may be inferred that the star is orbiting an unseen companion. Such systems, detected because of variations in position, are called **astrometric binaries.**

Simple brightness measurements can also tell us when a star which may appear to be single is actually part of a binary system. If our line of sight happens to lie close to the plane of the orbit, so that the two stars alternately pass in front of each other as we view the system, then the observed brightness will decrease each time one star is in front of the other (Fig. 18.17). This is called an **eclipsing binary.**

Finally, spectroscopic measurements can also be used to recognize binary systems, again even in a case in which a star appears single. If two stars are actually present, and if they have nearly equal brightnesses, spectral lines from both will appear in the spectrum. If we find a star with line patterns representing two different spectral classes, we can be sure that two different stars are present; such a system is called a **spectrum binary.** Even if it is not clear that two different spectra are present (that is, if the two stars have similar spectral classes, or if one is too faint to contribute noticeably to the spectrum), the Doppler effect may reveal the fact that the star is double. Because the stars are orbiting each other, they move back and forth along our line of sight (as long as we are not viewing the system face-on), and this motion produces alternating blueshifts and redshifts of the spectral lines (Figs. 18.18 and 18.19). Binaries in which these periodic Doppler shifts are seen are called **spectroscopic binaries,** and

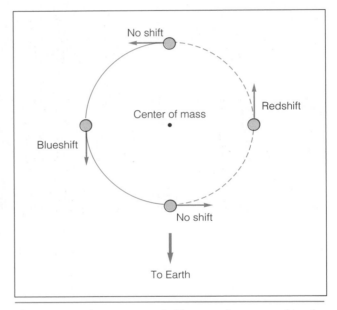

Figure 18.17 An eclipsing binary. *If our line of sight happens to be aligned with the orbital plane of a double star system, the two stars will alternately eclipse each other, and this causes brightness variations in the total light output from the system, as shown. In the upper figure, the smaller star is passing in front of the larger one, in the lower sketch, the smaller one is passing behind. The segmented graph in each case, called the light curve, illustrates how the total brightness of the system varies as these eclipses occur.*

Figure 18.18 A spectroscopic binary. *As a star orbits the center of mass of a binary system, its velocity relative to the Earth varies. This produces alternating redshifts and blueshifts in its spectrum, as the star recedes from and approaches us. For simplicity, only one of the stars is shown here to be moving, although in reality both do.*

they are more common than spectrum binaries.

A given binary system may fall into more than one of these categories. For example, a relatively nearby system seen edge-on could be a visual binary, if both stars can be seen in the telescope; it may also be an eclipsing binary, if the two stars alternately pass in front of each other; and it almost surely will be a spectroscopic binary, since the motion back and forth along our line of sight is maximized when we view the orbit edge-on.

Binary stars are important, not just because there are so many of them, but also because they provide a

Figure 18.19 Spectroscopic binaries.
Here are spectrograms of spectroscopic binaries, showing the periodic Doppler shifts due to orbital motion. The sequence of three spectra above shows a single-lined spectroscopic binary, in which lines of only one star are seen to shift back and forth. Below are spectra of a double-lined spectroscopic binary, in which lines of both stars are visible when the two are moving in opposite directions along the line of sight. The emission-line spectra at top and bottom of each set of stellar spectra are used to provide a reference for the wavelength scale. (Yerkes Observatory)

direct means of determining many basic stellar properties, as we will learn in the next chapter.

VARIABLE STARS

Although we tend to think of stars as eternal, steady beacons of light in the nighttime sky, close examination reveals that some of them are not so steady. There are stars that pulsate regularly, varying in brightness as they do so. A few of the brightest of these have been known since ancient times, but most have been found by modern astronomers who made systematic searches for them, usually by comparing photographs of the same region of sky, taken at different times. Such a comparison will reveal any stars that change dramatically in brightness from one photograph to the next. (It is quite possible that most or all stars oscillate in some manner, but usually the pulsations are too subtle to cause noticeable changes in brightness, as in the case of our own Sun; see Astronomical Insight 17.2.)

Like peculiar stars, variable stars have been classified according to their observed properties, primarily the period of pulsation but also the shape of the curve representing the variation in brightness (called a **light curve**). Most of these classes are named after the prototype, which is usually the first (and often the brightest) star discovered in that class. Examples include the **δ Cephei** (or **Cepheid**) stars, with periods of 1 to 100 days; the **RR Lyrae** stars, with pulsation times of less than 1 day; and the **Mira,** or long-period, variables, which can take up to 2 years to complete one cycle.

A pulsating variable star physically expands and contracts as it goes through its cycle (Fig. 18.20). This is known because the spectrum shows a periodic Doppler shift as the star's outer layers rise and fall in concert with the variations in brightness. The surface temperature also varies, and so does the spectral class. The brightness may change by as much as one magnitude (or much more, especially in the case of the long-period variables), or as little as a few hundredths of a magnitude.

As with most peculiar stars, the pulsating variables are normal stars passing through a particularly unstable

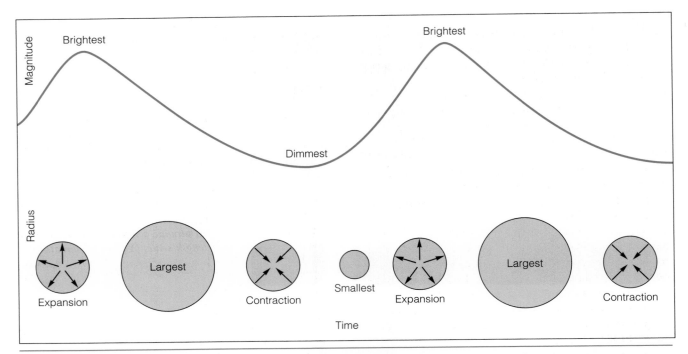

Figure 18.20 A pulsating variable star.   *The curved line shows how the brightness (in magnitude units) varies as the star expands and contracts. The sequence of sketches illustrates how the expansion and contraction phases are related in sequence to the variations in brightness. The surface temperature also varies.*

point in their evolutions. A star has a natural vibration frequency, just like a bell, so that if it is disturbed at all, it will pulsate at this frequency for a while before dying down to a steady state again. For certain stars, however, a kind of resonance is established, and once the vibrations start, they continue. These are the pulsating variables.

Some of these stars have certain uniform properties, such as average luminosities, that allow them to be used as distance indicators; these types of variables will be discussed in Chapter 24. In other chapters (particularly Chapter 23), we will discuss stars that vary in a highly irregular fashion rather than pulsating at a steady rate. Many of these irregular and eruptive variables are stars that are in late stages of their lifetimes. Thus, we will describe them along with other types of dying stars.

PERSPECTIVE

In this chapter we have learned the basics of stellar observations, by discussing the three fundamental types of observations and the characteristics of stars revealed

by each. The stellar properties we have discussed—namely, positions, brightnesses, and spectra—represent the superficial information that comes directly from the observations. We are now ready to see how much more can be learned about stars by combining these directly observed properties with our knowledge of the laws of physics.

SUMMARY

1. All that we know about stars is derived from the light we receive from them.
2. There are three basic types of stellar observation: measurements of position, brightness, and spectra.
3. Positional astronomy (that is, astrometry) has developed techniques capable of measuring position to an accuracy of about 0.01 arcseconds, which is sufficient to detect proper motions, binary motions in some cases, and stellar parallaxes for stars up to 100 parsecs away.
4. Astronomers generally carry out stellar brightness measurements by using the stellar magnitude system, in which a difference of 1 magnitude corresponds to

a brightness ratio of 2.512.

5. Magnitudes can be measured at different wavelengths, allowing the determination of color indices, or overall wavelengths, resulting in the determination of bolometric magnitudes.

6. Stellar spectra contain patterns of absorption lines that depend primarily on the surface temperatures of stars, and which therefore can be used to assign stars to spectral classes that represent a sequence of temperatures.

7. Peculiar stars are those that do not conform to the usual spectral types, most often because they have unusual surface compositions, but sometimes because they have emission lines.

8. Astronomers can detect binary-star systems on the basis of positional variations of one or both stars (astrometric binaries), brightness variations (eclipsing binaries), or composite or periodically Doppler-shifted spectral lines (spectrum or spectroscopic binaries).

9. Some stars are variable, in brightness or spectral appearance, in many cases because of physical pulsations.

REVIEW QUESTIONS

Thought Questions

1. Compare the techniques used today for positional measurements in astronomy with those used by the ancient Greeks. How much more accurate are today's techniques?

2. How can the proper motion and the parallax of a star be distinguished from one another?

3. Does the Sun have annual parallax motion, as seen from the Earth? Explain.

4. Briefly discuss the role of stellar parallax in the development of the heliocentric theory.

5. Describe how the ancient stellar magnitude system has evolved into the modern one. What has been retained, and what has changed?

6. Explain in your own words why the magnitude is smaller, the brighter a star is. Why do extremely bright objects have negative magnitudes?

7. Why is the difference between visible and bolometric magnitudes largest for stars whose temperatures are very different from that of the Sun? Why is the bolometric correction always negative?

8. Explain, in terms of what you learned about ionization in Chapter 6, why stars of different temperature have different absorption lines in their spectra.

9. What does the presence of emission lines in the spectrum of a star tell you about that star? Explain.

10. Summarize the different types of binary stars and how they are observed.

Problems

1. Convert the following distances from parsecs to light-years: 100 pc; 1.33 pc (the distance to the nearest star); 8,000 pc (the distance to the center of our galaxy); 46,000 pc (the distance to the large Magellanic Cloud, one of the closest external galaxies to our own); 700,000 pc (the distance to the nearest galaxy like our own).

2. What is the distance (in parsecs) to the following stars: (a) a star whose parallax is $p = 0\rlap{.}''25$; (b) a star whose parallax is $p = 0\rlap{.}''04$; (c) a star whose parallax is $p = 0\rlap{.}''005$?

3. How much fainter is a twelfth-magnitude star than an eleventh-magnitude star? How does a fourth-magnitude star compare with a fifth-magnitude star?

4. Compute the brightness ratio for: (a) a fourth-magnitude star and a seventh-magnitude star; (b) a star with magnitude $m = 17$ and a star with $m = 22$; (c) a star of $m = 6$ and a star of $m = -1$.

5. How much brighter is the Sun ($m = -26$) than the full Moon ($m = -12$)? How does the full Moon compare with Venus at its brightest ($m = -4$)?

6. How much fainter are the faintest stars detectable to modern telescopes ($m = 26$) than the brightest stars in the sky ($m = -1$)?

7. Use Wien's law to find the temperature where the wavelength of maximum emission moves from the V band to the B band (assume that the wavelength boundary between the bands occurs at 5,000 Å). Discuss your result in relation to the color index $B-V$.

8. Suppose a star has a bolometric correction of -2. If its visual magnitude is 13, what is its bolometric magnitude? What fraction of its total energy is emitted in the visual band?

9. In a spectroscopic binary system a spectral line of one of the stars is observed to undergo periodic Doppler shifts. The rest wavelength of the line is 4570.90 Å, and it shifts periodically from an observed wavelength of 4570.44 Å to 4571.36 Å. What is the orbital velocity of this star with respect to its companion? (Assume that the orbit is circular.)

10. In an eclipsing binary system, the duration of the eclipse of star B by star A is 100 hours, and the velocity of star A is 50 km/sec with respect to star B. What is the diameter of star B?

ADDITIONAL READINGS

Probably the best sources of additional information on stars and their observational properties are general astronomy textbooks, of which there are a number readily available in nearly any library. A few more specific resources are listed here as well.

Beyer, Steven L. 1986. *The star guide.* Boston, Mass.: Little, Brown, Inc.

Kaler, James B. 1987. The B stars: Beacons of the sky. *Sky and Telescope* 74(2):174.

_____. 1986. Cousins of our sun: The G stars. *Sky and Telescope* 72(5):450.

_____. 1986. "The K stars: Orange giants and dwarfs. *Sky and Telescope* 72(2):130.

_____. 1986 M stars: Supergiants to dwarfs. *Sky and Telescope* 71(5):450.

_____. 1986. The origins of the spectral sequence. *Sky and Telescope* 71(2):129.

_____. 1987. The temperate F stars. *Sky and Telescope* 73(2):131.

_____. 1987. White Sirian stars: Class A. *Sky and Telescope* 73(5):491.

McAlister, H. A. 1977. Binary star speckle interferometry. *Sky and Telescope* 53(5):346.

Mihalas, D. 1973. Interpreting early type stellar spectra. *Sky and Telescope* 46(2):79.

Rubin, Vera. 1986. Women's work: Women in modern astronomy. *Science 86* 7(6):58.

Upgren, A. 1980. New parallaxes for old: a coming improvement in the distance scale of the universe. *Mercury* 9 (6):143.

Welther, Barbara. 1984. Annie Jump Cannon: Classifier of the stars. *Mercury* (1):28.

MATHEMATICAL METHODS

Stellar Magnitudes

 In this chapter we have introduced the concept of stellar magnitudes, basing our discussion on the ratios of brightness represented by differences in magnitude. In the examples we discussed, we always dealt with magnitude differences that were whole numbers. The reason for this is that the process of converting from magnitude difference to brightness ratio (or vice versa) is a bit more complicated mathematically if fractions of magnitudes are involved.

It is possible, however, to derive a formula for making the conversion, using a mathematical tool known as the logarithm. The common logarithm is the power to which 10 must be raised to yield a given value. For example, the logarithm of 100 (10^2) is 2, since 10 must be raised to the second power to equal 100. Similarly, the logarithm of 1,000 (10^3) is 3, of 10,000 (10^4) is 4, and so on.

For numbers that are not simple powers of 10, the logarithm is fractional. For example, the logarithm of 65 is 1.81; the logarithm of 1,465 is 3.17; and so on. To find these values, you must use a table of logarithms (which is seldom done these days), a slide rule (also unusual today), or an electronic calculator having the logarithm function (by far the most common method).

From our discussion in the text, we know that a magnitude difference of 1 means that the brightness ratio is the fifth root of 100; a ratio of 2 corresponds to the fifth root of 100 times the fifth root of 100; and so on. We can generalize this to write the equation

$$\frac{b_1}{b_2} = 100^{1/5(m_2 - m_1)}, \tag{1}$$

where $m_2 - m_1$ is the magnitude difference corresponding to the brightness ratio b_1/b_2. By taking the logarithm of both sides, we can also write

$$\log\left(\frac{b_1}{b_2}\right) = \frac{1}{5}(m_2 - m_1) \log 100 \tag{2}$$

or

$$m_2 - m_1 = 2.5 \log\left(\frac{b_1}{b_2}\right).$$

(to see how this is done, you need to know that the logarithm of a number having an exponent is equal to that exponent multiplied times the logarithm of base number; that is $\log x^y = y \log x$. Also recall that $\log 100 = 2$).

Now we have equations that allow us to convert magnitude differences to brightness ratios, or vice versa, which work for any fractional values of the magnitude difference. For example, if we wish to know the brightness ratio of two stars whose magnitudes are $m_1 = 2.83$ and $m_2 = 6.14$, we use equation 1 to find

$$\frac{b_1}{b_2} = 100^{1/5(6.14 - 2.83)} = 100^{1/5(3.31)} = 100^{0.66} = 21.1.$$

If we want to find the magnitude difference between two stars that have a brightness ratio of $b_1/b_2 = 11,600$, we find from equation 2

$$m_2 - m_1 = 2.5 \log (11,600) = 2.5(4.06) = 10.16.$$

These expressions are quite general and work equally well for negative magnitudes (as for very bright objects) and either positive or negative magnitude differences. Further discussion of the application of these equations can be found in Appendix 10.

CHAPTER 19

Fundamental Stellar Properties and the H-R Diagram

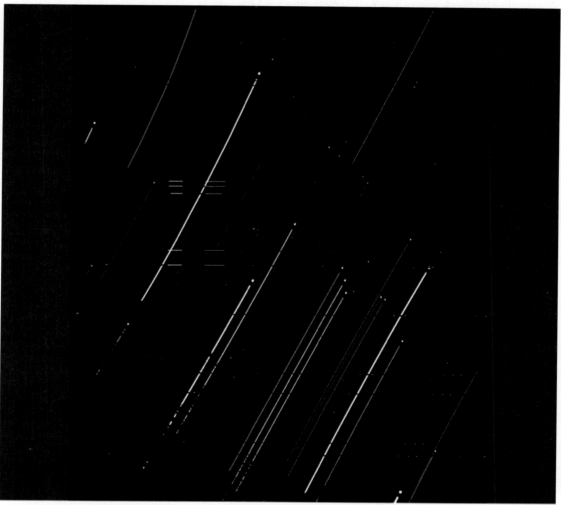

The constellation Orion rising, the colors of the stars indicating their temperatures. (National Optical Astronomy Observatories)

CHAPTER PREVIEW

Absolute Magnitudes and Stellar
  Luminosities
Surface Temperatures of Stars
The Hertzsprung-Russell Diagram
  Spectroscopic Parallax
Stellar Diameters
Stellar Rotation
Binary Stars and Stellar Masses
Stellar Composition
Magnetic Field

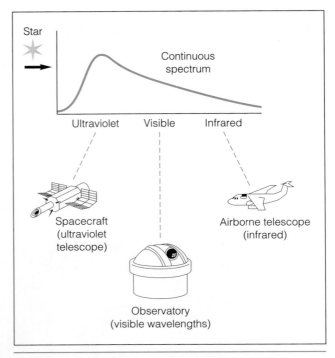

Figure 19.1 Measuring a star's luminosity. *A simple concept in principle, the determination of a stellar luminosity can be complex. The star's brightness must be measured at all wavelengths where it emits light (that is, its bolometric magnitude must be measured), and then allowance must be made for its distance, in order to calculate how much energy the star is emitting from its surface.*

Before being able to understand how stars work, astronomers needed to determine their physical properties and how they are related to each other. A number of fundamental quantities characterize a star, and in this chapter we will see how the observations described previously can be used to determine those quantities. These include the luminosities, the temperatures, radii, and, above all, the masses of stars. In this chapter a new, very powerful distance-determination method will also be described.

ABSOLUTE MAGNITUDES AND STELLAR LUMINOSITIES

The luminosity of a star is the total amount of energy it emits from its surface, in all wavelengths; it is usually measured in units of ergs per second (see Chapter 5). If the distance to a star is known, it seems an easy task to measure the apparent brightness and then allow for the distance by taking into account the inverse-square law of light propagation. It is, however, a very painstaking observational problem to measure the intensity of light from a star in absolute units, especially when it must be done over all wavelengths, including the ultraviolet and infrared (Fig. 19.1). Because of the complexities associated with such measurements, astronomers have done this only for a relatively small number of stars, and they have determined the luminosities of others by making comparisons with these standard stars.

The luminosity of a star cannot be determined unless its distance is known, because the brightness we measure directly is affected by distance (recall our discussion of the inverse square law in Chapter 6). Partly for this reason, partly because of familiarity with the magnitude system, and partly because it is easier to make comparative (rather than absolute) measurements, we speak in terms of how bright stars would look if they were all the same distance from us. To do this, we define the **absolute magnitude,** which is the magnitude a star would have if it were seen at a standard distance of 10 parsecs (Fig. 19.2). This distance was chosen arbitrarily, and is used by all astronomers. Thus a comparison of absolute magnitudes reveals differences in luminosities, because all the effects of distance have been canceled out. Table 19.1 lists absolute magnitudes for a variety of objects.

To understand how absolute magnitudes are determined, consider a specific example. Suppose a certain star is 100 parsecs away and has an apparent magnitude (the observed magnitude) of 7.3. To determine

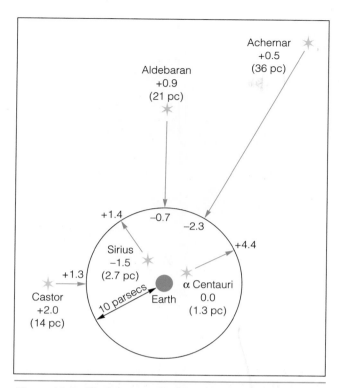

Figure 19.2 Absolute magnitudes. *A sample of stars are shown at their correct relative distances (actual distances given in parentheses). The arrows indicate the imaginary movement of these stars to a 10 parsec distance from the Earth. The numbers indicate the apparent (actual distance) and absolute (10 pc distance) magnitudes.*

Table 19.1 Visual absolute magnitudes of selected objects

Object	Absolute Magnitude
Typical bright quasar	−28
Brightest galaxies	−25
Milky Way galaxy	−20.5
Andromeda galaxy	−21.1
Type I supernova (maximum brightness)	−18.8
Large Magellanic Cloud	−18.7
Type II supernova (maximum brightness)	−17
Small Magellanic Cloud	−16.7
Typical globular cluster	−8
The most luminous stars	−8
Typical nova outburst	−7.7
Vega (bright star in summer sky)	+0.5
Sirius	+1.41
Alpha Centauri (nearest star)	+4.35
Sun	+4.83
Venus (greatest elongation)	+28.2
Full moon	+31.8
100-W bulb	+66.3

the absolute magnitude, we must find out what this star's magnitude would be if it were only 10 parsecs away. The inverse-square law tells us that since the star would be a factor of 10 closer, it would appear a factor of $10^2 = 100$ brighter. Thus, if it were only 10 parsecs away rather than 100, this star would be a factor of 100 brighter, corresponding to exactly 5 magnitudes brighter. Therefore, its absolute magnitude is $7.3 − 5 = 2.3$.

It is always possible to derive the absolute magnitude from the measured apparent magnitude and the distance, by following a similar line of thought. Of course, the answer generally will not involve simple factors of 100, as did the example just cited, but the logic is the same. If we are doing these conversions mentally, it helps to remember that a factor of 10 in distance always corresponds to a difference of 5 magnitudes. Thus, if in our example the star had been 1,000 parsecs away rather than 100, then to move it to a distance of 10

parsecs would have required bringing it a factor of 100 = 10^2 closer, so its absolute magnitude would have been $5 + 5 = 10$ magnitudes brighter than its apparent magnitude. Similarly, a star 10,000 parsecs away has an absolute magnitude 15 magnitudes brighter (smaller) than its apparent magnitude.

It should be clear that if we somehow knew the absolute magnitude of a star and measured its apparent magnitude, we could work these arguments backward to determine the distance to the star. This is a very important point, and we will discuss it later.

Let us now return to the question of luminosities. By determining the absolute magnitudes of stars, we can compare their luminosities, since the distance effect has been removed. Thus if one star's absolute magnitude is 5 magnitudes smaller than another's, then we know that its luminosity is a factor of 100 greater. To actually express the luminosity in terms of ergs per second, we make a comparison with one of the standard stars for which a direct measurement of the luminosity has been made. Strictly speaking, since the luminosity refers to energy emitted over all wavelengths, we must use bolometric absolute magnitudes in order to determine luminosities (recall that a bolometric magnitude is simply a magnitude measured over all wavelengths).

What astronomers find when they determine stellar luminosities is that the values from star to star can vary over an incredible range. There are stars with luminosities as small as 10^{-4} that of the Sun or as great as 10^6 that of the Sun, a range of 10 billion from the faintest to the most luminous! The luminosity has by far the broadest range of values of any parameter for stars; the others that we will discuss cover only ranges of a factor of 100 or so from one extreme to the other.

SURFACE TEMPERATURES OF STARS

We have already learned something about how the temperatures of stars are determined, and we need not add much to that. In Chapter 18 we saw that the color of a star depends on its surface temperature, as does the spectral type, so that by observing either, we can deduce the temperature. Recall that the color index, the difference between the blue *(B)* and visual *(V)* magnitudes, is a measured quantity that indicates temperature. A negative value of $B - V$ means that the star is brighter in blue than in visual light, and therefore is a hot star. A large positive value indicates a cool star. A specific correlation between color index and surface temperature has been established and is used to determine temperatures from observed values of $B - V$.

More refined estimates of temperature can be made from a detailed analysis of the degree of ionization, which is done by measuring the strengths of spectral lines formed by different ions. This is basically the same as simply estimating the temperature from the spectral type, since in either case the point is that the strengths of spectral lines of various ions depend on how abundant those ions are, which in turn depends on the temperature.

The temperature referred to here has been referred to as the surface temperature, although stars do not have solid surfaces. We are really referring to the outermost layers of gas, where the absorption lines form. This region is called the **photosphere** of a star, and it actually has some depth (although it is very thin compared with the radius of the star).

Stellar temperatures range from about 2,000 K for the coolest M stars to 50,000 K or more for the hottest O stars. The temperature of the Sun, a G2 star, is a little less than 6,000 K.

Figure 19.3 Henry Norris Russell. *One of the leading astrophysicists of the era when a physical understanding of stars was first emerging, Russell made many major contributions in a variety of areas. (Princeton University)*

THE HERTZSPRUNG-RUSSELL DIAGRAM

We have seen that temperature and spectral type are closely related, and we have learned how astronomers deduce the luminosities of stars. In the first decade of this century, the Danish astronomer Ejnar Hertzsprung and, independently, the American Henry Norris Russell (Fig. 19.3), began to consider how luminosity and spectral type might be related to each other. Each gathered data on stars whose luminosities (or absolute magnitudes) were known, and found a close link between spectral type (that is, surface temperature) and absolute magnitude (or luminosity). This relationship is best seen in the diagram constructed by Russell in 1913 (Fig. 19.4), now called the **Hertzsprung-Russell,** or **H-R, diagram.** In this plot of absolute magnitude (on the vertical scale) versus spectral class (on the horizontal axis), stars fall into narrowly defined regions rather than being randomly distributed. A star of a given spectral class cannot have just any absolute magnitude, and vice versa.

Most stars fall into a diagonal strip running from the upper left (high temperature, high luminosity) to the lower right (low temperature, low luminosity) (Fig. 19.5). This strip has been given the name the **main sequence.** A few stars are not in this sequence, but instead appear in the upper right (low temperature,

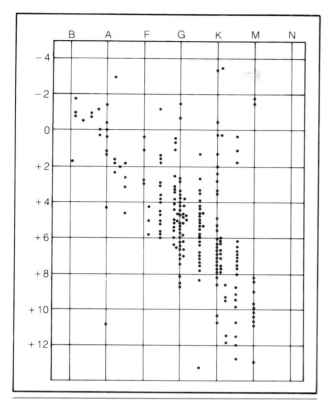

Figure 19.4 The first H-R diagram. *This plot showing absolute magnitude versus spectral class was constructed in 1913 by Henry Norris Russell. (Estate of Henry Norris Russell, reprinted with permission)*

high luminosity). Since the spectra of these stars indicate that they are relatively cool, their high luminosities cannot be a result of the fact that they have greater temperatures than the main-sequence stars of the same type. The only way one star can be much more luminous than another of the same temperature is if it has a much greater surface area; recall that the two stars will emit the same amount of energy per square centimeter of surface per second, as dictated by Stefan's law (Chapter 6). Hertzsprung and Russell independently realized that these extra-luminous stars located above the main sequence must be much larger than those on the main sequence, and they named these stars **giants** and **supergiants.**

The distinction among giants, supergiants, and main-sequence stars (known as **dwarfs**) has been incorporated into the spectral classification system used by modern astronomers. A **luminosity class** has been added to the spectral type with which we are already familiar.

The luminosity classes, designated by Roman numerals following the spectral type, are I for supergiants (further subdivided into classes Ia and Ib), II for extreme giants, III for giants, IV for stars just a bit above the main sequence, and V for main-sequence stars, or dwarfs (see Fig. 19.6). Thus the complete spectral classification for the bright summertime star Vega, for example, is A0V, meaning that it is an A0 main-sequence star. The red supergiant in the shoulder of Orion, Betelgeuse, has the full classification M2Iab (it is intermediate between luminosity classes Ia and Ib). It is usually possible to assign a star to the proper luminosity class from examination of subtle details of its spectrum.

Another group of stars has become known (mostly since the time Russell first plotted the H-R diagram), that do not fall into any of the standard luminosity classes, but which instead appear in the lower left (high-temperature, low-luminosity) corner of the diagram. Since these stars are hot but not very luminous, they must be very small, and they have been given the name **white dwarfs.** These objects have some very bizarre properties, which will be discussed in Chapter 23.

The H-R diagram is a fundamental tool for understanding stars and the relationships among their various properties, as well as how they evolve. We will be referring to it continually throughout this section of the book.

Spectroscopic Parallax

The H-R diagram can be used to find distances to stars, even stars that are very far away. The idea is really very simple: If we know how luminous a star is intrinsically (that is, how much energy it is actually emitting from its surface), and we measure how bright it appears to be, we can determine how far away it is, because we know that the difference between its luminosity and its observed brightness is caused by the distance. The only problem lies in knowing the intrinsic brightness of the star (its luminosity), and this is where the H-R diagram comes in.

Once we determine the spectral class of a star, we can place it on the H-R diagram (as long as we are sure we know the luminosity class, so we know whether it is on the main sequence or is a giant or supergiant). After we have placed it on the diagram, we simply read off of the vertical axis the absolute magnitude of the star, which is a measure of its luminosity (Fig. 19.7). A comparison of the absolute magnitude with the apparent magnitude, therefore, amounts to the same thing as a comparison of the intrinsic and apparent bright-

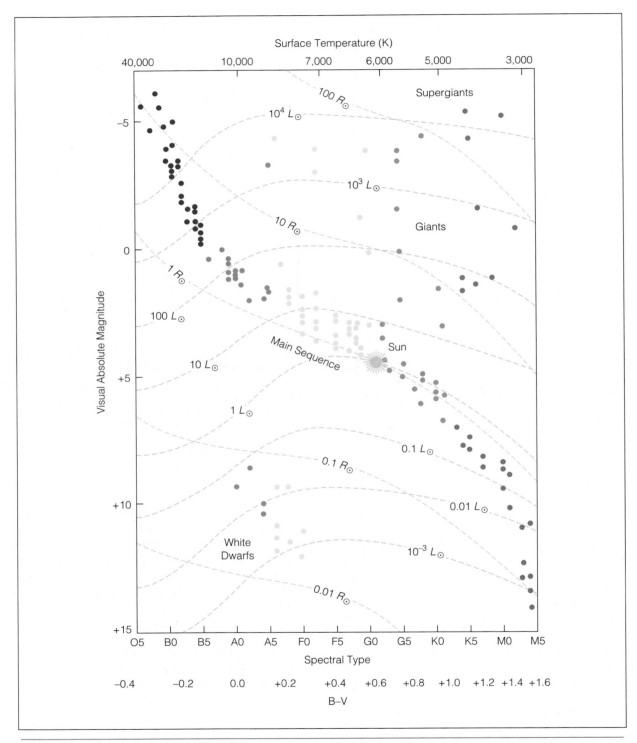

Figure 19.5 A modern H-R diagram. *This diagram shows the three commonly-used temperature indicators on the horizontal axis. Blue dashed lines indicate stellar radii in solar units, and red dashed lines indicate luminosities, also in solar units (both have been derived using bolometric corrections appropriate for each spectral type; see discussions in Chapter 18).*

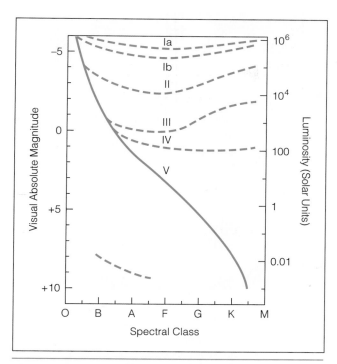

Figure 19.6 Luminosity classes. *This H-R diagram shows the locations of stars of the luminosity classes described in the text. A complete spectral classification for a star usually includes a luminosity class designation, if it has been determined.*

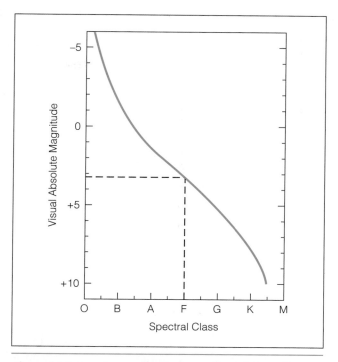

Figure 19.7 Spectroscopic parallax. *Knowledge of a star's spectral class (including the luminosity class) allows the absolute magnitude, and hence the distance, to be determined. This figure shows how the absolute magnitude (+3.3) of an F0 main-sequence star is read from the H-R diagram. The distance can then be found by comparing the absolute and apparent magnitudes of the star (as explained in the text and Appendix 10).*

nesses of the star, and from such a comparison the distance can be found.

Consider these examples. If the difference $m - M$ (that is, the apparent minus the absolute magnitude), which is called the **distance modulus,** is 5, then the star appears 5 magnitudes, or a factor of 100, fainter than it would at the standard distance of 10 parsecs. A factor of 100 in brightness is created by a factor of 10 change in distance, so this star must be 10 times farther away than it would be if it were at 10 parsecs distance; therefore, it is $10 \times 10 = 100$ parsecs away. Similarly, a star whose distance modulus $m - M$ is 10 is 1,000 parsecs away. If $m - M = 15$, then the distance is 10,000 parsecs. It should be obvious that if $m = M$ (that is, $m - M = 0$), then the distance to the star must be 10 parsecs, because this is the distance that defines the absolute magnitude. As a rule of thumb, it helps to remember that for every 5 magnitudes of difference between the apparent and absolute magnitudes, the distance increases by a factor of 10.

This method is very powerful, because it can be used for very large distances. All that is needed is to be able

to place a star on the H-R diagram so that its absolute magnitude can be determined, and to measure its apparent magnitude.* Because this distance-determination technique requires that the spectrum of a star be classified so that it can be placed on the H-R diagram, it is called the **spectroscopic parallax** method (the word *parallax* is used by astronomers as a general word for any distance measurement, even though, technically speaking, no parallax is measured in this case).

It should be noted that any time it is possible to determine the absolute magnitude of an object, even if it is by some means other than placing it on the H-R

*There is an additional bit of information needed: The interstellar medium contains a haze of fine dust particles that tends to make stars appear dimmer than they otherwise would, and allowance must be made for this effect. The interstellar dust is discussed in Chapter 25, and the method for correcting distance measurements is given in Appendix 10.

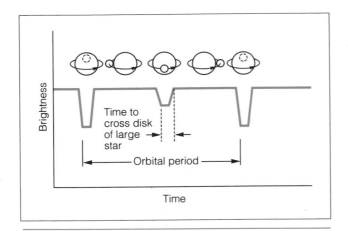

Figure 19.8 Measuring stellar diameters in an eclipsing binary. *The duration of each eclipse is combined with knowledge of the orbital speeds of the stars (from the Doppler effect) to yield the diameters of the stars.*

Figure 19.9 Surface features on Betelgeuse. *The large areas of contrasting color on this image of the red supergiant Betelgeuse (α Orionis), obtained by use of interferometry, probably represent convective cells on the star's surface. (National Optical Astronomy Observatories)*

diagram, its distance can be found by comparing the absolute and apparent magnitudes. Any object whose absolute magnitude can be determined so that it can be used as a distance indicator is called a **standard candle.** We will see in later sections how important these objects are in exploring the size scale both of the galaxy and of the universe.

STELLAR DIAMETERS

We have already seen that a star's position in the H-R diagram depends partly on its size, since the luminosity is related to the total surface area. If two stars have the same surface temperature (and therefore the same spectral type), but one is more luminous than the other, we know that the brighter star must also be larger. The Stefan-Boltzmann law (see Chapter 6) specifically relates luminosity, temperature, and radius; use of this law allows the radius to be determined if the other two quantities are known.

Eclipsing binaries provide another means of determining stellar radius, one that is independent of other properties. Recall that these are binary systems in which the two stars alternately pass in front of each other, as we view the orbit nearly edge-on. The eclipsing binary is very likely also to be a spectroscopic binary, so that the speeds of the two stars in the orbits can be measured from the Doppler effect. We therefore know how fast the stars are moving, and from the duration

of the eclipse we know how long it takes one to pass in front of the other; the simple formula "distance equals speed times time" thus gives us the diameter of the star that is being eclipsed (Fig. 19.8). Even if no information on the orbital velocity is available, the relative diameters of the two stars can be deduced from the relative durations of the alternating eclipses.

Eclipsing binaries provide the most direct means of measuring stellar sizes, but unfortunately, there are not many of them. In most cases the radii are estimated from the luminosity and temperature, as just described. In a few cases of nearby stars, stellar radii have been measured directly by use of interferometry, which overcomes the blurring effects of the Earth's atmosphere. Only relatively nearby or very large stars can be measured this way, however (Fig. 19.9).

Stars on the main sequence do not vary greatly in radius, ranging from perhaps 0.1 times the Sun's radius for the *M* stars at the lower right-hand end to 10 or 20 solar radii at the upper left. Of course, large variations in size occur as we go away from the main sequence, either toward the giants and supergiants, which may be more than 100 times the size of the Sun, or toward the white dwarfs, which are as small as 0.01 times the size of the Sun.

STELLAR ROTATION

To measure how rapidly a star is spinning is a fairly straightforward procedure, but there is one serious complication. Part of the light that reaches us from a spinning star is emitted from the portion of its surface that is moving toward us, and part is emitted from the portion that is moving away. Thus, because of the Doppler effect, spectral lines formed in the approaching portion are blueshifted, and those formed in the receding portion are redshifted. When we measure the spectrum of a star, the spectral lines are broadened, being made up of light from all parts of the surface. The more rapid the rotation, the greater the Doppler shift, and the broader the spectral lines (Fig. 19.10). Thus, in principle, a star's velocity of rotation can be determined from measurements of the width of its spectral lines.

The complication is that we do not know the orientation of the star's rotation axis. If we view a star from straight above its pole, there will be no Doppler shift or line broadening, no matter how fast it is spinning, because there will be no motion along our line of sight (recall that the Doppler effect occurs only for motions directed straight toward or away from the observer). If we view a star in the plane of its equator, on the other hand, then the Doppler shift does cause its spectral lines to be broadened, and we can make a direct measurement of its rotational speed. Unfortunately, we rarely know whether we are viewing a star pole-on, equator-on, or, as is more likely, at some random angle in between. The result is that we usually cannot precisely determine the true rotational speed of a particular star; instead, by sampling large numbers of stars, we can make general statements about the range of stellar rotation speeds that exist.

Many stars rotate slowly, with surface speeds of less than 10 km/sec (the Sun's rotational speed at the equator is about 2 km/sec), and rotation periods of at least several days or a few weeks. There are exceptions, particularly among the hot stars (which, as we will see, are young objects that may still be spinning rapidly as a result of their formation). In these cases, rotational velocities up to 450 km/sec are found, corresponding to rotation periods as short as 1 or 2 days.

BINARY STARS AND STELLAR MASSES

The mass of a star is the most important of all its fundamental properties, for it is the mass that governs

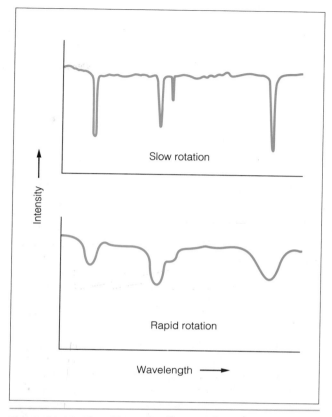

Figure 19.10 The effect of stellar rotation. *The Doppler effect causes spectral lines in a rapidly rotating star (bottom) to be broadened, because one edge of the star's disk is rapidly approaching us while the other edge is receding. Little broadening occurs in a slowly rotating star (top).*

most of the others. (This point will be discussed at some length in the next chapter.) Unfortunately, there is no direct way to see how much mass a star has. The only way to measure a star's mass is by observing its gravitational effect on other objects, and this is possible only in binary star systems, in which the two stars hold each other in orbit by their gravitational fields. The fact that binary systems are common provides us with many opportunities to determine masses by analyzing binary orbits.

The basic idea is rather simple, although the application may be quite complex, depending on the type of binary system. Kepler's third law is used, in the form derived by Newton. Remember that if the period P is measured in years, the average separation of the two stars (the semimajor axis a) is measured in astronomical units, and the masses m_1 and m_2 in units of solar

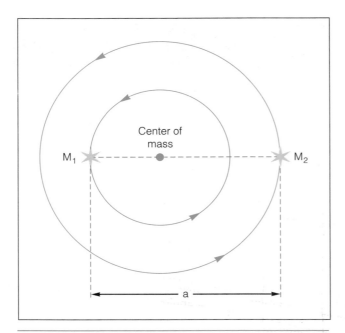

Figure 19.11 Binary star orbits. *This figure illustrates the terms used in Kepler's third law. Two stars of masses* m_1 *and* m_2 *(*m_1 *larger than* m_2 *in this case) orbit a common center of mass, each making one full orbit in period P. The semimajor axis a that appears in Kepler's third law is actually the sum of the semimajor axes of the two individual orbits about the center of mass; this sum corresponds to the average distance between the two stars.*

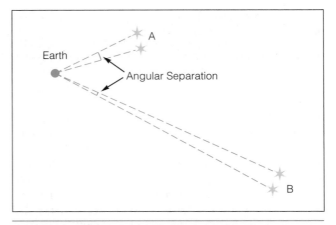

Figure 19.12 The effect of distance on measurements of binary star orbits. *Here two binary systems have the same actual separation, but from Earth the nearer system (A) appears more widely separated than the more distant system (B). Therefore the distance to a binary system must be known in order to analyze the stellar orbits.*

masses, then Kepler's third law is

$$(m_1 + m_2)P^2 = a^3$$

We need only observe the period and the semimajor axis in order to solve for the sum of the two masses. Careful observation of the sizes of the individual orbits (actually, of the relative distances of the two stars from the center of mass; see Fig. 19.11) also yields the ratio of the masses; when both the sum and the ratio are known, it is simple to solve for the individual masses.

Complications arise when some of the needed observational data are difficult to obtain. The period is almost always easy to measure with some precision, but not so the semimajor axis. One main problem is that the apparent size of the orbit is affected by our distance from the binary (Fig. 19.12), and so the distance must be well known if *a* is to be accurately determined. Another problem is that the orbital plane is inclined at a random, unknown angle to our line of

sight, so that the apparent size of the orbit is foreshortened by an unknown amount. In some cases it is possible to unravel these confusing effects by carefully analyzing the observations, and in other cases it is not. Even so, some information about stellar masses can be gained, but usually only in terms of broad ranges of possible values, rather than precise answers.

The masses of stars vary along the main sequence from the least massive stars in the lower right to the most massive in the upper left. The M stars on the main sequence have masses as low as 0.08 solar masses, whereas the O stars reach values as great as 60 solar masses. It is likely that stars occasionally form with even greater masses (perhaps up to 100 solar masses), but as we will see in the next chapter, such massive stars have very short lifetimes, so it is rare to find one.

The giants, supergiants, and white dwarfs have masses comparable to those of main-sequence stars. Hence, their obvious differences from main-sequence stars in other properties such as luminosity and radius have to be the result of something other than extreme or unusual mass. (This is discussed in the next chapter.)

For main-sequence stars, there is a smooth progression of all stellar properties from one end to the other (Table 19.2). The mass and the radius vary by similar factors, whereas the luminosity changes much more rapidly along the sequence. The mass of an O star is perhaps 100 times that of an M star, whereas the luminosity is greater by a factor of 10^9 or more.

ASTRONOMICAL INSIGHT 19.1

Fundamental Properties and Interferometry

In order for astronomers to gain any understanding of the physical processes that govern stars, they must have reliable information on the properties of the stars. We cannot hope to understand the relationships among basic properties such as luminosity, mass, diameter, and temperature until we can accurately measure these properties. This chapter is devoted to descriptions of how astronomers measure basic stellar parameters, emphasizing traditional techniques. Today major advances are being made using new methods,

particularly **interferometry.**

Interferometry, discussed briefly in Chapter 7, involves using the wave nature of light to improve resolution, the ability to detect and measure fine detail. Interferometry takes advantage of the manner in which light waves interfere with each other, and the fact that the interference depends sensitively on the relative pathlengths followed by separate beams of light. Separation of beams can be accomplished in at least two different ways: Separate telescopes can be used, so that precise timing of the arrival of light waves provides precise directional information; or the Earth's atmosphere can be allowed to do the separation. As light from a small source such as a star passes through the atmosphere, density fluctuations in the air cause the light to be split into separate bundles which

can interfere with each other. Analysis of the interference pattern can then lead to a reconstruction of the source image.

The basic principles of interferometry have been understood for several decades, but application has been very difficult because of the precision required, both in instruments and in measurements. The precision requirements vary with observed wavelength, and thus it has been in radio astronomy, which involves very long wavelengths, where the first widespread applications have occurred. Today, however, there is intense activity in visible-light interferometry, made possible by modern technology. Advances in computers have been particularly important, allowing the necessary high-speed capacity for analyzing

(continued on next page)

Table 19.2 Properties of main sequence stars

Spectral Type	Mass ($M_\odot$)	Temperature	$B - V$	Luminosity ($L_\odot$)	M_V	B.C.	Radius ($R_\odot$)
O5V	40	40,000	−0.35	5×10^5	−5.8	−4.0	18
B0V	18	28,000	−0.31	2×10^4	−4.1	−2.8	7.4
B5V	6.5	15,500	−0.16	800	−1.1	−1.5	3.8
A0V	3.2	9,900	0.00	80	+0.7	−0.4	2.5
A5V	2.1	8,500	+0.13	20	+2.0	−0.12	1.7
F0V	1.7	7,400	+0.27	6	+2.6	−0.06	1.4
F5V	1.3	6,580	+0.42	2.5	+3.4	0.00	1.2
G0V	1.1	6,030	+0.58	1.3	+4.4	−0.03	1.1
G5V	0.9	5,520	+0.70	0.8	+5.1	−0.07	0.9
K0V	0.8	4,900	+0.89	0.4	+5.9	−0.19	0.8
K5V	0.7	4,130	+1.18	0.2	+7.3	−0.60	0.7
M0V	0.5	3,480	+1.45	0.03	+9.0	−1.19	0.6
M5V	0.2	2,800	+1.63	0.008	+11.8	−2.3	0.3

Note: Masses are in units of $M_\odot$, the solar mass (whose value is 1.99×10^{33} gm); luminosities are in units of $L_\odot$, the solar luminosity (having a value of 3.83×10^{33} erg/sec), and radii are in units of $R_\odot$, the solar radius (whose value is 6.96×10^{10} cm). The temperatures are **effective temperatures,** which are a measure of surface temperature. The heading $B - V$ stands for the color index, M_V stands for visual absolute magnitude, and B.C. stands for bolometric correction.

ASTRONOMICAL INSIGHT 19.1
Fundamental Properties and Interferometry (continued)

interference patterns.

The principal stellar property that can be measured with interferometry is the diameter. What is actually measured is the *angular* diameter; to convert this to a physical diameter requires knowing the distance to the star. Angular diameters for stars within a few hundred parsecs of the Earth can be as small as 0.001 arcsecond (that is, a *milliarcsecond*) or less. Recall that there are 60 arcseconds in an arcminute, 60 arcminutes in a degree, and 360 degrees in a full circle; multiplying these together shows that there are 1,296,000 arcseconds in a full circle. Thus, an arcsecond is a very small angle (about the angular size of a dime seen at a distance of 2 miles), yet modern interferometry allows angular diameters 1,000 times smaller to be measured (try to imagine the angular size of a dime seen from a distance of 2,000 miles!).

Interferometry also leads to better knowledge of stellar masses, because the improvement in resolution offered by interferometry allows better determination of binary star orbits. As we learn in this chapter, the masses of stars are measured by the application of Kepler's third law to orbits of stars in binary systems. It is not usually possible for astronomers

using ordinary telescopes to directly measure the separation of the two stars, but interferometry allows this separation to be measured for many more systems. Thus, interferometry allows astronomers to measure stellar masses more accurately and for more stars, a great improvement in our knowledge of basic stellar properties.

Other stellar parameters can be determined from the ones that are directly measured using interferometry. If the diameter of a star is measured and the luminosity is known, then the surface temperature can be derived from the Stefan-Boltzmann equation, which relates surface temperature, luminosity, and radius (see Chapter 6). These properties, together with the mass of a star (also determined with the help of interferometry), are required in order to construct meaningful models of interior conditions (as described in the next chapter).

In addition to improving our knowledge of basic properties of stars, interferometry can also tell us much about other interesting aspects. For example, interferometric measurements may be the best method we have for detecting planets orbiting other stars. (Here infrared wavelengths are particularly useful, because planets are cool and emit mostly infrared radiation.) Interferometry can also tell us a great deal about star formation and young stars still surrounded by gas and dust. Closer to home, interferometry allows

solar system astronomers to measure planetary and satellite diameters, and to determine fundamental properties of interplanetary bodies such as asteroids.

Several new interferometry programs are being started or are underway all over the world. Most involve the use of multiple telescopes, usually mounted on tracks so that their separations can be changed (this allows greater accuracy and a greater range of angular diameters that can be measured). These projects are identical in concept to the *Very Large Array*, the huge radio interferometry observatory in New Mexico (Chapter 7). Other new projects are designed to do **speckle interferometry**, the technique where the image of a star is reconstructed after being broken up by passage of starlight through the Earth's atmosphere. There are even plans being made to put interferometric instruments into space, so that they can start with images that are free of the distortions caused by the Earth's atmosphere. If this can be done, then it may become possible to measure angular diameters as small as onemillionth of an arcsecond (that is, a *microarcsecond*). This would allow direct measures of stellar diameters for stars the size of the Sun up to distances of nearly 10,000 parsecs, equivalent to the distance from the Sun to the center of our galaxy. This will truly broaden the horizons of astronomers studying not only the properties of stars, but also the galaxy.

STELLAR COMPOSITION

The next property we should consider is the chemical makeup of a star. We have already learned that this does not vary much from one star to another on the main sequence; that the major observed differences among the spectral types are a result of temperature effects. It is important, however, to be able to measure the elemental abundances in stars, both to see what they are made of and to understand the variations that exist.

We learned previously that each chemical element has its own set of spectral lines. It would therefore seem simple to determine the composition of a star by seeing which elements are represented in its spectrum. This is complicated, however, by the fact that the temperature and density, which control the ionization and excitation, have a very profound, dominant influence on the spectrum.

To measure chemical abundances, therefore, requires a more subtle analysis. It is necessary to know the temperature and density fairly well to begin with, so that these effects may be taken into account, and then to analyze the strengths of the spectral lines to see what the relative abundances are. This process is carried out most accurately by using complex computer programs that can calculate a simulated spectrum for comparison with the observed spectrum (Fig. 19.13). In the calculation of the theoretical spectrum, the temperature, density, and chemical composition are all varied until the best match with the observed spectrum is found. This is a costly and time-consuming process, and is not done in such detail for many stars.

In general terms, the result of stellar composition studies shows that most stars consist almost entirely of hydrogen and helium, with everything else constituting, at most, a small percentage of the total mass. We have already learned that the Sun and indeed the entire solar system were formed out of material with similar composition; we now find this to be a nearly universal effect. This provides an important clue regarding how the universe itself formed, as we shall see in Chapter 31.

MAGNETIC FIELD

An often overlooked, but undeniably fundamental, property of a star is its magnetic field. We expect that stars should have fields; after all, the Sun and many of the planets do, and we suspect that magnetism in these bodies originates in fluid motions in their interiors. There is every reason to expect the same processes to be at work in other stars.

To determine whether a distant star has a magnetic field, and furthermore to measure its strength, is very difficult. It was found some years ago that some spectral lines of certain elements are split into two or more closely spaced lines in the presence of a magnetic field (Fig. 19.14). In a process called **Zeeman splitting**, electromagnetic forces split certain energy levels of an atom, so that an electron can be in one or two or more different levels, whereas there would only be one if there were no magnetic field. Thus a spectral line involving one of these split levels is itself split into two lines because the electron that absorbs a photon of light may be in either level. The amount of splitting is determined by the strength of the magnetic field, so in

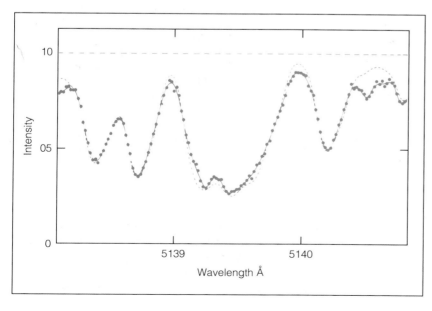

Figure 19.13 Determination of stellar composition. *This diagram illustrates a modern technique for measuring the chemical composition of a star. A theoretical spectrum (solid line) is compared with the observed spectrum (dotted and dashed line). The abundances of elements assumed present are varied in the computed spectrum until a good match is achieved. (D. L. Lambert)*

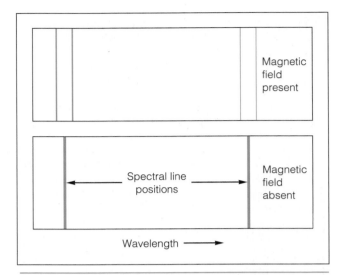

Figure 19.14 The measurement of a magnetic field. *The presence of a magnetic field splits certain absorption lines of some elements, in a process called Zeeman splitting. The amount of separation is a measure of the strength of the magnetic field.*

principle it is possible to determine the field strength by measuring how widely the lines are split.

In practice, such measurements are very difficult, however, for a number of reasons. The splitting of spectral lines is normally very small, so very precise measurements are needed. Furthermore, stellar rotation or gas motions in the outer layers can cause the spectral lines to be broadened by the Doppler effect, so that the separate lines are merged into one, and their separation cannot be measured. Because of these difficulties, we do not have much information on magnetic-field strengths in typical stars. The fields have been measured in a number of stars having especially strong ones, and values up to hundreds or thousands of times greater than the Sun's magnetic-field strength have been found.

The lack of more complete data on magnetic-field strengths in normal stars is frustrating, because it is likely that these fields play important roles in stellar formation and evolution.

PERSPECTIVE

We now know how all the basic properties of stars are derived from observational data. We can categorize stars, classify them, and describe them in any way we wish. Now we are ready to see how they work and why the various quantities are related the way they are and not in any other way.

SUMMARY

1. We determine stellar luminosity through knowledge of the distance to a star and the star's apparent magnitude. The absolute magnitude, a measure of luminosity, is the magnitude a star would have if it were seen from a distance of 10 parsecs.
2. Stellar temperatures can be inferred from the $B - V$ color index, estimated from the spectral class, or determined from the degree of ionization in the star's outer layers.
3. The Hertzsprung-Russell diagram shows that the luminosities and temperatures of stars are closely related, and that stars which do not fall on the main sequence are either larger (as with red giants) or smaller (white dwarfs) than those on the main sequence.
4. We can measure the distance to a star by first determining the star's spectral class, then using the H-R diagram to infer its absolute magnitude, and finally comparing the absolute magnitude with its apparent magnitude to yield the distance. This technique is called spectroscopic parallax.
5. We can determine stellar diameters directly in eclipsing binaries through knowledge of the orbital speed and the duration of the eclipses.
6. The rotational speeds and periods of stars are determined from the Doppler shifting of the spectral lines, which causes the lines to be broadened. However, these speeds are almost always ambiguous because of the unknown orientation of the rotation axis.
7. We can derive stellar masses in binary systems through the use of Kepler's third law, when the period and the orbital semimajor axis are known from observations.
8. The chemical composition of a star can be deduced from the strengths of the absorption lines in its spectrum, and is most commonly and accurately done by comparison with theoretically computed spectra.
9. The magnetic-field strengths of stars are probably important in their structure and evolution, and in some cases can be inferred from the Zeeman splitting of spectral lines. However, these strengths are not easily measured for most stars.

REVIEW QUESTIONS

Thought Questions

1. Why is it difficult to directly measure a star's brightness at all wavelengths?

2. Explain the relationship between the luminosity of a star and its absolute magnitude. Discuss both absolute visual magnitudes and absolute bolometric magnitudes.

3. Summarize all the methods that have been discussed so far for determining the surface temperatures of stars.

4. Explain the significance of the fact that stars in an H-R diagram (representing luminosity versus surface temperature) are not scattered randomly, but instead fall into well-defined regions.

5. Explain why a G2 star that lies above the main sequence must have a larger diameter than a G2 star that lies on the main sequence.

6. Explain the spectroscopic parallax method for finding the distances to stars. Why do you think the word "parallax" is used?

7. Discuss the role of standard candles in determining distances to astronomical objects.

8. Summarize the methods by which stellar diameters are measured.

9. Could Kepler's third law, in the form discovered by Kepler himself, have been used to determine the masses of stars? Explain.

10. Describe how the chemical composition of a star is determined. Explain how part of the method depends on observations, and part of it depends on theoretical calculations.

Problems

1. Determine the absolute magnitude of the following stars: (a) a star 1,000 pc from the Sun with apparent magnitude $m = 12.3$; (b) a star 10,000 pc from the Sun with $m = 12.3$; (c) a star 10 pc away with $m = 6.8$; (d) a star 1 pc away with $m = 2.1$.

2. If a star has absolute bolometric magnitude $M_{bol} = 1.5$, and another star has $M_{bol} = 4.5$, which is more luminous, and by how much?

3. If one star is 4 times hotter than another, and has twice as large a radius, which is more luminous, and by how much?

4. Using the spectroscopic parallax method, find the distances to the following stars: (a) a star with apparent magnitude $m = 7.2$ and absolute magnitude $M = -2.2$; (b) a star with $m = 18.6$ and $M = 8.6$; (c) a star with $m = 9.0$ and $M = 9.0$; and (d) a star with $m = 5.8$ and $M = -4.2$.

5. What is the rotational speed of a star whose 4500.9 Å line is broadened by the Doppler effect to a width of 0.90 Å?

6. What is the sum of the masses in a binary system whose orbital period is 1 year and whose semimajor axis is 2 AU? What are the individual masses of the two stars if the center of mass between them is one-fourth of the way from the more massive star (star A) to the less massive one (star B)?

ADDITIONAL READINGS

Abt, H. A. 1977. The companions of sun-like stars. *Scientific American* 236(4):96.

De Vorkin, D. 1978. Steps towards the Hertzsprung-Russell diagram. *Physics Today* 31(3):32.

De Vorkin, D. H. 1989. Henry Norris Russell. *Scientific American* 260(5):26

Evans, D. S., T. G. Barnes, and C. H. Lacy. 1979. Measuring diameters of stars. *Sky and Telescope* 58(2):130.

Getts, Judy. 1983. Decoding the Hertzsprung-Russell diagram. *Astronomy* 11(10):16.

Giampapa, M. S. 1987. The solar-stellar connection, *Sky and Telescope* 74:142.

Philip, A. G. D., and L. C. Green. 1978. The H-R diagram as an astronomical tool. *Sky and Telescope* 55(5):395.

MATHEMATICAL METHODS

The Distance Modulus

In this chapter, we have discussed the determination of distance from knowledge of the apparent and absolute magnitudes. We showed that the distance can be derived from the difference between the apparent and absolute magnitudes (called the *distance modulus*), but we were only able to demonstrate this with simple examples where the difference was a whole multiple of 5, because only these lead to results that can be estimated mentally. Here we show how to do these calculations in general cases.

In the last chapter, we showed an expression for the magnitude difference between two stars:

$$m_1 - m_2 = 2.5\log(b_2/b_1),$$

where m_1 and m_2 are the apparent magnitudes of stars that have a brightness ratio of b_2/b_1. We can use the same expression to compare the apparent and absolute magnitudes of the same star:

$$m - M = 2.5\log(b_{10}/b),$$

where m is the apparent magnitude, M is the absolute magnitude, and b_{10}/b is the ratio of the star's brightness at 10 pc distance to its actual brightness b. Now we invoke the inverse square law, which says that the apparent brightness is inversely proportional to the square of the distance, or

$$b_{10}/b = (d/10)^2,$$

where d is the actual distance to the star (in parsecs), and 10 represents the 10-parsec standard distance that defines the absolute magnitude. Finally we substitute $(d/10)^2$ for the brightness ratio b_{10}/b in the earlier expression, yielding

$$m - M = 2.5\log(d/10)^2 = 5\log(d/10)$$

(recall that the exponent in a logarithm can be brought outside as a multiplier).

The final expression just derived is the standard form in which the distance modulus is usually expressed. We can try a couple of examples to show how it works. First, let us take a simple one, such as a star having distance modulus $m = M = 10$. From our expression we find

$$m - M = 10 = 5\log(d/10),$$

which leads to

$$\log(d/10) = 2,$$

or

$$d/10 = 100,$$

and finally

$$d = 1,000 \text{ pc.}$$

This example is probably simpler if we just remember that a distance modulus of 10 always corresponds to a distance of 1,000 pc, as suggested in the text.

But now let's take a more complex example. Suppose a star is found to have apparent magnitude m of 16.39 and, from its location in the H-R diagram, an absolute magnitude M of 4.81. This example would be impossible to solve by memorizing a table of distance modulus versus distance, but is straightforward to do using our mathematical expression:

$$m - M = 16.39 - 4.81 = 11.58 = 5\log(d/10),$$

which yields

$$\log(d/10) = 2.32,$$

and

$$d/10 = 209,$$

and, finally,

$$d = 2,090 \text{ pc.}$$

Thus the mathematical expression for the distance modulus has general utility and can be applied to any situation in which the apparent and absolute magnitudes are known (i.e., to any standard candle). One correction is sometimes needed, however, which is discussed in the Mathematical Methods section at the end of Chapter 25.

CHAPTER 20

Stellar Structure: What Makes a Star Run?

Surface features on Betelgeuse, detected by interferometry (National Optical Astronomy Observatories)

We have learned how astronomers determine from observations the physical characteristics of stars. From this we have learned a great deal about their surface properties, but very little about what goes on inside them or how they evolve. To understand these secrets, we must apply the laws of physics and calculate theoretically the internal structure and evolution, requiring that the calculations reproduce the observable properties. If this is done successfully, then we can have some confidence that the calculations also accurately describe the internal conditions. We have already seen what can be learned, in our discussions of the Sun in Chapter 17. In this chapter, we will rely on our previous discussions, for it is possible to apply the same physical principles to other stars that were useful for the Sun.

WHAT IS A STAR, ANYWAY?

As we found for the Sun, a star is a spherical ball of hot gas. The outer layers are partially ionized, and the interior, where the temperature and pressure are much higher, are fully ionized, so that the gas there consists of bare atomic nuclei and free electrons. Continuous radiation is generated at the core of a star, and from there photons of light make their way slowly out to the surface, where they are emitted into space from a layer called the **photosphere,** just as in the Sun. Absorption lines are formed in the photosphere, so that the spectra of normal stars are absorption-line spectra, just as the Sun's is. The interior regions of a star are very dense because of gravitational compression, but they are still gaseous because of the high temperature. Temperatures in the cores of stars range from 10 to 100 million degrees absolute.

Like the Sun, most stars are made primarily of hydrogen, although, as we will see, this changes gradually over a star's lifetime because of nuclear reactions. (Recall that in the Sun's core, hydrogen is gradually being converted into helium.)

The internal structure of a star, like that of the Sun, is governed by the balance between gravity and pressure that we call **hydrostatic equilibrium.** The gas particles all exert gravitational forces on each other, so that the entire star is held together by its own gravity. This force is always directed towards the center; thus, the star is forced into a symmetric, spherical shape. The fact that gas is compressible explains how gravity creates a state of high density inside.

A gas that is compressed heats up, causing it to exert greater pressure on its surroundings. Thus, a star's interior is very hot and the internal pressure is high. If it were not for this pressure, gravity would cause a star to keep shrinking. A balance is struck between gravity, which is always trying to squeeze a star inward; and pressure, which pushes outward (Fig. 20.1). This balance, which can be stable for long periods of time only if there is a source of energy inside the star,

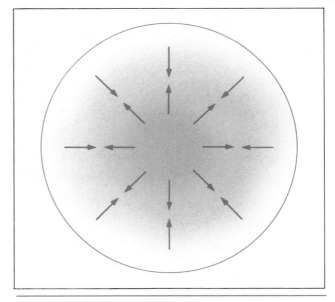

Figure 20.1 Hydrostatic equilibrium.   *This cutaway sketch of a star shows its spherical shape. The arrows represent the balanced forces of gravity (inward) and pressure (outward), and the shading indicates that the density increases greatly toward the center. Equilibrium is reached when the core becomes sufficiently hot to attain the pressure necessary to counterbalance gravity.*

plays a dominant role in determining the internal structure. Whenever the balance is disturbed, as it is during some phases of a star's lifetime, the internal structure changes (this is discussed in the next chapter).

It is the mass of a star that causes it to reach a certain state of internal compression and no other. The amount of gravitational force is set by the total mass, and this force in turn determines how much pressure is needed to balance gravity. The star will be compressed until this pressure is reached, and then it will become stable. the pressure required to balance gravity dictates the temperature inside the star. Hence a star's mass determines the internal density and temperature as well as the overall size of the star, since this is a function of the mass and the density.

Luminosity is also governed largely by the mass. The luminosity of a star is simply the amount of energy per second generated inside it that eventually reaches the surface and escapes into space, and it is determined by the temperature in the interior. As we have just seen, the temperature is set primarily by the mass. The dependence of luminosity on mass for main-sequence stars is demonstrated by making a graph of luminosity versus mass (Fig. 20.2). This is the **mass-luminosity relation,** and it was discovered observationally before there was a full theoretical explanation of it.

Thus, virtually all the observed properties of stars depend on their masses (Fig. 20.3). This explains why the main sequence represents a smooth run of masses increasing from the lower right to the upper left in the H-R diagram: The luminosity and temperature, which define a star's place in the diagram, both depend on mass.

Of course, there are stars that do not fall on the main sequence, yet their masses are not different from those of main-sequence stars. This tells us that there must be something other than mass that can influence

Figure 20.2 The mass-luminosity relation.   *This diagram, based on observed luminosities and masses for stars in the main sequence, shows how well the two quantities correlate with each other.*

a star's properties. This other factor is the chemical composition. As we shall see, this varies as a star ages. The supergiants, giants, and white dwarfs have different chemical makeups than main-sequence stars. In the next section we will see how these differences arise.

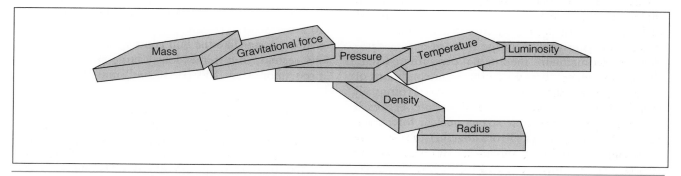

Figure 20.3 The importance of mass.   *A star's mass is the single quantity that governs all its other properties for a given composition. The sequence shown here is general, but the details vary for different chemical compositions.*

The amount of internal compression, and hence the temperature and luminosity of a star, depend on the average mass of the individual nuclei in the star's core. The heavier the particles that make up the gas, the more tightly they are compressed by gravity, the hotter it gets, and the greater the luminosity. When a star is formed, it consists mostly of hydrogen, but as it ages its core material is converted to helium at first, and later possibly to other, even heavier, elements. In the process, the core heats up and the star becomes more luminous.

The fact that all properties of a star depend on just its mass and composition was recognized several decades ago and is usually referred to as the **Russell-Vogt Theorem,** after the astrophysicists who first stated it. In modern times, it is known that other influences, such as magnetic fields and rotation, must play roles in governing a star's properties as well. Just what these roles are is not yet well understood, however.

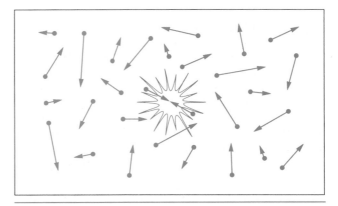

Figure 20.4 Nuclear fusion.   *The core of a star is a sea of rapidly moving atomic nuclei. Occasionally a pair of these nuclei, most of which are simple protons (hydrogen nuclei), collides with sufficient velocity to merge and form a new kind of nucleus (deuterium in this case, composed of one proton and one neutron). Energy is released in the process.*

NUCLEAR REACTIONS AND ENERGY TRANSPORT

We have seen that the core of a star must be very hot to maintain the pressure required to counterbalance gravity. Since energy flows outward from the core, there must be some source of heat in the interior. If there were not, then the star would gradually shrink.

Nuclear fusion reactions provide the only source of heat capable of maintaining the required temperature over a sufficiently long period of time. A particular type of reaction was described in Chapter 17, where we discussed the Sun's interior. Recall that in atomic fusion reactions, nuclei of light elements such as hydrogen combine to form nuclei of heavier elements. In the process, a small fraction of the mass is converted into energy according to Einstein's famous equation $E = mc^2$. It is this energy, in the form of heat and radiation, that maintains the internal pressure in a star.

Fusion reactions can occur only under conditions of extremely high temperature and density. Protons all carry positive electrical charges and therefore repel each other. The force (called the **strong nuclear force**) that holds the protons and neutrons together in a nucleus and which causes fusion to occur, can overcome this repulsive electric force only over very short distances. Therefore the nuclei must collide at very high velocities (Fig. 20.4) in order to combine, so that they can get close enough together despite their electromagnetic re-

pulsion for each other. The high temperature of a stellar core imparts high speeds to the nuclei, so that they collide with great energy, and the high density means that collisions will be frequent. Even so, only occasionally do two nuclei combine in a fusion reaction; a single particle may typically collide and bounce around inside a star for millions or even billions of years before it reacts with another. There are so many particles, however, that reactions are continually occurring despite the low probability of reaction for any individual particle.

The amount of energy released in a single reaction between two particles is small, about 10^{-5} erg. An erg (defined in Chapter 5) is itself a small quantity; a 100-watt light bulb radiates 10^9 ergs per second. Hence, a light bulb would require 10^{14} reactions per second to keep glowing, if nuclear reactions were its energy source. A star like the Sun emits more than 10^{33} ergs per second, so a tremendously large number of reactions (more than 10^{38} per second) must be occurring in its interior at all times.

In Chapter 17 we learned that the reaction that powers the Sun is the proton-proton chain, in which hydrogen is converted into helium, with 0.007 of the initial mass being converted into energy. The net result of this reaction is the combination of four hydrogen nuclei into one helium nucleus; two of the protons must be converted into neutrons in the process (see Chapter 17 or Appendix 11 for details of this reaction sequence).

There is another reaction sequence, called the **CNO cycle** (see Appendix 11), that has the same net effect, but operates primarily in stars hotter and more massive than the Sun. *All* stars on the main sequence produce their energy by one of the two processes that convert hydrogen into helium in their cores; the specific reaction that is dominant depends on the internal temperature, which in turn depends on the star's mass. The point on the main sequence that divides proton-proton chain stars from CNO stars corresponds roughly to spectral class F0; stars above that point produce most of their energy by the CNO cycle, and those below that point do so by the proton-proton chain.

Once it is produced in the core, energy must somehow make its way outward to the stellar surface. For most stars, two different energy-transport mechanisms are at work at different levels (Fig. 20.5). One of these is **radiative transport,** meaning that the energy is carried outward by photons of light. In the core, where it is very hot, the photons are primarily γ-rays, but as they slowly move outward, being continually absorbed and reemitted on the way, they are gradually converted to longer wavelengths. When the light emerges from the stellar surface, it is primarily in the visible wavelength region (or the ultraviolet or infrared, if the star is very hot or very cool). Radiative transport is a very slow process, as we learned in the case of the Sun,

where it takes thousands of years for a single photon to make its way from the core to the surface.

The second means of energy transport inside stars is **convection,** a process already discussed briefly in Chapter 17. Convection is an overturning of the gas, as heated material rises and cooled material sinks. The same process causes warm air to rise towards the ceiling of a room and is responsible for the overturning of water in a pot that is being heated from the bottom. When conditions are right for convection to occur, it is much more efficient than radiative transport. Energy travels outward much faster in convective zones than in zones were radiation is the chief means of energy transport.

Whether radiative transport or convection will be the dominant energy-transport mechanism in a star depends on the temperature structure (specifically, on how rapidly the temperature decreases with distance from the center). Model calculations show that most stars have both a convective zone and a radiative region. In stars like the Sun (i.e., for stars on the lower half of the main sequence, of spectral types F, G, K, and M), convection occurs in the outer layers, while radiative transport is the principal means of energy transport in the interior. For stars on the upper main sequence, the situation is reversed: There is convection in the central core but not in the rest of the star, where

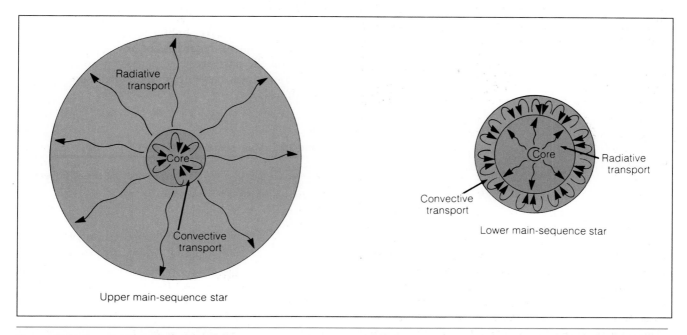

Radiative transport

Core

Convective transport

Upper main-sequence star

Core

Radiative transport

Convective transport

Lower main-sequence star

Figure 20.5 Energy transport inside of stars.   *In a star on the upper portion of the main sequence, energy is transported by convection in the inner zone and radiation in the outer regions. The opposite is true of lower main-sequence stars.*

radiative transport is responsible for conveying the energy to the surface.

STELLAR LIFE EXPECTANCIES

The lifetime of a star is determined by the amount of energy it can produce in nuclear reactions and the rate at which the energy is radiated away into space (Fig. 20.6). When all the available nuclear fuel has been used up, the star undergoes major changes in its structure and properties, and it leaves the main sequence (this will be discussed in detail in the next chapter). In Chapter 17 we found that the Sun has a life expectancy of about 10 billion years, assuming that 10 percent of its mass undergoes reactions, and that 0.007 of that mass is converted into energy. For other stars in which the CNO cycle is the dominant reaction, the calculation is made the same way, and the fraction of the initial mass converted into energy is the same. The calculation involves determining the total amount of energy available to the star during its lifetime (derived from the formulae $E = mc^2$), and dividing this by the luminosity, which is the rate at which energy is expended.

We can apply similar calculations to other stars (Table 20.1). A star near the top of the main sequence, for example, may have 50 times the mass of the Sun, but at the same time it uses its energy as much as 10^6 times more rapidly. (Recall how rapidly the luminosity of a star varies with its mass, as discussed in the last chapter.) These numbers lead to an estimated lifetime that is $50/10^6 = 5 \times 10^{-5}$ times that of the Sun. This star will last only half a million years! (Actually, in such a massive star, more than 10 percent of the total mass can undergo reactions, and the lifetime is accordingly longer than this simple estimate; it is a few million years.) By astronomical standards, such massive stars exist for only an instant before using up all their fuel and dying. This is one reason why stars of this type are very rare; only a few may be around at any given moment, regardless of how many may have formed in the past (as it happens, very massive stars also do not form very often, another reason for their rarity).

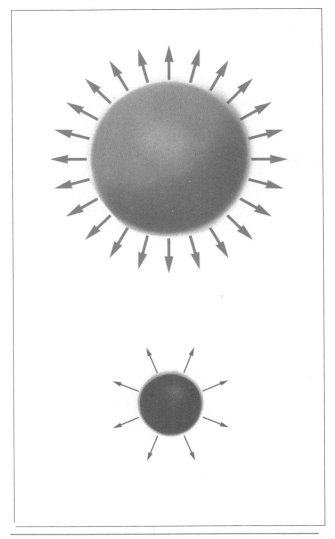

Figure 20.6 Stellar lifetimes. *Even though the more massive star (top) has a greater supply of nuclear fuel, it loses energy so much more rapidly than the lower-mass star (bottom), that its lifetime is far shorter.*

Table 20.1 Main-sequence lifetimes

Spectral Type	Hydrogen-Burning Lifetime (years)
O5V	5×10^5
B0V	9×10^5
B5V	8×10^7
A0V	4×10^8
A5V	1×10^9
F0V	3×10^9
F5V	5×10^9
G0V	9×10^9
G5V	1×10^{10}
K0V	2×10^{10}
K5V	4×10^{10}
M0V	2×10^{11}
M5V	3×10^{11}

Now let us consider a lower main-sequence star, for example an M star with mass 0.08 solar masses, and luminosity 10^{-4} times that of the Sun. Its lifetime is $0.08/10^{-4} = 800$ times *greater* than the Sun's lifetime, or about 8×10^{12} years. This is a very long time, greater than the age of the universe. No star ever formed with such a low mass has yet had time to use up all of its fuel; all such stars born are still on the main sequence. For this reason, the vast majority of all stars in existence today are low-mass stars. (It is also true that these stars form in greater numbers, adding further to their dominance in numbers.)

HEAVY ELEMENT ENRICHMENT

Clearly the nuclear reactions have an effect on the chemical composition of a star, because they change one element into another. We have stressed that all stars consist of about the same mixture of hydrogen and helium, with a trace amount of other elements, but now we find that, in the core of a star, hydrogen is gradually converted into helium. While this may not immediately affect the surface composition, which is what astronomers can measure directly from spectral analysis, it does change the internal composition. It is this change that causes a star to evolve, because the core density is altered as hydrogen nuclei combine into the heavier helium nuclei, and because when the hydrogen runs out, the star must make major structural adjustments as its primary source of internal pressure disappears.

These adjustment are discussed in Chapter 22; for now, let us consider only the change in abundances of the elements. There are other nuclear reactions that can take place in stars, after the hydrogen-burning stage has ended, if the core temperature reaches sufficiently high levels. One such reaction is the conversion of helium into carbon, by a sequence called the **triple-alpha reaction** (see Appendix 11; helium nuclei are called alpha particles, and in this reaction, three of these combine to form one carbon nucleus). When the triple-alpha reaction takes place, the core of a star changes its composition from helium to carbon.

Many additional reactions can occur, if in later stages the core temperature goes even higher (Fig. 20.7). These reactions produce ever-heavier products, so that as a star goes through successive stages of nuclear burning, its internal composition changes from what had originally been predominantly hydrogen to elements as heavy

as iron (which has twenty-six protons and thirty neutrons in its nucleus).

These reactions require ever-higher temperatures, because as the particles become more massive, it takes more energy to keep them moving fast enough to react. Furthermore, the heavier nuclei have greater electrical charges and therefore stronger repulsive forces tending to keep them apart. For both of these reasons, reactions involving the fusion of heavy nuclei require higher temperatures than those in which lightweight nuclei such as hydrogen undergo fusion.

It is the most massive stars that go through the greatest number of reaction stages and, as we have just seen, these stars live only a short time. In later chapters we will discuss the role played by these massive stars in the enrichment of chemical abundances in the galaxy.

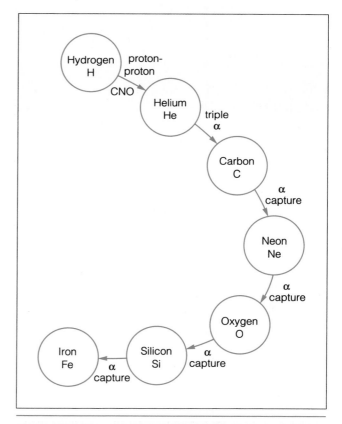

Figure 20.7 The enrichment of heavy elements. *As a star ages, each time one nuclear fuel is exhausted, another ignites if the temperature in the core rises sufficiently (which depends on the mass of the star). The net result is an enrichment of heavy elements.*

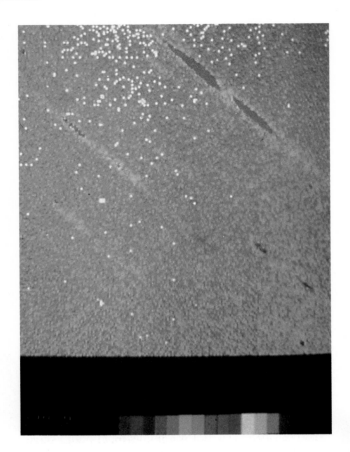

STELLAR CHROMOSPHERES AND CORONAE

Let us now consider other processes that may affect a star's structure and evolution. Some very important ones take place near the surface, rather than deep inside.

As we have seen, in most stars convection occurs at some level inside. For stars on the upper portion of the main sequence, convection takes place near the center and has no directly observable consequences. For stars on the lower portion of the main sequence, however, where convection occurs in the outer layers, there are important effects. Stars of spectral types F, G, K, and M, where surface convection is thought to occur, also are found to have chromospheres and coronae. (Recall from Chapter 17 that chromospheres and coronae produce ultraviolet emission lines and X-ray emission, and these signatures are generally found in *all* cool stars; Figs. 20.8 and 20.9.)

The presence of chromospheres and coronae seems to be linked with the presence of convection in the outer layers (Fig. 20.10). Stars on the lower portion of

Figure 20.8 Ultraviolet emission lines in a cool star. *At top is a spectrum obtained by the* International Ultraviolet Explorer, *showing, in several segments of ultraviolet spectrum, emission lines (red features). At bottom is a plot showing the intensity of these emission lines (in a different star). (top: NASA; bottom: J. Bennett)*

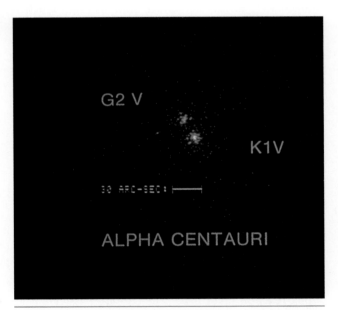

Figure 20.9 An X-ray image of cool stars. *This image, obtained by the* Einstein Observatory's *X-ray telescope, shows our nearest neighbor, Alpha Centauri. This is a multiple star system, and here we see that two of the stars, a G star and a K star, emit X-rays from their coronae. (Harvard-Smithsonian Center for Astrophysics)*

the main sequence generally have both phenomena. Somehow, the kinetic energy of the turbulent motions in the convective layer is transported into higher zones, where it causes heating. As we learned in our discussion of the Sun, the precise mechanism for converting the energy of convection into heat is not understood but almost certainly involves magnetic fields. Observations of flare activity and ultraviolet chromospheric and coronal emission lines show that all cool stars have activity cycles similar to that of the Sun (see discussion in Chapter 17). Detailed observations of brightness variations in some stars demonstrate the presence of "starspots," similar to sunspots. X-ray data show that virtually all cool stars have chromospheres and coronae. Thus it seems likely that magnetic fields are important in all these stars, playing key roles in transporting energy into high levels of the atmospheres.

What of the red giants and supergiants? These stars can have surface temperatures comparable to the lower main-sequence stars, but obviously their internal structure is rather different, since they are so much larger. These stars also have convection in their outer layers,

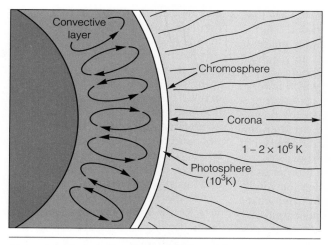

Figure 20.10 A stellar chromosphere and corona. *Stars in the lower portion of the main sequence all have chromosphere and coronae, analogous to the Sun (see Chapter 13 for more details). The photosphere is the region in which the star's continuous spectrum and absorption lines are formed; the chromosphere is a thin, somewhat hotter region just above; and the corona is a very hot, extended region outside of that. The source of heat for the chromosphere and corona is probably related to convective motions in the star's outer layers.*

however, and indeed they also have chromospheres, although it is not certain that the extremely high temperatures characteristic of coronae are present.

STELLAR WINDS AND MASS LOSS

So far we have said nothing about the possibility that the hot stars might have chromospheres or coronae. Some of the upper main-sequence stars are known to have emission lines in their spectra, indicating that they have some hot gas above their surfaces. Visible-wavelength emission lines are found in only a few extremely hot or luminous stars, however.

In the late 1960s, with the first observations of ultraviolet spectra, it was immediately found that many hot stars (nearly all the O stars and many of the hotter B stars) have ultraviolet emission lines. Furthermore, there are absorption features that show enormous Doppler shifts, indicating that the stars are ejecting material at speeds as great as 3,000 kilometers per second! The most extreme **stellar winds,** as these outflows are called (Fig. 20.11), occur in the O stars, but most cool giants and supergiants also have winds, usually with lower speeds.

The cause of the winds is not known, although there are indications of how the high velocities are reached. Once gas begins to move outward, light from the star can exert sufficient force to accelerate it to high speeds. This force, called **radiation pressure,** is very weak (you certainly can't feel the breeze from a light bulb, for example), but O and B stars are so luminous that strong acceleration of the wind is possible. For the most luminous of these stars, radiation pressure may be sufficiently strong to initiate the winds, but for the slightly less luminous O and B stars, calculations show that radiation pressure is not capable of creating the initial outflow but that it can accelerate the gas to high speed once the flow begins. Some other mechanism must start the outflow first. It is not known what this mechanism is, but recently it has been suggested that subtle pulsations (see Astronomical Insight 20.1) might be responsible.

Very high degrees of ionization are observed in the winds from O and B stars. Recently X-ray emission from stars with winds has been discovered, and it is possible that the X-rays are responsible for ionizing the gas (X-ray photons have high energies and cause ionization when they are absorbed by atoms). The X-rays

ASTRONOMICAL INSIGHT 20.1

Stellar Pulsations and Mass Loss

Stars in the upper portion of the H-R diagram all lose mass through stellar winds, but the nature of the winds in the hot stars is quite different from that in the cool stars. The hot stars have very rapid winds, with velocities up to several thousand kilometers per second, whereas the cool giants and supergiants have slower winds, with speeds of perhaps 10 to 20 km/sec. The rate at which mass is lost can be comparable, however, because the cool-star winds are much denser. It has generally been assumed that, since the winds in the two types of stars are so different, the causes of the winds must also be different. Recent evidence suggests, to the contrary, that the basic mechanisms may actually be very similar.

We have learned a little about pulsating variable stars, such as the Cepheid variables and the RR Lyrae stars. These are stars that physically expand and contract, growing brighter and dimmer in the process. Other kinds of pulsating variable stars have also been known to exist, such as the so-called long-period variables, red supergiants that take as long as 1 or 2 years to go through a single cycle of expansion and contraction. It happens that these stars are the ones in the upper right-hand portion of the H-R diagram that have the strongest stellar winds, suggesting a connection between the pulsations and the loss of mass. Specifically, it has been suggested that every time such a star expands to its largest

European Southern Observatory, D. Baade

size and then begins to contract, some of the gas from its outer layers continues to drift outward, because the surface gravity of such an extended star is very low. It is known from infrared observations that the escaping matter can condense to form tiny, solid grains of material. Photons of light from the star (which is very luminous then push the grains outward, creating a steady flow. The outward force exerted on the grains by photons from the star is called **radiation pressure.**

The idea that radiation pressure plays a role in creating the winds from cool stars is not new; what has not been well understood is the mechanism for getting the outflow started, and it now appears that stellar pulsations may provide the answer. Similarly, it has been known for some time that the very high velocities reached by winds from hot luminous stars are due to radiation pressure acting on gas atoms and ions, but there has been great difficulty in discovering how the outflow gets started in the first place,

so that radiation pressure can then take over and provide the acceleration. Now it has been found that many hot stars pulsate in a very subtle way, and that these pulsations may provide the missing mechanism for initiating hot stellar winds, in analogy with cool star pulsations and winds.

A star can pulsate **nonradially,** meaning that it does not simply expand and contract as a whole, but instead that its surface consists of separate regions that alternately rise and fall. A very simple example would be a star that alternately is elongated (a bit like a cigar) and squashed. Much more complex "modes" of pulsation are possible, however, with the result that the surface of a star may have many zones that rise and fall over small distances, but with high frequency. These pulsations may be caused by internal rotation (see Astronomical Insight 17.2).

It has been suggested very recently that the small velocities in rising regions on a star's surface may be sufficient to allow radiation pressure to act on that gas, accelerating it to very high velocities and creating the observed winds. The general phenomenon has been dubbed "mass-tossing," and has striking overall similarities to the mechanism described here for the winds in cool stars. Some details are quite different, such as the type of stellar pulsation, the velocity, and the types of particles that are pushed by radiation pressure, but the general mechanism is surprisingly similar. Detailed theoretical and observational studies of stellar pulsations and their possible relationship to stellar winds are among the most active areas of research in stellar structure today.

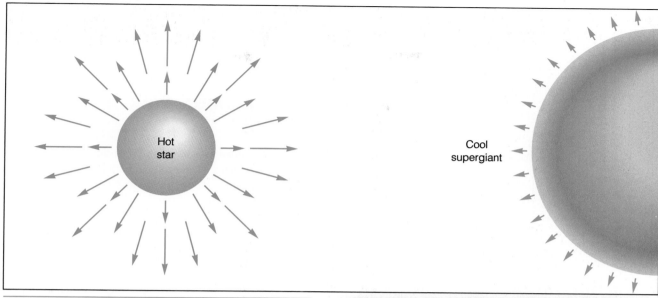

Figure 20.11 Stellar winds. *In the drawing at top a luminous hot star (left) ejects gas at a very high velocity; the lengths of the arrows indicate that the gas accelerates as it moves away from the star. A luminous cool star (especially a K or M supergiant) is so extended in size that the surface gravity is very low, and material drifts away at relatively low speeds (right). In both types of stars, radiation pressure probably helps accelerate the gas outward. At bottom is a photograph of a hot star with an extensive bubble of gas around it, created by a stellar wind. (© 1984 ROE/AAT Board)*

must be produced in some very hot region, either gas in the wind that is heated by turbulence, or a hot coronal region in the star's outer layers. It would be a surprise to theorists if these hot stars have coronae, because the O and B stars are not thought to have convection in the outer layers. Perhaps this is incorrect, and they do have convection zones near their surfaces, or perhaps some other process is creating the coronae in these stars.

Whatever the cause of the winds, they have important consequences. Analysis of the ultraviolet emission lines shows that, in some cases, stars are losing matter at such a great rate that a large fraction of the initial mass may be lost during their lifetimes. An O star might lose as much as one solar mass every 100,000 years. If such a star begins life with twenty or thirty solar masses and lives a million years, it could lose ten solar masses, a significant fraction of what it had to begin with. This has important effects on how such a star evolves, as we shall see.

The supergiants in the upper right of the H-R diagram also lose mass through stellar winds (Fig. 20.12), but these winds have a distinctly different nature. The

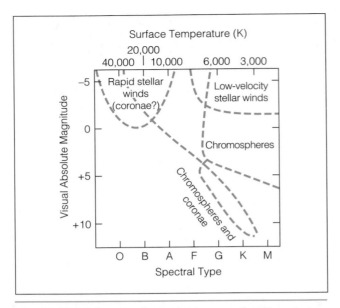

Figure 20.12 Chromospheres, coronae, and stellar winds in the H-R diagram. *Here we see that most stars of type F or cooler have chromospheres, although coronae may be confined to the main sequence. Very luminous stars, hot and cool alike, have sufficiently strong winds to result in significant loss of mass.*

red supergiants are so large that their surface gravity is very low, and the gas in the outer layers is not tightly bound to the star. Radiation pressure can easily push the gas outward, at the relatively low speed of 10 or 20 kilometers per second. The amount of mass lost can be just as great as in the hot stars, however, because these low-velocity winds are much denser than the high-speed winds from the O and B stars. It is thought that slow pulsations in red giant and supergiant stars may play a role in triggering mass loss, just as rapid pulsations may help create the high-speed winds in hot stars (see Astronomical Insight 20.1).

Because the red supergiants are relatively cool objects, the gas in their outer layers is not ionized, and even molecular species form there. In addition, small solid particles called dust grains can condense, so that the star becomes shrouded in a cloud. In extreme cases the dust cloud becomes so thick that little or no visible light escapes to the outside. The dust grains become heated, however, and emit infrared radiation, so the star can still be detected with the use of an infrared telescope.

Some of the dust that forms in the outer layers of red supergiants escapes into the surrounding void, and interstellar space gets contaminated in this way with a kind of interstellar haze (the interstellar dust is discussed at greater length in Chapter 25).

STELLAR MODELS: TYING IT ALL TOGETHER

Once astrophysicists have understood all the phenomena that govern a star's internal structure and its outward appearance, they must combine their knowledge of all these processes into a theoretical calculation (Fig. 20.13) that will tell them what goes on inside the star, and what will happen to it in time. When the calculations correctly reproduce the observable quantities, it is assumed that they also correctly illustrate the nonobservable aspects, such as the star's interior conditions. We can have reasonable confidence in such calculations, because it can be demonstrated that the resulting solutions are unique; that is, no other solution to the mathematical equations can exist. The most difficult question is whether we know all the correct equations and physical parameters to begin with.

The calculation of a model star consists of simultaneously solving a set of basic mathematical equations describing the physical processes, and solving them repeatedly for different depths inside the star. In this way, the calculations may proceed from the observed surface conditions to the center, or from assumed central conditions to the surface, with adjustments made until the calculations match the observed properties of the star. To carry out such a calculation requires a substantial amount of time on a large computer.

To see how a star evolves, it is necessary to compute the model many times over, each time taking into account changes in the star that took place in the previous step in the calculation. For example, a certain amount of hydrogen is converted into helium at each step, and this change in the chemical composition of the core must be taken into account in the next step. The conversion of hydrogen into helium uses up the hydrogen fuel, so that the star will eventually exhaust its supply, and it causes the density of the core to increase, because helium nuclei are more massive than hydrogen nuclei. These things cause the star's overall structure to change, and the calculations keep track of the changes, allowing the evolution to be traced. In the next chapter we will discuss the evolution of stars, and how it is determined from a combination of theory and observation.

ASTRONOMICAL INSIGHT 20.2

Stellar Models and Reality

In this chapter we discuss the physical processes that are believed to govern the internal conditions and external properties of stars. We will find that, despite the remoteness of stars and the fact that all of our observational information is derived only through the analysis of light emitted from their surfaces, we know quite a bit about their insides. It may seem surprising that we could have any confidence at all in our theories of stellar interiors and atmospheres, but the fact that we do derives from our faith that the same laws of physics apply to distant stars just as they do on Earth. Given that, stars are actually rather simple objects, in comparison with, for example, a solid planet or a biological system.

Stars are made entirely of gas, and this is the major reason for the simplicity of describing them theoretically. Gases obey certain well-defined laws, which are verified through experiment and theory. Add the fact that stars are basically symmetric objects and stable (not changing with time), and we can invoke additional physical laws in the form of equations for the forces acting on gas particles in a star and for the production and transport of energy. Within the context of the adopted physical laws, there is only one solution to the set of equations that describe a star's structure and properties. This is a very powerful statement, for it implies that all we need to do is be sure to include the correct laws of physics in our calculations, and we will get the correct solution.

Aside from the fact that our known laws of physics may be incomplete, the major difficulty in calculating theoretical models of stars lies in the fact that some of the processes occurring in stars are quite complex when examined in sufficient detail. For example, it is easy to treat the transport of energy from the core to the surface of a star when the energy is carried solely in the form of radiation, but very difficult when the energy is transported by flowing streams of heated gas, in the process we call convection. The calculations are simple when a star is spherically symmetric, but they become more complex when the star is rotating rapidly, so that its shape becomes somewhat flattened. Magnetic fields in stars are probably very complex, if our Sun is any example, and theory has not yet fully treated even the Sun's magnetic field. These are cases in which we know the laws of physics well enough, but their application is made complex because of the vast amount of detail that must be incorporated into our calculations.

Despite the complications, stellar model calculations are quite good. We know this because it is possible to achieve very good agreement between observed quantities (such as stellar luminosity, radius, surface temperature, and others) and those calculated from theory. The first test of any theoretical model for a star is to see whether the model successfully reproduces the observed surface conditions. If it does, then we have confidence that the model must also represent the interior fairly well.

Today's stellar structure theorists are busily refining their models, not so much because they are incorrect, but because it is becoming increasingly possible to add details to the calculations. Just as new telescope technology increases the ability of astronomers to observe astronomical objects, new advances in computer technology increase the capability of theorists to represent reality in ever greater detail. The first stellar models were calculated with slide rules or slow, mechanical desk calculators, and it took weeks or months to calculate a single model. In the past two or three decades, model complexity has increased in proportion to computer capacity and speed, and it is now possible to calculate many models (for different stages in the evolution of a star, for example) in hours. Today we are on the brink of a new generation of "supercomputers," machines with vastly greater speed and capacity for massive calculations, and no one will benefit from this more than the stellar structure theorist. As the capabilities of the computers increase, so will the ability of astrophysicists to represent reality in their calculations.

POINT	RADIUS	DENSITY	TEMP	PRESSURE	ENERGY GEN	LUMIN	OPACITY	DELAD	DELRAD	BETA
1	0.	2.8526E+00	3.5455E+07	1.7651E+16	1.9682E+05	0.	3.4602E-01	2.8868E-01	3.0013E+00	7.7421E-0
2	5.4736E+08	2.8526E+00	3.5456E+07	1.7651E+16	1.9680E+05	9.0509E+32	3.4602E-01	2.8868E-01	3.0013E+00	7.7421E-0
3	7.3124E+08	2.8525E+00	3.5455E+07	1.7650E+16	1.9679E+05	9.1397E+32	3.4602E-01	2.8868E-01	3.0011E+00	7.7421E-0
4	8.8736E+08	2.8525E+00	3.5455E+07	1.7650E+16	1.9677E+05	1.6376E+33	3.4602E-01	2.8868E-01	3.0009E+00	7.7421E-0
5	1.0385E+09	2.8525E+00	3.5455E+07	1.7650E+16	1.9675E+05	2.6208E+33	3.4602E-01	2.8868E-01	3.0008E+00	7.7421E-0
6	1.1927E+09	2.8524E+00	3.5455E+07	1.7649E+16	1.9673E+05	3.9644E+33	3.4602E-01	2.8869E-01	3.0005E+00	7.7421E-0
7	1.3548E+09	2.8524E+00	3.5454E+07	1.7649E+16	1.9671E+05	5.6460E+33	3.4602E-01	2.8869E-01	3.0001E+00	7.7421E-0
8	1.5283E+09	2.8523E+00	3.5454E+07	1.7648E+16	1.9668E+05	8.3930E+33	3.4602E-01	2.8869E-01	2.9998E+00	7.7421E-0
9	1.7159E+09	2.8523E+00	3.5457E+07	1.7648E+16	1.9664E+05	1.1801E+34	3.4602E-01	2.8869E-01	2.9990E+00	7.7421E-0
10	1.9206E+09	2.8522E+00	3.5457E+07	1.7647E+16	1.9660E+05	1.6659E+34	3.4602E-01	2.8869E-01	2.9994E+00	7.7421E-0
11	2.1449E+09	2.8521E+00	3.5455E+07	1.7646E+16	1.9654E+05	2.3202E+34	3.4602E-01	2.8868E-01	2.9989E+00	7.7422E-0
12	2.3910E+09	2.8519E+00	3.5455E+07	1.7645E+16	1.9647E+05	3.2162E+34	3.4602E-01	2.8868E-01	2.9983E+00	7.7422E-0
13	2.6639E+09	2.8518E+00	3.5455E+07	1.7643E+16	1.9639E+05	4.4432E+34	3.4602E-01	2.8869E-01	2.9976E+00	7.7422E-0
14	2.9647E+09	2.8515E+00	3.5453E+07	1.7641E+16	1.9629E+05	6.1231E+34	3.4602E-01	2.8869E-01	2.9967E+00	7.7423E-0
15	3.2977E+09	2.8513E+00	3.5455E+07	1.7639E+16	1.9615E+05	8.4230E+34	3.4602E-01	2.8869E-01	2.9956E+00	7.7423E-0
16	3.6666E+09	2.8510E+00	3.5454E+07	1.7636E+16	1.9600E+05	1.1571E+35	3.4602E-01	2.8869E-01	2.9942E+00	7.7424E-0
17	4.0755E+09	2.8506E+00	3.5446E+07	1.7632E+16	1.9580E+05	1.5806E+35	3.4602E-01	2.8869E-01	2.9925E+00	7.7424E-0
18	4.5291E+09	2.8501E+00	3.5452E+07	1.7628E+16	1.9554E+05	2.1477E+35	3.4602E-01	2.8893E-01	2.9903E+00	7.7425E-0
19	5.0324E+09	2.8496E+00	3.5451E+07	1.7623E+16	1.9521E+05	2.9979E+35	3.4602E-01	2.8870E-01	2.9876E+00	7.7426E-0
20	5.5910E+09	2.8480E+00	3.5445E+07	1.7616E+16	1.9491E+05	4.0871E+35	3.4602E-01	2.8870E-01	2.9847E+00	7.7427E-0
21	6.2112E+09	2.8480E+00	3.5434E+07	1.7608E+16	1.9446E+05	5.5949E+35	3.4602E-01	2.8870E-01	2.9816E+00	7.7429E-0
22	6.8990E+09	2.8469E+00	3.5420E+07	1.7598E+16	1.9391E+05	7.6580E+35	3.4602E-01	2.8870E-01	2.9760E+00	7.7431E-0
23	7.6660E+09	2.8455E+00	3.5421E+07	1.7585E+16	1.9324E+05	1.0468E+36	3.4602E-01	2.8871E-01	2.9701E+00	7.7434E-0

Figure 20.13 Building a model star.

Modern astrophysicists reconstruct what is happening inside stars with the use of complex computer programs that simultaneously solve large numbers of equations describing the physical conditions. Often the result is plotted, as a convenient means of showing the implications of these calculations. (C. Hansen, University of Colorado)

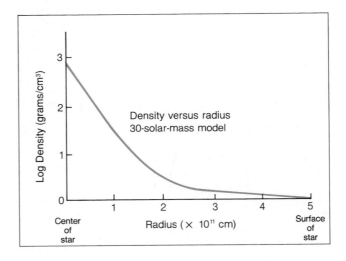

PERSPECTIVE

We now are acquainted with all the physical processes that affect a star; we know what a star is and what makes it run. We have seen that mass plays a dominant role in dictating all the other properties, although the composition, which changes with time, is also important. We are prepared to see what the life story of a star is—how it is born, how it lives, and how it ends its days.

SUMMARY

1. A star is gaseous throughout, with temperature and density increasing toward the center.

2. The balance between gravity and pressure within a star, called hydrostatic equilibrium, governs the internal structure.

3. The mass of a star, along with its chemical composition, governs its central temperature and pressure (through hydrostatic equilibrium), its luminosity, its radius, and its internal structure.

4. Nuclear fusion reactions occur in the core of a star, where the temperature and density are high enough to allow nuclei to collide with sufficient velocity and frequency.

5. In most stars (those on the main sequence), hydrogen is converted into helium by reactions in the core.

6. The energy inside a star is transported by convection or by radiation. On the lower main sequence (including the Sun), radiative transport dominates in the inner parts of the star, while convection operates in the outer layers. On the upper main sequence, convection is dominant in the core, while radiative transport dominates in the outer layers.

7. The lifetime of a star depends on its mass and how rapidly it uses up its nuclear fuel; the lifetime decreases

dramatically from the lower main sequence to the upper main sequence.

8. Nuclear reactions in stars, and the recycling of matter between stars and interstellar space, result in a gradual increase in the abundance of heavy elements in the universe.

9. Stars in the cool half of the H-R diagram have chromospheres and coronae, in analogy with the Sun.

10. Stars in the hot portion of the H-R diagram, and very luminous cool supergiants as well, have stellar winds that can cause the loss of large fractions of the initial stellar masses. The loss of mass can have significant effects on how the stars evolve.

11. To determine the interior conditions in a star, astronomers must compute them from known laws of physics and observations of the surface conditions.

REVIEW QUESTIONS

Thought Questions

1. How do we know that a star must be in a state of balance between gravity and pressure? What would happen if this balance were disturbed?

2. Explain in your own words why mass is a star's most fundamental property; that is, explain how the mass of a star is the most important factor in determining its other properties.

3. Why do nuclear reactions in stars occur only in the innermost region, that is, in the core?

4. Summarize the ways in which energy produced in a star's core is transported outward, to the surface. Point out ways in which the transport varies with the type of star.

5. Explain why the lifetime of a star is *shorter,* the more massive the star.

6. As a star ages and undergoes successive nuclear reaction stages, a higher temperature is required to start each new reaction. Why?

7. Compare the stellar winds from O stars with the solar wind.

8. Summarize the differences between an upper main-sequence star (an O or B star) and one on the lower main sequence (K or M), in structure, energy generation and transport, and outer layers.

9. If mass is exchanged between two stars in a binary system, how do you think it might affect the lifetimes of the two stars?

10. Discuss stellar model calculations. Why do we think they tell us anything realistic about conditions inside stars,

where we cannot observe? How does a mathematical model of a star stand as a scientific theory; that is, does it have the attributes of a theory?

Problems

1. If a single proton-proton chain reaction produces only 10^{-5} ergs, how many reactions per second must occur in order to supply the Sun with the 4×10^{33} erg/sec that it radiates? How many reactions (CNO cycle, which produces about the same amount of energy per reaction) must occur per second in an O star whose luminosity is 10^{38} erg/sec?

2. Estimate the hydrogen-burning lifetime of a star of fifteen solar masses and luminosity 10,000 times that of the Sun, and a star of 0.2 solar masses and luminosity 0.008 that of the Sun.

3. When the hydrogen in a star's core has all been converted to helium, there are only one-fourth as many nuclei in the core as there were before the reactions. If the average distance between nuclei is the same, by how much has the core's density increased?

4. Suppose the speed of the wind from an O supergiant star is 2,000 km/sec. What is the Doppler-shifted wavelength of the three-times ionized carbon line, whose rest wavelength is 1548.8 Å (in the ultraviolet)?

5. If a star has an initial mass of twenty solar masses and a lifetime of 5 million years, how much of its mass will it lose if the rate of mass loss is one solar mass every 500,000 years?

6. Suppose the Sun becomes a red giant star with a radius of 100 times its present radius. By how much would its surface gravity and escape velocity be decreased? Discuss the importance of your answer in view of the stellar winds from red supergiants.

ADDITIONAL READINGS

Aller, L. H. 1971. *Atoms, stars, and nebulae.* Cambridge, Mass.: Harvard University Press.

Boss, A. P. 1985. Collapse and formation of stars. *Scientific American* 252(1):40.

Cox, A. N., and J. P. Cox. 1967. Cepheid pulsations. *Sky and Telescope* May, p. 278.

Giampapa, Mark S. 1987. The solar-stellar connection. *Sky and Telescope* 74(2):142.

Graham, J. 1984. A new star becomes visible in Chile. *Mercury* 13(2):60.

Hoyle, F. 1975. *Astronomy and cosmology: A modern course.* San Francisco: W. H. Freeman.

Jones, B., D. Lin, and R. Hanson. 1984. T Tauri and TIRC: A star system in formation. *Mercury* 13 (3):71.

Perry, J. R. 1975. Pulsating stars. *Scientific American* 232(6):66.

Shu, F. 1982. *The physical universe.* San Francisco: W. H. Freeman.

Smith, E. P., and K. C. Jacobs. 1973. *Introductory astronomy and astrophysics.* Philadelphia: Saunders.

Snow, T. P. 1981. Dieting for stars, or how to lose 10^{25} grams per day (and feel better too!). *Griffith Observer* 45(5):2.

Weymann, R. J. 1978. Stellar winds. *Scientific American* 239(2):34.

MATHEMATICAL METHODS

Energy Production in Nuclear Reactions

In this chapter and in Chapter 17 we discussed the total energy released by nuclear reactions in a stellar core. In the conversion of hydrogen into helium, a small fraction (0.007) of the initial mass is converted into energy. Here we can show where this fraction comes from and can illustrate how the luminosity of a star is produced.

The net result of the reaction (either the proton-proton chain or the CNO cycle) is to convert four hydrogen nuclei into one helium nucleus. To find the amount of mass that is "lost"—that is, converted into energy—we can compare the mass of the four hydrogen nuclei with the mass of a single helium nucleus. Appendix 3 lists the elements and provides their "atomic weights," which are actually masses. These are expressed in atomic mass units (abbreviated simply by u; 1 u has the value of 1.660531×10^{-24} gram).

For hydrogen, the atomic weight is 1.0080 u; for helium it is 4.0026 u. If we add up the mass of four hydrogen nuclei, we find that 4×1.0080 $u = 4.0320$ u. Note that this is larger than the mass of the helium nucleus; the difference found by subtracting is 0.0294 u. This is the quantity of mass that disappears during the reaction. Note that it is 0.0073 of the original mass of the four hydrogen nuclei; now we see the origin of the factor of 0.007 that was given in this chapter and in Chapter 17.

To find how much energy is created by the conversion of this mass, we now switch to grams:

$(0.0294$ $u) \times (1.66 \times 10^{-24}$ grams$/u) = 4.88 \times 10^{-26}$ gram. Using Einstein's equation $E = mc^2$ to find the energy equivalent of this mass, we find

$$E =$$

$$(4.882 \times 10^{-26} \text{ gram}) \times (2.99793 \times 10^{10} \text{ cm/sec})^2$$

$$= 4.39 \times 10^{-5} \text{ erg.}$$

This tiny quantity of energy is the amount released by a single reaction (technically, by the sequence of reactions that converts a single group of four hydrogen nuclei into one helium nucleus).

This is a tiny quantity of energy, but very many reactions can take place throughout the core of a star, so the total energy that is potentially available is enormous. To see this, consider the core of the Sun, which we estimate to contain about one-tenth of the total solar mass, or about 2×10^{32} grams. If 73 percent of this mass was hydrogen when the Sun was young, then the mass in the form of hydrogen was about 1.4×10^{32} grams. Dividing this by the mass of the hydrogen nucleus (i.e., 1.0080 u, or 1.67×10^{-24} gram), we find that the Sun's core initially contained some 8.4×10^{55} hydrogen nuclei. For every four of these, one reaction can take place, so there can be 2.1×10^{55} reactions during the Sun's lifetime. Multiplying this number by the energy released per reaction yields a total lifetime energy output for the Sun of just about 10^{51} ergs. This yields a lifetime for the Sun of roughly 10^{10} years, given the luminosity of about 4×10^{33} erg/sec.

This result is consistent with the estimated lifetime we found in Chapter 17, but here we have explained its origin in more detail. For stars of other masses, we can estimate lifetimes as discussed in this chapter, using the ratio of the star's mass and luminosity to the solar values.

CHAPTER 21

Star Clusters and Observations of Stellar Evolution

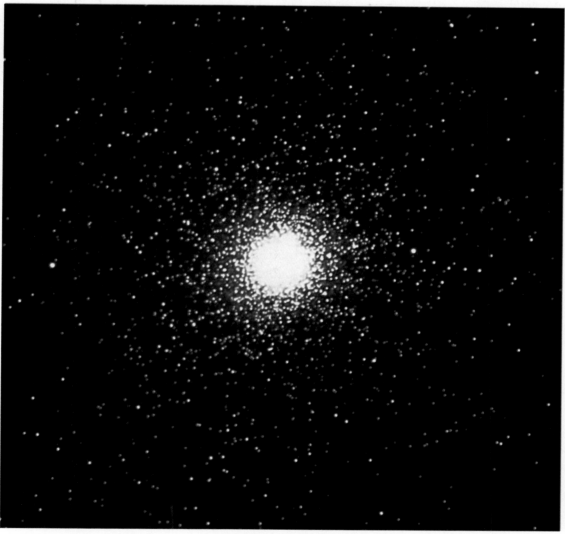

The globular cluster 47 Tucanae. (National Optical Astronomy Observatories)

CHAPTER PREVIEW

Properties of Clusters
The Assumption of a Common Distance
The Assumption of a Common Age
The Main-Sequence Turnoff and Cluster Ages
Young Associations and Stellar Infancy
Star Formation

How can we watch a star evolve? After all, even the most short-lived ones are around for hundreds of thousands of years, and human studies of astronomy date back only three or four millennia. Occasionally we catch a star in the act of change, as it makes a transition from one stage in its evolution to another, but clearly we cannot hope to see a star through its whole lifetime, watching it form, live, and die.

We can, however, piece together a story of stellar evolution by examining many stars of different ages, and deducing from this the sequence of events that occurs in a single star. An apt analogy is the deduction of the life story of a tree from the examination of a forest, where seeds, young saplings, mature trees, and dead logs are all found. For stars, matters are complicated by the fact that the manner of evolution depends on the mass of the star, whereas for a group of trees we expect all to have about the same properties. As we shall see, however, stars, particularly groups of them, provide certain clues for the detective trying to reconstruct their life stories.

PROPERTIES OF CLUSTERS

Within our galaxy, not all stars are uniformly distributed, but instead many are located in concentrated regions called clusters. A few clusters are sufficiently prominent to be visible to the unaided eye, and others can be viewed with a small telescope or a good pair of binoculars (Table 21.1). The Pleiades (Fig. 21.1), a bright group to the west of Orion, is perhaps the best-known cluster for Northern Hemisphere observers, being easily visible in the evening sky throughout the late fall and early winter. Other clusters easy to find with a small telescope are the Hyades, the double cluster in Perseus, and M13 (Fig. 21.2), a fantastic, spherical col-

Table 21.1 Selected clusters of stars

Cluster	Type	m_v	Distance (pc)	No. Stars	Diameter (pc)	Age (years)
47 Tuc	Globular	4.0	5,100	10^4	10	1.5×10^{10}
M103	Open	6.9	2,300	30	5	1.6×10^7
h Persei	Open	4.1	2,250	300	16	1.0×10^7
χ Persei	Open	4.3	2,400	240	14	1.0×10^7
M34	Open	5.6	440	60	30	1.3×10^8
Perseus	Open	2.2	167	80	12	1.0×10^7
Pleiades	Open	1.3	127	120	4	5.0×10^7
Hyades	Open	0.6	42	100	5	6.3×10^8
S Mon	OB	4.3	740	60	6	6.3×10^6
τ CMa	Open	3.9	1,500	30	3	5.0×10^6
Praesepe	Open	3.7	159	100	4	4.0×10^8
o Velorum	Open	2.6	157	15	2	2.5×10^7
M67	Open	6.5	830	80	4	4.0×10^9
θ Carinae	Open	1.7	155	25	3	1.3×10^7
Coma	Open	2.8	80	40	7	5.0×10^8
ω Cen	Globular	3.6	5,000	10^5	20	1.5×10^{10}
M13	Globular	5.9	7,700	10^5	11	1.5×10^{10}
M22	Globular	5.1	3,000	10^5	9	1.5×10^{10}
M15	Globular	6.4	14,000	10^6	11	1.5×10^{10}

Figure 21.1 The Pleiades. *This relatively young galactic cluster is a prominent object in the Northern Hemisphere during the late fall and winter. (© 1980 Anglo-Australian Telescope Board)*

lection of hundreds of thousands of stars. It is apparent that the stars in a cluster are gravitationally bound together.

There are at least three distinct types of star clusters. Those containing a modest number of stars (up to a few hundred), found in the disk of the galaxy, are referred to as **galactic,** or **open,** clusters (Figs. 21.3 and 21.4). The Pleiades (see Fig. 21.1) and the Hyades are examples. Loose groupings of hot, luminous stars are also found in the plane of the galaxy, and are called **OB associations,** being dominated by stars of spectral types O and B. These clusters, in contrast with the other types mentioned here, are probably not gravitationally bound together, but appear grouped simply because the stars have recently formed together. The giant **globular clusters,** such as M13 (Fig. 21.2), are found in a spherical volume about the galactic center and are not confined to the plane of the galaxy (Fig. 21.5). In later chapters, especially Chapters 24

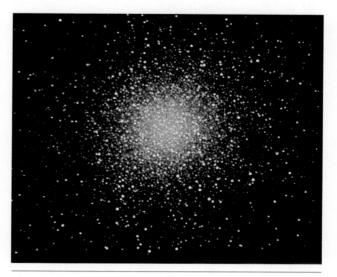

Figure 21.2 The globular cluster M13. *(U.S. Naval Observatory)*

Figure 21.3 The open cluster NGC3293. *(© 1977 Anglo-Australian Telescope Board)*

Figure 21.4 The open cluster NGC4755. *(National Optical Astronomy Observatories)*

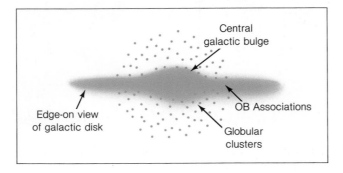

and 26, we will discuss how the various types of clusters fit into the story of the formation and evolution of the galaxy. For now we will exploit certain properties of the clusters that help us see how stars evolve.

There are two important assumptions astronomers make about a cluster of stars: (1) The stars in a cluster are all at the same distance from us; and (2) they formed together so that all are the same age and have the same composition initially.

The Assumption of a Common Distance

The first of these assumptions is, strictly speaking, only approximately true. Clearly, a cluster of stars has some spatial extent, so that the stars on the near side are closer to us than those on the far side. The diameter of a cluster, however, is rarely more than a few parsecs and is usually small compared with the distance from the Earth to the cluster. Therefore, for all practical purposes, the stars in a cluster are all at the same distance from us. This means that any observed differences from star to star within a cluster must be real differences and cannot be a false effect created by differing distances from us. Brightness differences, for example, have to be the result of variations in the luminosities of stars within the cluster.

Because distance effects are eliminated when we compare stars within a cluster, it is possible to plot an H-R diagram for a cluster without first determining the absolute magnitudes of the stars. We simply plot apparent magnitude against spectral type, or, more commonly, against the color index $B - V$, which is easier to measure for a large number of stars (Fig. 21.6). To do this requires only the determination of the visual *(V)* and blue *(B)* magnitudes of the cluster stars, a rather straightforward observational procedure, and then the construction of a plot of *V* versus $B - V$. The result, called a **color-magnitude** diagram, looks just like an H-R diagram, except that the vertical scale is an apparent-magnitude scale. This is because the differences in magnitude of the cluster stars are the same, whether we use apparent or absolute magnitudes.

Figure 21.5 The locations of clusters in the galaxy. *The open or galactic clusters are found in the disk of the Milky Way, as are the OB associations, which are always located in spiral arms (see Chapter 24). The globular clusters, to the contrary, inhabit a large spherical volume, and are not confined to the plane of the disk.*

Figure 21.6 H-R diagrams for star clusters.   *Here, plotted on the same axes, are H-R diagrams (using color index rather than spectral type) of several clusters. Different symbols are used in the right-hand portion to distinguish among the giant stars in different clusters. The solid lines indicate the main sequence turn-off points, which are related to the ages of the clusters.*

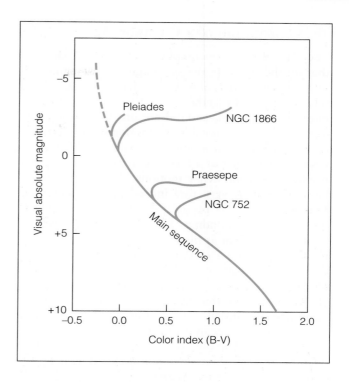

One advantage of plotting cluster H-R diagrams is that these diagrams can be used to determine the distances to clusters (Fig. 21.7). A comparison of a cluster H-R diagram with a standard one (that is, one with an absolute-magnitude scale) allows us to see what the absolute-magnitude scale for the cluster diagram should be. This in turn tells us the difference between the absolute and apparent magnitudes for the cluster, and from this difference we deduce the distance. This procedure, known as **main-sequence fitting**, depends only on the assumption that the cluster has a main sequence that is identical to that of the standard H-R diagram, a safe assumption as long as the chemical composition of the cluster is not unusual.

This technique is identical to the spectroscopic parallax method of distance measurement (Chapter 20), applied to a group of stars.

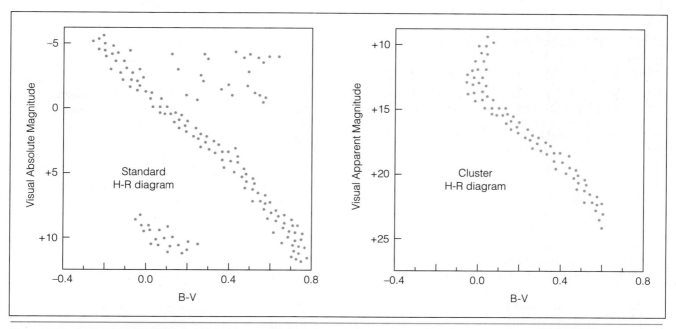

Figure 21.7 Main-sequence fitting.   *This is a standard H-R diagram, with a cluster H-R diagram next to it using apparent magnitude on the vertical axis. A comparison of the two diagrams leads to an estimate of the distance to the cluster.*

The Assumption of a Common Age

The second major assumption made about clusters, that all the stars are the same age, is very important in studying stellar evolution. The basic premise is that the stars formed together, out of a common cloud of interstellar material, and did not just happen to come together. The basis for this assumption is the extremely low probability that a number of stars could ever encounter each other and form a cluster by chance, as a result of random stellar motions. A close encounter between stars is very rare indeed, and furthermore, even when one does occur, it is very unlikely to result in gravitational bonding. Binary systems as well as clusters must form as binaries or clusters. A corollary is that all stars in a cluster must begin with the same chemical composition, since they form from a common source of material.

The knowledge that all stars in a cluster have the same age and initial composition gives astronomers tremendous leverage in deducing the evolutionary histories of stars. The only major parameter that differs from one star to another within a cluster is mass, so the observed differences between stars in a cluster give us clues as to how the properties of stars depend on mass. Comparison of H-R diagrams of clusters of different ages then gives us a picture of how these different masses evolve.

THE MAIN-SEQUENCE TURNOFF AND CLUSTER AGES

Striking differences are found between H-R diagrams of different clusters. The Pleiades, for example, have an almost complete main sequence, but no red giants. By contrast, a globular cluster has no stars on the upper portion of the main sequence, but it has a large number of red giants. A wide variety of intermediate cases can be found.

From considerations discussed in the last chapter, we expect massive stars to evolve most rapidly. These are the stars that start out near the top of the main sequence, and they have such tremendous luminosities that they use up their nuclear fuel rapidly. Tying this together with the observed differences between cluster H-R diagrams leads to the conclusion that the upper main-sequence stars must evolve into red giants when they have used up all their hydrogen fuel. In a very young cluster, even the most massive stars might still be on the main sequence, whereas in an older cluster,

these massive stars will have had time to use up their core hydrogen and become red giants. Hence, the Pleiades are relatively young, whereas a globular cluster is very old.

As a cluster ages, more and more of its stars use up their hydrogen and leave the main sequence. The upper main sequence gradually erodes away as its stars move to the right in the H-R diagram, toward the red-giant region (Figs. 21.8 and 21.9). This process creates a definite cutoff point on the main sequence, above which there are no stars. The location of this cutoff depends entirely on the age of the cluster. The position of this **main sequence turnoff,** as it is commonly called, indicates the age of a cluster. If, for example, we find a cluster whose main sequence stops at the position of stars like the Sun (that is, it has no stars above G2, or $B - V = 0.6$, on the main sequence), then we know that all stars more massive than one solar mass have burned their hydrogen and evolved toward the red-giant region. In the last chapter, we learned that the Sun's lifetime for hydrogen burning is about 10 billion years. We conclude, therefore, that such a cluster is just about 10 billion years old. If it were younger, its main sequence would extend farther up, and if it were older, even the Sun-like stars would have had time to evolve into red giants, and the main-sequence turnoff would be farther down.

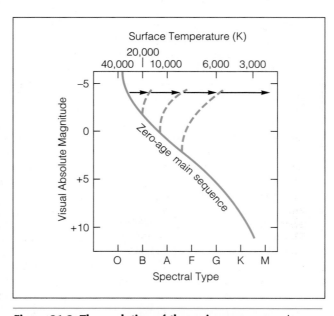

Figure 21.8 The evolution of the main sequence. *As a cluster ages, the stars on the upper main sequence move to the right. The farther down the main sequence this has happened, the older the cluster.*

ASTRONOMICAL INSIGHT 21.1

Star Tracks

We have seen that as a star evolves, its position in the H-R diagram changes. The correct deduction of how a star's position changes was not easily made, however. During the years following the development of the H-R diagram, great intellectual effort was concentrated on discovering how stars evolve, but because the necessary information was not yet complete, many of the first ideas were wrong. This resulted in an assortment of views on the direction of motion in the H-R diagram, and evolutionary tracks thought to be made by stars during their lifetimes changed directions a few times.

By the second decade of this century, it was well established that the red giants formed a distinct grouping of stars in the H-R diagram, above the main sequence. Nothing was known of the true energy source that powers stars, and little was known of stars' chemical composition. It was thought that stars glowed by virtue of energy released in a slow gravitational contraction, which caused the stars to shrink and heat, at least in their early stages. It was also suspected that within the main sequence, the hot stars were youngest and the cool stars the oldest. In 1913 Henry Norris Russell, in an address to the American Astronomical Society, put these ideas together when he suggested a comprehensive theory of stellar evolution. He proposed that a star begins life as a red giant, and for a while it contracts and heats, moving across the top of the H-R diagram from right to left. Eventually the temperature gets so hot inside the star that the gas there becomes highly ionized (now we know this is true from the start). Because ionized gas (it was supposed incorrectly) cannot compress and heat further, the star begins to cool while still contracting, and it moves down the main sequence from upper left to lower right. The evolutionary track of a star was therefore thought to resemble a backward figure seven, and the main sequence was thought to be just an age sequence.

In the decade following Russell's hypothesis, several astrophysicists, including most notably Sir Arthur Eddington and Sir James Jeans, developed theoretical models of stellar structure. These models, which of necessity incorporated many approximations, appeared to be consistent with the evolutionary scenario of Russell. But by the early 1920s, discrepancies began to arise. The chief problem was that there appeared to be no reason for the ionized gas in a star's interior to become incompressible, and therefore no reasonable justification for assuming that a contracting star should stop heating up and start cooling when its interior became ionized. The new theoretical work, along with developing evidence for a close relation between a star's mass and its luminosity, led to the first suggestion (again by Russell) that the main sequence represented a mass sequence, where stars are stable and spend a long time with no movement in the H-R diagram. Russell suggested that the difference between the main-sequence stars and the red giants was that the two groups had different internal energy sources, although his ideas of what those sources might be were rather vague. He still thought that the red giants were young stars, and that they moved down to the main sequence as their first energy source was exhausted and the second took over. Thus, in 1923 (when this was hypothesized), the evolutionary tracks attributed to stars were much like those known today, but with the motion in the exact opposite direction.

The proper sequence of events in a star's lifetime finally began to be sorted out in the late 1920s and early 1930s, with Russell once again leading the way. Two major clues were developed: (1) Cluster H-R diagrams (studied by Hertzsprung, among others) showed that as a cluster ages, it loses stars from its upper main sequence and gains red giants; and (2) the Russell-Vogt theorem was developed, which demonstrated that a star's properties are all determined specifically by its mass and composition. It finally became clear that a star moves from the main sequence to the red-giant region, and that this movement must be caused by a change in its internal composition. It was even suggested that the change of composition involves the conversion of hydrogen into helium, although it was not until the late 1930s that the exact nature of this process was understood.

The study of star tracks in the H-R diagram demonstrates how plausible, but incorrect, theories can be deduced from incomplete information, yet it also shows how patient and careful collection of data will eventually lead to the correct picture. The incorrect hypotheses developed along the way were important, for they helped inspire the continued efforts to solve the problem.

YOUNG ASSOCIATIONS AND STELLAR INFANCY

Earlier in this chapter we mentioned OB associations, loosely bound groups of O and B stars. We now know that these are very young clusters, probably the young-est in the galaxy, because they contain stars on the extreme upper portion of the main sequence. In many cases, the stars are moving away from the center and escaping the association, so within a few million years no concentration of stars will remain.

If the OB associations are the youngest ensembles of stars in the galaxy, we might expect to find star formation actually occurring in or near them. Some young clusters show remnants of the process of star birth, in the form of gas and dust surrounding the stars (Figs. 21.10, 21.11, and 21.12). The Pleiades provide a good example of this; photographs show bluish clouds of dust, shining by the reflected light of the stars in the cluster (see Fig. 21.1). In other cases, not only these **reflection nebulae**, but also hot, glowing gas clouds called **HII regions** may be found. The notation *HII*

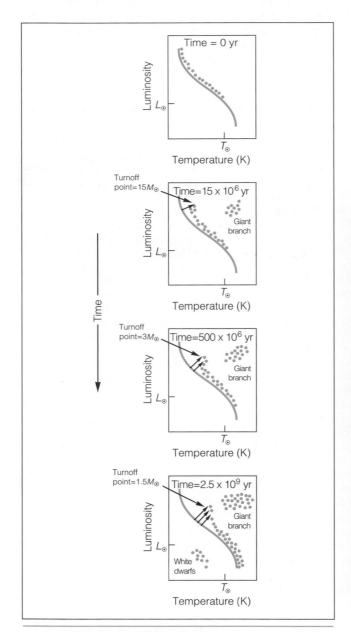

Figure 21.9 Evolution of the H-R diagram for a star cluster. *This shows how the H-R diagram for a cluster of stars changes as the cluster ages and the most massive stars evolve to become red giants.*

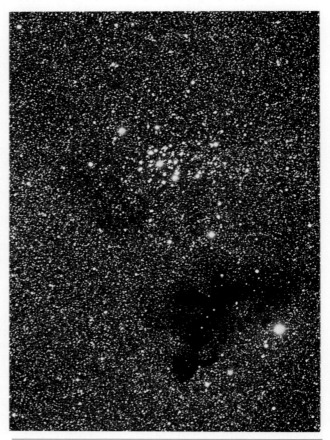

Figure 21.10 A very young cluster with nearby dust cloud. *The cluster, NGC6520, contains several hot, young blue stars. The nearby dust cloud is so dense that it blocks out the light from the stars behind it. (© 1980 Anglo-Australian Telescope Board)*

refers to hydrogen in its ionized form; these hot gas clouds are called HII regions because they glow most strongly in emission lines of hydrogen (these lines form when protons and electrons combine to form hydrogen atoms, which can occur only in a region that is ionized). HII regions appear red because hydrogen has a very strong emission line at a wavelength of 6,563 Å in the red portion of the spectrum (see Chapter 6).

Since several young clusters are permeated by gas and dust left over from star formation, it is natural to expect that in even younger, newly forming clusters, the gas and dust would be much denser, so dense, in fact, that the embryonic stars might be hidden from the view of observers. Hence the search for stars in the process of forming takes us into dense interstellar clouds.

How do we probe these dark and dusty regions? For a couple of reasons, this is best done by observing in infrared wavelengths. One reason is that infrared radiation penetrates much farther through clouds of gas and dust than does visible light. Another is that the dust around a newborn star is hot, and it glows at infrared wavelengths. Therefore, to see infant stars, we must look for infrared sources buried in dark clouds (Figs. 21.13 and 21.14).

We have found many regions in our galaxy in which star formation is apparently taking place. One of the best observed of these is in the sword of Orion, in the Orion nebula, a great, glowing HII region. A very young cluster of stars is located there, and infrared observations reveal a number of infrared sources embedded within the dark cloud associated with the cluster (see Fig. 21.13).

In other, similar regions, faint variable stars called **T Tauri** stars are found. These appear to be newborn stars just in the process of shedding the excess gas and dust from which they formed; in fact, spectroscopic measurements indicate that some material may still be falling into these stars, as though the collapse of the

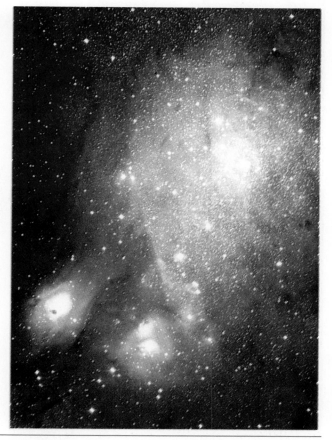

Figure 21.11 Regions of young stars and nebulosity. *Here are two regions populated by recently formed stars and their associated interstellar gas and dust. (© 1981, 1984 Anglo-Australian Telescope Board)*

Figure 21.12 A dark cloud. *This is the well-recognized Horsehead nebula. Many stars in the process of formation have been detected, by infrared techniques, within the dark and dusty regions seen here. (© 1984 Anglo-Australian Telescope Board)*

Figure 21.13 The Orion nebula. *At top is a normal photograph, showing the familiar nebula. Below is an infrared image of the same region. Here little of the glowing gas can be seen, but the young, hot stars that are embedded within the outer portions are clearly visible. Note that the Trapezium, a close group of four brilliant young stars, is visible in both images. (top: U.S. Naval Observatory; bottom: © 1984 Anglo-Australian Telescope Board)*

Figure 21.14 An embedded infrared source. *In this infrared photograph of a region within the Orion nebula, blue indicates the location of an intense, relatively warm source that is thought to be a newborn star. (R. D. Gehrz, J. Hackwell, and G. Grasdalen, University of Wyoming)*

cloud from which they are forming is not yet complete. We learned in Chapter 17 that the Sun, in its evolution, probably went through an early phase as a T Tauri star.

STAR FORMATION

The process of star formation deduced from the types of observations just described begins with the gravitational collapse of an interstellar cloud (Fig. 21.15). It is not certain what causes a cloud to begin to fall in on itself, but once it starts, gravity takes care of the rest (see Astronomical Insight 21.2). Calculations show that the innermost portion collapses most quickly, while the outer parts of the cloud are still slowly picking up speed. The temperature in the core builds up as the density increases, but for a long time the heat escapes as the dust grains in the interior radiate infrared light. Eventually, however, the cloud becomes opaque in the center, and radiation no longer can escape directly into space. After this, the heat builds up much more rapidly, and the resulting pressure causes the collapse to slow to a very gradual shrinking. There may be a second episode of rapid collapse, when hydrogen molecules in the central condensation begin to break apart due to the increasing temperature. When this happens, the

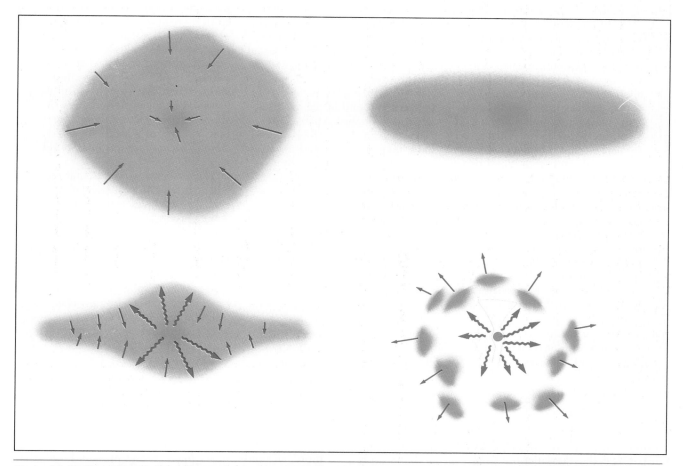

Figure 21.15 Steps in star formation.   *A rotating interstellar cloud collapses, most rapidly at the center. At first infrared radiation escapes easily and carries away heat, but eventually the central condensation becomes sufficiently dense that this radiation is unable to escape, and the core becomes hot. After a lengthy period of slow contraction, nuclear reactions begin in the protostar. At some stage, in a process not well understood, the young star develops a strong wind that helps clear away remaining debris.*

ASTRONOMICAL INSIGHT 21.2

Triggers of Star Formation

An interstellar cloud can collapse under its own gravity only if its density exceeds a certain value for its size, such that the inward gravitational force overcomes the ordinary gas pressure as a result of random motions of the cloud particles. There is a specific relationship among density, temperature, and cloud size called the **Jeans criterion** (named after Sir James Jeans, the British astrophysicist who derived it in the early decades of the twentieth century), which predicts the mass of cloud that can collapse under a given set of physical conditions. This criterion states that, as long as there are normal interstellar cloud densities (several thousand to perhaps a million atoms per cubic centimeter in dark clouds), only very massive clouds can collapse gravitationally. This poses the dilemma of explaining how stars with masses like that of the Sun or smaller can form.

One answer to this question has been to assume that as a massive cloud collapses, it breaks up into smaller fragments that can then individually collapse into stars. This is possible because as the overall density increases during collapse, the Jeans criterion allows smaller and smaller masses to condense individually. Furthermore, this picture provides a nice explanation of how a cluster of stars may form, starting with a very massive cloud that creates fragments as it collapses, the fragments in turn forming individual stars.

But what of individual, isolated stars? There is no evidence, for example, that the Sun was ever a member of a cluster (although if it had formed in a cluster, by now the cluster could have dissipated). Evidently it is possible for small-mass clouds to collapse and form stars, contrary to the Jeans criterion.

A hint at the answer to this problem is found in regions of active star formation, such as the Orion nebula. Here observations show that the interstellar material is chaotic; it is stirred up by violent shock fronts, probably themselves triggered by supernova explosions or strong stellar winds. The entire interstellar medium is in a constant state of turmoil (see Chapter 25), so that a cloud of gas and dust is likely to be ravaged sooner or later by a passing shock front. Theoretical calculations show that in some circumstances this event will cause the cloud to compress enough to make it collapse, even though the cloud's initial density may not have been high enough to satisfy the Jeans criterion.

Supernova explosions occur in massive stars when all the nuclear fuel in the core is depleted (see Chapter 22 for a more complete discussion). These are the same stars that have very short lifetimes. Hence, when a cluster of stars forms, its most massive members will evolve and explode rather quickly as supernovae. The shock waves from these explosions can then trigger the collapse of interstellar material remaining from the original formation of the cluster, creating a new generation of stars. The most massive of these new stars will also evolve quickly, themselves dying in supernova explosions and creating shock waves that can give birth to yet another generation of stars. A sort of chain reaction of star formation is set off, gradually consuming an entire interstellar cloud complex.

There is some evidence that the Sun's formation was associated with a nearby supernova explosion. Certain atomic isotopes found in meteorites show evidence of having formed out of an isolated interstellar cloud that was compressed by shocks created in a supernova outburst. This was not part of a chain-reaction process like that just described, presumably because the Sun's predecessor cloud was not in a dense region of gas and dust.

heat created by the infall is taken up by the molecules as they dissociate, instead of contributing to the pressure. Once the dissociation is complete, the central pressure builds again, and the collapse is replaced by another period of slow shrinking. Material still falling in crashes into the dense core, creating a violent shock front at its surface. The luminosity of the core, which by this time may have a temperature of 1,000 K or more, depends on its mass. If the mass is large, that is, several times the mass of the Sun, then the luminosity may be sufficient to blow away the remaining material by radiation pressure. For a lower-mass

protostar, as the central dense object is called, material continues to fall in. Eventually the density of this material becomes low enough that the protostar can be seen through it. Regardless of mass, a point is reached at which a starlike object becomes visible to an observer; this stage may be identified with the T Tauri stars mentioned earlier. If the obscuring matter is swept away violently, the object may appear to brighten up dramatically in a very short time. In a few cases, a previously rather faint star was found to have brightened suddenly by several magnitudes.

In any case, the protostar continues to shrink slowly, growing hotter in its core. The shrinking stops when the temperature becomes high enough for nuclear reactions to begin to occur in the center. This is a major landmark in the process of star formation, for once the reactions have started, the star is on its way to becoming stable (that is, it shrinks no further), and it lies on the main sequence, where it will spend most of its life converting hydrogen into helium in its core.

It is interesting to trace on the H-R diagram the path the star takes as it forms (Fig. 21.16). When it is a cold,

dark cloud, its position is far to the right and down, well off the scales of temperature and luminosity appropriate for stars. As the cloud heats up in the core and begins to glow in infrared wavelengths, it moves up and to the left, eventually attaining sufficient luminosity to fall onto the standard H-R diagram, but still off to the right of the main sequence. When it reaches the phase of slow contraction, it moves gradually to the left and down, finally reaching the main sequence when the reactions begin.

The entire process, from the beginning of cloud collapse to the point at which the newly formed star is on the main sequence (Table 21.2), takes many millions or even billions of years for the least massive stars, and only a few hundred thousand years for the most massive. In a cluster with stars of various masses, the most massive stars may form, live, and die before the least massive ones even reach the main sequence. The H-R diagram for such a cluster shows stars located to the right of the main sequence at the lower end (Fig. 21.17).

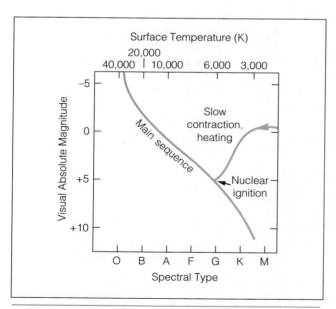

Figure 21.16 The path of a newly forming star on the H-R diagram. *As a protostar heats up, it eventually becomes hot and luminous enough to appear in the extreme lower right-hand corner of the H-R diagram. When the core becomes opaque, the rapid collapse slows to a gradual shrinking, and the protostar moves across to the left as it heats. Eventually nuclear reactions begin, and the star is on the main sequence.*

Table 21.2 Steps in star formation (one solar mass)

1. Gravitational collapse of interstellar cloud: Collapse proceeds most rapidly at the center, where the density grows most quickly. The cloud core remains cool because heat escapes in the form of infrared radiation.

2. Slow contraction of core: Once the core density is high enough so that the cloud becomes opaque to infrared radiation, heat is trapped, and the temperature and pressure in the core rise. The central portion of the cloud contracts very slowly for a while.

3. Infall: Material from the outer portions of the cloud continues to fall in, creating heating and shock waves. When the hydrogen molecules in the core begin to break up because of the additional heating, the breakup of molecules takes up heat from the gas, reducing the pressure, so that a new collapse phase occurs.

4. Final contraction: After the second collapse, the internal pressure again rises to balance gravity, and the central condensation enters a new phase of very slow contraction. At this point, the protostar is still surrounded by gas and dust, and is usually observable only as an infrared source.

5. Nuclear ignition: The protostar continues to heat gradually as the contraction goes on. Eventually, the core becomes hot enough to start nuclear fusion reactions, and the contraction stops because there is an internal source of energy to provide pressure that balances gravity fully. The star is now on the main sequence. Observationally, it may still be associated with gas and dust and may still undergo a T Tauri phase.

ASTRONOMICAL INSIGHT 21.3

Hyperactive Young Suns

For many years, a class of peculiar stars called **T Tauri** stars has been recognized and studied. These stars, named after the brightest and first-identified example, are characterized by cool-star spectra (most are G stars), and by prominent emission lines. The T Tauri objects are often associated with dense clouds of interstellar matter and are therefore generally best observed at infrared wavelengths, which penetrate the interstellar dust far better than visible light.

The presence of emission lines has been taken as an indicator that these stars have very active chromospheres, a supposition borne out by recent ultraviolet and X-ray observations. Whereas emission lines in visible wavelengths do not provide much information on the temperature of the gas, there are many good diagnostics of hot gas in ultraviolet and X-ray wavelengths. Data obtained with the *International Ultraviolet Explorer* have revealed a number of bright ultraviolet emission lines indicative of a very hot gas. X-ray emission from many T Tauri stars has been detected as well, particularly by the *Einstein Observatory* satellite (see Chapter 7). There is no doubt that the T Tauri stars have chromospheres, probably heated by energy transported from the outer convective zones of the stars to higher levels. This is analogous to the Sun's chromosphere, except that the amount of energy in-

volved in the T Tauri chromospheric heating is apparently greater than in the Sun.

There are indications, in the Doppler shifts that sometimes are seen in the emission and absorption lines in T Tauri star spectra, that gas is flowing outwards at high velocities. Careful study of the Doppler shifts has shown that the situation is not simple, however. Several cases have been found in which T Tauri stars are apparently *receiving* gas that is falling in, and certain stars have even shown both infall and outflow at the same time. Probably a more accurate picture would be to assume that, rather than having an overall wind flowing one way or the other, a typical T Tauri star has gas simultaneously rising and falling, in analogy with the solar prominences whose motions are governed by the Sun's magnetic field.

There are other indirect indications that magnetic fields play important roles in T Tauri stars. From observations of bright patches called **Herbig-Haro objects** that are sometimes seen amid the nebulosity near T Tauri stars, it is inferred that gas is ejected at great velocities from the stars and is beamed along very specific directions. The Herbig-Haro objects appear to be "hot spots" where the high-velocity beams strike surrounding material. Very recent radio observations have directly revealed a linear jet or beam of hot gas emanating from one suspected T Tauri star, helping to clarify this interpretation of the Herbig-Haro objects.

The general consensus on T Tauri stars is that they are very young objects on their way to be-

coming stars. This conclusion is based on their locations within dense regions of interstellar material, on their apparent instability (the spectra and the gas motions they reveal are often variable with time), and on comparisons of their positions in the H-R diagram with theoretical calculations of the properties of newly forming stars. The "T Tauri wind," a phase thought to be undergone by young stars in general, is usually invoked to explain how the early solar system was cleared of leftover debris from the formation of the planets (see Chapter 16).

Continued radio observations of T Tauri stars has shown that the outflows often associated with them are directed away from the poles of the stars. Thus, a T Tauri star may have a disk of solid particles in its equatorial plane, a pair of opposing jets of hot gas emanating from its poles, and matter falling in at intermediate latitudes. The so-called bipolar flows, the opposing beams of ionized gas flowing away from the stars along their rotation axes, are now known to be quite common among young stars. It is interesting that these beams have quite a bit in common with the jets of hot gas that are often observed to be emitted from the nuclei of galaxies (Chapter 30); perhaps there are similarities in the formation mechanisms.

The study of T Tauri stars has added a lot to our picture of the formation of the Sun and of the solar system. New observational tools soon to be available will do even more, by providing X-ray, ultraviolet, visible, and infrared data on these infant suns.

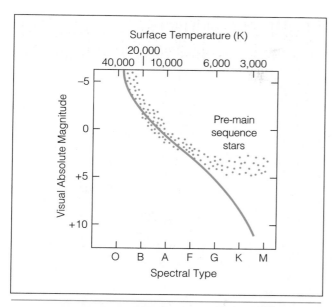

Figure 21.17 Pre-main sequence stars. *H-R diagrams of very young clusters often show stars that have not yet reached the main sequence, but are still moving toward it. At the same time, some of the massive, short-lived stars at the top of the main sequence may have already evolved away from it, having used up their hydrogen fuel.*

PERSPECTIVE

The observational evidence tells us that stars must form out of interstellar material, and that they spend most of their lives on the main sequence. H-R diagrams of clusters show that, upon completion of the main-sequence lifetime, a star moves up and to the right, becoming a red giant. By combining these observational deductions with the results of stellar model calculations, we can deduce the detailed life stories of individual stars.

SUMMARY

1. Steps in stellar evolution are deduced from observations of stars in various stages, from newly formed ones to those in the late stages of their lives.

2. Clusters are groups of stars that formed together out of the same material.

3. The relative brightnesses of stars in a cluster reflect real differences in luminosity, because all the stars in a

cluster are approximately at the same distance from us.

4. The distances to clusters can be determined by plotting a color-magnitude diagram and comparing the location of the main sequence with that on a standard H-R diagram.

5. The fact that all stars in a cluster have the same age and initial composition means that the observed differences from star to star within the cluster are entirely a result of differences in stellar mass. This allows us to determine how stars of different mass evolve.

6. As stars evolve, they move off of the main sequence, becoming cooler and more luminous, and move into the red-giant region of the H-R diagram.

7. The age of a cluster can be inferred from the location of its main-sequence turnoff point.

8. In associations of young, hot stars, interstellar material is often left over from the process of star formation.

9. Stars in the process of forming are often embedded within dense clouds and are best observed in infrared wavelengths.

10. A star forms from the gravitational collapse of an interstellar cloud, which condenses most rapidly at the center. The new star begins to heat up only after the gas in the core of the cloud becomes opaque, and the resulting protostar continues to slowly shrink until its interior becomes hot enough for nuclear reactions to begin.

REVIEW QUESTIONS

Thought Question

1. Summarize the importance of star clusters in the study of stellar evolution.

2. Describe the various types of star clusters mentioned in the text. Include an indication of where they are likely to be found in the galaxy.

3. If a cluster is observed to have O stars in it, how do we know that it must be a young cluster? Is it possible for the same cluster to also contain M stars?

4. Suppose two stars in a cluster have identical $B - V$ color indices and apparent magnitudes. How do their masses and other properties compare? Could you make the same assumption for a pair of stars not in the same cluster, but which also have identical color indices and apparent magnitudes?

5. Describe the main sequence fitting technique for finding distances to star clusters. On what assumptions about the stars in a cluster does this method depend?

6. Explain, in your own words, how the main-sequence turnoff for a cluster H-R diagram is related to the age of the cluster.

7. Explain why infrared observations are an important method for observing star formation.

8. Summarize the sequence of events thought to take place as a star forms.

9. Is a young star in hydrostatic equilibrium when it is in the slow gravitational contraction stage? Explain.

10. What determines the location on the main sequence where a star will lie after it has formed?

Problems

1. Comparison of the main sequence of a cluster color-magnitude diagram with a standard H-R diagram shows that the apparent magnitudes of the stars on the cluster diagram main sequence are 5 magnitudes larger than the absolute magnitudes of the corresponding stars on the standard H-R diagram. How far away is the cluster?

2. Using what you learned about stellar lifetimes and about the properties of main-sequence stars (Table 20.2), calculate the age of a cluster whose brightest main-sequence stars belong to spectral type F5. (Hint: Use Table 20.2 to find the mass and luminosity of an F5V star, then calculate its lifetime.) What is the $B - V$ color index for the main sequence turnoff for this cluster (again, use Table 20.2)?

3. When a star leaves the main sequence, it becomes a red giant, having a larger radius and lower surface temperature. Suppose the Sun becomes a red giant, with a radius 20 times greater and a surface temperature one-half its present value. By how much does its luminosity increase? How many magnitudes brighter (in bolometric magnitudes) is it?

ADDITIONAL READINGS

Allen, David A. 1987. Star formation and the IRAS galaxies, *Sky and Telescope* 73(4):372.

Bok, B. J. 1972. The birth of stars. *Scientific American* 227(2):48

———. 1981. The early phases of star formation. *Sky and Telescope* 227(2):48.

Boss, Alan P. 1985. Collapse and formation of stars. *Scientific American* 252(1):40.

Cohen, M. 1975. Star formation and early evolution. *Mercury* 4(5):10.

Fienberg, R. T. 1989. Brown dwarfs coming and going. *Sky and Telescope* 78:482.

Gehrz, R. D., D. C. Black, and P. M. Solomon. 1984. The formation of stellar systems from interstellar molecular clouds. *Science* 224(4651):823.

Graham, John. 1984. A new star becomes visible in Chile. *Mercury* 13(2):60.

Herbst, W., and G. E. Assousa. 1979. Supernovas and star formation. *Scientific American* 241(2):138.

Jones, B., D. Lin, and R. Hanson. 1984. T Tauri and TIRC: A star system in formation. *Mercury* 13(3):71.

Lada, Charles J. 1986. A star in the making. *Sky and Telescope* 72(4):334.

Lada, C. J. 1982. Energetic outflows from young stars. *Scientific American* 247(1):82.

Loren, R. B., and F. J. Vrba. 1979. Starmaking with colliding molecular clouds. *Sky and Telescope* 57(6):521.

Scoville, N., and J. S. Young. 1984. Molecular clouds, star formation, and galactic structure. *Scientific American* 250(4):42.

Spitzer, L. 1983. Interstellar matter and the birth and death of stars. *Mercury* 12(5):142.

Waldrop, M. Mitchell. 1983. Stellar nurseries. *Science 83* 4(4):40.

Werner, M. W., E. E. Becklin, and G. Neugebauer. 1977. Infrared studies of star formation. *Science* 197:723.

Zeilik, M. 1978. The birth of massive stars. *Scientific American* 238(4):110.

MATHEMATICAL METHODS

The Jeans Length and Stellar Masses

The formation of a star begins when an interstellar cloud becomes unstable and begins to fall in on itself. The collapse may be precipitated by an outside force, such as the impact of a passing shock wave, but it may also happen spontaneously. In the early years of this century, the British astrophysicist Sir James Jeans analyzed the conditions under which spontaneous collapse would occur.

Jeans reasoned that collapse would take place if the inward force of gravity exceeded the outward force of gas pressure within a cloud. By setting these forces equal, he was able to derive an equation for the radius of a cloud that would collapse. The size of the cloud depends only on the temperature of the gas (which governs the pressure) and the density (which determines the gravitational force). The so-called Jeans Length R_J is

$$R_J = 3.5 \times 10^7 \sqrt{\frac{T}{\rho}},$$

where R_J is the radius (in centimeters) of the cloud that is just about to begin to fall in on itself; T is the temperature of the gas in the cloud (on the absolute scale), and ρ is the density (in grams/cm^3).

A typical interstellar cloud has a temperature of about 50 K and density about 10^{-21} grams/cm^3 (about 1,000 atoms per cubic centimeter). Substituting these values yields

$$R_J = 3.5 \times 10^7 \sqrt{\frac{50}{10^{-24}}} = 8 \times 10^{18} \text{ cm},$$

or about 2.5 parsecs (approximately 8 light-years).

It is interesting to consider the mass contained in a cloud such as this. The cloud mass is equal to its volume times its density. If we assume that it is spherical in shape, so that the volume is $\frac{4}{3}\pi R^3$, then the mass is

$$M = \frac{4}{3}\pi R^3 \rho = 2 \times 10^{36} \text{ grams.}$$

To get this value, we substituted $R_J = 8 \times 10^{18}$ cm for R and 10^{-21} grams/cm^3 for ρ in the equation.

Note that a mass of 2×10^{36} grams is about 1,000 times the mass of the Sun. This is far more massive than any known star (in fact, calculations show that such a massive star could not be stable), and certainly much more massive than the vast majority of stars. Hence we need to explain how stars of ordinary mass can be formed.

While the Jeans equations are thought to be essentially correct, there are some important considerations that are not included. For example, it is possible for a cloud to collapse due to an outside force, as already mentioned, and this could happen in a much smaller cloud than calculated from the Jeans Length. It has also been considered possible that turbulence or other kinds of gas motions could cause a collapsing cloud to fragment, separating into smaller, denser clouds that can then individually collapse to form stars of normal mass (this process may well explain the formation of star clusters). Finally, recent suggestions have been made that stars eject a majority of the initial cloud mass during the late stages of collapse. It has been known for some time that very young stars (such as T Tauri stars) undergo phases where they eject matter. If enough is cast off in this process, then the final mass of the star that is formed can be far less than that of the original cloud before collapse.

CHAPTER 22

Life Stories of Stars

The Helix nebula, a shell of gas emitted by a dying star. (© 1979 Anglo-Australian Telescope Board; photograph by D. Malin)

CHAPTER PREVIEW

Stars Like the Sun
The Helium Flash
Planetary Nebula to White Dwarf
The Evolution of Massive Stars
Nuclear Evolution of a Massive Star
The End of the Reactions
Formation of the Heaviest Elements
Supernova 1987A
Evolution in Binary Systems

Combining our knowledge of stellar structure with observations showing how stars evolve, we can develop
a picture of the story of an individual star, tracing it
from its birth to its death. Because so much depends
on the mass of a star, we will discuss three separate
examples: a star like the Sun, a star with a mass a few
times that of the Sun, and a very massive star. In the
process we will see many important differences among
the three cases.

We have already discussed the formation of stars,
finding that the process is rather similar for all masses,
except for the time required. We will not reiterate this
discussion here; we will begin our story for each star
after it has reached the main sequence.

STARS LIKE THE SUN

A star with the mass of the Sun, when it arrives on the
main sequence, is similar to the Sun in all its properties.
Thus it is a G2 star, with a surface temperature around
6,000 K and a luminosity of about 10^{33} erg/sec. Its
composition initially is about 73 percent hydrogen by
mass, and roughly 25 percent helium, with only about
2 percent of the mass composed of other elements. It
has convection in the outer layers, and it has a chromosphere and a corona. The star must change with
time, because its core composition changes. Table 22.1
summarizes the stages in the life of a star like the Sun.

In this star's core, the proton-proton chain converts
hydrogen into helium. As we saw previously (in
Chapters 17 and 20), this process lasts some 10 billion
years for a star of 1 solar mass. During this time, the
core gradually shrinks as the hydrogen nuclei are re-

Table 22.1 Evolution of a 1-$M_\odot$ star

1. *Hydrogen burning:* During the 10^{10} years the star
spends on the main sequence, hydrogen is converted to
helium through the proton-proton chain. The luminosity
gradually increases as the core becomes denser and hotter.

2. *Development of degenerate core and evolution to red
giant:* When hydrogen in the core is used up, the core
contracts until it is degenerate. A hydrogen shell source
outside the core still undergoes reactions, causing the outer
layers to expand and cool. The star becomes a red giant.

3. *Helium flash:* The core eventually gets hot enough to
ignite the triple-alpha reaction, in which helium is
converted to carbon. Because the core is degenerate and
cannot expand and cool to counteract the new source of
heat, the reaction rapidly takes place throughout the core.
So much heat energy is added that degeneracy is quickly
destroyed, and the reactions become stable.

4. *Stable helium burning:* Once reactions are taking place
in the core again, the major energy production is there,
and the shell source becomes less important. The outer
layers contract, and the surface heats up; the star moves
back to the left on the H-R diagram.

5. *Second red giant stage:* When the core helium is used
up, the energy again comes from a shell (an inner shell,
where helium is converted to carbon, and an outer shell,
where hydrogen is still converted to helium). The outer
layers expand again, and the star becomes a red giant for
the second time.

6. *Mass loss:* Through a stellar wind and through
pulsational instabilities, the star loses its outer layers,
exposing the hot core. The core continues shrinking but
never again is hot enough for nuclear reactions. The star
becomes a planetary nebula periodically as the outer layers
are shed.

7. *White dwarf:* The core again becomes degenerate as it
shrinks. Eventually the outer layers are gone altogether,
leaving only the hot, degenerate core, which is a white
dwarf.

placed by a smaller number of the heavier helium nuclei, and the internal temperature rises. This causes an
increase in luminosity, and the star gradually moves
up in the H-R diagram. The main sequence, therefore,
is not a perfectly narrow strip but has some breadth,
since stars on it slowly move upward as their luminosities increase (Fig. 22.1). The starting point, the lower
edge of the main sequence, is called the **zero-age main
sequence (ZAMS)**, because this is where newly formed
stars are found.

When the hydrogen in the core of the star is gone,
reactions there cease. By this time, however, the temperature in the zone just outside has nearly reached the
point at which reactions can take place, and with a

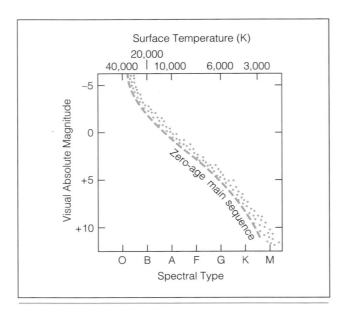

Figure 22.1 The width of the main sequence. *As stars on the main sequence gradually convert hydrogen into helium in their interiors, the cores shrink a little, and get hotter. This increases the luminosity, and the stars gradually move upward on the H-R diagram. As a result, the main sequence, consisting of stars of a variety of ages, is not a narrow strip, but has some breadth.*

Figure 22.2 The hydrogen shell source. *After the hydrogen in the core of a star is completely used up, the core, now composed of helium, continues to shrink and heat. This causes a layer of gas outside the core to reach the temperature required for nuclear reactions. A shell source, in which hydrogen is converted into helium, is ignited. The star's outer layers expand and cool, and the star becomes a red giant.*

little more shrinking and heating of the core, the proton-proton chain begins again in a spherical shell surrounding the core (Fig. 22.2). At this point, the star has an inert helium core, and it is producing all its energy in the hydrogen-burning shell, which steadily moves outward inside the star, eating its way through the available hydrogen.

The core continues to shrink and heat. This heating enhances the nuclear reactions in the shell, causing it to produce more and more energy. As the source of energy is heated and moves toward the surface, the outer layers of the star are forced to expand, and as this occurs, the surface cools. The luminosity of the star increases because of its increased surface area, so it moves upward on the H-R diagram (Fig. 22.3). It also moves to the right, because the surface temperature is decreasing. The star rapidly becomes a red giant, reaching a size of 10 to 100 times its main-sequence radius (Fig. 22.4). The outer layers are constantly overturning as convection occurs to a great depth (Fig. 22.5).

The helium core becomes extremely dense. At first the temperature is not high enough for any new nuclear reactions to start there. (The triple-alpha reaction, in

which helium nuclei combine to form carbon, requires a temperature of about 100 million degrees.) As the core continues to shrink, the matter there takes on a very strange form. Electrons have a property that prevents them from being squeezed too close together, and this creates a new kind of pressure, which becomes the principal force supporting the star against the inward force of gravity. The gas in the stellar core contains many free electrons, and eventually their resistance to being compressed becomes the dominant pressure that supports the core against further collapse. When this happens, the gas is said to be **degenerate.**

A degenerate gas has many unusual properties. One of them is that the pressure no longer depends on the temperature, and vice versa. If the gas is heated further, it will not expand to compensate, as an ordinary gas would. As we will soon learn, there are important consequences if nuclear reactions start in the degenerate region of a star.

We now have a red giant star, with highly expanded outer layers. Near the center is a spherical shell in which hydrogen is burning in nuclear reactions, and inside of this is a degenerate core containing helium nuclei and degenerate neutrons. The core is small, but it may contain as much as a third of the star's total mass.

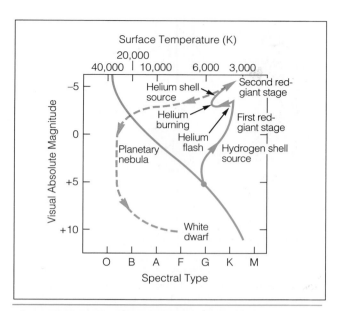

Figure 22.3 Evolutionary track of a star like the Sun.
This H-R diagram illustrates the path a star of one solar mass is thought to follow as it completes its evolution. The dashed portion, following the second red giant stage, is less certain than the earlier stages.

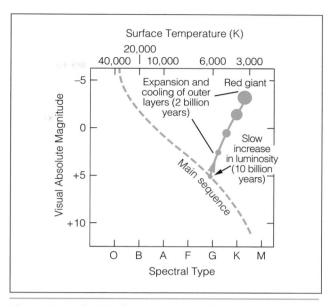

Figure 22.4 The development of a red giant. *As a star's outer layers expand because of the shell hydrogen source, they cool. At the same time the surface area increases, raising the star's luminosity. The star therefore moves up and to the right on the H-R diagram.*

The Helium Flash

During the red giant phase, the helium core continues to be heated by the reactions going on around it. Eventually the temperature becomes sufficiently high for the triple-alpha reaction to begin. When it does, there are spectacular consequences.

Ordinarily, when a gas is heated, it expands, and this limits how hot it can get, because an expanding gas tends to cool. In a degenerate gas, however, no such expansion occurs, because pressure does not depend on temperature in the usual way. Hence, when the reactions begin in the degenerate core of a red giant, the temperature goes up quickly, and the core retains the same density and pressure. The increased temperature speeds up the reactions, producing more heat, in turn accelerating the reactions even further. There is a rapid snowball effect, and in an instant (literally seconds), a large fraction of the core is involved in the reaction. This spontaneous runaway reaction is called the **helium flash.** Although it creates dramatic consequences for the star's interior, calculations show that there is little or no immediate effect visible to an observer. The overall direction of the star's evolution is changed, however, which would be apparent after thousands of years.

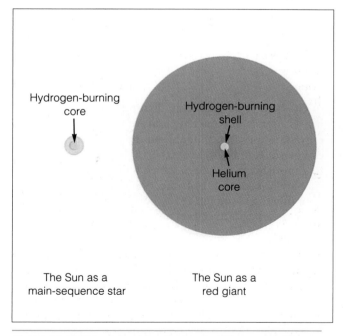

Figure 22.5 Structure of a red giant. *When a star like the Sun becomes a red giant its diameter grows much larger, but most of its mass remains concentrated in a small inner region. Energy is supplied by nuclear reactions in the hydrogen shell source.*

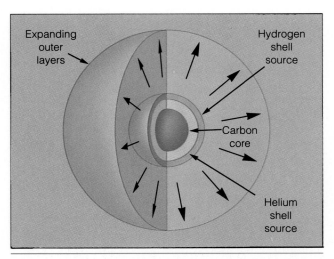

Figure 22.7 Internal structure late in a star's life. *The core of this star has undergone hydrogen burning and helium burning and is composed of carbon. Hydrogen and helium shell sources are still undergoing reactions and producing energy.*

Figure 22.6 The H-R diagram for a globular cluster.
These star clusters are very old, and all the upper main sequence stars have evolved to later stages. Here we see a number of stars in or approaching the red giant stage, and a number on the horizontal branch, where they move following the helium flash. Only very old clusters have a horizontal branch. (Besides their great ages, globular clusters also have a relatively low content of heavy elements, and this affects how the stars evolve. This point is discussed more fully in Chapter 26.)

The helium flash quickly disrupts the core, destroying its degeneracy (as temperature increases, more energy states become available to the electrons, and they can be squeezed closer together). The core then returns to a more normal state in which temperature and pressure are linked, and the triple-alpha reaction continues, now in a more stable, steady fashion. The outer layers of the star begin to retract as the star reverts to a more uniform internal structure, and as they do so, they become hotter. The star moves back to the left on the H-R diagram (see Fig. 22.3). Old clusters, particularly globular clusters, are found to have a number of stars in this stage of evolution, forming a sequence called the **horizontal branch** (Fig. 22.6). This is seen only in clusters old enough for stars as small as 1 solar mass to have evolved this far; as we learned in previous chapters, this requires an age of 10 billion years or more.

In due course, the helium in the core becomes exhausted, leaving an inner core of carbon (Fig. 22.7). Helium still burns in a shell around the core; there could even be an active hydrogen-burning shell farther out in the star at the same time. The star expands and cools, becoming a red giant for the second time. This second red giant stage is short-lived, however, lasting perhaps 1 million years. A degenerate core again develops, but there is not another dramatic flare-up in its interior. In fact, no more nuclear reactions ever occur in the core of this low-mass star, because the temperature never reaches the extremely high levels needed to cause heavy elements like carbon to react.

Planetary Nebula to White Dwarf

The star in its second red giant stage may be viewed as having two distinct zones: the relatively dense interior, consisting of a carbon core inside a helium-burning shell which in turn is surrounded by a hydrogen-burning shell; and the very extended, diffuse outer layers. The inner portion may contain up to 70 percent of the mass but occupies only a small fraction of the star's volume.

As the nuclear fuel in the interior runs out, the core shrinks. The details of what happens to the outer layers are not clear, but apparently the star ejects portions of this material in a series of minor outbursts that could be likened to blowing smoke rings. The star gently gets rid of its outer layers.

Figure 22.8 Planetary nebulae.   *At left is the Dumbbell Nebula; at right, the Helix Nebula. In both, the blue-green color is produced by emission lines of twice-ionized oxygen. The central stars, which are responsible for ejecting the gas of the nebulae, are visible at the center of each. (Left: Lick Observatory photograph; right: © 1979 Anglo-Australian Telescope Board.)*

The evidence that this occurs is primarily observational, since a successful theory of how it happens has not yet been developed. A number of objects have been observed that seem to consist of a hot, compact star surrounded by a shell of expanding gas (Fig. 22.8). These shells often have a bluish green color because of the emission lines (primarily those of twice-ionized oxygen) by which they glow, and through a telescope or on a photograph they bear some resemblance to the planets Uranus and Neptune. Because of this resemblance, they are called **planetary nebulae.** Some, such as the Dumbbell Nebula (see Fig. 22.8) or the Ring Nebula (Fig. 22.9), are well-known objects visible with a small telescope.

The star that remains in the center of a planetary nebula lies far to the left in the H-R diagram (see Fig. 22.3), having a surface temperature that may be as high as 100,000 K or more, far hotter than any main-sequence star. When the shell is being ejected, the star's surface temperature increases as hot inner gas becomes exposed on the surface. In essence, such a star is a naked core, left behind when the outer layers are cast off.

Some nuclear reactions may still be going on in a shell inside the star, but they do not last much longer, since the fuel in the core is depleted. The density, already very high, increases as the stellar remnant con-

Figure 22.9 The Ring Nebula.   *This is one of the best-known and most beautiful planetary nebulae, the Ring Nebula in the constellation Lyra. (Lick Observatory photograph)*

denses under the force of gravity. A larger and larger fraction of the star's interior becomes degenerate. As the star shrinks, it stays hot, but its luminosity decreases because of its diminishing surface area, and it moves downward on the H-R diagram (see Fig. 22.3). Eventually the pressure of the degenerate electron gas stops the collapse, and the star becomes stable again, but now it is a very small object, so dim that it lies well below the main sequence, in the lower left-hand corner of the H-R diagram. It is called a **white dwarf.**

A white dwarf is a bizarre object in many ways. It is supported by degenerate electrons whose peculiar properties govern its internal structure. It has a mass as great as that of the Sun (some white dwarfs are even slightly more massive), yet it is approximately the size of the Earth. This means that its density is incredibly high, roughly a million times that of water. A cubic centimeter of white-dwarf material would weigh a ton at the surface of the Earth!

A white dwarf does not do much. It undergoes no more nuclear reactions, so it does not evolve further. It just sits there, radiating away the heat energy still contained inside and slowly cooling off. A white dwarf can be stirred into action under certain circumstances (as described in the next chapter), but in most cases, this is the end of the line for a star like the Sun.

THE EVOLUTION OF MASSIVE STARS

Stars more massive than the Sun follow the same general course of events as the Sun, but with significant differences. The major similarity is that the majority of the star's lifetime is spent on the main sequence, while nuclear reactions in the core convert hydrogen to helium, producing the energy necessary to balance gravity and supply the luminosity of the star. The principal contrasts with the evolution of a lower-mass star are that everything happens more rapidly in a massive star, and that many more stages of nuclear reactions can occur. The final result, or remnant, can be quite different as well.

Nuclear Evolution of a Massive Star

Even though the first nuclear-burning stage is the same as that of the Sun, the details are different. Whereas the low-mass stars convert hydrogen to helium by the proton-proton chain, in stars more than about twice as massive as the Sun another reaction, the CNO cycle, is more efficient and is therefore dominant. (This reaction sequence is described in Chapter 20 and in Appendix 11.)

The luminosity and surface temperature of a star, hence its location on the main sequence, depend on its initial mass. A five solar-mass star would have surface temperature roughly 12,000 K and spectral type B8,

for example; a 10 solar-mass star would have surface temperature about 20,000 K and would be type B2; and a 20 solar-mass star would have surface temperature roughly 30,000 and would be type O9.

The core temperature inside a star governs the rate of nuclear reactions, and this is why massive stars evolve more rapidly than low-mass ones. The higher temperature in massive star cores (itself a direct result of the high mass; see discussion in Chapter 20) results in rapid reaction rates and quick exhaustion of the nuclear fuel. A five solar-mass star will consume all the hydrogen in its core in only about 100 million years (about 1 percent of the main-sequence lifetime of the Sun), and a 20 solar-mass star will use up its hydrogen in even less time, perhaps 1 million years.

The high core temperatures of massive stars have another important impact on the reactions: they allow further reaction stages to occur. After each form of nuclear fuel is exhausted, the core contracts and the temperature increases, reaching the point where a new reaction can begin. In the massive stars, helium will combine by the triple-alpha reaction to form carbon (while hydrogen continues to burn in a shell outside the core), and when a significant amount of carbon is present, it may react with helium to form oxygen through a process called an **alpha-capture reaction.** In alpha-capture reactions, helium nuclei (which are called alpha particles) are added to larger nuclei, increasing the mass of the nucleus by the addition of two neutrons and two protons, and changing its identity. Carbon, with 6 neutrons and 6 protons, becomes oxygen, with 8 neutrons and 8 protons; oxygen can then add another alpha particle to become neon, with 10 neutrons and 10 protons, and so on. At sufficiently high temperatures, even more complex reactions can occur, involving the capture of free neutrons by large nuclei, forming ever-heavier nuclei. In this way, elements as heavy as iron can be produced in the core of a massive star.

After each nuclear reaction stage, the reaction continues in a shell around the core. Hence, a star whose core has finished hydrogen burning will still have a shell outside the core where hydrogen continues to react. Later, when helium in the core is used up, there will be a helium-burning shell outside the core, and a hydrogen-burning shell will remain outside that. As successive reaction stages occur in the core, a series of concentric shells develops, each undergoing a different reaction, leading to an onionlike internal structure (Fig. 22.10). The result is a star made of layers, each of a different composition. Ultimately a core made of iron will form, after which the orderly progression of re-

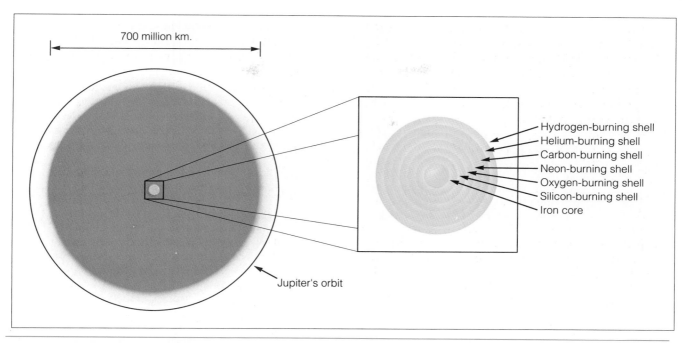

Figure 22.10 A massive star at the end of its life. *This star has undergone all possible reaction stages in its core. The core consists of multiple layers of different composition, including six active shell sources and an inert iron core. This star is about to undergo collapse and supernova explosion.*

action stages must end, with consequences described in the next section.

As the successive nuclear reaction stages occur in the core of a star, the star's position in the H-R diagram changes in a fashion similar to that of a one solar-mass star as it goes from main sequence to red giant. A massive star also makes this transition, although no helium flash occurs because the higher core temperature prevents the gas from becoming degenerate. Once the star becomes a red giant, it soon turns to the left and down in the H-R diagram as helium burning begins (Fig. 22.11). When the helium is used up, a second red

Figure 22.11 The evolution of a massive star. *This evolutionary track in the H-R diagram shows several steps followed by a star in the 5–10 solar mass range. Each time the nuclear fuel in the core is exhausted, a shell burning source becomes dominant, causing the outer layers to expand, moving the star into the red giant region. When a new core source ignites, the star moves back to the left. The more massive the star, the greater the number of burning stages.*

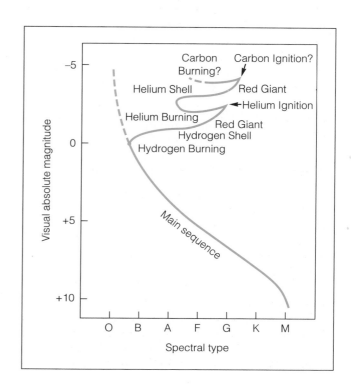

ASTRONOMICAL INSIGHT 22.1

The Missing Hypergiants

 In this chapter, we have seen that a star usually moves up and to the right when its main-sequence lifetime is over. Examination of the H-R diagram shows no stars in the extreme upper right-hand corner, however. There are no red supergiants as luminous as the brightest hot main-sequence stars (see Fig. 19.5). This is where we might have expected to find extreme red supergiants or "hypergiants," which should form when stars from the uppermost portion of the main sequence exhaust their core hydrogen and their outer layers expand. Now astronomers may have solved the mystery of the missing hypergiants.

One way to reduce the luminosity of a star is to take away some of its mass. We discussed (in Chapter 20) the relationship between mass and luminosity; from that discussion it is easy to see that a reduced mass would reduce the luminosity of a star. Hence we expect that mass loss might be the answer to our question. As we have seen, mass loss does occur for stars in the upper portion of the H-R diagram. The hot stars in the upper left have high-speed stellar winds that can result in the loss of as much as one solar mass every 100,000 years.

Calculations of stellar evolution that include the effects of mass loss do not solve our problem, however. The stars from the upper main sequence are now predicted to move straight across to the right instead of up and to the right, but observations show that even this does not occur. There would still be red hypergiants in the upper right corner of the H-R diagram.

The very unusual blue supergiant P Cygni, along with a handful of other stars like it, appears to provide the answer. P Cygni was first noticed more than 300 years ago, when it flared up in brightness by several magnitudes, becoming for a while a third-magnitude star. After about 10 years it faded back to its normal magnitude of 5 or 6, roughly its brightness today. Modern observations of P Cygni show that it has a strong, rapid stellar wind; analysis of the ultraviolet spectral lines formed in the wind tell us that the current rate of mass loss is about 1 solar mass every 100,000 years, similar to other stars in the upper left region of the H-R diagram. But occasionally the mass-loss rate increases temporarily (for a few months), reaching values closer to 1 solar mass every 10,000 years (the star

---

giant phase follows, as the energy is again supplied primarily by a shell source and the outer layers expand and cool for a second time. Each time one form of fuel is exhausted, a new red giant phase is begun, and each time a new fuel is ignited, the star will contract its outer layers and move to the left and down in the diagram. The number of these loops a star will undergo depends on how much mass it has, because this governs the internal temperature the star can ultimately reach, and this in turn determines how many reaction stages can occur.

Predicting exactly what happens to a star of a given mass is made difficult by the fact that massive stars lose matter at several points in their lifetimes (see Chapter 20), and this can alter the internal conditions. Very massive stars eject matter in high-speed winds while on the main sequence, and stars of all masses lose matter through gradual winds when they are in the red giant stage. Through these processes, a significant fraction of the initial mass of a star can be lost during its lifetime.

The End of the Reactions

We have seen that a star like the Sun will become a white dwarf when all possible nuclear reactions have run their course. The star becomes stable, with the pressure of degenerate electron gas supporting it against the force of gravity, and it remains a white dwarf perpetually (except under certain circumstances discussed in the next chapter). Some massive stars also become white dwarfs, although with different compositions than the white dwarfs left by low-mass stars (Fig. 22.12). It is not certain that the Sun will undergo helium burn-

ASTRONOMICAL INSIGHT 22.1

entered an outburst phase like this in early 1990).

These periods of enhanced mass loss raise the possibility that perhaps all upper-main-sequence stars lose mass much faster on average than they appear to be doing most of the time. Even an average mass-loss rate of 1 solar mass every 10,000 years is not enough to explain why upper-main-sequence stars do not make it to the upper right corner of the H-R diagram, however. But calculations show that if the average rate of mass loss were as high as 1 solar mass every 1,000 years, this could explain the discrepancy; upper-main-sequence stars would then lose their entire outer envelope of hydrogen before having time to expand and become red hypergiants.

So what about P Cygni? We have seen that its normal mass-loss rate is 1 solar mass every 100,000 years, and that it goes

through occasional episodes when the rate goes up to 1 solar mass every 10,000 years. But recall that in the 1600s the star underwent an outburst of much greater magnitude than the smaller-scale episodes of enhanced winds. Now theoretical considerations, combined with observations of gaseous nebulae surrounding the cores of highly evolved, massive stars, suggest that during such an outburst the star ejects an entire solar mass at once! If stars like P Cygni have these outbursts every 1,000 years or so, as is suggested by current observations, then the average mass-loss rate is 1 solar mass every 1,000 years, just what is needed to explain the lack of red hypergiants.

Astronomers now think that it is a general phenomenon for stars from the uppermost portion of the main sequence to undergo periods of instability and eruptive outbursts as they are leaving the

main sequence, so that their average mass-loss rates are as high as 1 solar mass every 1,000 years. This results in the loss of the outer envelope, preventing the star from becoming a red hypergiant. The hot, hydrogen-deficient core of the star becomes exposed, revealing an object known to astronomers as a Wolf-Rayet star (many of these are seen). The reason it took so long for astronomers to recognize this phenomenon is that stars such as P Cygni are extremely rare; only about half a dozen have been found in our galaxy and several neighboring galaxies combined. These objects are so uncommon because there are very few massive stars in the first place, and because the extreme mass-loss period of their lifetimes is very short. As happens often in astronomy, the rare, peculiar stars have told us very much about how all stars evolve.

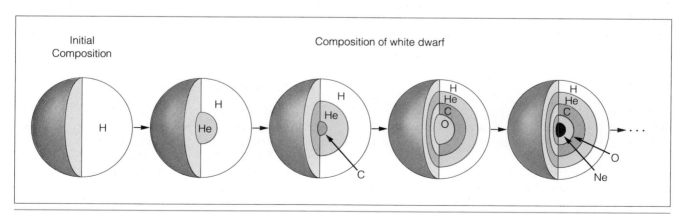

Figure 22.12 White dwarf composition. *The more nuclear burning stages a star goes through, the greater the variety of elements produced in its interior. The white dwarf that results, therefore, has a composition that depends on the number of reaction stages, and this, in turn, depends on the mass of the star.*

ing; if not, the Sun will become a white dwarf composed chiefly of helium. If nuclear reactions proceed to helium burning, then the resulting white dwarf will be made of carbon; if the next stage occurs, in which the alpha-capture reaction of carbon with helium proceeds to completion, the resulting white dwarf will be made primarily of oxygen.

For stars above a certain mass, however, a white dwarf becomes an unattainable endpoint for a star, because there is a firm limit on how massive a white dwarf can be. This limit occurs because above a certain mass degenerate electron gas pressure is no longer sufficient to support the star against gravity, and the star

cannot exist as a stable white dwarf. The mass limit for white dwarfs, called the **Chandrasekhar limit,** is about 1.4 solar masses. Stars that start out with considerably more mass than this can end up as white dwarfs if they lose enough mass during their lives; it is the *final* mass of a star that determines whether or not it can become a white dwarf. There are uncertainties in the calculations (mostly due to poorly known rates of mass loss), but it is now thought that stars with initial masses up to 6 or 8 solar masses can end as white dwarfs. Stars that begin with more mass than this apparently do not lose enough to become white dwarfs, and something else must happen.

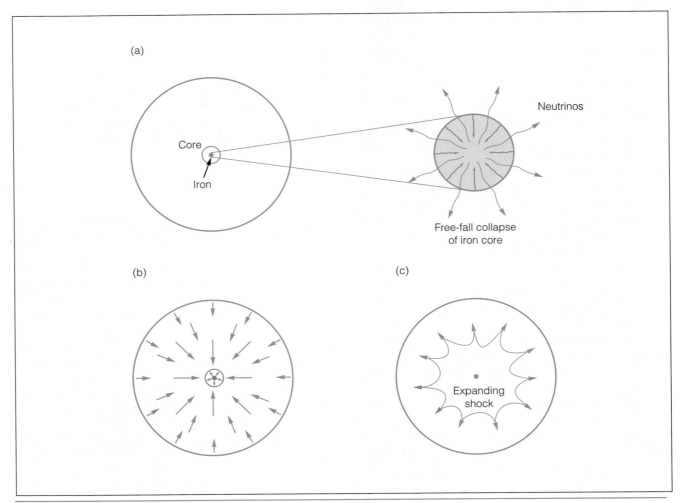

Figure 22.13 Type II supernova.   *The tiny iron core at the star's center undergoes rapid collapse because the only nuclear reactions that can occur there remove heat energy (i.e., they are endothermic). The compression converts the core to pure neutrons, releasing copious quantities of neutrinos and becoming rigid in the process (a). The outer layer of the star falls in onto the neutron star core, and rebounds (b). This creates a shock wave which travels outward (c) and, with the help of the outflowing neutrinos, blows off the star's outer layers in a supernova explosion, leaving behind a neutron star remnant.*

Stars that initially have between 8 and about 20 solar masses probably end up in another very compact state, even smaller and more dense than a white dwarf. Many of them reach this state by a very violent process called a **supernova** explosion (Fig. 22.13). Once iron has formed in the core of such a star, no further energy-producing reactions can take place as the core continues to contract and heat. Elements more massive than iron can be formed only by reactions that use up more energy than they create; these are called **endothermic** reactions. As reactions of this type begin under the incredibly high temperature (over 10^9 K) and density (nearly 10^{10} grams/cm^3) conditions in the iron core, heat energy is removed from the surroundings by the reactions. This accelerates the contraction of the core, because the pressure is diminished, and the contraction becomes a free-fall collapse.

Very rapidly (within a few ten-thousandths of a second) the core reaches sufficient density that protons and electrons are forced together in a reaction called an **inverse beta decay,** forming a sea of neutrons (Fig. 22.14). A vast quantity of neutrinos, the tiny, massless particles discussed in Chapter 17, is released, carrying away enormous quantities of energy and further enhancing the collapse. The neutrons left behind, like electrons, cannot be packed closer together beyond a certain limiting density, and they create a new kind of degenerate gas pressure. This **degenerate neutron gas pressure** is sufficient to halt the collapse of the core very abruptly. What was an effective vacuum into which the core was falling on itself suddenly becomes a very hard object, and material still falling in crashes onto its surface. A violent rebound, or shock wave, is formed and moves rapidly outward through the star. This shock may become stalled as it encounters the material of the outer layers, but the enormous burst of neutrinos from the core soon (within less than a second) pushes the shock on its way again, and the entire outer portion of the star is blown away. This is the mechanism for a form of supernova called a **Type II supernova,** and is basically what happened in the Large Magellanic Cloud in February of 1987, when Supernova 1987A occurred (this is described in more detail in the next section). A Type II supernova is defined on the basis of the appearance of spectral emission lines due to hydrogen, which indicates that the exploded star still had an outer envelope of "normal" gas when its core collapsed. (In the next chapter we will discuss Type I supernovae, in which no hydrogen lines are present.)

The core that is left behind by the collapse and rebound of a star of 8 to 20 solar masses is made almost

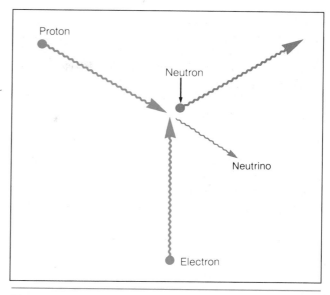

Figure 22.14 Inverse beta decay. *In this process, which occurs at a significant rate only under conditions of very high pressure and temperature, a proton and an electron are forced together, forming a neutron and a neutrino. Energy is absorbed in the process, rather than being produced.*

entirely of neutrons and is incredibly dense. A mass between 1 and 2 solar masses is compressed into a spherical volume whose diameter is only a few miles (typically 10 km, or about 6 miles), roughly the size of a modest city. This object, called a **neutron star,** is as dense as 10^{14} to 10^{15} grams/cm^3, comparable to the density of an atomic nucleus. Needless to say, matter in this form has bizarre properties that are not reproducible on the Earth. For years after the possibility of neutron stars was raised by theorists, no observational evidence for them was known, and they were seldom discussed. However, as we shall see in the next chapter, there is now compelling evidence that neutron stars are indeed very numerous in our galaxy.

What of the stars that start life with more than 20 solar masses? These stars go through the same succession of nuclear reaction stages, although they do this so rapidly that their outer layers do not have sufficient time to respond to all the internal changes, and they may not actually go through the red giant part of the H-R diagram. In a matter of several hundred thousand to a million years, all the possible reactions may be completed and an iron core formed. In the meantime, these stars stay in the upper portion of the H-R diagram, remaining hot at their surfaces while losing their outer layers through dense, high-speed winds. A class of "peculiar" stars called **Wolf Rayet stars,** characterized by high masses and unusual surface compositions along with rapid mass loss, are thought to represent an intermediate stage in the lives of very massive stars, a stage in which products of the CNO cycle are exposed on the surface after the outermost layers have blown away.

In the cores of the most massive stars, it is thought that a neutron star forms just as it does in the intermediate-mass stars. The formation of a neutron star is a virtually automatic result of the rapid collapse of an iron core, always happening because as soon as the density is high enough, the core "hardens" and further infall is stopped by the outward shock that forms because of rebound. In the most massive stars, however, the shock becomes stalled and neutrino pressure from within is not sufficient to get it going again. Meanwhile, more material from the iron-producing layer in the star continues to fall slowly onto the core, adding to its mass. Soon the mass limit for a neutron star (not known exactly, but probably between 1.6 and 2.4 solar masses) is exceeded, and the stellar core collapses anew. This time nothing can stop it; there is no form of pressure that can counteract gravity, and the collapse continues without limit. Within less than one-thousandth of a second, the core becomes a **black hole.** The name derives from the fact that the force of gravity is so strong that even photons of light cannot escape; there is no direct means of communication with or observation of the inside of a black hole. Properties of these most bizarre objects are discussed in the next chapter.

Note that in the process of forming a black hole, a neutron star is created temporarily and a shock rebounds from its surface. The shock becomes stalled, however, by the large amount of matter above it, and the pressure of neutrinos streaming outward from the core is insufficient to push the shock outward. Therefore a supernova explosion probably does not take place in these circumstances. It appears that black hole formation by this mechanism is not accompanied by any spectacular, explosive event.

The mass limit dividing stars that become neutron stars from those that become black holes is not accurately known and may not even be a firm limit, depending instead on details such as the initial rotation rate of the star or minor aspects of its composition. Certainly the rate of mass loss during and after its main-sequence lifetime has an important effect on the end result, and as yet mass-loss rates are not predictable to a very high degree of accuracy. In the foregoing discussion, we have placed the cutoff between stars that form neutron stars and those that form black holes at about 20 solar masses; some think that it may be as high as 30 or 40 solar masses.

Formation of the Heaviest Elements

It was pointed out earlier in this discussion that the formation of iron is the natural endpoint of nuclear reactions in the core of even the most massive stars, because further reactions are endothermic and precipitate the collapse of the core. There are a variety of reactions that can and do take place during the supernova outburst, however, and these are responsible for the formation of most of the elements heavier than iron. All of these reactions involve the capture of free neutrons by nuclei, thereby increasing the nuclear masses. There are two general classes of neutron-capture reactions; those that proceed slowly (these are called **s-process reactions**), and those that proceed rapidly (**r-process reactions**). The slow ones, s-process reactions, occur during the latter stages of evolution before the collapse of the iron core, and produce some heavy elements then. The r-process reactions occur during the explosion and produce greater quantities of heavy elements. Whether formed before the supernova or during it, all of these heavy elements (as well as other products of stellar nuclear reactions, such as carbon, nitrogen, and oxygen) are dispersed into the interstellar medium as a result of the explosion and are thereby made available for subsequent generations of stars. Therefore we say that all the heavy elements crucial to life and technology are the result of supernova explosions.

Supernova 1987A

One of the most spectacular and scientifically beautiful events in the history of astronomy was observed on February 23, 1987 (Fig. 22.15). A supernova bright enough to be easily seen with the naked eye was discovered in the Large Magellanic Cloud, providing astronomers the first opportunity in nearly 400 years to study so bright a supernova. The new supernova, the first to be seen in 1987 and the first since the invention of the telescope, was named Supernova 1987A, following a tradition used by astronomers to designate supernovae in other galaxies.

Figure 22.15 The region of the supernova before and after the explosion. *At left is a photograph made before the supernova occurred; the photograph at right was made afterward. At this point, the supernova was a fourth-magnitude object, on its way to an eventual peak brightness of magnitude 2.9. The large size of its image compared to normal stars is due to blurring on the photographic emulsion caused by the brilliance of the supernova; it was actually a point source, no bigger than the images of other stars. (© 1987 REO/AAT Board)*

Astronomers around the world were galvanized by the discovery as the news traveled through the International Astronomical Union's telegraph service and by telephone. NASA's *International Ultraviolet Explorer* (*IUE;* see Chapter 7) was making ultraviolet spectroscopic observations within 14 hours, and ground-based observers everywhere in the Southern Hemisphere trained their telescopes on it. Radio telescopes observed the supernova within hours of the discovery as well. Unfortunately, the event occurred at a time when space astronomy was in a lull between major missions (the *Hubble Space Telescope,* for example, was still sitting on the ground because of delays in the Shuttle program), so it was not possible to obtain all the data that would have been desirable. A Japanese X-ray telescope called *Ginga* had just been launched, however, and was soon able to make X-ray observations. NASA moved quickly to mobilize a number of sounding-rocket and high-altitude aircraft and balloon experiments as well, so there was substantial coverage of the supernova in wavelengths that do not reach the ground. Perhaps the most spectacular observations of all were not in any wavelength of electromagnetic radiation, however, but consisted instead of the detection of neutrinos from the explosion.

Spectroscopic data quickly showed that the supernova was of Type II, because hydrogen lines were seen in its spectrum. The expansion velocity implied by the Doppler shifts of these lines was at least 20,000 km/sec.

Recall that standard models assume that Type II supernovae result from the core collapse of massive stars which still have normal (that is, hydrogen-rich) gas in their outer layers.

Of the many unique opportunities the supernova has provided astronomers, one of the most important was the chance to analyze the star that blew up. Many photographic plates of the region exist in observatory files around the world, so it was possible to examine them to determine what kind of star the progenitor was. After some confusion caused by the crowded field of stars, the one that blew up was identified as a blue supergiant (type B3I) listed in catalogs as Sk-69°202. This was a major surprise for many astronomers, because until then it had been thought that only red supergiants exploded in supernovae of Type II (see previous section).

The unusual nature of the progenitor, once it was established, helped explain some other peculiar features of the explosion. For example, the supernova never reached the luminosity normally expected, and it reached its peak brightness (just brighter than third magnitude) on May 20, some 85 days after its discovery, an unusually long time. Radio data, on the other hand, showed a rapid decline almost from the beginning, whereas other Type II supernovae typically do not reach their peak radio emission until hundreds of days after the outburst. The *Ginga* satellite showed no X-rays coming from Supernova 1987A, although this may not be unusual. (There is very little information on X-ray emission in the early stages of other supernovae.)

All of these peculiarities fit well with supernova models involving core collapse in a massive star that is not as large as a red supergiant. The small size of the blue supergiant (compared to a red supergiant) accounts for the low luminosity, since relatively little surface area was available to radiate light. Another very important measure of the size of the star came from the detection of neutrinos, which escape directly from the core as the star collapses. The observed light is emitted later, when the shock front caused by the collapse reaches the star's surface. Thus, the neutrinos reached the Earth before the supernova was visible, but the short delay (equivalent to the time it took for the shock to travel from the star's core to its surface) indicates that the star could not have been as large as a red supergiant. The best fit to all the observations suggests that Sk-69°202 had a mass of about 20 solar masses and a diameter of roughly 43 times the diameter of the Sun.

As already mentioned, it was a surprise to most astronomers that such a star should explode, although some model calculations had been made suggesting that it could happen. It is still not clear what stage of evolution the star was in, but most astronomers think it had been a red supergiant that then contracted and moved toward the left in the H-R diagram before it exploded. Whether this scenario is correct or not, it is generally believed that the explosion itself was the result of iron core collapse, by the mechanism described in the preceding sections. A complicating factor in determining the exact evolutionary stage the star was in is that the Large Magellanic Cloud has lower overall abundances of heavy elements than "normal" spiral galaxies like our own and most others where Type II supernovae are observed, and lower abundances of heavy elements can alter the evolution of a star.

Since Supernova 1987A failed (by about a factor of 5) to reach the magnitude expected for a Type II supernova, many at first assumed that this was a freak event and not any kind of common occurrence. Further analysis showed, however, that many supernovae of brightness comparable to this one could occur in other galaxies without being detected, because they would not stand out as well. Therefore, supernovae like 1987A might actually occur fairly frequently. There have been a small number of other "peculiar" supernovae with some similarities to 1987A.

The detection of neutrinos from Supernova 1987A was a spectacular event for astronomy and for physics.

A burst of neutrinos was detected simultaneously by two separate experiments, one in Japan and one in Ohio. This marked the first time that extraterrestrial neutrinos had been detected except for those from the Sun (see Chapter 17), and is regarded by some as establishing a new branch of astronomy. As already explained, the timing of the arrival of the neutrinos provided valuable information on the size of the star. The neutrinos represent something far more important in terms of understanding the supernovae, however; the energy they carried away from the explosion was greater than that in all other forms by roughly a factor of 100. The visible light emitted over the first several months amounted to some 10^{49} ergs of energy; the kinetic energy of the expanding gas cloud entailed perhaps a few times 10^{51} ergs; and the energy released in neutrinos was about 3 or 4 times 10^{53} ergs. These numbers are thought to be typical of Type II supernovae, according to model calculations, but Supernova 1987A provided the first confirmation of the important role of neutrinos. What we see of a supernova, as brilliant as it is, is only a small fraction of the energy produced. Furthermore, even though Supernova 1987A was dimmer than many other supernova in visible wavelengths of light, its total energy output was comparable to the brightest of supernovae.

Physicists who study elementary particles were as excited by the detection of neutrinos from Supernova 1987A as the astronomers were. Detailed analysis of the energy distribution of the neutrinos that were detected, along with the exact timing of their arrival, provided information about their basic properties. As mentioned elsewhere in this text, there has been some debate on the question of neutrino mass. The standard theories of neutrinos say that they are massless; yet controversial experimental data, along with a host of astronomical problems that would be solved, suggest that neutrinos might have mass. The data from the supernova neutrinos are not as complete as would have been desirable, but the information that was obtained still allows something to be learned about neutrino masses. Analysis of the data indicates that the mass must be zero or very small, too small to solve the astronomical problems alluded to here (such as the total mass content of the universe and the low observed neutrino emission from the Sun).

After the initial outburst had diminished, during which the brightness of the supernova was caused by the high temperature of its expanding atmosphere, the main source of its luminosity was energy released by the radioactive decay of a form of nickel that was produced

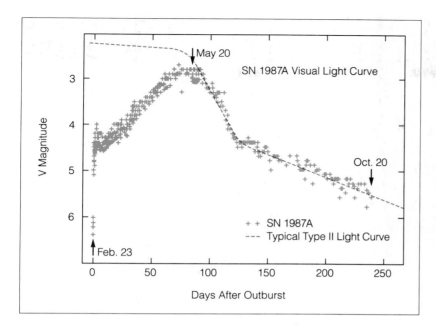

Figure 22.16 The light curve of Supernova 1987A. *This graph shows the variation in magnitude of Supernova 1987A since its discovery up to late fall, 1987. The early part of the light curve is very unusual compared to other known supernovae, but once this one reached its peak it began to behave very much like the light curves of other Type II Supernovae. The reason for the difference in the early period is that the star that exploded was smaller than is thought typical, so that it had less surface area to emit light. The later part of the light curve is consistent with energy production by the decay of radioactive nickel (^{56}Ni). (J. Doggett and R. Fesen, University of Colorado)*

in large quantities (about 0.09 solar mass) during the explosion. This form of nickel decays to another radioactive substance, an isotope of cobalt, and this then decays to iron. Thus the end product of the supernova is the production of great quantities of iron, but ironically this is not the iron from the core of the progenitor star, but rather iron produced indirectly by nuclear reactions resulting from the explosion.

One of the most important clues to the nature of the energy production in the supernova's expanding shell is the manner in which the brightness decreases with time. A graph showing magnitude versus time, called a *light curve*, is shown in Fig. 22.16. Radioactive decay causes the brightness to drop off in a specific fashion, and until recently, this is just what the observed light curve was doing. Then the brightness began to dip below the expected curve, probably because some of the energy from radioactive decay began to escape directly into space in the form of gamma rays, rather than being absorbed and converted into visible light, an indication that the expanding cloud of debris is starting to thin.

It is considered a virtual certainty that a neutron star formed during the collapse and explosion of Sk-69°202, because no other mechanism is known for producing the huge quantity of neutrinos that was detected. If so, then eventually energy radiated by the neutron star should begin to dominate its luminosity, causing the light to stop diminishing so rapidly. Therefore astron-

omers are waiting for the first signs of energy from the neutron star, which could show up in X-ray wavelengths or in visible light, representing X-ray emission from the neutron-star that has been absorbed and reemitted.

The effects of the supernova on the interstellar gas and dust around it and along the line of sight between it and the Earth are also expected to be very interesting and informative. For example, it is suspected that much of the interstellar dust that pervades the space between the stars (see Chapter 25) forms through the condensation of gas in the material ejected from supernovae. Thus astronomers are monitoring the infrared emission from the vicinity of the supernova, looking for the glow that is expected from warm dust, if large quantities of dust are formed.

Previously existing interstellar dust has already played an important role in observations of the supernova. *Light echoes* from the supernova have been seen (Fig. 22.17). These are pale circles of light centered on the supernova, caused by scattering due to layers of dust in the foreground (Fig. 22.18). The dust particles reflect light most efficiently at a certain angle, so that we see the reflected light only from the narrow ring of dust that lies at that angle relative to the direction of the supernova (a very similar effect due to ice particles in the Earth's upper atmosphere sometimes creates pale rings around the Moon). The light echoes from Supernova 1987A consist of light that was emitted during

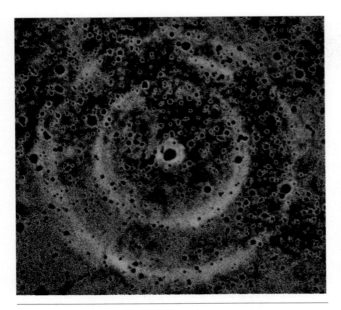

Figure 22.17 Light echoes from Supernova 1987A.
These rings of light from Supernova 1987A are reflected from sheets of gas and dust in the foreground. (© 1989 Anglo-Australian Telescope Board. Photo by David Malin)

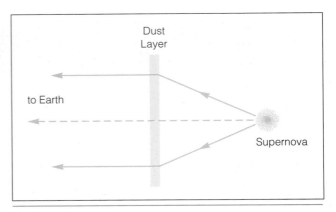

Figure 22.18 The formation of light echoes.   *The light emitted straight toward the Earth got here first; its arrival first signaled the occurrence of the supernova. Light that struck the dust layer at just the right angle to be reflected (scattered) toward Earth arrived here several months after the initial outburst was observed, due to the longer path that it traveled.*

the initial outburst of the supernova, but was delayed in reaching the Earth, because it took an indirect route. The fact that two distinct echoes were seen indicates that there are two separate layers of interstellar dust lying on the near side of the supernova.

The occurrence of Supernova 1987A has already had a major impact on nearly every field of astronomy. The technical journals are dominated by articles on theory or observations of the supernova, and we may expect this to continue for months or years. Eventually, as the expanding shell evolves into a normal supernova remnant, it is expected to become one of the brightest radio and X-ray sources in the sky, as the ionized gas disperses and emits energy by synchrotron and other nonthermal processes. In due course, perhaps in months or years, the gaseous remnant may become thin enough to allow astronomers to detect the neutron star, particularly if it is a pulsar (discussed in the next chapter). For decades or centuries, the gaseous remnant will be a strong source of visible and ultraviolet emission lines; for centuries it will remain an intense X-ray source; and for millennia it will be a strong radio emitter. We may expect many generations of astronomers to take part in the study of this supernova, and many future astronomy classes to learn about it.

EVOLUTION IN BINARY SYSTEMS

So far our discussion of stellar evolution has been concerned with single stars. About half the stars in the sky are double, however, and in some cases this can affect the lives of the stars. If the two stars are sufficiently far apart—perhaps several astronomical units or more—then most likely each evolves on its own, as though the companion were not there. If, however, they are very close together, significant interactions can drastically alter the normal course of events.

If either star is massive enough to have a strong wind, matter can be transferred to the other star (Fig. 22.19). Even if no strong wind is present initially, as soon as either star reaches the red giant stage, it can swell up sufficiently to dump material onto the other through the point between them where their gravitational forces are equal. If one of the two is a compact stellar remnant, such as a white dwarf, neutron star, or black hole, the result can be spectacular, as discussed in the next chapter.

Regardless of how it happens, once mass begins to transfer from one star to the other, the evolution of one or both is modified. The one that gains mass speeds up its evolution, while the one that loses mass may slow down. This process was discovered because certain binary systems were found in which the *less* massive star was the one that had evolved the farthest,

ASTRONOMICAL INSIGHT 22.2

The Mysteries of Algol

The bright double star known as Algol (β Persei) has played a leading role in the discovery of two important astronomical phenomena. Both of these phenomena, the existence of binary stars and the process of mass transfer in close binaries, went unrecognized for some time because they seemed too radical for most astronomers to accept.

The first report that Algol's brightness varies came as long ago as 1667, when the Italian astronomer Geminiano Montanari noted that the star occasionally appeared fainter than normal. At that time a few variable stars were known, but no explanation for them had been found. It was not until 1782 that anyone rediscovered the variability of Algol; in that year the English astronomer John Goodricke found that the light variations were periodic, with a brief decrease in brightness occurring every 69 hours. Goodricke made the then-novel suggestion that the dimmings could be due to the presence of a companion star that orbited Algol, eclipsing it each time it passed in front. At that time there was controversy over earlier suggestions that some stars appeared to be double, and the prevailing view (including that of the eminent English astronomer William Herschel) was one of skepticism.

In independent studies of Goodricke's idea, the Swiss astronomer Daniel Huber and an obscure English clergyman named William Sewall soon analyzed the light variations of Algol to deduce crudely the properties of the suspected companion. The writings of both men, published originally in 1787 and 1791, were ignored by their contemporaries and then lost, not being rediscovered until well into the twentieth century. Meanwhile, the notion that stars could be double gradually became accepted, as additional Algol-type stars were found, and observations of visual binaries became common. Interestingly, it was William Herschel who, in 1803, conclusively proved the existence of double stars when he was able to actually measure orbital motion in a visual binary system. Finally, in 1889, spectroscopic observations of Algol itself revealed the orbital motion of the bright star about its companion, establishing that stars can be double even though they present a single-point image when viewed in a telescope or when photographed.

Much more recently, Algol played a similar leading role in the discovery of mass transfer in binary systems. During the twentieth century, Algol has been widely observed, using photometric and spectroscopic techniques, and a paradox was revealed as an understanding of stellar evolution arose. Analysis of the data on Algol using Kepler's third law and the shape of the light curve (a graph showing the brightness of the star versus time) showed that the visible star is a B main-sequence star having a mass of 3.7 times that of the Sun, whereas the fainter star is a G giant star having a mass of only 0.81 times the Sun's mass. From observations of cluster H-R diagrams and stellar evolution theory, it was known that more massive stars evolve faster than less massive ones. Thus the fact that the *less* massive member of the Algol system was already a red giant, while the *more* massive star was still on the main sequence presented a major mystery, which became known as the Algol paradox.

The solution to this puzzle was found only in the past 20 years or so, as ultraviolet observations have revealed massive stellar winds from some kinds of stars, and other evidence (mostly from telescopes in space) for mass transfer between stars in close binary systems has been found. Evidently the G star in the Algol system originally contained more mass and was once an upper main-sequence star, whereas its companion started out on the lower main sequence. As the first star, being more massive, became a red giant, it swelled up to such a large size that matter began to flow across to the companion. The first star became less massive and fainter than the companion, which in turn became hotter and brighter as its mass grew. Today the original situation is reversed, with the star that started as the less massive one now being the more massive. At some future time, there may be another reversal, as the B star becomes a red giant and swells up, spilling mass back onto its companion.

The existence of binary stars and the process of mass transfer in close binaries were both profound discoveries in astronomy. It is ironic that hints of both were provided by Algol, but in both cases the evidence was found long before astronomers were ready to understand and accept the implications.

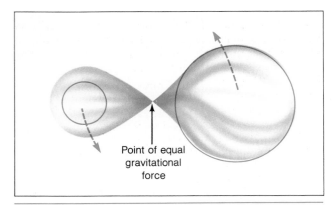

Point of equal
gravitational
force

Figure 22.19 Mass transfer in a binary system. *When two stars in a binary system are close together, material from one may be transferred to the other. There is a point in space between the two stars at which material would feel equal gravitational forces from each star; if one star has a wind or expands as a red giant so that its outer layers reach this point, then matter can flow from one star to the other. This can alter the evolutions of both.*

quite contrary to what is expected. The most famous such system is the eclipsing binary known as Algol, or β Persei. This binary system consists of a red giant and a B main-sequence star. The B star is more massive, but is still on the main sequence, whereas the red giant has already left the main sequence. The "Algol paradox" was solved only when it was realized that the red giant must have been the more massive star initially but later lost mass to its companion (See Astronomical Insight 22.2).

The most spectacular observational effects in mass-transfer binary systems occur when one star has gone all the way through its evolution and become a white dwarf, neutron star, or black hole and then gains new material from the companion star. Special effects, often including X-ray emission, are observed, and it is under these circumstances that stellar remnants are most easily detected. This will be discussed at some length in the next chapter.

PERSPECTIVE

We have now seen how stars are born, live their lives, and die, and we know how the sequence of events depends on mass. The least massive stars, those with

much less than a solar mass, have been seldom discussed, simply because they evolve so slowly that none has yet had time, since the universe began, to complete its main-sequence lifetime. The evolution of these stars has had little effect on the rest of the galaxy, while the much more rare heavyweight stars are profoundly important because their rapid evolution has enriched the galaxy with heavy elements.

Before we finish discussing stars and turn our attention to larger-scale objects in the universe, we must consider the properties of the stellar remnants left behind as stars die.

SUMMARY

1. A star with the mass of the Sun spends about 10^{10} years on the main sequence, producing its energy by nuclear reactions that convert hydrogen to helium in its core.

2. When the core hydrogen is gone, the reactions stop there but continue in a spherical shell outside the core. The outer layers expand and cool, and the star becomes a red giant.

3. The core continues to shrink and heat until it becomes degenerate. When it eventually gets hot enough, a helium flash occurs, after which the star is powered by the triple-alpha reaction, converting helium into carbon.

4. Following the helium-burning phase, the star may undergo instabilities that cause the ejection of one or more planetary nebulae.

5. The star eventually becomes degenerate throughout and ends its evolution as a white dwarf.

6. A more massive star, in the 5-to-10-solar-mass range, undergoes many of the same evolutionary steps as the 1-solar-mass star, but it does so much more quickly.

7. A massive star may go through several red giant phases, as successive nuclear reaction stages produce ever-heavier elements in the stellar core.

8. This star will become a white dwarf only if it loses enough mass, through stellar wind or other processes, to bring its mass below the 1.4-solar-mass limit for white dwarfs.

9. If the star is too massive at the end of its evolution to become a white dwarf, it may explode in a supernova, leaving a neutron star.

10. When all possible reaction stages are completed, the massive star will collapse and may explode in a

supernova, or perhaps more likely will continue to collapse, forming a black hole.

11. In close binary systems, mass can be transferred from one star to the other, speeding the evolution of the one that gains mass.

REVIEW QUESTIONS

Thought Questions

1. Explain why the main sequence is somewhat broadened rather than being a thin strip in the H-R diagram. Where do you think the Sun lies within the broad strip of the main sequence?

2. We have learned that when helium-burning reactions start in a star like the Sun, they begin in an instantaneous "helium flash." Why is there not a "hydrogen flash" earlier in a star's lifetime, when it forms, and nuclear reactions begin for the first time?

3. Explain why a star like the Sun becomes a red giant when its hydrogen fuel is gone.

4. White dwarfs have a small range of masses; that is, most have masses that are very similar. Despite this, white dwarfs can have a wide variety of different compositions. Explain how this can be so.

5. Summarize the contrasts between the evolution of a 1-solar-mass star and a star of 5 solar masses.

6. When a star undergoes successive nuclear reaction stages, each stage is shorter than the previous one. Explain why. (*Hint:* You need to recall what you have learned about the temperatures required for successive reactions, and the dependence of luminosity on temperature.)

7. Discuss the role of stellar winds in the evolution of massive stars.

8. Compare and contrast the evolution of a 5-solar-mass star with that of a star of 15 solar masses.

9. The relative abundances of the elements in the galaxy (Appendix 3) show that the abundances generally decline with increasing atomic weight, with hydrogen the most abundant, helium next, and so on. Contrary to this trend, iron is relatively abundant, compared to other elements near it in atomic weight. Based on what you have learned about stellar evolution, explain why this is so.

10. Discuss the effects of mass exchange in binary systems on the evolution of the two stars.

Problems

1. A star like the Sun may become as large as 20 times its original radius when it is a red giant, and as small as 0.01 times its original radius when it is a white dwarf.

Compute its density, surface gravity, and escape velocity for the red giant and white dwarf stages, relative to their values when the star is on the main sequence. Assume that the mass remains constant throughout.

2. If a star loses one-tenth of a solar mass each time it ejects a planetary nebula, how many times would a star of initial mass of 2.4 solar masses have to eject planetary nebulae in order to become a white dwarf?

3. What is the wavelength of maximum emission for the central star of a planetary nebula whose surface temperature is 100,000 K? What kind of telescope would be best for observing this object?

4. If a neutron star has a radius that is 1/500 times that of a white dwarf of half the mass of the neutron star, compare the average densities of the two.

5. A supernova may have an absolute magnitude as small (bright) as -19. How much more luminous is such a supernova than the Sun, whose absolute magnitude is $+5$?

ADDITIONAL READINGS

Bethe, Hans A. 1985. How a supernova explodes. *Scientific American* 252(5):60.

Flannery, B. P. 1977. Stellar evolution in double stars. *American Scientist* 65:737.

Greenstein, George. 1985. Neutron stars and the discovery of pulsars. Part 1, 2. *Mercury* 14(2):34, 14(3):66.

Kaler, J. 1981. Planetary nebulae and stellar evolution. *Mercury* 10(4):114.

Kawaler, S. D., and D. E. Winget. 1987. White dwarfs: Fossil stars. *Sky and Telescope* 74(2):132.

Malin, D. and N. Allen. 1990. Echoes of the Supernova. *Sky and Telescope* 79:22.

Rosa, M., and S. Harrington. 1988. Light echoes from Supernova 1987A. *Mercury* 17:63.

Schaefer, Bradley E. 1985. Gamma-ray bursters. *Scientific American* 252(2):52.

Seward, Fredrick. 1986. Neutron stars in supernova remnants. *Sky and Telescope* 71(1):6.

Sweigart, A. V. 1976. The evolution of red giant stars. *Physics Today* 29(1):25.

Tomkin, J., and D. L. Lambert. 1987. The strange case of Beta Lyrae. *Sky and Telescope* 74:354.

Wheeler, J. A., and R. P. Harkness. 1987. Helium-rich supernovas. *Scientific American* 257(5):50.

Wyckoff, 1979. Red giants: The inside scoop. *Mercury* 8(1):7.

The following articles are all related to Supernova 1987A:

Helfand, David. 1987. Bang: The supernova of 1987. *Physics Today* 40(8):24.

Lattimer, J. M., and A. S. Burns. 1988. Neutrinos from Supernova 1987A. *Sky and Telescope* 76:348.

Murdin, Paul, and Lesley Murdin. 1985. *Supernova.* Cambridge Mass.: Cambridge University Press.

Schorn, Ronald A., 1987. Neutrinos from hell. *Sky and Telescope* 73(5):477.

————. 1987. SN1987A: Watching and waiting. *Sky and Telescope* 74(1):14.

————. 1987. A supernova in our backyard. *Sky and Telescope* 73(4):382.

————. 1987. Supernova 1987A's fading glory. *Sky and Telescope* 74(3):258.

————. 1987. Supernova shines on. *Sky and Telescope* 73(5):470.

————. 1987. A surprising supernova. *Sky and Telescope* 73(6):582.

Time. March 23, 1987. 129(12):60.

Webbink, R. F. 1989. Cataclysmic variable stars. *American Scientist* 77:248.

Woosley, S., and T. Weaver. 1989. The great Supernova of 1987. *Scientific American* 261(2):32.

MATHEMATICAL METHODS

The Densities of Compact Stellar Remnants

One of the basic properties of matter is its density. The density can affect the manner in which atoms interact with each other, such as the manner in which chemical bonds form and the nature of the resulting substance (e.g., whether it is gas, liquid, or solid, and in a gas, whether the particles are molecules or free atoms). Along with density, the temperature of matter helps control its behavior; for example, a gas of a certain density may be either molecular or atomic, or even ionized, depending on the temperature.

It is interesting to consider the densities of the compact remnants that stars become at the ends of the nuclear burning lifetimes. In this discussion, we consider the average density, which is the mass of an object divided by its volume, which for a sphere is

$$\frac{4}{3}\pi R^3.$$

(where R is the radius). If ρ is the density of a spherical object, then we have

$$\rho = \frac{3M}{4\pi R^3}.$$

First let us consider the Sun. Its mass is 2×10^{33} grams, and its radius is 7×10^{10} cm, so its average density is

$$\rho = \frac{3(2 \times 10^{33} \text{ grams})}{4\pi(7 \times 10^{10} \text{ cm})^3}$$

$$= 1.4 \text{ gram/cm}^3.$$

On average, the Sun is only about 40 percent denser than water. The physical conditions in the Sun allow its matter to be in gaseous form throughout, and gases can be compressed, so the density is far higher at the center than at the surface. The central density is more than 100 times greater than the average density, and more than a billion times denser than the gas in the photosphere.

Now let us consider the stellar remnants. A white dwarf has a typical mass about equal to that of the Sun, but its radius is 100 times smaller than the solar radius. Hence the average density of a white dwarf is

$$\rho = \frac{3(2 \times 10^{33} \text{ grams})}{4\pi(7 \times 10^{8} \text{ cm})^3}$$

$$= 1.4 \times 10^6 \text{ grams/cm}^3.$$

As discussed in the text, at such high densities, matter takes on a form that is quite unlike anything we are familiar with on Earth. The free electrons in the stellar core are squeezed so close together that fundamental quantum mechanical forces keep them from further compression, and the gas acts in some ways more like solid matter. The density is more nearly uniform throughout, rather than varying significantly from the surface to the center.

A neutron star may have a mass about twice that of the Sun, and a radius of only 10 km, or 10^6 cm. Substituting these values into the equation for density yields

$$\rho = \frac{3(4 \times 10^{33} \text{ grams})}{4\pi(1 \times 10^{6} \text{ cm})^3}$$

$$= 1 \times 10^{15} \text{ grams/cm}^3.$$

This value is so high that matter takes on another completely unfamiliar form, in which protons and electrons cannot exist but are instead forced together to form neutrons. The star has become, in effect, a giant atomic nucleus. Again, this strange material is quite rigid, and the density is nearly constant throughout.

CHAPTER 23

Stellar Remnants

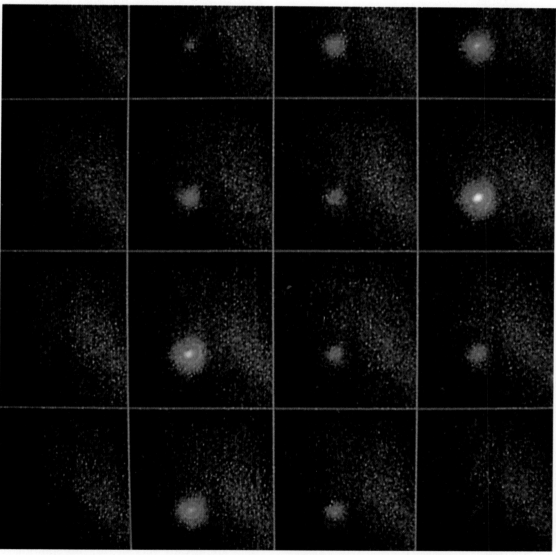

X rays from the Crab pulsar, turning on and off (Harvard-Smithsonian Center for Astrophysics)

CHAPTER PREVIEW

White Dwarfs, Black Dwarfs
  White Dwarfs, Novae, and Supernovae
Supernova Remnants
Neutron Stars
  Pulsars: Cosmic Clocks
  Neutron Stars in Binary Systems
Black Holes: Gravity's Final Victory
  Do Black Holes Exist?

Our story of stellar evolution is almost complete. We have seen how the properties of stars are measured, and we have learned how they form, live, and die. The only missing link is what is left of a star after its life cycle is completed.

As we saw in the previous chapter, the form of remnant that remains at the end of a star's life depends on the mass the star had when it finally ran out of nuclear fuel. Three possibilities were mentioned: white dwarf, neutron star, and black hole. In this chapter, each of these three bizarre objects is discussed, with emphasis on the observational properties.

WHITE DWARFS, BLACK DWARFS

Table 23.1 summarizes the properties of a typical white dwarf. An isolated white dwarf is not a very exciting object, from an observational point of view. It is rather dim (Fig. 23.1), and its spectrum is nearly featureless, except for a few very broad absorption lines created by the limited range of chemical elements it contains (Fig. 23.2). The great width of the lines is caused by

Table 23.1 Properties of a typical white dwarf

Mass: 1.0 $M_\odot$
Surface temperature: 10,000 K
Diameter: 0.008 $D_\odot$ (1.0 $D_\oplus$)
Density: 5×10^5 g/cm³
Surface gravity: 1.3×10^5 g
Luminosity: 2×10^{31} erg/sec (0.005 $L_\odot$)
Visual absolute magnitude: 11

Figure 23.1 Sirius B.   *The dim companion to Sirius (to left of large image of Sirius A) was the first white dwarf to be discovered; analysis of its mass, temperature, and luminosity led to the realization that it is very small and dense. (Palomar Observatory, California Institute of Technology)*

the immense pressure in the star's outer atmosphere; pressure tends to smear out the atomic energy levels, because of collisions between atoms. In effect, the energy levels become broadened, and this results in broadened spectral lines when transitions of electrons occur between the levels.

Additional broadening or shifting of spectral lines may occur because of magnetic effects. If the star had a magnetic field before its contraction to the white dwarf state, that field is not only preserved but is intensified in the process of contraction. Thus a white dwarf may have a very strong magnetic field, and as we have seen previously (in Chapters 17 and 19), a magnetic field causes certain spectral lines to be split into two or more parts. In a white dwarf, where the lines are already very broad, this splitting (called the Zeeman effect) usually just makes the lines appear even broader.

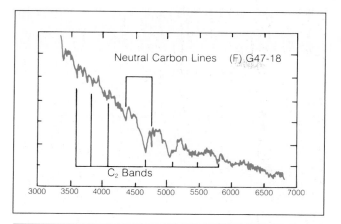

Figure 23.2 A white dwarf spectrum. *Here we see that white dwarf spectra are without many strong features, and that the spectral lines tend to be rather broad. This white dwarf has a large abundance of carbon, as a result of the triple-alpha reaction. (G. A. Wegner)*

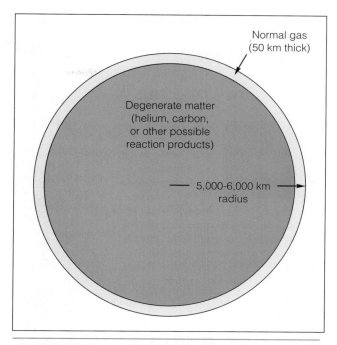

Figure 23.3 The internal structure of a white dwarf. *The star's interior is degenerate and would cool very rapidly, except that the outer layer of normal gas acts as a very effective insulator, trapping radiation so that heat escapes only very slowly.*

The lines in the spectrum of a white dwarf are also shifted, always toward longer wavelengths. This is not caused by motion of the star; if it were, it would be telling us that somehow all the white dwarfs in the sky are receding from us. This is instead a different type of redshift, one caused by gravitational forces. One prediction of Einstein's theory of general relativity is that photons of light should be affected by gravitational fields. White dwarfs, with their immensely strong surface gravities, provide a confirmation of this prediction. In later sections, we will see even more extreme examples of these **gravitational redshifts.**

Left to its own devices, a white dwarf will simply cool off, eventually becoming so cold that it no longer emits visible light. At this point, it cannot be seen and simply persists as a burnt-out cinder that may be referred to as a black dwarf.

The cooling process takes a long time, however; several billion years may go by before the star becomes cold and dark. This may seem surprising, since the white dwarf has no source of energy, staying hot only as long as it can retain the heat that it contained when it was formed. The outer skin of the white dwarf acts like a very efficient thermal blanket, however, since it consists of gas that is nearly opaque. Radiation cannot penetrate it easily, and because there is no other way of transporting heat away from the interior into space, it takes a long time for the heat to filter out.

Inside the white dwarf, energy is transported very efficiently by a process called *conduction*, which keeps the temperature nearly constant throughout the interior. In conduction, the energy is transported by electrons which are free to move rapidly.

The thermal blanket created by the outer atmosphere is very thin; most of the volume of the star is filled with degenerate electron gas (Fig. 23.3), as discussed in the last chapter. The atmosphere, consisting of ordinary gas, may be only about 50 km thick (remember, the white dwarf itself is about the size of the Earth, with a radius of some 5,000 to 6,000 km). The composition of the interior may be primarily helium, if the star never progressed beyond the hydrogen-burning stage; it may be carbon, if the triple-alpha reaction was as far as the star got in its nuclear evolution before dying; or it may be some heavier element, if the original star was sufficiently massive to have undergone several reaction stages. In any case, however, the mass of the white dwarf must be less than the limit of 1.4 solar masses.

ASTRONOMICAL INSIGHT 23.1

The Story of Sirius B

Sirius, also known as α Canis Majoris, is the brightest star visible in the heavens. In 1844 it was discovered to be a binary, because of its wobbling motion as it moves through space (recall the discussion of astrometric binaries in Chapter 18). Its period is about 50 years, and the orbital size about 20 astronomical units; analysis of these data using Kepler's third law leads to the conclusion that the companion must have a mass near that of the Sun. At the distance of Sirius (only 2.7 parsecs), a star like the Sun should be easily visible, yet the companion to Sirius defied detection for some time.

Part of the difficulty lay in the overwhelming brightness of Sirius itself, whose light would tend to drown out that of a lesser, very close star. Astronomers were so intrigued by the mysterious nature of the companion, however, that intensive efforts were made to find it, and finally, in 1863, it was seen with an exceptionally fine,

newly developed telescope.

It was immediately confirmed that this star was unusually dim for its mass, having an apparent magnitude of 8.7 (the apparent magnitude of the primary star is -1.4) and an absolute magnitude of 11.5, nearly 7 magnitudes fainter than the Sun. Spectra obtained with great difficulty (again because of the brightness of the companion) showed that this object, called Sirius B, has a high temperature, similar to that of an O star.

When Henry Norris Russell constructed his first H-R diagram in 1913, he included Sirius B, finding it to lie all by itself in the lower lefthand corner. The only way for the star to have such a low luminosity while its temperature was so high was for it to have a very small radius, and it was deduced at that time that Sirius B was about the size of the Earth. The incredible density was estimated immediately, and it was realized that a totally new form of matter was involved. The term *white dwarf* was coined to describe Sirius B and other stars of this type, which were discovered soon in other binary systems.

The physics of this new kind of matter took some time to be

understood, partly because the theory of relativity had to be applied (the electrons in a degenerate gas move at speeds near that of light). The Indian astrophysicist S. Chandrasekhar was the first to solve the problem of the behavior of this kind of gas, and by the 1930s he was able to construct realistic models of the structure of a white dwarf. From this work came the 1.4 solar mass limit we have referred to in the text.

Modern observational techniques offer the possibility of detecting white dwarfs more easily. For example, even though Sirius B is about 10 magnitudes (or a factor of 10,000) fainter than Sirius A in visible wavelengths, its greater temperature makes it relatively brighter in the ultraviolet. There is even a wavelength, about 1,100 Å, below which the white dwarf is actually brighter than its companion. Hence Sirius B and other white dwarfs have been detected with ultraviolet telescopes, and with the development of more sophisticated space instruments, it will become possible to study the properties of many white dwarfs in this way.

White Dwarfs, Novae, and Supernovae

Ordinarily, a white dwarf cools off and is never heard from again. There are circumstances, however, in which it can be resurrected briefly for another role in the cosmic drama. It has recently become apparent that white dwarfs are central characters in producing spectacular events.

In the previous chapter, we discussed the evolution of stars in close binary systems, where mass can transfer from one star to the other. Consider such a system,

in which the star that was originally more massive has gone far enough in its evolution to become a white dwarf. Now the companion begins to transfer matter to the white dwarf (Fig. 23.4), probably because the companion has expanded to become a red giant. The white dwarf may respond to the addition of the new material in a variety of ways, depending on its composition and on the rate at which mass is added. All of the possible responses lead, sooner or later, to an explosive outburst, or to many repeated outbursts, because of the degenerate nature of the white dwarf.

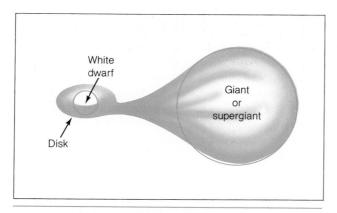

Figure 23.4 Mass transfer to a white dwarf. *Matter is transferred onto a white dwarf by a red giant companion. Theoretical calculations show that the new material will orbit the white dwarf in a disk but will eventually become unstable and fall down onto the white dwarf. When it does so, nuclear reactions flare up, creating a nova outburst.*

and cool when heated (recall the discussion of the helium flash in the previous chapter). In other cases, mass may accrete onto the white dwarf for some time, perhaps decades or centuries, before the new material becomes hot enough to undergo nuclear reactions. The intensity of the outburst depends on how much matter is involved. If enough material is consumed in this explosion, the formerly very dim white dwarf may become more luminous than any ordinary star (Fig. 23.5). This is a **nova.** Its luminosity may be 20,000 to 600,000 times that of the Sun (corresponding to an absolute magnitude in the range -6 to -9). A star that becomes a nova may do so repeatedly, usually with separations of decades between outbursts.

In some binary systems, rather small amounts of matter reach the white dwarf at regular intervals, creating relatively minor flare-ups. A general name for such systems is **cataclysmic variables,** and there are a number of specific types, each named after a particular prototype. The frequency of the outbursts can be as high as several times a day, but typically it is once every few days.

If the transferred gas reaches the white dwarf with a high velocity, so that heating occurs because of the energy of impact on the white dwarf's surface, there may be an instant nuclear explosion, enhanced by the fact that the degenerate dwarf material cannot expand

If material trickles onto the white dwarf at a slow rate, so that not much heating is caused by the impact, then the white dwarf can gradually gain mass, without the infalling material causing any explosions. In this way, a white dwarf can approach the 1.4 solar mass

Figure 23.5 Nova Cygni 1975. *This was one of the brightest novae in recent years. The photograph at left shows the region without the nova; at right is the same region near the time of peak brightness, when the apparent magnitude of the object was near $+1$. (E. E. Barnard Observatory, photograph by G. Emerson)*

limit beyond which degenerate electron gas pressure can no longer support it. As a white dwarf gains mass, it becomes *smaller;* this is another peculiar property of degenerate matter. Thus, the white dwarf increases in density. What happens as the white dwarf approaches the mass limit depends on its composition; it may, more or less quietly, suddenly contract, becoming either a neutron star or black hole. If its composition consists of carbon, however, theory shows that nuclear reactions will begin instead, and the degeneracy will cause these reactions to very rapidly consume much of the white dwarf's mass, creating an immense explosion. This is another mechanism for producing a supernova explosion.

In the previous chapter, we discussed supernovae caused by the collapse and rebound of a massive star at the end of its nuclear lifetime; now we see that supernovae can also be caused by the addition of mass to a white dwarf. Supernovae created in these two different ways are generally similar in their maximum brightness and in the length of time required to become dim again. But they are rather different in many details, such as the types of spectral lines observed and the shape of the **light curve,** a graph showing the brightness variations with time (Fig. 23.6). Supernovae caused by the explosion of a massive star are usually Type II supernovae, because hydrogen lines appear in the spectrum, while those produced in the explosion of a white dwarf are Type I supernovae, having no hydrogen features.

White dwarfs, then, are most prominent in binary systems in which mass transfer from a normal companion takes place. We will see later in this chapter that other forms of stellar remnants are also best observed in these circumstances.

SUPERNOVA REMNANTS

We have learned that stars can explode as supernovae in two different ways, in some cases leaving a stellar remnant (a neutron star or black hole). In all cases there will be another form of matter left over—an expanding gaseous cloud called a **supernova remnant.**

Several supernova remnants are known to astronomers (Table 23.2). A few, like the prominent Crab Nebula (Fig. 23.7), are detectable in visible light, but many are most easily observed at radio wavelengths. This is partly because visible light is affected by the interstellar dust that can hide our view of a remnant, whereas radio waves are largely unaffected. But it is primarily because, as a supernova remnant ages, its emission in visible light fades out long before it stops emitting radio waves. Thus there is an extended time period (10,000 to 20,000 years) when the remnant emits radio radiation and little else. The radio emission

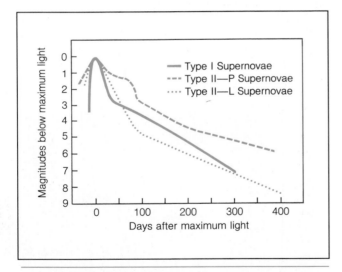

Figure 23.6 Supernova light curves. *The two types of supernovae are readily distinguished by their spectra and their light curves, which show marked differences. Type I supernovae usually have no hydrogen lines in their spectra and are thought to be explosions of carbon white dwarfs that have accreted matter. Type II supernovae have hydrogen lines in their spectra and are the explosions that occur in massive stars when all nuclear fuel is gone. Peak brightnesses for the two types are slightly different and, as seen here, the rates of decline are also quite different. Type II supernovae have been subclassified into types P and L. (J. Doggett and D. R. Branch)*

Table 23.2 Prominent supernova remnants

Remnant	Age (yr)	Distance* (l.y.)
Vela X	10,000	1,600
Cygnus Loop	20,000	2,300
Lupus	975	4,000
IC 443	60,000	4,900
Crab Nebula	936	6,520
Puppis A	4,000	7,500
Cassiopeia A	200	9,130
Tycho's Supernova	400	7,800
Kepler's Supernova	370	14,340

*Distance data are from a modern re-analysis by Saken, J., Fesen, R. A. and Shull, M. J. 1991. *Astrophys. J., Suppl., (in press)*

Figure 23.7 Supernova remnants. *At upper left is the Crab Nebula, the remnant of a star that was seen to explode in 1054* A.D. *Below is a photograph of part of the Veil Nebula, showing wispy, filamentary structure. At upper right is a radio image of the remnant of Tycho's supernova. (Top left: Lick Observatory photograph; top right: National Radio Astronomy Observatory; bottom: © 1980 ROE/AAT Board)*

Figure 23.8 The synchrotron process. *Rapidly moving electrons emit photons as they spiral along magnetic field lines. The radiation that results is polarized, and its spectrum is continuous but lacks the peaked shape of a thermal spectrum. The electrons must be moving very rapidly (at speeds near that of light), so whenever synchrotron radiation is detected, a source of large quantities of energy must be present.*

is not a result of the temperature of the gas, as in thermal emission, but is instead produced by a different mechanism, called **synchrotron emission.**

Synchrotron emission, given its name by the particle accelerators in which it is produced on Earth, occurs when electrons move rapidly through a magnetic field. The electrons must have speeds near that of light; as they travel through the magnetic field, they are forced to move in a spiraling path, and they emit photons as they do so (Fig. 23.8). The emission occurs over a very broad range of wavelengths, including some visible, ultraviolet, and even X-ray radiation (Fig. 23.9), but these other wavelengths are often not as easily detected as radio, because the radio emission is the strongest and most persistent. Synchrotron radiation can be distinguished from thermal radiation (i.e., radiation due to the temperature of the object, as discussed in Chapter 6) in two ways: (1) Synchrotron radiation oc-

Figure 23.9 Emission from a supernova remnant. *This shows three different images of the same supernova remnant, Cassiopeia A. At left is a radio map, at upper right an X-ray image, and at lower right a visible-light image. The remnant is less obvious in visible light because there are many other, brighter sources in the field of view. (Left: National Radio Astronomy Observatory; top right: Harvard-Smithsonian Center for Astrophysics; bottom right: K. Kamper and S. van den Bergh)*

curs more uniformly over a broad range of wavelengths, rather than peaking sharply at a particular wavelength; and (2) synchrotron radiation is usually polarized (see Chapter 6).

Supernova remnants that are detected in visible wavelengths usually glow in the light of several strong emission lines, most notably the bright line of atomic hydrogen at a wavelength of 6,563 Å, in the red portion of the spectrum. These lines often show large Doppler shifts, indicating that the gas is still moving rapidly as the entire remnant expands outward from the site of the explosion that gave it birth. Often a filamentary structure is seen, suggestive of turbulence, but probably modified in shape by magnetic fields.

Remnants are visible today at the locations of several famous historical supernova explosions. The Crab Nebula is the most prominent, having been created in the supernova observed by Chinese astronomers in A.D. 1054. Other supernovae seen by Tycho Brahe (in 1572) and by Kepler and Galileo (1604) also left detectable remnants, although neither is as bright in visible wavelengths as the Crab. Apparently a supernova remnant can persist for 10,000 years or more before becoming too dissipated to be recognizable any longer. One of the best-studied remnants, Cassiopeia A (Fig. 23.9), is apparently only about 300 years old, but the supernova that created it was quite dim and not widely observed. This supernova may have been obscured by interstellar dust, or it may have been an unusually dim one intrinsically (see Astronomical Insight 23.2).

The energy of a supernova explosion is immense, comparable to the total amount of radiant energy the Sun will emit over its entire lifetime. Some of this energy takes the form of mass motions as the remnant expands. This kinetic energy heats the expanding gas to very high temperatures. The entire process has a profound effect on the interstellar gas and dust that permeates the galaxy, as we shall see in Chapter 25.

NEUTRON STARS

We have referred to a neutron star as a stellar remnant composed entirely of neutrons that are in a degenerate state, similar to that of the electrons in a white dwarf. The pressure created by the degenerate neutron gas is greater than that of a degenerate electron gas, so a star slightly too massive to become a white dwarf can be supported by the neutrons. Again, however, there is a limit on the mass one of these objects can have. This

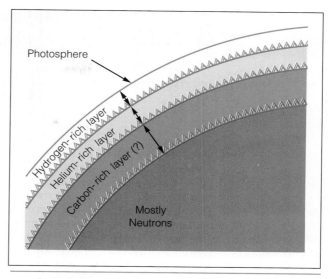

Figure 23.10 The structure of a neutron star. *This diagram shows that the outer regions of a neutron star may consist of thin layers of various elements that were produced by nuclear reactions during the star's lifetime. These outer layers are thought to have a rigid crystalline structure because of the intense gravitational field of the neutron star.*

limit, which depends on other factors such as the rate of rotation, is between 2 and 3 solar masses. Thus, there is a class of stars, whose masses at the end of their nuclear lifetimes are between 1.4 and about 2 or 3 solar masses, that become neutron stars.

The structure of a neutron star (Table 23.3 and Fig. 23.10) is even more extreme than that of a white dwarf. All of the mass is compressed into an even smaller volume (now the radius is about 10 km), and the gravitational field at the surface is immensely strong. The layer of normal gas that constitutes its atmosphere is only centimeters thick, and beneath it may be zones

Table 23.3 Properties of a typical neutron star

Mass: 1.5 $M_{\odot}$
Radius: 10 km
Density: 10^{14} g/cm³
Surface gravity: 7×10^9 g
Magnetic field: 10^{12} Earth's field
Rotation period: 0.001–100 sec

ASTRONOMICAL INSIGHT 23.2

Reconstructing an Exploded Star*

 Supernova explosions are rare events. We seldom get to observe one that is nearby. (Recall that the one in 1987 in the Large Magellanic Cloud was the first in nearly 400 years bright enough to be visible to the unaided eye.) When we find them in distant galaxies, we usually know nothing about the star before it exploded. Most of what we know about which kinds of stars blow up is based on theoretical understandings of nuclear reactions and stellar structure. Our knowledge would be more complete if we knew more about the progenitors of supernovae. It has been possible to reconstruct some information in the case of one recent supernova in addition to the great supernova of 1987.

The most recent supernova explosion known to have occurred in our galaxy took place some 300 years ago. The explosion itself was not noted in historical records, which implies that it may not have reached as great a brightness as those usually seen in distant galaxies. We know about the 300-year-old event because the remnant of expanding gas that was left behind is very easily observed today. Known as Cassiopeia A, it is one of the brightest objects in the radio sky and is observed in visible and X-ray wavelengths as well. Images of this remnant are shown in Figure 23.9.

Scientists believe that as a massive star ages and undergoes many successive nuclear reaction stages, it develops a layered internal structure, often likened to an onion. At the core is the material currently undergoing reactions, outside the core is a layer of whatever element was produced by the previous reaction stage, and so on. When such a star proceeds all the way to iron formation in the core and explodes, the layered structure should be preserved in the expanding supernova remnant. The surface material will fly outward most rapidly, the next layer a little slower, and so on. Thus it may be possible, by measuring the properties of debris at different distances from a supernova explosion, to reconstruct the internal structure of the progenitor. This has recently been done successfully for the star that formed Cassiopeia A.

Very careful visible-light observations have revealed fragmentary wisps of gas far outside the main part of the supernova remnant. According to the scenario just described, these filaments of gas, which have been flying away at higher speed than the rest of the remnant, must be fragments of the original surface of the star. This provides astronomers an opportunity to compare the surface material of the exploded star with observed surface properties of known types of stars, to see what kind of star the progenitor was.

Spectroscopic analysis has shown that the gas in these outlying fragments contains large quantities of nitrogen. This leads to the conclusion that the progenitor star was a nitrogen-rich **Wolf-Rayet star,** a hot star of high mass but only moderate diameter. These stars are observed here and there in the galaxy, and are believed to be the remains of large, massive stars whose outer layers have been stripped away by stellar winds. (See the discussions of Wolf-Rayet stars in Chapter 22 and of stellar winds in Chapter 20.) This discovery was made before the supernova of 1987, so the progenitor of Cassiopeia A was the first to be identified.

It is not surprising that a Wolf-Rayet star should explode when its nuclear lifetime is over. These stars, despite having lost a fraction of their original mass because of stellar winds, still have masses 20 or more times that of the Sun. Furthermore, the supernova that formed the Cassiopeia A remnant was apparently not as luminous as many that are observed in other galaxies, and this can be explained if the progenitor had only a moderately large radius. (A Wolf-Rayet star, while much larger than the Sun, is still considerably smaller than a massive supergiant or a hot upper main-sequence star.) Possibly many supernovae occur in relatively small stars and go unobserved because of their modest luminosities (modest only by comparison with the very brilliant supernovae that are seen in the most faraway galaxies).

Unfortunately, the technique used to reconstruct the progenitor of the Cassiopeia A supernova remnant cannot be widely applied to other remnants, because in a few hundred years the gaseous fragments cool to the point where they no longer can be found and analyzed spectroscopically. Thus, the number of instances in which we know what kind of star blew up will remain very small.

*Based on information provided by Dr. Robert Fesen, whose work is described here.

of different chemical composition resulting from previous shell-burning episodes, each zone only a few meters thick at the most. Inside these surface layers is the incredibly dense neutron gas core, which takes the form of a crystalline lattice. The temperature throughout is very high, but because the surface area is so small, a neutron star is very dim indeed. It may also have a magnetic field that is much more intense than even that of a white dwarf, resulting from the compression of the star's original field.

Pulsars: Cosmic Clocks

The properties of neutron stars were predicted theoretically several decades ago, but until 1967 it was not expected that they could be observed, because of their low luminosities. In that year, however, an accidental discovery by a radio astronomer established a new class of objects called **pulsars,** which are now believed to be neutron stars. As it turned out, pulsars are only one of two distinct ways in which neutron stars can be detected; the other is described in the next section.

Pulsars are radio sources that flash on and off very regularly, and with a high frequency (Fig. 23.11). One of the most rapid of them, the one located in the Crab Nebula, repeats itself 30 times a second, and several others are now known to flash even more rapidly than the Crab pulsar; one of these pulses at the incredible frequency of 885 times per second! The slowest known pulsars have cycle times of more than a minute, on the other hand. In every case, the pulsar is only "on" for a small fraction of each cycle.

The discovery of pulsars produced a great deal of excitement and speculation, including the suggestion that they were beacons operated by an alien civilization. Once the initial shock of discovery wore off, however, a number of more natural explanations were offered by astronomers who sought to establish the identity of the pulsars. It was well known that a variety of stars pulsate regularly, alternately expanding and contracting, but none were known to do it so rapidly. Theoretical studies showed that such quick variations as those exhibited by the pulsars should occur only in very dense objects, denser even than white dwarfs. This led astronomers to think of neutron stars.

Enough was known from theoretical calculations, however, to rule out rapid expansion and contraction of neutron stars as the cause of the observed pulsations. The calculations showed that the vibration period of such an object would be even shorter than the observed periods of the pulsars.

Figure 23.11 The radio emission from a pulsar. *Here, plotted on a time scale, is the radio intensity from a pulsar. The radio emission is weak or nonexistent except for a very brief flash once each cycle (in some cases, there is a weaker flash between each adjacent pair of strong flashes).*

A second possibility—that the rapid periods were produced by rotation of the objects—was considered. As in physical contraction and expansion, the rotation hypothesis ruled out ordinary stars and white dwarfs. To rotate several times per second, a normal star or white dwarf would have to have a surface speed in excess of the speed of light, a physical impossibility. Such an object would be torn apart by rotational forces before approaching such a rotational speed. A neutron star, on the other hand, could rotate several times per second and remain intact.

When the Crab pulsar was discovered, a great deal of additional information became available. This pulsar was found to be gradually slowing its pulse rate, something that could best be explained if the pulses were linked to the rotation of the object. The rotation could slow as the pulsar gave up some of its energy to its surroundings.

This discovery cleared up another mystery. The source of energy that powers the synchrotron emission from the Crab nebula had been unknown. Now it was suggested that the slowdown of the rotating neutron star could provide the necessary energy, either by magnetic forces exerted on the surrounding ionized gas or by transfer of energy from the pulsar to the surrounding material through the emission radio waves.

The remaining question was how the pulses were created by the rotation. Evidently a pulsar acts like a lighthouse, with a beam of radiation sweeping through space as it spins. The question was why a rotating neutron star should emit beams from just two points on its surface.

The most probable explanation has to do with the strong magnetic fields that neutron stars are likely to have. If a neutron star has a strong field, then electrons from the surrounding gas are forced to follow the lines of the field, hitting the surface only at the magnetic

ASTRONOMICAL INSIGHT 23.3

Very Rapid Pulsars

 Since the discovery of the first pulsars in 1967, many more have been found. The typical period of pulsation, thought to represent the rotation period of the neutron star responsible for the pulses, is 0.03 second or longer (the pulsar in the Crab supernova remnant was for many years the most rapid known pulsar, spinning 30 times per second). Recently, however, some pulsars have been found to have periods far shorter than 0.03 second. These pulsars, dubbed "millisecond pulsars" because their periods are measured in units of thousandths of a second, probably have different origins than the majority of pulsars.

The frequency of pulses from the very rapid pulsars was at first difficult to understand. As a pulsar ages, its rotation rate should either stay constant or slow (slowing is caused by the drag between the magnetic field lines of the neutron star and ionized gas that may exist around the pulsar). The pulsation rate of the Crab pulsar has been observed to be slowing, and similar slowing is expected for any neutron star formed in an environment where magnetic forces can act on surrounding gas. Yet the millisecond pulsars are apparently rotating at rates much faster than any reasonably old pulsar should be capable of. The contradiction between expectation and

observation became far more severe recently, when a millisecond pulsar was discovered in a globular cluster. Recall (from Chapter 21) that globular clusters are very old conglomerations of stars, not places where young, massive stars are expected to form and then die in supernova explosions creating neutron stars in the process. Therefore the existence of very young pulsars in globular clusters was difficult to understand.

The leading explanation of very rapid pulsars today is that they are neutron stars whose spin rate has actually *increased* since they formed, quite in contradiction with the normal expectation. The increased spin rate is presumed to have come about because new matter has been added to these neutron stars, increasing their angular momentum, hence increasing their spin rate. For example, in a binary system where one star is a neutron star and the other a normal star undergoing expansion to a red giant phase, it is likely that matter would transfer from the normal star to the neutron star (see discussion of evolution in binary systems, Chapter 20). As matter is transferred to the neutron star, it follows a spiral path because of the relative orbital motion of the two stars, and this spiral path causes the new material to fall onto the neutron star with a large amount of angular momentum This in turn causes the neutron star to spin more rapidly when it gains matter from its companion, turning the normal neutron star into a very rapidly spinning one. Thus, a once-slow

pulsar can become a very rapid, millisecond pulsar.

Very recently a new millisecond pulsar was discovered in a binary system. Analysis of the orbital properties of the pulsar revealed that its companion star has a mass of only 2 percent of the mass of the Sun. This is far too little mass for a normal star, and it is being suggested that this star originally had much more mass, but that most of its material has somehow been lost. The leading hypothesis is that the neutron star has caused its companion to evaporate, by heating the companion's outer layers. The neutron star in this system has been dubbed the "black widow" star, because it is "eating" its mate.

Since the discovery of the first millisecond pulsar, several more have been found, including a few in globular clusters. Too many have been discovered to allow an explanation in terms of residual rotation from the time of formation; it is now widely accepted that these rapid pulsars must have been spun up to high rotation rates by the addition of material from binary companions. It is quite possible that many of the rapid pulsars have in turn vaporized their companion stars, in a bizarre form of stellar cannibalism. The black widow star may not be the only one of its kind.

poles. The result is an intense beam of electrons traveling along field lines, especially concentrated near the magnetic poles of the neutron star, where the fields are crowded together. The rapidly moving electrons emit synchrotron radiation as they travel along the field lines, creating narrow beams of radiation from both magnetic poles of the star. If the magnetic axis of the star is not aligned with the rotation axis (Fig. 23.12), then these beams will sweep across the sky in a conical pattern as the star rotates. If the Earth happens to lie in the direction intersected by one of these beams, then we see a flash of radiation every time the beam sweeps by us.

Synchrotron radiation is normally emitted over a very broad range of wavelengths, suggesting that pulsars may "pulse" in parts of the spectrum other than the radio. Indeed this is the case: The Crab pulsar has been identified as a visible-light and X-ray pulsar (Fig. 23.13). Although the details are still somewhat vague, the rotating neutron star model seems to be the best explanation of the pulsars. Since special conditions (i.e., nonalignment of the magnetic and rotation axes, and the need for the beam to cross the direction towards the Earth) are required for a neutron star to be seen from Earth as a pulsar, it follows that there should be many neutron stars that do not manifest themselves as radio pulsars. In the next section, we will see how some of these neutron stars are detected.

Neutron Stars in Binary Systems

Earlier we made a general statement that stellar remnants are often most easily observed when they are in binary systems. We have already seen that a white dwarf in a binary can flare up violently if it receives new matter from its companion star. A neutron star reacts similarly in the same circumstances.

If a neutron star is in a binary system in which the companion object is either a hot star with a rapid wind or a cool giant losing matter because of its active chromosphere and low surface gravity, some of the ejected mass will reach the surface of the neutron star. Very little of it falls directly down onto the surface; instead, much of it swirls around the neutron star, forming a disk of gas called an **accretion disk** (Fig. 23.14). The individual gas particles orbit the neutron star like microscopic planets and fall inward only as they lose energy in collisions with other particles. The accretion disk acts as a reservoir of material, slowly feeding it inward, towards the neutron star.

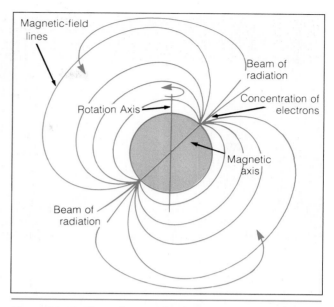

Figure 23.12 The pulsar mechanism. *Here we see a rapidly rotating neutron star, with its magnetic axis out of alignment with its rotation axis. Synchrotron radiation is emitted in narrow beams from above the magnetic poles, where charged particles, constrained to move along the magnetic field lines, are concentrated. These beams sweep the sky as the star rotates, and if the Earth happens to lie in a direction covered by one of the beams, we observe the star as a pulsar.*

The disk is very hot because of the immense gravitational field close to the neutron star. As gas in the accretion disk slowly falls in closer to the neutron star, gravitational potential energy is converted into heat. The gas in the disk reaches temperatures of several million degrees, hot enough to emit X-rays. Hence a neutron star in a mass-exchange binary system is likely to be an X-ray source, and a number of such systems have been found in the past fifteen years (Fig. 23.15), since the advent of X-ray telescopes launched on rockets or satellites. Often it is known that an X-ray object is part of a binary system, because the X-rays are periodically eclipsed by the companion star. Eclipses are made likely by the proximity of the two stars (mass exchange would not occur unless it were a close binary) and the fact that the mass-losing companion is likely to be a large star, either an upper main-sequence star or a supergiant.

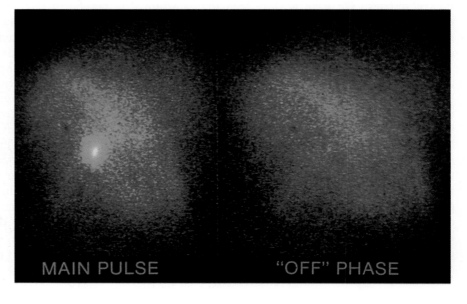

Figure 23.13 Visible and X-ray flashes from the Crab pulsar. *At top is a sequence of visible-light photographs accompanied by a light curve, showing how the pulsar in the Crab Nebula flashes on and off during its 0.03-second cycle. Below is a pair of X-ray images, showing the pulsar "on" and "off" in different parts of its cycle. (Top: National Optical Astronomy Observatories; bottom: Harvard-Smithsonian Center for Astrophysics)*

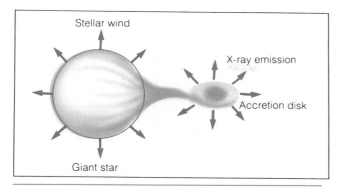

Figure 23.14 An accretion disk. *Here a giant star, losing mass through a stellar wind, has a neutron star companion. Material that is trapped by the neutron star's gravitational field swirls around it in a disk, which is so hot from compression that it glows at X-ray wavelengths. A nearly identical situation can arise when the compact companion is a black hole.*

The so-called **binary X-ray sources** (a few are listed in Table 23.4) are among the strongest X-ray-emitting objects known. Many of them are, most likely, neutron stars, although the possibility exists that some are black holes, which would similarly produce X-ray emission as material fell inward.

There is a slightly different way in which neutron stars can emit X-rays, which also occurs in mass-exchange binaries. If the infalling material trickles down onto the neutron star in a steady fashion, then there is continuous radiation of X-rays, as we have just seen. If, however, the material falling in does so sporadically, and arrives in substantial quantities every now and then, a major nuclear outburst occurs each time (Fig. 23.16), in close analogy to the nova process involving a white dwarf. This apparently happens in some cases, producing random but frequent X-ray outbursts. The intensity of the outburst, as in the case of a nova, depends on how much matter has fallen in and been consumed in the reactions. The neutron star binary systems where this occurs are called **bursters.** One important contrast with novae is that the flare-up of a burster occurs much more rapidly, lasting only a few seconds. Another is that most of the emission occurs only in very energetic X-rays, so these outbursts do not show up in visible light.

BLACK HOLES:
GRAVITY'S FINAL VICTORY

In the last chapter we saw what happens to a massive star at the end of its life: It falls in on itself, and there

Figure 23.15 An X-ray map. *This shows the locations in our galaxy of X-ray sources cataloged by the* High Energy Astronomical Observatory 1, *an orbiting X-ray satellite that operated in the late 1970s. Most of the sources here are binary systems in which one member is a compact stellar remnant such as a neutron star or a black hole. The concentration of sources along the galactic equator indicates that the massive stars that can end up as neutron stars or black holes tend to form, evolve, and die close to the plane of our disklike galaxy (see discussion in Chapter 26). (Smithsonian Astrophysical Observatory)*

Table 23.4 Selected binary X-ray sources

Source	Period (days)	Mass of System ($M_\odot$)	Nature of Stars
Cygnus X-1	5.6	40	O9Ib supergiant; probable black hole
Centaurus X-3	2.09	20–25	B0Ib-III giant; pulsar (neutron star)
Small Magellanic Cloud X-1	3.89	15–25	B0Ib supergiant; pulsar (neutron star)
Hercules X-1	8.95	20–30	B0.5Ib super-giant; probable neutron star
Vela X-1	1.70	2–5	HZ Hercules (A star); pulsar (neutron star)
A 0620-00	0.32	11	K dwarf; probable black hole (companion mass $\simeq 10\ M_\odot$)
Large Magellanic Cloud X-3	1.70	15	B3 main-sequence star; probable black hole (mass $\simeq 9M_\odot$)

is no barrier—not electron degeneracy nor neutron degeneracy—that can stop it.

The gravitational field near the surface of a collapsing star grows in strength as the mass of the star becomes concentrated in an ever-smaller volume. This gravitational field has important effects in the near vicinity of the star, although at a distance it remains unchanged from what it was before the collapse. Close to the star, though, the structure of space itself is strongly distorted, according to Einstein's theory of general relativity. Einstein discovered that accelerations caused by changing motion and those caused by gravitational fields are equivalent, and from this it follows that space must be curved in the presence of a gravitational field, so that moving particles follow the same path that they would follow if they were being accelerated. This applies to photons of light, just as it does to other kinds of particles.

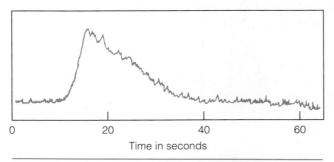

Figure 23.16 The X-ray light curve of a burster. *This schematic illustration of the X-ray intensity from one of these sources shows how rapidly the emission flares up and then drops off. Like the nova process, these outbursts are thought to be caused by material falling onto the surface of a neutron star and igniting brief episodes of nuclear reactions.*

Normally, the effects of the enhanced curvature of space are not noticeable, except under very careful observation, or when considering very large distances. Later in this text, when we discuss the universe as a whole and its overall structure, we will consider the latter situation. For now, we confine ourselves to local regions where space can be distorted by very strong gravitational fields (Fig. 23.17).

Let us consider a photon emitted from the surface of a star as it falls in on itself (Fig. 23.18). If the photon is emitted at any angle away from the vertical, its path will be bent over farther. If the gravitational field is strong enough, the photon's path may be bent over so far that it falls back onto the stellar surface. Photons emitted straight upward follow a straight path, but they lose energy to the gravitational field, causing their wavelengths to be redshifted (we discussed this phenomenon in the case of white dwarfs). When the gravitational field has become strong enough, the photon loses all of its energy and cannot escape. Another way of stating this is that the speed required for escape from the star exceeds that of light. To an outside observer, at the point when this happens the star becomes invisible, because no light from the star can reach the Earth.

The radius of the star at the time when its gravitational field becomes strong enough to trap photons is called the **Schwarzschild radius**, after the German astrophysicist who first calculated its properties some 60 years ago. The Schwarzschild radius depends only on the mass of the star that has collapsed. For a star of 10 solar masses, it is 30 km; a star of twice that mass would have twice the Schwarzschild radius, and so on (Table 23.5).

When a collapsing star has shrunk inside its Schwarzschild radius, it is said to have crossed its **event horizon**, because an outside observer cannot see it or anything that happens to it after that point. We can

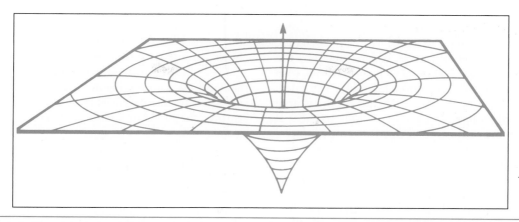

Figure 23.17 Geometry of space near a black hole. *Einstein's theory of general relativity may be interpreted in terms of a curvature of space in the presence of a gravitational field. Here we see a representation of how this curvature varies near a black hole (but this is only a representation, not a realistic illustration, because the curvature of space near a black hole is actually a distortion of three-dimensional space, not two-dimensional space as shown here).*

have no hope of ever seeing what happens inside the event horizon, but since no force is known that could stop the collapse, we assume that it continues. The mass becomes concentrated in an infinitesimally small region at the center, which is called a **singularity**, because mathematically it is a single point.

What happens to the matter that falls into a singularity is a subject for speculation, but we can be more concrete in discussing what happens just outside the

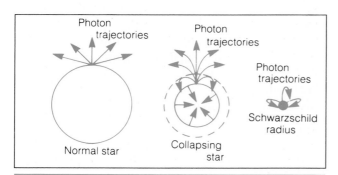

Figure 23.18 Photon trajectories from a collapsing star. *Light escapes in essentially straight lines in all directions from a normal star (left), whose gravitational field is not sufficient to cause large deflections. At an intermediate stage of collapse (center), photons emitted in a cone nearly perpendicular to the surface can escape, while others cannot. Those emitted at just the right angle go into orbit around the star, while those emitted at greater angles fall back onto the stellar surface. After collapse has proceeded to within the Schwarzschild radius, no photons can escape.*

event horizon. From the outsider's point of view, the time sense is distorted by the extreme gravitational field and acceleration in such a way that the collapse would seem to slow gradually, taking an infinite time to reach the event horizon. This slowdown, however, is most significant in the last moments before the star's disappearance, and by then the star would be essentially invisible anyway, as most escaping photons are redshifted into the infrared or beyond. The star would seem to disappear rather quickly, despite the stretching of the collapse in time caused by relativistic effects. To an observer unfortunate enough to be falling into the black hole, the fall would go very quickly; the sense of time would not be distorted by relativistic effects in the same way as that of the distant observer. (The infalling observer would suffer serious discomfort from being stretched out by tidal forces.)

Table 23.5 Schwarzschild radii for various objects

Object	Schwarzschild Radius
Sun	3 km
Earth	0.9 cm
150-lb person	5×10^{-23} cm
Jupiter	2.9 m
Star of 50 $M_\odot$	150 km
Typical globular cluster	5×10^4 km
Nucleus of a galaxy	$10^6 - 10^8$ km ($10^{-7} - 10^{-5}$ pc)
Massive cluster of galaxies	10^{15} km (100 pc)
The universe	10^{26} km (10^{13} pc)

Mathematically, a black hole can be described completely by three quantities: its mass, its electrical charge, and its spin. The mass, of course, is determined by the amount of matter that collapsed to form the hole, plus any additional material that may have fallen in later. The electrical charge, similarly, depends on the charge of the material from which the black hole formed; if it contained more protons than electrons, for example, it would end up with a net positive charge. Because particles with opposite charges attract each other, and those with like charges repel, it is thought that electrical forces would maintain a fairly even mixture of particles during and after the formation of the black hole, so that the overall charge would be nearly zero. To illustrate this, imagine that a black hole was formed with a net negative charge. If there were ionized gas around it afterwards (as there likely would be, with some of the matter from the original star still drifting inward), the negative charge of the hole would repel additional electrons, preventing them from falling in, while protons would be accelerated inward. In time, enough protons would be gobbled up to neutralize the negative charge.

The spin of a black hole is not so easily dismissed, however. It stands to reason that if the star were spinning before its collapse, and most likely it would have been, then conservation of angular momentum would cause the rotation to speed up greatly as the star shrank. A high spin rate actually shrinks the event horizon, allowing an outside observer to see closer in to the singularity residing at the center. It is even possible mathematically—although it presents a physical dilemma—to have sufficient spin that there is no event horizon, so that whatever is in the center is exposed to view. A **naked singularity,** as this has been dubbed, would not produce the usual gravitational effects of a black hole, and it would be possible to blunder into one without any forewarning.

For the most part, it is assumed that the spin rate is never so large that a naked singularity can form, and in fact black hole properties are usually specified by mass alone, neglecting the effects of charge and spin. It is assumed that our main hope of detecting a black hole is by its gravitational effects, determined entirely by the mass.

Before turning to a discussion of how to find a black hole, we should mention that some black holes may be formed by processes other than the collapse of individual massive stars. The others include the so-called mini black holes, very small ones postulated to have formed under conditions of extreme density early in the history of the universe; and supermassive black holes, thought possibly to inhabit the cores of large galaxies, having formed from thousands or millions of stellar masses coalescing in the center. These other possibilities are discussed in later chapters; for now, we turn to the hunt for stellar black holes.

Do Black Holes Exist?

The mass that goes into a black hole during its formation still exists there, hiding inside its event horizon. Even though no light can escape, its gravitational effects persist. The gravitational force of the star is exactly the same as it was before the collapse, except at points so close that they would have been inside the original star.

Our best chance of detecting a black hole, then, is to look for an invisible object whose mass is too great to be anything else. Even if we do find such a thing, it is really only a circumstantial argument, because the conclusion that it is a black hole relies on the theory that says neither a white dwarf nor a neutron star can survive if the mass is sufficiently great.

The best chance of determining the mass of an object is when it is in orbit around a companion star, where Kepler's third law can be applied to find the masses. The search for black holes, then, leads us to examine binary systems, looking for invisible, but massive, objects.

This search has been facilitated by another property of black holes; if they find new material to pull into themselves, the trapped matter forms an accretion disk, just as we described in the case of a neutron star in a mass-transfer binary. We have already seen that such a disk becomes so hot that it emits X-rays. Probably some of the binary X-ray sources, therefore, are due to black holes in binary systems rather than to neutron stars. The best way to distinguish between the two possible types of remnants in these systems is to determine the mass of the invisible companion, to see whether it is too great to be a neutron star.

The determination of the masses in a binary system is difficult, if not impossible, if only one of the two stars can be seen, because the information on their speeds and on the inclination angle of the orbit is incomplete. The fact that many X-ray binaries are eclipsing systems (at least the X-ray source is eclipsed; Fig. 23.19) helps to determine the tilt of the orbit in

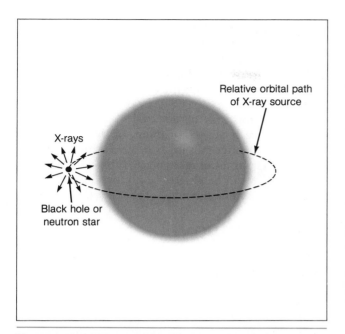

Figure 23.19 Eclipses of an X-ray source in a binary system. *Because the separation of the two stars in a mass-transfer system is always very small, and because the mass-losing star is usually very large (a red giant or a hot giant or supergiant with a stellar wind), eclipses occur easily. If the mass-receiving star is a neutron star or black hole, and therefore an X-ray emitter, the X-rays are likely to be eclipsed periodically by the companion. Hence many such systems are eclipsing X-ray binaries.*

Figure 23.20 Cygnus X-1. *This X-ray image obtained by the orbiting* Einstein Observatory *shows the intense source of X-rays at the location of a dim, hot star which is thought to be a normal star that is losing mass to its invisible black hole companion. (Harvard-Smithsonian Center for Astrophysics)*

some cases. The orbital period is usually easily determined from the eclipse frequency as well, but the orbital velocities, needed to deduce the sizes of the orbits, are often very difficult or impossible to measure. The unseen star, of course, emits little or no light, so there is no hope of measuring Doppler shifts in its spectral lines, and the normal companion star is often a hot giant or supergiant, and these stars usually have very broad spectral lines, whose velocities cannot be measured accurately. Nevertheless, sometimes enough information can be derived to at least place limits on the mass of the invisible companion, however, and in one such case, the X-ray binary called Cygnus X-1 (Fig. 23.20), the collapsed object appears to contain at least 8 solar masses. This object is one of three leading black hole candidates, the other two being the sources LMC X-3 and A0620-00 (included in Table 23.4).

Although the jury is still out on the question of the existence of black holes, the circumstantial evidence in favor is strong. Furthermore, it is easier to accept the existence of black holes than to find a way for a collapsing, massive star to avoid becoming one. As we have stressed, no force is known that can resist gravity in this circumstance. To believe that black holes do not form, one has to invent some way of preventing collapse, and also some way of explaining the observational evidence for invisible massive objects in some binary systems. It is far less complicated to accept that black holes do form and exist in these binaries, because to do so requires the fewest unproven assumptions. The principle called **Occam's Razor,** a guiding philosophy for scientists, states that the explanation of any phenomenon that requires the fewest arbitrary assumptions is most likely to be the correct one.

Most astronomers today have adopted the concept of black holes, and not only believe that they exist, but consider them an integral part of our universe, playing numerous important roles.

PERSPECTIVE

We have followed the stars into their graves. We have examined their corpses to see how they decay and have come away with our minds filled with wonder at the novel forms that matter can take. We have seen that the nature of stellar remnants depends entirely on how much mass is left when stars run out of nuclear fuel and collapse, the three possibilities being white dwarfs, neutron stars, and black holes. Stars that die alone, no matter what final form they take, are not likely to be detected again (except for the closest white dwarfs and the neutron stars that happen to appear to us as pulsars), while those that exist in binary systems may be reincarnated in spectacular fashion if mass exchange takes place.

Having learned all we can about stars as individuals, we are now ready to move on to larger scales in the universe, to examine galaxies, and ultimately the universe itself.

SUMMARY

1. A white dwarf gradually cools off, taking billions of years to become a cold cinder.

2. If new matter falls onto the surface of a white dwarf—for example in a binary system in which the companion star loses mass—then the white dwarf may become a cataclysmic variable if the matter arrives in small amounts with high energy, or it may become a nova if matter builds up slowly to the point where a larger quantity is involved in the explosion. If it exceeds the white dwarf mass limit, it may become a neutron star or black hole or may explode as a Type I supernova.

3. Massive stars are likely to explode as Type II supernovae when all possible nuclear reaction stages have ceased. The supernova explosion creates an expanding cloud of hot, chemically enriched gas known as a supernova remnant.

4. In some cases, a remnant of 2 to 3 solar masses is left behind, in the form of a neutron star, consisting of degenerate neutron gas.

5. A neutron star is too dim to be seen directly in most cases but may be observed as a pulsar (depending on the alignment of its magnetic and rotation axes, and our line of sight), and in a close binary system it may become a source of X-ray emission.

6. Some neutron stars that receive new material in clumps flare up occasionally as X-ray sources called bursters.

7. If the final mass of a star exceeds 2 or 3 solar masses, it will become a black hole at the end of its nuclear reaction lifetime.

8. The immensely strong gravitational field near a black hole traps photons of light, rendering the black hole invisible.

9. A black hole may be detected by its gravitational influence on a binary companion, or by the X-rays it emits if new matter falls in, as in a close binary system in which mass transfer takes place. A black hole binary X-ray source can be detected only by analysis of the orbits to determine the mass of the unseen object.

REVIEW QUESTIONS

Thought Questions

1. Why is the final mass of a star, rather than the initial mass, the factor that determines what form of remnant the star will leave? How can the final mass be different from the initial mass?

2. Summarize the techniques by which white dwarfs can be detected from the Earth.

3. Explain why white dwarfs, with no source of energy, can stay hot for as long as a billion years.

4. Describe the differences between a nova, a supernova of Type I, and a supernova of Type II.

5. Compare the properties of a neutron star with those of a white dwarf. Include a description of the form of pressure force that counteracts gravity in the two types of remnants.

6. How are neutron stars detected from the Earth?

7. Explain why the pulses from pulsars were finally judged to be the result of rapid rotation of neutron stars, rather than rapid expansion and contraction or rotation of some other type of star. What was the role of the Crab Nebula and its pulsar in this discovery?

8. Both neutron stars and black holes can be X-ray sources in binary systems where mass exchange occurs. How can astronomers tell which kind of remnant is present in a given binary X-ray source?

9. Explain why some massive stars become black holes, while others become neutron stars or blow up completely, leaving no remnant.

10. Why would a neutron star or a black hole be expected to have a very rapid rotation rate?

Problems

1. Compare the average density, surface gravity, and escape velocity of a neutron star (2 solar masses, radius 10^{-4} solar radii) with the same properties of the Sun.

2. Suppose mass is added to a white dwarf and undergoes nuclear reactions, creating a nova and releasing energy according to $E = mc^2$. How much energy is released if the mass that undergoes reactions is 10^{-4} solar masses? In Type I supernova, the mass that undergoes reactions is about 1 solar mass. What is the energy released in such a supernova? In both cases, assume that only 1 percent of the mass that undergoes reactions is converted into energy.

3. If the rotation period of the Crab pulsar is 0.033 sec and its radius is 10 km, what is the speed of a point on its surface at the equator? What fraction of the speed of light is this?

4. Apart from the beams that create the pulsar effect, a neutron star glows from its surface because of its high temperature. Suppose a neutron star has a surface temperature of 1 million degrees (10^6 K). What is its wavelength of maximum emission? What kind of telescope could be used to observe it?

5. The surface area of a neutron star is so small that the luminosity is low, despite the high temperature. To show this, calculate the luminosity of a neutron star with surface temperature 10^6 K and radius 10 km, relative to the luminosity of the Sun.

6. The formula for the Schwarzschild radius of an object of mass M and radius R is $R_s = 2\,GM/c^2$. Use this formula to find the Schwarzschild radii for the following: (a) the Sun; (b) a star of 20 solar masses; and (c) the Milky Way galaxy (mass approximately 10^{45} grams).

7. Suppose an X-ray binary system is found in which the visible star is a red giant of a spectral type thought to have a mass of about 12 solar masses. The orbital period is 3.65 days, and the semimajor axis is 0.12 AU. Calculate the sum of the masses from Kepler's third law, then try to decide whether the compact companion star is a neutron star or a black hole.

ADDITIONAL READINGS

Anderson, L. 1976. X-rays from degenerate stars. *Mercury* 5(5):2.

Backer, D. C., and S. Kulkarni. 1990. A new class of pulsars. *Physics Today* 43(3):26.

Bethe, H. A. 1985. How a supernova explodes. *Scientific American* 252(5):60.

————— . 1990. Supernovae. *Physics Today* 43(9):24.

Black, D. L. 1976. Black holes and their astrophysical implications. *Sky and Telescope* 50(1):20 (Part I); 50(2):87 (Part 2).

Clark, David H. 1985. *The quest for SS433*. New York: Penguin Books.

Greenstein, G. 1985. Neutron stars and the discovery of pulsars. *Mercury* (2):34 (Part I); (3):66 (Part 2).

Gursky, H., and E. P. J. van den Heuvel. 1975. X-ray emitting double stars. *Scientific American* 232(3):24.

Helfand, D. J. 1978. Recent observations of pulsars. *American Scientist* 66:332.

Kafatos, M., and A. G. Michalitsianos. 1984. Symbiotic stars. *Scientific American* 251(1):84.

Kaufmann, W. 1974. Black holes, worm holes, and white holes. *Mercury* 3(3):26.

Kawaler, S. D., and D. E. Winget. 1987. White dwarfs: fossil stars. *Sky and Telescope* 74:132.

Lewin, W. H. G. 1981. The sources of celestial X-ray bursts. *Scientific American* 244(5):72.

McClintock, J. 1988. Do black holes exist? *Sky and Telescope* 75:28.

Price, R. H., and K. S. Thorne. 1988. The membrane paradigm for black holes. *Scientific American* 258(4):69.

Schaefer, B. E. 1985. Gamma-ray bursters. *Scientific American* 252(2):52.

Schramm, D., and W. Arnett. 1975. Supernovae. *Mercury* 4(3):16.

Seward, F. 1986. Neutron stars in supernova remnants. *Sky and Telescope* 71(1):6.

Shaham, Jacob. 1987. The oldest pulsars in the universe. *Scientific American* 256(2):50.

Smarr, L., and W. H. Press. 1978. Spacetime: Black holes and gravitational waves. *American Scientist* 66:72.

Thorne, K. S. 1974. The search for black holes. *Scientific American* 233(6):32.

Van der Klis, M. 1988. Quasi-periodic oscillations in celestial X-ray sources. *Scientific American* 259(5):50.

Van Horn, H. M. 1973. The physics of white dwarfs. *Physics Today* 32(1):23.

Webbink, R. F. 1989. Cataclysmic variable stars. *American Scientist* 77:248.

Wheeler, J. C. 1973. After the supernova, what? *American Scientist* 61:42.

MATHEMATICAL METHODS

Escape Speed and the Schwarzschild Radius

In this chapter we refer to the Schwarzschild radius, which we define as the radius of an object whose surface gravity is great enough so that photons of light cannot escape. Here we show how the Schwarzschild radius is calculated.

To do this, we must remind ourselves of the concept of escape speed, introduced in Chapter 5 and discussed at some length in the Mathematical Methods section at the end of Chapter 10. The escape speed is the upward speed a particle must have in order to break away from the gravitational field of another body such as a planet or a star. To find the escape speed, we compare the kinetic energy of the moving body to the gravitational potential energy due to the other body. If the kinetic energy of the moving body is greater than the gravitational potential energy due to the other body, the moving body can escape. If it has less, then it is trapped.

The kinetic energy of a body of mass m moving at speed v is $\frac{1}{2}mv^2$. The gravitational potential energy of a body of mass m in the gravitational field of another body of mass M is GmM/R, where M is the mass of the other body, R is the separation between the two bodies (equal to the radius of the other body if body m is at the surface of body M), and G is the gravitational constant. Setting these two expression equal, we find

$$\frac{mv^2}{2} = \frac{GmM}{R}.$$

Cancelling m from both sides and rearranging, we find

$$v = \sqrt{\frac{2GM}{R}}.$$

This is the familiar formula for the escape velocity, as given in Chapter 5.

Now suppose the particle trying to escape body M is a photon of light, whose speed is the speed of

light, c. If we substitute c for v and solve for R, the radius of the object with a gravitational field that is trapping the particle (the photon) whose speed is c, we find

$$R_S = \frac{2GM}{c^2}.$$

This value for R_S is called the Schwarzschild radius (usually designated R_S), because it was first derived by Karl Schwarzschild soon after Albert Einstein developed his theory of special relativity.

Schwarzschild realized that if the speed of light is the limiting speed for any object in any frame of reference, then it is possible to envision objects so massive and small that their escape speeds equaled the speed of light. Objects with greater surface gravities would prevent the escape of any particles, even photons (such objects became known as black holes much later, when physicist John A. Wheeler coined the term).

We can calculate the Schwarzschild radii for any spherical body; all we need to know is the mass. For the Sun, whose mass is 2×10^{33} grams, the formula yields

$$R_S = 2(6.67 \times 10^{-7}) \, (2 \times 10^{33})/(3 \times 10^{10})^2$$
$$= 3 \times 10^5 \text{ cm}.$$

This is equal to 3 km, the value given in the text for the Schwarzschild radius of the Sun.

We could calculate Schwarzschild radii for other objects such as planets or stars by use of the same formula, but it is easier if we note that in the equation for the Schwarzschild radius the only value that changes from one object to another is the mass, and that the change is proportional to the mass. Thus an object having twice the mass of the Sun also has twice the Schwarzschild radius; an object 10 times more massive has 10 times the Schwarzschild radius, and so on. Thus it is easy to derive the Schwarzschild radius for any object, once we have its mass in units of the solar mass.

(Note: The derivation given here is a bit oversimplified, although it gives the correct result. To do this properly requires the use of general relativity, and the calculation of the gravitational redshift at which photons are shifted to infinite wavelengths, at which point they have zero energy and cannot escape.)

The Formation of Suns and Solar Systems

BY STEPHEN E. STROM AND SUZAN EDWARDS

The two authors of this guest editorial, Stephen Strom and Suzan Edwards, are leaders in the understanding of young stars and the gas and dust associated with them. Dr. Strom is chair of the Five College Astronomy Department, an association of five universities in Massachusetts that have joined together to share faculty and research facilities in astronomy. From his home location at the University of Massachusetts, Dr. Strom finds time to continue his prolific research programs despite his administrative duties (in addition to chairing a major astronomy department, Dr. Strom has been prominent in national astronomical affairs, having served on numerous advisory committees to NASA, the National Science Foundation, and the National Academy of Sciences, and having held offices as high as vice president of the American Astronomical Society). Since completing his doctorate at Harvard in 1962, Dr. Strom has pursued a number of different fields in astronomy, quickly rising to a leadership position in each, as evidenced by the numerous awards and lectureships he has won. Among his areas of work are stellar atmospheres theory,

galactic structure and evolution, the formation and growth of interstellar dust grains, and the early stages in the lifetimes of stars, where most of his effort goes today. In studying this latter topic, Dr. Strom has joined forces with Dr. Edwards, and together they have reached profound new understandings of the processes associated with star formation.

Dr. Edwards, also a member of the Five Colleges Astronomy department, is based at Smith College. Since receiving her Ph.D. from the University of Hawaii in 1980, she has worked almost exclusively in research on young stars and prestellar objects. This has led her into new territory in probing the role of mass loss during what is essentially a mass-accretion process. Her most recent work with Dr. Strom has led to a

comprehensive picture of star formation and the evolution of the associated gas and dust, and along with that, a new understanding of how planetary systems are formed.

In this article, Drs. Strom and Edwards present a beautiful account of the processes by which low-mass stars like the Sun are formed, and along with them, the story of planetary formation. In the process of summarizing these early stages in star formation, Drs. Strom and Edwards also clear up some longstanding problems in understanding how stars as small as the Sun can form from interstellar clouds initially having much more mass.

How did the solar system form? If we could be transported back in time to the epoch of planet formation, what would we see? Are the Earth and its sister planets examples of a rare cosmic accident or an inevitable consequence of star formation? These questions are modern expressions of timeless and universal human curiosity about how the Earth came to be, and whether we are alone in the universe. Over the past centuries, women and men driven by a passion to answer these questions have developed new conceptual tools and built powerful machines. Millenia ago, our ancestors throughout the world recorded the paths of visible planets across the sky and strove to develop algorithms for predicting their motions, and paradigms for understanding why they followed such regular paths. Over the past centuries, we have discovered and catalogued many of the denizens of our solar system—the nine planets and their satellites, thousands of asteroids, and billions of comets—and have come to understand the forces that govern their motion. Astronomers working with Earth-based telescopes, chemists analyzing fragments of meteors and tiny grains of cosmic dust, and mathematicians and physicists working with pen, paper, and powerful computers have evolved models of the size, mass, chemical composition and structure of these bodies. We have developed a variety of working hypotheses and trial models to explain their origin and evolution, and that of the solar system as a whole. We have sent robotic observers into space to visit planets, their satellites, and comets in order to deepen our understanding and to test and challenge our

models. We have set foot on the Moon and brought back pieces of that world to our own. The intellectual context provided by our predecessors and the tools available now or foreseeable in the near term promise that, by early in twenty-first century, we will learn whether our planetary system is one of many or is unique.

Our optimism that we will soon come to know how Earth came into being, and whether our home planet (and implicitly, life) are singular events, derives from the revolutionary progress of the past half century toward understanding how stars and planets are assembled from interstellar gas and dust. We now believe that the life of a star like the Sun begins in darkness, deep within a cloud of molecular hydrogen gas and micron-sized interstellar dust grains. Shielded by dust from incoming starlight, and by magnetic fields from cosmic rays launched from exploding supernovae, material in such clouds can cool and form dense cores of gas and dust approximately 0.3 light-year in radius. These cores move randomly through the larger cloud, occasionally encountering other cores at distances close enough to raise gravitational tides that set each core spinning. These nowrotating bodies cool still further, until the pressure generated by the motion of gas molecules and dust can no longer balance the pull of gravity exerted by the ensemble of such molecules and dust comprising the core. The core then begins to collapse, forming at its center a protosun surrounded by a flattened, rotating disk of gas and dust comparable in extent to that of our own solar system; material in the core continues to rain onto the disk and protosun and to in-

crease their mass. These early stages of stellar and disk formation are hidden from view by the obscuring dust of the protostellar core.

Gas and dust in the disk orbit the protosun in approximate Keplerian motion. Friction between adjacent annuli of material orbiting around the protosun causes material to migrate inwards. Because this inward-flowing material carries with it the angular momentum of orbital motion, its accretion by the protosun causes the protosun to spin more rapidly. Adding too much high-angular-momentum material to the protosun would eventually force it to spin so rapidly that centrifugal forces exceed gravity, at which stage the protosun would break up. Nature apparently intercedes to prevent this catastrophe by redirecting some of the accreting spinning material into an outward-flowing wind. These winds carry a great deal of momentum and energy, sufficient in fact to dissipate the protostellar core, thereby ending the epoch of infall from the core and fixing the mass of the disk and star. The process of core dissipation can be seen in radio observations, which reveal large-scale (up to 3 light-years in size), biconical regions accelerated by highly collimated winds that are directed outward along the north and south poles of the star-disk system. During the last stages of core dissipation, these bipolar, collimated winds are traced by spectacular jets of ionized plasma (see figure).

The loss of matter during the period of bipolar outflow provides an explanation for a longstanding problem: how to make low-mass stars out of interstellar clouds that initially have far more mass than

A jet from a young star. *Energetic mass outflows seem to be an intrinsic part of the star formation process. Part of the outflow occurs in narrow "jets" of ionized gas, which can extend to distances as great as 100,000 times the extent of the solar system. Here the faint, red star in the upper portion of the figure is the source of the jet, which is characterized by a series of knots of shocked, glowing gas moving outward at hundreds of km/sec. Jets such as this are believed to emerge perpendicular to the plane of the disk of gas and dust that forms as the new star condenses (the disk, about the size of the solar system, is too small to be seen here). (R. Mundt, Max Planck Institut fur Astronomie; photo obtained at the Calar Alto 3.5-m telescope, in Spain)*

a single sun. As described elsewhere in this text (Chapter 21), a low-mass cloud is unlikely to collapse spontaneously to form a star, whereas more massive clouds can become gravitationally unstable much more easily. But then the cloud must fragment or give up some of its material substance if the end product is to be a solar-type star. The rapid streams of gas flowing from the polar regions in young stars provide a mechanism for curtailing the accretion of mass, thus allowing small stars to form.

As the shroud of protostellar material is lifted by the wind, the protosun and its surrounding disk become visible and accessible to study with ground- and space-based optical telescopes. Material in the disk continues to accrete onto the protosun until the disk becomes so tenuous that it is transparent to visible and infrared light. At his point, accretion stops and the disk becomes quiescent, so much so that the motions of dust and gas grains perpendicular to the disk plane are no longer coupled. Much as cream separates

from milk, the dust grains settle toward the midplane of the disk, forming a thin, opaque layer surrounded by a much thicker atmosphere of gas molecules. The orbits of individual dust grains are not quite circular and not perfectly centered about the protosun. Consequently, the orbit of one dust grain typically crosses that of another. Occasionally, they meet at these crossings and collide. In many cases, the colliding grains adhere to one another, forming larger grains, and eventually kilometer-size "planetesimals." Comets and possibly some asteroids may be examples of relic planetesimals. Collisions between planetesimals may lead ultimately to the building of planetary-size bodies. Observational evidence of frequent collisions between planetesimals and larger entities is provided by the oldest craters observed on the Moon, and on other bodies (planets and satellites) that have suffered little or no surface reprocessing. The Earth-Moon system itself, and the displacement of planetary rotation axes with respect to the rotation axis of the solar system also can be explained if collisions between lunar-size and larger bodies occurred during the epoch of planet assembly.

Our own work is aimed at testing this picture by observing large numbers of solar-like stars over a wide range of ages: from their first appearance following dissipation of their natal cores by winds (approximately 1 million years after birth), to ages of about 1 billion years (or about 20% of the age of the Sun). During these "early stages" of solar system formation, it is possible to observe either visible light scattered by, or infrared radiation emitted by the micron-size dust grains from which planetesimals (and later, planets) were assembled. By measuring the radiation from such grains surrounding many solar-type stars of differing ages, we can infer the disk structure for star-disk systems in varying stages of evolution. These snapshots allow us to assemble a statistical picture of the changes in disk structure as a function of time and the range of times required to effect these evolutionary changes. Over the past 5 years we have assembled infrared and optical ob-

servations of a large sample of very young, solarlike stars and found evidence of disks of solar system size and mass approximately the same size as that required to build a planetary system similar to our own: At least half of all solar mass stars younger than 3 million years are surrounded by such disks. If these disks all build solar systems, then it is likely that there will be many planets similar to Earth orbiting other stars, a few of them perhaps populated by sentient beings.

Although we don't know with certainty whether planets have formed in any disk other than the one from which our solar system was assembled, we can say confidently that disks around other stars must evolve: Stars 10 million years and older no longer show the strong infrared emission from the disk, and the powerful winds associated with their younger progenitors. We interpret our observations to mean that dust settling and planetesimal growth likely take place on a time scale of 10 million years or less. Recently, we have detected very tenuous disks surrounding older stars. The remarkable edge-on disk discovered by Smith and Terile to surround the nearby 100 million-year-old star β Pictoris (Fig. 16.22, page 375) provides a vivid example of a tenuous disk. The dust in these disks cannot be left over from the protosun/disk formation epoch. Forces acting on such dust would quickly cause it to spiral into the central star on a time scale far shorter than the age of the star. Rather, we believe it most likely that the dust grains we observe now are byproducts of the era when planetesimal collisions led to the growth of plane-

tary cores; collisions that build larger bodies leave behind debris, including micron-size dust that absorbs radiation from the parent star and glows at a temperature that admits detection via infrared measurements with Earth-based telescopes. If our guess is correct, then by observing tenuous disks around stars of a wide range of ages, we can monitor the rate of dust production as a function of stellar/disk age and can thereby estimate the planetesimal collision rate. The clues we piece together by observing disks around many stars can then be compared with the evidence we have assembled from close-up observation of our own planetary system.

Most of the our evidence about disks is indirect. With the exception of β Pic, we have no pictures of disks; rather we infer their presence from a variety of observations that find (we believe) most economical interpretation in the context of the picture just described. To image disks requires far higher angular resolution than has been achieved to date with Earth-bound telescopes; the candidate primordial solar systems around the nearest forming solarlike stars have an angular size of less than 1 in. Images of stars viewed through the turbulent atmosphere above ground-based telescopes are blurred, thus preventing direct observation of disks. By making observations from space (for example with the *Hubble Space Telescope*), we expect to resolve disks and winds emanating from young stars and to track their evolution by imaging large numbers of solarlike stars of differing ages. The 1990s will also witness the construction of a new generation of large telescopes.

These telescopes will be equipped with sophisticated optical systems capable of rapidly compensating for image distortion wrought by atmospheric turbulence and producing sharp images of stars, disks, and winds. If they achieve their promise, these new telescopes may permit us to follow the evolution of the gas and dust content of disks and to diagnose the chemical composition of different parts of the disk. It may even be possible to observe forming planets in these disks, if they are massive enough and located at favorable distances from the star.

Ultimately, we would like to know whether evolving disks around solarlike stars actually form planets. The direct detection of planets around other stars is extremely difficult, because the light from the parent overwhelms the scattered or thermal radiation from the planet. Instead, detection of planetary systems must rest on the measurement of extremely subtle motions in the position or velocity of their parent stars. Planets orbiting a sun produce small reflex motion in the sun about the planet-sun center of mass (see Chapter 16). Their presence can be inferred from precise measurements of the motion (both radial and tangential) of their parent stars. Even though these measurements are difficult, promising efforts to search for other planetary systems are currently underway at several ground-based observatories. However, characterization of these systems—and in particular the detection of Earth-mass planets—probably will require space-based measurements.

Sometime over the next 10 years, we will image disks around a large sample of stars covering a wide range of ages. These images, combined with ground- and space-based infrared measurements, will allow us to chart their evolutionary paths and place astronomical constraints on the events that might lead to the assembly of planetesimals and later, planets. During the next half century, we will develop the tools and techniques that may allow us to learn how many of these evolving disks actually form planets, and to learn with certainty whether other earths orbit other suns. Possibly a reader of this text will experience the joy and deep satisfaction of being the first man or woman to examine these images of our past or to uncover the evidence of a twin to our solar system. At that moment, she or he should listen for the voices of all the ancients who wanted to know. They will be chanting in silent celebration.

THE MILKY WAY

In this section we explore the great conglomeration of stars to which our Sun belongs. The Milky Way, a spiral galaxy, is itself a dynamic entity, with its own structure and evolution governed by the complex interactions of stars with each other and with the interstellar gas and dust. Although this system of some 10^{11} individual stars is infinitely more complicated than a single star and its life story is correspondingly more difficult to unravel, great progress has nevertheless been made in piecing together the puzzle.

In the first of the three chapters of this section, we examine the overall properties of our galaxy. We learn how its size and shape, and the Sun's location within it, were deduced from observations carried out from our position within the great disk, where our view is obscured by interstellar haze. We discuss how the various parameters describing the galaxy were derived from observation, and we learn of some immense and fascinating mysteries that remain.

Chapter 25 focuses on the diffuse material between the stars. The interstellar medium is so rarefied that a superficial consideration might lead us to deem it unworthy of our time and effort. But we find, to the contrary, that the interstellar medium is a vital part of the galactic life story. Stars form from this material, and in evolving and dying they return their substance to it. The interstellar medium itself is a dynamic, active component of the galaxy, one that has in recent years provided astronomers with several major surprises. In this chapter we survey the general properties of the gas and dust that pervade space, and pay particular attention to their role in the evolution of the galaxy.

Finally, Chapter 26 ties together the disparate data gathered and examined in the preceding two chapters, and tells the story of the formation and evolution of our galaxy. We see how the major features of the Milky Way arose from a sequence of developments dictated by simple laws of physics previously learned. Although there remain some mysteries, such as the quantity of mass in the far reaches of the galactic halo and the traces of violent activity at the core of the galaxy, most of the story can be told.

Having probed the Milky Way, we will then be ready to move outward, to examine and understand the countless galaxies in the universe beyond our own.

CHAPTER 24

Structure and Organization
of the Galaxy

A 21-cm radio map of the Milky Way. (Bonn Radio Observatory)

We have discussed stars as individuals, as though they existed in a vacuum, isolated from the rest of the universe. For the purpose of analyzing the structure and evolution of stars, this is a suitable approach, but if we want to understand the full stellar ecology, the properties of individuals must be discussed in the context of the larger environment.

Even a casual glance at the nighttime sky shows that the stars tend to be grouped rather than randomly distributed. The most obvious concentration is the Milky Way, the diffuse band of light stretching from horizon to horizon, clearly visible only in areas well away from city lights (Fig. 24.1). To the ancients, the Milky Way was merely a cloud; Galileo with his primitive telescope recognized it as a region of great concentration of stars. In modern times we know the Milky Way as a galaxy,

the great pinwheel of billions of stars to which the Sun belongs (Table 24.1). The hazy streak across our sky is a cross-sectional, or edge-on, view of the galaxy, seen from a point within its disk.

The overall structure of the Milky Way is a bit like a phonograph record, except that it has a large central bulge, the center of which is called the **nucleus** (Figs. 24.2 and 24.3). Nearly all the visible light is emitted by stars in the plane of the disk, although the galaxy also has a **halo,** a distribution of stars and star clusters centered on the nucleus but extending well above and below the disk. The most prominent objects in the halo are the **globular clusters,** very old, dense clusters of stars characterized by their distinctly spherical shape.

To envision the size of the Milky Way requires a new unit of distance. In Chapter 19 we discussed stellar distances in terms of parsecs, and we found that the nearest star is about 1.3 parsecs from the Sun. To expand to the scale of the galaxy, we speak in terms of **kiloparsecs,** or thousands of parsecs. The visible disk of the Milky Way is roughly 30 kiloparsecs (abbreviated kpc) in diameter, and the disk is a few hundred parsecs thick. Light from one edge of the galaxy takes about 100,000 years to travel across to the far edge.

Our galaxy is so large that we must discuss new methods of measuring distance, for even the main-sequence-fitting technique—the most powerful we have mentioned so far—fails when star clusters are too distant and faint to allow determination of the individual stellar magnitudes and spectral types or color indices.

Table 24.1 Properties of the Milky Way

Mass: 1.4×10^{11} M$_\odot$ (interior to Sun's position)
Diameter of disk: 30 kpc
Diameter of central bulge: 10 kpc
Sun's distance from center: 8.7 kpc
Diameter of halo: 100 kpc (very uncertain)
Thickness of disk: 1 kpc (at Sun's position)
Number of stars: 4×10^{11}
Typical density of stars: 20 stars/pc^3 (solar neighborhood)
Average density of interstellar matter: 10^{-24} g/cm^3 (roughly equal to one hydrogen atom/cm^3)
Luminosity: 2×10^{10} L$_\odot$
Absolute magnitude: -20.5
Orbital period at Sun's distance: 2.5×10^8 years

VARIABLE STARS AS DISTANCE INDICATORS

It was stressed in Chapter 19 that it is always possible to determine the distance to an object if both its absolute and apparent magnitudes are known. In that chapter we talked about spectroscopic parallaxes, in which the absolute magnitude of a star is derived from its spectral class, and we discussed main-sequence fitting, where a cluster of stars is observed and its main sequence determined, so that it can be fitted to the main sequence in the standard H-R diagram. Now we will learn about a special type of star whose absolute magnitude can be determined from its other properties, allowing it to be used to measure distances.

In Chapter 18 we briefly mentioned variable stars, including those which pulsate regularly. These stars

Figure 24.1 The Milky Way.   *This mosaic (top) shows a major portion of the cross-sectional view of our galaxy that we see from Earth. The bottom image is an infrared view of the Milky Way, obtained by the IRAS satellite. (Top: Mt. Wilson and Las Campanas Observatories, Carnegie Institution of Washington; bottom: NASA)*

physically expand and contract, changing in brightness as they do so. One of the first such stars to be discovered was δ Cephei, which is sufficiently bright to be seen with the unaided eye. Following the discovery of the variability of δ Cephei in the mid-1700s, other stars were found to have similar brightness variations, and these as a class became known as **Cepheid variables,** named after the prototype. Several other types

of variable stars have subsequently been identified (Table 24.2), each with its own distinctive properties.

The usefulness of pulsating variables as distance indicators was discovered in 1912, when an astronomer named Henrietta Leavitt (Fig. 24.4) at the Harvard College Observatory carried out a study of variable stars located in the Magellanic Clouds, the two small satellite galaxies of the Milky Way. Leavitt noticed that

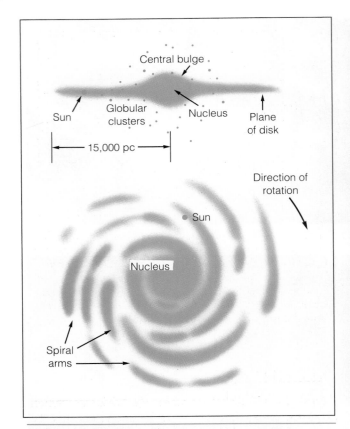

Figure 24.2 The structure of our galaxy. *These sketches illustrate the modern view of the Milky Way.*

Figure 24.3 A spiral galaxy similar to the Milky Way. *This galaxy (NGC 7331) probably resembles our own, as seen from afar. (Palomar Observatory, California Institute of Technology)*

Table 24.2 Types of pulsating variables

Type of Variable	Spectral Type	Period (days)	Absolute Magnitude	Change of Magnitude
δ Cephei (Type I)	F,G supergiants	3–50	−2 to −6	0.1–2.0
W Virginis (Type II Cepheids)	F,G supergiants	5–30	0 to −4	0.1–2.0
RR Lyrae	A, F giants	0.4–1	0.5 to 1.2	0.6–1.3
RV Tauri	G, K giants	30–150	−2 to −3	Up to 3
δ Scuti	F giants	0.1–0.2	2	0.1
β Cephei (β Canis Majoris)	B giants	0.1–0.2	−3 to −5	0.1
Long-period variables (Mira)	M supergiants	80–600	+2 to −3	3–7

Figure 24.4 Henrietta Leavitt. *Leavitt was part of the group of astronomers at Harvard who pioneered the development of spectral classification. In addition, she discovered the period-luminosity relationship for Cepheid variables in the course of studying stars in the Magellanic Clouds. (Harvard College Observatory)*

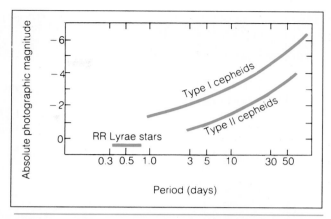

Figure 24.5 The period-luminosity relationship for variable stars. *This diagram shows how the pulsation periods for Cepheid and RR Lyrae variables are related to their absolute magnitudes. The fact that there are two types of Cepheids, with somewhat different relationships, was not recognized at first, and this led to some early confusion about distance scales.*

the average brightnesses of these stars were correlated with the period of pulsation; that is, the longer the time between brightenings, the brighter the star. The Magellanic Clouds are so far away (about 55 kpc) relative to their own dimensions that all the stars are, for practical purposes, at the same distance from us. This meant that the correlation of period with apparent magnitude was actually a correlation with absolute magnitude, suggesting the possibility that the absolute magnitude could be determined from the period of pulsation.

Subsequent analysis established the numerical relationship between the period and absolute magnitude (Fig. 24.5), so that the measurement of distance to a variable star becomes a simple matter of determining the period of pulsation by counting the days between times of maximum brightness, then using the established correlation to determine the absolute magnitude. The distance is then found by comparing the absolute and apparent magnitudes, using the standard technique

discussed in Chapter 19. Because the Cepheid variables are giant stars, they are quite luminous and can be observed at great distance, even beyond the Milky Way. Therefore these stars are very useful in measuring the scale of our galaxy.

It was discovered eventually that there are two distinct types of Cepheid variables, with different period-luminosity relationships (see Fig. 24.5). Before this distinction was recognized, there was some difficulty in establishing the correct relationship, hence distances based on Cepheid variables eventually had to be revised. This is discussed further in a subsequent section of this chapter.

Another type of pulsating star, the **RR Lyrae variables,** also were found to be reliable distance indicators. These stars have periods of only a few hours, and they all have the same absolute magnitude. Any time we identify an RR Lyrae star by measuring its period, we immediately know its absolute magnitude and hence its distance. The RR Lyrae stars are not as luminous as the Cepheids, but they are still bright enough to be observed at large distances. These stars are often found in globular clusters and are therefore sometimes called **cluster variables.** They played a very important role in the early measurements of the size and shape of our galaxy, as we will learn later in this chapter.

ASTRONOMICAL INSIGHT 24.1

The Invisible Milky Way

One of the most revealing facets of modern observational technology for astronomers has been the ability to see how the universe looks in wavelengths of radiation not visible to the eye. We have made frequent mention of this in earlier chapters (for example, the discussion of the Sun in Chapter 17). Our own galaxy, the Milky Way, provides another informative illustration of the power of observations in invisible wavelengths.

The series of galactic maps shown here are portraits of our galaxy in six different wavelengths, ranging from the longest (Map 1) to the shortest (Map 6). Each is organized on the same coordinate system, showing the galaxy as seen edge-on from the Earth. The direction toward the center of the galaxy is in the middle of each map, so that we are scanning along the plane of the disk (that is, in the direction of galactic longitude) when we look right or left of the center, and we are looking out of the plane (along the direction of increasing galactic latitude) when we look up or down from the center line. It is immediately apparent from all six maps that our galaxy has a disklike form, because all six show strong concentration about the central horizontal line.

Map 1 was made by surveying the galaxy at a wavelength of 21 cm, so that emission due to interstellar hydrogen atoms was measured. The relative intensities

MAP 1

MAP 2

MAP 3

Top: A. Stark and C. Heiles; middle: NASA; bottom: Lund Observatory

ASTRONOMICAL INSIGHT 24.1

MAP 4

○ AGN
○ SNR
○ BINARY
○ CLUSTER
○ STELLAR CORONAL
○ UNIDENTIFIED

MAP 5

MAP 6

Top: From McCammon, et al. 1983, Astrophysical Journal 269:107, image provided by W. T. Sanders, University of Wisconsin; middle: Smithsonian Astrophysical Observatory; bottom: French National Research Council, courtesy C. Laurent

of this emission are color-coded, with pale violet representing the maximum intensity. We see from this that the interstellar hydrogen is concentrated in the plane of the galactic disk. Because the 21-cm emission is able to penetrate interstellar gas and dust, this map shows us material out to the farthest reaches of the galaxy. Careful analysis of the Doppler shifts of the 21-cm emission reveals the spiral structure of the disk, something not at all easy to see in a simple edge-on view such as this one.

Map 2, obtained by the *IRAS* satellite, shows the infrared emission characteristic of cold matter, particularly interstellar dust. Again we see that this material is highly concentrated in the galactic plane, even more so than the hydrogen gas represented in the first map. By tracing the distribution of cold dust in this way, we are locating the places in the galaxy where stars are forming, a very important thing to know if we are to understand the evolution of the galaxy. As in the 21-cm map, the infrared sky survey shows us material out to great distances, most of the way across the galaxy.

Map 3, representing the visible appearance of the galaxy, shows in broad scope what we have known of our galaxy through millennia. Again, we see that most of the stars, as well as the dark blotches of interstellar dust clouds, are concentrated in the plane of the disk. In this picture, we are seeing out to a limited distance, particularly in the plane, because of obscuration by gas and dust. In the plane of the Milky Way, the stars we can see lie within a few hundred parsecs from us; out of the plane, where there is less interstellar obscuration, we see stars as far as a few thousand parsecs away.

Map 4, representing low-energy X-rays, is the most limited of all in the distance it reaches. The X-ray emission shown in this map arises in interstellar gas that is very close to the Sun, so we are seeing only the "local" neighborhood. The interstellar gas emits X-rays because it is very hot;

ASTRONOMICAL INSIGHT 24.1

The Invisible Milky Way
(continued)

temperatures of nearly 1 million degrees are implied by the presence of this emission (the origin of this hot gas, in supernovae and stellar winds, is discussed in the next chapter).

Map 5 shows what the galaxy looks like when it is surveyed in high-energy X-ray wavelengths. Now instead of seeing the diffuse, hot gas in our neighborhood, we see point sources of energetic activity. Included are X-ray binary systems (discussed in Chapter 24), X-ray burst sources, and various kinds of explosive stars in our

own galaxy, as well as the active nuclei of other galaxies and quasars (Chapter 30). The stellar sources of X-rays in our galaxy tend to be concentrated in the plane of the disk, where most of the stars are, whereas the extragalactic sources are more readily seen at high galactic latitudes, well above or below the plane of the disk.

Finally, Map 6 shows what the galaxy looks like when observed in gamma rays. These extremely energetic photons are produced primarily in nuclear reactions. The general emission of gamma rays shown in this map is highly confined to the plane of the galactic disk, and is thought to arise from collisions between very rapid atomic nuclei (protons) with molecules of hydrogen in interstellar

clouds. When these collisions take place at sufficiently high energy, nuclear reactions occur, resulting in the production of elementary particles and gamma rays. The rapidly-moving protons (called cosmic rays; see Astronomical Insight 25.2) are probably produced in supernova explosions.

There are many other wavelengths in which it would be useful to map the Milky Way. To do so is helpful for astronomers seeking to understand the distribution of matter and the physical state of the galaxy, and very important in studies of galactic evolution and for comparisons of our galaxy with others. Perhaps in future textbooks additions may be made to the series of maps shown here.

THE STRUCTURE OF THE GALAXY AND THE LOCATION OF THE SUN

Because the solar system is located within the disk of the Milky Way, we have no easy way of getting a clear view of where we are in relation to the rest of the galaxy. All we see is a band of stars across the sky, which tells us that we are in the plane of the disk. It is not so easy to determine where we are within the disk with respect to its edge and center.

The first serious attempts to solve this problem were based on determinations of the density of stars in space around the Sun. The method was to choose random regions of the sky and count the number of stars of each magnitude in these regions. The number of stars increases with decreasing brightness, because faraway stars appear faint, and in a given direction there are many more faraway stars than nearby ones. Therefore we can determine the distribution of stars in space by seeing how rapidly their number increases with decreasing brightness (that is, with increasing distance).

In the first decade of the twentieth century, the Dutch astronomer J. C. Kapteyn employed this method and got a surprising result. The density of stars appeared to fall off in all directions from the Sun, implying that

our solar system is in the densest portion of the Milky Way (Fig. 24.6). The most logical interpretation was that the Sun is at the very center of the galaxy, for it seemed likely that it should be densest at the core. The thought that the Sun was at the center of the known universe made many astronomers uneasy, because in

Figure 24.6 The structure of our galaxy. *This sketch illustrates the view of our galaxy developed by J. C. Kapteyn. The shaded region indicates the rough extent and shape of the galaxy as he perceived it, with the Sun at the center.*

the past, every assumption that the Earth held a special place in the universe had been proved incorrect. It turns out there was indeed a flaw in the work of Kapteyn, although he could hardly have been aware of it. Before the error in Kapteyn's method was found, however, two other studies showed that the Sun was probably not at the center of the galaxy.

One of these studies, carried out by Harlow Shapley (Fig. 24.7), made use of the globular clusters, the spherical star clusters found to lie outside the confines of the galactic disk. As noted earlier, these clusters tend to contain RR Lyrae variables, so it was possible for Shapley to determine their distances and hence their locations with respect to the Sun and the disk of the Milky Way (Fig. 24.8). He found that the globular clusters are arranged in a spherical volume centered on a point several thousand parsecs from the Sun, and he argued that this point must represent the center of the galaxy. It would not make physical sense for the globular clusters to be concentrated around any location other than the center of the entire galactic system.

Shapley's conclusion, first published in 1917, was not widely accepted initially, but other supporting evidence came to the fore in the 1920s. Two scientists, Jan Oort of Holland (already mentioned in connection with the origin of comets; see Chapter 15), and the Swede Bertil Lindblad, carried out careful studies of the motions of stars in the vicinity of the Sun. What they found was that these motions could best be understood if the Sun and the stars around it were assumed to be orbiting a distant point. That is, there are systematic, small-velocity differences between stars, similar to those between runners on a track who are in the inside and outside lanes (Fig. 24.9). It appeared from these studies that the Sun is following a more-or-less circular path about a point several thousand parsecs away, indicating that the center of the galaxy is located at that distant center of rotation. This supported Shapley's view of the galaxy, although there were still uncertainties about how far the Sun was from the center.

Let us now return to the work of Kapteyn, for the next major discovery was the one that revealed the flaw in his conclusions. In 1930, a study performed by the Swiss-American astronomer R. Trumpler showed that the galaxy is filled with an interstellar haze that makes stars appear fainter than they otherwise would. Trumpler made this discovery by examining the apparent brightnesses of star clusters whose distances he knew from their apparent sizes. What he found was that distant clusters appear much fainter than they

Figure 24.7 Harlow Shapley. *Shapley's work on the distances to globular clusters was a key step in the determination of the size of the Milky Way. Shapley also played a prominent role in the discovery of galaxies outside our own (see Chapter 27). Much of his observational work was done before 1920, when Shapley was a staff member at the Mount Wilson Observatory. He later became director of the Harvard College Observatory. (Harvard College Observatory)*

Figure 24.8 Shapley's measurements of globular clusters. *This is one of Shapley's original figures illustrating the distribution of globular clusters in the galaxy. This is a projection down onto the galactic plane. The Sun is at the point where the straight lines intersect, and each circle centered on that point represents an increase in distance of 10,000 parsecs. Note that the distribution of globulars is centered some 10 to 20 kiloparsecs from the Sun. (Estate of H. Shapley, reprinted with permission.)*

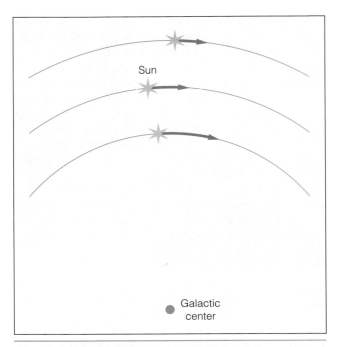

Figure 24.9 Stellar motions near the Sun. *Stars just inside the Sun's orbit move faster than the Sun, whereas those farther out move more slowly. Analysis of the relative speeds (as inferred from measurements of Doppler shifts) and distances of stars like these led to the realization that the Sun and stars near it are orbiting a distant galactic center.*

should, and that the effect increases with increasing distance. This was the first direct evidence that there is a pervasive interstellar medium in the space between the stars; until then, it was known only that interstellar clouds and nebulae existed, as seen on photographs showing dark regions and brightly glowing gas clouds.

The discovery of the general interstellar dust medium in 1930 laid to rest the apparent conflict between the findings of Kapteyn and those of Shapley, Oort, and Lindblad, and the only major question remaining was to find the true size of the galaxy. This required refinement of the period-luminosity relation for variable stars, which, as noted earlier, was confused for a while by the fact that there are two types of Cepheids, which had not been recognized as distinct. Eventually a consensus was reached that the Sun is about 10 kiloparsecs from the center of a disklike galaxy whose total diameter is about 30 kiloparsecs.

More recently, evidence has emerged that in fact the Sun is a little closer to the center, perhaps only 7 or 8

kiloparsecs out. We see that the problem of determining where we are in our galaxy is not necessarily completely solved even today. There are also significant uncertainties in determining the extent of the galaxy beyond the Sun's position, as we will see in later sections of this chapter.

GALACTIC ROTATION AND STELLAR MOTIONS

An important result of the work of Lindblad and Oort was the development of an understanding of the overall motions in the galaxy. Oort's analysis especially was useful in this regard, for he showed not only that the Sun and the stars near it are orbiting the distant galactic center, but also that the rotation of the galaxy is differential, meaning that the galaxy does not spin like a rigid disk, but that each star follows its own orbit at its own speed. Thus the galaxy in the vicinity of the Sun does not act like a rigid disk, but like a fluid, with each star moving as an independent particle.

This is not true of the inner portion of the galaxy. There the entire system rotates more like a rigid object, like a record on a turntable. In the region where rigid-body rotation is the rule, the speeds of individual stars increase with distance from the center, whereas in the outer portion they decrease with distance (Fig. 24.10). In the outer reaches of the galaxy, each star orbits the central portion of the galaxy with little influence from its neighbors, and the stellar orbits are approximately described by Kepler's laws. This is why the orbital speeds decrease with distance from the center in this part of the galaxy. There is an intermediate distance, just inside the Sun's orbit, where a transition between rigid-body and Keplerian orbits occurs, and it is here that stars have the greatest orbital velocities. The Sun, near this peak position, travels at about 250 kilometers per second, taking roughly 250 million years to make one complete circuit about the galaxy. In its 4- to 5-billion-year lifetime, the Sun has thus completed 15 to 20 orbits.

Individual stellar motions do not necessarily follow precise circular orbits about the galactic center. What we have described so far is the overall picture that develops from looking at the composite motions of large numbers of stars. If we look at the individuals instead, we find that each star in the great disk has its own particular motion, which may deviate slightly from the ideal circular orbit. These individual motions are

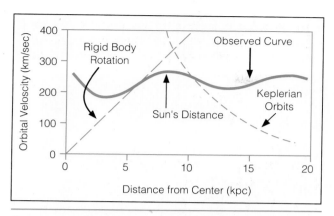

Figure 24.10 The rotation curve for the Milky Way. *This diagram shows how the stellar orbital velocities vary with distance from the center of the galaxy. The fact that the curve does not simply drop off to lower and lower velocities beyond the Sun's orbit, as it would if all the mass of the galaxy were concentrated at its center, indicates that there is a lot of mass in the outer portions of the galaxy. (Data from M. Fich)*

comparable to the paths of cars on a freeway, where the overall direction of motion is uniform, but a bit of lane-changing occurs here and there.

In the galaxy, a star's deviation in motion from a perfect circular orbit is called the **peculiar velocity.** The Sun's motion also deviates from perfect circular motion about the galactic center. This deviation, called the **solar motion,** is estimated from the Sun's average velocity with respect to other stars in its vicinity, along with the assumption that, on average, these nearby stars follow a circular orbit about the galaxy. The Sun has a velocity of about 20 kilometers per second with respect to the standard established by the average motion of its neighbors, that is, with respect to what it would be if it moved precisely in a circular orbit. The solar motion is in a direction about 45 degrees from the galactic center and slightly out of the plane of the disk. Most peculiar velocities of stars near the Sun are comparable, amounting to only minor departures from the overall orbital velocity of about 250 kilometers per second. In Chapter 26 we will discuss the **high-velocity** stars, which greatly deviate from the circular orbits followed by most stars in the Sun's vicinity.

THE MASS OF THE GALAXY

Once the true size of the Milky Way was determined,

it became possible to estimate its total mass. This could be achieved by measuring the star density in the vicinity of the Sun and then assuming that the entire galaxy has about the same average density, but a much simpler and more accurate technique became possible with the discovery that the stars in our region of the galaxy obey Kepler's laws. We note, however, that the mass-determination method about to be described yields only the mass *interior* to the Sun's orbital distance from the galactic center. Mass further out is difficult to measure, as discussed later in this chapter.

Kepler's third law, in the more complete form developed by Newton, expresses a relationship among the period, the size of the orbit, and the sum of the masses of the two objects in orbit about each other:

$$(m_1 + m_2)P^2 = a^3,$$

where m_1 and m_2 are the two masses (in units of the Sun's mass), P is the orbital period (in years), and a is the semimajor axis (in astronomical units). If we consider the Sun to be one of the two objects, and the galaxy itself to be the other, we can use this equation to determine the mass of the galaxy. As we have already seen, the orbital period of the Sun is roughly 250 million years; that is, $P = 2.5 \times 10^8$ years. The orbit is nearly circular, with the galactic nucleus at the center, so the semimajor axis is approximately equal to the orbital radius; $a = 10$ kpc $= 2 \times 10^9$ AU. Now we can solve Kepler's third law for the sum of the masses:

$$m_1 + m_2 = \frac{a^3}{P^2} = 1.3 \times 10^{11} \text{ solar masses.}$$

Since the mass of the Sun (1 solar mass) is inconsequential compared with this total, we can say that the mass of the galaxy itself is about 1.3×10^{11} solar masses. The Sun is slightly above average in terms of mass, so we conclude that the total number of stars in the galaxy must be $3-4 \times 10^{11}$, that is, several hundred billion.

This method, based on Kepler's third law, refers only to the mass inside the orbit of the Sun; the matter that is farther out has no effect on the Sun's orbit. Thus when we estimate the mass of the galaxy by applying Kepler's third law in this way, we neglect all the mass that lies farther out. To overcome this problem, radio measurements of interstellar gas in the outer portions of the galaxy have recently been used to determine the orbital velocity (and therefore the orbital period) of material in the outer reaches of the galaxy. It was found that the velocity does not decrease as rapidly with distance as had previously been thought, and this in turn

ASTRONOMICAL INSIGHT 24.2

The Shape of the Milky Way

It is more difficult in many ways to study our own galaxy than it is to observe other galaxies. The reason, of course, is that we can observe the Milky Way only from an interior position, where we see a cross-sectional view of the disk of the galaxy. We do not get to step back and see the entire system of stars, and worse yet, interstellar gas and dust (discussed in the next chapter) limit the distance to which we can see even in our cross-sectional view. For these reasons, some of the most basic properties of our galaxy are still poorly known.

One of these properties is the shape of the galaxy. Galaxies outside our own have been classified according to their shapes (as discussed in Chapter 27), since it appears that shape has much to tell us about the overall properties and evolution of a galaxy. Thus it is of interest to know the shape of our own galaxy, so that we can compare it with others. Two recent developments have added to our knowledge of the shape of the Milky Way.

One such development was the launch, in late 1989, of the *Cosmic Background Explorer (COBE)*, a satellite designed to measure infrared and short-wavelength radio emission from space (the reasons for doing this are discussed in Chapter 29). *COBE* has obtained data at the infrared wavelengths where the

NASA: courtesy of the COBE Science Working Group

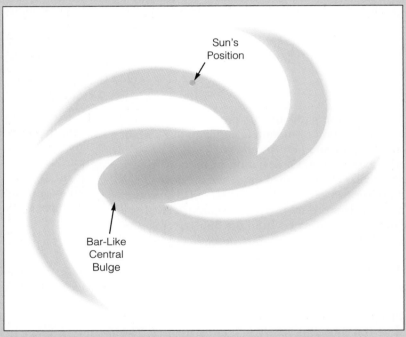

Sun's Position

Bar-Like Central Bulge

L. Blitz, University of Maryland, and D. Spergel, Princeton University

coolest stars emit, and a composite cross-sectional view of the Milky Way in these wavelengths has been assembled (see photo, above). In this view, we see much farther than in similar visible-light images, because the infrared radiation is not as strongly blocked out by interstellar dust. This image provides the best information

yet on the relative size and shape of the galactic bulge, which contains mostly old, dim, red stars, as compared with the thickness of the galactic disk. For the first time, astronomers have a picture of our galaxy that can be compared with similar pictures of other spiral galaxies that are seen edge-on.

The second recent development has to do with the shape of the central bulge of the galaxy, as it would appear if we could see it face-on. Obviously we cannot do that, so we must use indirect methods to infer the shape. One of the most common techniques is to measure the motions of hydrogen gas, observed through its 21-cm emission. As discussed in the text, this emission can be observed over enormous distances, and it allows astronomers to map the spiral structure of the galaxy. It also allows detailed studies of the motions of the gas at different distances from the galactic center.

Several observers have found evidence that the central bulge of the galaxy is elongated; that is, it is apparently not circular, but instead has a barlike structure (see sketch, facing page). Most of the observations leading to this conclusion were made at positions close to the center of the galaxy, but now even gas farther out has been shown to be moving in a way that is consistent with a central bar. The gravitational effects of a bar are different than those of a circular object, and the result is that gas orbiting the center follows ever-more-elliptical orbits, the closer in you observe it.

Many other spiral galaxies have barlike central bulges; about half of all spirals are actually classified as barred spirals. Most are more highly elongated than the bar in our galaxy, but there seem to be many different degrees of elongation. As discussed in Chapter 27, the presence or absence of a bar may be linked to the mass and extent of the galaxy's halo.

---

led to the conclusion that quite a bit of mass lies beyond the Sun's orbit (recall Fig. 24.10). In a later section, we will discuss the possibility that most of the mass of the galaxy lies in the halo and that the actual mass of the galaxy may exceed 10^{12} solar masses.

SPIRAL STRUCTURE AND THE 21-CENTIMETER LINE

So far we have spoken of the galactic disk as though it were a uniform, featureless object, but we know that this is not a completely accurate picture. The Milky Way is a spiral galaxy, and if we could see it face-on, we would see the characteristic pinwheel shape normally found in galaxies of this type (Fig. 24.11).

There is a common misconception about the spiral structure in the Milky Way (and other spiral galaxies)—namely, that there are few stars between the visible spiral arms. In reality, the density of stars between the arms is nearly the same as it is in the arms. The most luminous stars, however, the young, hot O and B stars, tend to be found almost exclusively in the arms. This is because the interstellar gas and dust tends to

Figure 24.11 Spiral structure. *This galaxy has prominent spiral arms, as ours does, because hot, luminous young stars tend to be concentrated in the arms. Even the barlike central region may resemble our galaxy, according to recent reports (see Astronomical Insight 24.2). (© 1977 ROE/AAT Board)*

be concentrated in the arms, so young stars tend to be concentrated there also. Because these are the brightest stars, their presence in the arms makes the spiral structure stand out.

The fact that we live in a spiral galaxy was not easily discovered, again because we are located within it and see only a cross-sectional view. It was not until 1951 that investigations of the distribution of luminous stars revealed traces of spiral structure in the Milky Way, and even that technique was limited to a small portion of the galaxy. The obscuration caused by interstellar material limits our view of even the brightest stars to a local region, about a thousand parsecs from the Sun at most.

A major advance in measuring the structure of the galaxy occurred in the same year, when the 21-centimeter radio emission of interstellar hydrogen atoms was first detected. It had been predicted that hydrogen should emit radiation at this wavelength, and the Americans E. M. Purcell and H. I. Ewen were the first to detect this emission, using a specially built radio telescope.

One great advantage of the hydrogen 21-centimeter emission for measuring galactic structure is its ability to penetrate the interstellar medium to great distances. Whereas the brightest stars can be detected at distances of at most a few hundred parsecs, hydrogen clouds in space can be "seen" by radio telescopes from all the way across the galaxy, at distances of several thousand parsecs. Because hydrogen gas is the principal component of the interstellar medium and tends to be concentrated along the spiral arms, observations of the 21-centimeter radiation can be used to trace the spiral structure throughout the entire Milky Way (Fig. 24.12).

When a radio telescope is pointed in a given direction in the plane of the galaxy, it receives 21-centimeter emission from each segment of spiral arm in that direction. Because of differential rotation, each arm has a distinct velocity from the others that are closer to the center or farther out. Therefore, instead of a single emission peak at a wavelength of exactly 21.1 centimeters, what we see is a cluster of emission lines near this wavelength but separated from each other by the Doppler effect. By combining measurements such as these with Oort's mathematical analysis of differential rotation, it was possible to reconstruct the spiral pattern of the entire galaxy. Radio emission from the carbon monoxide (CO) molecule has also been used to map the distribution and velocities of relatively dense interstellar clouds, providing further information on galactic rotation and spiral structure (Fig. 24.13).

The pattern is much more complex than in some galaxies, where two arms are seen elegantly spiraling out from the nucleus. Instead the Milky Way consists of bits and pieces of a large number of arms, giving it

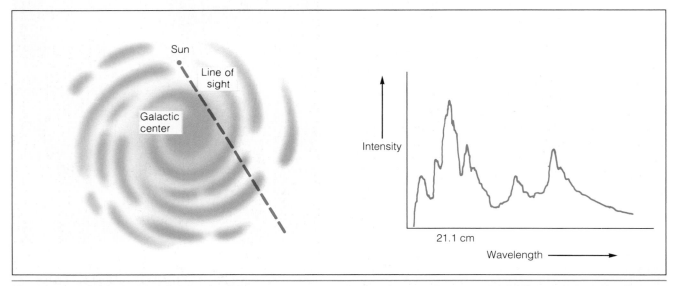

Figure 24.12 21-centimeter observations of spiral arms. *At left is a schematic diagram of the galaxy, showing the direction in which a radio telescope might be pointed to produce a 21-centimeter emission-line profile like the one at right. There are many components of the 21-centimeter line, each corresponding to a distinct spiral arm, and each at a wavelength reflecting the Doppler shift between the velocity of that arm and the Earth's velocity.*

Figure 24.13 Galactic rotation from carbon monoxide observations. *This is a map of CO radial velocities observed with a radio telescope. It shows that interstellar gas on one side of the galactic center (which lies at galactic longitude L = 0°) is approaching us, while material on the other side of the center is receding. Data like these are used to construct the galactic rotation curve such as the one shown here (From Robinson et al., 1984, Astrophysical Journal 282:131; image processing at the Remote Sensing Center, University of Massachusetts)*

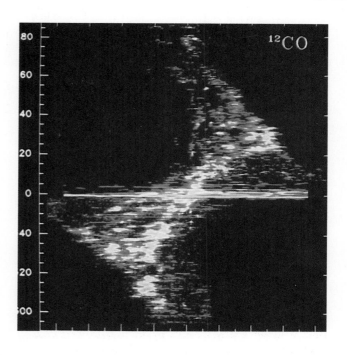

a definite overall spiral form, but not a smoothly coherent one. As we will learn in Chapter 27, the differences between our type of spiral structure and the regular appearance of some other spiral galaxies may reflect different origins for the arms.

THE GALACTIC CENTER: WHERE THE ACTION IS

The center of our galaxy is a mysterious region, forever blocked from our view by the intervening interstellar medium (Fig. 24.14). From Shapley's work on the distribution of the globular clusters, as well as Oort's analysis of stellar motions, it was known by the 1920s that the center of our great pinwheel lies in the direction of the constellation Sagittarius. There we find immense clouds of interstellar matter and a great concentration of stars. It is one of the richest regions of the sky to photograph, although the best observations can be made only from the Southern Hemisphere.

The central portion of the galaxy consists of a more-or-less spherical bulge, populated primarily by relatively cool stars. The absence of hot, young stars implies that the central bulge has had relatively little recent star formation, something we will have to deal with in our discussion of the history of the galaxy (Chapter 26).

Figure 24.14 The galactic center. *The dark regions are dust clouds in nearby spiral arms. Because of obscuration, photographs such as this do not reveal the true galactic center, but only nearby stars and interstellar matter in the plane of the disk and the outer portions of the central bulge of the galaxy. (© 1980 ROE/AAT Board)*

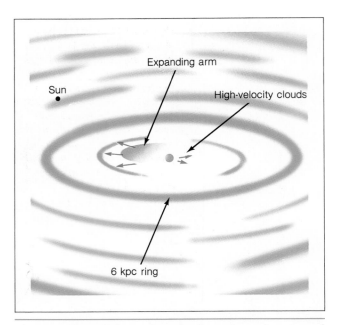

Figure 24.15 Activity at the galactic center. *A variety of evidence for energetic motions associated with the center of our galaxy is indicated here.*

At the center of the bulge lies the nucleus of the galaxy, a small region best located by radio observations. While we cannot see into the central portion of the nucleus in visible light, we can probe the region at longer wavelengths where the interstellar extinction is not such a problem. Radio and, more recently, infrared observations have revealed some interesting features of the galactic core (Fig. 24.15).

The first clue that something unusual was taking place there came from 21-centimeter observations of hydrogen, which revealed a turbulent mixture of clouds moving about at high speed. They also showed that one of the inner spiral arms, about 3 kiloparsecs from the center, is expanding outward at a velocity of more than 100 kilometers per second. This is suggestive of some sort of explosive activity at the center that has forced matter to move outward.

More recent infrared observations of emission lines from hot gas clouds have shown that interstellar clouds near the center of the nucleus are moving quite rapidly in their orbits, which in turn implies (through the use of Kepler's third law) that they are orbiting a very massive central object. Radio observations made with the *Very Large Array* reveal a remarkable arc of gas extending about 200 parsecs above the galactic plane,

apparently held in place by a magnetic field (Fig. 24.16). The origin of this magnetic field is mysterious; it has been suggested that a dynamo created by the enigmatic central object may be responsible.

Space observations show that a very small object precisely at the center of the galaxy is emitting an enormous amount of energy in the form of X-rays. This object is less than 1 parsec in diameter, yet it appears to be responsible for all the violent and energetic activity described. From the observed stellar velocities, the mass of the central object has been estimated at roughly 10^6 solar masses. This is only a small fraction of the total mass of the galaxy, but it is much greater than the mass of any known object within it and must therefore represent some new kind of astronomical entity. One explanation that has been offered by astronomers is that a very massive black hole resides at the core of our galaxy. (Some astronomers argue that a very massive, yet compact, cluster of ordinary stars could account for the observed motions, but the black hole hypothesis seems to be more widely accepted.) The gravitational influence of a black hole would be responsible for the rapid orbital motions of nearby objects, and matter falling in would be compressed and heated, which would account for the X-ray emission. How such an object formed is unknown, but the evidence indicates that at some time during the formation and evolution of the Milky Way, great quantities of matter were compressed into a very small volume at its center. Such an intense buildup of matter may have resulted from frequent collisions among stars that caused vast numbers of them gradually to coalesce into a single, massive object at the center. The prospect that such a beast inhabits the core of our galaxy is a bizarre one, yet it may tie in with some of the strange happenings that have been observed in the nuclei of other galaxies (see Chapter 30).

GLOBULAR CLUSTERS REVISITED

We have not yet said much about the globular clusters, the gigantic spherical conglomerations of stars that orbit the galaxy. A large globular cluster can be an impressive sight when viewed with a small telescope, and a number of them are popular objects for astronomical photography (Fig. 24.17). Over 130 of these clusters have been cataloged; no doubt many more are obscured from our view by the disk of the Milky Way.

A single cluster may contain several hundreds of

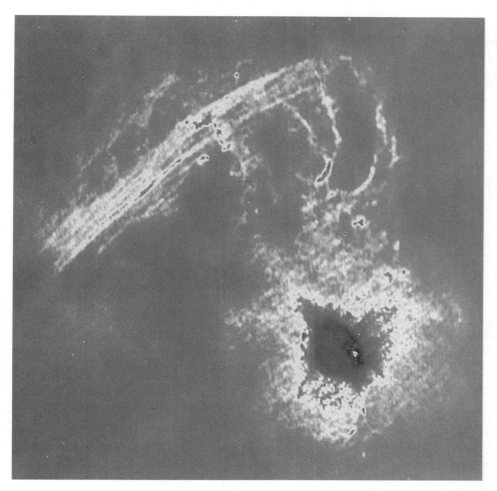

Figure 24.16 A radio image of the galactic center. *This image, made at a wavelength of 6 centimeters (where thermal continuum emission is measured, rather than a spectral feature such as the 21-centimeter line), shows an arc of gas extending some 200 parsecs above the plane of the galactic disk. (F. Yusef-Zadeh, M. Morris, and D. Chance)*

thousands of stars and have a diameter of 10 to 20 parsecs. The mass is usually in the range of several hundred thousand to a few million solar masses. To contain so many stars in a relatively small volume, globular clusters must be very dense compared to the galactic disk. The average distance between stars in a globular cluster is only about 0.1 parsec (recall that the nearest star to the Sun is more than 1 parsec away). If the Earth orbited a star in a globular cluster, the nighttime sky would be a spectacular sight, with hundreds of stars brighter than the first magnitude.

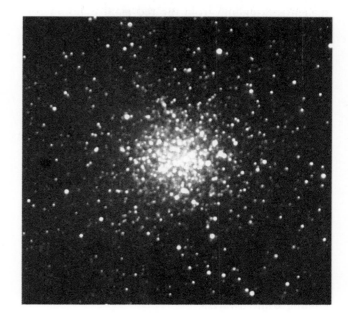

Figure 24.17 A globular cluster. *This cluster, designated Hodge 11, actually resides in one of the Magellanic Clouds, which are small galaxies just outside the Milky Way. (© 1984 ROE/AAT Board)*

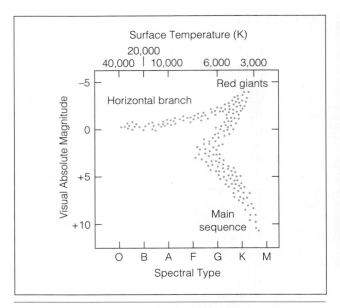

Figure 24.18 The H-R diagram for a typical globular cluster. *The main sequence branches off at a point near the bottom, indicating the great age of the cluster.*

Figure 24.19 A globular cluster X-ray source. *Several globular clusters have been found to contain intense X-ray sources at their centers. This X-ray image obtained with the* Einstein Observatory *clearly shows the central source of a globular cluster. (Harvard-Smithsonian Center for Astrophysics)*

An H-R diagram for a globular cluster (Fig. 24.18) is rather peculiar looking compared to the familiar diagram for stars in the galactic disk. The main sequence is almost nonexistent, having stars only on the extreme lower portion. On the other hand, many red giants and a number of blue stars lie on a horizontal sequence extending from the red giant region across to the left in the diagram. Putting all these facts together points to a very great age for globular clusters. As we saw in Chapter 21, in a cluster H-R diagram the point where the main sequence turns off toward the red giant region indicates the age of the cluster; the lower the turnoff point, the older the cluster. Using this technique to date globular clusters leads typically to age estimates of 14 to 16 billion years, comparable to the accepted age of the galaxy itself. Thus globular clusters are among the oldest objects in the galaxy. As we shall see, their presence in the galactic halo provides important data on the early history of the galaxy.

The sequence of stars extending across the globular cluster H-R diagram from the red giant region to the left is called the **horizontal branch.** It is most likely that these are stars that have completed their red giant stages (as discussed in Chapter 21) and are moving to the left on the diagram, perhaps on their way to becoming white dwarfs.

Some globular clusters have been found to contain X-ray sources in their centers (Fig. 24.19). In a few cases these are X-ray "bursters," described in Chapter 17 as being neutron stars that are slowly accreting new matter on their surfaces, probably in binary systems where the companion star is losing mass. The other globular cluster X-ray sources may be binaries with mass transfer onto a compact stellar remnant, but without the outbursts typical of bursters.

A MASSIVE HALO?

The halo of the Milky Way galaxy has traditionally been envisioned as a very diffuse region, populated by a scattering of dim stars and dominated by the giant globular clusters. There was little or no evidence of any substantial amount of interstellar material, and the halo was assumed not to contain a significant fraction of the galaxy's mass.

Some of these ideas are changing as a result of very recent discoveries, and now it is believed that the halo of the galaxy may be much more extensive and massive than previously thought (Fig. 24.20). We mentioned

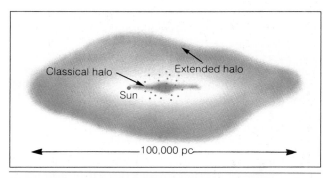

Figure 24.20 The extensive halo of the Milky Way. *This illustrates the shape and scale of the very extended galactic halo, whose presence has been inferred from the shape of the galactic rotation curve (see Fig. 20.10 and from absorption lines formed by interstellar gas.*

in an earlier section of this chapter that radio observations have revealed an unexpectedly high amount of mass in the outer portions of the disk. This indicates that the galaxy is more extensive than previously believed. At the same time, other evidence (to be described in later chapters) led to a general expectation that many galaxies may have large quantities of matter in their halos. Finally, in the early 1980s, ultraviolet observations revealed a large amount of very hot, rarefied interstellar gas in the halo of our galaxy, probably extending to distances of several thousand parsecs above and below the plane of the disk. This gas is very turbulent, with clouds traveling at speeds of several hundred kilometers per second, and is highly ionized, indicating a temperature of at least 100,000 K or more.

At present we cannot say much about the origin or total quantity of material in the halo of our galaxy. It is quite possible that, in addition to the gas now known to be present, there are enough stars, too dim to be detected, to add significantly to the total mass of the galaxy. Many of these stars could be white dwarfs or neutron star remnants of an early generation of massive stars, according to recent suggestions that the Milky Way might have had a large number of massive stars early in its history (discussed in Chapter 26). Some astronomers believe that as much as 90 percent of the galaxy's mass may be in the halo. This means that the total mass of the galaxy is more like 10^{12} solar masses than the 10^{11} or so that we estimated from applying Kepler's third law to stars in the Sun's vicinity.

The *Hubble Space Telescope* should be able to settle this question by making more sensitive measurements of both the gas and the stars in the halo.

PERSPECTIVE

We have discussed the anatomy of the galaxy, particularly its overall structure and motions. The solar system is located nearly 10 kpc from the center of a flattened, rotating disk containing a few hundred billion stars. With the help of radio and infrared observations, astronomers have unraveled the structure of the disk, which consists of spiral arms (delineated by bright, hot stars and interstellar gas), and a central nucleus that contains relatively dim, red stars along with a mysterious, energetic gremlin that stirs up the core region. Many questions remain about the structure and extent of our galaxy; some will be answered by the *Hubble Space Telescope*.

We have yet to discuss the workings of the galaxy; we have seen what it is like but have said little about why. Before we endeavor to do so, we will discuss the interstellar medium, a vital part of the galactic ecology.

SUMMARY

1. The Milky Way is a spiral galaxy consisting of a disk with a central nucleus and a spherical halo, where the globular clusters reside. The disk is about 30 kpc in diameter.

2. Distances within the Milky Way can be determined from the measurement of variable-star periods and application of the period-luminosity relation.

3. The true size of the Milky Way and the Sun's location within it were difficult to determine because the view from our location within the disk is obscured by interstellar matter.

4. Star counts were once interpreted as indicating that the Sun is in the densest part of the Milky Way, but measurements of the distribution of globular clusters and analysis of stellar motions show that the Sun is 7 or 8 kpc from the center. The discrepancy was resolved when it was discovered that interstellar extinction affects the star counts.

5. The inner part of the galactic disk rotates rigidly, whereas the outer part rotates differentially, with each star following its own individual orbit, approximately described by Kepler's laws.

6. Individual stellar orbits in the Sun's vicinity are generally in the plane of the disk, and are nearly circular. Deviations from perfect circular motion by stars are called peculiar velocities, and in the case of the Sun, the solar motion.

7. The spiral structure of our galaxy is most easily and directly measured through radio observations of the 21-cm line of hydrogen atoms in space. The spiral pattern is complex, with many segments of spiral arms.

8. The mass of the galaxy, determined by the application of Kepler's third law to the Sun's orbit, is roughly 10^{11} solar masses. This technique does not take into account any mass that resides in the halo.

9. A variety of evidence indicates there is chaotic, energetic activity associated with the central core of our galaxy. The data show a compact, massive object existing there, and the best explanation is that it is a massive black hole.

10. The globular clusters that inhabit the halo of the galaxy are very old and therefore provide information about the early history of the galaxy.

11. There may be large quantities of mass in the galactic halo, in the form of interstellar gas or dim, cool stars, or both. It appears possible that as much as 90 percent of the mass of the galaxy is in the halo.

REVIEW QUESTIONS

Thought Questions

1. Explain how the use of pulsating variable stars to find distances is similar to the spectroscopic parallax method described in Chapter 19.

2. What would Kapteyn have found if there had not been any interstellar dust obscuration to modify his star-count method for determining the structure of the galaxy?

3. Compare the reasoning of Shapley in deciding that the globular clusters must be concentrated around the galactic center with that of Aristarchus who, more than 2,000 years earlier, had deduced that the Sun is at the center of the solar system.

4. Summarize the nature of stellar orbits about the galaxy. Why are the highest orbital speeds attained by stars at the position where rigid-body rotation gives way to fluid motion (Keplerian orbits)?

5. Explain why the 21-cm line of atomic hydrogen is a powerful tool for mapping the structure of the galaxy.

6. Suppose the mass of the galaxy were determined from the application of Kepler's third law to the orbit of a star 20 kpc from the galactic center. How might the result compare with that found from analysis of the Sun's orbit?

7. Summarize the evidence that a massive black hole might exist at the core of the galaxy.

8. How is it known that globular clusters are very old, relative to the rest of the galaxy?

9. Given the great age of the globular clusters, do you think their chemical composition might differ from that of younger stars?

10. What is the evidence that our galaxy might have a very massive halo, containing most of the galaxy's mass? Can you think of any observational methods for learning more about this halo?

Problems

1. Calculate the distance to a star whose parallax is $p = 0.005$ (nearly the smallest parallax that can be measured), using the method described in Chapter 18. How does your answer compare with the overall diameter of the galaxy?

2. Suppose a Cepheid variable is found to have a period of 10 days, and an average apparent magnitude of $m = +16$. From the period-luminosity relation, the average absolute magnitude for such a star is $M = -4$. What is the distance to this star? Compare your answer with the maximum distance that can be measured with the stellar parallax technique.

3. Suppose that interstellar obscuration makes a star appear 5 magnitudes fainter than it would otherwise. If the star has apparent magnitude $+17$, what would its apparent magnitude be if it were not obscured by dust? If the star's absolute magnitude is $+2$, how far away is it? Compare your answer with what you would find if you did not correct the apparent magnitude for the effects of dust.

4. Confirm the orbital period of the Sun about the galaxy by calculating how long it would take to travel the circumference of a circle whose radius is 10,000 pc at a speed of 250 km/sec. (You will have to convert to consistent units.)

5. If the rest wavelength of the hydrogen emission line is 21.1 cm, what would be the observed wavelength from an interstellar cloud whose velocity is 150 km/sec away from the Earth?

6. Suppose that in a distant spiral galaxy, the orbital speed at a distance of 5 kpc from the center is found to be 150 km/sec. What is the orbital period for stars at that point in the galaxy? What is the mass of the galaxy interior to that point?

7. If there is an object at the galactic center with a mass equal to 1 million solar masses, what is the orbital period for a star 10 parsecs from this object? If this star's orbit is circular, what is its orbital speed? Comment on the relationship of your result to observations suggesting the presence of a massive black hole at the galactic core.

8. To appreciate what the nighttime sky might be like if the Earth orbited a globular cluster star, calculate the apparent magnitudes of the first five stars in the list of the brightest stars (Appendix 8) if each star were only 0.1 pc away (the average distance between stars in a globular cluster).

ADDITIONAL READINGS

Bok, B. J., and P. Bok. 1981. *The Milky Way.* Cambridge, Mass.: Harvard University Press.

Bok, B. J. 1984. A bigger and better Milky Way. *Astronomy* 12(1):6.

Clark, Gail O. 1985. Ancients of the universe: Globular clusters. *Astronomy* 13(5):6.

de Boer, K. S., and B. D. Savage. 1982. The coronas of galaxies. *Scientific American* 247(2):54.

Geballe, T. R. 1979. The central parsec of the galaxy. *Scientific American,* 241(1):52.

Gingerich, O., and B. Welther. 1985. Harlow Shapley and the Cepheids. *Sky and Telescope* 70(6):540.

Hodge, P. 1986. *Galaxies.* Cambridge, Mass.: Harvard University Press.

———. 1987. The local group: Our galactic neighborhood. *Mercury* 6(1):2.

King, I. R. 1985. Globular clusters. *Scientific American* 252(6):78.

Mulholland, D. 1985. The beast at the center of the galaxy. *Science 85* 6(7):50.

Sanders, R. H., and G. T. Winkon. 1974. The center of the galaxy. *Scientific American* 230(4):66.

Townes, C. H., and R. Genzel. 1990. What is happening in the center of our galaxy? *Scientific American* 262(4):46.

Tucker, W., and K. Tucker. 1989. Dark matter in our galaxy. *Mercury* 18(1):2 (Part I): *Mercury* 18(2):51 (Part II).

Weaver, H. 1975 and 1976. Steps towards understanding the large-scale structure of the Milky Way. *Mercury* 4(5):18 (Part I); 5(6):19 (Part II).

MATHEMATICAL METHODS

The 21-cm Emission Line of Hydrogen

In this chapter, we have discussed the profound effect that radio observations of the 21-cm emission line of hydrogen have had on our understanding of the structure of the galaxy we live in. It is useful to gain a better understanding of how hydrogen atoms create emission at this wavelength.

In Chapter 6, where we discussed atomic structure and the properties of emission and absorption lines, we stressed the fact that the electrons orbiting an atomic nucleus can exist only in specific orbits or energy levels. When an electron moves from one energy level to another, it absorbs or emits a quantity of energy equal to the energy difference between the two levels, by absorbing or emitting a photon whose energy (i.e., wavelength) corresponds to this energy difference. In Chapter 6, we discussed the major energy levels in an atom, whose differences usually correspond to photon wavelengths in the ultraviolet, visible, or infrared portions of the electromagnetic spectrum.

Some atoms, however, have electron energy levels that are very close together, so that the energy dif-ference between levels corresponds to photons with very long wavelengths. Hydrogen is such an atom; the lowest principal energy level of hydrogen is split into two very closely spaced levels. The reason for the splitting is that both the proton and the electron, which comprise the entire structure of a hydrogen atom, spin. The rotation axes of the two particles can be aligned in the same direction or in opposite directions, and the total energy of the atom is slightly different in each case (the energy is a bit lower when the spin axes are in opposite directions).

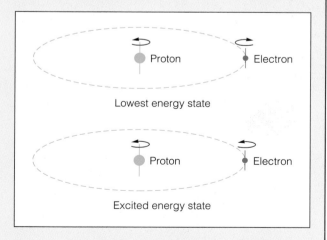

Lowest energy state

Excited energy state

(continued on next page)

MATHEMATICAL METHODS

The 21-cm Emission Line of Hydrogen (continued)

The energy difference between the two states is about 9.42×10^{-18} erg. Using the formula from Chapter 6 equating photon energy to wavelength, we find that

$$\lambda = \frac{hc}{E},$$

where λ is the wavelength (in centimeters), h is the Planck constant (whose value is 6.626×10^{-27} erg-sec), c is the speed of light (3×10^{10} cm/sec), and E is the energy level difference (in erg). Substituting 9.42×10^{-18} erg for E, we find

$$\lambda = \frac{(6.626 \times 10^{-17} \text{ erg-sec}) (3 \times 10^{10} \text{ cm/sec})}{(9.42 \times 10^{18} \text{ erg})}$$

$$= 21.1 \text{ cm}.$$

This is the wavelength a photon has that is emitted or absorbed when a hydrogen atom changes between these two energy states (such changes are called "spin-flip" transitions, because the electron reverses the direction of its spin axis).

Ordinarily a hydrogen atom will be in its lowest possible energy state—that is, in the state where the proton and electron spins are in opposite directions. But collisions between atoms can cause the electron to gain the tiny amount of energy needed to jump to the upper state, with parallel proton and electron spin axes. The excited atom will remain in this state for a very long time (about 100 million years, on average) but eventually it will spontaneously jump back to the lower state, emitting a 21-cm photon in the process. Only a tiny fraction of all hydrogen atoms in interstellar space are in the excited state at any one moment, but hydrogen is so abundant that the galaxy is permeated by 21-cm photons all the time anyway.

CHAPTER 25

The Interstellar Medium

A Hubble Space Telescope *image of a portion of the Orion nebula, showing details as small as the solar system (NASA/JPL: Space Telescope Science Institute)*

The very rarefied matter between the stars, diffuse though it is, constitutes one of the most important elements in the galactic environment. Stars form from it, and as they age and die, they return their substance to it. As we have seen, the interstellar haze limits our observations of the galactic disk to the nearest few hundred parsecs, allowing us to view only a very small fraction of the entire galactic volume.

Despite its obvious importance, the interstellar medium, by ordinary earthly standards, is so tenuous that it verges on nonexistence. Even in the densest interstellar clouds, the density of particles is less than one-trillionth of the Earth's sea-level atmospheric density. Only the most sophisticated artificial vacuum pumps can even approach the natural vacuum of space.

The medium is so pervasive, however, and the volume of space so large, that a significant fraction of the total mass of the galactic disk is contained in this form. There are two distinct types of interstellar material, which are thoroughly mixed together throughout space: tiny solid particles called interstellar dust grains; and gas particles, which may be atoms, ions, or molecules, depending on the temperature and density. The grains are relatively few in number, and constitute only about 1 percent of the total mass in the interstellar medium.

THE INTERSTELLAR DUST

The fact that there are interstellar dust grains was recognized long ago. Photographs of the Milky Way show extensive, irregular dark regions (Figs. 25.1–25.4) where dense, dust-bearing clouds completely hide the stars behind them. Knowledge of the presence of these clouds is ancient, but it was not recognized until Trumpler's work in 1930 that there is a general distribution of dust grains throughout the spaces in between the obvious clouds. Today the dust is studied by a number of techniques, and a great deal has been learned about its nature. The *IRAS* satellite, with its infrared sensors highly sensitive for detecting thermal emission from cold objects, found numerous thin, wispy dust clouds in space. It is not yet clear how these interstellar "cirrus" clouds, as they have been called, are related to other types of interstellar clouds (described later in this chapter).

Even though the dust grains constitute only a tiny fraction of the total mass in the interstellar medium, they have a very important influence on the starlight that passes through space. There are two basic effects: (1) the obscuration of light from distant stars, com-

Figure 25.1 Dust clouds along the Milky Way. *This is a portion of the Milky Way in Sagittarius, showing multitudes of stars as well as extensive patches of interstellar clouds. (© 1980 Anglo-Australian Telescope Board)*

Figure 25.2 A gaseous nebula. *Clouds of gas and tiny dust particles such as this are the birthplaces of stars. (© 1981 Anglo-Australian Telescope Board)*

Figure 25.3 An extensive region of nebulosity.   *This is a well-known region of dark clouds, including one known as the Horsehead nebula. Contained within these clouds are numerous infrared-emitting objects thought to be young stars.* (© 1980 Royal Observatory, Edinburgh)

monly referred to as **interstellar extinction;** and (2) the polarization of starlight. Both effects provide important information about the nature of the dust grains themselves.

The extinction of starlight by interstellar dust particles is the result of a combination of absorption and **scattering,** a process in which the light essentially bounces off of the dust grains. When a photon is scattered, its wavelength remains fixed, but its direction is altered. The scattering process is more effective for short wavelengths of light than for longer wavelengths, so a dis-

tant star appears redder than it otherwise would, because the red light that it emits reaches us more easily than the blue (Fig. 25.5). For a similar reason, the Sun appears red when it is near the horizon so that its light must traverse a long pathlength through the Earth's atmosphere; the fine particles in the air scatter the blue light but allow the red to come through relatively unhindered. Because the extinction of starlight is so much more severe at short wavelengths, ultraviolet telescopes cannot probe distances as great as visible-light telescopes can.

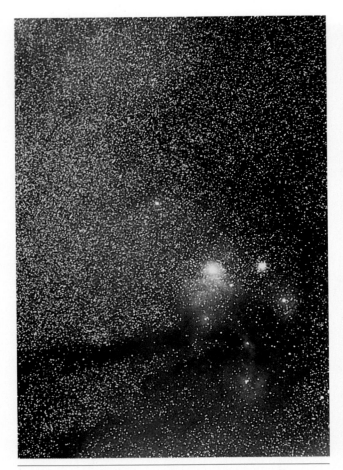

Figure 25.4 A dark cloud. *The dark streaks in the lower portion of this photograph are regions where dark clouds hide newly forming stars. This is part of a well-studied cloud complex near the bright red supergiant star Antares. (Fr. R. E. Royer)*

It is possible to determine how much dust lies in the direction of a given star by measuring how much redder the star appears than it should. Recall the color index $B-V$, the difference between the blue and visual magnitudes of a star (see Chapter 18). The redder a star is, the greater the value of $B-V$. (Do not forget: If a star is red, it is brighter in visual than in blue light, so the V magnitude is smaller than the B magnitude). The effect of interstellar extinction is to make the value of $B-V$ greater than it would otherwise be.

To estimate the amount of dust in the line of sight to a distant star, therefore, simply requires a comparison of the star's observed $B-V$ color index with the value it would have without extinction. The latter quantity is determined from the spectral type of the star. Remember that $B-V$ and spectral class both measure the same stellar property, namely the temperature; therefore all stars of a given spectral type have the same intrinsic value of $B-V$.

Theoretical considerations allow astronomers to determine the average size of the grains, from the variation of extinction with wavelength (known as the extinction curve; Fig. 25.6). The fact that the grains scatter blue light more effectively than red, and the way in which the scattering efficiency varies as a function of wavelength leads to the conclusion that there are two distinct types of grains, both very small: (1) those with an average diameter of about 5×10^{-5} cm (or about 5,000 Å, comparable in size to the wavelength of visible light); and (2) a population of even smaller grains, whose diameters are only about 10^{-6} cm (or about 100 Å), which create extremely strong ultraviolet extinction.

The polarization of starlight tells us something about the shape of the grains. In Chapter 6 we learned that when light is polarized, the electric and magnetic fields of the waves have a preferred orientation. The interstellar dust creates this effect by selectively absorbing light of one orientation, so that what gets through to us is the light with the orientation perpendicular to that (Fig. 25.7). The dust has the same effect as the polarizing lenses commonly used in sunglasses.

The only way the interstellar dust grains can produce such an effect is if they are not spherical, but elongated (Fig. 25.8). They must also be aligned in some organized fashion, so that a majority of the grains between the Earth and a given star are parallel to each other. Their alignment is probably caused by an interstellar magnetic field, which has a constant orientation over large regions of space. Thus the fact that the grains produce polarization tells us something not only about their properties, but also about the galaxy—namely, that it has an overall magnetic field. It appears that the field tends to parallel the spiral arms, at least in the vicinity of the Sun.

Little has been learned about the chemical composition of the grains. Theoretical work indicates that a variety of common substances, including silicates, oxides, graphite, and iron-bearing minerals, could cause the observed extinction. There is a pronounced peak of extinction near the ultraviolet wavelength of 2,200 Å, which is best explained as being a result of absorption by a form of carbon, so it is probable that some of the grains contain carbon.

Figure 25.5 Scattering of light by interstellar grains.   *The grains tend to absorb or deflect blue light more efficiently than red, so a star seen through interstellar material appears red.*

The origin of the interstellar grains is not well understood; there seem to be several possible processes, all of which may contribute to their existence. Tiny solid particles can form from a gas through condensation, if the proper combination of pressure and temperature is reached. As we learned in Chapter 16, this probably occurred early in the formation of the solar system, and therefore we expect that some interstellar grains are born in the vicinity of newly formed stars, then expelled into space, perhaps by a stellar wind. A more important source of grains appears to be the extended outer atmospheres of red giant and supergiant stars, where the proper conditions apparently exist, and radiation pressure from these very luminous stars expels the grains into interstellar space. In addition, infrared observations show that grains can form in the expanding material around novae and in planetary nebulae. The formation of interstellar grains is a by-product of the formation and evolution of stars, and is therefore an important part of the large-scale recycling of matter between stellar and interstellar forms.

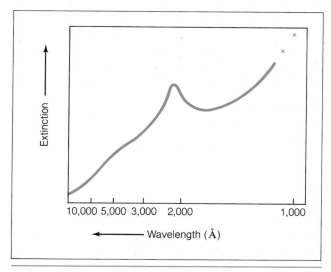

Figure 25.6 Interstellar extinction.   *The extinction of starlight by interstellar dust is greatest at short wavelengths, especially in the ultraviolet, and decreases toward longer wavelengths, being very small in the infrared and virtually nonexistent in radio wavelengths.*

OBSERVATION OF INTERSTELLAR GAS

Interstellar gas can be detected by several different techniques. As in the case of dust, one manifestation

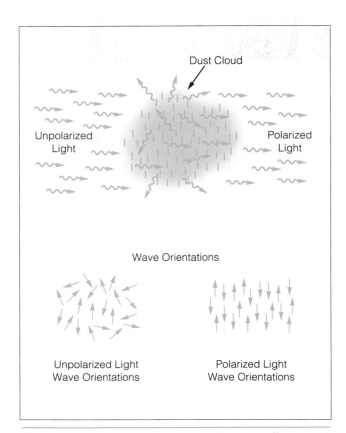

Figure 25.7 Polarization of starlight by interstellar grains.   *Starlight passing through the interstellar medium becomes weakly polarized, indicating that the dust grains in space are elongated in shape and tend to be aligned, probably by the galactic magnetic field.*

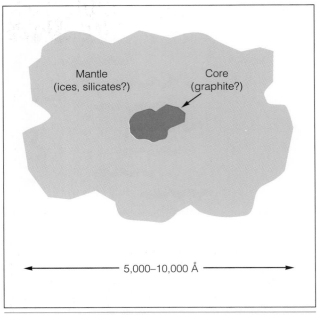

Figure 25.8 An interstellar grain?   *No direct information is yet available on the composition and the detailed shapes of interstellar grains. This sketch illustrates a popular model, but for a sobering look at what the grain shapes might really be like, see the photo of an interplanetary grain in Chapter 15 (Fig. 15.28).*

is in the form of rather obvious clouds that can be seen on photographs of the sky. Rather than being dark clouds, however, the visible gas regions are bright concentrations of material that is hot enough to glow. Interstellar gas is also observed in the radio portion of the spectrum, as we learned in the previous chapter when discussing the 21-cm line of hydrogen. As we will see, there are other types of radio emission lines from the interstellar gas as well.

Perhaps the best method for measuring the properties of the interstellar gas, however, is to observe the absorption lines this gas forms in the spectra of stars (Fig. 25.9). As light from a distant star travels through space on its way to the Earth, it encounters atoms, ions, and molecules along the way, each of which can absorb light at specific wavelengths. The result is that the spectrum of the star has extra absorption lines in

it, in addition to the ones created in its own atmosphere. It is possible to distinguish the interstellar from the stellar absorption lines, by several criteria. The interstellar lines are usually much narrower than those formed in the star's atmosphere; they may occur at a different Doppler shift; and they may represent a different degree of ionization than occurs in the gas in the stellar atmosphere. The first interstellar absorption lines were discovered accidentally just after the turn of the century, but it was not until the 1920s that their interstellar origin was firmly established.

Because of the low density of the interstellar gas, collisions between particles are very rare, and as a result, nearly all the ions and atoms have their electrons in the lowest possible energy level. For most elements, the absorption lines that can be formed when the electrons are in this lowest state lie at ultraviolet wavelengths; only a few relatively rare elements, such as calcium and sodium, have interstellar absorption lines in visible wavelengths. Therefore it was not until the 1970s, when ultraviolet telescopes were launched on satellites, that the most common species could be di-

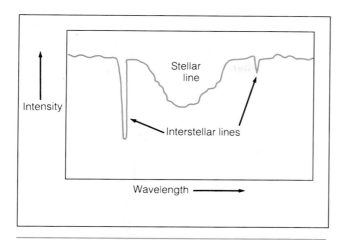

Figure 25.9 Interstellar absorption lines. *Gas atoms and ions in space absorb photons from distant stars, creating absorption lines in the spectra of the stars. Here we see some interstellar absorption lines in ultraviolet wavelengths, where most of these lines are found. The interstellar lines can be distinguished from the star's own spectral lines because they are narrower, usually have a different Doppler shift (because the star and the cloud producing the line move at different velocities), and represent different states of ionization.*

rectly observed in space (Fig. 25.10).

The chemical composition of the interstellar medium is nearly identical to that of the Sun and other stars, which is no surprise, since the stars form from this material. As we will see in the next section, however, some elements tend to be in the form of grains rather than gas, so the interstellar gas itself has a somewhat different distribution of the elements than does the Sun.

Clouds and Nebulae

The interstellar material is not uniformly distributed, but tends to be patchy, with most of the matter contained in relatively dense concentrations called **interstellar clouds.** In the line of sight to a faraway star, there are likely to be several clouds, their presence being revealed to us by the extra redness they impart to the color of the star and by the interstellar absorption lines they form in its spectrum. These most common of interstellar clouds are not usually hot enough to glow nor thick enough to show up as dark patches in the sky like the dense clouds found in some regions, and for this reason they are called **diffuse clouds.**

In comparing the properties of different types of

interstellar clouds (Table 25.1), we find it convenient to think in terms of the density: the number of particles per cubic centimeter. At sea level on the Earth, the atmospheric density is about 2×10^{19} molecules per cubic centimeter. In a typical diffuse interstellar cloud, the number is more like 10 to 1000/cm³, a factor of 10^{16} to 10^{18} less! The densest clouds, the dark ones that allow no light to pass through, have densities of about 10^6 particles/cm³ at most, still a factor of 10^{13} less than the density of the Earth's atmosphere. The temperature of the gas in a diffuse cloud is typically in the range of 50 to 150 K and is as low as 20 to 50 K in dark clouds.

Diffuse clouds are thought to be ordinarily less than 1 parsec thick, although in many cases they may stretch for many parsecs in one dimension, as though they were thin sheets of matter, rather than being more or less spherical. As mentioned earlier in this chapter, very thin, wispy dust clouds, which may be identified with diffuse clouds, have been detected at infrared wavelengths with the *IRAS* satellite (Fig. 25.11). The total mass of a diffuse cloud is difficult to assess because of its unknown extent but is often comparable to the mass of the Sun. Thus we conclude that these clouds are not the type from which massive stars or star clusters are formed.

The composition of diffuse clouds is dominated by hydrogen, with other elements present in much smaller quantities. The hydrogen is primarily in atomic form, except in the central portions of the clouds, where the density may be sufficient to allow the formation of

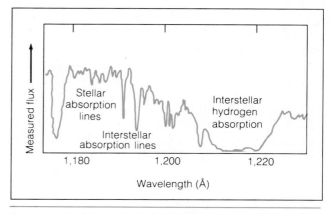

Figure 25.10 Ultraviolet interstellar absorption lines. *This is a portion of a spectrum obtained with the* Copernicus *satellite. Several stellar lines and a few interstellar lines are seen here. The interstellar hydrogen feature is so wide because of the great abundance of this element in space.*

Table 25.1 Interstellar medium conditions

Component	Temperature (K)	Density (per cm³)	State of Gas	How Observed
Coronal gas (intercloud)	10^5-10^6	10^{-4}	Highly ionized	X-ray emission, UV absorption lines
Warm intercloud gas	1,000	0.01	Partially ionized	21-cm emission, UV absorption lines
Diffuse clouds	50–150	1–1000	Hydrogen in atomic form, others ionized	Visible, UV absorption lines, 21-cm emission
Dark clouds	20–50	10^3-10^5	Molecular	Radio emission lines, IR emission and absorption lines

Figure 25.11 Infrared cirrus. *This image, obtained by the IRAS satellite, reveals extensive regions of diffuse, patchy clouds in interstellar space (shown here as orange against the blue background). The faint, wispy clouds seen in the upper portion here have been dubbed "infrared cirrus" because of their resemblance to cirrus clouds in the Earth's atmosphere. (NASA)*

Figure 25.12 An emission nebula. *The red color of this cloud of hot gas is due to emission by hydrogen atoms. This is a portion of the Rosette nebula. (© 1984 Anglo-Australian Telescope Board)*

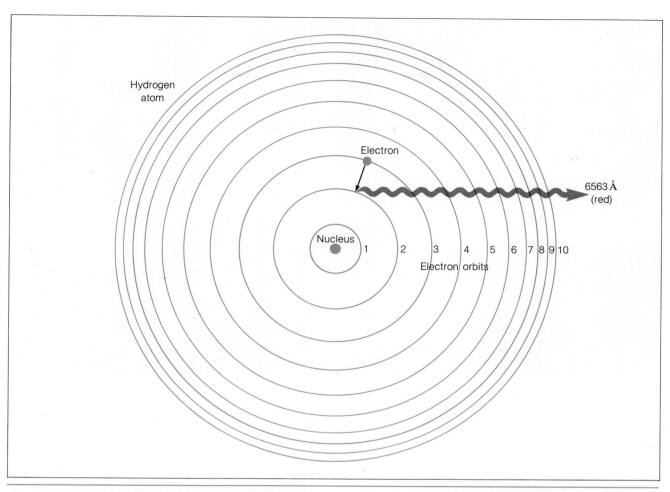

Figure 25.13 Hydrogen lines from an emission nebula. *In heated interstellar gas, electrons and protons combine to form hydrogen atoms, usually with the electron initially in an excited state. It then drops down to the lowest state, emitting photons at each step. The strongest emission in visible wavelengths corresponds to the jump from level 3 to level 2, and has a wavelength of 6,563 Å. This is why emission nebulae appear red.*

molecular hydrogen (H_2). The abundances of some of the elements are lower than the accepted "cosmic" values. It is as though some of the elements, among them iron, calcium, titanium, and other metallic species, are missing from the interstellar gas. The likely explanation is that these materials are preferentially in the solid dust grains, which gives us some indirect indication of what the grains are made of.

As we have pointed out, diffuse clouds are essentially transparent, but there are conditions under which they show up in photographs. If a hot star is embedded within a cloud, its radiation ionizes and heats the gas to the point where it glows. Such a cloud is known as an **emission nebula** (Fig. 25.12), nebula being a general word for any interstellar cloud that is dense or bright

enough to show up in photographs. In the case of a hot star embedded within a diffuse cloud, the emission occurs when atoms are ionized and then combine again with electrons to form new atoms. The electron usually starts out in a high-energy state, and then quickly drops down to the lowest level, emitting a photon of light each time it makes a downward jump. Because most of the light from an emission nebula is from regions where hydrogen is ionized, these nebulae are called **HII regions,** the H standing for hydrogen, and the Roman numeral II indicating the ionized state (astronomers generally use this type of notation to indicate the degree of ionization of an element). The strongest line of hydrogen has a wavelength of 6,563 Å in the red portion of the spectrum (Fig. 25.13), so HII regions stand out

Figure 25.14 A dark cloud. *The dark, patchy region here is a dense, cold interstellar cloud, the kind of region where stars can form. The interior temperature may be as low as 20 K, and the density roughly 10^5 particles per cubic centimeter, a high vacuum by earthly standards, but quite dense for the interstellar medium. Some reflection nebulosity is present in this region also. (© 1984 Anglo-Australian Telescope Board)*

especially well in photographs taken with red-sensitive film. A particularly prominent HII region surrounds the stars of the Trapezium, in the sword of Orion, and is commonly called the Orion Nebula (see Fig. 25.2).

If a diffuse cloud happens to lie just behind or to one side of a hot star, it may be visible to us as a **reflection nebula** (Fig. 25.14). In this case it is the dust, rather than the gas, that causes the cloud to glow. What we see is the light that has been scattered in our direction by dust grains (Fig. 25.15). This light is always

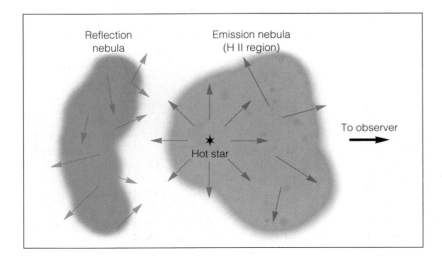

Figure 25.15 Bright nebulae. *This shows how reflection and emission nebulae are formed. Light from a hot star is reflected off of a background cloud containing dust. The cloud has a bluish color, because short-wavelength photons are scattered most effectively. The stellar radiation also ionizes gas that is very close to the star, creating a region around it that glows at the wavelengths of certain emission lines. This type of emission nebula is also known as an HII region.*

Figure 25.16 The Pleiades. *This relatively young galactic cluster is a prominent object in the Northern Hemisphere during the late fall and winter. (© 1980 Anglo-Australian Telescope Board)*

blue, for the same reason the daytime sky is blue: The tiny particles scatter blue light more effectively than longer wavelengths. The best-known reflection nebula is the one in the young star cluster called the Pleiades, and it lends photographs of this group of stars an ethereal quality (Fig. 25.16).

DARK CLOUDS AND THE MOLECULAR ZOO

In many ways the most fascinating of the interstellar objects are the dark clouds, whose densities and dimensions are sufficient to block out all the light of stars behind them. Dark-cloud regions are highly concentrated in the plane of the Milky Way and have a variety of amorphous sizes and shapes (see Figs. 25.1–25.4). Often bright HII regions are located within and among them, for the regions where dark clouds are common are also the sites of star formation, and therefore many young, hot stars are mixed in.

The masses of dark clouds are comparable to, or, in the cases of large cloud complexes, far in excess of a single star mass, ranging up to hundreds of thousands of solar masses in some instances. Thus it is quite possible for massive stars and star clusters to form from material in this type of cloud, and it is in dark-cloud regions that the most active star formation occurs.

As we have already noted, the temperature in the interior of a dark cloud is extremely low, perhaps only 10 to 20 K, and the density in the range of 10^3 to 10^6 particles/cm^3. Under these conditions, atoms can combine into molecular form, and a wide variety of species have been detected. The discovery of the molecular nature of the gas in dark clouds was relatively recent, the first species (the simple molecule OH) having been detected in 1963. The reason these observations were not made sooner is that sophisticated new radio astronomy techniques had to be developed.

Because these clouds are so dense, it is not possible to observe them in the same manner as the diffuse clouds. No stars can be seen through the dark clouds, so there is no way to measure absorption lines formed there. Molecules have special properties, however, that produce radio emission lines, and these lines can be observed.

A molecule is a combination of atoms in a single particle, held together by chemical bonds. These bonds may be viewed as a merging of energy states, so that some of the electrons orbit the entire combinations of atoms, acting as the glue that holds them together. The electrons still have fixed energy levels, and there is more complex structure as well. The atoms act as though they are held together with springs, in the sense that they can vibrate, and different frequencies of vibration correspond to different energy states. A molecule can also rotate on its axis, and different rotation speeds also correspond to different energy levels. The vibration and rotation states are close together, so the wavelengths of light corresponding to transitions between them are long. In most cases, the rotational levels are so closely spaced that the differences between them correspond to radio wavelengths. Thus if a molecule jumps from one rotational state to a lower one (that is, if it slows its rotation), it emits a photon at a radio wavelength. Each kind of molecule has its own characteristic set of rotational energy levels, so each has its own particular set of radio emission lines. Molecules have the same sort of fingerprints in the radio spectrum that atoms and ions do in visible and ultraviolet light. (Molecules, having such complex structure, also have spectral lines at visible and ultraviolet wavelengths, but these cannot be easily observed in dark clouds, because it is usually not possible to observe stars lying behind such clouds in order to analyze absorption lines.)

In a dark cloud, the density is sufficiently high that molecules collide every now and then, and when they do they can be excited to high rotational energy levels. Once a molecule has been kicked into a rapid rotation state, it will soon slow (in a single, instantaneous jump) to a lower state, emitting a radio photon in the process. Thus a dark cloud is constantly releasing radio emission at a variety of fixed wavelengths corresponding to the types of molecules inside. As we learned in discussing the 21-cm line, radio lines are not strongly affected by interstellar extinction, so the emission from a dark cloud escapes easily into space.

Most of the molecules that exist in dark clouds emit at wavelengths of a few millimeters or centimeters. Since the emission from a distant cloud is greatly attenuated (because of the inverse-square law) by the time it reaches the Earth, very sensitive receivers are required to detect it, and the technology for doing this has been developed only in the last two decades. Every time a new advance is made in receiver sophistication, new molecules are immediately detected.

The number of molecules discovered in space has grown to more than seventy (see Appendix 12). Most are rather simple species, the most common being diatomic molecules, which consist of only two atoms. Of these, the dominant one is molecular hydrogen which, for a subtle reason, does not emit radio waves, and has been detected in dark clouds only in active regions where it is excited to emit infrared radiation. The next most common molecule is carbon monoxide (CO), and following this the abundant species include OH (called the hydroxyl radical), water vapor (H_2O), and an assortment of other simple combinations of abundant elements.

Some rather complex large molecules have begun to be detected as well, the current grand champion being $HC_{11}N$, a species consisting of a total of thirteen atoms, eleven of them carbon. There may be even larger molecules in space, most likely other species containing long chains of carbon atoms, which form easily under the conditions that prevail inside the dark clouds.

All the molecules detected so far are made up of combinations of just a few of the most common elements, primarily hydrogen, carbon, nitrogen, oxygen, and sulfur. There probably are molecules containing other elements, but because of their lower abundances, these species are difficult to detect. The second most common element after hydrogen is helium, but this is an inert element, meaning that it cannot easily combine with other atoms to form molecules. Helium is certainly present in interstellar clouds, but only in atomic form.

Perhaps the most interesting of all the species detected in dark clouds are the **organic molecules,** those containing certain combinations of carbon atoms that are also found in living material. Their presence shows that when stars form in dark clouds, the ingredients for life are already present. We have seen (in Chapter 15) that amino acids have been found in meteorites, indicating that these molecules were present before the planets formed. We can speculate that perhaps these relatively complex organic molecules formed even before the first solid matter in the solar system.

ASTRONOMICAL INSIGHT 25.1

Giant Molecules in Space

Even the densest interstellar clouds are very good vacuums by laboratory standards. The formation of molecules, which requires frequent collisions between atoms, proceeds best in dense gases. The Earth's atmosphere is almost entirely molecular, and the density of particles (at sea level) is about 2×10^{19} per cubic centimeter. The densest interstellar clouds have only 10^5 to 10^6 particles per cubic centimeter. Even so, most of the particles in these clouds are molecules.

Generally, interstellar molecules are very small and simple species, such as two-atom molecules like H_2 (hydrogen), CO (carbon monoxide), CN (cyanogen), and OH (hydroxyl). More complex species are found, however, but only in the darkest and densest clouds. Appendix 12 lists all known species; there we see that the largest have about a dozen atoms (and these largest ones have been seen in only one place, a complex of dark clouds in the Taurus region). Now astronomers have found evidence of the existence of much larger species, having 50 or more atoms.

The evidence consists of infrared emission lines that are seen coming from gas (called circumstellar gas) surrounding some hot stars. For a long time, these emission lines were unidentified, but recently it has been found that they match in wavelength with features formed by a class of molecules called polycyclic aromatic hydrocarbons (PAHs). These are enormous, flat molecules consisting of a hexagonal pattern of carbon atoms and having a few other species (primarily hydrogen, but also including nitrogen, oxygen, and other common elements) attached around the edges. On the Earth, PAHs are not naturally occurring species, but were developed for use as lubricants and solvents. The sketch below shows the structure of a few of the more common PAHs.

At first, the idea that molecules as large and complex as PAHs could exist in interstellar (or circumstellar) space seemed hopelessly unrealistic. But the hypothesis has borne up well under intense scrutiny, and today astronomers are coming around to the point of view that, at least in some environments, these species do exist. The primary arguments are, first, the excellent wavelength matches between the laboratory and astronomical measurements of the emission lines; and, second, the fact that the infrared emission lines are associated only with carbon-rich environments (such as the gas ejected from stars whose carbon abundance has been enriched through the triple-alpha reaction). This latter argument strongly suggests that whatever forms the circumstellar emission lines is made of carbon, as the PAHs are.

The formation mechanism for species such as PAHs must be understood before astronomers can accept their reality without reservation. But here, too, progress is being made. It is already known that small dust particles can form in the winds that flow outward from cool giant and supergiant stars, so it seems possible that large molecules could develop there as well (in fact, beyond a certain size of molecule, the distinction between grain and molecule becomes inconsequential). Model calculations have shown that PAHs could form in the winds from carbon-rich red giants, given certain conditions for temperature, density, and wind speed that are not unrealistic.

An alternative formation mechanism has also been suggested: the fragmentation of larger grains made of graphite. Graphite is a form of pure carbon whose structure consists of layers of carbon molecules in hexagonal patterns. These layers are themselves quite strong, but are not tightly bonded to each other (this is why graphite feels slippery). In effect, a graphite grain consists of layers of PAH molecules; if such a grain is broken apart, perhaps by turbulence or shock waves in a stellar wind, the pieces would be PAHs. It is widely suspected that graphite grains are common in the interstellar medium and that they form in the winds coming from carbon-

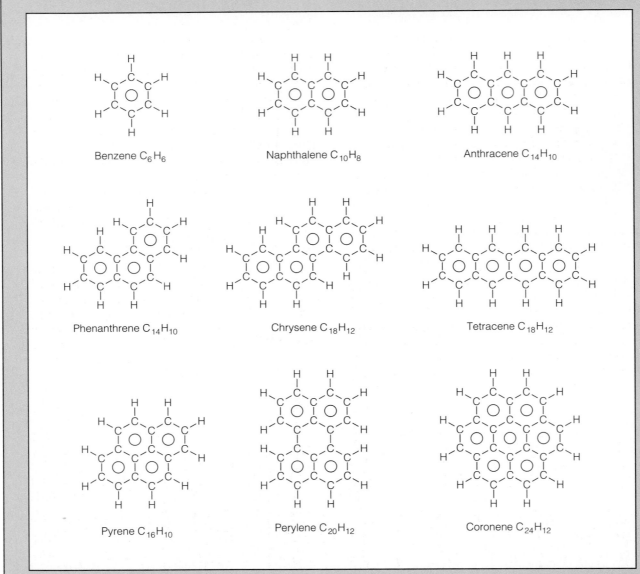

Structure of some typical PAH molelcues.

ASTRONOMICAL INSIGHT 25.1
Giant Molecules in Space (continued)

rich stars. Thus, the idea of giant PAH molecules in interstellar space may not be as crazy as it once seemed.

Many questions remain about how common these species really are, and how important they are in the diffuse interstellar medium, away from mass-losing stars. To-day many scientists are directing their research toward answering these questions.

INTERSTELLAR VIOLENCE AND THE ROLE OF SUPERNOVAE

Now that we have examined the principal components of the interstellar medium, let us take another look at its overall properties. We have found a diverse collection of ingredients, which can be characterized by a wide variety of temperatures and densities.

When the first spectroscopic observations of the interstellar gas were made more than 50 years ago, it was noticed that some diffuse clouds are moving through space at rather large speeds, up to 100 kilometers per second or even more (Fig. 25.17). Since then it has been found that the motions are often directed away from groups of bright stars, as though the gas were being expelled by some force originating in the stars.

When 21-cm observations of hydrogen gas became possible, similar motions were found. A variety of rapidly moving clouds was discovered, and in some regions there appeared to be expanding loops or shells of gas (Fig. 25.18). Some of these regions were associated with known supernova remnants, where it was clear that the gas was still expanding from the original explosion. In other cases, however, nothing was found at the center of expansion except perhaps a group of young stars. Whatever the cause, the picture developed by these observations was one of an interstellar medium in turmoil, with random, high-speed motions throughout.

In the 1970s new evidence of energetic phenomena in space was uncovered, in two forms. Data obtained from X-ray satellite and rocket experiments showed that the interstellar medium in the plane of the galaxy emits X-rays. At about the same time, ultraviolet spectroscopic observations revealed previously undetected ions in space that form only at very high temperatures. The combination of these X-ray and ultraviolet data led to the conclusion that the space between cool clouds is filled with a very hot gas, whose temperature is 1 million degrees or more. The density of the hot gas between the clouds is very low even by interstellar standards; it is 10^{-4} to 10^{-3} particles per cubic centimeter, the equivalent of one particle in every cubic volume 10 to 20 centimeters on a side!

We now see that the interstellar medium not only is diversified, but also is being constantly disturbed, which creates the observed high-speed motions and high temperatures. All this activity requires a substantial amount of energy input; a quiet pond does not spontaneously form ripples and eddies unless something stirs it up.

The agitation of the interstellar medium, it is now understood, is primarily the result of supernova ex-

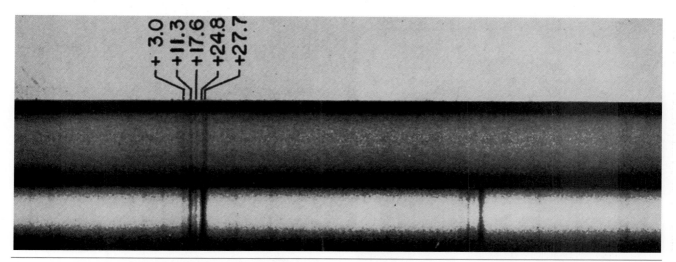

Figure 25.17 The effect of cloud motions on interstellar lines. *This is a spectrogram of a distant star that has several interstellar clouds along its line of sight. The upper spectrum shows a line due to ionized calcium; the lower spectrum shows atomic sodium (two major lines). Each line consists of several closely spaced features representing individual clouds in the line of sight. Each cloud has a slightly different velocity, so that the lines are separated by the Doppler effect. Velocities (in km/sec) of the individual clouds with respect to the Sun are indicated. (G. Munch)*

Figure 25.18 An extensive supernova remnant. *This is a large portion of the Vela supernova remnant, which covers a large region of sky in the Southern Hemisphere. (© 1980 Royal Observatory, Edinburgh)*

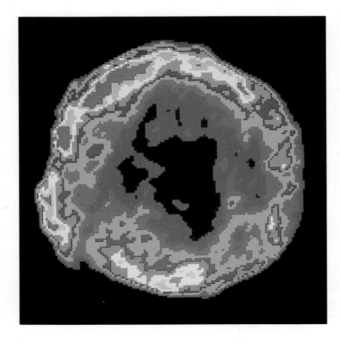

plosions. These occur in the galaxy at a frequency of one every few decades, and each time, a vast amount of energy is released into space. Material expands outward from the site of the explosion, sweeping up a shell of gas that continues to expand, and creating a cavity of very rarefied, extremely hot gas inside the shell (Fig. 25.19). As we have seen, interstellar clouds can be compressed by the outburst, triggering star formation, and clouds can be accelerated to high speeds as well. The net effect of repeated supernova explosions in the galaxy is that a large fraction of its volume has become filled with the hot gas that is left behind

Figure 25.19 The remnant of Tycho's supernova. *In 1572 a prominent supernova occurred, and was observed by Tycho Brahe. (The reputation his analysis earned him led to his obtaining royal funding for his observatory.) This is a radio map of the remnant of this supernova, showing how the expanding gas has created a huge, spherical bubble. (National Radio Astronomy Observatory)*

ASTRONOMICAL INSIGHT 25.2

The Mysterious Cosmic Rays

We have known for a long time that the Earth is constantly being bombarded by rapidly moving subatomic particles from space, which have been given the name **cosmic rays.** Consisting of electrons or atomic nuclei, these particles can be detected in several ways, the most common of which is to trace their tracks in special liquid-filled containers called bubble chambers. The cosmic ray particles are electrically charged, and their passage through such a container creates a trail of fine bubbles. From the length and shape of this track, we can infer the charge and speed of the particles. Cosmic ray velocities are usually near the speed of light.

The Earth's magnetic field deflects most cosmic rays, particularly the less massive and energetic ones, so that they reach the ground most easily near the poles.

From the point of view of life on Earth, this is probably a good thing, for cosmic rays are suspected of causing genetic mutations and could be harmful in other ways if allowed to reach the surface unimpeded. (On the other hand, mutations caused by cosmic rays are thought to have assisted evolutionary processes in the development of life on Earth.)

Extensive observations of cosmic rays show that most are either electrons or nuclei of atoms, with all of the electrons stripped free. Most of the latter are single protons, the nuclei of hydrogen atoms, but some are nuclei of heavy elements, including metallic species known to be formed only in supernova explosions. The majority of the cosmic rays that reach the Earth originate in the Sun, but many, particularly the more massive and energetic ones, come from all directions in the plane of the galaxy and seem to fill its volume more or less uniformly. Apparently they are trapped by the galactic magnetic field and cannot escape into intergalactic space. Because their paths

are curved and deflected by the magnetic field, it is impossible to tell where they come from. Therefore the source of the galactic cosmic rays is not understood, but a strong possibility is that they are released in supernova explosions. The total energy contained in cosmic rays darting about the galaxy is very large, and only supernovae seem capable of producing the necessary energy.

The distribution of cosmic-ray nuclei of various elements is generally similar to the chemical abundances found in stars and the interstellar material, although there are some differences, most of which can be attributed to nuclear reactions that can occur when the rapidly moving nuclei strike interstellar dust grains. Such reactions are responsible for the production of certain elements in space, such as the lightweight species beryllium.

Cosmic rays have little effect on our everyday lives, but they have the potential of telling us quite a bit about the energetic activities that take place in interstellar space.

---

by the expanding shells. Much of the cloud material has been accelerated, so that high-velocity clouds are seen here and there, particularly in loops or fragments of spherical shells.

Another mechanism that has the same effect, and which contributes to this picture, is the expansion of stellar winds. We learned in Chapter 20 that hot, blue stars as a rule have high-speed winds; these are streams of gas flowing outward at speeds as great as 3,000 kilometers per second. Such a wind will act in some ways like a supernova explosion, sweeping up material around the star and creating an expanding shell

(Figs. 25.20 and 25.21). Although a supernova releases far more energy in a brief moment, a stellar wind may last several million years, eventually injecting a comparable amount of energy into the surrounding gas. In a cluster of young stars, the winds can combine to evacuate a large bubble around the entire cluster, closely simulating the effect of a supernova explosion.

In one recently discovered case, a gigantic shell was found surrounding a vast volume in the constellation Cygnus (Fig. 25.22); the total energy required to create that structure is estimated to be comparable to that of several hundreds or even thousands of supernovae. It

Figure 25.20 Formation of an interstellar bubble. *A supernova explosion or a strong stellar wind occurs in the midst of a cloudy interstellar medium (left). As the resultant shock waves expand, a spherical cavity is formed, with a concentration of swept-up gas surrounding it. The cavity, or bubble, is filled with very hot, rarefied gas. In real situations, the medium is probably not so uniform initially, and the resulting cavity is not so symmetrically shaped as shown here.*

is not clear how this "superbubble" formed. Possibly a very large star cluster at its center experienced a number of supernovae whose expanding shells combined to form the single huge structure seen today. Several other examples of "superbubbles" are now known.

The hot, tenuous interstellar gas is known to extend well above and below the plane of the galactic disk. Perhaps the galaxy as a whole has a wind emanating outward, a composite of the expansion of all the supernova explosions and stellar winds that have occurred within it. In any case, we can no longer think of the interstellar void as a cold, desolate place where nothing ever happens.

PERSPECTIVE

If this chapter had been written 20 years ago, it would have had a much different flavor. The interstellar medium would have been viewed as a quiescent back-

ground against which all the activities of stellar evolution are played out. Instead, we now have a picture of a dynamic role for interstellar matter in the affairs of the galaxy, with the stars providing the necessary energy as they live and die, and the interstellar material in turn creating the conditions necessary for the births of new generations. The rich interplay between stars and interstellar matter is the dominant theme of galactic ecology, and its role will be brought to the fore again in the next chapter, as we discuss the formation and evolution of the Milky Way.

SUMMARY

1. The interstellar medium, although extremely tenuous, contains a significant fraction of the mass of the galactic disk.
2. Interstellar grains cause extinction and reddening of starlight, because they scatter and absorb photons, and do so most efficiently at short wavelengths. The

Figure 25.21 A bubble created by a stellar wind.
This nebula, called NGC2359, contains a roughly spherical cavity that has been evacuated by the high-velocity wind from a hot, luminous star. (© 1979 Anglo-Australian Telescope Board)

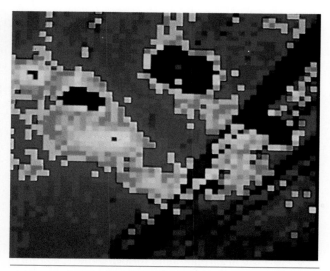

Figure 25.22 The Cygnus superbubble.   *This diagram shows the composite X-ray emission from a huge region in the constellation Cygnus. The yellow arclike structure outlines a gigantic bubble of superheated gas that has been swept up by a series of supernova explosions. The dimensions of the bubble are roughly 600 by 450 parsecs. (W. C. Cash)*

grains also cause polarization of starlight, indicating that they are elongated and preferentially aligned, probably by a galactic magnetic field.

3. Analysis of the interstellar extinction created by dust grains leads to the conclusion that many of the grains are about 5,000 Å in diameter, and that there is a second group of dust grains, responsible for the strong extinction of ultraviolet wavelengths, that are much smaller in size, with diameters of about 100 Å.

4. The grains also cause polarization of starlight, implying that they are elongated in shape and that they are aligned by a galactic magnetic field.

5. The gas in space was first detected through absorption lines in the spectra of distant stars. Most of these lines lie at ultraviolet wavelengths.

6. The interstellar medium is patchy, containing cold clouds embedded in a hot, more rarefied, intercloud medium. Most of the mass in space is in the clouds, and most of the volume is filled with the hot gas.

7. Interstellar clouds have a variety of densities and temperatures. The so-called diffuse clouds have densities of about 1 to 100 particles per cubic centimeter and temperatures of 50 to 100 K. We can observe these clouds in absorption-line measurements. Also, if near a hot star, they may appear as emission or reflection nebulae.

8. The dark clouds, with temperatures of 10 to 20 K and densities up to 10^6 particles per cubic centimeter, consist of gas that is primarily molecular in form. The most common molecules are hydrogen (H_2) and carbon monoxide (CO). A large number of different species have been detected, primarily through their radio emission lines.

9. The interstellar medium displays various forms of energetic activity: There are random, often rapid, cloud motions; the gas between clouds is extremely hot; and here and there are gigantic loop and ring structures suggestive of expanding shells. All of this activity can be attributed to energy injected into the interstellar medium by supernova explosions and stellar winds.

REVIEW QUESTIONS

Thought Questions

1. Summarize the conditions in the different types of interstellar environments discussed in this chapter, and say how each type of region is observed.

2. List all the different ways in which astronomers detect and study interstellar dust.

3. Suppose a certain star lies behind a large quantity of interstellar dust. Why is it much more difficult to observe this star in ultraviolet wavelengths than in infrared wavelengths?

4. How can astronomers distinguish between a cool star that is intrinsically red and a hotter star that appears just as red because of interstellar dust?

5. How can absorption lines formed in the spectrum of a star by interstellar gas be distinguished from absorption lines formed in the atmosphere of the star itself?

6. Explain why the interstellar absorption lines for most elements lie in ultraviolet wavelengths.

7. How can most of the mass in interstellar space be in the form of clouds when the clouds fill only a small fraction of the volume of interstellar space?

8. Explain why the densest interstellar clouds are best observed at radio wavelengths.

9. What is the evidence that the interstellar medium is an active, energetic place?

10. Summarize the role of the interstellar medium in the evolution of stars.

Problems

1. If interstellar dust causes a star to appear only 1/100 as bright as it would otherwise, how many magnitudes of extinction is the dust creating?

2. Suppose a B0 star is observed to have a color index of $B-V = 0.20$. It is known from data on other B0 stars that the normal color index for a star of this type is $B-V = -0.30$. What is the color excess for this star? (See Appendix 10.)

3. If 15 percent of the mass of the galaxy is in the form of interstellar matter, estimate the average density of the interstellar medium by dividing the interstellar mass by the volume of the galactic disk. (Assume the mass of the galaxy is 2×10^{11} solar masses. Assume that the galactic disk has a radius of $r = 15,000$ pc and a thickness of $z = 200$ pc; its volume V is equal to its area of πr^2 times its thickness, or $V = \pi r^2 z$. To get an answer in g/cm^3, convert the mass to grams and all distances to centimeters before you calculate the density.)

ADDITIONAL READINGS

Blitz, L. 1982. Giant molecular cloud complexes in the galaxy. *Scientific American* 246(4):84.

Buhl, D. 1972. Light molecules and dark clouds. *Mercury* 1(5):4.

Chaisson, E. J. 1978. Gaseous nebulas. *Scientific American* 239(6):164.

Chevalier, R. A. 1978. Supernova remnants. *American Scientist* 66:712.

Chevalier, R. A., and C. L. Sarazin. 1987. Hot gas in the universe. *American Scientist* 75:609.

Dame, T. M. 1988. The molecular Milky Way. *Sky and Telescope* 76:22.

Eicher, David J., and David Higgins. 1987. The secret world of dark nebulae. *Astronomy* 15(9):46.

Gehrz, R. D., D. C. Black, and P. M. Soloman. 1984. The formation of stellar systems from interstellar molecular clouds. *Science* 224(4651):823.

Greenberg, J. M. 1984. The structure and evolution of interstellar grains. *Scientific American* 250(6):124.

Heiles, C. 1978. The structure of the interstellar medium. *Scientific American* 238(1):74.

Herbig, G. H. 1974. Interstellar smog. *American Scientist* 62:200.

Herbst, E., and W. Klemperer. 1976. The formation of interstellar molecules. *Physics Today* 29(6):32.

Jura, M. 1977. Interstellar clouds and molecular hydrogen. *American Scientist* 65:446.

Kaler, J. B. 1982. Bubbles from dying stars. *Sky and Telescope* 63(2):129.

Knacke, R. F. 1979. Solid particles in space. *Sky and Telescope* 57(4):347.

Lewis, R. S., and E. Anders. 1983. Interstellar matter in meteorites. *Scientific American* 249(2):54.

Malin, David. 1987. In the shadow of the horsehead. *Sky and Telescope* 74(3):253.

Miller, J. S. 1974. The structure of emission nebulas. *Scientific American* 231(4):34.

Scoville, N., and J. S. Young. 1984. Molecular clouds, star formation, and galactic structure. *Scientific American* 250(4):42.

Smith, D. H. 1990. Seeking the origins of cosmic rays. *Sky and Telescope* 79:479.

Snow, T. P. 1976. The interstellar medium: Much ado about nothing? *Griffith Observer* 40(9):2.

Verschuur, Gerritt L. 1987. Molecules between the stars. *Mercury* 16(3):66.

MATHEMATICAL METHODS

The Effect of Extinction on Distance Measurements

In the "Mathematical Methods" section at the end of Chapter 19, we developed and used the following expression for the distance modulus:

$$m - M = 5\log\left(\frac{d}{10}\right),$$

where m and M are the apparent and absolute magnitudes of a star whose distance is d parsecs. This expression is useful in finding the distance to an object whose apparent magnitude is measured and whose absolute magnitude is estimated by some indirect technique, such as the star's position in the H-R diagram or its period, if it is a Cepheid variable, for example. This expression for the distance modulus makes no allowance for the presence of interstellar dust, which will cause a distant star to appear fainter than it otherwise would. Here we show how the effect of interstellar extinction is taken into account.

Since dust extinction makes a star appear dimmer, the effect is to make the distance modulus larger than it should be. If we continue to define m to be the measured apparent magnitude, then the "true" apparent magnitude (i.e., the apparent magnitude the star would have if there were no extinction) is smaller than the measured value m. The correction factor, which we call A_v, is the amount (in magnitude units) by which the star appears dimmer than it should. Thus, in the distance modulus expression, we write

$$(m - A_v) - M = 5\log\left(\frac{d}{10}\right),$$

which is usually rearranged to read

$$m - M = 5\log\left(\frac{d}{10}\right) + A_v.$$

Now to find the distance to the star we need not only its apparent and absolute magnitudes, but also an estimate of how much extinction is present. This cannot usually be measured directly, but it can be estimated from the amount of reddening the star suffers; that is, the degree to which the star's color (as measured by the color index $B - V$) is redder than it should be for a star of its spectral type (this is explained further in Appendix 10).

Let us show by an example how important the effect of extinction can be in determining the distance to a star. Consider the example we give in the "Mathematical Methods" section at the end of Chapter 19. There we had a star whose apparent magnitude was $m = 16.39$ and whose absolute magnitude was $+4.81$, leading to a distance of 2,090 parsecs. Now suppose that this star is dimmed by 2.4 magnitudes due to interstellar dust; that is, $A_v = 2.40$. Now we have

$$5\log\left(\frac{d}{10}\right) = m - M - A_v$$
$$= 16.39 - 4.81 - 2.40 = 19.18,$$

or

$$\log(d/10) = 19.18/5 = 1.84.$$

This yields

$$d/10 = 68.55 \text{ pc},$$

or, finally,

$$d = 686 \text{ pc}.$$

This is about one third of the distance we found before, when we assumed that no dust extinction was present. Thus we see that in determining stellar distances, it is very important to estimate the amount of extinction that is present and allow for its effects.

CHAPTER 26

The Formation and Evolution of the Galaxy

A spiral galaxy (NGC253), showing its stellar populations. (© 1980 Anglo-Australian Telescope Board; photo by D. Malin)

The galaxy we observe today is the end product of some 10 to 20 billion years of evolution. A variety of processes have shaped it, and many have left unmistakable evidence of their action. In this chapter we will discuss the major influences on the galaxy, along with the developments they have brought about.

We have laid the groundwork that will enable us to understand one of the most elegant concepts in the study of astronomy—the interplay between stars and interstellar matter and the long-term effects of this cosmic recycling on the evolution of the galaxy.

Figure 26.1 A spiral galaxy similar to the Milky Way. *This vast conglomeration of stars and interstellar matter is following its own evolutionary course, just as individual stars do. Note that the spiral arms tend to be blue, while the central bulge is yellow. These colors indicate the types of stars that dominate in the different parts of the galaxy. (© 1980 Anglo-Australian Telescope Board; photo by David Malin)*

STELLAR POPULATIONS AND ELEMENTAL GRADIENTS

To begin this story, we focus first on the overall structure of the galaxy (Fig. 26.1) and the types of stars found in various regions. We have already seen that most of the young stars in the galaxy tend to lie along the spiral arms. Indeed, it is the brilliance of the hot, massive O and B stars and their associated HII regions that delineates the arms, making them stand out from the rest of the galactic disk. We have also noted that in the nuclear bulge of the galaxy, as well as in the globular clusters and the halo in general, there are few young stars and relatively little interstellar material.

Careful scrutiny has revealed a number of distinctions between the stars that lie in the disk, particularly those in spiral arms, and the stars in the nucleus and halo. Analysis of the chemical compositions of the stars has shown that those in the halo and nucleus tend to have relatively low abundances of heavy elements such as metals, while in the vicinity of the Sun (and in spiral arms in general), these species are present in greater quantity. Nearly all stars are dominated by hydrogen, of course, but in the halo stars the heavy elements represent an even smaller trace than in the spiral arm

stars. The relative abundance of heavy elements in a halo star may typically be a factor of 100 below that found in the Sun; if iron, for example, is 10^{-5} as abundant as hydrogen in the Sun, it may be only 10^{-7} as abundant as hydrogen in a halo star.

A picture has emerged in which the galaxy has two distinct groups of stars, those in the halo and nuclear bulge on one hand, and those in the spiral arms on the other (Fig. 26.2). These two groups have been designated **Population II** for the stars in the halo and **Population I** for those in spiral arms; (Table 26.1). The Sun is thought to be typical of Population I, and its chemical composition is usually adopted as representative of the entire group.

Almost every time an attempt is made to classify astronomical objects into distinct groups, the boundaries between them turn out to be a little indistinct. There are usually intermediate objects whose properties fall between categories, and the stellar populations are no exception to this. For example, the stars in the

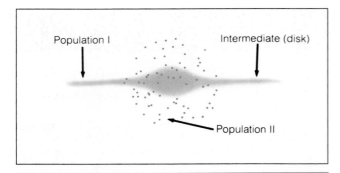

Figure 26.2 The distribution of Populations I and II. *This cross-section of the galaxy shows that Population II stars lie in the halo and central bulge, whereas Population I stars inhabit the spiral arms in the plane of the disk. Intermediate population stars are distributed throughout the disk.*

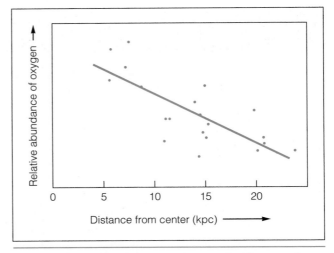

Figure 26.3 An abundance gradient. *This diagram shows how the abundance of oxygen (relative to hydrogen) varies with distance from the center of the Andromeda galaxy. More stellar generations have lived and died in the dense regions near the center, so nuclear processing is further advanced in the central region than it is farther out. (Data from Blair, W. P., Kirschner, R. P., and Chevalier, R. A. 1982, Astrophysical Journal 254:50)*

disk of the galaxy that do not fall strictly within the spiral arms have intermediate properties between those of Population I and Population II, and are often called simply the **disk population.** Furthermore, there are gradations within the two principal population groups, and we speak therefore of "extreme" or "intermediate" Population I or Population II objects. It is much more accurate to view the stellar populations as a smooth sequence of stellar properties represented by very low heavy element abundances at one end and solarlike abundances at the other.

The differences between chemical compositions from one part of the galaxy to another provide astronomers with very important clues in their efforts to piece together the history of the galaxy. Changes that occur gradually over substantial distances are referred to as

gradients, and the variations of stellar composition from one part of the galaxy to another are therefore referred to as **abundance gradients.** There is a gradient of increasing heavy-element abundances from the halo to the disk (Fig. 26.3). Within the disk itself, there is a similar gradient from the outer to the inner portions (this does not include the stars in the nuclear bulge, which, as we have pointed out, tend to have low abundances of heavy elements).

Table 26.1 Stellar populations

	Population I	Population II
Age	Young to intermediate	Old
Distribution	Disk, spiral arms	Halo, central bulge
Composition	Normal metals	Low heavy elements
Constituents	Disk, arm stars	Stars low in heavy elements
	O, B stars	High-velocity stars
	Interstellar matter	Globular clusters
	Type I Cepheids	Type II Cepheids
	Type II supernovae	RR Lyrae stars
	The Sun	Type I supernovae
		Planetary nebulae

A War Benefit

The discovery of stellar populations was actually based on observations of another galaxy, rather than the Milky Way, and the circumstances that led to the discovery were rather unusual.

During World War II, following the Japanese attack on Pearl Harbor in late 1941, cities on the West Coast were often kept dark at night to reduce the danger of bombardment. This was accomplished by enforced blackouts, in which all outside lights were kept turned off, and windows were covered so that no light got out.

As a result of these blackouts, major cities on the West Coast experienced some of the darkest nighttime skies of this century. This was a boon to local astronomers.

Many scientists were preoccupied with war-related research efforts and were not available to take advantage of the excellent observing opportunities. At the Mt. Wilson Observatory in the mountains overlooking Pasadena, however, there was a German-born astronomer named Walter Baade, a naturalized United States citizen restricted to nonsensitive research because of missing citizenship papers. Baade made extensive use of the 100-inch telescope at Mt. Wilson during the blackouts in Los Angeles, and was able to obtain extraordinarily fine photographs of galaxies neighboring the Milky Way.

One of these is the Andromeda nebula, a large spiral thought to be similar in many respects to our own galaxy. Baade's images of Andromeda were so fine and clear that he was able to make out individual stars, and to assess their properties. From colors and locations of the stars, he realized that there were systematic differences between different parts of the galaxy, and from these differences he developed the notion of two distinct stellar populations. Only later was the idea applied to the Milky Way, after careful observation revealed the same distribution of stellar properties.

Most of the stars near the Sun belong to Population I; they are disk stars orbiting the galactic center in approximately circular paths, like the Sun. There are a few nearby Population II stars here and there, and they are distinguished by several properties in addition to their compositions. The most easily recognized is that their motions depart drastically from those of Population I stars. As a rule, Population II stars do not follow circular paths in the plane of the disk, but instead follow highly eccentric orbits that are randomly oriented (Fig. 26.4), much like cometary orbits in the solar system. Thus, most of the Population II stars that are seen near the Sun are just passing through the disk from above or below it. The Sun and the other Population I stars in its vicinity move in their orbits at speeds around 250 km/sec, so the Sun moves very rapidly with respect to the Population II stars that pass through its neighborhood in a perpendicular direction. From our perspective here on the Earth, these Population II objects therefore are classified as **high-velocity stars.**

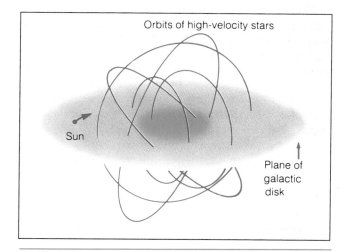

Figure 26.4 The orbits of high-velocity stars. *These stars follow orbits that intersect the plane of the galaxy. When such a star passes near the Sun, it has a high velocity relative to us, because of the Sun's rapid motion along its own orbit.*

Historically, high-velocity stars were discovered and recognized as a distinct class even before the differences between Population I and Population II were enumerated. It was only later that these objects were equated with Population II stars following randomly oriented orbits in the galactic halo.

The systematic differences between the stellar populations arose from differing conditions at the times when they formed. In this regard it is important to recall that globular clusters, representative of Population II, are among the oldest objects in the galaxy, while Population I, on the other hand, includes young, newly formed stars. We will follow this discussion further in a later section.

STELLAR CYCLES AND CHEMICAL ENRICHMENT

The fact that stars in different parts of the Milky Way have distinctly different chemical makeups is readily explained in terms of stellar evolution. Recall (from discussions in Chapter 20) that as a star lives its life, nuclear reactions gradually convert light elements into heavier ones. The first step occurs while the star is on the main sequence, during which hydrogen nuclei deep in its interior are fused into helium. Later stages, depending on the mass of the star, may include the fusion of helium into carbon, and possibly the formation of even heavier elements by the addition of further helium nuclei. The most massive stars, which undergo the greatest number of reaction stages, also form heavy elements in the fiery instants of their deaths in supernova explosions.

All of these processes work in the same direction: They act together to gradually enrich the heavy-element abundances in the galaxy. Material is cycled back and forth between stars and the interstellar medium, and with each passing generation a greater supply of heavy elements is available. As a result, stars formed where there has been a lot of stellar cycling in the past are born with higher quantities of heavy elements than those formed where little previous cycling has occurred. Stars formed before much cycling occurred therefore tend to have low abundances of heavy elements. This explains why old stars belong to Population II; they formed out of material that had not yet been chemically enriched.

The Population I stars, such as the Sun, are those that condensed from interstellar material that had pre-viously been processed in stellar interiors and in supernova explosions. Therefore, as a rule, they formed at moderate to recent times in the history of the galaxy, and are not as ancient as Population II objects.

Besides elemental abundance variations with age, there can also be variations with location. As noted in the preceding section, there is a distinct abundance gradient within the disk of the galaxy, the stars nearer the center having higher abundances of heavy elements than those farther out. This is a reflection of enhanced stellar cycling near the center, rather than an age difference. The central portion of the disk is where the density of stars and interstellar material is highest, and therefore in this region there has been relatively active stellar processing, because star formation has proceeded at a greater rate than in the less dense outer portions of the disk. Over the lifetime of the galaxy, more generations of stars have lived and died in the central region than in the rarefied outer reaches.

THE CARE AND FEEDING OF SPIRAL ARMS

In Chapter 24 we described the structure of the galaxy and how its spiral nature was discovered and mapped, but we said nothing about how the spiral arms formed, nor why they persist. Both questions are important, for it is clear that if the arms were simply streamers trailing along behind the galaxy as it rotates, they should wrap themselves up tightly around the nucleus (Fig. 26.5), like string being wound into a ball. Because the galaxy is old enough to have rotated at least 40 to 50 times since it formed, something must be preventing the arms from winding up, or they would have done so long ago.

Maintaining the spiral arms—a distinct question from that of forming them in the first place—is a very complex business, and it is not yet fully understood how this is done. The first successful theory on this appeared in 1960 and is still being refined. The essence of this theory is that the large-scale organization of the galaxy is imposed on it by wave motions. We have understood waves as oscillatory motions created by disturbances, and we know that waves can be transmitted through a medium over long distances while individual particles in the medium move very little. In the case of water waves, for example, a floating object simply bobs up and down as a wave passes by, whereas the wave itself may travel a treat distance. Here we are distinguishing

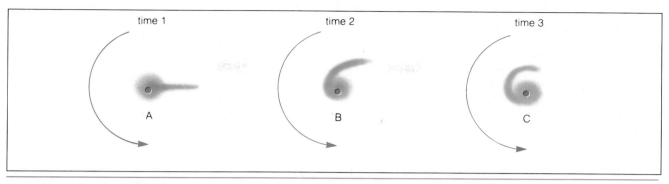

Figure 26.5 The windup of spiral arms.   *If spiral arms were simply streamers of material attached to a rotating galaxy, they would, within a few hundred million years, wind tightly around the nucleus. The galaxy is much older than that, so some other explanation of the arms is needed.*

between the wave and the medium through which it moves.

The waves that apparently govern the spiral structure in our galaxy are not transverse waves, like those in water, but are compressional waves, similar to sound waves and to certain seismic waves (the P waves; see Chapter 9 for an elaboration on the different types of waves). In this case the wave pattern consists of alternating regions of high and low density. When a compressional wave passes through a medium, the individual particles vibrate back and forth along the direction of the wave motion. There is no motion perpendicular to that direction, in contrast with water waves.

The theory of galactic spiral structure that invokes waves as a means of maintaining the spiral arms is called the **density wave theory.** This theory supposes that there is a spiral-shaped wave pattern centered on the galactic nucleus, creating a pinwheel shape of alternating dense and relatively empty regions (Fig. 26.6). The density waves have more effect on the interstellar medium than on stars, so the spiral arms are characterized primarily by concentrations of gas and dust. These lead, in turn, to concentrations of young stars, because star formation is enhanced in regions where the interstellar material is compressed (Fig. 26.7).

In its simplest form, the wave pattern is double; that is, there are just two spiral arms emanating from the nucleus, on opposite sides. There are galaxies that have such a simple spiral structure, but many, including the Milky Way, are more complicated. The density wave theory allows the possibility of having more arms, if the waves have shorter "wavelength," or distance between them (there is an alternative explanation for the somewhat disorganized spiral structure of galaxies like ours, which is discussed shortly).

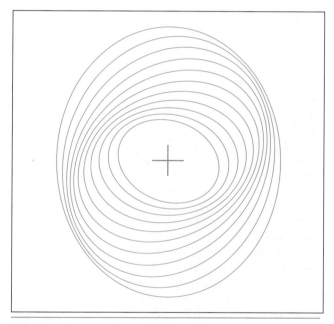

Figure 26.6 The density wave theory.   *This drawing depicts the manner in which circular orbits are deformed into slightly elliptical ones by an outside gravitational force. The nested elliptical orbits are aligned in such a way that there are density enhancements in a spiral pattern. This pattern rotates at a steady rate, and does not wind up more tightly.*

The waves rotate about the galaxy at a fixed rate that is constant from inner portions to the outer edge. Thus, while the outer portions of the arms appear to trail the rotation, in fact they move all the way around

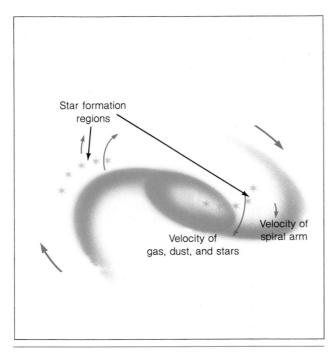

Figure 26.7 The effect of a spiral density wave on the interstellar medium. *As the wave moves through the medium, material is compressed, leading to enhanced star formation. The young stars, HII regions, and nebulosities along the density wave are what we see as a spiral arm.*

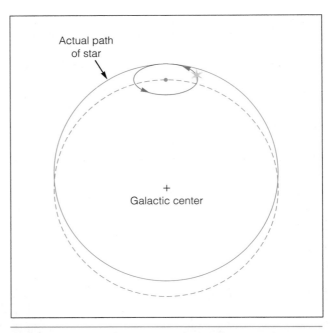

Figure 26.8 Epicyclic orbits. *The oscillatory motions created by spiral density waves cause individual stars to have orbits about the galaxy that are not perfect circles but are oval instead. The shape of the orbit may be viewed as a combination of a large circle and a small circle (epicycle). (Adapted from Shu, F. 1982, The Physical Universe [San Francisco: University Science Press], Fig. 12.12, p. 267.)*

the galaxy in the same time period that the inner portions do. The waves are essentially rigid, in strong contrast with the motions of the stars. Just as in the case of water waves, the motion of the spiral density waves is quite distinct from the motions of individual particles (i.e., stars) in the medium through which the waves travel. In the Milky Way, individual stars orbit the galaxy at their own speed. In the inner part of the galactic disk, out to the Sun's position, the stars move faster than the density waves. This means that as a star circles the galaxy, every so often it will overtake and pass through a region of high density as it penetrates a spiral arm. Because their motion is slowed slightly when they are in a density wave, stars tend to become concentrated in the arms, just as cars on a highway become jammed together at a point where traffic flow is constricted. Farther out in the galaxy, the rotation speed of the spiral density waves is faster than that of the individual stars.

The motion of an individual star about the galaxy is not precisely circular, although it is close. The spiral density waves impose slight oscillatory motions on the

stars as they pass through the waves, with the result that the orbit of a star can be represented as an epicyclic motion (Fig. 26.8).

We have noted that the main reason the spiral arms stand out is that hot, luminous young stars are to be found exclusively in the arms because star formation is enhanced there, where the interstellar gas is compressed. According to the density wave theory, the most active stellar nurseries should be located on the inside edge of the arms, where stars and gas catch up with and enter the compressed region. This is difficult to check observationally for our own galaxy, but it seems to be true of others.

The most luminous stars have such short lifetimes that they evolve and die before having sufficient time to pull ahead of the arm where they were born; therefore, we find no bright, blue stars between the arms. Less luminous stars—those that live billions instead of only millions of years—can survive long enough to orbit the galaxy several times, and these stars become spread almost uniformly throughout the galactic disk, between the arms as well as in them. The Sun, for example, is old enough to have circled the galaxy some 15 to 20 times, and therefore has passed through spiral arms and the intervening gaps on several occasions. It is only coincidence that the Sun is presently in a spiral arm; it is not necessarily the one in which it formed.

The initial formation of the density wave that is the cause of the spiral arms is a separate question. A rotating disk of particles, such as the disk of a galaxy like ours, will naturally form spiral density waves if disturbed by an outside force (Fig. 26.9). This suggests

that the spiral density waves in our galaxy were triggered by the gravitational influence of a neighbor galaxy, most likely the Magellanic Clouds, a pair of small galaxies that orbit the Milky Way.

Another possible mechanism for creating spiral density waves, with no help from neighboring galaxies, occurs in certain galaxies that have noncircular central bulges. In these **barred spiral galaxies** (see Chapter 27), the asymmetric shape of the central region can create the gravitational disturbance needed to initiate spiral

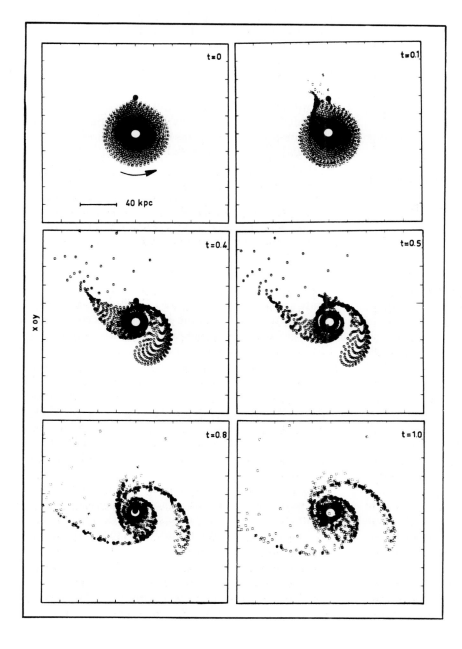

Figure 26.9 The creation of spiral structure. *This sequence of computer-generated models shows how a rotating disk forms spiral density waves when subjected to a gravitational force. (After T. M. Eneev, N. H. Kuzlov, and R. A. Sunyaev, 1973,* Astronomy and Astrophysics *22:41)*

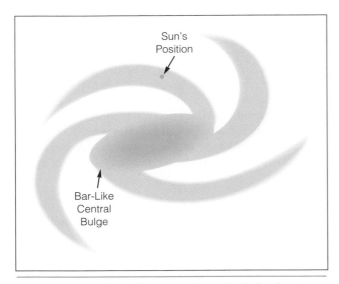

Figure 26.10 Is the Milky Way a barred spiral galaxy?
This sketch illustrates a new picture of our home galaxy that has recently emerged, as a result of the analysis of gas motions observed through the Doppler shifts in radio emission lines. As we will learn in the next chapter, about half of all spiral galaxies have bar-like nuclei, so it may not be surprising that our own galaxy shows this trait. (data from L. Blitz, University of Maryland, and D. Spergel, Princeton University)

density waves. Some calculations suggest that the barred spiral structure is the more natural form for a disklike galaxy to take, and that the presence of a massive, extended halo is what prevented our own galaxy from becoming a barred spiral instead of the "normal" spiral it is. If so, this suggests that other normal spiral galaxies may also have massive halos, whereas the barred spirals may not. In this context, it is very interesting that a recent study shows that the Milky Way may actually be a barred spiral (Fig. 26.10). A new analysis of gas motions in the outer galaxy indicates that the gas is orbiting a central bulge of stars that is elongated, rather than being round. The shape of the bulge has been likened to that of a watermelon. Thus our galaxy may be a mild form of barred spiral.

It is possible that in some spiral galaxies the structure is not governed by spiral density waves at all. The more chaotic and disorganized spirals, possibly including the Milky Way, consist of many fragments of spiral arms, and have no simple overall spiral pattern. The fragmentary spiral arms in these galaxies may be gigantic star-formation regions that have been stretched into arclike shapes by the differential rotation of the galaxy.

If so, then in such galaxies the spiral structure must be regarded as constantly changing and evolving rather than the longstanding pattern that is located in other galaxies by density waves.

GALACTIC HISTORY

We are now in a position to tie together all the diverse information on the nature of the Milky Way, and to develop from that a picture of its formation and evolution. There are still many important uncertainties in this picture, however, and some of these will be pointed out.

The pertinent facts that must be explained include the size and shape of the galaxy, its rotation, the distribution of interstellar material, elemental abundance gradients, and the dichotomy between Population I and Population II stars in terms of composition, distribution, and motions. The task of fitting all the pieces of the puzzle together is made easier by the fact that we can reconstruct the time sequence, knowing that stars with high abundances of heavy elements were probably formed more recently than those with lower abundances.

The oldest objects in the galaxy are in the halo, which is dominated by the globular clusters but contains a large (but unknown) number of isolated, dim, red stars as well. Estimates based on the main sequence turnoff in the H-R diagrams for globular clusters indicate ages of between 14 and 16 billion years, and we conclude that the age of the galaxy itself is comparable (some globular clusters appear to be somewhat younger; this is discussed in the next section).

The spherical distribution of the halo objects about the galactic center demonstrates that when they formed, the galaxy itself was round. Evidently the progenitor of the Milky Way was a gigantic spherical gas cloud, consisting almost exclusively of hydrogen and helium. Very early, perhaps even before the cloud began to contract (Fig. 26.11), the first stars and globular clusters formed in regions where localized condensations occurred. These stars contained few heavy elements and were distributed throughout a spherical volume, with randomly oriented orbits about the galactic center. Eventually the entire cloud began to fall in on itself, and as it did so, star formation continued to occur, so that many stars were born with motions directed toward the galactic center. These stars assumed highly eccentric orbits, accounting for the motions observed today in Population II stars.

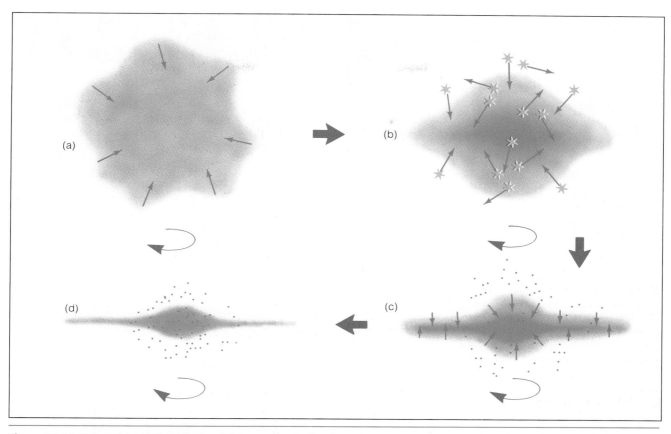

Figure 26.11 Formation of the galaxy.   *The pregalactic cloud, rotating slowly, begins to collapse (a). The stars formed before and during the collapse have a spherical distribution and noncircular, randomly oriented orbits (b). The collapse leads to a disk with a large central bulge, surrounded by a spherical halo of old stars and clusters (c). The disk flattens further, and eventually forms spiral arms (d). Stars in the disk have relatively high heavy element abundances because they formed from material that had been through stellar nuclear processing.*

Apparently the pregalactic cloud was originally rotating, for we know that as the collapse proceeded, a disklike shape resulted. Rotation forced this to happen, just as it did in the contracting cloud that was to form the solar system (see Chapter 16). Rotational forces slowed the contraction in the equatorial plane, but not in the polar regions, so material continued to fall in there. The result was a highly flattened disk. Stars that had already formed before the disk took shape retained the orbits in which they were born; stars are unaffected by the fluid forces (e.g., viscosity) that caused the gas to continue collapsing to form a disk.

Stars that were formed after the disk had developed had characteristics unlike those of their predecessors in at least two respects: They contained greater abundances of heavy elements, because by this time some

stellar cycling had occurred, enriching the interstellar gas; and they were born with circular orbits lying in the plane of the disk. These are the primary traits of Population I stars.

Since the time of the formation of the disk, there have been additional, but relatively gradual, changes. Further generations of stars have lived and died, continuing the chemical enrichment process, particularly in the inner region of the disk, creating the chemical abundance gradient mentioned earlier. Apparently the enrichment process was once more rapid than it is today, because the present rate of star formation is too slow to have built up the quantities of heavy elements that are observed in Population I stars. At some time in the past, probably just when the disk was forming, there must have been a period of intense star formation,

ASTRONOMICAL INSIGHT 26.2

The Missing Population III

It was explained in the text that the two populations of stars should be viewed as having a range of stellar characteristics such as age and metal content, rather than a strict separation of two completely distinct classes of stars. The most extreme Population II stars, those with the lowest metal abundances and the greatest ages, represent one end of the range, whereas the youngest stars in spiral arms, containing the highest metal abundances, represent the other end of the range.

The fact that the extreme Population II stars contain *some* heavy elements, even though it may be only 1 percent of the metal content of Population I stars, has proven difficult to explain. In the picture of galactic formation and evolution presented in this chapter, the Population II stars are portrayed as the first stars to form from the collapsing cloud of gas that was to become the galaxy. This cloud of gas is assumed to be primordial—that is, never to have been involved in stellar formation and processing. Therefore, the cloud should have had a composition consisting almost entirely of hydrogen and helium, with *no* heavy elements. The

question, then, is why the oldest stars we observe, those belonging to extreme Population II, have the metal content that they do. Even these stars must have formed from material that had previously been processed through stars, becoming somewhat enriched with heavy elements.

The missing population of stars formed from completely unprocessed material and having virtually zero heavy-element abundances has become known as **Population III.** Several astronomers have spent considerable time and effort in the attempt to find such stars, with no success.

No widely accepted explanation for the lack of Population III stars has been devised yet, but there have been suggestions. One is that before the pregalactic cloud collapsed, there was a brief episode of star formation and nuclear processing, in which a number of very massive, short-lived stars were formed and quickly evolved to explode in supernovae, creating heavy elements. These heavy elements then became part of the gas that collapsed to form the disk, so that stars formed during and after the collapse—the oldest stars we now observe—contained some heavy elements, and represent what we know today as Population II. None of the original Population III stars remain, because they were all too massive and short-lived to survive for long.

The chief difficulty in this and other pictures in which an early

star-formation episode occurred is to explain how star formation became so active at a time when the pregalactic material was still widely dispersed in a giant, tenuous cloud. Despite our lack of understanding of *how* this occurred, it seems probable that it did, since we find no Population III stars. Furthermore, there is independent evidence that galaxies do undergo periods of intense star formation and processing; such episodes are observed in some other galaxies. Therefore, it is quite possible that our galaxy went through such a time early in its history. Several studies are under way today in an effort to understand how and why this may have occurred.

Another possibility has recently been suggested, however. Contrary to longstanding assumptions about nuclear reactions in the earliest moments of the formation of the universe, it is now thought possible that some heavy elements were formed then. If so, the "primordial" material from which our galaxy formed may have contained some heavy elements, thus explaining how the oldest stars could have been born with the observed composition of Population II stars. In this view, there never was a Population III. The recent developments on the theory of the universe that lead to this suggestion are discussed in Chapter 31.

during which the abundances of heavy elements in the galaxy jumped from almost none to nearly the present level. A large fraction of the galactic mass must have been cooked in stellar interiors and returned to space in a brief episode of stellar cycling whose intensity has not been matched since. (Today some galaxies are observed that seem to be in such a state of rapid stellar processing; these are called **starburst galaxies,** and they are discussed in Chapter 27.)

Apparently during this early phase of intense star formation, massive stars were created at a greater rate than they are today. Studies of star formation have shown that isolated dark clouds tend to form low-mass stars, while large star-formation regions, characterized by intense ultraviolet radiation and warm temperatures, tend to form stars of greater mass. Thus when intense star formation occurred early in the history of the galaxy, many massive stars were born, which quickly produced heavy elements and dispersed them into the interstellar medium.

If large numbers of massive stars did form early in the galaxy's history, then today there may be many compact stellar remnants such as neutron stars and black holes. As mentioned in Chapter 24, there could even be enough of these remnants left over to account for the large amount of invisible mass believed to exist in our galaxy.

We have just about recounted all the events that led up to the present-day Milky Way, except for the formation of the spiral arms. It is not known when this took place, but it was probably soon after the formation of the disk itself. We see few gas-rich, disk-shaped galaxies without spiral structure, which leads to the conclusion that a disk galaxy containing large quantities of interstellar matter does not exist long in an armless state.

It is difficult to guess exactly what may become of the galaxy in the future. The interstellar gas and dust is being consumed by star formation, but only gradually, so the general appearance of the galaxy may not change substantially for many billions of years. Perhaps in the very distant future, when most of the ingredients for new stars have been used up, our galaxy will become an armless disk containing only old stars and stellar remnants.

THE AGE OF THE GALAXY

The scenario we have described for the formation of the galaxy is thought to be generally correct, but there remain some substantial uncertainties. Most of these have to do with the timescale on which the galaxy formed, and the duration of certain stages along the way.

There have been discrepancies in the age of the galaxy as estimated from different techniques. We have already mentioned the use of globular cluster ages as an indicator of the age of the galaxy. The general idea is that the galaxy is probably comparable in age to the oldest systems within it, the globular clusters. This would not be correct, however, if there was some delay between the time the galaxy formed and the time the globular clusters developed. Now there is some hint that the ages of the globular clusters are not so well understood.

The age of a cluster is determined from the main-sequence turnoff point, as described in Chapter 21. We did not emphasize, however, that while this method provides reliable *relative* ages, it relies on stellar evolution models to yield actual ages in years. The accuracy of the model calculations is always a bit uncertain, and is especially so in cases of old clusters whose chemical composition is deficient in heavy elements.

Recent analyses of globular clusters in our galaxy reveal that they are probably not all the same age. Some appear to be as "young" as 11 billion years, while others are as old as 16 billion years. If the globular clusters really have such a wide range of ages, this suggests that the formation of the galaxy was not as simple as outlined in the previous section. There must have been a rather extended period of time (perhaps 5 billion years) before the galaxy completed its contraction to a disklike shape. During this time the galaxy must have retained the roughly spherical shape of the halo, while the globular clusters formed here and there at different times.

It is not easy to explain why the galaxy would have undergone such a drawn-out early phase, prior to collapsing to form a disk. One suggestion is that the galaxy formed from the conglomeration of several smaller systems of stars, which themselves formed earlier (and at different times, accounting for the spread in ages now observed). In this picture, our galaxy was created by the merger of a number of subgalaxies, systems of stars much like globular clusters. Each of these subsystems would have formed separately from the collapse of a cloud of gas, in contrast with the picture in which the entire galaxy formed from a single cloud that collapsed. As we will see in the coming chapters, there is growing evidence that galaxies in general formed in this way; that the earliest stellar systems in the universe were smaller than galaxies, with galaxies forming later as the result of mergers of these smaller systems.

There are at least two other ways of estimating the age of the Milky Way, independent of the globular cluster method. In one method, white dwarfs are used as age indicators. It is possible, through a statistical analysis of the luminosities of white dwarfs scattered around our part of the galaxy, to determine the age of the oldest ones (this depends on knowledge of the time required for white dwarfs to cool to the point of being too dim to be observed). This method yields a rather low estimate of the age of the galaxy—around 9 billion years.

The other method of age estimation also depends in part on theoretical calculations. From knowledge of the rates of the nuclear reactions that produce heavy elements, it is possible to estimate how old the galaxy must be in order to contain the quantities of these elements that exist today. There are substantial uncertainties in this method, so that it does not provide very precise answers. It has been shown, however, that the current abundances of heavy elements are consistent with an age for the galactic disk of 11 to 15 billion years, in keeping with the age estimated from the oldest globular clusters.

It is not understood why the white-dwarf method and the globular-cluster method of age determination provide such different answers. Certainly the picture we have developed of the formation and evolution of our galaxy is subject to further revision as this question is resolved and as the wide range of globular cluster ages is further explored.

PERSPECTIVE

We have now seen our galaxy in a new light, as an active, dynamic entity. Individual stars orbit its center in individual paths, yet the overall machinery is systematically organized. The majestic spiral arms rotate at a stately pace, while stars and interstellar matter pass through them. We have come to understand the active processes of stellar cycling and chemical enrichment, which still occur today. In a constant turnover of stellar generations, the deaths of the old give rise to the births of the new, while the violence of their death throes energizes a chaotic interstellar medium.

The lessons we have learned by examining our own Milky Way will be remembered as we move out into the void, probing the distant galaxies.

SUMMARY

1. Stars in the galactic halo have low abundances of heavy elements and are referred to as Population II stars, whereas those in the spiral arms of the disk have "normal" compositions and are called Population I stars.
2. There are gradients of increasing heavy-element abundance from halo to disk, and from the outer to the inner portions within the disk.
3. Population II stars have randomly oriented, highly eccentric orbits, whereas Population I stars have nearly circular orbits that lie in the plane of the disk. Population II stars, when passing through the disk near the Sun's location, are seen as high-velocity stars.
4. The variations in heavy-element abundance from place to place within the galaxy reflect variations in age: Very old stars, formed before stellar nuclear reactions had produced a significant quantity of heavy elements, have low abundances of these elements, whereas younger stars were formed after the galactic composition had been enriched by stellar evolution.
5. Spiral arms are probably density enhancements produced by spiral density waves, which rotate about the galaxy while stars and interstellar material pass through them. Because the interstellar gas is compressed in these density waves, star formation tends to occur there, and this in turn explains why young stars are found predominantly in the spiral arms.
6. The existence and characteristics of Population I and II stars can be explained in a picture of galactic evolution that begins with a spherical cloud of gas that has little or no heavy elements at first. Population II stars are those that formed while the cloud was still spherical or just beginning to collapse, but when it still had no significant quantities of heavy elements. Population I stars are those that formed later, after the galaxy had collapsed to a disk, and after stellar evolution had produced some heavy elements.
7. The spiral arms formed after the disk had been created by the collapse of the original cloud. The spiral density waves that maintain the arms probably started as a result of gravitational disturbances, possibly from nearby galaxies.

8. Different methods for estimating the age of the galaxy provide contradictory values; furthermore, there is a wide spread in the ages of globular clusters, suggesting that the galaxy may have formed from the merger of several subsystems of different ages.

REVIEW QUESTIONS

Thought Questions

1. Summarize the properties of Population I and Population II objects in the galaxy.

2. Explain in your own words what a gradient is.

3. Explain why the Sun is thought to be a Population I object.

4. How are high-velocity stars and comets similar?

5. How are the spiral arms of the galaxy similar to the rings of Saturn?

6. Explain the distinction between the orbital periods of stars in the galaxy and the rotation period of the spiral arms.

7. Why do stars form primarily in the spiral arms of a galaxy like ours?

8. Why do the globular clusters have the lowest heavy-element abundances of any objects in the galaxy?

9. Explain why the evolution of massive stars, rather than that of the much more common low-mass stars, has been the principal contributor to the enrichment of heavy elements in the galaxy.

10. Summarize the various roles played by supernovae in the evolution of the galaxy.

Problems

1. Suppose the spectrum of a star is analyzed, and iron is found to be present in its outer layers in the ratio of 1:40,000 compared to hydrogen. Is this a Population I or a Population II star? Explain how you decided.

2. The strong line of ionized calcium, whose rest wavelength is 3,933.8 Å, is observed at a wavelength of 3,932.0 Å in the spectrum of a star. To which population does this star belong? Explain how you decided.

3. The Sun moves faster in its orbit than the rotation speed of the galaxy's spiral arms, so the Sun repeatedly overtakes any given arm. If the Sun passes through the arm it is now in every 5×10^8 years, how many times has it done so in the 15-billion-year lifetime of the galaxy?

4. If the mass of the galaxy is 1.5×10^{11} solar masses, what is the orbital period of a globular cluster whose semimajor axis is 20 kpc? (Remember to convert this into astronomical units.)

ADDITIONAL READINGS

Bok, B. J. 1981. Our bigger and better galaxy. *Mercury* 10(5):130.

Bok, B. J., and P. Bok. 1981. *The Milky Way*. Cambridge, Mass.: Harvard University Press.

Burbidge, G., and E. M. Burbidge. 1958. Stellar populations. *Scientific American* 199(2):44.

Hodge, Paul. 1986. *Galaxies*. Cambridge, Mass.: Harvard University Press.

Iben, I. 1970. Globular cluster stars. *Scientific American* 223(1):26.

Larson, R. B. 1979. The formation of galaxies. *Mercury* 8(3):53.

Mathewson, Don. 1985. The clouds of Magellan. *Scientific American* 252(4):106.

Scoville, N., and J. S. Young. 1984. Molecular clouds, star formation, and galactic structure. *Scientific American* 250(4):42.

Shu, F. H. 1973. Spiral structure, dust clouds, and star formation. *American Scientist* 61:524.

————. 1982. *The physical universe*. San Francisco: W. H. Freeman.

Silk, Joseph L. 1986. Formation of the galaxies. *Sky and Telescope* 72(6):582.

Weaver, H. 1975 and 1976. Steps towards understanding the spiral structure of the Milky Way. *Mercury* 4(5):18 (Pt. I); 4(6):18 (Pt. II); 5(1):19 (Pt. III).

MATHEMATICAL METHODS

Orbital Velocities of Stars in the Galaxy

 In discussing the rotation of the Milky Way, we have said that the inner portion rotates like a rigid body, while in the outer portion the stars follow individual orbits according to Kepler's third law. Let us explore what these statements mean.

In a spinning rigid body, by definition, all parts complete a rotation in the same period of time. The farther from the center a particular portion is, the greater the distance it must travel to complete one rotation. The circumference of a circle is $2\pi R$, where R is the radius; thus, as R increases, the distance traveled must increase in proportion to R. This means that the speed must also increase in proportion to R. So when we say the inner portion of the galaxy rotates like a rigid body, we mean that the orbital speeds of the individual stars are proportional to their distances from the center of the galaxy.

But how does the orbital speed vary with distance from the center in the region where stars follow Keplerian orbits? We can find out by writing Kepler's third law:

$$MP^2 = R^3,$$

where we have substituted the total mass of the galaxy M for the sum of the masses $m_1 + m_2$ (since the mass of an individual star is negligible), and we have substituted the orbital radius R for the semimajor axis a, on the assumption that the orbits are circular. P is the orbital period. (In this form of Kepler's third law, M is in solar masses, P is in years, and R is in astronomical units.)

During one orbital period, a star must travel a distance equal to the orbital circumference. If the orbital speed is V, then application of the formula speed multiplied by time equals distance gives us

$$VP = 2\pi R,$$

which can be rearranged to yield

$$P = \frac{2\pi R}{V}.$$

Substituting this into the expression for Kepler's third law leads to

$$\frac{4\pi^2 R^2 M}{V^2} = R^3,$$

which can rearranged to yield

$$V = 2\pi\sqrt{\frac{M}{R}}.$$

Since M represents the mass of the galaxy, which is constant, this result shows that the orbital speed is inversely proportional to the square root of the distance from the galactic center. Thus in a region where the orbits are Keplerian, the orbital speed decreases with increasing distance from the galactic center, whereas in a rigid-body region the orbital speed increases with distance from the center.

The dashed line in Figure 24.10 shows the expected drop-off in orbital speed with increasing distance in the outer part of the galaxy, and also shows that the actual orbital speeds do not drop off as expected. As discussed in the text, this indicates that the stars in the outer regions of the galaxy are not following Keplerian orbits. That is, their motions are not as expected if these stars were influenced only by the gravity of the central galaxy. Instead it is apparent that motions in the outer galaxy are dominated by the gravitational effects of other matter farther out.

Globular Clusters and the Evolution of the Milky Way

BY RICHARD B. LARSON

Richard Larson is one of today's preeminent theorists in the area of star formation and the formation and evolution of stellar systems. He has made fundamental contributions to our knowledge of these processes; much of what is written about star formation in Chapters 12 and 16 of this book is based on his work, for example. Following his education at the University of Toronto and the California Institute of Technology, Professor Larson joined the faculty at Yale in 1968 (where your author was an undergraduate student in over his head in one of Professor Larson's early graduate-level classes). During his 20 years at Yale, Professor Larson has developed a reputation as a sound and innovative researcher into dynamic processes involving the gravitational collapse of prestellar and pregalactic clouds, and the systems that form as a result. In his work, he has sought particularly to understand the roles of turbulence, fragmentation, accretion, and disk formation as he has explored the real world of random and nonuniform collapse, as opposed to the more orderly processes that have traditionally been envisioned. In this article, Professor Larson describes how recent observational studies by others,

combined with his own theoretical analyses, are leading to a radically different picture of the formation of our own galaxy than the one depicted in this text.

It has usually been assumed that the globular clusters, the gigantic spherical conglomerations of stars that populate the galactic halo, all formed at about the same time early in the history of the Milky Way galaxy. The standard scenario for the creation of our galaxy calls for formation of the globular scenario for the creation of our galaxy calls for formation of the globular clusters when the pregalactic cloud was still more or less spherical in shape, followed by rapid collapse of the remaining gas to the disklike appearance

that the galaxy has today. The globular clusters remained in a spherical distribution, forming the halo of the galaxy that is now observed.

One of the main points of debate in modern theories of the formation of the Milky Way is the length of time required for the collapse from a spherical cloud to a disklike galaxy to occur. In the original models of this process, it was proposed that the collapse was quite rapid—that is, that it took place in less than 1 billion years. In this case, the ages of the globular clusters should all be within a billion years of each other. More recently, a competing point of view has emerged, in which it is thought that the galaxy may have formed in a less orderly fashion from the merger of a number of smaller, galaxylike systems of stars; the time for a formation of a disk might then have been much longer than a billion years, and we would not necessarily expect that all the globular clusters formed at the same time.

Thus, the ages of globular clusters can provide key information on which model for the formation of the galaxy is correct. Traditionally, age estimates for globulars have been based on rather crude measures of stellar magnitudes and colors derived from photographs. A cluster H-R diagram would be constructed and the main sequence turnoff compared with computed models of stellar evolution, but the results were restricted in accuracy by limitations in the sensitivity of photographic film. In recent years, new devices for detecting and measuring star images have been developed, leading to much more accurate measurements of much fainter stars than previously possible. These new devices are electronic cameras called **charge-coupled devices** or **CCDs**, and they take the place of film in obtaining pictures of objects such as globular clusters. CCDs allow very faint stars to be detected and their brightnesses to be measured, and therefore allow astronomers to obtain much more accurate cluster H-R diagrams than they could using film.

Several recent studies of the ages of globular clusters based on these new data have yielded a wide range of values, from 11 to perhaps 17 billion years. Because different groups of astronomers have made independent measurements in several cases and have found similar results, this range of ages seems likely to be real. Particular cases support this conclusion: For example, the two clusters NGC288 and NGC363 lie in similar locations and have similar composition, yet they have different main-sequence turnoff points and different horizontal branch structures and are thought, therefore, to be 3 to 5 billion years apart in age.

If these results are basically correct, we can now begin to piece together a chronology for the formation of the halo and disk of our galaxy. The formation of halo clusters appears to have continued for a period of perhaps 6 billion years, or one-third of the age of the universe (see Chapter 21), and was not completed until about 11 billion years ago. The galactic disk in the Sun's vicinity then apparently began to form at about the same time that halo formation ended, and with about the same content of heavy elements. Various indicators of the age of the local galactic disk all yield age estimates between 9 and 12 billion years, comparable to the ages of the youngest globular clusters in the halo. Thus it appears that the formation of the local part of the galactic disk took up where halo formation left off approximately 10 to 12 billion years ago.

What was happening during the 6 billion years preceding the formation of a disk in the solar vicinity? The continuing formation of globular clusters in the halo and the failure of an extended disk to form over this long period suggests that conditions remained too chaotic during this time for gas to settle permanently into a disk. Understanding this phase of galactic evolution clearly requires understanding the origin of the globular clusters themselves. Our present knowledge of cluster systems in galaxies does not indicate any sharp distinction between globular and open clusters, so there is no reason to suppose that the globular clusters formed in a fundamentally different way from open clusters whose formation we can see taking place at the present. Together with others, I have argued that the globular clusters did not form in isolation, but rather, like open clusters, as parts of much larger star-forming complexes or galactic subsystems. These systems may have resem-

bled compact galaxies such as some that are observed outside our own galaxy; one such galaxy, discovered by Fritz Zwicky and designated I Zw 18, has a central star formation region that may be forming a globular cluster at the present time. If our galaxy contained a number of subordinate galaxylike objects at early times, some remnants of this early population of subsystems may survive at present in the form of dwarf galaxies near the Milky Way, such as the Fornax system, which contains several apparently normal globular clusters.

A strong piece of evidence favoring this view is the fact that the heavy-element abundances in globular clusters vary greatly from cluster to cluster, yet are very similar for stars within each cluster. This could be explained if the globular clusters formed in a number of protogalactic subsystems that for a time evolved separately but remained internally well-mixed as they continued to form stars and clusters. In other words, these systems may have evolved as separate small galaxies for up to several billion years before finally being merged together into the galactic halo.

The observational results show that there is a continuity of ages and chemical composition from the halo to the disk of our galaxy. This suggests that the disk formed from gas that was left over from, and had already been chemically enriched by, the subsystems of stars that formed the early halo. This entire picture supports the recent suggestions that large galaxies form from collections of smaller galaxies that merge together but retain their own identities, in some cases, for an extended period, as stars form and evolve within them. In this picture, the simple idea of a disk forming rapidly from the collapse of a spherical cloud is invalid. Instead, we must view the galaxy formation process as being more chaotic, involving a merging together of subsystems, star formation and evolution within them, and then their disruption by collisions, with the leftover gas settling finally into a disk.

A question not directly answered by the globular cluster evidence concerns the number and typical size of the subsystems or subordinate galaxies from which the Milky Way formed: Were there many very small ones, or only a relatively few large ones? The fact that the galactic halo contains almost exclusively objects with low abundances of heavy elements could mean that the subunits were mostly not very large, since large ones would have undergone more rapid star forma-

tion and hence would have built up greater abundances of heavy elements. There is still a wide range of possibilities in theoretical models, so it will be important to gain further information on the kinds of objects from which our galaxy formed. In any event, it appears almost certain that the early origins of the Milky Way were less orderly than envisioned in traditional models such as the one discussed in this book.

SECTION VI

EXTRAGALACTIC ASTRONOMY

We are ready now to move out of the confines of our own galaxy, to explore the universe beyond. The Milky Way is just one of billions of galaxies in the cosmos, and we will find that although our own system is typical of a certain class of these objects, there is a wide variety of shapes, sizes, and peculiarities associated with other galaxies.

The first chapter of this section describes both the history of the discovery of galaxies beyond the Milky Way and the observational properties of these galaxies. Next, chapter 28 begins to bridge the gap between our discussion of individual objects (that is, galaxies) and the structure and organization of the universe as a whole. In this chapter we examine the distribution of galaxies, paying particular attention to groupings and clusters and the possibility that these are in turn organized into larger associations.

Chapter 29 describes two major aspects of the universe, the expansion of the universe and the cosmic background radiation that fills it. Both are legacies of the fiery origin of the universe, and their observed properties teach us something of its early history.

With the perspective gained from Chapter 29, in the next chapter we turn our attention back to individual objects, ones whose properties can best be appreciated in the context of an expanding universe. A variety of peculiar galaxies, and especially the quasistellar objects, have fantastic properties that were only discovered by astronomers because the nature of the universal expansion was already known. These objects, on the frontiers of the observable universe, can potentially tell us a great deal about its history.

Finally, the last chapter of this section describes the current state of cosmology, the science of the universe as a whole. Here we tie together all the diverse information gained from the preceding chapters and assess the overall structure of the universe, and we especially emphasize the question of its future. If there is a single premier challenge to modern astronomy, it is this, and we are on the verge of knowing the answer. Having completed this section, we will be nearly finished with our introduction to astonomy.

CHAPTER 27

The Nature of the Nebulae

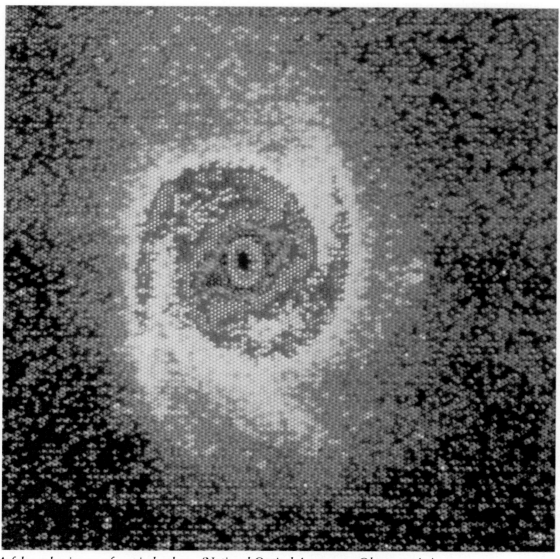

A false color image of a spiral galaxy. (National Optical Astronomy Observatories)

CHAPTER PREVIEW

The Hubble Classification System
Galactic Distances
The Masses of Galaxies
Luminosities, Colors, and Diameters
The Origins of Spirals and Ellipticals

We have spoken of our galaxy as one of many, a single member of a vast population that fills the universe. Given all that we have learned about our ordinary position in the cosmos, this is no surprise. Having traced our painful progression from the geocentric view to the realization that we occupy an insignificant planet orbiting an ordinary star in an obscure corner of the galaxy, we should be surprised if we found that our galaxy held any kind of unique status in the larger environment of the universe as a whole. It doesn't.

Despite the seeming inevitability of this idea, the actual proof that our galaxy is not alone was some time in coming, arriving only after considerable debate and controversy. The so-called **nebulae**, dim, fuzzy objects scattered throughout the sky, have been known since the early days of astronomical photography, but it was not until the mid-1920s that they were demonstrated to be galaxies, rather than more nearby objects such as gas clouds or star clusters. The proof of their galactic nature was announced in 1924, when the American astronomer Edwin Hubble (Fig. 27.1) reported that he had found Cepheid variables in the prominent Andromeda galaxy (until then known as the Andromeda nebula, since its true nature was not known), and used the period-luminosity relation to show that this nebula was entirely too distant to be within our own galaxy.

Figure 27.1 Edwin Hubble. *Hubble's discovery of Cepheid variables in the Andromeda Nebula led to the unambiguous conclusion that this object lies well beyond the limits of the Milky Way and must therefore be a separate galaxy. Hubble later made important discoveries about the properties of galaxies and what they tell us about the universe as a whole (see Chapter 29). (Mt. Wilson and Las Campanas Observatories, Carnegie Institution of Washington)*

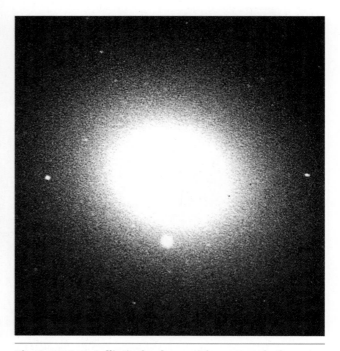

Figure 27.2 An elliptical galaxy. *These smooth, featureless galaxies are probably more common than spirals. (National Optical Astronomy Observatories)*

Figure 27.3 A spiral galaxy.   *At left is NGC 4321, a beautiful example of a face-on spiral galaxy. During the early years of the twentieth century, there was considerable controversy among astronomers over the true nature of objects like this. The view at right shows how thin the disk of a spiral galaxy is, relative to its diameter. (Left: J. D. Wray, University of Texas; right: U.S. Naval Observatory)*

THE HUBBLE CLASSIFICATION SYSTEM

Having established that nebulae are truly extragalactic objects, Hubble began a systematic study of their properties. The most obvious basis for establishing patterns among the various types was to categorize them according to shape. Hubble did this, designating the sphe-

roidal nebulae **elliptical galaxies** (Fig. 27.2), as distinct from the **spiral galaxies** (Fig. 27.3). Within each of the two general types, Hubble established subcategories on the basis of less dramatic gradations in appearance (Table 27.1). The ellipticals displayed varying degrees of flattening and were sorted out according to the ratio of the long axis to the short axis, with designations from E0 (spherical) to E7 (the most highly flattened). The number following the letter E is determined from

Table 27.1 Types of galaxies

Type	Designation	Characteristics
Elliptical	E0–E7	Spheroidal shape; subtype determined by the formula $10(1 - b/a)$, where a is the long axis and b is the short axis of the galaxy.
Dwarf elliptical	dw E	Spheroidal shape; very low mass and luminosity.
S0	S0	Disk-like galaxies with no spiral structure.
Spiral	Sa–Sc	Disk-like galaxies with spiral arms; subtype determined by relative size of nucleus and openness of spiral structure; *Sa* refers to large nucleus and tightly wound arms, whereas *Sc* refers to small nucleus and open spiral arms.
Barred spiral	SBa–SBc	Spiral galaxies with elongated, barlike nuclei; subtypes determined the same way as in spiral galaxies.
Irregular I	Ir I	Disk-like galaxies with evidence of spiral structure, but not well organized.
Irregular II	Ir II	Galaxies that do not fit into any of the other types; some Ir II galaxies have been found to be normal spirals heavily obscured by interstellar gas and dust.

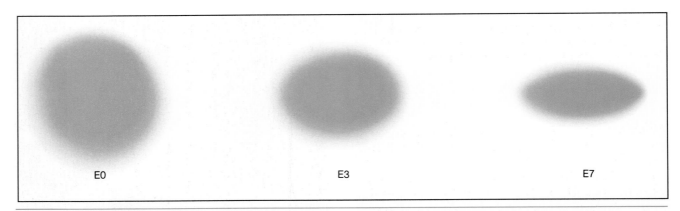

Figure 27.4 The shapes of ellipticals. *The numerical designation (following the letter E) is given by the formula $(1 - b/a) \times 10$, where b is the short axis and a the long axis of the galaxy image. Here are three examples: the E0 galaxy has $a = b$ (that is, it is circular); the E3 galaxy has $a = 1.43b$; and the E7 has $a = 3.3b$. The E7 galaxy is the most highly elongated of the ellipticals.*

Figure 27.5 Spiral galaxies. *Here are four typical examples of spiral galaxies. At top left is M64, top right is M66 (these designations refer to the Messier Catalog, see Appendix 15); at bottom left is NGC6946, bottom right is NGC4321 (these designations are from the* New General Catalog*). (Top left, top right, and bottom left: U.S. Naval Observatory; bottom right: J. D. Wray, University of Texas)*

Figure 27.6 An assortment of spirals. *This sequence shows spiral galaxies of several subclasses. (Palomar Observatory, California Institute of Technology)*

the formula $10(1 - b/a)$, where a is the long axis and b is the short axis, as measured on a photograph (Fig. 27.4).

Hubble's classification of the spirals was based on the tightness of the arms and the compactness of the nucleus (Figs. 27.5 and 27.6). The types ranged from Sa (tight spiral, large nucleus) to Sc (open arms, small nucleus). The Milky Way is probably an Sc in this system, although it is difficult to determine, because we cannot get an outsider's view of our galaxy. Some have classified the Milky Way as an Sb galaxy.

Hubble recognized a variation of the spiral galaxies in which the nucleus has extensions on opposing sides, with the spiral arms emanating from the ends of the extensions. These he called **barred spirals,** and assigned them the designations SBa through SBc, using the same criteria as before in establishing the a, b, and c subclasses (Figs. 27.7 and 27.8). Following Hubble's original work on galaxy classification, astronomers recognized an intermediate class called the S0 galaxies.

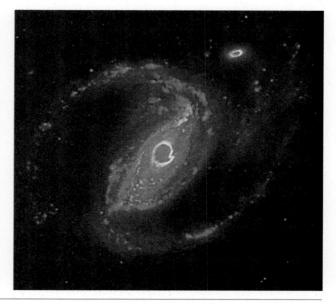

Figure 27.7 Barred spirals. *At left is M83, a well-known example of a spiral galaxy with a barlike structure through the nucleus; at right is a computer-enhanced view of another barred spiral. (left: © 1980 Anglo-Australian Telescope Board; right: photo by H. C. Arp, image processing by J. J. Lorre, Image Processing Laboratory, JPL)*

Figure 27.8 Barred spirals. *Roughly half of all spiral galaxies have central elongations, or bars, from which the spiral arms emanate. (Palomar Observatory, California Institute of Technology)*

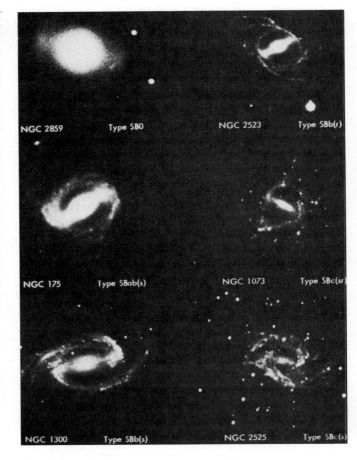

NGC 2859   Type SB0

NGC 2523   Type SBb(r)

NGC 175   Type SBab(s)

NGC 1073   Type SBc(sr)

NGC 1300   Type SBb(s)

NGC 2525   Type SBc(s)

These appear to have a disk shape, but no trace of spiral arms.

Although Hubble himself did not draw an evolutionary connection among the various types of galaxies, he assembled an organizational chart that was for a time thought by many astronomers to represent an age sequence. Because there are two types of spirals, Hubble chose not to force all the types into a single sequence, but instead split the sequence into two branches, and the result became known as the **tuning fork diagram** (Fig. 27.9). It was the prevailing belief that elliptical galaxies evolve into spirals, the originally spherical distribution of gas gradually flattening into a disk (as in S0 galaxies) and then developing spiral arms. This view was later reversed, after the discovery of stellar populations (see Chapter 26). Since elliptical galaxies consist primarily of old stars, with little interstellar matter and therefore little active star formation, and since spirals contain prominent young stars

and interstellar gas and dust, it seemed reasonable to assume that spirals evolve into ellipticals. This was the accepted view as recently as the early 1960s. It was eventually found, however, that even spiral galaxies contain old populations of stars and clusters and are just as old as the ellipticals. The tuning fork diagram is not an age sequence after all, and the differences in galactic type must be explained in some other way.

Most of the galaxies listed in catalogs are spirals, which are about evenly divided between normal and barred spirals. Only about 15 percent of the listed galaxies are ellipticals, and a comparable number are S0 galaxies. The remaining few are called irregular galaxies, because they do not fit into the normal classification scheme. Apparently there are many small, dim, elliptical galaxies that are usually not sufficiently prominent to appear in catalogs, and it seems likely that ellipticals actually outnumber spirals in the universe. This certainly is the case in dense clusters of

ASTRONOMICAL INSIGHT 27.1

The Shapley-Curtis Debate

Formal debate traditionally has been a respected means of determining right in cases of dispute. Debating is a time-honored practice, undertaken by some for entertainment, and by others for more serious reasons. Our legal system, for example, is based on something akin to debates that take place in court rooms, with a judge or jury to decide who is right. Occasionally, scientific questions are argued in the format of a debate, the judge being the test of time and the acquisition of new data to test the disputed theories, the jury being the community of scientists. A famous debate on the nature of the so-called nebulae was one of the singular events in early twentieth-century astronomy.

At the opening of this century, it was not known that there were galaxies beyond our own. (In fact, as we saw in Chapter 24, it was not even known where we were located in our own galaxy, nor was there any good information on our galaxy's size and shape.) By 1920, large numbers of indistinct, hazy objects had been found sprinkled throughout the sky. Some astronomers believed these to be other galaxies; others thought they were clouds of gas within our own galaxy. Both sides marshalled evidence to support their views.

Adherents of the "island universe" hypothesis, the idea that

the so-called nebulae were distant galaxies like our own, tried to show that the nebulae lay outside of our galaxy. In doing so, they argued that our galaxy was smaller than thought by some, and this in turn hinged on disagreements among astronomers over distance scales, particularly those based on pulsating variable stars in clusters. The distribution of nebulae was mapped and found to avoid the plane of the Milky Way, which indicated to the proponents of the island universe hypothesis that the nebulae lay far outside our galaxy, so that the galactic dust layer could block our view of the ones lying near the plane of the disk. Observations of novae in some of the nebulae (particularly our neighbor, the Andromeda nebula) were used to show that they must be outside of our galaxy, if they were as luminous as novae seen within the Milky Way.

Proponents of the local hypothesis for the nebulae argued that the Milky Way was large enough to include them. This is where the major disagreement on distance scales came into the argument, one side believing the galaxy to be some 10 times larger than did those on the other side of the issue. The problem was exacerbated by the assumption (on both sides) that there was no significant interstellar absorption, so that distances could be estimated simply from apparent magnitudes. Some "novae" were seen in other galaxies that, if comparable in luminosity to novae in our galaxy, would have to be close by, within the galaxy. (These were later identified as supernovae, far more luminous than ordinary novae.) The

most telling argument favoring the local hypothesis was the observation, later shown to be incorrect, that some of the spiral nebulae were actually seen to be rotating. An object as large as the Milky Way would have to be rotating faster than the speed of light in order for the rotation to be observed from a great distance over a time of only several months, so this was taken as proof that the nebulae could not be as large as galaxies, and therefore must be close by.

The chief proponents of the opposing points of view met in formal debate before the National Academy of Sciences, in Washington, D.C., on April 26, 1920. Representing the island universe hypothesis was H. D. Curtis, who had amassed most of the arguments cited above; opposing him was the formidable Harlow Shapley, who was already famous for his work on the size scale of the galaxy (see Chapter 24). The first part of the debate was devoted wholly to a discussion of the size of the Milky Way, and the second part of the debate was on the nature of the nebulae.

Shapley's arguments were considered more persuasive, and in fact he was more nearly correct on the question of the size scale of the galaxy (except for the omission of interstellar absorption). The judgment of time and the clues of new data found later by the jury ended up showing that Curtis was correct on the question of the nature of the nebulae, however. His arguments about the distribution of the nebulae, the existence of a class of "super" novae, galaxies, as we will see. Table 27.2 lists a few prominent galaxies whose properties indicate the typical values for galaxies of different types.

Although the irregular galaxies are, by definition, misfits, it has proven possible to find some systematic characteristics even in their case. Most have a hint of spiral structure, although they lack a clear overall pattern, and these have been designated as **Type I Irregulars.** The rest, a small group, simply do not conform in any way to the normal standards, and are assigned

the classification of **Type II Irregulars,** or peculiar, galaxies (Fig. 27.10). The Magellanic Clouds (Fig. 27.11) are irregulars of Type I.

Some of the Type II irregulars have the appearance of pairs of galaxies in collision (Fig. 27.12). In these cases, very unusual overall shapes are seen, but careful examination reveals hints of two galaxies in the process of merging, or at least passing through each other (Fig. 27.13). Since galaxies are mostly empty space, in such a collision few direct impacts of stars occur. The

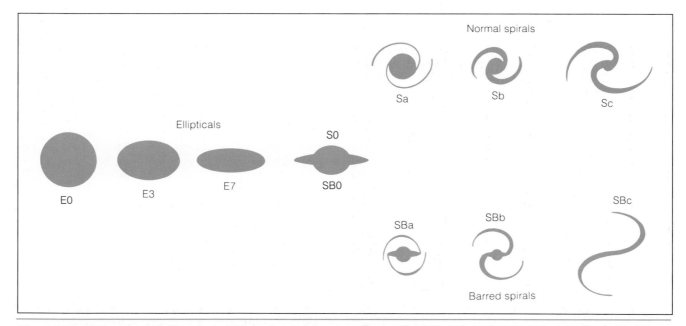

Figure 27.9 The tuning-fork diagram.   *This is the traditional manner of displaying the galaxy types originally devised by Hubble. For quite some time this was thought to be an evolutionary sequence, although the imagined direction of evolution was reversed at least once. Now it is known that in the normal course of events galaxies do not evolve from one type to another.*

Table 27.2 Selected bright galaxies

Galaxy	Type	Angular Diameter (arcmin)	Diameter (kpc)	Distance (kpc)	Apparent Magnitude[a]	Absolute Magnitude[a]	Mass ($M_\odot$)
Large Magellanic Cloud	Ir I	460	7	55	0.1	−18.7	10^{10}
Small Magellanic Cloud	Ir I	150	3	63	2.4	−16.7	2×10^{9}
Andromeda (M31)	Sb	100	16	700	3.5	−21.1	3×10^{11}
M33	Sc	35	6	730	5.7	−18.8	1×10^{10}
M81	Sb	20	16	3,200	6.9	−20.9	2×10^{11}
Centaurus A	EOp	14	15	4,400	7	−20	2×10^{11}
Sculptor system	dw E	30	1	85	7	−12	3×10^{6}
Fornax system	dw E	40	2	170	7	−13	2×10^{7}
M83	SBc	10	12	3,200	7.2	−20.6	—
Pinwheel (M101)	Sc	20	23	3,800	7.5	−20.3	2×10^{11}
Sombrero (M104)	Sa	6	8	1,200	8.1	−22	5×10^{11}
M106	Sb	15	17	4,000	8.2	−20.1	1×10^{11}
M94	Sb	7	10	4,500	8.2	−20.4	1×10^{11}
M82	Ir II	8	7	3,000	8.2	−19.6	3×10^{10}
M32	E2	5	1	700	8.2	−16.3	3×10^{9}
Whirlpool (M51)	Sc	9	9	3,800	8.4	−19.7	8×10^{10}
Virgo A (M87)	E1	4	13	13,000	8.7	−21.7	4×10^{12}

[a]The apparent and absolute magnitudes represent the light from the entire galaxy. Because the light from a galaxy comes from a large area in the sky, a galaxy of a given apparent magnitude is not as easily seen by the eye as a star of the same magnitude. Even though several galaxies in the list are brighter than sixth magnitude, only the two Magellanic Clouds and the Andromeda galaxy are easily visible to the unaided eye.

Figure 27.10 A peculiar galaxy.   *Not all galaxies fit the standard classifications. This is M82, a well-known example of a galaxy with an unusual appearance. For a time, it was thought that the nucleus of this galaxy was exploding, but more recent analysis has indicated that the odd appearance is due instead to a very extensive region of interstellar gas and dust surrounding the center of a spiral galaxy that is undergoing an episode of intense star formation. The image at right uses computer-enhanced colors to highlight the gas and dust. (Left: Lick Observatory photograph; right: photo by H. C. Arp, Image Processing by J. J. Lorre, Image Processing Laboratory, JPL)*

ASTRONOMICAL INSIGHT 27.2

Infrared Galaxies

 Until very recently, nearly all that we knew about the properties of galaxies and about their numbers was based on observations made in visible wavelengths of light. We now know that limiting ourselves to this small portion of the electromagnetic spectrum has severely limited our knowledge of galaxies and their evolution.

The *Infrared Astronomical Satellite (IRAS)* mapped the entire sky at four wavelengths far into the infrared portion of the spectrum. As noted in other chapters in this text, *IRAS* data have brought about major changes in our understanding of many objects in the universe, ranging from planets to newborn stars to interstellar clouds of gas and dust. It also produced new understandings of the galaxies that populate the universe.

The *IRAS* sky survey revealed huge numbers of faint, fuzzy sources of infrared emission not associated with any objects in our galaxy. These objects are now thought to be distant galaxies, and many of them are invisible in all wavelengths except the infrared. This means that most of the energy that they emit comes from very cold material. These infrared objects are now believed to

be spiral galaxies so obscured by dust that they are seen only in the far infrared wavelengths where the dust emits. Most of the light from the stars in these galaxies is absorbed by dust, which is heated in the process to temperatures of several tens of degrees and then emits infrared radiation.

The discovery of the infrared galaxies has revealed a whole new class of galaxies containing far more interstellar matter than normal spirals. The interpretation of these galaxies is uncertain: Are they aging galaxies undergoing a new phase of evolution in which stars emit vast quantities of material into space; or are they newly forming galaxies undergoing rapid star formation? The second point of view, that these are young galaxies undergoing massive star formation episodes, is becoming more widely accepted. If this is correct, it has important implications for galactic evolution. It implies that intervals of very intense and rapid star formation may be normal stages in the early evolution of spiral galaxies. Visible-light observations, as well as infrared data, have already led astronomers to define a class of galaxies called **starburst galaxies,** galaxies thought to be in stages of rapid star formation. The new infrared data from *IRAS* appears to have established the widespread existence of starburst galaxies.

If a starburst stage is normal in the early evolution of spiral galaxies, then one problem confound-

ing studies of the history of our own galaxy may have been solved. As noted in Chapter 26, the abundance of heavy elements in the Milky Way is higher than would be expected if current star formation rates were the norm throughout the history of the galaxy. This has led to speculation that our galaxy was either formed from material that had already undergone elemental enrichment by nuclear reactions in some earlier generation of stars, or that the galaxy underwent a much more intense episode of star formation sometime in the past than it is undergoing now. The discovery by *IRAS* that starburst phases of galactic evolution may be common leads astronomers to believe that our own galaxy was once a starburst galaxy. If so, the Milky Way may once have been virtually invisible in all wavelengths except the far infrared. Perhaps alien astronomers several billion light-years away are at this moment discovering the Milky Way as a source of far infrared radiation, to be included in their equivalent of our *IRAS* catalogs.

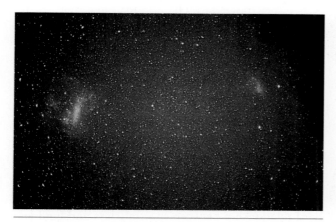

Figure 27.11 The Magellanic Clouds. *These two irregularly shaped, small galaxies lie just outside the Milky Way galaxy and orbit it, taking hundreds of millions of years for each complete orbit. (Fr. R. E. Royer)*

Figure 27.12 Galaxies in collision? *This unusual object is thought to be a pair of galaxies running into each other. (Palomar Observatory, California Institute of Technology)*

galaxies can pass through one another like ghosts and go their separate ways, except that their gravitational effects cause major distortions of their shapes. Even the Milky Way suffers from such an effect; the gravitational forces of the Magellanic Clouds have caused a slight warping of the galactic disk. As we will see in the next chapter, collisions among galaxies, particularly in clusters, can have very important effects on the evolution of both the individual galaxies and the clusters in which they are found.

When it became possible to observe galaxies in wavelengths other than visible light, new kinds of galaxies that do not fit the standard classification scheme were discovered. Infrared sky maps in particular revealed new types of galaxies, some not detectable in visible light. The *IRAS* satellite (Chapter 7) found many infrared galaxies that are so dim in visible wavelengths that they had not been previously noticed. These appear to be galaxies that contain very large quantities of interstellar gas and dust, so that much of the starlight is obscured by dust and reemitted in infrared wavelengths. Some of these galaxies seem to be undergoing very intense star formation and are called **starburst galaxies.** Perhaps the best-known example, although only recently recognized as a starburst galaxy, is M82, a peculiar-looking galaxy thought for a long time to be undergoing some kind of explosion (see Fig. 27.10). Starburst galaxies may be normal galaxies undergoing very intense episodes of star formation; recall from

Chapter 26 that our own galaxy is thought to have had at least one early period when stars formed much more rapidly than they do now. Perhaps at some time in the past, the Milky Way was a starburst galaxy.

GALACTIC DISTANCES

To probe the physical nature of the galaxies, we must first know galactic distances, because without this information we cannot deduce such fundamental parameters as masses and luminosities. We have already mentioned one technique for distance determination that can be applied to some galaxies: the use of Cepheid variables. These stars are sufficiently luminous to be identified as far away as a few million parsecs (that is, a few **megaparsecs,** abbreviated Mpc), which is sufficient to reach the Andromeda Nebula and several other neighbors of the Milky Way. This technique is not adequate, however, for probing the distances of most galaxies; other techniques had to be developed.

Recall, once again, from our discussions of stellar distance determinations (Chapter 19), that we can always find the distance to an object if we know both its apparent and absolute magnitudes. This is the basis of the spectroscopic-parallax method, the main-sequence-fitting technique, and even the Cepheid variable period-luminosity relation. A general term for any

Figure 27.13 The effect of a near-collision between galaxies. *This is M51 (also known as the Whirlpool galaxy), with its smaller companion. Below is a computer simulation of the correct orientation of these two galaxies as seen from different angles. At left the orientation is as we see it; at right is a side view. (Top: © 1975 Royal Greenwich Observatory; bottom: A. Toomre and J. Toomre)*

object whose absolute magnitude is known from its observed characteristics is **standard candle,** and an assortment of these are used in extending the distance scale to faraway galaxies.

The most luminous of stars are the red and blue supergiants, those which occupy the extreme upper regions of the H-R diagram. These can be seen at much greater distances than Cepheid variables, and therefore are important links to distant galaxies. The absolute magnitudes of these stars are inferred from their spectral types, just as in the spectroscopic-parallax technique, although they are so rare that there is substantial uncertainty in assuming that they conform to a standard relationship between spectral class and luminosity. In a variation on this technique, it is simply assumed that there is a fundamental limit on how luminous a star can be, and that in a collection of stars as large as a galaxy, there will always be at least one star at this limit. Then to determine the distance to a galaxy, we need only to measure the apparent magnitude of the brightest star in it, and to assume that its absolute magnitude is at the limit, which is about $M = -8$. This technique extends the distance scale by about a factor of 10 beyond what is possible with Cepheid variables—that is, to 10 or more megaparsecs.

Other standard candles come into play at greater distances. These include supernovae, which, at peak

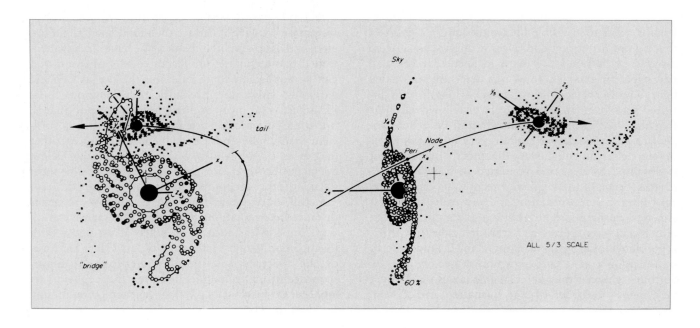

brightness, are thought to reach the same absolute magnitude. (Care must be exercised however, because the two types of supernovae have different absolute magnitudes, and because the recent supernova in the Large Magellanic Cloud has shown that some supernovae have unusual peak brightnesses.) A supernova is fantastically bright and can even outshine the entire galaxy in which it resides, reaching an absolute magnitude as bright as $M = -19$. By adopting supernovae as standard candles, we can measure distances of hundreds of millions of parsecs. One drawback is that we can use this technique only when a supernova happens to occur in a galaxy, which may be only once every 10 to 100 years. Therefore, if we wish to estimate the distance to a specific galaxy, and we do not want to wait that long, we must use some other standard candle.

Other objects useful for this purpose include certain types of star clusters and bright HII regions. In the latter case, it is not the absolute magnitude but the diameter of the HII region that is assumed to be known.

Then the apparent diameter (that is, the angular diameter) can be used to infer the distance. The ultimate standard candle is the brightest galaxy in a cluster of galaxies, in which case we adopt the assumption that this object should have the same absolute magnitude in every instance. This assumption makes it possible to measure distances to objects up to several billion parsecs away.

A relatively new technique, called the **Tully-Fisher method,** shows great promise for spiral galaxies. This method is based on a relationship that has been discovered between the masses of spiral galaxies and their luminosities. The more massive a spiral galaxy is, the more luminous it is; hence the luminosity and therefore the distance to a galaxy can be estimated by measuring the mass. The mass is not measured directly but is inferred from the rotational speed of the galaxy, as indicated by the width of the 21-cm emission line of hydrogen (Fig. 27.14). The more rapidly the outer portions of the galaxy are rotating, the wider the 21-cm line, because of the Doppler effect. The width of the

Figure 27.14 Rotation velocity and the width of the 21-centimeter line. *This shows how the rotation velocity of a spiral galaxy is related to the width of the 21-centimeter emission line from hydrogen gas in the galaxy. Because of the Doppler effect, 21-centimeter photons from one side of the galaxy are shifted toward shorter wavelengths, those from the other side are shifted toward longer wavelengths, and those from the central regions are not shifted. The result is a 21-centimeter line that is broadened according to how rapidly the galaxy is rotating. Because the rotation speed is governed by the central mass of a galaxy, the width of the line is related to the mass of the galaxy, and can be used to estimate its absolute magnitude for distance determination.*

21-cm line leads directly to an estimate of the galaxy's absolute magnitude; a comparison of this with the apparent magnitude yields a distance estimate. This technique works best when infrared magnitudes are used because infrared light is relatively unaffected by interstellar dust within the galaxy being observed. The Tully-Fisher method, along with a similar relation between luminosity and velocity dispersion for elliptical galaxies, is now being used in efforts to improve the intergalactic distance scale. Better knowledge of the distances to galaxies has important applications to our understanding of the size scale of the universe itself, as discussed in Chapters 29 and 31.

It is important to keep in mind how uncertain these techniques are. To assume that all objects in a given class are identical in basic properties such as luminosity is always risky, especially when such assumptions are applied to objects as distant as external galaxies or as rare as the brightest star in a galaxy. There is little else that can be done, however, so we must simply recognize the inherent limitations in accuracy and take them into account.

Figure 27.15 Measuring a galactic rotation curve. *This sequence of images illustrates how radio observations of atomic hydrogen are used to derive the rotation curve for a spiral galaxy. At upper left is a visible-light photograph of the nearly edge-on galaxy; at upper right is a radio map obtained in the 21-cm line of hydrogen. In the two lower panels the radio data have been color-coded, on the left to indicate relative intensities of the hydrogen emission; and on the right to indicate relative line-of-sight velocities. In this last panel we can see that one side of the galaxy (red) is receding and the other side (blue) is approaching. Comparison of the velocities leads to a determination of the rotational speed as a function of distance from the galactic center. (National Radio Astronomy Observatory)*

THE MASSES OF GALAXIES

The masses of nearby galaxies can be measured in the same manner as the mass of our own galaxy, by applying Kepler's third law to the orbital motions of stars or gas clouds in the outer portions. All that is required is to measure the orbital speed at some point well out from the center, and to determine how far from the center that point is (which in turn requires knowledge of the distance to the galaxy). Then Kepler's third law leads to

$$M = \frac{a^3}{P^2}$$

where M is the mass of the galaxy (in solar masses), and a and P are the semimajor axis (in AU) and the period (in years) of the orbiting material at the observed point (see Chapter 24).

The difficulties of this technique are that it is hard to isolate individual stars in distant galaxies and then to measure their speeds (using the Doppler shift), and that both the distance and the orientation of the galaxy must be known before the true orbital speed and semimajor axis can be determined. The orientation can usually be deduced for a spiral galaxy, since it has a disk shape whose tilt can be seen, and distance can be estimated through the use of one of the methods just outlined. By allowing light from many stars to enter the telescope and spectrograph, and then measuring the aggregate velocity of a large area of the galaxy, we can avoid the problem of isolating individual stars. This is somewhat imprecise, because there is always some internal motion within the portion of the galaxy that is observed, but it leads to useful estimates of the orbital speed in the outer regions.

Radio astronomy techniques are especially useful, because the gas that produces the radio emission (usually the 21-cm line of hydrogen is used in this case) often extends farther out in the galaxy than the visible stars (Fig. 27.15), and it is important to know the orbital velocities at the greatest possible distance from the galactic center. Furthermore, velocity measurements can be made very accurately with radio emission-line data, and interstellar extinction is not a problem.

In most cases the orbital velocities are measured at several points within a galaxy, from the center out as far as possible, and the data are plotted on a **rotation curve**, which is simply a diagram showing the variation of orbital velocity with distance from the center

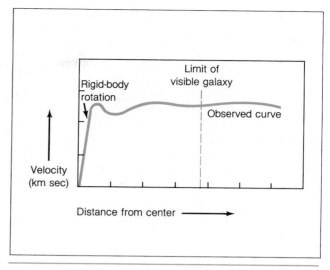

Figure 27.16 A schematic rotation curve for a galaxy.
This illustrates the fact that in most spiral galaxies, the rotation velocity does not drop off in the outer regions, as it would if most of the mass were concentrated at the center of the galaxy. The data on rotation speed in the outermost regions comes from radio observations.

(Fig. 27.16). This is useful for several reasons: It provides a means of checking whether the observations go sufficiently far out in the galaxy to reach the region where the orbiting material follows Kepler's third law; it helps in determining the orientation of the galaxy; and it ensures that the velocities have been measured as far out as possible from the center.

The technique of measuring rotation curves as just described can best be applied to spiral galaxies, where there is a disk with stars and interstellar gas orbiting in a coherent fashion. In elliptical galaxies there is no such clear-cut overall motion, and a slightly different technique must be used. The individual stellar orbits are randomly oriented within an elliptical galaxy, so there is a significant range of velocities within any portion of the galaxy's volume. This range of velocities, called the **velocity dispersion** (Fig. 27.17), is greatest near the center of the galaxy, where the stars move fastest in their orbits, and is smaller in the outer regions. Furthermore, the greater the mass of the galaxy, the greater the velocity dispersion (at any distance from the center), so a measurement of this parameter can lead to an estimate of the mass of a galaxy. The velocity dispersion is deduced from the widths of spectral lines

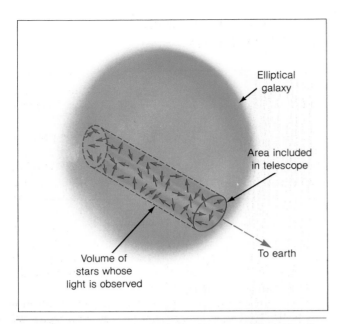

Figure 27.17 The velocity dispersion for an elliptical galaxy. *There is no easily measured overall rotation of an elliptical galaxy, so the mass cannot be estimated from a rotation curve. Instead the average velocities of stars at a known distance from the center are used. Light from a small area of the galaxy's image is measured spectroscopically. The random motions of the stars in the observed part of the galaxy broaden the spectral lines by the Doppler effect, and the amount of broadening is a measure of the average velocity of the stars in the observed region. At a given distance from the center of the galaxy, the higher the average velocity, the greater the mass contained inside that distance.*

formed by groups of stars in different portions of a galaxy; the greater the internal motion within the region that is observed, the greater the widths of the spectral lines, owing to the Doppler effect (caused by the fact that some stars are moving toward the Earth and others away from it).

Kepler's third law can sometimes be used in an entirely different way in determining galactic masses. There are double galaxies here and there in the cosmos, orbiting each other exactly like stars in binary systems. In these cases, Kepler's third law can be applied, leading to an estimate of the combined mass of the two galaxies. The uncertainties are even more severe than in double stars, however, because the orbital period of

a pair of galaxies is measured in hundreds of millions of years. Thus the usual problems of not knowing the orbital inclination or the distance to the system are compounded by inaccuracies in estimating the orbital period, something that is usually very well known for a double star. Still, this technique is useful and has one major advantage: It takes into account all the mass of a galaxy, including whatever part of it is in the outer portions, beyond the reach of the standard velocity curve or velocity-dispersion measurements. Galactic masses estimated from double systems are generally much larger than those based on measurements of internal motions within galaxies, possibly indicating that most galaxies have extensive halos containing large quantities of matter. As noted in Chapter 24, there is independent evidence that our own galaxy has a massive halo, perhaps containing as much as 90 percent of the total mass.

LUMINOSITIES, COLORS, AND DIAMETERS

Once the distance to a galaxy has been established, its luminosity and size can be deduced directly from the apparent magnitude and apparent diameter. Both quantities are found to vary over wide ranges, with luminosities as low as 10^6 and as high as 10^{12} times that of the Sun, and diameters ranging from about 1 to 100 kpc.

Generally the elliptical galaxies display a wider range of luminosities and sizes than the spirals. The latter tend to be more uniform, with luminosities usually between 10^{10} and 10^{12} solar luminosities and diameters between 10 and 100 kpc. The smallest elliptical galaxies are called **dwarf ellipticals.** These may be very common, but because they are too dim to be seen at great distances, we can only say for sure that there are many of them near our own galaxy.

It appears that galaxies of a given type tend to be fairly uniform in other properties, just as stars of a given spectral type are the same in other ways. One quantity often used by astronomers to characterize galaxies is the **mass-to-light** (M/L) ratio, which is simply the mass of a galaxy divided by its luminosity, in solar units. Values of the mass-to-light ratio typically range from 10 to over 100. Any value larger than 1 means that the galaxy emits less light per solar mass than the

Sun; that is, such a galaxy is dominated in mass by stars that are dimmer than the Sun or contains substantial quantities of mass in forms other than stars (such as interstellar matter). Even the smaller values of M/L that are observed for galaxies are much larger than 1; a value of M/L = 10, for example, means that 10 solar masses are required to produce the luminosity of one Sun.

Elliptical galaxies tend to have the largest mass-to-light ratios, consistent with our earlier statement that these galaxies are relatively deficient in hot, luminous stars. Spirals, on the other hand, which contain some of these stars, have the lower M/L values. It is worth stressing again, however, that even the low values of M/L for these galaxies are much greater than 1, indicating that they too are dominated in mass by low-luminosity stars. Hence a spiral galaxy, with all its glorious bright blue stars, actually has far more dim red ones.

The colors of galaxies also can be measured, with the use of filters to determine the brightness at different wavelengths. In general, the spirals are not as red as the ellipticals, again indicating that the latter galaxies contain a higher fraction of cool, red stars. There are also color variations within galaxies; for example, in a spiral, the central bulge is usually redder than the outer portions of the disk where most of the young, hot stars reside.

Both the mass-to-light ratios and the colors of galaxies are indicators of the relative content of Population I and Population II stars. Recall that Pop. I stars tend to be younger, and included in this group are all the bright blue O and B stars. Population II, by contrast, includes only old, red, relatively dim stars. Therefore, a red overall color, along with a high mass-to-light ratio, implies that Pop. II stars are dominant, whereas a low value of M/L and a bluer color means that some Pop. I stars are mixed in. Thus, elliptical galaxies seem to consist almost entirely of Pop. II objects, whereas spirals contain a mixture of the two populations.

This dichotomy between the two types of galaxies is found also when we compare the interstellar matter content of spirals and ellipticals. Photographs of spirals, especially edge-on views, often clearly show the presence of dark dust clouds, and face-on views typically reveal a number of bright HII regions. Neither shows up on photographs of elliptical galaxies. Radio observations of the 21-cm line of hydrogen bear this out; emission is usually strong in spirals and weak in ellipticals.

THE ORIGINS OF SPIRALS AND ELLIPTICALS

In attempting to explain the differences between spiral and elliptical galaxies, astronomers have suggested several types of theories, none of which is fully developed at present.

The weight of all the evidence cited in the previous sections is that spiral galaxies are still dynamic, evolving entities, with active star formation and recycling of material between stellar and interstellar forms, whereas elliptical galaxies have reached some sort of equilibrium in which these processes are not taking place at a significant rate. Because both types of galaxies contain old stars, we know that there are no significant age differences between the two types; we cannot simply conclude that the ellipticals are older and have run out of gas.

One of the earliest suggestions was that rotation is responsible for the division between the two types. We learned in the previous chapter that our own galaxy is thought to have formed from a rotating cloud of gas that flattened into a disk as it contracted. The rotation of the cloud was the cause of the disk formation, so perhaps elliptical galaxies are the result of contracting gas clouds that did not rotate rapidly enough to form disks. It is not clear how this would account for the lack of interstellar material and star formation, however.

One problem with this idea is that elliptical galaxies can and do rotate, yet they have not formed disks. Therefore rotation cannot be the full explanation. Apparently the key is whether a disk forms before most of the gas in the contracting cloud has been consumed by star formation (Fig. 27.18). If all the gas is converted into stars before the collapse has proceeded very far, the result is an elliptical galaxy. If the initial rate of star formation is not so great, the cloud has time to form a disk while it still contains a large quantity of interstellar gas and dust. The reason that the timing is so important is that stars, once formed, act as individual particles and continue to orbit the galaxy without forming a disk. In contrast, gas acts as a fluid and settles into a disk. (The essence of this is whether there is viscosity that will dissipate energy and allow the material to sink into the plane of a disk; stars don't encounter each other frequently enough to create a fluid viscosity, but gas particles do.)

We may ask why star formation occurs quickly in some cases, consuming all the gas and leading to formation of an elliptical galaxy; and more slowly in other stars. Why are spiral galaxies thought to be as old as elliptical galaxies?

10. Explain how the theories described in the text for galaxy formation can account for all the observed differences between spiral and elliptical galaxies.

Problems

1. To help demonstrate the difficulty of observing individual stars in other galaxies, calculate the apparent magnitude of a supergiant star whose absolute magnitude is $M = -8$, if such a star were observed in a galaxy that is 100,000 pc away (that is, a bit beyond the Magellanic Clouds), and in a galaxy 1,000,000 pc away (a little farther than the Andromeda galaxy).

2. Suppose an elliptical galaxy is found to be twice as long as it is wide. What type of elliptical is it?

3. A supernova of Type I is observed in a distant galaxy. The absolute magnitude of such a supernova is assumed to be -19. If the supernova has an apparent magnitude of $+24$ (near the limit that can be observed with the largest telescopes), how far away is the galaxy in which it lies?

4. Suppose that in a spiral galaxy, the rotation curve shows that at a distance of 15 kpc from the center, the orbital velocity is 150 km/sec. Assuming that the orbital motions at this distance from the center obey Kepler's laws, what is the mass of this galaxy?

ADDITIONAL READINGS

Allen, D. A. 1987. Star formation and *IRAS* galaxies. *Sky and Telescope* 73:372.

Bok, B. J. 1964. The large cloud of Magellan. *Scientific American* 210(1):32.

Davies, J., M. Disney, and S. Phillipps. 1990. Are spiral galaxies heavy smokers? *Sky and Telescope* 80:37.

de Boer, K. S., and B. D. Savage, 1982. The coronas of galaxies. *Scientific American* 247(2):54.

Hirshfeld, A. 1980. Inside dwarf galaxies. *Sky and Telescope* 59(4):287.

Hodge, Paul. 1981. The Andromeda galaxy. *Scientific American* 244(1):88.

_____ . 1986. *Galaxies*. Cambridge, Mass.: Harvard University Press.

_____ . 1964. The Sculptor and Fornax dwarf galaxies. *Sky and Telescope*, Dec., p. 336.

Larson, R. B. 1977. The origin of galaxies. *American Scientist* 65:188.

Mathewson, D. 1984. The clouds of Magellan. *Scientific American* 252(4):106.

Rubin, V. C. 1983. Dark matter in spiral galaxies. *Scientific American* 248(6):96.

Strom, S. E., and K. M. Strom. 1977. The evolution of disk galaxies. *Scientific American* 240(4):56.

Talbot, R. J., E. B. Jensen, and R. J. Dufour. 1980. Anatomy of a spiral galaxy. *Sky and Telescope* 60(1):23.

Toomre, A., and J. Toomre. 1973. Violent tides between galaxies. *Scientific American* 229(6):38.

MATHEMATICAL METHODS

The Stellar Luminosity Function and Galactic Mass-to-Light Ratios

 In this chapter, the point was made that the mass-to-light ratio (M/L) of a galaxy is dominated by the most luminous stars in the galaxy, which are far from being the most abundant stars. Here we show this in more quantitative terms.

To see how rare the hot, luminous stars are in a galaxy, consider the table below:

Absolute Magnitude	No. of Stars	Mass	Luminosity
−5	0.006	$25M_\odot$	$10,000L_\odot$
0	1	5	100
+5	34	1	1
+10	78	0.4	0.01
+15	125	0.1	0.0001

This is a partial listing of the *luminosity function,* which is the astronomer's term for a galactic census showing the relative numbers of stars having different luminosities. We see that there is a general trend toward increasing numbers of stars with increasing absolute magnitude (i.e., with decreasing luminosity). The number appears to drop off abruptly around absolute magnitude +17 or +18, which corresponds to the point where stellar masses are so small that nuclear core reactions never start.

We can use the luminosity function tabulated above to see the effect of the few very luminous stars on the mass-to-light ratio (M/L) of a galaxy. First, consider what M/L would be if we include only the stars comparable to the Sun and those that are dimmer—that is, the stars in the table having absolute magnitude +5 or greater. If we add up the masses and luminosities of these stars, using the relative numbers as given in the table, we get $M = 77.7 \ M_\odot$. The sum of the luminosities for the same stars is $L = 34.8 \ L_\odot$. The mass-to-light ratio is $M/L = 77.7/34.8 = 2.2$. One solar luminosity is produced by every 2.2 solar masses.

Now we compute the value of *M/L* for the entire list, so that we can see the effect of the small number of stars more luminous than the Sun. Now the total mass is $M = 82.9 \ M_\odot$, and the total luminosity is $L = 194.8 \ L_\odot$. The mass-to-light ratio is $M/L = 0.43$. Only 0.43 solar masses are required to produce a solar luminosity. The hot, luminous stars, few in number, have radically changed *M/L* for the galaxy.

We note that both of the values for *M/L* calculated here are far smaller than the actual values measured for galaxies. This discrepancy is caused in part by the incompleteness of our adopted luminosity function (which uses a somewhat arbitrary sampling of stellar absolute magnitudes), but it also illustrates the effect of unseen mass, a factor not included in our calculation. The unseen mass in galaxies has been discussed in the text; some of it is in the form of interstellar material; some is in stellar remnants such as white dwarfs, neutron stars, and black holes; and some is in the form of the mysterious "dark matter" that seems to permeate the universe.

CHAPTER 28

Clusters and Superclusters: The Distribution of Galaxies

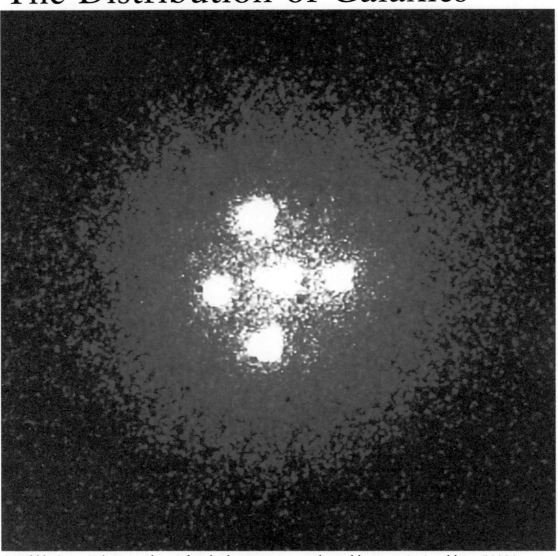

A Hubble Space Telescope photo of multiple quasar images formed by a gravitational lens. (NASA European Space Agency: Space Telescope Science Institute)

Although galaxies may be considered the largest single objects in the universe (if indeed an assemblage of stars orbiting a common center can be viewed as a single object), there are yet larger scales on which matter is organized. Galaxies tend to be located in clusters (Fig. 28.1) rather than being distributed uniformly throughout the cosmos, and these in turn have an uneven distribution, with concentrations of clusters referred to as **superclusters.** It may be that superclusters are the largest things in the universe.

Clusters of galaxies range in membership from a few (perhaps half a dozen) to many hundreds or even thousands. Just as stars in a cluster orbit a common center, so are galaxies in a group or cluster gravitationally bound together, following orbital paths about a central orbit. In a large, rich cluster, the frequent encounters between galaxies have, over time, caused the members to take on a smooth, spherical distribution; in a small group like the one to which the Milky Way belongs, the arrangement of individual galaxies is more haphazard, creating an amorphous overall appearance.

THE LOCAL GROUP

We can begin by discussing the **Local Group** of galaxies, the cluster that we live in. This group consists of about thirty members arranged in a random distribution (Appendix 13 and Fig. 28.2). Despite the relative proximity of these galaxies to us, it has been difficult to ascertain their properties in some cases, because of obscuration by our own galactic disk. It is not even possible to say with certainty how many members belong to the Local Group.

Among the member galaxies are three spirals, two of which, the Andromeda galaxy and the Milky Way, are rather large and luminous. These are probably the brightest and most massive galaxies in the entire cluster. Most of the other members are ellipticals, many of them dwarf ellipticals (Fig. 28.3). There may be additional members of this type, undetected so far because of their faintness. Four irregular galaxies, two of them being the Large and Small Magellanic Clouds (Fig. 28.4), are also found within the Local Group. In addition, there are globular clusters, which are probably distant members of our galaxy, but which lie so far away from the main body of the Milky Way that they appear isolated.

Figure 28.1 A cluster of galaxies. *Several galaxies are visible in this photograph of a portion of a large cluster of galaxies, called the Fornax cluster. (© 1984 Royal Observatory, Edinburgh)*

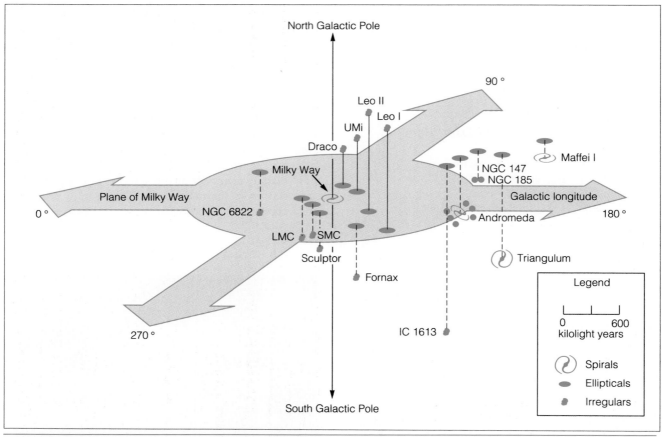

Figure 28.2 The Local Group. *This diagram schematically illustrates the arrangement of galaxies in the Local Group, with respect to the Milky Way.*

The Local Group is about 1,000 kpc in diameter and has a roughly disklike overall shape, with the Milky Way located a little off center. Beyond the outermost portions of the Local Group, there are no conspicuous external galaxies for a distance of some 1,100 kpc.

The Magellanic Clouds and the Andromeda galaxy (also commonly known as M31, its designation in the widely cited Messier Catalog; Appendix 15) have been particularly well studied, because of their proximity and prominence, and because of what they can tell us about galactic evolution and stellar processing.

The Large and Small Magellanic Clouds appear to the unaided eye as fuzzy patches, easily visible only on dark, moonless nights. They lie near the south celestial pole and can therefore be easily seen only from the Southern Hemisphere. Their name originated from the fact that the first Europeans to see them were Ferdinand

Magellan and his crew, who made the first voyage around the world in the early sixteenth century.

The Magellanic Clouds are satellites of the Milky Way, having lesser masses and following orbits about it, taking several hundred million years to make each circuit. Lying between 50 and 60 kpc from the Sun, both are irregulars of Type I. They contain substantial quantities of interstellar matter and are quite obviously the sites of active star formation, with many bright nebulae and clusters of hot, young stars (Fig. 28.5). Measurements of the colors of these galaxies and of the spectra of some of their brighter stars indicate that they have somewhat lower heavy-element abundances than Pop. I stars in our galaxy. This seems to indicate that the Magellanic Clouds have not undergone as much stellar cycling and recycling as the Milky Way. These and other Type I irregular galaxies may generally be

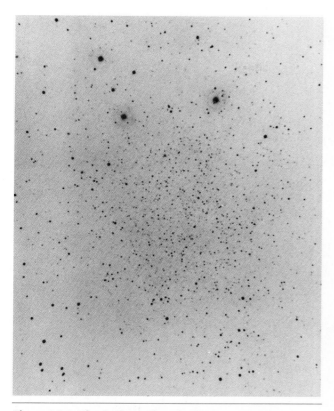

Figure 28.3 The Sculptor dwarf galaxy. *This loose conglomeration of stars lies about 83 kiloparsecs from the Sun. Such a small, dim galaxy would not be noticed if it were much farther away, but the number of systems of this type in the Local Group leads to the assumption that they may be very common in the universe. (P. W. Hodge, photo taken at the Boyden Observatory)*

Figure 28.4 The Magellanic Clouds. *The Large Cloud is shown in the upper photo, and the Small Cloud in the lower picture. Our nearest neighbors, these galaxies orbit as satellites of the Milky Way. (Top:© 1984 Royal Observatory, Edinburgh; bottom: R. J. Dufour, Rice University)*

Figure 28.5 The Tarantula nebula. *This spectacular region of ionized gas and hot dust is an active site of star formation in the Large Magellanic Cloud. The great supernova that occurred in early 1987 lies very near this nebula (but outside the field of view of this photograph; see discussion of Supernova 1987A in Chapter 22). (© 1984 Anglo-Australian Telescope Board)*

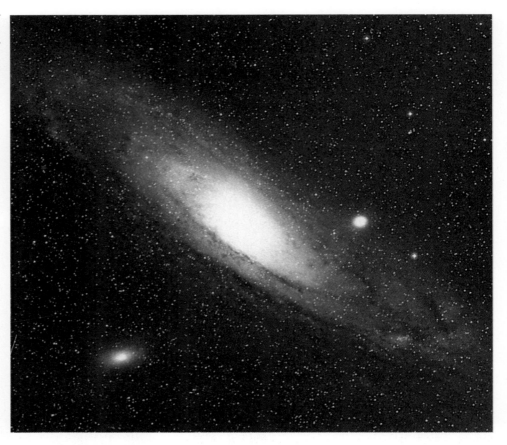

Figure 28.6 The Andromeda galaxy. *At a distance of over 2 million light-years, this galaxy is the most distant object visible to the unaided eye. The naked eye sees only an extended, fuzzy patch of light, rather than the detailed view shown here, which was obtained with a telescope and a time-exposure. (Palomar Observatory, California Institute of Technology).*

viewed as galaxies in extended adolescence, not yet having settled down into mature disks. In the Magellanic Clouds, the reason for the unrest is probably the gravitational tidal forces exerted by the Milky Way.

The great spiral M31, the Andromeda galaxy (Fig. 28.6), is the most distant object visible to the unaided eye, lying some 700 kpc from our position in the Milky Way. All that the eye can see is a fuzzy patch of light, even if a telescope is used; but when a time-exposure photograph is taken, then the awesome disk and spiral arms stand out. The Andromeda galaxy is so large, extending more than one degree across the sky, that full portraits can be obtained only by using relatively wide-angle telescope optics (most large telescopes have extremely narrow fields of view).

The Andromeda galaxy is probably very similar to our own galaxy, and thus has taught us quite a bit about the nature of the Milky Way. The two stellar populations were first discovered through studies of stars in Andromeda. In fact, the effects of stellar pro-

cessing on the chemical makeup of different portions of a galaxy are better determined for Andromeda than for the Milky Way. Observations of the Andromeda galaxy in invisible wavelengths (Fig. 28.7) have revealed similarities in the distribution of interstellar matter and X-ray sources as well.

Another very important by-product of observations of the Andromeda galaxy, as well as those of some other members of the Local Group, is the calibration of our distance scales for measuring objects farther away. Knowledge of distances to galaxies within the Local Group, measured through the use of the period-luminosity relationship for Cepheids, has allowed astronomers to establish the absolute magnitudes of various standard candles, which, in turn, are used to find distances to more faraway galaxies. Each step in extending the distance scale must be built upon previous steps, and the Local Group galaxies represent an important stage in the process of reaching out to the greater distances of other clusters of galaxies. It would

Figure 28.7 New views of the Andromeda galaxy. *Here are an X-ray image (left) and an infrared picture (right) of our neighbor spiral galaxy. The X-ray view reveals associations of stars with hot coronae and binary X-ray sources; the infrared image shows the locations of cold dust clouds. (Left: Harvard-Smithsonian Center for Astrophysics; right: NASA)*

have been very difficult indeed to establish the distances of extragalactic objects if we did not live within a cluster of galaxies where this calibration process is possible.

Before leaving the Local Group, we return to the question of its membership, and the rank of the Andromeda galaxy and the Milky Way within it. New galaxies are occasionally discovered and found to belong to the Local Group. These are most often dwarf ellipticals. In the early 1970s, however, infrared observations revealed two new galaxies (Fig. 28.8), both lying behind the disk of the Milky Way and therefore obscured from our direct view. These galaxies, called Maffei 1 and Maffei 2 after their discoverer, are both rather large, one being an elliptical and the other a

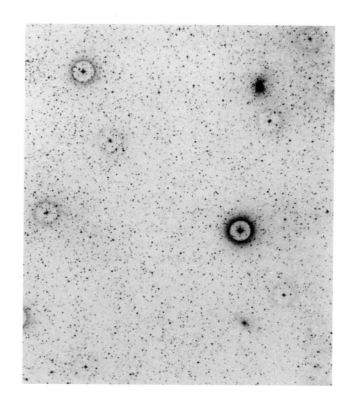

Figure 28.8 Maffei 1 and 2. *This infrared photograph reveals two large galaxies (the fuzzy images, upper right and lower right) that were for a while considered possible members of the Local Group. More extensive analysis has shown, however, that they probably are not. (The circular marks are overexposed images of closer stars.) (H. Spinrad)*

Figure 28.9 A portion of a rich cluster of galaxies. *Most of the objects in this photograph are galaxies. This cluster lies in the constellation Hercules. (National Optical Astronomy Observatories)*

spiral. Their distances were not well known for a while, and it appeared possible that they were members of the Local Group. If this had been the case, they would have displaced the Andromeda galaxy and the Milky Way as the dominant members, and this would have altered our perception of the cluster as a whole. Recent evidence has shown, however, that the two galaxies are probably too distant to be members of the Local Group.

RICH CLUSTERS: DOMINANT ELLIPTICALS AND GALACTIC MERGERS

In contrast with small clusters such as the Local Group, many clusters of galaxies are much larger and contain hundreds or even thousands of members (Fig. 28.9). The distribution of galaxies in one of these rich clusters tends to be smooth, so that the density of galaxies increases gradually from the outermost regions to the center, where the density of galaxies is highest. By contrast, in a small cluster like the Local Group, the galaxies are unevenly distributed, with no central dense region.

Because the density of galaxies in a rich cluster is relatively high, there are numerous close encounters between galaxies as they follow their individual orbits about the center. When two galaxies pass close by each other, they exert mutual gravitational forces that can have profound effects. Apparently these effects include

the conversion of some galaxies from one type to another.

The central regions of large, rich clusters are highly dominated by elliptical and S0 galaxies, more so than in small groups or isolated galaxies. Near the center of a rich cluster, some 90 percent of the galaxies may be ellipticals or S0s (Fig. 28.10), whereas among noncluster galaxies about 60 percent are spirals. This contrast is probably partially a result of the frequent near-collisions between galaxies in dense clusters. When two galaxies have a close encounter, the tidal forces they exert on each other stretch and distort them (Fig. 28.11). Under some circumstances, any interstellar matter in them can be pulled out and dispersed. The outer regions, such as the halos, can also be stripped away. Intracluster gas (see next section) is even more effective at converting spirals to ellipticals. Pressure created by motion through this gas can sweep out a galaxy's interstellar matter and halt star formation. A spiral galaxy subjected to these cosmic upheavals may assume the form of an elliptical. In time, most of the spirals in an entire cluster, particularly those in the dense central region, may be converted into ellipticals and S0s.

The frequent gravitational encounters between galaxies in a rich cluster have another interesting effect: They cause a buildup of galaxies right at the center of

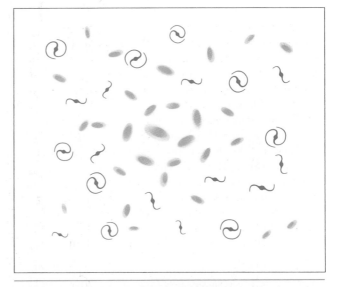

Figure 28.10 Galaxy types in a rich cluster. *In dense, highly populous clusters of galaxies, nearly all the members in the central portion are ellipticals, and there is often a giant, dominant elliptical at the center. Only in the outer portion are many spirals seen.*

Figure 28.11 A collision between galaxies. *The photo above shows a pair of galaxies undergoing a near-collision. At right is a computer simulation of their interaction, showing how the present appearance was created. This type of interaction between galaxies is believed responsible in some cases for the removal of interstellar matter, converting spiral galaxies to elliptical or S0 galaxies. (photograph from Palomar Observatory, California Institute of Technology; computer simulation from A. Toomre and J. Toomre)*

the cluster. When two galaxies orbiting within a cluster come together, one will always gain speed and move to a larger orbit, while the other (always the more massive of the two) will lose speed and drop closer to the center. Thus a gradual sifting process, analogous to the differentiation that occurs inside some planets as the heavy elements sink toward the core, gradually builds up a dense central conglomeration of galaxies at the heart of the cluster. There these galaxies may actually merge, the end result being a single gigantic elliptical galaxy, which continues to grow larger as new galaxies fall in. In at least one case, a radio image shows a galaxy with two nuclei, apparently the result of a merger still in progress (Fig. 28.12).

The scenario just described was developed rather recently, as astronomers sought to explain why elliptical galaxies are so prevalent in many rich clusters, and why a single, giant elliptical is found at the center of many of these clusters (Fig. 28.13). These giants, designated cD galaxies, are at the extreme upper limit of known galactic sizes, masses, and luminosities, and are often peculiar in certain ways, as we will learn in

Chapter 30. Interestingly, many galaxies outside of rich clusters also appear to have undergone mergers.

MASSES OF CLUSTERS

There are two methods for measuring the mass of a cluster of galaxies, and both are quite uncertain. This is a crucial problem, as we will see in Chapter 31, because of the importance of knowing how much mass the universe contains.

The simpler and more straightforward of the two methods is to estimate the masses of the individual galaxies in a cluster (using techniques described in the previous chapter) and then to add up these masses. In many cases, particularly for distant clusters in which it is impossible to measure the rotation curves or velocity dispersions of individual galaxies, the only way to estimate their masses is to measure their brightnesses and then use a standard mass-to-light ratio to derive their masses. This technique is inaccurate because it

Figure 28.12 A radio image of a galactic merger. *This is a radio map of a giant elliptical galaxy with two nuclei (the bright points at lower center, each having a pair of wispy gaseous jets emanating from it; the jets are discussed in Chapter 30). Evidently this galaxy is in the process of forming from the merger of two galaxies whose centers have not yet completely merged. (National Radio Astronomy Observatory)*

depends on our knowing the distance, and because it assumes that the galaxies adhere to the usual mass-to-light ratios for their types. It also does not take into account any matter in the cluster that may lie between the galaxies.

The second method is similar to the velocity-dispersion technique used to estimate masses of elliptical galaxies (Fig. 28.14). The mass of a cluster is estimated from the orbital speeds of galaxies in its outer portions: The faster they move, the greater the mass of the cluster. This method has the advantage that it measures all the mass of the cluster, whether the mass is in galaxies or between them. But it has the disadvantage that the velocity measurements, particularly for a very distant cluster, are difficult, and it is necessary to know the cluster distance in order to determine how far from the center the observed galaxies are. Furthermore, the technique is only valid if the galaxies are in stable orbits about the cluster; if the cluster is expanding or some of the galaxies are not really gravitationally bound to it, then the results are incorrect. In rich clusters, at least, the smooth overall shape and distribution give the appearance of a bound system, and this technique is probably valid. It may not be valid for small clusters such as the Local Group, however.

Figure 28.13 A giant elliptical galaxy. *This is M87, a well-known example of a dominant central galaxy in a rich cluster. This particular galaxy is discussed further in Chapter 30. (National Optical Astronomy Observatories)*

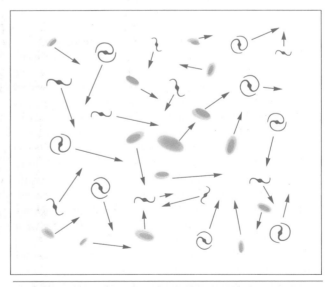

Figure 28.14 Galaxy motions within a cluster. *Galaxies move randomly within their parent cluster. In a large cluster, where the overall distribution of galaxies is uniform, analysis of the average galaxy velocities is used to estimate total mass of the cluster, just as the velocity dispersion method is used to measure the masses of elliptical galaxies.*

Whatever the reason, this method always leads to an estimated cluster mass that is much greater than that based on standard mass-to-light ratios. It is as though a large fraction of the cluster's mass is either in extensive galactic halos, which are ignored when the individual galaxy masses are estimated, or in intergalactic space.

GRAVITATIONAL LENSES AND DARK MATTER IN CLUSTERS

In our discussion of the derivation of masses of clusters of galaxies, we found that there seems to be much more mass present in many clusters than can be accounted for by the visible galaxies in those clusters. This implies that there is more mass present in these clusters than we can see in the form of stars and galaxies. The invisible matter inferred to be present has been dubbed **dark matter.** Whatever this dark matter is, it has profound implications for the overall distribution of mass in the universe, and plays a key role in determining the ultimate fate of the universe, as discussed in Chapter 31.

Very recent observations of clusters of galaxies have lent new insights into the distribution of the dark matter, while leaving its form and origin as mysterious as ever. Several clusters have been found to contain huge glowing arcs or rings of light around them (Fig. 28.15). These so-called **luminous arcs** are truly gigantic, if they lie at the same distance from us as the clusters in which they are seen. They appear to be millions of light-years in length, comparable in size to the largest structures associated with individual galaxies (such as the enormous radio-emitting regions surrounding radio galaxies; see Chapter 30).

It is now thought that the luminous arcs are not material structures at all, but instead that they are extended, magnified, and distorted images of galaxies farther away. It was shown theoretically by Albert Einstein, and later confirmed by experiment, that a gravitational field deflects photons of light (this was discussed in Chapter 23). Because of this deflection, under certain circumstances a gravitational field can act like a lens, creating a magnified or distorted image of an object in the distance (Fig. 28.16). If there is perfect alignment among the source of light, the massive object in between, and the observer, the image formed by the gravitational lens will be a circle. If the alignment is not quite perfect, the image can consist of one or more arcs of light, or of multiple images. Several years ago,

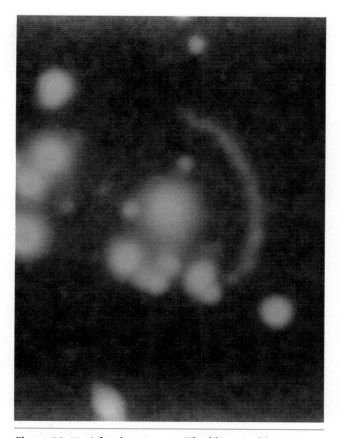

Figure 28.15 A luminous arc. *The blue, ringlike structure shown here is gigantic in scale, dwarfing the individual galaxies in this cluster. (F. Hammer and J. H. Jones, Canada-France Hawaii Telescope)*

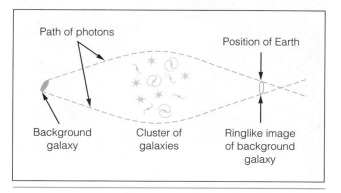

Figure 28.16 Gravitational lens creating luminous areas. *The gravitational field of a foreground cluster of galaxies can deflect the light from a distant object, creating a ringlike image if the alignment is perfect as shown here, or arclike images if the alignment is not quite perfect.*

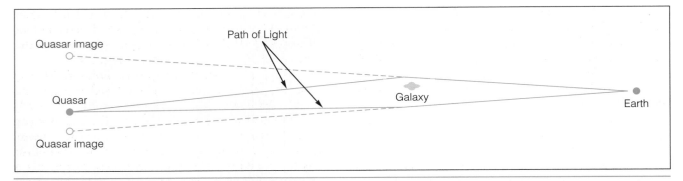

Figure 28.17 A double quasar image. *This diagram illustrates how an intervening galaxy can act as a gravitational lens, forming two images of a distant quasar.*

a closely spaced pair of very distant objects called quasars (see Chapter 30) was observed to have virtually identical properties, and it was deduced that this was two images of the same quasar, formed by the gravitational lensing effect of a galaxy in between the Earth and the quasar (Fig. 28.17). Since then many other examples of multiple images of quasars have been found.

Now it is thought that the luminous arcs seen around some galaxy clusters are also the result of gravitational lenses.

If a distant galaxy lies behind a massive cluster of galaxies, then depending on the precision of the alignment as seen from the Earth, the gravitational field of the cluster can form one or more arclike, magnified

Figure 28.18 Dark matter in a cluster of galaxies. *It is possible to deduce the distribution of dark matter in a cluster of galaxies by analyzing the luminous arcs created by the gravitational lensing effect of the dark matter. The left panel shows the optical image of a cluster of galaxies; close examination reveals a number of short, blue arcs. The right panel shows, on the same scale, the distribution of dark matter required to produce the blue arcs gravitationally. The dark matter is concentrated toward the center of the cluster but has a smoother distribution than the galaxies. (T. A. Tyson, Bell Laboratories)*

images of the background galaxy. From the size and shape of the image, it is possible to calculate the mass contained in the cluster of galaxies, because the properties of the image depend on the strength of the gravitational field that forms the lens. Thus we have a new method of determining the masses of galaxy clusters, one that is quite independent of the usual assumptions that have to be made about the orbits of individual galaxies within the cluster.

This new method yields results that agree very well with those determined from other methods. The masses of clusters are still found to be much larger than the sum of the masses of the visible galaxies. Thus this new tool is helping to confirm the growing belief among astronomers that the universe contains vast quantities of invisible, dark matter. At the same time, it is also providing a new and dramatic confirmation of one of the predictions made by Einstein's theory of general relativity.

It has actually been proven possible, through the analysis of the luminous arcs observed in and around a cluster, to determine the distribution of the dark matter in the cluster (Fig. 28.18). In some clusters, a large number of luminous arcs are seen. Using the equations that describe the deflection of light by a gravitational field, astronomers have been able to deduce the mass distribution that would be needed to create the observed pattern of arcs. In this way, it has proven possible to actually map the dark matter in several clusters of galaxies. It is almost as if the dark matter can be seen. It is found that the dark matter traces the distribution of visible galaxies in each cluster, but without much fine-scale variation from place to place. So whatever the dark matter is, it tends to be associated with the glowing matter that makes up the visible stars and galaxies in the universe.

INTERCLUSTER GAS AND X-RAY EMISSION

For a variety of reasons , astronomers have sought to know whether the space between galaxies is truly empty or whether there is an intergalactic medium analogous to the interstellar material that exists within galaxies. The quantity of intergalactic matter could play a vital role in the future evolution of the universe (see Chapter 31) and could have a profound effect on the fate of individual galaxies, in terms of their motions and their shapes. If galaxies move through a gaseous

medium, drag will be created, which will help them congregate toward the centers of rich clusters and aid in stripping off their halos and interstellar gas; in this way spirals will be converted to ellipticals.

Searches for intergalactic gas that utilize techniques similar to those traditionally employed to study the interstellar matter in our own galaxy have generally yielded negative or ambiguous results. The lack of strong evidence for an intergalactic medium provides useful information, however. It tells us that there is very little gas there, and what is there must be so hot that it is essentially invisible, because a very highly ionized gas does not create visible-wavelength absorption lines.

The possibility that there is a very hot intergalactic medium appeared to be supported by recent X-ray observations of clusters of galaxies, which showed that many clusters are filled with a diffuse gas that glows at X-ray wavelengths (Fig. 28.19). The spectrum of the X-ray emission implies that this gas is indeed very hot;

Figure 28.19 Intracluster gas in the Virgo cluster. *This is an X-ray image from the* Einstein Observatory, *showing emission from the entire central portion of this cluster of galaxies. The X-rays are being emitted by very hot gas that fills the space between galaxies. The X-ray emission is shown in blue; radio images of galaxies in the cluster are shown in red. (X-ray data from the Harvard-Smithsonian Center for Astrophysics; radio data and image processing by the National Radio Astronomy Observatory)*

Figure 28.20 The Supercluster. *This is an artist's concept of the cluster of galaxy clusters to which the Local Group belongs. The Local Group is at position (0,0) in this diagram. (Illustration from* Sky and Telescope Magazine *by Rob Hess. © 1982 by the Sky Publishing Corporation; also with the permission of R. B. Tully)*

its temperature is estimated to be as high as 100 million degrees, much hotter than even the highly ionized gas in the interstellar medium in our galaxy. This observation raised the possibility that the general intergalactic void is filled with such gas, although if it is, the density outside the clusters must still be rather low, or the X-ray data would have revealed its presence.

A different type of X-ray measurement, however, indicated that the hot, intracluster gas is not part of a general intergalactic medium, but instead originates in the galaxies themselves. Spectroscopic measurements made at X-ray wavelengths revealed that the gas contains iron, a heavy element, in nearly the same quantity (relative to hydrogen) as in the Sun and other Pop. I stars. Such a high abundance of iron could only have been produced in nuclear reactions inside of stars. Therefore this intracluster gas must once have been involved in part of the cosmic recycling that goes on in galaxies, as stars gradually enrich matter with heavy elements before returning it to space. How the gas was then expelled into the regions between galaxies is not clear, but it may have been swept out during near-collisions between galaxies, or it may have been ejected in galactic winds created by the cumulative effect of supernova explosions and stellar winds.

Estimates of the quantity of mass contained in the hot intracluster gas show that this material falls far short of the mass contained in the clusters in the form of dark matter. Hence we are still left with the mystery of what form most of the mass in galaxy clusters is in.

SUPERCLUSTERS, VOIDS, AND WALLS

We turn now to consider the overall organization of the matter in the universe. We noted earlier that clusters of galaxies may represent the largest scale on which matter in the universe is clumped, but that there is some evidence for a higher-order organization, even in the case of the Local Group (Fig. 28.20). It appears that clusters of galaxies are themselves concentrated in certain regions, commonly referred to as superclusters.

The reality of superclusters has been difficult to establish, and for a while many astronomers were not convinced. Today, however, there are few doubts that clusters of galaxies tend to be grouped, although there is still disagreement on the significance of this. The best evidence lies in the distribution of rich clusters (some

1,500 of these have been cataloged), which clearly tend to congregate in certain regions, with relatively empty space in between.

The uncertainty that remains has to do with whether the groupings are random occurrences or whether they reflect a fundamental unevenness in the distribution of matter in the universe. This is usually tested by comparing the apparent clustering of clusters with what would be expected from coincidence if they were actually distributed in a random fashion (after all, in any random scattering of objects, there will always be occasional accidental groupings). Mathematical experiments of this type appear to show that the observed superclusters could be random concentrations of clusters and therefore may not have profound significance for the universe as a whole. On the other hand, very recent observations appear to show that a large number of clusters of galaxies may be linked into an enormous filamentary supercluster spanning a major portion of the sky. Previously unrecognized because it crosses the plane of the Milky Way, where our view of distant galaxies is obscured, this supercluster, if real, is far too large to have been the result of random groupings of clusters. This may show that there is a basic unevenness in the distribution of matter in the universe.

Other signs of an uneven distribution of matter in the universe have begun to show up in the observations. As astronomers perform larger and larger surveys of galaxy positions and distances, probing more and more deeply into the far reaches of the universe, they find increasing evidence that the galaxies are not distributed randomly. The structures being discovered are larger even than superclusters, yet still much smaller than the overall size scale of the universe itself. Today it still is not clear what the term "supercluster" means. While the reality of enormous concentrations of clusters of galaxies is unquestioned, these structures appear to be segments of gigantic sheets and arcs, rather than isolated groupings. Between the sheets and arcs are huge volumes that appear to be relatively devoid of visible matter.

Several years ago, a group of astronomers carrying out a large-scale survey of galaxies discovered a huge region that appeared to be empty, devoid of galaxies. This cavity or void was some hundreds of millions of parsecs across. More recently, this void and others like it have been found to be part of a cell-like structure in the universe, where sheets of galaxies and clusters of galaxies separate large, relatively empty regions from each other (Fig. 28.21). The structure of the universe may be likened to a gigantic Swiss cheese or bubble bath. Very recently, a huge wall-like structure was found, consisting of galaxies and clusters arranged in an enormous sheet so big that its true dimensions are yet to be uncovered.

The result of all these surveys and findings is that the universe appears to be far lumpier and more uneven than previously envisioned. The question now is how

Figure 28.21 Arclike superclusters. *This shows the distribution of galaxies in a region of the sky that has recently been mapped completely, showing that superclusters are arranged in curved filamentary structures. This is evidence that the distribution of matter in the universe is not uniform, but that there is a large-scale structure. (M. Geller, J. Huchra, M. Kurtz, and V. de Lapparent, Harvard-Smithsonian Center for Astrophysics).*

it got this way; did the universe start out in a uniform state which then became lumpy, or was it somehow formed in an uneven and clumpy state? The possible answers to this question are discussed in the next section and in Chapter 31.

THE ORIGINS OF CLUSTERS AND SUPERCLUSTERS

The problem of explaining the formation of clusters of galaxies and larger structures involves fundamental questions about the universe itself. We must ask how and why (and when) the universe became clumpy enough for condensations of matter to collapse gravitationally and form galaxies and clusters. In some respects, this brings up issues similar to those confronted in explaining the formation of stars, because the probability that a cloud will collapse depends on its density and temperature (see the discussion of star formation in Chapter 21). For a given gas density, the cloud mass that will spontaneously collapse is determined by the temperature. The higher the temperature, the greater the internal energy of the gas, and the larger the region that must become self-gravitating if collapse is to occur. If we believe, as most astronomers now do, that 90 percent or more of the mass of the universe is in the form of dark matter rather than visible stars (see Chapter 31), then it is the temperature of the dark matter that determined how galaxy formation proceeded.

Thus, to understand galaxy and cluster formation, we need to invoke the properties of the mysterious, pervasive dark matter that apparently fills the universe but defies detection. The dark matter content of the universe is a central topic in modern discussions of cosmology; this is described in some detail in Chapter 31. For now, the essential point is that many lines of evidence suggest that the majority of the mass in the universe is in forms that we cannot see. If this dark matter is hot, then the first condensations would have had masses comparable to or even greater than the masses of superclusters, around 10^{15} solar masses. If these were the first "objects" to form, then the resulting superclusters must have fragmented sequentially, forming clusters and then individual galaxies. This view, sometimes called the "top-down" theory of galaxy formation because the largest structures formed first, would support the notion that superclusters represent a fundamental unevenness in the distribution of matter, rather than having formed by the random aggregation of galaxies and clusters.

The opposing theory, called the "bottom-up" model, suggests that individual galaxies, or perhaps even smaller objects, formed first and then collected together to form clusters and superclusters. This would have happened if the dark matter was cold, because then the masses of the first condensations would have been around 10^5 or 10^6 solar masses, comparable to globular clusters. Some support for this view is found in the shapes of many galaxies, particularly those seen very far away and hence as they were very long ago, which appear to be the result of the merger of star clouds having masses smaller than a normal galaxy (as discussed in the Guest Editorial following Chapter 26, it is possible that our own Milky Way galaxy may have formed through the merger of smaller subsystems).

A crucial distinction between the "top-down" and "bottom-up" theories is the time when galaxy formation occurred. In the cold dark matter (or bottom-up) picture, galaxy formation occurred relatively recently (that is, about 10 billion years ago), whereas in the hot dark matter (or top-down) theory, the first galaxies would have appeared much earlier, within the first billion years after the universe began. It is possible to estimate when galaxy formation took place, because we can look back into the past by looking sufficiently far away. The light from distant objects takes time to reach us; for example, if we look at a galaxy 1 billion light-years away, we see it as it was 1 billion years ago. If we can look back far enough to see galaxies in the process of formation, their distance from us will tell us how long ago the formation occurred. A possible protogalaxy, a huge amorphous gas and star-formation system some 12 billion light-years away, has been tentatively identified as a galaxy just in the process of formation. If this is correct, then galaxy formation apparently took place at about the right time to be consistent with the bottom-up theory.

The enormous extent of some superclusters becomes a problem for the bottom-up theory, however. There appears to be an overall bubblelike network of filamentary superclusters (see Fig. 28.21), which is very difficult to explain as a result of aggregation of clusters. Furthermore, indirect evidence (discussed in the next chapter) suggests the existence of huge concentrations of mass comparable to the 10^{15} solar masses expected from the top-down, or hot dark matter, theory. Thus, at present we do not have a clear basis for adopting either picture over the other; perhaps in the end we

ASTRONOMICAL INSIGHT 28.1

Galaxies at the Edge of the Universe

Astronomers are forever probing the most distant reaches of the universe that can be reached with telescopes. This is important because of the new kinds of objects that might be revealed, because of the new information that can be found on the overall nature of the universe and the distribution of matter in it, and because the farther away we look, the farther back in time we are seeing (because of light-travel time; this is discussed further in the next three chapters). One of the primary motivations for building the *Hubble Space Telescope* is to detect fainter objects than can be seen with telescopes that look up through the Earth's atmosphere.

Now it appears that ground-based telescopes may have "scooped" the *Space Telescope* by observing the most distant objects it will ever be possible to see. Using modern electronic detectors in place of photographic film, astronomers have obtained images of regions of the sky that, on existing photographs, appear to be completely dark and devoid of objects. The superb sensitivity of the new detectors revealed a host of faint blue objects rather than a dark sky. These ethereal images represent objects as faint as magnitude 30, exceeding the expected limiting sensitivity of the *Space Telescope*.

It is not easy to demonstrate conclusively what the objects are, but the most likely explanation is that they are very distant galaxies.

Very Faint Galaxies?   *The faint blue objects in this image, obtained through the use of a very sensitive electronic detector with a large telescope, are thought to be galaxies. If so, they are being seen as they were many billions of years ago, when the universe was young. (A. Tyson, Bell Laboratories).*

If so, the light from them has taken several billion years to reach us, which means that these galaxies must have been very young at the time the light was emitted.

It is possible that these objects are galaxies seen as they were during or shortly after their formation. If so, then a search for even fainter galaxies might reveal nothing. If there were no galaxies at earlier times, as some theories of cosmology suggest, then a search for more distant galaxies would be the same as a search for galaxies at times earlier than the era when galaxies formed. If the theories are correct, there should have been no galaxies at these early times.

In most theories for the evolution of the universe (Chapter 31), it is believed that at the earliest times no stars and galaxies existed. It is thought that all the material of the universe was in the form of atoms and ions, and possibly huge quantities of "dark"

matter, which may still be the dominant form of matter in the universe (see discussions in this chapter and in Chapter 31). At some time, several billion years ago, stars and galaxies came into being through the gravitational condensation of atomic matter, and the universe began to look something like it does today. The very young galaxies now being detected may represent the time when galaxy formation was just beginning. If so, then no matter how much farther we probe, nor how faint the objects we seek, we will find nothing but dark skies beyond these objects.

This finding, if confirmed, has profound implications. For one thing, it explains a longstanding riddle: If the universe were really infinite, then in any direction from the Earth our line of sight should eventually intersect the surface of a star, and the night-time sky should be bright instead

(continued on next page)

ASTRONOMICAL INSIGHT 28.1
Galaxies at the Edge of the Universe (continued)

of dark. Now we see that in effect the universe is not infinite because there is a "horizon" in time before which no luminous objects such as stars and galaxies existed (see the discussion of Olber's paradox in Chapter 31).

Even though no stars and galaxies can be seen farther away (i.e., longer ago) than some limiting distance (or some early time),

we do have a way of observing the universe as it was at times even before there were stars and galaxies. The universe is filled with radiation that is left over from the earliest moments of its creation, and this background radiation is observed today (this is discussed in the next chapter).

The conclusions discussed here concerning the faint objects seen in very sensitive images of the sky are quite new and must be regarded as tentative. When and if these results are widely accepted, astronomers will still continue pushing toward ever larger and more sensitive telescopes, because

there is much yet to be learned about the universe. There is always more to be discovered, and making discoveries often requires gathering more light. For example, to definitely show whether the newfound faint objects really are recently formed galaxies, it will be necessary to obtain spectra of them, and this requires more light than is needed for simply detecting the objects. Obtaining their spectra is difficult or impossible with existing telescopes. It is safe to assume that astronomers will never outgrow their need for larger telescopes!

---

will find that both are wrong, or that some combination of the two prevails. In any event, the extent and nature of dark matter in the universe, as well as the manner in which it became lumpy and formed galaxies and clusters, are very important to our understanding of the universe as a whole.

Despite the apparent difficulty in explaining the enormous observed structures as due to random gravitational encounters among galaxies, recent model calculations show that this could have happened. Using larger computer models than previously available, a group of astronomers has recently shown that such structures as the huge filaments, voids, and "walls" of galaxies that have been observed could have formed by the gradual process of gravitational clumping, starting from an early universe in which the matter was uniformly and evenly distributed. If so, then the "bottom-up" theory, in which the dark matter is cold and therefore dominated by gravitational forces, may prove successful in explaining the observed overall distribution of matter in the universe.

PERSPECTIVE

We have, at last, finished our tour of the universe, having discussed nearly all the forms and types of organization that matter can take. We still must deal with a few specific kinds of objects that have not yet been described, but the overall picture of the universe in its

present state is now more or less complete. We are ready to tackle questions having to do with the nature of the universe itself, its overall properties, and its dynamic nature, and we will do so before examining some of the peculiar objects that are clues to the past.

SUMMARY

1. Many galaxies are members of clusters, rather than being randomly distributed throughout the universe.
2. The Milky Way is a member of a cluster called the Local Group, which contains about thirty galaxies.
3. The nearest neighbors to the Milky Way are the Magellanic Clouds, both Type I Irregulars.
4. The Andromeda galaxy, a huge spiral of type Sb, is similar in size and general properties to the Milky Way, and the two are among the most prominent members of the Local Group.
5. Rich clusters of galaxies may contain thousands of members. In contrast to the general situation for isolated galaxies and small groups, in rich clusters elliptical and S0 galaxies are the most common, and there is a relatively small fraction of spirals.
6. The dominance of elliptical galaxies in rich clusters is probably the result of the conversion of spirals into ellipticals by tidal forces exerted by other galaxies and by drag created by intracluster gas.
7. In many rich clusters, there is a giant elliptical galaxy at the center that probably formed from the

merger of several galaxies that settled there as a result of collisions.

8. Masses of clusters can be determined from the sum of the masses of individual galaxies (measured by techniques described in Chapter 27), or from the internal velocity dispersion of the galaxies in the cluster.

9. Observations of luminous arcs in clusters of galaxies confirm the presence of dark matter, and allow the distribution of the dark matter to be mapped.

10. Although there is no evidence for a general intergalactic medium, there is a very hot, rarefied gas filling the space between galaxies in some rich clusters. Observed best at X-ray wavelengths, this gas appears to have been created by the galaxies themselves (through mass loss in galactic winds that result from stellar winds and supernova explosions).

11. Clusters of galaxies tend to be grouped into aggregates called superclusters, but these may be random concentrations, rather than fundamental inhomogeneities of the universe. Superclusters appear to be sheetlike, with an overall distribution resembling a series of huge arclike structures, and this may tell us something about the early distribution of matter in the universe.

12. Clusters of galaxies formed because of an uneven distribution of matter at some point early in the history of the universe, but it is not known how this clumpiness originally developed. One of the two competing possibilities is that the early universe was hot, so that only very massive objects could collapse; this is the "top-down" theory, in which superclusters of galaxies formed first and then fragmented to form galaxies. The other scenario, called the "bottom-up" theory, assumes that the matter in the universe was cold, so that objects as small as galaxies could condense and then aggregate into clusters.

REVIEW QUESTIONS

Thought Questions

1. Summarize the membership of the Local Group of galaxies.

2. Why does a small cluster such as the Local Group have an amorphous shape rather than a spherical, smooth shape such as is seen in rich clusters of galaxies?

3. How does our location in a cluster (the Local Group) help us to measure distances to galaxies far away, outside our cluster?

4. Explain why the relative numbers of galaxies of different types may not be the same in a rich cluster of galaxies as it is among galaxies not belonging to such clusters.

5. Why would it be surprising if the Local Group included a giant elliptical galaxy among its members?

6. Summarize the techniques that are used for finding the masses of clusters of galaxies. For each method discussed, explain any difficulties or shortcomings.

7. Why is it thought that the intracluster gas observed in rich clusters of galaxies has been expelled from the galaxies in the cluster, rather than being truly intergalactic gas?

8. Explain why there has been some question about the significance of superclusters, even though their existence is well established.

9. What explanations have been offered by astronomers for the apparent concentration of superclusters in arclike or sheetlike structures?

10. How is the problem of explaining the formation of individual galaxies similar to the problem of explaining the formation of individual stars within a galaxy?

Problems

1. If a dwarf elliptical galaxy has an absolute magnitude of $M = -15$, and could be detected with an apparent magnitude as faint as $m = +20$, how far away can these galaxies be found? Compare this distance with the diameter of the Local Group and with the distance of the Virgo cluster of galaxies, a moderately large cluster some 15 Mpc away.

2. What is the travel time for light to reach us from (a) the Magellanic Clouds; (b) the Andromeda galaxy; (c) the Virgo cluster; (d) a distant cluster that is 500 Mpc away? What do these times imply for our ability to understand the evolution of galaxies and of the universe?

3. Suppose the temperature of intracluster gas in a rich cluster of galaxies is 100 million degrees (10^8 K). At what wavelength does this gas emit most strongly? What kind of telescope is required to observe it?

ADDITIONAL READINGS

Allen, D. A. 1987. Star formation and *IRAS* galaxies. *Sky and Telescope* 73:372.

Burns, Jack O. 1986. Very large structures in the universe. *Scientific American* 255(1):38.

Chaffee, F. H. 1980. Discovery of a gravitational lens. *Scientific American* 243(5):60.

Chevalier, R. A., and C. L. Sarazin. 1987. Hot gas in the universe. *American Scientist* 75:609.

de Vaucouleurs, G. 1987. Discovering M31's spiral shape. *Sky and Telescope* 74:595.

Geller, M. J. 1978. Large-scale structure of the universe. *American Scientist* 66:176.

Gorenstein, M. V. 1983. Charting paths through gravity's lens. *Sky and Telescope* 66(5):390.

Gorenstein, P., and W. Tucker. 1978. Rich clusters of galaxies. *Scientific American* 239(5):98.

Hirshfeld, A. 1980. Inside dwarf galaxies. *Sky and Telescope* 59(4):287.

Hodge, Paul. 1981. The Andromeda galaxy. *Scientific American* 244(1):88.

———. 1986. *Galaxies.* Cambridge, Mass.: Harvard University Press.

Kaufman, M. 1987. Tracing M31's spiral arms. *Sky and Telescope* 73:135.

Lawrence, J. 1980. Gravitational lenses and the double quasar. *Mercury* 9(3):66.

Mathewson, D. 1984. The clouds of Magellan. *Scientific American* 252(4):106.

Rubin, V. C. 1983. Dark matter in spiral galaxies. *Scientific American* 248(6):96.

Silk, J., A. S. Szaley, and Y. B. Zel-dovich. 1983. The large-scale structure of the universe. *Scientific American* 249(4):56.

Struble, M. F., and H. J. Rood. 1988. Diversity among galaxy clusters. *Sky and Telescope* 75:16.

Tucker, W., and K. Tucker. 1989. Dark matter in our galaxy. *Mercury* 18(1):2 (Part I): *Mercury* 18(2):51 (Part II).

Turner, E. L. 1988. Gravitational lenses. *Scientific American* 259(1):54.

MATHEMATICAL METHODS

Cluster Masses

In the text, we have referred to determinations of cluster masses based on measurements of the velocity dispersion. Here we can explain more fully what this means.

It can be shown, from Newton's laws of motion and gravitation, that in a system having a large number of orbiting bodies in which all the bodies are gravitationally bound together, there is a specific relationship between the total gravitational potential energy of the system and the average kinetic energy of the individual bodies. This relationship is called the *virial theorem,* and it provides the basis for deriving cluster masses based on velocity dispersions. From the virial theorem it is possible to derive the following relationship:

$$M = \frac{2Rv^2}{G},$$

where R is the characteristic radius of the cluster (there is a more specific definition, which we will not go into here), v^2 is the velocity dispersion, and G is the gravitational constant.

The velocity dispersion is the square of the average velocity of the individual galaxies in the cluster at distance R from the center. This term must be derived from observations of the Doppler shifts of the spectra of individual galaxies. Recall that when we measure a Doppler shift, we are measuring only the effect of motion directly toward or away from us (i.e., the radial velocity), which is not the true three-dimensional velocity. If the directions of the motions are random, it can be shown that the true velocity dispersion is equal to 3 times the square of the average radial velocity. Therefore, we can substitute for v^2 the value $3V^2$, where V is the measured average radial velocity at distance R. This yields

$$M = \frac{6RV^2}{G}.$$

Let us try an example. Suppose a cluster is observed to have characteristic radius $R = 10$ Mpc and average radial velocity of 1,000 km/sec for galaxies at this distance from the center. To solve for M, we must first put all quantities in a common system of units; we convert R to $R = 3.1 \times 10^{25}$ cm and V to 10^8 cm/sec. Then we find

$$M = \frac{6(3.1 \times 10^{25})(10^8)^2}{(6.67 \times 10^{-8})} = 2.8 \times 10^{49} \text{ grams}.$$

This is a fairly meaningless number until we convert it into solar masses; division by the Sun's mass of 2 $\times$ 10^{33} grams yields

$$M = 1.4 \times 10^{16} \text{ solar masses}.$$

This is rather large, and would correspond to a very rich cluster of galaxies.

In using this technique to measure cluster masses, astronomers generally have the greatest difficulty in accurately determining the average radial velocity, since the objects are usually quite faint, and the spectral lines (created by the composite spectra of all the stars in the galaxies) are quite broad and diffuse, without sharp edges. As emphasized in the text, this method usually leads to cluster masses that are much greater than the sum of the masses of the visible galaxies, thus lending support to the idea that large quantities of dark matter populate the universe.

CHAPTER 29

Universal Expansion and the Cosmic Background

The microwave receiver used to measure the anisotropy of the cosmic background radiation. (George Smoot, Lawrence Berkeley Laboratory)

A recurrent theme throughout our study of astronomy has been the dynamic nature of celestial objects. Everything we have studied in detail, from planets to stars to galaxies, has turned out to be in a constant state of change. Now that we are prepared to examine the entire universe as a single entity, we should not be surprised to see this theme maintained. That the universe itself is evolving, with a life story of its own, has been accepted by most astronomers, but until recently, some still doubted it. In this chapter, we will study the evidence that the universe is evolving.

HUBBLE'S GREAT DISCOVERY

Even before it was established unambiguously that the nebulae were really galaxies, a great deal of effort went into observing them. Their shapes were studied carefully, and their spectra were analyzed in detail. Typically the spectrum of a galaxy resembles that of a moderately cool star, with many absorption lines (Fig. 29.1). Because the spectrum represents the light from a huge number of stars, each moving at its own velocity, the Doppler effect causes the spectral lines to be rather broad and indistinct. Nevertheless, it is possible to analyze these lines in some detail, and one of the early workers in this field, V. M. Slipher, discovered during the decade before the Shapley-Curtis debate in 1920 that the nebulae tend to have large velocities away from the Earth. The spectral lines nearly always were found to be shifted toward the red, and this tendency was most pronounced for the faintest of the nebulae. Slipher measured velocities as great as 1,800 km/sec.

Following Slipher's work, others continued to study spectra of nebulae, with heightened interest after Hubble's demonstration in 1924 that these objects were undoubtedly distant galaxies, comparable to the Milky Way in size and complexity. Hubble and others estimated distances to as many nebulae as possible by use of the Cepheid variable period-luminosity relation for the nearby ones, and other standard candles such as novae and bright stars for more distant nebulae.

In 1929 Hubble made a dramatic announcement: The speed with which a galaxy moves away from the Earth is directly proportional to its distance. If one galaxy is twice as far away as another, for example, its velocity is twice as great. If it is 10 times farther away, it is moving 10 times faster.

The implications of this relationship between distance and velocity are enormous. It means that the universe itself is expanding, its contents rushing outward at a fantastic pace. All the galaxies in the cosmos are moving away from all the others (Fig. 29.2).

To envision why the velocity increases with increasing distance, it may be helpful to resort to a commonly used analogy: Imagine a loaf of bread that is rising. If the dough has raisins sprinkled uniformly through it, the raisins will move farther apart as the dough expands. Suppose that the raisins are 1 centimeter apart before the dough begins to rise, and consider what has happened 1 hour later, after it rises to the point at which adjacent raisins are 2 centimeters apart. A given raisin is now 2 cm from its nearest neighbor, but 4 cm away from the next one over, 6 cm from the next one, and so on. The distance between any pair of raisins has doubled. From the point of view of any one raisin, its nearest neighbor had to move away from it at a speed of 1 cm/hr, the next raisin farther away had to move at 2 cm/hr, the next one had a speed of 3 cm/hr, and so on. If we let the dough continue to rise to the point where the adjacent raisins are 3 cm apart, then we find that all the distances between pairs have tripled over what they were to start with, and the speeds needed to accomplish this are again directly proportional to the distance between a given pair. The farther away a raisin is to begin with, the farther it must move to maintain the regular spacing during the expansion, and therefore the greater is its velocity. In the same way, the galaxies in the universe must increase their separations from each other at a rate proportional to the distance between them (Fig. 29.3), or else they would bunch up. By observing the velocities of galaxies, astronomers are keeping watch on the raisins to see how the loaf of bread is coming along.

CLUSTER NEBULA IN	DISTANCE IN LIGHT-YEARS	RED-SHIFTS
VIRGO	78,000,000	1,200 KM/SEC
URSA MAJOR	1,000,000,000	15,000 KM/SEC
CORONA BOREALIS	1,400,000,000	22,000 KM/SEC
BOOTES	2,500,000,000	39,000 KM/SEC
HYDRA	3,960,000,000	61,000 KM/SEC

Figure 29.1 Spectra of galaxies. *These examples illustrate the broad absorption lines characteristic of spectra of large groups of stars such as galaxies. It was noticed before 1920 that the spectral lines tend to be shifted toward the red in galaxy spectra, which indicates that galaxies as a rule are receding from us. (Palomar Observatory, California Institute of Technology)*

It is important to realize that it doesn't matter which raisin we choose to watch; from any point in the loaf, all other raisins appear to move away with speeds proportional to their distances from that point. Thus we do not conclude that our galaxy is at the center of the universe; from *any* galaxy, it would appear that all others are receding. It is also important to realize that the raisin-bread analogy has serious shortcomings: It suggests that the universe can be viewed as a simple three-dimensional object, with an edge and a center, which is inappropriate (this is discussed further in Chapter 31). It is better to try to envision a universe with no boundaries and no center; space itself is getting bigger. There is no location within it that can today be identified as the center, and there is no "outside" from which the universe can be viewed.

Following Hubble, several other astronomers have extended the study of galaxy motions to greater and greater distances, with the same result: As far as the telescopes can probe, the galaxies are moving away from each other at speeds proportional to their distances. Expansion is a major feature of the universe we live in, one that must be taken into account as we seek to understand its origins and its fate.

It is often convenient to display the data showing the universal expansion in a graph of velocity versus distance (as Hubble originally did; see Fig. 29.4), or on a plot of redshift versus apparent magnitude (Fig. 29.5), since increasing apparent magnitude (i.e., decreasing brightness) indicates increasing distance.

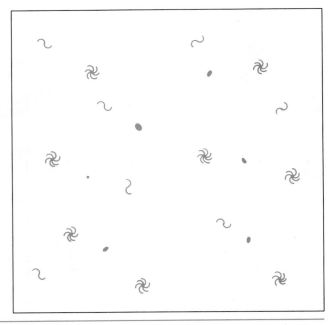

Figure 29.2 The expanding universe. *These drawings show a number of galaxies at one time (left), and again at a later time (right). All of the spacings between galaxies have increased because the universe is expanding.*

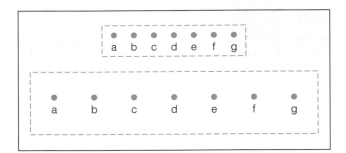

Figure 29.3 Velocity of expansion as a function of distance. *At upper left is a row of galaxies, and at lower right is the same row 1 billion years later, when the universe (represented by the rectangular box) has expanded so that the distances between adjacent galaxies have doubled from 1 to 2 megaparsecs. From the viewpoint of an observer in any galaxy in the row, the recession velocity of its nearest neighbor is 1 megaparsec per billion years; of its next nearest neighbor, 2 megaparsecs per billion years; of the next, 3 megaparsecs per billion years; and so on. The velocity is proportional to the distance in all cases, for any pair of galaxies.*

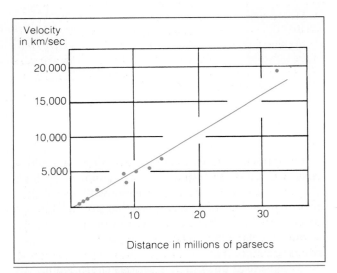

Figure 29.4 The Hubble law. *This early diagram prepared by E. Hubble and M. Humason shows the relationship between galaxy velocity and distance. The slope of the relation has been altered since, with addition of data for more galaxies and improvements in the measurement of distances, but the general appearance of this figure is the same today. (Estate of E. Hubble and M. Humason. Reprinted with permission)*

Figure 29.5 A modern version of the Hubble law. *Often, apparent magnitude is used on the horizontal axis instead of distance, because the two quantities are related, particularly if the diagram is limited to galaxies of the same type, as this one is. At lower left the small rectangle indicates the extent of the relationship as Hubble first discovered it; today many more galaxies, much dimmer, have been included. The most rapid galaxies are not actually traveling at the speed of light as this diagram implies, because relativistic corrections have not been applied (see discussion in the next chapter). (Adapted from J. Silk 1980, The Big Bang [San Francisco: W. H. Freeman])*

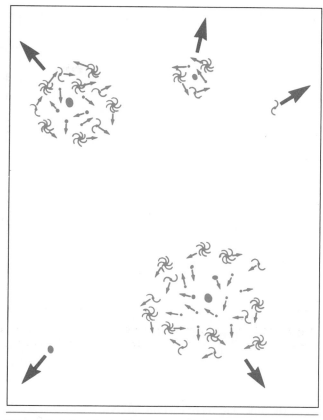

Figure 29.6 Local motions. *Although there is a systematic overall expansion of the universe, individual galaxies within clusters, and even clusters within superclusters, have random individual motions. Hence within the Local Group the galaxies are not uniformly receding from each other.*

The galaxies within a cluster have their own orbital motions, which are separate from the overall expansion (Fig. 29.6). As the universe expands, the clusters move apart from each other, while the individual galaxies within a cluster do not. The motions of galaxies within a cluster are relatively unimportant for very distant galaxies that are moving away from us at speeds of thousands of kilometers per second, but for nearby ones, the orbital motion can rival or exceed the motion caused by the expansion of the universe. The Andromeda galaxy, for example, is actually moving *toward* the Milky Way at a speed of about 100 km/sec, whereas at its distance of 700 kpc from us, the universal expansion should give it a recession velocity of about 40 km/sec.

HUBBLE'S CONSTANT AND THE AGE OF THE UNIVERSE

The relation discovered by Hubble can be written in simple mathematical form:

$$v = Hd,$$

where v is a galaxy's velocity of recession and d is its distance, in megaparsecs. The Hubble constant, H, is then given in units of km/sec/Mpc, if v is in units of km/sec.

The value of H is difficult to establish (Table 29.1). It is done by collecting as large a body of data as possible on galactic velocities and distances, and then

Table 29.1 Example measurements of the Hubble constant

Value of H (km/sec/mpc)	Observer	Date	Implied Age of Universe (years)
540	Hubble	1930	1.8×10^9
260	Baade	1949	3.8×10^9
180	Humason, et al.	1955	5.4×10^9
75	Sandage	1956	1.3×10^{10}
56	Sandage and Tammann	1974	1.8×10^{10}
100	van den Bergh	1975	1.0×10^{10}
80	Tully and Fisher	1977	1.2×10^{10}
100	de Vaucouleurs, et al.	1979	1.0×10^{10}
95	Aaronson	1980	1.0×10^{10}
65	Mould, et al.	1980	1.5×10^{10}
50	Sandage and Tammann	1982	2.0×10^{10}
82	Aaronson and Mould	1983	1.2×10^{10}

Note: Information from M. Rowan-Robinson, *The Cosmological Distance Ladder*, 1985 (New York: W. H. Freeman).

deducing the value of *H* that best represents the relationship between distance and velocity. Hubble did this first, finding that $H = 540$ km/sec/Mpc, meaning that for every megaparsec of distance, the velocity increased by 540 km/sec. The standard candles were not very well established in Hubble's day, when it was still news that the nebulae were distant galaxies, and this value of H turned out to be an overestimate. The best modern values for *H* are between 50 and 100 km/sec/Mpc. Until recently, a consensus had been developing that the best value is 55 km/sec/Mpc, but a new distance determination for a number of galaxies has now suggested a value close to 90 km/sec/Mpc. The precise value of *H* has extremely important implications for our understanding of the universe and its expansion, and intense research effort is being devoted to refining the estimates.

If the universe is expanding, it follows that all the matter in it used to be closer together than it is today (Fig. 29.7). If we follow this logic to its obvious conclusion, then we find that the universe was once concentrated in a single point, from which it has been expanding ever since.

From the rate of expansion, we can calculate how long ago the galaxies were all together in a single point. From the simple expression

Figure 29.7 Backtracking the expansion. *If the universe began with a small size, the present rate of expansion provides information on how long ago the expansion began. With considerable uncertainty arising from the fact that the expansion rate has probably not been constant, this leads to an estimated age for the universe of between 10 and 20 billion years.*

$$\text{Time} = \frac{\text{Distance}}{\text{Velocity}}.$$

we find

$$\text{age} = \frac{d}{Hd} = \frac{1}{H}.$$

The age of the universe is equal to $1/H$, if the expansion has been proceeding at a constant rate since it began. As we will see in Chapter 31, this assumption of a constant expansion rate is not strictly true (it was more rapid in the beginning), but it gives a useful estimate of the age for now.

If we adopt, for the sake of discussion, a value for H of 75 km/sec/Mpc, then the age of the universe is 4.1×10^{17} sec $= 1.3 \times 10^{10}$ years. (To do this calculation, first you must convert the value of a megaparsec into kilometers.)

A simple calculation based on the observed expansion rate has led to a profound conclusion: Our universe is about 13 billion years old. If this value is correct, it just fits in with what we have learned about galactic ages. The Sun, for example, is thought to be about 4.5 billion years old, and the most aged globular clusters are apparently some 14 billion years old, acceptably close, within the uncertainties of the measurements. On the other hand, a much higher value of H would lead to an age for the universe that is much smaller than the estimated ages of globular clusters, and those who favor a value nearer to 100 km/sec/Mpc are facing just such a dilemma.

We can easily see what happens to our estimate of the age of the universe if we choose other values of H. For example, if Hubble's value of 540 km/sec/Mpc had been correct, then the age would have been calculated as less than 2 billion years, and if the recently suggested value of about 90 km/sec/Mpc is found to be correct, then the age is 11 billion years. In either case, there is a conflict with the estimated ages of globular clusters; it will be interesting to see how this conflict is resolved if indeed a value for H of 90 km/sec/Mpc becomes widely accepted. The conflict worsens when we realize that age estimates based on $1/H$ are probably too large, because they do not take into account the fact that the expansion rate at the earliest times was somewhat more rapid than it is today. This further complicates things, if a value of 90 km/sec/Mpc is found to be best. Whatever the correct ages are, this situation serves to illustrate how uncertain some of the most important measurements in astronomy can be.

There are many other important questions about the universe that depend on the value of H, and this is the reason so much effort is being expended to define its value as accurately as possible. One of the principal reasons for development of the *Hubble Space Telescope* is to probe the most distant possible galaxies, to determine their distances and velocities so that this information can be used to help find the correct value of H.

REDSHIFTS AS YARDSTICKS

The expansion of the universe has a very practical benefit for astronomers concerned with the properties of distant galaxies: It provides a means of determining their distances (Fig. 29.8). If we know the value of H, then we can find the distance to a galaxy simply by measuring its velocity as indicated by the Doppler shift of its spectral lines. The distance is given by

$$d = \frac{v}{H}.$$

Thus, if we adopt $H = 75$ km/sec/Mpc, then a galaxy

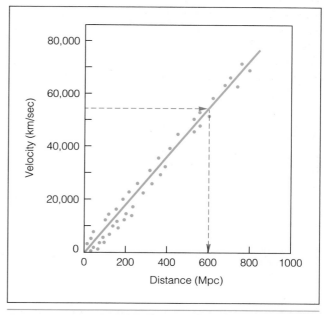

Figure 29.8 Finding distance from the Hubble law. *This diagram shows how the distance to a galaxy can be deduced directly from the Hubble law, if the redshift, hence the velocity, is measured.*

ASTRONOMICAL INSIGHT 29.1

The Distance Pyramid

Having reached the ultimate distance scales, we can pause to look back at the progression of steps that got us here. As noted in the text, the use of Hubble's relation to estimate distances to the farthest galaxies depends critically on the value of H, which can be determined only from knowledge of the distances to a representative sample of galaxies. Those distances in turn depend on various standard candles, themselves calibrated through the use of other methods, such as the Cepheid variable period-luminosity relation, which are applicable only to the most nearby galaxies. Our knowledge of these close extragalactic distances rests on the known distances to objects within the Milky Way, such as star clusters containing variable stars, which are used to calibrate the period-luminosity relation. The distances to these clusters, in turn, are known from more local techniques, such as spectroscopic parallaxes and main-sequence fitting, and these ultimately depend on the distances to the most nearby stars, derived from trigonometric parallaxes.

Thus, the entire progression of distance scales, all the way out to the known limits of the universe, depends on our knowledge of the distance from the Earth to the Sun, the basis of trigonometric

Distance (pc)	Key Object	Method
10^{-6}	Sun	↓ Radar, planetary motions
10^{-5}		
10^{-4}		
10^{-3}		
10^{-2}		↓ Trigonometric parallax
10^{-1}		
1	α Centauri	
10^1		
10^2	Hyades cluster	↓ Moving cluster method
10^3		Main-sequence fitting
10^4	Limits of Milky Way	(spectroscopic parallax)
10^5	Magellanic Clouds	
10^6	Andromeda galaxy	↓ Cepheid variables
10^7		Brightest stars
10^8	Virgo cluster	
10^9		
10^{10}		↓ Brightest galaxies

parallax measurements. At every step of the way outward, our ability to measure distances rests on the previous step. If we revise our measurement of local distances, we must accordingly alter all our estimates of the larger scales, affecting our perception of the universe as a whole. This elaborate and complex interdependence of distance determination methods is known as the **distance pyramid,** and it is sometimes depicted in graphic form (see table).

A very important step in the sequence shown above is represented by the Hyades cluster, a galactic star cluster some 43 parsecs away. The distance to this cluster is determined from the **moving cluster method,** in which a combination of Doppler shift measurements and apparent angular motion measurements reveal

the true direction of cluster motion, so that a comparison of angular motion and true velocity then gives the distance to the cluster. Because much of the rest of the distance pyramid depends on the application of this technique to the Hyades cluster, great care has been taken in the analysis, and every so often it is redone. Whenever the distance to the Hyades is altered slightly, the impact spreads as astronomers judge how this affects other distance scales in the universe. One of the things the *Hubble Space Telescope* may accomplish is to make direct parallax measurements of stars in the Hyades, providing a new and important measurement of the distance to the cluster and thus revising, once again, our understanding of the distance scale of the universe.

found to be receding at 7,500 km/sec, for example, has the distance

$$d = \frac{7,500}{75} = 100 \; Mpc.$$

This technique can be applied to the most distant and faint galaxies, where it is nearly impossible to observe any standard candles. It is important to remember, however, that the use of Hubble's relation to find distances is only as accurate as our value of H, and the value that we use is derived in turn from distance determinations that depend on standard candles. Furthermore, because the expansion rate of the universe has not been constant, for very distant galaxies it is necessary to know what the rate was in the past to accurately determine the distance.

A COSMIC ARTIFACT: THE MICROWAVE BACKGROUND

Since Hubble's discovery of the universal expansion, many astronomers and physicists have explored its significance, both for what it may tell us about the future and for the information it provides about the past. A number of scientists, beginning in the 1940s with George Gamow and his associates, have done extensive studies of the nature of the universe in its earliest days. It was quickly realized that, if all the matter were originally compressed into a small volume, it must have been very hot, for the same reason that any gas heats up when compressed. The fiery conditions inferred for the early stages of the expansion have led to the term **big bang,** which is the commonly accepted title for all theories of the universal origin that start with an expansion from a state of high density and temperature.

Gamow's primary interest was in analyzing nuclear reactions and element production during the big bang (the modern understanding of this is described in Chapter 31). But in the course of his work he came to another important realization: The universe must have been filled with radiation when it was highly compressed and very hot, and this radiation should still be with us. At first, when the temperature was in the billions of degrees, it was γ-ray radiation, but later, as the universe cooled, next X-rays, and then ultraviolet light filled it. (Recall the relationship between temperature and wavelength of maximum emission, as stated in Wien's law; see Chapter 6). During the first million

Figure 29.9 The cooling of the universal radiation. *At left is a box representing the early universe, filled with intense gamma-ray radiation. At right is the same box, greatly expanded. The radiation is still there, but its wavelength has increased, and the temperature represented by its thermal spectrum has decreased.*

years or so after the expansion started, the radiation was constantly being absorbed and reemitted (i.e., scattered) by the matter in the universe, but eventually (when the temperature had dropped to around 4,000 K), protons and electrons combined to form hydrogen atoms, and from that time on there was little interaction between the matter and the radiation. The matter, in due course, organized itself into galaxies and stars, while the radiation has simply continued to cool as the universe has continued to expand (Fig. 29.9). The intensity and spectrum of the radiation is dependent only on the temperature of the universe and is described by the laws of thermal radiation (see Chapter 6).

Gamow and others made rough calculations showing that the present temperature of the radiation filling the universe should be very low indeed, about 25 K. More recently, R. H. Dicke and colleagues, quite independently, carried out new calculations, predicting an even lower temperature, about 5 K. Using Wien's law, it is easy to calculate the most intense wavelength of such radiation: If the temperature is 5 K, then the value of λ_{max} is 0.06 cm, in the microwave part of the radio spectrum.

Gamow's principal interest was in element production during the big bang, rather than in the remnant radiation. Furthermore, in the late 1940s the technology needed to detect the radiation had not yet been developed. Hence no attempt was made in Gamow's

ASTRONOMICAL INSIGHT 29.2

Overlooking the Cosmic Background

Sometimes significant discoveries are delayed because no one expects them, or because few people take unconventional predictions seriously. The latter appears to have delayed the discovery of the cosmic microwave background radiation by nearly two decades.

When George Gamow predicted the existence of the cosmic background radiation in 1947, the prediction constituted a minor postscript to a major paper about the formation of the elements. In the paper, Gamow and his colleagues showed that the present-day abundances of helium and other light elements were produced during the early stages of the universal expansion, while other, heavier elements must have been produced in stellar cores. No one seems to have paid much attention to the prediction about the cosmic background, despite its

significance. No one made an attempt to look for the background radiation for the next 18 years.

One of the reasons for the delay was technological. Thermal radiation corresponding to a temperature of 25 K (the value predicted by Gamow and colleagues) has its peak intensity at a wavelength of 0.01 cm, or 0.1 mm, in a portion of the microwave spectrum where little work had been done by 1947. The technology to do so was just becoming available, as a result of studies of microwave emissions from the Sun, and as a response to radar developments during World War II. Thus, one reason nobody looked for the predicted radiation was that few scientists had the technology to do so.

The other reason was more subtle. Apparently few scientists were motivated to undertake a major experimental project such as the search for background radiation. The cost and time required for developing an experiment to search for the background radiation were high, and in the late 1940s there were many areas of research that ap-

peared more promising. The basic answer to the question of why the background was not discovered before 1965 is that no one looked for it before then. At nearly the same time as Gamow's prediction, astronomers learned that observations at other radio wavelengths—particularly the 21.1 cm wavelength where atomic hydrogen should emit—would be very productive scientifically, and early radio astronomy followed these paths.

Even when the cosmic background was first detected the discovery was accidental, although there is little doubt that Robert Dicke and his colleagues would have found it soon if Penzias and Wilson had not. Ironically, the discovery could have been made at least 15 years earlier. The history of twentieth-century cosmology might have been quite different if astronomers had taken Gamow's prediction more seriously.

time to search for the radiation, and the idea was forgotten until Dicke's work. The remnant radiation was eventually detected in 1965 by accident, when radio astronomers Arno Penzias and Robert Wilson found a persistent source of background noise while testing a new antenna (Fig. 29.10).

The implication soon became clear: The noise was the universal radiation that the theorists had predicted should exist. This was one of the most significant astronomical observations ever made, for it provided a direct link to the origins of the universe. More than a piece of circumstantial evidence, this was a real arti-

fact, the kind of hard evidence that carries weight in a court of law. Penzias and Wilson later were awarded the Nobel Prize for their discovery.

THE CRUCIAL QUESTION OF THE SPECTRUM

There remained questions about the interpretation of the radiation, however, and doubt in the minds of some who did not accept the big bang theory of the origin

Figure 29.10 The discoverers of the universal background. *Robert Wilson, left, and Arno Penzias, in front of the horn-reflector antenna at Bell Laboratories with which they found a persistent noise that turned out to be the remnant radiation from the big bang. (Courtesy of AT&T)*

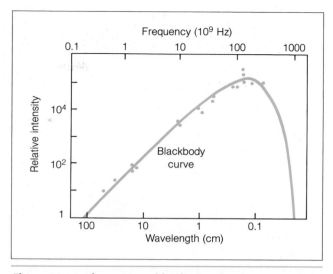

Figure 29.11 The measured background spectrum. *This curve shows the spectrum of the cosmic background radiation, plotted as a function of wavelength (a frequency scale is given at the top as well). The solid line represents a thermal spectrum for a temperature of just under 3 K.*

of the universe. If the radiation was really the remnant of the primeval fireball, then it was expected to fulfill certain conditions, and intensive efforts were made to find out whether it did.

One important prediction was that the spectrum should be that of a simple glowing object whose emission of radiation is the result only of temperature (that is, it should be a thermal spectrum, as explained in Chapter 6). If, on the other hand, the source of the radiation was something other than the remnant of the initial universe, such as distant galaxies or intergalactic gas, it could have a different spectrum. Thus, the shape of the spectrum, its intensity as a function of wavelength, was a very important test of the interpretation of its origin.

Some of the early results were confusing, because of unforeseen complications, but now the situation seems to have been resolved. The spectrum does indeed follow the shape expected (Fig. 29.11), and the radiation is considered to be a relic of the big bang. The most difficult part of measuring the spectrum was observing the intensity at short wavelengths, around 1 millimeter or less, where the Earth's atmosphere is especially impenetrable, but this was finally accomplished with high-altitude balloon payloads.

The net result of all the effort expended in determining the spectrum is that the temperature of the radiation is 2.7 K, and its peak emission occurs at a wavelength of 1.1 mm. The radiation is commonly referred to as the **microwave background,** or the **3-degree background radiation.**

The care that went into establishing the true nature of the background radiation reflects the importance of what it has to tell us about the universe. It is now widely accepted that the cosmic background radiation is indeed the result of a highly compressed and hot early universe, and this has proved to be very important in ruling out alternative theories about the earliest times.

ISOTROPY AND DAILY VARIATIONS

Another expectation of the background radiation, in addition to the nature of its spectrum, is that it should fill the universe uniformly, with no preference for any particular location or direction. A medium or radiation field that has no preferred orientation, but instead looks the same in all directions, is said to be **isotropic.** Some of the early observations of the 3-degree background were designed to test its isotropy, for again, failure to meet this expectation could imply some origin other than the big bang.

Figure 29.12 A microwave receiver used to measured the cosmic background. *The pair of receiving horns are used to compare the radiation intensity in different directions. (G. Smoot, Lawrence Berkeley Laboratory)*

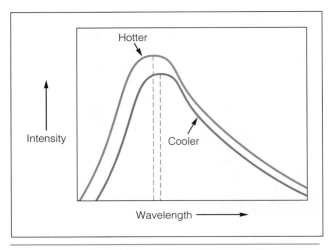

Figure 29.13 The effect of a Doppler shift on the cosmic background radiation. *If the observer on the Earth has a velocity with respect to the radiation, this shifts the peak of the spectrum a little, and affects the temperature that observer deduces from the measurements. Motion of the Earth results in a daily cycle of tiny fluctuations in the observed temperature, known as the 24-hour anisotropy.*

A test of isotropy, in simple terms, is just a measurement of the radiation's intensity in different directions, to see whether or not it varies (Fig. 29.12). From the time it was discovered, the microwave background showed a high degree of isotropy, as expected. The issue has been pressed, however, for a variety of reasons. One possible alternative explanation of the radiation was that it arises from a vast number of individual objects, such as very distant galaxies, so closely spaced in the sky that their combined emission appears to come uniformly from all over. To test this idea, astronomers have attempted to find out whether the radiation is patchy on very fine scales, as it would be if it came from a number of point sources. So far, no evidence of any clumpiness has been found, and the big bang origin for the radiation has not been threatened.

Some clumpiness is to be expected, even if the radiation is the remnant of the early expansion of the universe. It is known that galaxies formed from density enhancements in the universe at an early time, and these condensations may have affected the radiation locally, leading today to small fluctuations in the temperature of the radiation. These fluctuations, expected to occur over angular sizes on today's sky of around 1 arcminute, have not been detected, even with the

Cosmic Background Explorer, a satellite observatory designed for the purpose (see further discussion in the next section).

In addition to any subtle localized nonuniformities expected in the big bang theory, a broad asymmetry is expected, because of the motion of the Earth with respect to the frame of reference of the radiation. Because of the Doppler shift, the observed intensity of the radiation as seen by a moving observer will vary with direction, depending on whether the observer is looking toward or away from the direction of motion. If the observer looks ahead, toward the direction of motion, then a blueshift occurs, the peak of the spectrum being shifted slightly toward shorter wavelengths (Fig. 29.13). In terms of temperature, this means that the radiation looks a little hotter when viewed in this direction. If the observer looks the other way, there is a slight redshift, and the measured temperature is cooler.

The Earth is moving in its orbit about the Sun. The Sun, in addition, is orbiting the galaxy, the galaxy is moving along its own path through the Local Group, and apparently the Local Group is moving with respect to the local supercluster. All of these motions combined represent a velocity of the Earth with respect to the frame of reference established by the background radiation, a velocity that should produce slight differences in the radiation temperature if it is viewed in different directions. Because of the Earth's rotation, if we simply point our radio telescope straight up, we should alternatively see high and low temperatures, as our telescope points toward and then (12 hours later) away from the direction of motion. This means that there should be a daily variation cycle in the radiation, referred to as the **24-hour anisotropy** or the **dipole anisotropy** (*anisotropic* is the term for a nonuniform medium).

The actual change in temperature from one direction to the other is very small (though not as small as the localized fluctuations discussed above), and was not successfully measured until a few years ago. The observed temperature difference from the forward to the rear direction is only a few thousandths of a degree, and very sophisticated technology was required for this to be measured. The data show that the Earth is moving with respect to the background radiation at a speed of about 360 km/sec. When the Sun's known orbital velocity about the galaxy is taken into account, along with the galaxy's motion within the Local Group, this implies that the Local Group is "falling" toward the very large Virgo cluster of galaxies with a speed of about 570 km/sec. Some of this motion is caused by the gravitational influence of the Virgo cluster, and apparently some is from the effects of distant superclusters. Very recently it has been found that the motion toward the Virgo cluster is a component of a much larger motion, part of a general streaming of all the clusters in our part of the universe. This streaming motion may be caused by the gravitational attraction of something extraordinarily massive—even compared to clusters of galaxies—yet invisible. This "Great Attractor," the subject of current excitement and controversy, may be related to the dark matter thought to contain most of the mass of the universe (see Chapter 31).

The use of the 3-degree background radiation as a tool for deducing the local motions with respect to the radiation is predicated upon the assumption that the radiation itself is fundamentally isotropic on large scales (as distinguished from the localized fluctuations discussed earlier). At present there is no reason to doubt this, but the possibility remains that it is not—that the universe itself is somehow lopsided. If it is, it will be very difficult indeed to measure its asymmetry, because of the complexity of the motions that our observing platform, the Earth, undergoes.

RESULTS FROM THE COSMIC BACKGROUND EXPLORER

In November, 1989, NASA launched a new satellite called the *Cosmic Background Explorer (COBE;* Fig. 29.14). The mission of *COBE* is to make more refined and detailed observations of the 3 K background radiation than was previously possible from the ground. One of the major advances of *COBE* is its ability to avoid contamination due to the Earth's atmosphere, which interferes strongly with ground-based observations because it absorbs and emits radiation in the same wavelength region where the cosmic background is strongest. The *COBE* satellite carries a container of liquid helium that is used to maintain extremely low temperatures in the electronic devices that detect the background radiation, helping to reduce extraneous emission that would interfere with the measurements of the cosmic radiation.

Figure 29.14 The cosmic background explorer. *This satellite, to be launched in the 1990s, will make the most precise and extensive measurements yet of the microwave background radiation (NASA)*

Figure 29.15 The revised cosmic background spectrum from *COBE*. *The* Cosmic Background Explorer (COBE) *has measured the background radiation at many wavelengths, as indicated by the points. The result is the most accurate measurement yet of the temperature of the radiation, which is found to be 2.735 K, derived by fitting a theoretical thermal radiation spectrum (solid line) through the points. (NASA, courtesy of the COBE Science Working Group)*

One of the initial tasks of *COBE* was to measure the spectrum of the cosmic background more accurately than had been done previously from the ground. *COBE* succeeded very well in this, confirming that the spectrum of the radiation is thermal and finding that its temperature is 2.735 K (with a measurement uncertainty of only 0.06 K; Fig. 29.15). This result further solidifies the interpretation of the background radiation as due to the universe itself, rather than to some other kind of source.

Another goal of *COBE* is to establish clearly whether the background radiation is clumpy or uneven. The initial results strongly rule out variations from place to place on an angular scale of one degree or a bit smaller (Fig. 29.16). This has important consequences for our understanding of the early universe and the process by which it became clumpy and formed galaxies and clusters of galaxies. If the universe had been clumpy at the time when the background radiation was emitted (i.e., the time of the "decoupling" of matter and radiation; see Chapter 31), then there should have been detectable variations in the strength of the background radiation from place to place. The fact that the background radiation appears to be extremely smooth and invariant with direction indicates that the early universe was very uniform.

If this conclusion is confirmed by further *COBE* observations, it strongly supports the suggestion, already discussed in the previous chapter, that the clusters and superclusters of galaxies observed today are

ASTRONOMICAL INSIGHT 29.3

Falling Galaxies

In our discussion of the cosmic microwave background radiation, we pointed out the uniformity (that is, isotropy) of the radiation. At the same time, we discussed the fact that there is a daily variation in the radiation temperature that is measured from the Earth, because the Earth is moving through space relative to the reference frame of the radiation, so that as the Earth rotates, we observe alternately in the forward and rearward directions. This daily variation, variously called the 24-hour anisotropy or the dipole anisotropy, has been analyzed to determine the direction and speed of the Earth, the Milky Way galaxy, the Local Group of galaxies, and even the local supercluster of galaxies.

Carrying out this analysis is very complex, because all the different motions add together. To determine the overall motion of the Local Group or the local supercluster, first it is necessary to make allowance for the Earth's orbital motion around the Sun, the Sun's motion about the galaxy, and the galaxy's motion within the Local Group. When all this has been done, the result is not very definite, and little can be said about the largest-scale motions.

There is, however, another approach to determining the overall motion of our galactic system within the framework established by the expanding universe. Careful statistical analyses of the observed motions of galaxies seen in different directions can indirectly reveal the motion of our own galaxy or group of galaxies. If a sample of galaxies is chosen for the best possible uniformity of properties and for the greatest possible extent in distance, then that set of galaxies forms its own reference frame against which the motion of the Local Group or even the local supercluster can be measured. The observations consist of measuring distances for a large number of galaxies, and obtaining spectra so that Doppler shifts can be measured, yielding line-of-sight velocities (i.e., radial velocities) for all the galaxies observed. If the Earth were at rest within the framework of this large collection of galaxies, then the average of all the velocities would be zero, because by random chance as many galaxies would be approaching us as receding from us. If, on the other hand, the Earth is moving in some steady direction with respect to the sample of galaxies, then we would see a trend toward redshifts in one direction and blueshifts in the other direction. This is analogous to measuring the anisotropy in the cosmic background radiation by determining its temperature (i.e., its Doppler shift) in opposite directions.

Many astronomers have attacked the problem of determining the Earth's motion relative to large populations of distant galaxies, with differing results. The difficulties lie in measuring the distances to the galaxies, which must be known in order to interpret the motions, and in obtaining sufficiently complete coverage of the sky to be sure that the distant galaxies really do represent a reference frame that is moving with the expansion of the universe (that is, a frame of reference against which we can measure our own motion with respect to the expansion).

Despite the difficulties, one of the most ambitious and complete of these studies has led to a fascinating conclusion: Our region of the universe, including the Local Group as well as other clusters of galaxies nearby, appears to be in motion with respect to the reference frame set by the expanding universe. That is, the Local Group and the other galaxy clusters in our portion of the universe are all moving together with a speed of about 570 kilometers per second. The motion is directed toward a large cluster of galaxies (known as the Virgo cluster), so it is tempting to conclude that this cluster is the cause. Its mass is not sufficient, however, and it appears that a far more massive object of some kind, lying about 3 times farther away (or about 50 Mpc from us) and having a mass of roughly 10^{16} solar masses (comparable to the largest known superclusters), is providing the gravitational attraction needed to explain the inferred motion of our cluster of galaxies. This unseen, mysterious object toward which we are falling has been dubbed the "Great Attractor."

When the presence of the Great Attractor was first proposed, astronomers were unable to find any galaxies in the region of the sky where it should be, and they speculated that perhaps the Great Attractor consists of dark matter. More diligent searches have since revealed some galaxies in the region, but it is still not known

(continued on next page)

ASTRONOMICAL INSIGHT 29.3

Falling Galaxies (continued)

whether there are enough there to explain the enormous gravitational attraction that is causing the motion of our part of the universe. Several astronomers are at work on this problem, and we may expect new developments on a continual basis for some time to come. Astronomers are especially intrigued by the Great Attractor because of what it may tell us about the structure of the universe, and because of the possibility that it is made of dark matter, the invisible material that may dominate the content of the universe.

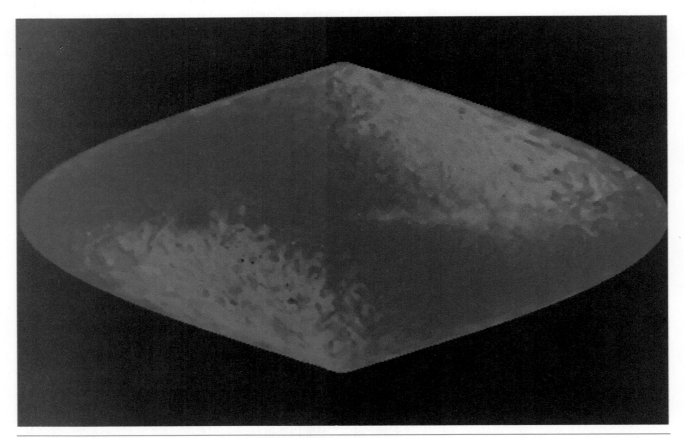

Figure 29.16 *COBE* data on anisotropy in the cosmic background. *This figure shows, through false-color representation, the survey obtained by COBE of anisotropy (patchiness) in the cosmic background radiation. In galactic coordinates, this shows two important aspects of the radiation: (1) There is no detected patchiness or lumpiness on small, angular scales; and (2) the dipole anisotropy (also known as the 24-hour anisotropy) is clearly present, represented here by blue color in the forward direction of the Earth's motion and pink in the rearward (opposite) direction. (NASA, courtesy of the COBE Science Working Group)*

the result of random gravitational encounters. It appears that the universe was formed in a very uniform state, and that whatever clumpiness it has today has developed since then. Thus the early *COBE* results support the concept of a universe that was originally very smooth and that is clumpy and lumpy today as a result of random gravitational aggregation, and not as a by-product of some intrinsic imbalance in its structure that was present from the start. It will be very interesting to see whether further results from *COBE* continue to support this agreement in conclusions reached by way of very different methods.

PERSPECTIVE

We have learned now that the universe, like all the subordinate objects within it, is a dynamic, evolving entity. One of the grandest stories in astronomy has been the unfolding of the concept of universal expansion. An idea forced on astronomers by the evidence of the redshifts, it has led directly to the big bang picture, which in turn tells us the age of the universe, explains the origin of some of the elements, and is verified through the presence of the microwave background radiation.

Before we finish the story by studying modern theories of the universe and their implications for its future, we must take a closer look at some of the objects that inhabit it. There is a breed of strange and bizarre entities whose very nature seems to be tied up with the evolution of the universe, and which therefore will provide us with a few additional bits of evidence for our final discussion.

SUMMARY

1. It was discovered early in this century that galaxies tend to have redshifted spectral lines, implying that they are moving away from us.
2. Hubble discovered in 1929 that the velocity of recession is proportional to distance, demonstrating that the universe is expanding.
3. The rate of expansion, expressed in terms of Hubble's constant H, is probably between 50 km/sec/Mpc and 100 km/sec/Mpc.
4. The fact that the universe is expanding implies that it originated from a single point, and the rate of expansion, if constant, tells us that it began some 10 to 20 billion years ago. This may be considered to be the age of the universe.
5. The relationship between speed and distance allows the distance to a galaxy to be estimated from its velocity, through the Doppler effect. This method of determining distance extends to the farthest limits of the observable universe.
6. The fact that the universe began in a single point implies that it was very hot and dense initially, and this in turn implies that the early universe was filled with radiation. As the expansion has proceeded, this radiation has been transformed into microwave radiation, with a thermal spectrum corresponding to a temperature of about 3 K. The cosmic background radiation was discovered in 1965.
7. To distinguish a primordial origin for the background radiation from other possible origins (such as many galaxies spread throughout the universe), it is necessary to measure the spectrum to see whether it truly is a thermal spectrum. Observations to date are consistent with the assumption that it is.
8. Another important test is to determine whether the radiation is isotropic, or whether it might be unevenly intense in different directions. Careful observations have revealed no evidence of patchiness or unevenness that might imply that the radiation arises in a large number of individual sources, or that the early universe was nonuniform in any way.
9. There is a 24-hour (or dipole) anisotropy in the radiation, however, created by the Doppler effect resulting from the Earth's motion. By comparing the radiation intensity in the forward and backward directions, it has been possible to determine the Earth's velocity with respect to the reference frame of the radiation. The Earth's motion is a combination of its orbital motion about the Sun, the Sun's orbital motion about the galaxy, the galaxy's motion within the Local Group, and the Local Group's motion within the local supercluster.
10. The early results from the *COBE* satellite confirm the conclusion that the spectrum of the cosmic background radiation is thermal, and support the suggestion that the early universe was very smooth and free of clumpiness.

REVIEW QUESTIONS

Thought Questions

1. In your own words, explain how all galaxies can be receding from our own, yet our galaxy is not at the center of the universe.

2. Explain the distinction between the motions of galaxies within a cluster of galaxies and motions resulting from the expansion of the universe.

3. How is the value of the Hubble constant measured? Why is it difficult to obtain a precise value?

4. If the universal expansion was once more rapid than it is today, how does that affect estimates of the age of the universe that are based on the present-day expansion rate?

5. Explain how the distances we measure to the most faraway galaxies depend on how accurately we know the Sun–Earth distance.

6. Explain why the assumption that the early universe was hot and dense leads to the conclusion that it must have been filled with radiation.

7. Why was measuring the spectrum of the background radiation so important in determining whether it was truly remnant radiation from the early expansion of the universe? Why is measuring the spectrum difficult?

8. Explain why the isotropy of the background radiation is important in establishing whether it is remnant radiation from the early expansion of the universe.

9. If the background radiation is isotropic, why are there small shifts in its temperature as observed in opposite directions from the Earth?

10. What might subtle variations in the intensity of the background radiation tell us about the fundamental structure of the universe?

Problems

1. What is the speed of recession of a galaxy whose ionized calcium line is observed at a wavelength of 3,937.61 Å? (The rest wavelength of the line is 3,933.68 Å.)

2. Suppose three galaxies are situated so that the distance from galaxy A to galaxy C is twice the distance from A to B. After a period of time, the distance between galaxies A and B has doubled. How has the distance between A and C changed? How does the distance between A and C now compare with the distance between A and B?

3. The age of a supernova remnant can be determined from a technique analogous to determining the age of the universe from the Hubble constant. To see the analogy, calculate the age of a supernova remnant whose radius is 10 pc and whose outer portions are expanding at a speed of 1,000 km/sec. (*Note:* You'll have to convert parsecs to kilometers.) Express your answer in years.

4. Suppose the current value of the Hubble constant is 75 km/sec/Mpc, but because the initial expansion was more rapid, the average value over the history of the universe is 80 km/sec/Mpc. What is the true age of the universe, and how does it compare with the value derived by assuming the expansion has been steady at the present rate?

5. Suppose three galaxies are observed, and in their spectra the position of the ionized calcium line is observed to be 3,936 Å for galaxy A, 4,028 Å for galaxy B, and 3,942 Å for galaxy C (the rest wavelength is 3,933 Å). Find the velocities and distances to the galaxies, assuming a Hubble constant of $H = 75$ km/sec/Mpc.

ADDITIONAL READINGS

Abell, G. 1978. Cosmology—the origin and evolution of the universe. *Mercury* 7(3):45.

Barrow, J. D., and J. Silk. 1980. The structure of the early universe. *Scientific American* 242(2):118.

Dressler, A. 1987. The large-scale streaming of galaxies. *Scientific American* 257(3):46.

Ferris, Timothy. 1984. The radio sky and the echo of creation. *Mercury* 13(1):2.

Gulkis, S., P. M. Lubin, S. S. Meyer, and R. F. Silverberg. 1990. The *Cosmic Background Explorer. Scientific American* 262(1): 132.

Hodge, P. 1984. The cosmic distance scale. *American Scientist* 72(5):474.

Kippenhahn, R. 1987. Light from the depths of time. *Sky and Telescope* 73(2):140.

Layzer, D. 1975. The arrow of time. *Scientific American* 233(6):56.

Muller, R. A. 1978. The cosmic background radiation and the new aether drift. *Scientific American* 238(5):64.

Parker, Barry. 1986. The discovery of the expanding universe. *Sky and Telescope* 72(3):227.

Shu, F. H. 1982. *The physical universe.* San Francisco: W. H. Freeman.

Silk, J. 1980. *The big bang: The creation and evolution of the universe.* San Francisco: W. H. Freeman.

———. 1989. Probing the primeval fireball. *Sky and Telescope.* 79:600.

Smith, D. H. 1989. Cosmic fire, terrestrial ice. *Sky and Telescope* 78:471.

Webster, A. 1974. The cosmic background radiation. *Scientific American* 231(2):26.

Weinberg, S. 1977. *The first three minutes.* New York: Basic Books.

MATHEMATICAL METHODS

Doppler Shifting the Cosmic Background

 In the text, we noted that the motion of the Earth relative to the rest frame of the cosmic background radiation causes a small Doppler shift in the radiation as seen from the Earth. The resulting shift appears as a small difference in the temperature of the radiation as seen in opposing directions. We can gain some appreciation of the difficulty of measuring this small shift by calculating the shift in the wavelength of maximum emission and the temperature difference that is involved.

The velocity of the Earth with respect to the radiation is about 360 km/sec. From the Doppler formula, we see that this results in a wavelength shift of

$$\frac{\Delta\lambda}{\lambda} = \frac{v}{c},$$

or

$$\Delta\lambda = \lambda\left(\frac{v}{c}\right),$$

where $\Delta\lambda$ is the shift in wavelength, λ is the rest wavelength, v is the speed of the Earth relative to the background radiation, and c is the speed of light.

For the rest wavelength, we take the wavelength of maximum emission of the radiation, which is 1 mm = 0.1 cm. Substituting, we find

$$\Delta\lambda = 0.1 \text{ cm}\left(\frac{360 \text{ km/sec}}{300,000 \text{ km/sec}}\right)$$

$$= 0.00012 \text{ cm}.$$

This is the amount by which the peak emission is shifted. When we observe in the forward direction of the Earth's motion, the shift is toward shorter wavelengths; that is, the wavelength of maximum emission is 0.1 cm − 0.00012 cm = 0.09988. When we look in the backward direction, the shift is toward longer wavelengths, so the wavelength of maximum emission is 0.10012 cm.

Now we can calculate the temperature difference this wavelength shift corresponds to by using Wien's law. Recall (from Chapter 6) that this law says that the wavelength of maximum emission is inversely proportional to the absolute temperature. If we use λ_1 and λ_2 to designate the two wavelengths, and T_1 and T_2 for the corresponding temperatures, then we can write

$$\frac{T_1}{T_2} = \frac{\lambda_2}{\lambda_1}.$$

Now we substitute 0.10012 for λ_1 and 0.09988 for λ_2, finding

$$\frac{T_1}{T_2} = \frac{0.09988}{0.10012} = 0.9976.$$

If we take 3 K for T_2, then $T_1 = 0.9976(3 \text{ K}) = 2.9928$ K, and the difference in temperature between the forward and backward directions is 3 K − 2.9928 K = 0.0072 K. This is a very small temperature difference, and it is very difficult to measure. It has been detected, however, with a value comparable to the one calculated here (in fact, what really happened is that the temperature difference was detected first, and the above calculations were performed in reverse order to find the velocity of the Earth relative to the radiation).

CHAPTER 30

Peculiar Galaxies, Active Nuclei, and Quasars

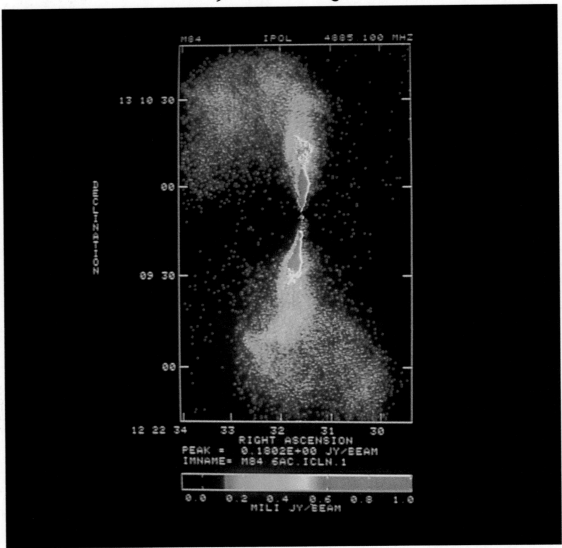

Radio jets emitted from a galaxy with an active nucleus. (National Radio Astronomy Observatory)

In discussing the characteristics of galaxies in the preceding chapters, we have overlooked a variety of objects, some of them galaxies and some possibly not, that have unusual traits. As in many other situations in astronomy, the so-called peculiar objects, once understood, have quite a bit to tell us about the more normal ones.

To fully appreciate the bizarre nature of some of these astronomical oddities, it was best to delay their introduction until this point, when we have essentially completed our survey of the universe. We know about the expansion and the big bang, and we are just in the process of trying it all together. In this chapter we will uncover a number of vital clues in the cosmic puzzle.

THE RADIO GALAXIES

Most of the first astronomical sources of radio emission to be discovered, other than the Sun, are galaxies. Most of these, when examined optically, have turned out to be large ellipticals, often with some unusual-appearing structure (Fig. 30.1). The first of these objects to be detected, and one of the brightest, is called Cygnus A (named under a preliminary cataloging system in which the ranking radio sources in constellation are listed

Figure 30.1 Centaurus A. *This is a giant elliptical radio galaxy, showing a dense lane of interstellar gas and dust across the central region. At upper right is a visible-light photograph; at lower left is a radio image; and at lower right is an X-ray image. Note that the radio and X-ray emission comes from the lobes above and below the visible galaxy. (Top: © 1980 Anglo-Australian Telescope Board, photo by D. Malin; lower right: National Radio Astronomy Observatory; lower left: Harvard-Smithsonian Center for Astrophysics)*

Figure 30.2 A radio image of a radio galaxy. *Here the lighter areas represent maximum radio brightness. The galaxy as seen in visible wavelengths does not even appear here; this image is completely dominated by the double-side lobes. (National Radio Astronomy Observatory)*

Figure 30.3 A map of a radio galaxy. *The intensity contours here represent radio emission from the two side lobes of the radio galaxy Cygnus A. The blurred image at the center illustrates the size of the visible galaxy compared to the gigantic side lobes. (Mullard Radio Astronomy Observatory, University of Cambridge)*

Figure 30.4 Details of radio lobes. *This image of Cygnus A was obtained with the* Very Large Array *and shows fine detail suggesting turbulence and flows in the lobes. (National Radio Astronomy Observatory)*

alphabetically). This galaxy has a strange double appearance, and it was eventually found, after sufficient refinement of radio observing techniques (see the discussion of interferometry in Chapter 7) that the radio emission comes from two locations on opposite sides of the visible galaxy and well separated from it (Figs. 30.2 and 30.3). High-resolution observations with the *Very Large Array* have revealed remarkable detail in some radio galaxies (Fig. 30.4).

This double-lobed structure is a common feature of the so-called **radio galaxies,** those with unusually strong radio emission. Most, if not all, ordinary galaxies emit radio radiation, but the term *radio galaxy* is applied only to the special cases in which the radio intensity is many times greater than the norm. The core of the Milky Way is a strong radio source as viewed from the Earth, but it is not in a league with the true radio galaxies. Typical properties of radio galaxies are listed in Table 30.1.

Although the double-lobed structure is standard among elliptical radio galaxies, the visual appearance varies quite a bit. We have already noted the double appearance of Cygnus A; another bright source, the giant elliptical Centaurus A (the object shown in

Fig. 30.1), appears to have a dense band of interstellar matter bisecting it, and it has not one, but two, pairs of radio lobes, one much farther out from the visible galaxy than the other. Other giant elliptical radio galaxies have other kinds of strange appearances. One of the most famous is M87, also known as Virgo A, which has a jet protruding from one side that is aligned with one of the radio lobes (Fig. 30.5). Careful examination shows that this jet contains a series of blobs or knots that appear to have been ejected from the core of the

Table 30.1 Properties of radio galaxies

Galaxy type: Most often elliptical or giant elliptical.

Radio luminosity compared to optical luminosity: 0.1 to 10.

Radio source shape: Either double-lobed or compact central source; often jets are seen that emit throughout the spectrum.

Variability: Intensity variations in times as short as days may occur in the compact radio sources.

Nature of spectrum: Usually synchrotron or inverse Compton spectrum, usually polarized.

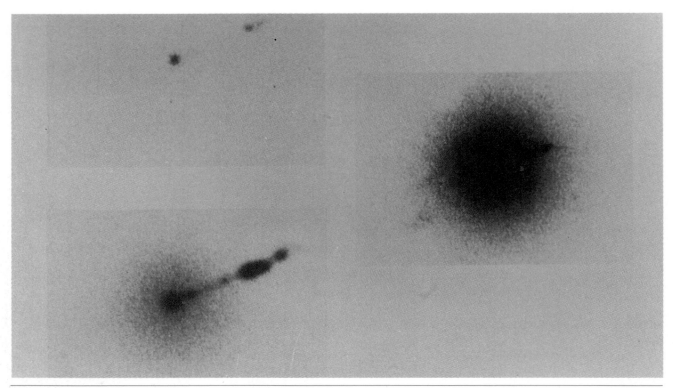

Figure 30.5 A giant elliptical radio galaxy with a jet. *This is M87, also known as Virgo A, showing its remarkable linear jet of hot gas. This is a short-exposure photograph, specially processed to maximize the visibility of the jet and its lumpy structure. On a longer-exposure photograph, this galaxy looks like a normal elliptical, the light from the jet being drowned out by the intensity of light from the galaxy. (H. C. Arp)*

galaxy in sequence. Many galaxies have been discovered to have radio-emitting jets. These include galaxies already known to be peculiar, such as Centaurus A (see Fig. 30.1), and at least one spiral galaxy (Fig. 30.6). In some cases double jets are seen on opposite sides of a galaxy (Fig. 30.7).

The size scale of the radio galaxies can be enormous. In some cases, the radio lobes or jets extend as far as 5 million parsecs from the central galaxy. Recall that the diameter of a large galaxy is only about 2 percent of this distance, and that the Andromeda galaxy is less than 1 million parsecs from our position in the Milky Way.

The first radio galaxies were detected in the 1940s, and by the 1950s studies of the radio spectra of these objects were in progress. It was discovered that they are emitting by the synchrotron process (already mentioned in Chapter 23), which requires a strong magnetic field and a supply of rapidly moving electrons.

The electrons are forced to follow spiral paths around the magnetic field lines, and as they do so, they emit radiation over a broad range of wavelengths. The characteristic signature of synchrotron radiation, in contrast with the thermal radiation from hot objects such as stars, is a sloped spectrum with no strong peak at any particular wavelength. The radiation from a synchrotron source is also polarized (see Chapter 6). Whenever an object is found to be emitting by the synchrotron process, it is immediately concluded that some highly energetic activity is taking place, producing the rapid electrons that are required to create the emission.

In some cases, it appears that a different process is responsible for the emission, particularly at wavelengths shorter than radio. If there is ordinary thermal radio emission from a galaxy, then the presence of electrons moving near the speed of light can alter the spectrum of the emission, as energy is transferred from

Figure 30.6 Jets in an unusual galaxy.   *Here are three views of the barred spiral galaxy NGC1097, showing its two remarkable linear jets. At top left is a black-and-white photograph that has been processed to enhance the visibility of the jets. At lower left is a false-color version. An enlarged view of the inner part of the galaxy is shown at right. (H. C. Arp)*

the electrons to the radiation. This process is called **inverse Compton scattering,** and it has the same implication as synchrotron radiation: There must be a source of vast amounts of energy, to produce the rapidly moving electrons required to create the radiation.

The source of this energy in radio galaxies is a mystery, although some interesting speculation has been fueled by certain characteristics of the galaxies. The double-lobed structure indicates that two clouds of fast

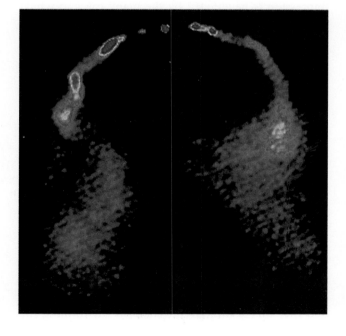

Figure 30.7 Dual radio jets.   *While visible jets are rare, radio images often reveal jetlike structures. Here is a radio image of a galaxy with a pair of jets emanating from opposite sides, and curving away from the galaxy, apparently because the galaxy is moving through an intergalactic gas medium. (National Radio Astronomy Observatory)*

Figure 30.8 A Seyfert galaxy. *This photograph of NGC 4151, a well-studied example, shows the enormous intensity of light from the nucleus compared to the rest of the galaxy. (Palomar Observatory, California Institute of Technology)*

double radio lobes. The fact that the jets are not *observed* in all radio galaxies is thought to be an effect of how the jets are aligned with respect to our line of sight. The emission from rapidly moving particles is confined to a narrow angle; therefore, an observer in the wrong direction from a jet will see little of its radiation.

The theory that massive black holes are the energy sources in radio galaxies has received some observational support. Recently it was reported that evidence for such a central, massive object has been found in the heart of M87, the radio galaxy with the well-known visible jet. Based on measurements of the velocity dispersion near the center of the galaxy, it was deduced that a large amount of mass must be confined in a very small volume at the core, and a black hole seemed the best explanation. Subsequent observations, however, have failed to confirm the reportedly high speeds of stars in the inner portions of the galaxy, and the case for a black hole in M87 has been weakened. Further efforts to find direct evidence of supermassive black holes are being made.

electrons are located on either side of the visible galaxy, and this leads naturally to the impression that the clouds of hot gas may have been ejected from the galactic core, perhaps by an explosive event that expelled material symmetrically in opposite directions. This impression is heightened by the many cases in which jets have been seen protruding from the core (Figs. 30.5, 30.6, and 30.7), and by instances in which more than one pair of lobes, aligned in the same direction but with one farther out than the other, is seen. In these galaxies, it seems that superheated, highly ionized gas is continually being accelerated from the core in opposing directions.

One method envisioned by astronomers for producing prodigious amounts of energy from a small volume such as the core of a galaxy is to have a giant black hole there, surrounded by an accretion disk, and this has naturally been a suggested source for the energy in the radio galaxies. The general idea is that superheated gas that has been compressed by the immense gravitational field of the black hole escapes along the rotation axes and is channeled into two opposing jets. It is probably better to view this process not as explosive, but as a more or less steady expulsion of gas. It has been suggested that *all* radio galaxies have jets, and that this is the only mechanism for forming the

SEYFERT GALAXIES AND ACTIVE NUCLEI

Although it is true that the strongest radio emitters among galaxies are giant ellipticals, they are by no means the only ones with evidence for explosive events in their cores. Some spiral galaxies also display such behavior. We learned in Chapter 24 that even the Milky Way is not immune; the evidence was summarized there for the existence of a compact, massive object at its center. There are other spiral galaxies with much more pronounced violence in their nuclei. These galaxies as a class are called **Seyfert galaxies** (Fig. 30.8 and Table 30.2), after Carl Seyfert, an American astronomer who discovered and cataloged many of them in the 1930s.

Seyfert galaxies have the appearance of ordinary spirals, except that the nucleus is unusually bright and blue, in contrast with the yellow color of most normal spiral galaxy nuclei. About 10 percent of them are radio emitters, with spectra indicating that the synchrotron process is at work. The radio emission usually comes from the nucleus, rather than from double side lobes. The spectrum of the visible light from a Seyfert nucleus typically shows emission lines, something completely out of character for normal galaxies. The emission lines, formed in an ionized gas, may be broad,

Table 30.2 Properties of Seyfert galaxies

Spiral galaxies.
Luminosities comparable to brightest normal spirals.
Bright, compact blue nuclei.
Nuclei show emission lines of highly ionized gas.
About 10% are radio sources.
Most are variable in times of days or weeks.
Emission from the nucleus is synchrotron radiation.
Some have radio jets.

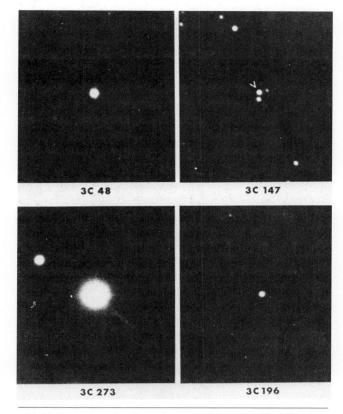

3C 48

3C 147

3C 273

3C 196

Figure 30.9 Quasi-stellar objects. *These starlike objects are quasars, whose spectra are quite unlike those of normal stars. As explained in the text, these objects are probably more distant, and therefore more luminous, than normal galaxies. (Palomar Observatory, California Institute of Technology)*

indicating speeds of several thousand km/sec in the gas that produces the emission, or they may be more narrow. The narrow-line Seyfert galaxy nuclei usually have emission lines characteristic of low-density regions, while the broad-line Seyferts typically indicate higher-density gas that is more tightly confined in the galactic core. Evidently the cores of these galaxies are in extreme turmoil, with hot gas swirling about and tremendous amounts of energy being generated.

Few spirals show such effects, and it is not clear whether the Seyfert behavior is a phase they all pass through at some time, or whether only a few act up in this manner. Later in this chapter we will discuss evidence supporting the first of these possibilities.

THE DISCOVERY OF QUASARS

In 1960, spectra were obtained of two starlike, bluish objects that had been found to be sources of radio emission (Fig. 30.9). No radio stars were known, and astronomers were very interested in these two objects, called 3C48 and 3C273 (their designations in a catalog of radio sources that had recently been compiled by the radio observatory of Cambridge University, in England). The spectra of the two objects, which had several strong emission lines, defied understanding for some 3 years. The objects became known as **quasars** (short for "quasistellar radio sources"), or **quasistellar objects** (sometimes simply **QSOs**) because of their starlike appearance but nonstellar spectra.

It was eventually realized that the strong emission lines in the two quasars were the well-known Balmer series of hydrogen, but shifted far toward longer wavelengths than those measured in the laboratory. If these shifts are due to the Doppler effect, then tremendous speeds are implied. The redshift in the spectrum of

3C273 is 16 percent, indicating that this object is moving away from us at 16 percent of the speed of light, or 48,000 km/sec! In 3C48, the shift was even greater, about 37 percent, corresponding to a speed of 111,000 km/sec (these speeds actually have to be corrected for relativistic effects, as described in Appendix 14; the correct values are 44,000 km/sec for 3C273 and 91,000 km/sec for 3C48).

Since the early 1960s, hundreds of additional quasars have been discovered. Only a small fraction are radio sources, and in other ways they may differ from the first two, but they invariably have highly redshifted emission lines, and in very many cases weak absorption lines are also seen. In many quasars the redshift is so huge that spectral lines with rest wavelengths in the ultraviolet portion of the spectrum are shifted all the way into the visible region. The grand champion today is a quasar with redshift of 473 percent, so that the strongest of all the hydrogen lines, with a rest wavelength of 1,216 Å, is shifted all the way to 6968 Å. This quasar is *not* traveling away from us at 4.73 times the speed of light, however; for very large velocities, the Doppler shift formula has to be modified in accordance with relativistic effects (Appendix 14). The velocity of this quasar is 94.1 percent of the speed of light, still an enormous speed.

THE ORIGIN OF THE REDSHIFTS

To understand the physical properties of the quasars, we must first discover the reason for the high redshifts. We have already tacitly adopted the most obvious explanation, that they are a result of the Doppler effect in objects that are moving very rapidly; but we still must ascertain the nature of the motions. Furthermore, at least one alternative explanation was suggested that has nothing to do with motions at all.

One consequence of Einstein's theory of general relativity, verified by experiment, is that light can be redshifted by a gravitational field. Photons struggling to escape an intense field lose some of their energy in the process, and as this happens, their wavelengths are shifted towards the red. We have already been exposed to this concept in Chapter 23, when we discussed the behavior of light near black holes, which have such strong gravitational fields that no light can escape at all. For a while it was considered possible that quasars are stationary objects sufficiently massive and compact to have large gravitational redshifts.

This suggestion has been largely ruled out, for two reasons. One is that there is no known way for an object to be compressed enough to produce such strong gravitational redshifts without falling in on itself and becoming a black hole. If enough matter is squeezed into such a small volume that its gravity produced redshifts as large as those in quasars, no known force could prevent this object from collapsing further. A neutron star does not have as large a gravitational redshift as those found in quasars, and we already know that the only possible object with a stronger gravitational field than that of a neutron star is a black hole.

The second objection is that, even if such a massive, yet compact, object could exist, its spectral lines would be very much broader than those observed in quasar spectra. The high pressure would distort electron en-

ergy levels so that the spectral lines would be smeared out (as in a white dwarf, but more extreme), and light emitted from slightly different levels in the object would have different gravitational redshifts, again causing spectral lines to be broadened.

For both of these reasons, we are forced to accept the Doppler shift explanation for the redshifts. The problem then is to somehow explain how such large velocities can arise. One possibility is that the quasars are relatively nearby (by intergalactic standards), and are simply moving away from us at very high speeds, perhaps as the result of some explosive event. This "local" explanation requires some care: To accept it, we must explain why no quasars have ever been found to have blueshifts. In other words, if quasars are nearby objects moving very rapidly, it is not easy to see why none happen to be approaching us but are instead all receding. It might be possible to argue that they originated in a nearby explosion (at the galactic center, perhaps) a long time ago, so any that happened to be aimed toward us have had time enough to pass by and are now receding (Fig. 30.10). There are serious difficulties with this picture, though, primarily in the amount of energy that would be required to get all these objects moving at the observed velocities, an amount that would dwarf the total light output of the galaxy over its entire lifetime.

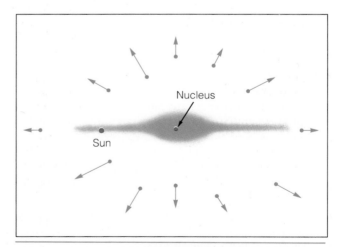

Figure 30.10 A "local" hypothesis for quasar redshifts. *If quasars were very rapidly moving objects ejected from our galactic nucleus some time ago, they could all be receding from our position in the disk by now. This would explain the fact that only redshifts are found, but there is no known mechanism for providing the energy required to produce such great speeds.*

ASTRONOMICAL INSIGHT 30.1

The Redshift Controversy

In the text we present a standard set of arguments demonstrating that quasar redshifts are cosmological, resulting from the expansion of the universe, and implying that the quasars must be very distant objects. Although it is true that most astronomers accept this viewpoint, it is not universally adopted, and there are those who favor other explanations of the redshifts.

The evidence cited by the opponents of the cosmological interpretation consists primarily of cases where quasars are found apparently associated with galaxies or clusters of galaxies, but do not have the same redshift as the galaxies. It is argued in these cases that the quasar is physically associated with the galaxy or cluster of galaxies, and is therefore at the same distance, so that its redshift cannot be cosmological.

The evidence favoring these arguments can be quite striking. When a quasar is found within a cluster of galaxies, there is a natural tendency to think of it as a physical member of the group. It can be argued from statistical grounds that the chances of accidental alignments between galaxies and quasars are so low that the associations between these objects that are observed are unlikely to be coincidental. Long-exposure photographs have even appeared to show gaseous filaments connecting a galaxy with a quasar that has a different redshift. This would seem to prove that the redshift cannot be cosmological.

This radical view of quasars leaves many important questions open. The most difficult of these is how to explain the redshifts, if they really are not cosmological, for the arguments cited in the text against gravitational and local Doppler redshifts are still valid. No satisfactory explanation of the quasar redshifts has been offered by the opponents of the cosmological viewpoint.

The counterarguments center on statistical calculations. One point is that the calculations showing that the quasar-galaxy associations have a low probability of occurring by chance are made after the fact. That is, after a few such associations were found, arguments were made that this was a very unlikely thing to occur by chance alignments of objects randomly distributed over the sky. This is somewhat akin to arguing that the chances of being dealt a certain combination of cards in a game are very low. This is true before the deal, but meaningless after the fact. Based on this and other analogies, many scientists think it incorrect to argue that chance associations of quasars and galaxies are improbable. More to the point, they argue, quasars associated with galaxies are more likely to be discovered than those that are not; therefore, it is natural to find a disproportionately high number of quasars that happen to be aligned on the sky with galaxies. This is an example of a **selection effect,** because the process by which quasars are found is not perfectly random and thorough; the attention of astronomers is naturally concentrated on regions where there are many galaxies, so those regions are more thoroughly searched for quasars.

Probably the most telling argument against the noncosmological redshifts for quasars arises from the fact (noted in the text) that a number of quasars are embedded in galaxies or within clusters of galaxies that share the same redshift. Recent observations show, in fact, that *every* quasar that is sufficiently nearby for an associated galaxy to be detected is indeed embedded within a galaxy that has the same redshift as the quasar. There has not yet been an exception to this, and the data amount to virtual proof that quasars are the cores of galaxies and that their redshifts are cosmological.

The controversy is by no means over, however. Adherents of the noncosmological interpretation of the redshifts continue to find remarkable cases of apparent galaxy-quasar connections where the redshifts do not match. Perhaps observations made in the near future, with the *Space Telescope* or other planned instruments, will provide unequivocal evidence supporting one side or the other.

We are left with one other alternative, one that still poses problems, but perhaps is more acceptable than the others. Let us consider the possibility that the quasars are very distant objects, moving away from us with the expansion of the universe. In this view they are said to be at "cosmological" distances, meaning that they obey Hubble's relation between distance and velocity, just as galaxies do. We find, if we adopt this assumption, that 3C273 is some 580 Mpc away, and 3C48 is more than 1300 Mpc away. (These distances are based on the assumption that the value of the Hubble constant is H = 75 km/sec/Mpc and that the rate of expansion of the universe has been constant, something that is probably not true; this will be discussed in Chapter 31.)

The best evidence that the quasars are at cosmological distances is the fact that some have been found in clusters of galaxies (Fig. 30.11), with the same redshift as the galaxies. This was an extremely difficult observation to make, because at the distances at which the quasars exist, the much fainter galaxies are very hard to see, even with the largest telescopes.

Supporting evidence for the cosmological redshift interpretation is provided by the **BL Lac objects.** Named for the prototype, an object called BL Lacertae that

was once thought to be a variable star, these are elliptical galaxies with very bright central cores. The nucleus of a BL Lac object displays many of the properties of a quasar, with radio synchrotron emission, variable brightness, and enormous luminosity. The surrounding galaxy shows an almost featureless spectrum with a redshift that is consistent with its apparent distance and which is undoubtedly cosmological. Generally, the redshifts of BL Lac objects are relatively small (by quasar standards), and these apparently are nearby objects very closely related to quasars. The implications of the fact that they are embedded in galaxies will be discussed later in this chapter.

Distances of hundreds or thousands of Mpc, inferred for the quasars with very large redshifts, have enormous implications. One is that the light that we receive from them has traveled on its way to us for many billions of years. When we look at a quasar, we are looking far into the history of the universe, and we must keep this in mind as we attempt to interpret them. The fact that quasars are seen only at high redshifts, meaning that they exist only at great distances, indicates that they were more common in the early days of the universe than they are now. This will prove to be a very important clue to their true nature. It has been speculated that there is a limit to quasar redshifts; that is, no quasars can be found with a redshift greater than this limit. The implication is fantastic, for it means that we are looking back to a time before the universe had organized itself into such objects as galaxies and quasars.

Another important implication of the great distances to quasars is that they are the most luminous objects known. Their apparent magnitudes (the brightest are thirteenth-magnitude objects) combined with their tremendous distances, imply that their luminosities are far greater than even the brightest galaxies, by factors of hundreds or even thousands. As we will see, explaining this astounding energy output is one of the central problems of modern astronomy.

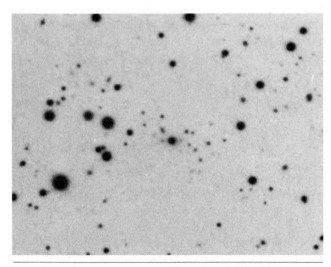

Figure 30.11 A quasar in a cluster of galaxies. *The image at the center of this very long exposure photograph is a quasar. The much fainter objects around it are galaxies, forming a cluster of which the quasar is apparently a member. The galaxies and the quasar have the same redshift, indicating that they lie together at the same distance. (H. Spinrad)*

THE PROPERTIES OF QUASARS

More than 1,500 quasars have been cataloged, and more are being discovered continually. Many are being studied with care, primarily by means of spectroscopic observations. Within the past few years, it has even become possible to observe quasars at infrared, ultraviolet, and X-ray wavelengths, with the use of satellite

Table 30.3 Properties of quasars

Characterized by large redshifts.

Spectra dominated by emission lines of highly ionized gas.

Optical luminosities 100 to 1,000 times those of normal galaxies.[a]

About 10% are radio sources.

Appear as compact, blue objects.

Many are variable in times of days or weeks.

Emission due to synchrotron radiation.

Some have radio or optical jets.

[a]Assuming that quasars are at cosmological distances.

Figure 30.12 A quasar with a jet.   *This is 3C273, one of the first two quasars discovered. This visible-light photograph reveals a linear jet very much like those seen in many radio galaxies. The radio structure of quasars is usually double-lobed, also similar to radio galaxies. (National Optical Astronomy Observatories)*

observatories such as the *Infrared Astonomy Satellite (IRAS), International Ultraviolet Explorer (IUE),* and the *Einstein Observatory.* The faintness of the quasars makes all these observations difficult, but their importance makes the effort worthwhile. The result of all the intensive work being done on quasars is that a great deal is now known about their external properties (Table 30.3), even though their origin and especially their source of energy remain mysterious.

The blue color that characterized the first quasars discovered is a general property of all quasars (except for extremely redshifted ones, where the blue light has been shifted all the way into the red part of the spectrum), but the radio emission is not. Only about 10 percent are radio sources, contrary to the early impression that they all are. This type of misunderstanding is known as a **selection effect,** because the process of selecting quasars was based at first on a certain assumption about their characteristics. For a while, the only method used to look for quasars was to search for objects with radio emission, so naturally, all those that were found were radio sources. The radio emission, and usually the continuous radiation of visible light as well, shows the characteristic synchrotron spectrum.

Apparently *all* quasars emit X-rays, again by the synchrotron process, according to the data collected by the *Einstein Observatory,* (new information on this will soon be available from the U.S.-German X-ray observatory *ROSAT,* which has just begun mapping the sky as this book goes to press). Hence X-ray observations may be a reliable technique for finding new quasars, particularly those of relatively low redshift, which are relatively nearby and hence have stronger X-ray emission than the more distant ones.

In some cases, photographs of quasars reveal some evidence of structure, instead of a single point of light. The most notable of these is 3C273, which played such a key role in the initial discovery of quasars. This object shows a linear jet extending from one side (Fig. 30.12), closely resembling the one emanating from the giant radio galaxy M87 (see Fig. 30.5). Perhaps this means that similar processes are occurring in these two rather different object.

Many quasars vary in brightness, usually over times of several days to months or years (although in at least one case, variations were seen over just a few hours). This variability is very important, for it provides information on the size of the region in the quasar that is emitting the light. This region cannot be any larger than the distance light travels in the time over which the intensity varies. There is no way an entire object that might be hundreds or thousands of light-years across can vary in brightness in a time of a month or so. The light-travel time across the object would guarantee that we would see changes only over times of hundreds or thousands of years, as the light from different parts of the object reached us (Fig. 30.13). Therefore the observed short-term variations in quasars tell us that the fantastic energy emitted by these objects is produced within a volume no more than a

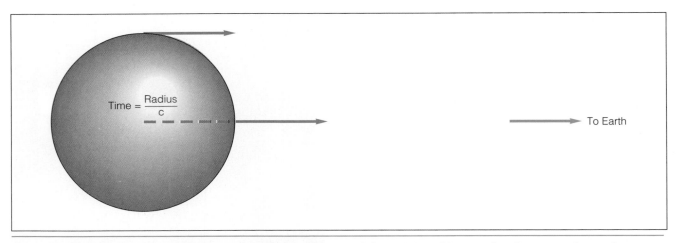

Figure 30.13 The implication of the time variability for the size of the emitting object. *This illustrates why an object cannot appear to vary in less time than it takes for light to travel across it. As a simple case, this spherical object is assumed to change its luminosity instantaneously. On Earth we observe the first hint of this change when light from the nearest part of the object reaches us, but we continue to see the brightness changing gradually as light from more distant portions reaches us.*

light-month, or about 0.03 parsecs, in diameter. This obviously places stringent limitations on the nature of the emitting object.

The spectra of quasars have already been described in broad outline, but there is a great deal of detail as well (Fig. 30.14). All quasars have emission lines, generally of common elements such as hydrogen, helium, and often carbon, nitrogen, and oxygen (some of the latter elements only have strong lines in the ultraviolet, and therefore are best observed when the redshift is sufficiently large to move these lines into the visible portion of the spectrum). In many cases, the emission lines are very broad, showing that the gas that forms them has internal motions of thousands of km/sec. The degree of ionization tells us that the gas is subjected to an intense radiation field, which continually ionizes the gas by the absorption of energetic photons.

Many quasars have absorption lines in addition to the emission features. Strangely enough, the absorption lines are usually at a different redshift (almost always smaller) than the emission lines, indicating that the gas that creates the absorption is not moving away from us as rapidly as the gas that produces the emission. Further confounding the issue is the fact that many quasars have multiple absorption redshifts; that is, they have several distinct sets of absorption lines, each with its own redshift, and each therefore representing a distinct velocity. This shows that there are several ab-

sorbing clouds in the line of sight. There are two possible origins for these clouds (Fig. 30.15). One is that the quasar itself, which is moving at the high speed indicated by the redshift of the emission lines, is ejecting clouds of gas, some of which happen to be aimed toward the Earth. These clouds therefore have lower velocities of recession than the quasar itself, so the redshift of the absorption lines they form is less than the redshift of the emission lines formed in the quasar.

The alternative view is that the absorption lines arise in the gaseous halos or disks of galaxies that happen to lie between us and the more distant quasar. Each galaxy has a recession velocity determined by the expansion of the universe, and because the intervening galaxies are not as far away as the quasars, they are not moving as fast and therefore have lower redshifts. If this interpretation of the absorption lines is correct, as is thought to be true in at least some cases, then their analysis should provide important information about the nature of the intervening galaxies. Of particular interest is the information quasar absorption lines might provide on the extensive halos that many galaxies are thought to have.

In some cases the absorption spectra of quasars show only lines formed by hydrogen, and none due to heavier elements. These lines apparently arise in gas that has not been enriched at all by stellar processing, and it is thought that these might be intergalactic clouds that

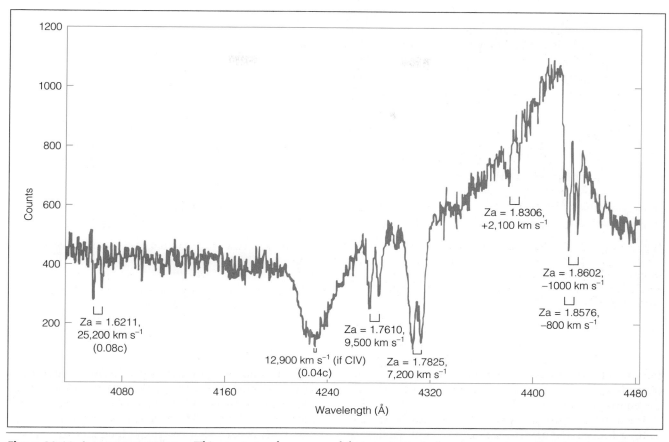

Figure 30.14 A quasar spectrum.   *This spectrum shows several features typical of quasar spectra: a broad emission line centered near 4,400 Å; absorption near 4,240 Å; and narrow absorption lines at several wavelengths. All features seen here are due to carbon that has lost three electrons (designated CIV). Redshifts are indicated (Z_a indicates the redshift of an absorption-line system). (Data from Weymann, R. J., R. F. Carswell, and M. G. Smith 1981, Annual Reviews of Astronomy and Astrophysics, 19:41.)*

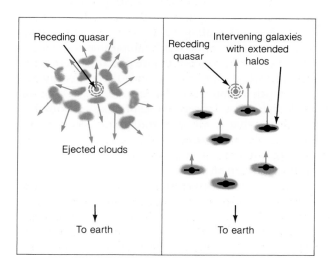

Figure 30.15 Two explanations of quasar absorption lines.   *In one scenario (left), the absorption lines are formed by clouds of gas ejected from the quasar. Clouds ejected directly toward the Earth have lower velocities of recession than the quasar itself. In the other more likely view (right), the absorption lines are formed in the extended halos of galaxies that happen to lie between us and the quasar. The galaxies are closer to us than the quasar, and therefore have lower velocities of recession and smaller redshifts.*

have never formed into galaxies, so that no stars have formed in them. These may be material remnants of the big bang; if so, they can tell us much about the exact amount of nuclear processing that occurred in the early stages of the expansion.

GALAXIES IN INFANCY?

A variety of theories, some of them rather fanciful, have been proposed to explain the quasars. One idea has gradually become widely accepted, however, and we will restrict ourselves to discussing only that one hypothesis, keeping in mind that there are other suggestions.

The prevailing interpretation of the nature of quasars was inspired by the fact that they existed only in the long-ago past, by their similarities to the BL Lac objects (which clearly are galaxies), and by their resemblance to the nuclei of Seyfert galaxies. The statement that they existed only in the past is based on the fact that all are very far away, so that the light-travel time ensures that we are seeing things only as they were billions of years ago, not as they are today. The resemblance to Seyfert nuclei is striking: Both are blue; both are radio sources in about 10 percent of the cases; both vary on similar time scales; and both have very similar emission line spectra. Seyferts lack the complex absorption lines often seen in the spectra of quasars, but this would be expected if the quasar absorption lines are formed in the halos of intervening galaxies, since Seyfert galaxies are not so far away that there are likely to be many other galaxies along their lines of sight. The nuclei of Seyferts differ from quasars also in the amount of energy they emit, being considerably less luminous.

The picture that is developing is that quasars are young galaxies, with some sort of youthful activity taking place in their centers. In this view, the Seyfert galaxies are descendants of quasars, still showing activity in their nuclei, but with diminished intensity. If we carry this idea another step, we are led to the suggestion that normal galaxies like the Milky Way are later stages of the same phenomenon; recall the mildly energetic (by quasar standards) activity in the core of our galaxy.

The notion that quasars may be infant galaxies is supported by some rather direct evidence. As noted earlier, some quasars have been found associated with clusters of galaxies, showing that they can be physically

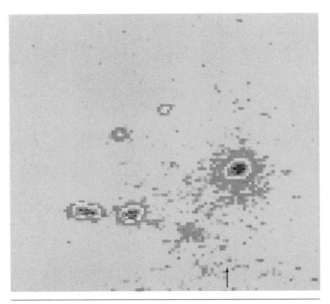

Figure 30.16 A quasar embedded in a galaxy. *This photograph, obtained with an electronic detector, contains several star images (top, sides), a galaxy (slightly elongated object just to the right of center), and a quasar (lower right). The quasar image is extended, just as the galaxy image is, and the fuzzy region surrounding it is indeed a galaxy. In every case where a quasar is close enough to us that a galaxy would be bright enough to be detected, one has been, indicating that all quasars lie at the centers of galaxies. (M. Malkan)*

located in the same region of space at the same time. In addition, there are a few examples known in which careful observation has revealed a fuzzy, dimly glowing region surrounding a quasar (Fig. 30.16). This is probably a galaxy, at the center of which a quasar is embedded. It is likely that all quasars are located in the nuclei of galaxies, which are so much fainter that they usually cannot be seen. A short-exposure photograph of a Seyfert galaxy (Fig. 30.17) shows only the nucleus, which is very much brighter than the surrounding galaxy, and it is clear that if we observed one of these galaxies from so far away that it was near the limit of detectability, all we would see would be a blue, starlike object resembling a quasar.

Although we can make a strong case that quasars are young galaxies, we are still far from understanding all of their properties. The main mystery remaining is the source of the tremendous energy of quasars and active galactic nuclei. Several possibilities have been raised, but today only one seems viable.

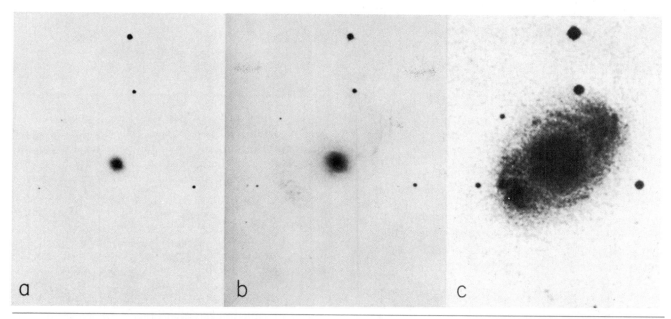

Figure 30.17 Three exposures of a Seyfert galaxy. *At left is a short-exposure photograph of the Seyfert galaxy NGC 4151, in which only the nucleus is seen, resembling the image of a star. The center and right-hand images are longer exposures of the same object, revealing more of its outer, fainter structure. This sequence illustrates how a quasar, which is even more luminous than a Seyfert galaxy nucleus, can appear as a starlike image even though it is embedded in a galaxy. (Photographs from the Mt. Wilson Observatory, Carnegie Institution of Washington, composition by W. W. Morgan)*

This possibility, one that keeps coming up in situations where large amounts of energy seem to be coming from small volumes of space, is that quasars are powered by massive black holes (Fig. 30.18). In the chaotic early days of collapse, when a galaxy was just forming out of a condensing cloud of primordial gas, a great deal of material may have collected at the center. If many stars formed there, frequent collisions could have caused them to settle in more and more tightly, until they coalesced to form a black hole. The gravitational influence of such a massive object would have then stirred up the surroundings, creating the high electron velocities required to fuel the synchrotron emission. Infalling matter would have formed an accretion disk around the black hole, from which X-rays would be emitted, in analogy with stellar black holes and neutron stars in binary systems. In this picture the source of energy in quasars is the same as in radio galaxies, described earlier in this chapter. Even the existence of jets of hot gas is the same; such jets have been observed emanating from several quasars.

Infrared data from the *IRAS* spacecraft have led to a suggested mechanism for building up sufficient densities in the cores of young galaxies to form supermassive black holes. The properties of some of the very luminous infrared galaxies indicate that they may be pairs of galaxies that are colliding and merging. The infrared properties of quasars are very similar to those of luminous infrared galaxies, suggesting that quasars may form as the result of galactic mergers early in the history of the universe. Recent intensive observations of a number of quasars show that they may indeed be the result of mergers of galaxies, at the time when galaxies were just forming. There is growing evidence, in fact, that *all* galaxies formed through the merger of smaller systems of stars (this was discussed in Chapters 26 and 27).

Whatever the origin and power source of the quasars, we will certainly learn a lot about the nature of matter and the early history of the universe when we are able to answer all the questions about them. The *Hubble Space Telescope*, with its broad wavelength coverage and great sensitivity, will provide invaluable information on these fundamentally important objects.

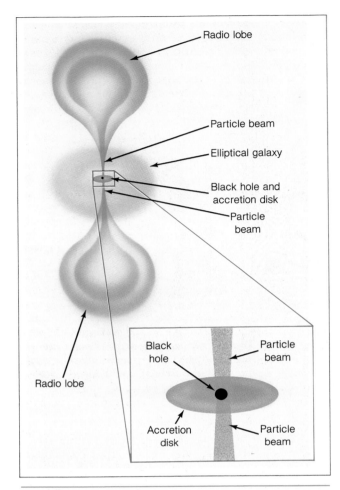

Figure 30.18 A geometrical model for a quasar, Seyfert galaxy, or a radio galaxy. *All of the objects have certain features in common that fit the picture shown here. A central object (most likely a supermassive black hole with an accretion disk) ejects opposing jets of energetic charged particles. These jets produce synchrotron radiation, and build up double radio-emitting lobes on either side of the central object. These lobes extend well beyond the confines of the galaxy in which the central object is embedded.*

PERSPECTIVE

In this chapter we have learned of new and wondrous things, the mighty radio galaxies and the enigmatic quasars. Along the way we have gained several important bits of information on the universe itself. It is time to tackle the fundamental question of the origin and fate of the cosmos.

SUMMARY

1. Radio galaxies emit vast amounts of energy in the radio portion of the spectrum. These galaxies, often giant ellipticals, commonly show structural peculiarities. The radio emission is nonthermal (usually synchrotron radiation), and in most cases is produced from two lobes on opposite sides of the visible galaxy.

2. The source of energy in a radio galaxy is unknown but appears to be concentrated at the core.

3. Seyfert galaxies are spiral galaxies with compact, bright blue nuclei which produce emission lines characteristic of high temperatures and rapid motions.

4. Several point sources of radio emission were found in the early 1960s to look like blue stars, and were therefore called quasistellar objects, or quasars.

5. Quasars have emission-line spectra with very large redshifts, corresponding to speeds that are a significant fraction of the speed of light. If these redshifts are cosmological, then quasars are on the frontier of the universe, are extremely luminous, and are seen as they were billions of years ago.

6. Quasars are sometimes radio sources, usually emit X-rays, are compact and blue, and sometimes vary over times of only a few days. This implies that their tremendous energy output arises in a volume only a few light-days across.

7. Many quasars have absorption lines at a variety of redshifts (almost always smaller than the redshift of the emission lines), which may be created by matter being ejected from the quasars, or, more likely, by intervening galactic disks and halos.

8. The most satisfactory explanation of the quasars is that they are the cores of very young galaxies. This interpretation is supported by the fact that they are only seen at great distances (that is, they existed only long ago), and that they resemble the nuclei of Seyfert galaxies. In several cases, photographs have revealed galaxies surrounding quasars, supporting this suggestion.

9. The source of the energy that powers quasars is a premier mystery of modern astronomy. The most probable explanation is that a quasar has a massive black hole at its core; this can produce enough energy (from infalling matter) to account for the great luminosity of quasars, as well as their time variability. By inference, similar objects may exist in the nuclei of galaxies such as Seyfert galaxies, radio galaxies, and the Milky Way.

REVIEW QUESTIONS

Thought Questions

1. Summarize the properties of radio galaxies, pointing out the ways in which they are distinguished from normal galaxies.

2. Why do astronomers suggest that massive black holes may be responsible for the vast amounts of energy being emitted from the cores of radio galaxies, Seyfert galaxies, and quasars?

3. Summarize the arguments for and against the hypothesis that the redshifts of quasars are cosmological, that is, that they result from the expansion of the universe rather than from some "local" phenomenon.

4. Explain, in your own words and using your own sketch, why the time scale of variations in light from a source such as a quasar places a limit on the diameter of the emitting region.

5. Compile a list of similarities and differences between quasars and Seyfert galaxies.

6. Why are quasar absorption lines almost always observed at smaller redshifts than the emission lines? Would this necessarily be true if the redshifts of the emission lines were not cosmological?

7. From what you have learned about quasars in this chapter and about the *Hubble Space Telescope* in Chapter 7, describe some important observations of quasars that might be made with the *Space Telescope*.

8. If quasars are young galaxies, how does this help demonstrate the difficulty of using faraway galaxies as standard candles in estimating large distances in the universe?

9. Explain why nearly all of the first quasars to be discovered are radio emitters, but only about 10 percent of all quasars are.

10. How can the material in intergalactic gas clouds give us information about conditions very early in the history of the universe?

Problems

1. Suppose that the red hydrogen line (rest wavelength 6,563 Å) is observed as a broad emission line from the nucleus of a Seyfert galaxy. The width of the line is 200 Å (that is, it extends for 100 Å in either direction from the rest wavelength). What is the maximum speed of the gas in the region where this line is emitted?

2. A quasar is found to have a redshift of $z = \Delta\lambda/\lambda = 2.45$. What is its recession velocity (use the relativistic Doppler shift formula given in Appendix 14)? How far away is it if the Hubble constant is $H = 75$ km/sec/Mpc? How long has the light from this quasar been on its way to us? How luminous is it if its apparent magnitude is $m = +18$?

3. If the brightness of a quasar changes in a time of one day, how large can the diameter of the light-emitting region be at most? How does your answer compare with the size of the solar system?

ADDITIONAL READINGS

Bechtold, J. 1988. High-resolution spectroscopy of quasar absorption lines. *Mercury* 17:188.

Blandford, R. D., M. C. Begelman, and M. J. Rees. 1982. Cosmic jets. *Scientific American* 246(5):124.

Burbidge, G. 1988. Quasars in the balance. *Mercury* 17:136.

Burns. J. O., and R. Marcus. 1983. Centaurus A: The nearest active galaxy. *Scientific American* 249(5):50.

Disney, M. J., and P. Veron. 1977. BL Lac objects. *Scientific American* 237(2):32.

Downes, Ann. 1986. Radio galaxies. *Mercury* 15(2):34.

Feigelson, E. D., and E. J. Schreier. 1983. The x-ray jets of Centaurus A and M87. *Sky and Telescope* 65(1):6.

Ferris, T. 1984. The radio sky and the echo of creation. *Mercury* XIII(1):2.

Gregory, S. 1988. Active galaxies and quasars: a unified view. *Mercury* 17:111.

McCarthy, P. 1988. Measuring distances to remote galaxies and quasars. *Mercury* 17:19.

Osmer, P. S. 1982. Quasars as probes of the distant and early universe. *Scientific American* 246(2):126.

Overbye, Dennis. 1982. Exploring the edge of the universe. *Discover* 3(12):22.

Posey, C. 1988. Three recent snapshots of how quasars can be triggered. *Mercury* 17:22.

Preston, R. 1988. Beacons in time: Maarten Schmidt and the discovery of quasars. *Mercury* 17:2.

Shipman, H. L. 1976. *Black holes, quasars, and the universe.* Boston: Houghton-Mifflin.

Silk, J. 1980. *The big bang: The creation and evolution of the universe.* San Francisco: W. H. Freeman.

Strom, R. G., G. M. Miley, and J. Oort. 1975. Giant radio galaxies. *Scientific American* 233(2):26.

Tesch-Fienberg, R. 1987. *IRAS* and the Quasars. *Sky and Telescope* 73:13.

Tyson, T., and M. Gorenstein. 1985. Resolving the nearest gravitational lens. *Sky and Telescope* 70(4):319.

Weedman, D. 1988. Quasars: A progress report. *Mercury* 17:12.

Wyckoff, S., and P. Wehinger, 1981. Are quasars luminous nuclei of galaxies? *Sky and Telescope* 61(3):200.

MATHEMATICAL METHODS

Quasar Luminosities

 In the text, we made the point that, if quasar redshifts are cosmological (i.e., due to the expansion of the universe), then quasars are very distant. If they are really so far away, then they must have enormous luminosities to appear as bright as they do. Here we can show how this latter conclusion was reached.

The brightest quasars have apparent magnitudes of approximately m = 13, and values of m = 16 are typical. Suppose we find a sixteenth-magnitude quasar whose redshift is $\Delta\lambda/\lambda$ = 3.00. From the relativistic Doppler formula (Appendix 14), we find that the velocity of recession of this quasar is 0.882c = 265,000 km/sec. Using this for v in the Hubble law and assuming the Hubble constant has the value H = 75 km/sec/Mpc, we find for the distance d

$$d = v/H$$
$$= (265,000 \text{ km/sec})/(75 \text{ km/sec/Mpc})$$
$$= 3,530 \text{ Mpc}.$$

Now we can find the absolute magnitude M of this quasar, using the distance modulus formula (see the Mathematical Methods section at the end of Chapter 18, or Appendix 10):

$$m - M = 5\log(d/10)$$

or

$$M = m - 5\log(d/10).$$

Substituting 16 for m and 3530 Mpc = 3.53 × 10⁹ pc for d, we find

$$M = 16 - 5\log(3.53 \times 10^9/10)$$
$$= 16 - 5(8.55) = -26.74.$$

The absolute magnitude of this quasar is −26.74. To find its luminosity, we can compare this with the Sun, which has a visual absolute magnitude of approximately +4.83; the quasar is 31.57 magnitudes more luminous. We want to know what ratio of luminosities is required to produce this difference in magnitudes.

From the definition of the magnitude system (Appendix 10 or the Mathematical Methods section at the end of Chapter 17), we find

$$L_q/L_\odot = 10^{-0.4(M_\odot - M_q)},$$

where L_q is the luminosity of the quasar, $L_\odot$ is the luminosity of the Sun, and M_q and $M_\odot$ are their respective absolute magnitudes. Substituting, we find

$$L_q/L_\odot = 10^{0.4[4.83-(-26.74)]} = 10^{12.63} = 4.24 \times 10^{12}.$$

This quasar is about 4 trillion times more luminous than the Sun, and roughly 40 times more luminous than the biggest and brightest galaxies. This line of reasoning shows how the tremendous luminosities of quasars were established.

(In doing this, we have conveniently overlooked one complicating factor, namely that the peak emission for a quasar is shifted because of the Doppler effect, so that it is technically not correct to use the visual apparent magnitude to compute the visual absolute magnitude; the shift is so large that the observed visual brightness corresponds to ultraviolet emission from the quasar. Interestingly, X-ray and ultraviolet observations of quasars show that even greater amounts of energy are emitted in these short wavelengths.)

CHAPTER 31

Cosmology: Past, Present, and Future of the Universe

The Cosmic Background Explorer *gathering data on the universal radiation field. (NASA)*
FOR PAGE 271

Given the time and opportunity, humans have, through the ages, devoted themselves to speculation on the grandest scale of all. Poets, philosophers, and theologians have approached the question of the origin and future of the universe in a countless variety of ways. So have astronomers, with the exception that they follow a somewhat restrictive set of rules; the answers they accept must not violate known laws of physics. By retaining this requirement, scientists attempt to approach problems in an objective, verifiable manner.

There are difficulties in maintaining this idealized posture, however, and we shall try to point them out. Because the universe in which we live is, as far as we know, unique, we have no opportunity to check our hypotheses by comparison with other examples. Furthermore, no matter how thoroughly and rigorously we can trace the evolution of the universe by application of known physical laws, there will always be fundamental questions that are beyond the scope of physics. As a result, even the most careful and objective scientists reach a point at which they have to make certain unverifiable assumptions, and at that point, they become philosophers or theologians. In this chapter we shall restrict ourselves to the questions that can in principle have objective, verifiable answers.

Technically, the study of the universe as it now appears is **cosmology**; that is really what the preceding thirty chapters have been all about. The study of its origin is **cosmogony**, and this word applies to the big bang as well as to the earlier theories on the origin of the solar system, once thought to be the entire universe. In practice, the general subject of the nature of the universe and its evolution is lumped under the heading of cosmology, and so it is on a pursuit of this subject that we now embark.

UNDERLYING ASSUMPTIONS

To even begin to study the universe as a whole, we must make certain assumptions. These can, in principle, be tested, although it is not clear whether there is a practical way to do so. One of the most important assumptions we make is that the known laws of physics can be applied to the universe as a whole. This is a bold assumption, since we are just now beginning to be able to observe significantly large distances (compared to the size of the universe), and it is possible that the laws that we know to apply to our own times and our own region are merely subsets of some large-scale set of laws, as yet undiscovered. Nevertheless, for now

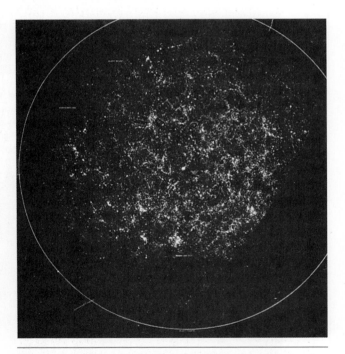

Figure 31.1 The cosmological principle. *The universe is assumed to be homogeneous and isotropic, meaning that it looks the same to all observers in all directions. Here is a segment of the universe, filled with galaxies. Their distribution, while not identical from place to place, is similar throughout. (© 1978, 1979 by The Co-Evolution Quarterly, Box 428, Sausalito, California)*

ASTRONOMICAL INSIGHT 31.1

The Mystery of the Nighttime Sky

A very simple question, asked over two centuries ago and discussed at length in the early 1800s by the German astronomer Wilhelm Olbers, leads to profound consequences. The question is: Why is the sky dark at night?

Olbers argued that in an infinite universe filled with stars, every line of sight, regardless of direction, should intersect a stellar surface. The nighttime sky should therefore be uniformly bright, not dark, as the star images in the sky would literally overlap each other. The fact that this is not so presents a paradox, one worth pondering. The dilemma created by asking why the sky is dark is called **Olbers' paradox.**

One suggested solution is that interstellar extinction can diminish the light from distant stars sufficiently to make the sky appear black. Careful consideration shows, however, that this is not so. Light that is absorbed by dust in space causes the grains to heat up. If the universe were really filled with starlight, as suggested by Olbers' paradox, then the grains would become so hot that they would glow, and we would still have a uniformly bright nighttime sky.

Another suggested explanation of Olbers' paradox has to do with the expansion of the universe. The redshift due to the expansion, if we look sufficiently far away, becomes so great that the light never gets here in visible form. It becomes redshifted to extremely large wavelengths, losing nearly all of its energy. This solution to Olbers' paradox also works in an infinitely old universe, so that even if there is no beginning to the universe, there is still a horizon.

While the expansion of the universe partially explains Olbers' paradox, it does not completely account for the darkness of the sky between galaxies. Yet another solution to the paradox turns out to be more important. To understand this solution, we must realize that the universe is not really infinite, at least from a practical point of view. At sufficiently great distances, we expect to see no galaxies or stars, because we will be looking back to a time before the universe began. Thus, even though the universe has no edge, there is a horizon beyond which stars and galaxies cannot be seen. Therefore, this argument goes, all lines of sight do not have to intersect stellar surfaces; many reach the darkness beyond the horizon of the universe, and we have a dark nighttime sky.

There is another factor that accounts for the darkness of the nighttime sky, which also has to do with the finite size of the universe. The source of energy available to produce light—that is, all the stars in the universe—is limited. If we calculate the total energy that possibly can be produced by all stars everywhere, we find that it is too little to make our sky appear bright. The universe simply does not contain enough energy.

The simple question of why the sky is dark at night, if analyzed fully, could have led to the conclusion that the universe has a finite age or a finite size a full century before this was learned from other observational and theoretical developments.

---

the only way we can proceed is to assume that our known laws can be applied.

A central rule traditionally set forth by cosmologists is that the universe must look the same at all points within it. This does not mean that the appearance of the heavens should be literally identical everywhere; it means that the general structure, the density and distribution of galaxies and clusters of galaxies (Fig. 31.1) should be constant. This assumption states that the universe is **homogeneous.**

A related, but slightly different assumption, has to do with the appearance of the universe when viewed in different directions. The assumed property in this case is **isotropy**; that is, the universe must look the same to an observer, no matter in which direction the telescope points. We have already encountered this concept, in connection with the cosmic microwave background radiation. Again, the assumption of isotropy does not imply identical constellations of galaxies and stars in all directions, just comparable ones.

Figure 31.2 Albert Einstein. *Among Einstein's great contributions was the development of a mathematical formalism to describe the interaction of gravity, matter, and energy in the universe with the nature of space-time. This framework, general relativity, has withstood all observational and experimental tests applied to it so far. (The Granger Collection)*

EINSTEIN: Einheitliche Feldtheorie von Gravitation und Elektrizität   415

Jnabhängig von diesem affinen Zusammenhang führen wir eine kontravariante Tensordichte $\mathfrak{g}^{\mu\nu}$ ein, deren Symmetrieeigenschaften wir ebenfalls offen lassen. Aus beiden bilden wir die skalare Dichte

$$\mathfrak{H} = \mathfrak{g}^{\mu\nu} R_{\mu\nu} \qquad (3)$$

und postulieren, daß sämtliche Variationen des Integrals

$$\mathfrak{J} = \int \mathfrak{H} dx_1 dx_2 dx_3 dx_4$$

nach den $\mathfrak{g}^{\mu\nu}$ und $\Gamma^{\alpha}_{\mu\nu}$ als unabhängigen (an den Grenzen nicht varierten) Variabeln verschwinden.

Die Variation nach den $\mathfrak{g}^{\mu\nu}$ liefert die 16 Gleichungen

$$R_{\mu\nu} = 0, \qquad (4)$$

die Variation nach den $\Gamma^{\alpha}_{\mu\nu}$ zunächst die 64 Gleichungen

$$\frac{\partial \mathfrak{g}^{\mu\nu}}{\partial x_\alpha} + \mathfrak{g}^{\sigma\nu}\Gamma^{\mu}_{\sigma\alpha} + \mathfrak{g}^{\mu\sigma}\Gamma^{\nu}_{\alpha\sigma} - \delta^{\nu}_{\alpha}\left(\frac{\partial \mathfrak{g}^{\mu\sigma}}{\partial x_\sigma} + \mathfrak{g}^{\mu\tau}\Gamma^{\sigma}_{\tau\sigma}\right) - \mathfrak{g}^{\mu\nu}\Gamma^{\sigma}_{\alpha\sigma} = 0. \qquad (5)$$

Wir wollen nun einige Betrachtungen anstellen, die uns die Gleichungen (5) durch einfachere zu ersetzen gestatten. Verjüngen wir die linke Seite von (5) nach den Indizes ν, α bzw. μ, α, so erhalten wir die Gleichungen

Figure 31.3 The field equations. *Here are portions of the equations that describe the physical state of the universe. To a large extent, the science of cosmology involves finding solutions to these equations. (The Granger Collection)*

Both of these assumptions are thought to apply only on the largest scales, larger than any obvious clumpiness in the universe, such as clusters or superclusters of galaxies. It should also be pointed out here, although it is not discussed until later in the chapter, that both assumptions have been questioned in some modern theories and observations of the structure of the universe. It is even thought surprising in some current theories that the universe should be as homogeneous as it is. Hence the assumption of homogeneity and isotropy should now be viewed more as an observation that must be explained, rather than as a postulate of cosmological theory.

The statement that the universe is homogeneous and isotropic is often referred to as the **Cosmological Principle,** and is often stated in terms of how the universe looks to observers; that is, the universe looks the same to all observers everywhere. It is difficult to verify the Cosmological Principle. We can test the assumption of isotropy by looking in all directions from Earth, and indeed we do not find any deviations. On the other hand, we cannot test the homogeneity of the universe by traveling to various other locations to see whether things look any different. What we can do is count very dim, distant galaxies, to see whether their density appears to be any greater or less in some regions than in others, but even this is complicated by the fact that

we are looking back in time to an era when the universe was more compressed and dense than it is now. The 3-degree background radiation gives us another tool for testing both homogeneity and isotropy, and it also appears to satisfy the Cosmological Principle. The best we can say for now is that the available evidence supports this principle.

EINSTEIN'S RELATIVITY: MATHEMATICAL DESCRIPTION OF THE UNIVERSE

The simple act of making assumptions about the general nature of the universe by itself tells us very little about its past or its future. It serves to elucidate certain properties of the universe as it is today, but to tie this into a quantitative description, to develop a theory capable of making predictions that can be tested, requires a mathematical framework. In developing such a mechanism for describing the universe, the astronomer seeks to reduce the universe to a set of equations and then to solve them, much as a stellar structure theorist studies the structure and evolution of stars by constructing numerical models.

The most powerful mathematical tool for describing the universe was developed by Albert Einstein (Fig. 31.2). His theory of general relativity represents the properties of matter and its relationship to gravitational fields. Within the context of his theory, Einstein developed a set of relations, called the **field equations** (Fig. 31.3), that express in mathematical terms the interaction of matter, radiation, and gravitational forces in the universe. Although alternatives to general relativity have been developed and their consequences explored, most research in cosmology today involves finding solutions to Einstein's field equations and testing these solutions with observational data.

The basic premise of general relativity is that acceleration created by a gravitational field is indistinguishable from acceleration due to a changing rate or direction of motion. One way to visualize this is to imagine that you are inside a compartment with no windows (Fig. 31.4). If this compartment is on the Earth's surface, your weight feels normal because of the Earth's gravitational attraction. If, however, you are in space and the compartment is being accelerated at a rate equivalent to 1 Earth gravity, your weight will also feel normal. There is no experimental way to tell the difference between the two situations, short of opening the door and looking out.

One consequence of the equivalence of gravity and acceleration is that an object passing near a source of gravitational field (that is, any other object having mass) undergoes acceleration and therefore follows a curved path. In a universe containing matter, this means that all trajectories of moving objects are curved, and it is often said that space itself is curved. The degree of curvature is especially high close to massive objects (Fig. 31.5), but there is also an overall curvature of the

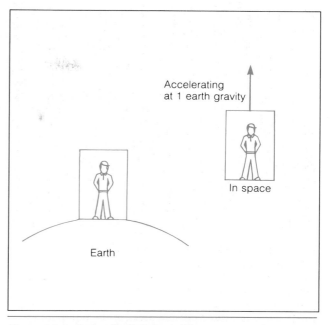

Figure 31.4 General relativity. *The person in the enclosed room has no experimental or intuitive means of distinguishing whether he is motionless on the surface of the Earth, or in space accelerating at a rate equivalent to 1 Earth gravity. The implication of this is that the acceleration due to motion and that due to gravity are equivalent, which in turn implies that space is curved in the presence of a gravitational field.*

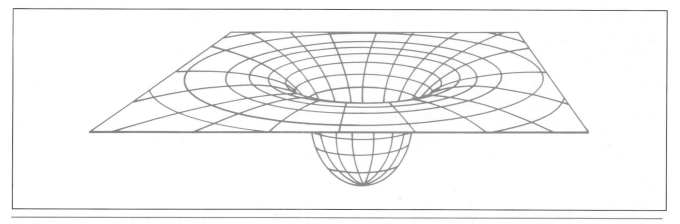

Figure 31.5 Curvature of space near a massive star. *This curved surface represents the shape of space very close to a massive star. The best way to envision what it means to say that space is curved in this way is to imagine photons of light as marbles rolling on a surface of this shape.*

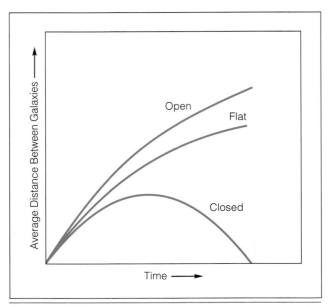

Figure 31.6 The three possible fates of the universe.
This diagram shows the manner in which the average distance between galaxies will change with time for the open, flat, and closed universe. In the first case, galaxies will continue to separate forever, although the rate of separation will slow. In the second case, the rate of separation will, in an infinite time, slow to a halt but will not reverse. In the third case, the galaxies eventually begin to approach each other, and the universe returns to a single point.

By the 1920s, solutions for an expanding universe were found, primarily by the Soviet physicist A. Friedmann and later by the Belgian G. LeMaître, who went so far as to propose an origin for the universe in a hot, dense state from which it has been expanding ever since. This was the true beginning of the big bang idea, and it was developed some 3 years before Hubble's observational discovery of the expansion. Of course, once it was found that the universe is not static but is actually in a dynamic state, the original need for Einstein's cosmological constant disappeared. Nevertheless, it is usually included by modern cosmologists in the field equations, but its value is assumed to be zero.

The general relativistic field equations allow for three possibilities regarding the curvature of the universe and three possible futures (Fig. 31.6). A central question of modern studies of cosmology has to do with deciding which possibility is correct. One is referred to as "negative curvature." An analogy to this is a saddle-shaped surface (Fig. 31.7), which is curved everywhere, has no boundaries, and is infinite in extent. This type of curvature corresponds to what is called an **open universe**, a solution to the field equations in which the expansion continues forever, never stopping. (It does slow down, however, because the gravitational pull of all the matter in it tends to hold back the expansion.)

universe, a result of its total mass content. The solutions to the field equations specify, among other things, the degree and type of curvature. We will return to this point shortly.

Einstein's solution to the equations, developed in 1917, had a serious flaw, in his view: It did not allow for a static, nonexpanding universe. In what he later admitted was the biggest mistake he ever made, Einstein added an arbitrary term called the **cosmological constant** to the field equations, solely for the purpose of allowing the universe to be stationary, neither expanding nor contracting.

Others developed different solutions, always by making certain assumptions about the universe. Some of these assumptions were necessary to simplify the field equations so that they could be solved. For example, W. de Sitter in 1917 developed a solution that corresponded to an empty universe, one with no matter in it.

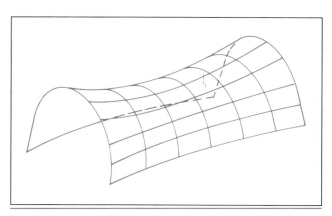

Figure 31.7 A saddle surface.   *This is a representation of the geometry of an open universe, which has negative curvature, is infinite in extent, has no center in space, and has no boundaries.*

A second possibility is that the curvature is "positive," corresponding to the surface of a sphere (Fig. 31.8), which is curved everywhere, has no boundaries, but is finite in extent. This possible solution to the field equations is called the **closed universe,** and it implies that the expansion will eventually be halted by gravitational forces and will reverse itself, leading to a contraction back to a single point.

The third and last possibility is that the universe is a **flat universe,** with no curvature. This corresponds to outward expansion being precisely balanced by the inward gravitational pull of the matter in the universe, so that the expansion will eventually come to a stop but will not reverse itself. This balanced, static state requires a perfect coincidence between the energy of expansion and the inward gravitational pull of the combined total mass of the universe, and may therefore seem very unlikely. But we do not know what set of rules governed the amount of matter the universe was born with, so there is no logical reason to rule out this possibility. Most astronomers, however, generally pose the question in terms of whether the universe is open or closed, without specifically mentioning the third possibility that it is flat. As we will see in the next section, if a perfect balance has been achieved, it will be very difficult to determine this from the observational evidence, which has large uncertainties.

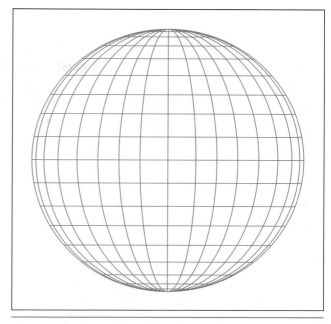

Figure 31.8 The surface of a sphere. *This surface represents a closed universe, which has positive curvature and a finite extent but no boundaries and no center.*

eral observational approaches to answering this: Determine whether there is enough matter in the universe to produce sufficient gravitational attraction to close it; or measure the rate of deceleration of the expansion to see whether it is slowing rapidly enough to eventually stop and reverse itself.

OPEN OR CLOSED: THE OBSERVATIONAL EVIDENCE

Substantial intellectual and technological resources are devoted today to the question of determining the fate of the universe. Theorists are at work developing and refining the solutions to the field equations, or seeking alternatives to general relativity that might provide equally valid or more valid representations. Observers are busily attempting to test the theoretical possibilities by finding situations in which competing theories should lead to different observational consequences. This is a difficult job, because most of the differences will show up only on the largest scales, and therefore to detect them requires observations of the farthest reaches of the universe.

Within the context of general relativity, the premier question traditionally has been whether the universe is open or closed. Historically, there have been two gen-

Total Mass Content

The total mass in the universe is the quantity that determines whether or not gravity will halt the expansion. The field equations express the mass content of the universe in terms of the density, the amount of mass per cubic centimeter. This is convenient for observers, because it is obviously simpler to measure the density in our vicinity than to try to observe the total mass everywhere in the universe. The field equations can be solved for the value of the density that would produce an exact balance between expansion and gravitational attraction (the density corresponding to a flat universe). If the actual density is greater than this critical density, then there is sufficient mass to close the universe, and the expansion will stop. The critical density depends on the value of Hubble's constant H; for

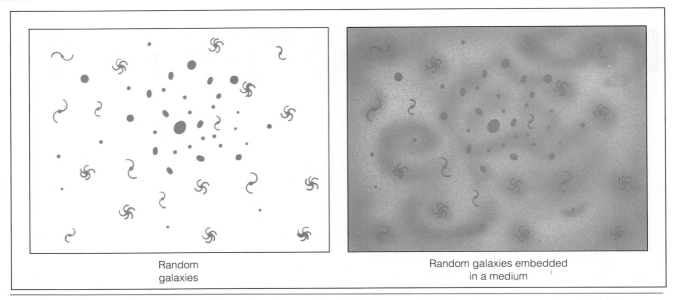

Random
galaxies

Random galaxies embedded
in a medium

Figure 31.9 Dark matter in the universe.   *Are the isolated galaxies that we see all that there is in the universe? Or are they merely points that happen to glow, embedded in a universal sea of unknown substance, whose mass density overwhelms that represented by the galaxies?*

a value near 55 km/sec/Mpc, it is calculated to be roughly 6×10^{-29} g/cm³, or about 3 protons per cubic meter, a very low value by any earthly standards.

The most straightforward way to measure the density of the universe is to simply count galaxies in some randomly selected volume of space, add up their masses, and divide the total by the volume. Care must be taken to choose a very large sample volume, so that clumpiness from clusters of galaxies is not important. This technique yields values for the density of around 10^{-31} to 10^{-30} grams/cm³, only a small percentage of the critical density. It may seem that we have already answered the question, and that the universe is open, but this method may overlook substantial quantities of mass.

One clue to this arises from determinations of cluster masses based on the velocity dispersion of the galaxies in them (see discussion in Chapter 28). These mass measurements, for reasons still not entirely clear, always yield much higher values than the estimated total mass of the visible galaxies in the cluster. The disparity can be as great as a factor of 10, so if the larger values are correct, the average density of the universe comes closer to the critical value. Even the larger values based on velocity-dispersion measurements, however, fall short (by a factor of ten!) of the amount needed to close the universe. If the mass density is to exceed the critical value, there must be large quantities of matter in some

form that has not yet been detected (Fig. 31.9). The hidden matter cannot be concentrated inside clusters of galaxies, because its presence would have been detected by the velocity-dispersion measurements.

If there is a lot of dark matter in the universe, it could take several forms. One possibility that is obvious to us by now is that there may be many black holes in the space between clusters. The idea that there may be black holes in intergalactic space has been seriously suggested by the astrophysicist Stephen Hawking, who hypothesized the existence of countless numbers of "mini" black holes, formed during the early stages of the big bang. These would have very small masses, much smaller even than the mass of the Earth, but would be so numerous that they could easily exceed the critical density. Current observational evidence argues against the existence of such objects in great numbers, however.

Another possibility is that neutrinos, the elusive subatomic particles produced in nuclear reactions, have mass. Recall from Chapter 17 that these particles permeate space, freely traveling through matter and vacuum alike, but that standard theory says they are massless. A recent controversial experiment indicates that this last assertion may not be correct, that they may contain minuscule quantities of mass after all. If so, they are sufficiently plentiful to provide more than the

ASTRONOMICAL INSIGHT 31.2

Dark Matter in the Universe

 Throughout the latter portions of this book, we have made many references to "dark matter," in our galaxy, in the halos of other galaxies, or in intergalactic space. The general trend has been that in all these environments, astronomers are finding evidence that the majority of the matter that is present is invisible. Perhaps it is useful at this point to summarize the evidence for dark matter in the universe, and to review some of the suggestions that have been made for its composition.

Within our galaxy there are several lines of argument favoring the conclusion that a substantial fraction, perhaps as much as 90 percent, of the mass is in some form other than normal stars. The motions of star clusters and gas clouds in the outer portions of the galaxy, as evidenced by the flat rotation curve beyond the Sun's position (Chapter 24), show that there must be large amounts of mass in the outer galaxy. Ninety percent or more of the mass of the Milky Way galaxy may reside in the form of invisible matter, most of it in the halo. Suggestions for this galactic dark matter include dim stars, which may be evolved stellar remnants such as black holes, neutron stars, and white dwarfs; or very low-mass normal stars that are simply too underluminous to be seen at any substantial distance. Both of these possibilities have run into difficulties, because of the enormous number of stars that would have had to evolve to stellar remnants already, or because of the huge quantity of low-mass stars that would have had to form in contradiction to what is known of the distribution of stellar masses that normally forms today. Even the so-called "brown dwarfs," objects too low in mass to become true stars, are probably not abundant enough to account for the dark matter in our galaxy. If all these stellar (or prestellar) forms of normal matter can be ruled out, then the best remaining suggestions are those that invoke elementary particles whose properties are not yet known even theoretically or from laboratory experiments.

On scales beyond our galaxy, one of the most direct lines of evidence for large quantities of dark matter arises in clusters of galaxies. Measurements of the overall gravitational effect of such a cluster, either from the orbital motions of the individual galaxies or from the masses required to create the observed gravitational lens effect (Chapter 28), invariably show the presence of much more mass than can be accounted for by the visible galaxies in the cluster. In this case, no quantity of dim stars within galaxies can explain the discrepancy, because such forms of matter are included in the estimates of the masses of individual galaxies. The most striking evidence that the dark matter is not within the galaxies in clusters comes from the analysis of the so-called "luminous blue arcs" that have been observed in many clusters (see Chapter 28). These arcs are distorted images of background galaxies, and the analysis of their shapes and distribution can yield information on the distribution of mass within the foreground cluster of galaxies; that is, by studying the images of background galaxies, it is possible to ascertain the shape of the lens that is forming them. In the clusters where this analysis has been carried out, it is found that the matter creating the gravitational lens is more spread out and smoothly distributed than the galaxies themselves. Furthermore, the mass required to create the lens effect is far greater than the mass known to be present in the form of hot, intracluster gas.

On scales beyond clusters of galaxies, the evidence for dark matter becomes more indirect. Other than the local streaming motion of our galaxy and many others toward the "Great Attractor" (Chapter 29), there is little opportunity to analyze motions of clusters of galaxies to look for anomalous effects that might require the presence of large amounts of unseen mass. The composition of the Great Attractor itself is still unknown, but recent arguments have been made that it may be an ordinary (but very massive) supercluster containing the usual allotment of luminous galaxies. On the other hand, some have suggested that it is indeed a concentration of dark matter. We may soon know more about this, in view of the intensive observational studies of the Great Attractor now taking place.

Perhaps the most compelling argument for a universal medium of dark matter comes from studies, both observational and theoretical, of the evolution of the universe itself. In this chapter, we have explored the current status of these theories and observations, and have found strong theoretical arguments that the universe is flat. That is, some modern theories call for a balance between the momentum of expansion of the universe and the self-gravitation that is slowing the expansion. Furthermore, observational estimates of the masses of galaxy clusters based on mass-to-light ratios (see Chapter 27) always fall short of the masses inferred from the motions of galaxies within the clusters, implying that at least 90 percent of cluster masses are invisible.

In order for the universe to be flat, there must be at least 10 times more mass in it than is observed, and the actual discrepancy is probably more like a factor of 100. This means that in the universe as a whole, dark matter may constitute as much as 80 to 99 percent of all the mass (the uncertainty arises because it is so difficult to measure precisely how much mass is present in the form of visible matter; current estimates are that the greatest quantity of visible mass allowed by observations is only one-fifth of the amount needed to make the universe flat, and that the most likely amount of visible matter present is only about 1 percent of the amount needed).

Clearly the issue of dark matter in the universe represents one of the most important questions astronomers are asking today. Much of the effort in observational astronomy, making use of the biggest telescopes available today and through the future use of new instruments such as the *Hubble Space Telescope*, will be focused on this problem. Similarly, particle physicists are working through theory and experiment to discover the existence and nature of exotic particles which may account for the dark matter. In the coming years, we may expect news from many fronts, but it remains to be seen how much light will be shed on this dark and dim problem.

critical density, thereby closing the universe. Further experiments designed to determine whether the neutrino has mass have been performed, all of them so far being more consistent with the standard notion that these particles are massless.

Significant information on the possibility that neutrinos have mass came from the recent supernova (Supernova 1987A) in the Large Magellanic Cloud. Neutrinos from reactions that took place during the explosion were detected at the Earth, and the time of arrival and the energies of these neutrinos allowed scientists to place a limit on how massive they can be. It was found that neutrinos probably cannot be massive enough to bring the average density of the universe up to the point where it is closed. Therefore, if the universe is closed, the dark, unseen matter must be in some form other than neutrinos.

There are other kinds of elementary particles that have been postulated to exist, on the basis of elementary particle theory. These particles have not been detected experimentally (the energies required far exceed those possible in any existing particle accelerators), but it is possible that they are sufficiently numerous and massive to close the universe. We must await new generations of particle accelerators before we will have an answer to the question of the significance of such particles.

If we base our current thinking only on the density of matter that has actually been observed, we are forced to conclude from the available observational evidence that the mass density of the universe is less than the critical density, and therefore that the universe is open. It is worth stressing, however, that many scientists believe there are substantial quantities of dark matter in the universe, and that it might be closed or flat.

The Deceleration of the Expansion

Another approach to answering the question of whether the universe is open or closed is perhaps being pursued more vigorously today. The objective is to compare the present expansion rate with what it was early in the history of the big bang, to see how much slowing, or deceleration, has occurred (Fig. 31.10). The expansion has certainly slowed in any case; the question is how much. If it has only decelerated a little, then we infer that is not going to slow down enough to stop and reverse itself; but if it has decelerated quite a bit, then we conclude that the expansion is coming to a halt, and that the universe is closed.

Establishing the deceleration rate requires knowing both the present expansion rate and the expansion rate at a time long ago, shortly after the big bang began. Neither quantity is easy to determine: We already learned that the value of the Hubble constant H, which tells us the present expansion rate, is uncertain to begin with. To measure the expansion rate at early times in the history of the universe is even more difficult, for it involves determining the distances and velocities of very faraway galaxies, so distant that we see them as they were when the universe was young. Such objects are now at the very limits of detectability, and it is hoped that the *Hubble Space Telescope* will push the frontier back far enough to permit observations of velocities in the early days of the expansion.

We still must rely on standard candles to establish the distance scale, and for such distant objects this procedure becomes even more uncertain than usual. When we look so far back in time, we are seeing galaxies that are much younger than those near us and therefore may have different properties than mature galaxies. For example, their content of stellar populations may be rather different than those in nearby galaxies, their brightest stars and nebulae may be different, and even their total galactic luminosities may be different. It may not be valid to assume that the brightest galaxy in a cluster of galaxies has the same absolute magnitude as in more nearby clusters.

The present evidence on deceleration is that the universe has not slowed very much and therefore is not on its way to stopping and beginning to contract. As we have seen, however, the observational uncertainties are great, and this conclusion is not very firm.

An indirect technique for measuring the deceleration has been recently employed and avoids many of the difficulties of observing extremely distant galaxies. Some light elements were created in the early stages of the expansion, and the amounts that were produced depend on the fraction of the universal energy that was in the form of mass rather than radiation at the time of element formation. The primordial mass density is in turn related to the present-day deceleration of the expansion, so if the primordial density is deduced from the abundances of elements that were created, we can infer the deceleration rate. The strategy, therefore, is to measure the abundance of some element that was produced in the big bang, and from that to derive the deceleration (but this method says nothing about mass that might be present in the form of elementary particles other than those that constitute ordinary atoms;

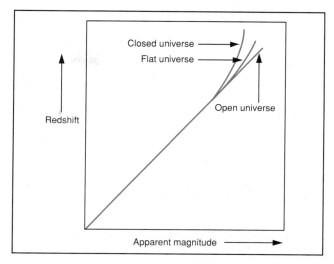

Figure 31.10 The effect of deceleration in the Hubble diagram. *This figure shows the relationship between velocity of recession and distance for the three possible cases: closed (upper curve), flat (middle), or open (lower). Present data seem to favor the open universe, although there is substantial uncertainty, largely because of problems in applying standard candle techniques to galaxies so far away that they are being seen as they were at a young age. The distinction is also made difficult by the subtlety of the contrast between the shapes of the curves.*

therefore use of deceleration measures to infer the mass density of the universe does not help in determining the dark matter content).

The most abundant element (other than hydrogen) produced in the big bang is helium, so if we could measure how much helium was created, this might be a good indicator of the deceleration. One problem with this element, however, is that it is also produced in stellar interiors, so it is difficult to determine how much of what we see in the universe today is really left over from the big bang. Another problem with using helium as a probe of the early universal expansion rate is that the quantity produced is not expected to vary much with different expansion rates; that is, the primordial helium abundance, even if we knew it well, is not a very sensitive indicator of the early expansion rate.

Better candidates are ^{7}Li, an isotope of the element lithium, and deuterium, the form of hydrogen that has

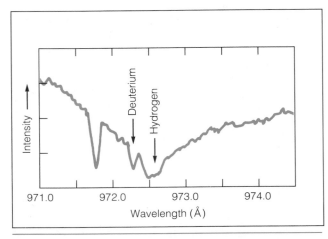

Figure 31.11   An ultraviolet absorption line of interstellar deuterium.   *The abundance of deuterium in space is determined from the analysis of absorption lines it forms in the spectra of background stars. Here we see a weak deuterium line close to a strong absorption feature due to normal hydrogen. The spectrum shown here was obtained by the* Copernicus *satellite. (J. B. Rogerson and D. G. York)*

one proton and one neutron in the nucleus. These species were also produced in the big bang, and, as far as is presently known, are not made in any other way. To date, most efforts have been concentrated on measuring the present-day abundance of deuterium, although in principle ^{7}Li can be measured as well. While deuterium can be destroyed by nuclear processing in stars, it is not produced in that way. (Even if it were, it would not survive the high temperatures of stellar interiors without undergoing further reactions.) The present deuterium abundance in the universe should therefore represent an upper limit on the quantity created in the big bang. Direct measurements of the amount of deuterium in space became possible in the 1970s, with the launch of the *Copernicus* satellite, which made ultraviolet spectroscopic measurements and was able to observe absorption lines of interstellar deuterium atoms (Fig. 31.11). The abundance of deuterium that was found is sufficiently high that it implies a low density, in turn pointing to a small amount of deceleration. Thus this test, in agreement with others, indicates that the universe is open. There is some uncertainty, though, because the *Copernicus* data seem to indicate an uneven distribution of interstellar deuterium throughout the galaxy, and this would not be expected if the deuterium formed exclusively in the big bang. A clumpy distribution of deuterium in the galaxy

could indicate that some of it is somehow produced by stars, in which case its abundance is not a strict test of deceleration. A recent measurement of the ^{7}Li abundance in the galaxy is consistent with the deuterium results; that is, the observed quantity of ^{7}Li is also large enough to imply that the universe is open.

Another technique, the direct measurement of the expansion rate at different times, based on counting galaxies with different redshifts, may be viewed as a modification of the attempts to infer the deceleration of the expansion. The result of the first application of this technique is that the universe appears to be flat; that is, the expansion appears to have just the rate needed to eventually stop (in an infinite time) but never to contract again. Certainly this technique will be further exploited, and this result further scrutinized in the coming years.

It is worth noting that all the observational techniques tried so far to discover whether the universe is open or closed have pointed towards a fairly close balance. That is, the data do not point to a universe that is closed or open by a wide margin. Whatever the correct answer is, we can say that the universe is close to being flat. This is very significant, for such a close balance is not necessarily expected in the big bang theory, and must be regarded in this theory as a coincidence. There are other models, however, in which a flat or nearly flat universe is expected; one of these is described in the next section.

THE INFLATIONARY UNIVERSE

The big bang cosmology is widely accepted, but there remain some nagging problems. One is the improbability for a universe starting from a singularity as envisioned in the big bang to be as symmetric as the observed universe. It is unlikely that the universe would appear as homogeneous and isotropic as observations tell us it is. A great deal of effort has been put into testing this property of the universe, because any departures from homogeneity would provide information on imbalances or asymmetries in the early epochs of the expansion. It is also thought surprising that the universe should be so closely balanced between being open and closed.

A new kind of expansion model has been developed in which the close balance and the homogeneity of the universe are easily understood. Calculations have shown that, at a certain point very early in the history of the universe, when the temperature was around 10^{27} K, conditions were right for small regions to separate themselves from the rest of the universe and then to expand very rapidly to much larger sizes. A reasonable analogy would be the creation of bubbles in a liquid, with the bubbles then growing much larger almost instantaneously. In this cosmological model, each "bubble" becomes a universe in its own right, with no possibility of communication with other bubbles.

The major advantage of this so-called **inflationary universe** model is that the universe born of a tiny cell in the early expansion would be expected to remain symmetric and homogeneous as it grew. The reasons for this are somewhat abstract but have to do with the fact that the various portions of the tiny region that was to expand into the present universe were so close together originally that they were in a form of equilibrium, with little or no variation in physical conditions being possible. In a larger region, such as in the standard big bang cosmology, no such equilibrium would have existed, and there would be no reason to expect such uniformity throughout the resulting universe.

Another advantage of the inflationary model is that in such a universe, the expansion would be expected to go on forever but would continue to slow, approaching a stationary state; that is, a flat universe is the natural result of rapid expansion from a tiny bubble as envisioned in this model. As we have seen, observations tell us that our universe is nearly flat, and may be precisely flat. If it were not so closely balanced, it would not be so difficult to determine whether it is open or closed.

The very success of the inflationary model in explaining the uniformity of the universe is also a problem: it fails to explain how the matter in the universe became clumpy enough to form galaxies, clusters, and superclusters. This problem faces the standard big bang model as well, though, and probably should not be considered a fatal flaw in the inflationary model. Considerable progress is being made on the question of galaxy formation, with much emphasis on the state of dark matter in the universe, which affects the typical sizes of condensations that could form. This is discussed in some detail at the end of Chapter 28.

The inflationary universe theory has reached a high degree of acceptance among cosmologists. This does not mean that the big bang model is wrong, but rather that it is a subordinate part of a much bigger picture, much like Newton's concept of gravity is now seen as a subset of Einstein's more complete theory, general relativity.

One of the most pleasing aspects of the inflationary model is also one that is very difficult to understand intuitively. In the standard big bang, we are forced to accept that the state of the universe as it began to expand was a given, something that just was. In other words, the big bang theory can only explain what happened after the beginning. In the inflationary model, the entire universe (and the infinite number of others that are possible) was literally created out of a vacuum. This means that no untidy, inexplicable initial conditions have to be invoked.

In a vacuum, energy may be present even though mass is not, and quantum mechanics tells us that particles can appear out of nothing. These occurrences may be viewed as random events, reflecting the fact that on the smallest of scales the fabric of the universe undergoes fluctuations. The spontaneous appearance of particles out of vacuum has been observed in laboratory experiments. General relativity theory tells us that if the energy of the vacuum is high enough, a strange state called "false vacuum" is created, in which a peculiar form of "negative" pressure acts as a repulsive force, causing the local region of space-time to expand, if matter is present. Another strange property of false vacuum is that during expansion the density remains constant, which means that additional matter is created out of the vacuum. In a very brief moment (about 10^{-30} seconds) the early universe would have expanded in size by a factor of 10^{25} or more and would have grown enormously in mass. This was the inflation phase already mentioned, which forms the basis of the inflationary model of the universe. The cooling that occurred during the expansion eventually (that is, after 10^{-30} seconds) would have broken the false vacuum, and the universe would then have been transformed into the more normal form of space we are familiar with, containing matter and having enough residual energy to power the expansion that has been proceeding ever since.

In the inflationary theory, then, the origin of our universe was a random quantum mechanical event, something now believed to occur naturally and inevitably, requiring nothing more than a vacuum with sufficiently high energy. It is now being speculated that perhaps new universes can begin within existing ones, that generation after generation of new bubbles of false vacuum occur randomly and expand. In this view it seems likely that our universe was born of another earlier one; it would be pretentious to think that, of all the possible universes, ours was the first.

If a new universe formed from a random fluctuation

in our own, there might be no way for us to know it, because the expansion could take place in different dimensions of space-time than those that we occupy and can sense. In effect, a tiny bit of space-time, if given enough energy, could pinch off and disappear from our universe, spawning a whole new universe that we cannot perceive. For this to happen would require an enormous concentration of energy in a tiny space, and thus is a very low-probability event. But in a sufficiently large universe, even a low-probability event can occur somewhere at some time.

As already noted, once the false vacuum was broken and the universe reverted to normal vacuum filled with radiation and matter, the rest of the expansion is described very well by the standard big bang theory. Hence, the long legacy of big bang cosmologies, beginning in the 1940s with the work of George Gamow and his associates, still provides a useful description of the evolution of our universe. This description is summarized in the next section.

THE HISTORY OF EVERYTHING

It is impossible to physically describe the universe at the precise moment that the expansion (or the inflation of our "bubble") began; it is physical nonsense to deal with infinitely high temperature and density. It is possible, however, to calculate the conditions immediately after the expansion started, and at any later time. Many of the most interesting events in the early history of the universe, and the ones currently under the most active investigation, occurred very early, before even a ten thousandth of a second had passed. Under the conditions of density and temperature that existed then, matter and the forces that act on it were quite different from anything we can experience, even in the most advanced laboratory experiments. Even the familiar subatomic particles such as protons and neutrons could not exist, but were replaced by *their* constituent particles.

Current particle physics theory holds that the most fundamental particles are **leptons,** which include electrons and positrons, and **quarks.** Modern theory also provides a basis for believing that all four of the fundamental forces in nature (see Astronomical Insight 5.1) may really be manifestations of the same phenomenon. So far it has been possible to show that three of the fundamental forces (the electromagnetic force and the weak and strong nuclear forces) are manifestations

of the same basic interaction. Under physical conditions that we are used to, these forces behave very differently, but when the density and temperature are very high, as they were early in the expansion of the universe, the forces are indistinguishable. Some aspects of the theory that shows this have been confirmed by laboratory experiments. The theoretical framework connecting the three forces is called **Grand Unified Theory** (**GUT,** for short). This name may be somewhat exaggerated, because the fourth force, gravity, has not yet been shown to be unified with the other three, and some theories of gravity indicate that it may not be.

The unification of the other three forces implies that, in the very early moments following the beginning of the expansion, until 10^{-35} seconds had passed, the universe contained only leptons and quarks and related particles and radiation, and the only forces operating in it were gravity and the unified force that was later to become recognizable as the electromagnetic, strong, and weak forces.

As the universe expanded and cooled, the three forces sequentially separated and became distinguishable from each other. Meanwhile a rich stew of particles was brewing, because particles can appear spontaneously from photons of radiation, and in its earliest times the universe was filled with very high energy photons. Recall that mass and energy are interchangeable according to the special relativity relation $E = mc^2$; particles (mass) can appear from photons (energy) if sufficient energy is present in the photons. This was the case for a brief time during the early expansion for the universe.

When particles appear in this way, they do so in pairs with opposing properties. Thus a single electron would not appear, but instead an electron and its oppositely charged counterpart, the positron, would appear. Protons appeared paired with "antiprotons," particles identical in mass but having opposite charges. For each type of particle, there is an antiparticle, opposite in key properties. (In some cases of neutral particles, the antiparticle and the normal particle are the same.) Pair production of particles from energetic photons (or from high-energy particle collisions) has been observed in particle accelerators.

Once a particle-antiparticle pair formed, each of the two particles moved freely through the young universe, until it encountered by random chance its antiparticle. When this happened, the two particles annihilated each other, converting their combined mass into energy, creating a new photon. Thus there was an equilibrium in which energy and matter were continually being con-

ASTRONOMICAL INSIGHT 31.3

Particle Physics and Cosmology

There is some irony in the fact that the science of the smallest scales, particle physics, is finding its greatest opportunities in cosmology, the science of the largest scales. One of the most rapidly advancing areas in modern particle physics coincides with one of the brightest fields in modern astrophysics, bringing full circle one of the themes of astronomy, the relationship between the science of the atom and that of the cosmos.

In particle physics, scientists are concerned with discovering the nature of matter and energy at its most fundamental level. We now know that the atom, and the major subatomic particles such as protons and neutrons, are actually composed of more basic particles, known in general as elementary particles. The most modern theories of elementary particles postulate that all matter is made up of two types of basic particles. One type, which includes electrons, are the **leptons**; the others, which can have several different combinations of properties, are called **quarks**. Quarks and other particles known only from theory are sought experimentally in particle accelerators, enormous machines that can accelerate particles to such high kinetic energies that when they collide they are broken down into their constituent elementary particles. Because there is

an equivalence between matter and energy (as expressed in Einstein's theory of relativity), elementary particles can be created from pure energy in sufficient quantities. Thus, the objective in building particle accelerators to probe elementary particles is to inject enough energy into the accelerators to create the particles being sought. Progress is made every time new levels of energy are reached.

The planned Superconducting Super Collider (SSC), having recently been approved for construction by the United States government, will provide higher energies than any existing accelerators, and will give scientists new capabilities for producing elementary particles. This instrument will consist of a huge circular tunnel (53 miles in circumference) in which particle beams will be accelerated by gigantic magnets and then allowed to collide. The collision energies will be comparable to particle energies in the early universe, only a small fraction of a second after the expansion began.

Even the SSC, however, will not answer questions about the very earliest times. Some of the particles that are predicted by theory will require energies so high that it may never be practical to create them in accelerators. This is where astrophysics, specifically cosmology, may come to the rescue. Energies high enough to create *all* possible particles certainly existed during the earliest stages of the universal expansion. Thus, elementary particle physicists may find some of the answers they

seek by studying what is known about the big bang.

We cannot actually observe the composition of the universe at the time of the early big bang, but we know that today's universe is the product of that era. Thus, by examining the detailed atomic composition of the universe today, particle physicists hope to deduce the mixture of particles that inhabited the universe during its initial moments. A remarkable degree of success has already been achieved in just this way, as particle theorists have been able to derive, with considerable certainty, the physical conditions that must have prevailed in early times. We are confident in our knowledge about the formation of the elements in the early universe because the theory of elementary particles is capable of reproducing the observed abundances of the elements believed to be present a fraction of a second after the start of the expansion. Now the hope is that the process can be reversed; that particle physicists can learn about elementary particle processes by observing the state of today's universe.

verted back and forth, as photons created particle pairs and particle pairs created photons.

This could have continued forever if not for the expansion and cooling of the universe. The cooling meant, according to Wien's law (Chapter 6), that the energy of the photons decreased steadily. Soon the photon energies were insufficient to produce particle pairs, and pair production ceased. This occurred at different times for particles of different masses; thus as the universe expanded, a threshold was passed for each type of particle, after which that particle was no longer produced.

Because particles and their antiparticles were both created, eventually they found each other and annihilated. If exactly equal numbers of particles and their antiparticles were formed, then in due time all would have annihilated, and there would be no matter in the universe today. The fact that there is matter tells us that particles and antiparticles must have been produced in slightly unequal numbers. The current understanding of how this came to be is that at the time when the four forces were separating from each other,

there were very slight asymmetries that favored normal particles over their antiparticles. It is estimated that normal particles had to outnumber antiparticles by only one part in 10^{16} to leave all the matter in the universe that we see today. It is thought possible that in other universes that may exist (see preceding section), different physical laws could prevail, and perhaps antimatter can dominate over normal matter.

Let us now consider an epoch in the early universe when matter was beginning to take on familiar forms, and the four fundamental forces were already acting as four distinct forces. In the inflationary theory, this followed the phase of rapid expansion, which was finished by 10^{-30} seconds after the beginning. The temperature dropped to 10 billion degrees (10^{10} K) by 1.09 seconds after the start, and by then protons and neutrons were appearing (Fig. 31.12). At this point, most of the energy of the universe was still in the form of radiation. Within a few minutes, conditions became better suited for nuclear reactions to take place efficiently, and the most active stage of element creation began. The principal products, in addition to the he-

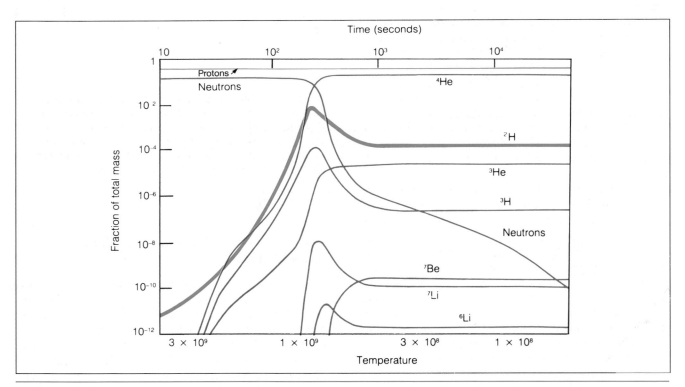

Figure 31.12 Element formation in the big bang. *This diagram illustrates the relative abundances of various light elements formed by nuclear reactions during the early stages of the big bang expansion. The production rates are shown as functions of time and temperature. (Data from Wagoner, R. V. 1973, The Astrophysical Journal 179:343)*

lium and deuterium mentioned in an earlier section, were tritium (another form of hydrogen, with one proton and two neutrons in the nucleus) and the elements lithium (three protons and four neutrons) and to a minor extent beryllium (four protons and five neutrons). Nearly all the available neutrons combined with protons to form helium nuclei, a process that was complete within 4 minutes of the beginning of the expansion. At this point, with some 22 to 28 percent of the mass in the form of helium, the reactions were essentially over, except for some production of lithium and beryllium over the next half-hour (see Fig. 31.12).

The expansion and cooling continued, but nothing significant happened for a long time after the nuclear reactions stopped. Eventually (at least 500,000 years after the beginning), the density of the radiation had reduced sufficiently that its energy was less than that contained in the mass (that is, the energy derived from $E = mc^2$ became greater than that contained in the radiation). At that point, it is said, the universe became matter-dominated, rather than radiation-dominated.

The matter and radiation continued to interact, however, because free electrons scatter photons of light very efficiently. The strong interplay between matter and radiation finally ended nearly a million years after the start of the big bang, when the temperature became low enough to allow electrons and protons to combine into hydrogen atoms. The atoms still absorbed and reemitted light, but much less effectively, because they could do so only at a few specific wavelengths. From this time on, the matter and the radiation went separate ways. The radiation simply continued to cool as the universe expanded, reaching its present temperature of 2.7 K some 10 to 20 billion years after the expansion began.

Sometime in the first billion years or so, the matter in the expanding universe became clumpy and fragmented into clouds and groups of clouds that eventually collapsed to form galaxies and clusters of galaxies (Fig. 31.13). As we noted in Chapter 28, the cause of the clumpiness and fragmentation is as yet unknown, but recent studies suggest that random gravitational encounters between bodies in a uniform early universe could have created today's observed uneven distribution of matter. Once galaxies began to form, the subsequent evolution of matter followed steps already described in the preceding chapters.

Because there is so much uncertainty about how the initial stages of fragmentation took place, there is a great deal of interest in probing into this process. To do so will be another goal of the *Hubble Space Telescope*.

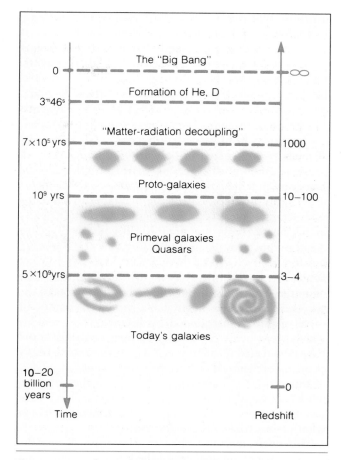

Figure 31.13 The evolution of the universe. *This diagram shows, as a function of time since the start of the big bang, significant stages in the evolution of the universe. It also shows the types of objects that existed at each stage, and the redshift at which they are (or would be) observed at the present time. (J. M. Shull)*

Possible Heavy Element Formation in the Big Bang

The standard model of the expansion of the universe shows that only a few of the lightest-weight elements could have been formed, because the expansion and cooling were so rapid that there was insufficient time for heavier elements to build up. Furthermore, the next heavier elements after helium and lithium tend to be unstable, immediately decaying by radioactive processes after being formed, so that even if the reactions had formed elements beyond lithium, they would not have lasted. For these reasons, it has generally been assumed that the universe contained essentially only hydrogen, helium, and a little lithium by the time it was a few minutes old, and that all other elements present today had to be formed by reactions in stars. One difficulty has been that no "pure" hydrogen and helium stars have ever been found; even the oldest stars observed today contain some traces of heavy elements. The origin of these traces has been difficult to unravel.

Now it has been suggested that perhaps some heavy elements were created in the early big bang after all. There is a family of reactions, called **neutron-capture reactions,** that are capable of producing heavy elements if a supply of free neutrons is available to combine with existing nuclei. Under conditions of sufficiently high density, these reactions could conceivably have created some heavy elements during the early big bang, but for this to happen, there would have had to exist neutron-rich regions. Recent theories indicate that quantum variations during the earliest moments of the universal expansion might have produced clumps of dense material surrounded by lower-density regions which were neutron rich. If so, then neutrons could have penetrated the dense regions, creating some heavy elements by neutron-capture reactions. Detailed models of this process, which depend in turn on the quantum models of early fluctuations in the universe, have been successful in producing enough heavy elements to match what is observed in the oldest stars.

This new picture of element formation during the big bang is not yet well established, but it is generating considerable interest because of its potential for explaining a longstanding problem with the composition of the oldest stars. We may expect further developments in the coming years.

WHAT NEXT?

To discuss the future of the universe is obviously a speculative venture. We cannot even answer, with absolute certainty, the basic question whether it is open or closed. There have been attempts to calculate future conditions in the universe, in analogy to the theoretical work on the early stages discussed in the last section. In the case of the future of the universe, however, the uncertainties are much greater, and the following descriptions should be regarded as very speculative.

If the universe is open or flat, as in the inflationary model, then it will have no definite end; it will just gradually run down. The radiation background will continue to decline in temperature, approaching absolute zero. As stellar processing continues in galaxies, the fraction of matter that is in the form of heavy elements will continue to grow, and the supply of hydrogen, the basic nuclear fuel, will diminish. It is predicted that all the hydrogen will be gone by about 10^{14} years after the birth of the universe, so the universe in which stars dominate has now lived approximately one ten-thousandth of its lifetime. The recycling process between stars and the interstellar medium will continue until this time, but gradually the matter will become locked up in black holes, neutron stars, white dwarfs, and black dwarfs. Dead and dying stars will continue to interact gravitationally, eventually colliding often enough in their wanderings that all planets will be lost (by about 10^{17} years) and galaxies will dissipate as their constituent stars are lost to intergalactic space (by about 10^{18} years). Further speculation shows that new physical processes will take over at later times. The new GUT theory tells us that the proton, a basically stable particle, may disintegrate in a very low probability process that occurs, on average, once in 10^{32} years for a given proton. When the universe reaches an age of about 10^{20} years, enough protons will begin to evaporate here and there that the energy produced will keep the remnant stars heated, although only to the modest temperature of perhaps 100 K. At 10^{32} years, most protons will have decayed, and the universe will consist largely of free positrons, electrons, black holes, and radiation (the extremely cold remnant of the big bang). The final stage that has been foreseen occurs at an age of 10^{100} years, when sufficient time has passed for all black holes to evaporate, and nothing is left but a sea of positrons, electrons, and radiation (theory says that black holes can disintegrate with very low probability, meaning that if we wait long enough, eventually they will do so).

If the universe is closed, then someday, perhaps some 50 billion years from now, the expansion will stop and will be replaced by contraction. The deterioration described above will still take place, until the final moments when the universe once again becomes hot and compressed, entering a new singularity. In some views, purely conjectural and without possibility of verification, such a contraction would be followed by a new big bang, and the universe would be reborn. This concept of an oscillating universe, pleasing to the minds of many, will not occur unless the present weight of the evidence favoring an open or flat universe is found to be in error.

PERSPECTIVE

Our story is essentially complete. We have now discussed the universe and all of the various objects and structures in it, and we have described what is known of its evolution and its fate.

We have, however, omitted consideration of what is perhaps the most important ingredient of all: life. Although much of what we might say about this is beyond the scope of astronomy, it is appropriate to assess what can be learned from objective scientific consideration. We will do so in the next chapter.

SUMMARY

1. In cosmological studies, astronomers usually adopt the Cosmological Principle, which states that the universe is both homogeneous and isotropic. Existing observational data tend to support this assumption.

2. Einstein's theory of general relativity, which describes gravitation and its equivalence to acceleration, is used to mathematically describe the universe as a whole. In the context of this theory, field equations are written that represent the interaction of matter, radiation, and energy in the universe. Solutions of the field equations define the properties of the universe.

3. There are three general solutions to the field equations that are considered by modern cosmologists. They correspond to a closed universe (positive curvature), an open universe (negative curvature), and a flat universe (no curvature).

4. A major question in modern astrophysics is whether the universe is closed (expansion eventually to be reversed) or open (expansion never to stop).

5. There are two general types of tests of whether the universe is open or closed: (1) to ascertain whether there is enough mass in the universe to gravitationally halt the expansion, or (2) to determine the rate of deceleration of the expansion, to see whether it is slowing enough to eventually stop.

6. The total mass content is measured in terms of the average density of the universe, and the matter that is visible in the form of galaxies is not sufficient to close the universe. Various suggestions have been made concerning other forms in which the necessary mass could exist.

7. The deceleration is measured in two ways: (1) by comparing past and present expansion rates, through observations of very distant galaxies; and (2) by inferring the early expansion rate from the present-day abundances of elements that formed only during the big bang.

8. Both the total mass content that is observed and the inferred deceleration of the expansion indicate that the universe is open or flat. This conclusion is not universally accepted, and observational tests are continuing.

9. Recent inflationary models, in which our universe formed through rapid expansion of an initially tiny bubble, show great promise. In these models a homogeneous, flat universe is expected, consistent with observations.

10. The early stages of the universal expansion, up to the time when matter and radiation decoupled, can be described quite precisely and certainly by modern physics. Following the initial moment of infinite density and temperature, the first atomic nuclei formed just over 1 second later, and all of the early element production was finished within a few minutes. Matter and radiation decoupled almost a million years later; it is not so well understood how the universe subsequently organized itself into stars and galaxies.

11. The future of the universe appears to have two possibilities. If it is open or flat, it will gradually become cold and disorganized. If it is closed, it will eventually contract, perhaps to a new beginning in another big bang.

REVIEW QUESTIONS

Thought Questions

1. Explain how the universe can be homogeneous even though galaxies are concentrated in clusters and superclusters.

2. Recall from the previous chapter that there is a daily anisotropy in the cosmic background radiation. Does this violate the assumption that the universe is isotropic? Explain.

3. Explain in your own words why general relativity implies that space is curved near massive objects.

4. Summarize the future of the universe under the three possibilities discussed in the text; that is, what will happen if the universe is open, closed, or flat?

5. Explain why it is difficult to determine the average mass density of the universe from observations.

6. It has been suggested (as discussed in Chapter 30) that most or perhaps all galaxies have very massive black holes in the cores. If so, would this change observational

estimates of the mass density of the universe, and therefore help determine whether the universe is open or closed? Explain.

7. Why is the deuterium abundance a better indicator of the early expansion rate of the universe than helium?

8. What are the advantages of the inflationary universe theory over the conventional big bang model?

9. Why did element production occur only during the first few minutes of the universal expansion?

10. Summarize the ways in which the *Hubble Space Telescope* will help to answer questions about the future of the universe.

Problems

1. The interstellar medium in the disk of our galaxy has an average density of about one hydrogen atom for every 10 cm^3 of volume. Convert this density into units of g/cm^3, assuming that the mass of a hydrogen atom is 1.7×10^{-24} g. How does your answer compare with the critical density for closing the universe? What implications does this have for whether the universe is open or closed?

2. Suppose that, on average, the universe contained one galaxy like ours (mass about 10^{11} solar masses) for every 10 Mpc^3. What is this density in terms of grams/cm^3, and how does it compare with the critical density for closing the universe? (*Hint:* A volume of 10 Mpc^3 corresponds to a cube that is 2.2 Mpc on each side; convert this to centimeters, and then compute the volume of the cube in cubic centimeters.)

ADDITIONAL READINGS

Abbott, L. 1988. The mystery of the cosmological constant. *Scientific American* 258(5):106.

Adair, R. K. 1988. A flaw in a universal mirror. *Scientific American* 258(2):50.

Barrow, J. D., and J. Silk. 1980. The structure of the early universe. *Scientific American* 242(4):118.

Bartusiak, Marcia. 1985. Sensing the ripples in spacetime. *Science 85* 6(3):58.

————. 1986 *Thursday's universe*. New York: Times Books.

Boslough, John. 1981. The unfettered mind: Stephen Hawking. *Science 81* 2(9):66.

Chevalier, R A. and C. L. Sarazin. 1987. Hot gas in the universe. *American Scientist* 75:609.

Davies, P. 1987. Particle physics for everybody. *Sky and Telescope* 74:582.

————. 1990. Matter-antimatter. *Sky and Telescope* 79:257.

Dicus, D. A., J. R. Letaw, D. C. Teplitz, and V. L. Teplitz. 1983. The future of the universe. *Scientific American* 248(3):90.

Ferris, T. 1977. *The red limit: The search for the edge of the universe*. New York: William Morrow.

————. 1984. The radio sky and the echo of creation. *Mercury* 13(1):2.

Field, G. B. 1982. The hidden mass in galaxies. *Mercury* 11(3):74.

Gaillard, M. K. 1982. Toward a unified picture of elementary particle interactions. *American Scientist* 70(5):506.

Gott, J. R., III, J. E. Gunn, D. N. Schramm, and B. Tinsley. 1976. Will the universe expand forever? *Scientific American* 234(3):62.

Gribbin, John. 1986. *In search of the big bang*. Bantam Books.

Guth, A. H., and P. J. Steinhardt. 1984. The inflationary universe. *Scientific American* 250(5):90.

Hawking, S. W. 1984. The edge of spacetime. *American Scientist* 72(4):355.

Malaney, R. A., and W. A. Fowler. 1988. The transformation of matter after the big bang. *American Scientist* 76:472.

Mallove, E. F. 1988. The self-reproducing universe. *Sky and Telescope* 76:253.

Overbye, Dennis. 1985. The shadow universe: Dark matter. *Discover* 6(5):12.

————. 1983. The universe according to Guth. *Discover* 4(6):92.

Page, D. N., and M. R. McKee, 1983. The future of the universe. *Mercury* 12(1):17.

Parker, B. 1988. The cosmic cookbook: The discovery of how the elements came to be. *Mercury* 17:171.

Penzias, A. A. 1978. The riddle of cosmic deuterium. *American Scientist* 66:291.

Schramm, D. N. 1974. The age of the elements. *Scientific American* 230(1):69.

Schwinger, Julian. 1986. *Einstein's legacy*. New York: W. H. Freeman.

Shu, F. 1983. The expanding universe and the large-scale geometry of spacetime. *Mercury* 12(6):162.

Silbar, M. L. 1982. Neutrinos: Rulers of the universe? *Griffith Observer* 46(1):9.

Silk, J., A. S. Szalay, and Y. B. Zel'dovich. The large-scale structure of the universe. *Scientific American* 249(4):56.

Trefil, J. S. 1978. Einstein's theory of general relativity is put to the test. *Smithsonian* 11(1):74.

————. 1983a. How the universe began. *Smithsonian* 14(2):32.

————. 1983b. How the universe will end. *Smithsonian* 14(3):72.

————. James S. 1983. *The moment of creation*. New York: Scribners.

Tucker, W., and K. Tucker, 1982. A question of galaxies. *Mercury* 11(5):151.

————. 1989. Dark matter in our galaxy. *Mercury* 18(1):2 (Part I); *Mercury* 18(2):51 (Part II).

Vilenkin, A. 1987. Cosmic strings. *Scientific American* 257(6):94.

Weinberg, S. 1977. *The first three minutes*. New York:

Basic Books.

Will, C. M. 1983. Testing general relativity: 20 years of progress. *Sky and Telescope* 66(4):294.

MATHEMATICAL METHODS

Making Particles from Radiation

 In the earliest times, particles are thought to have appeared spontaneously from the radiation that filled the universe. This is possible because of the equivalence of mass and energy, which is expressed by Einstein's formula $E = mc^2$. If a photon has enough energy, it can be converted into a particle whose mass is the equivalent of that amount of energy (actually a particle-antiparticle pair is always produced, because such quantities as total charge must still equal zero). Pair production of particles has been observed in laboratory experiments.

It is not difficult to see what conditions are needed to form given types of particles. Let us consider the electron and its antiparticle, the positron. These particles have masses of $m = 9.11 \times 10^{-28}$ gram, so a pair of them has a total mass m = 1.82×10^{-27} gram. Using Einstein's formula, we find the energy equivalent of this mass:

$$E = (1.82 \times 10^{-27}g) \, (3.00 \times 10^{10} \text{ cm/sec})^2$$

$$= 1.64 \times 10^{-6} \text{ erg.}$$

A photon having at least this much energy could spontaneously form an electron-positron pair.

Assuming that the radiation filling the universe as it expands is thermal radiation (as is confirmed by observations of today's cosmic background radiation), we can use Wien's law to find how hot the universe would have been when photons in it had sufficient energy to form electrons and positrons. First we convert the needed photon energy into a wavelength (see Chapter 6):

$$\lambda = \frac{hc}{E},$$

where λ is the wavelength, h is the Planck constant

(having the value $h = 6.626 \times 10^{-27}$ in the erg-cm-sec system of units), c is the speed of light, and E is the photon energy (in ergs). Substituting, we find

$$\lambda = \frac{(6.626 \times 10^{-27})(3 \times 10^{10})}{(1.64 \times 10^{-6})}$$

$$= 1.21 \times 10^{-10} \text{ cm.}$$

This is very short wavelength, in the gamma-ray portion of the spectrum.

Now we can use Wien's law to find the temperature for which this would be the wavelength of maximum emission. Wien's law states

$$T = \frac{0.29}{\lambda_{max}}$$

where T is the absolute temperature. Substituting for λ_{max} yields

$$T = \frac{0.29}{(1.21 \times 10^{-10})} = 2.4 \times 10^9 \text{ K.}$$

Thus in early times, when the temperature of the universe was 2.4 billion degrees or hotter, electrons and positrons formed spontaneously. Once the expansion had proceeded to the point where the temperature was lower than 2.4 billion degrees, these particles could no longer form.

Interestingly, electrons and positrons will annihilate each other if they meet, producing a photon having the energy calculated above (indeed such photons are occasionally detected coming from the interstellar medium). So in the early universe, these particles were disappearing as well as appearing. If exactly equal numbers of both particles had been formed, in time all would eventually annihilate, leaving the universe devoid of electrons and positrons (or other kinds of particle-antiparticle pairs). For reasons not yet fully understood, however, more electrons than positrons were formed, so that when the universe cooled to the point where these particles stopped forming, there were excess electrons. These have survived (as have other kinds of normal matter particles), which explains why today's universe contains matter, instead of being completely empty.

Patterns in the Universe

BY MARGARET J. GELLER

Dr. Geller is one of the world's foremost authorities on the structure of the universe. Starting with her graduate work at Princeton, her research has consistently focused on questions of the largest scope and the grandest scale imaginable as she has probed the overall distribution and motions of galaxies and what they tell us about the shape of the universe. Since receiving her doctorate at Princeton in 1975, Dr Geller has been associated during most of her career with the astronomy programs at Harvard and the Smithsonian Astrophysical Observatory, where she is now a Professor of Astronomy, Harvard University. Besides being a leader in cosmological research, Dr. Geller has also served the national astronomical community in several capacities, including her present position as a councilor of the American Astronomical Society and her activities on several advisory committees to both NASA and the National Science Foundation. The work described in this article has elevated Dr. Geller to the greatest heights of recognition and honor in our field; in 1990 she was awarded the prestigious MacArthur Fellowship, one of the most coveted prizes in any area of intellectual endeavor, and in 1991 she added the Newcomb Cleve-

land prize of The American Association for the Advancement of Science. In this article, Dr. Geller shares some of her thoughts on recent discoveries about the distribution of galaxies in the universe.

For more than 4,500 years we human beings have been making maps. We have mapped the surface of the Earth in almost excruciating detail. The concept of map making extends to regimes we cannot explore directly—we make maps of the positions of atoms in solid materials and in DNA, the key to life itself. From the smallest to the largest physical systems, maps reflect the human drive to know. Today one of the grand frontiers is the exploration of the universe over distances of hundreds of million, even billions, of light-years. With these maps, we hope to come to understand the formation and evolution of

the very large patterns in the distribution of galaxies.

Science and map making are closely related. The development of the theory of continental drift is a historical example. During the early twentieth century, clear knowledge of the large-scale geographic features of the Earth (particularly the shapes of the continents) led to the idea. Before 1912, Alfred Wegener noticed that the continents can be fitted together like pieces of a jigsaw puzzle, and he made the creative leap to suggest they once were a single giant land mass that later fragmented. The continents drifted apart to their present locations. The idea of a dynamic Earth was greeted with more than a bit of skepticism; it was even ridiculed. Now plate tectonics, which explains continental drift, is a full-fledged physical theory with abundant observational support.

The daunting enormity of the universe presents challenges to cosmic map makers that differ from those faced by map makers on the Earth. When we talk about "mapping the universe," we mean that we chart the positions of galaxies like our own Milky Way in a three-dimensional space we call "redshift space." Practical limitations dictate that a feasible study of the distribution of galaxies could include measurements for

10,000 or perhaps 100,000 galaxies. All surveys taken together include 40,000 thousand galaxies. In other words, we have mapped out a fraction of the universe comparable with the fraction of the Earth covered by the state of Rhode Island—about 1/100,000.

Although we are limited to mapping out a small fraction of the volume of the universe, we would like to try to answer a well-defined, perhaps deceptively simple, question: How big are the structures—the patterns (if any) in the distribution of galaxies? Once we focus on the question, the challenge is to obtain the best possible limits on the properties of big structures (if they exist).

The first step in making a map is the construction of a catalog of the positions of galaxies on the sky. Fortunately, during the 1960s, Fritz Zwicky and his collaborators cataloged the positions of more than 30,000 galaxies brighter than a limiting magnitude.

Until the 1930s extragalactic astronomers were limited to the "flatland" of maps of positions on the sky. In 1929, Hubble's discovery of the universal expansion opened the way for exploration of the three-dimensional structure of the universe. We get the third dimension by measuring the redshifts of galaxies which, in turn, give us estimates of their distances from us.

Modern detector technology makes it possible to use a 1.5-meter telescope (small by today's standards) to measure redshifts for galaxies within about 150 megaparsecs of the Earth. The Smithsonian Astrophysical Observatory operates a 1.5-meter telescope located on Mt. Hopkins, just south of Tucson, Arizona. Every clear

moonless night, John Huchra and I use this telescope to measure redshifts of galaxies. To measure a typical redshift in our survey, we collect photons from each galaxy for about half an hour. Within the next 5 to 10 years (depending on the weather in Tucson!) we plan to measure redshifts for some 15,500 galaxies. During the spring of 1985, Valérie de Lapparent, John Huchra, and I completed the first sample in this project. We chose the geometry of the sample for sensitivity to large patterns in the distribution of galaxies—patterns that persist over distances of a hundred million light-years or more. We measured redshifts for the approximately 1,100 galaxies in a 6″ wide strip stretching for 120° across the sky (figure A). Why this particular strip? The number of galaxies included is about what we could measure in one spring of observing time on the 1.5-meter telescope. The strip is roughly overhead in Tucson and seemed a convenient starting place.

In this map the Earth is at the vertex of the cone. The map shows only two of the three dimensions of the survey; the declination coordinate is perpendicular to the page. In the azimuthal direction we plot the right ascension—the coordinate on the sky that runs along the long dimension of the strip. Along the radial direction we plot the velocity derived from the redshift. Each point is a galaxy comparable with the Milky Way. Perhaps the map has a subliminal appeal because the pattern of galaxies looks like a human figure. In any case, the pattern is striking!

We designed a survey to look for big structures and we found some. Nearly all the galaxies lie in

thin, sharp structures that loop cross the map. If you treat the pattern of galaxies as a connect-the-dots exercise, your pencil outlines the vast dark regions we call voids. These regions contain few, if any, bright galaxies. The largest of these voids (centered at 15^h and 7500 km s^{-1}) has a diameter of 5,000 km s^{-1} or 150 million light-years. The pattern is reminiscent of a slice through the suds in the kitchen sink—the galaxies outline the voids, the interiors of the cosmic bubbles.

The map of recession velocities is not exactly a map of distances. There are distortions because of the motions of galaxies internal to systems like the Coma cluster. The torso of the "human figure" is the Coma cluster. It is stretched out along a line pointing toward us at the origin. The length of this "finger" is a measure of the typical motions of galaxies within the cluster. From the width and length of this finger we can estimate the amount of matter in the cluster by applying Newton's laws. When we perform this exercise, we are faced with one of the profound problems in extragalactic astronomy—the dark matter problem. If we measure the amount of light from the galaxies in the cluster and figure out the mass of stars responsible for the light, the total falls far short of the estimate based on Newton's laws. We come up short by a factor of 10. In other words, we see only about 10 percent of the mass; the rest is dark.

Other distortions of the velocity map may occur if all the galaxies in some region have a motion relative to the expansion of the universe. In general, we expect these velocities to be small compared

(continued on next page)

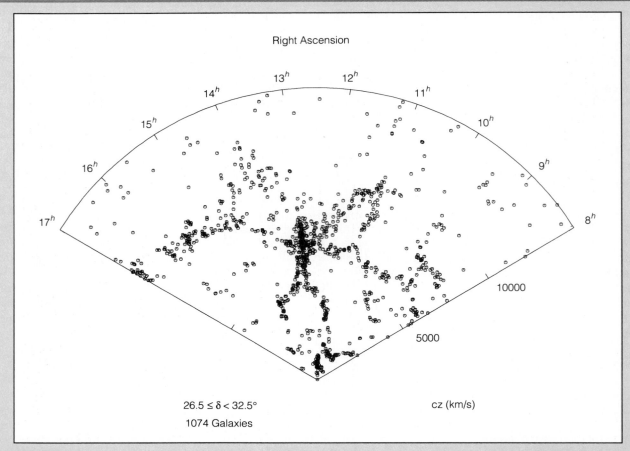

Right Ascension

$26.5 \leq \delta < 32.5°$

1074 Galaxies

cz (km/s)

Figure A

with the extent of the voids we see. Except for the fingers associated with rich clusters, we expect that the big patterns in our velocity map are a reasonable representation of the patterns we would see if we could measure distances directly. From here on we treat the map for the most part as though the "velocity" label read "distance."

We have now mapped the distribution of galaxies in three additional slices. Two are adjacent to the one in figure a and cover the declination range from 32.5° to 44.5°; the fourth is separated by

12° from these first three and covers the declination range from 8.5° to 14.5°. Figure B shows the distribution of galaxies in all four of these slices. The yellow points are galaxies in the three adjacent slices, and the orange points are the galaxies in the fourth slice. This map that contains some 4,000 galaxies confirms our original suggestion that galaxies are located primarily in thin surfaces. The most remarkable of these is the "Great Wall," which extends nearly horizontally across the entire map. The wall is about one-half to two-thirds of the way be-

tween the Earth and the outer (curved) boundary of the survey. This Great Wall is the largest structure we know and it probably extends outside the boundaries of the survey. It contains thousands of galaxies like the Milky Way. Our current estimates are that it extends for about 500 by 200 million light-years but is less than 15 million light-years thick. The Great Wall is probably composed of the surfaces surrounding several adjacent bubbles.

Surveys of other regions of the nearby universe support the view that large structures are the rule

Figure B

rather than the exception. Probes that reach more than 10 times as deep into the universe as the maps in figures a and b reveal a sequence of voids and thin "walls" perhaps similar to the Great Wall.

The pattern we observe in the distribution of galaxies has profound implications. One of the simplest and perhaps most disconcerting is that the largest voids we see are the largest we can detect within the limits set by the dimensions of our survey. There are, at the moment, no direct limits on even larger structures. These enormous dark, but not necessarily

empty, regions are common. Every survey big enough to contain one does so. These results make it clear that if we could sample the universe in volumes 100 million light-years on a side, the distribution of galaxies would differ substantially from one volume to the next. Some volumes ("patches") might be devoid of bright galaxies; others might contain lots of galaxies in one or several thin, sheetlike structures. This last statement reflects a profound change in our picture of the universe.

The development of the large

voids and sheets in the universe depends upon many poorly known or even unknown properties of the universe and its contents. However, the action of gravity on large scales must be an integral component of any explanation of the formation of structure we see. Galaxies and clusters of galaxies are held or "bound" together by the force of gravity. These large systems develop from smaller concentrations of matter in the early universe. The origin and properties of the small lumps and bumps are one of the puzzles

(continued on next page)

we have to solve.

We also have to understand the role of the ubiquitous dark stuff. Because we do not see the dark matter, we don't know what it is or where it is. If we could look directly at the way matter, rather than light-emitting matter, is distributed in the universe, we might see that it is actually very uniform and that galaxies form preferentially where the density of matter is somewhat higher than the average. Because the higher density regions are rare, the galaxy-formation process automatically leaves large hole in the distribution without having to move galaxies across enormous distances. Although computer simulations based on these ideas look much like the data, it is worrisome that we do not know the physics that determines where galaxies form.

If these models are right, the striking contrast between the voids and sheets in our maps is misleading. Galaxies are then the merest flotsam and jetsam a uniform sea of dark matter. Although the phenomenology of galaxies might be interesting in itself, they would be poor guides to the dynamics of the universe as a whole. The dark voids would be full of stuff—dark stuff and perhaps faint galaxies that would not be included in our surveys.

Physics other than the action of gravity may be important in making the structure we see. Perhaps explosions occurred when the universe was less than a billion years old and drove the hydrogen into thin shells that fragmented to form galaxies. This suggestion is appealing because it is easy to imagine how it produces a bubble-like pattern like the one we observe. Here too, the voids could be full of dark stuff. But these models have their drawbacks. Although there are many explosive, energetic phenomena in the universe ranging from supernovae to the sources of energy in the nuclei of galaxies, there are none that release sufficient energy over a short enough period to make the largest observed shells. Even worse, if there were such enormous explosions, we probably should have detected their effect as irregularities in the very smooth microwave radiation background.

The unanswered questions in cosmology are profound. I often feel that we are missing some fundamental element in our attempts to understand the large-scale structure of the universe. Advances in particle physics may eventually lead to the identification of the major constituents of the mysterious dark matter, which appears to dominate the large-scale dynamics of the universe. Detection of one of these particles would be a revolution in both particle physics and cosmology.

It is both sobering and exciting to think once again about the analogy between mapping the Earth and mapping the universe. It took centuries to learn the shapes of continents well enough to prompt the intuitive suggestion of continental drift. Then it took another half century to substantiate the theory.

Perhaps our approach to mapping the universe is more sophisticated than the early approaches to mapping the Earth. But the universe is immense; we are encumbered because we can only admire it from afar. The techniques required to uncover the fundamental physics governing the largest realm we can explore may be well beyond our current technology.

For observational cosmologists one challenge is to see how big a region we have to survey before one region looks essentially the same as another. The diameters of the largest voids we have seen so far are about 1 percent of the radius of the visible universe—still small. If we find similarly striking voids with much larger diameters, we might have to abandon one of the fundamental ideas in cosmology, the idea that the universe is homogeneous on some sufficiently large scale. We have to make deeper surveys to answer this question.

Very much deeper surveys of galaxies at large redshift can also tell us directly how the maps looked billions of years ago. Today we may not be smart enough to make a model that explains where the patterns we observe came from. In the future, large telescopes on the ground will enable us to directly observe the history of these patterns in the universe.

SECTION VII

LIFE IN THE UNIVERSE

We finish our survey of the dynamic universe with a discussion of living organisms and the prospects for finding that we are not alone in the cosmos. To many, this is the most central question we can ask. In the sole chapter of this section, we will examine the astronomer's attempt to answer it.

In discussing the question of life elsewhere in the universe, we will begin by looking into the origins of life here on Earth. In doing so, we will learn of biological evolution and the evidence that has been unearthed to tell us of our own beginnings. To do this, we will step outside the normal arena of astronomy, but we will then return to it in assessing whether the conditions that led to life on Earth might exist on other planets in our solar system and in the galaxy at large. We will discuss the probability that technological civilizations might be thriving here and there in the Milky Way.

We will complete our brief treatment of life in the universe by discussing the strategy for searching for other civilizations. It seems most likely that radio communications will be our initial means of contact, and there is an interesting exercise in logic involved with choosing the wavelength at which to search for or to send signals.

CHAPTER 32

The Chances of Companionship

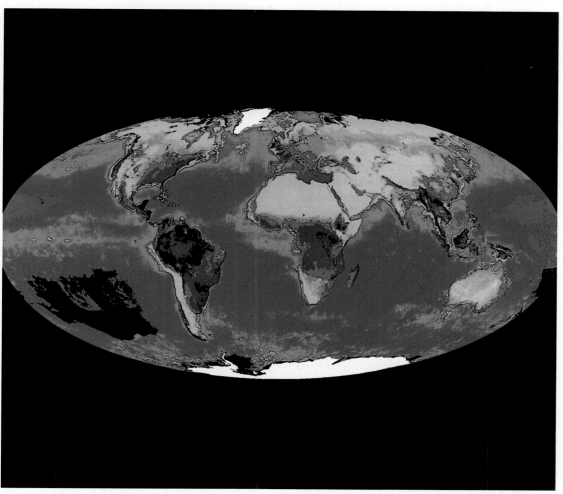

Life on Earth. Colors (green on continents; red and yellows in the seas) indicate relative concentrations of plants. (NASA)

Life on Earth
Could Life Develop Elsewhere?
The Probability of Detection
The Strategy for Searching

We have attempted to answer all of the fundamental questions about the physical universe that can be treated scientifically. Having done this, we know our place in the cosmos: We know something of its scale and age, and we realize how insignificant our habitat is.

In this chapter we contemplate whether we as living creatures are unique in the universe, or whether even that distinction must be shared. Nothing that we have learned so far leads us to rule out the possibility that other life forms, some of them intelligent, may exist. We believe that the Earth and the other planets in our solar system are a natural by-product of the formation of the Sun, and we have evidence that some of the essential ingredients for life were present on the Earth from the time it formed. Similar conditions must have been met countless times in the history of the universe, and will occur countless more times in the future.

Science cannot yet tell the full story of how life began, however, and we have not found any evidence that it actually does exist elsewhere. The mystery remains.

LIFE ON EARTH

We can start our discussion of possible extraterrestrial life by discussing the origins of the only life we know. Besides giving us some insight into the processes thought to have been at work on the Earth, this will help us later, when we are speculating about whether the same processes have occurred elsewhere.

Before the time of Charles Darwin (Fig. 32.1), in the mid-1800s, the view was widely held that life could arise spontaneously from nonliving matter. Darwin's work in the study of evolution, showing how species develop gradually as a result of environmental pressures, made such an idea seem improbable.

An alternative to the spontaneous formation of life was proposed in 1907 by the Swedish chemist S. Ar-

Figure 32.1 Charles Darwin. *Scientific inquiries by Darwin led to an understanding of evolution, one of the most profound concepts of human intellectual development. (The Granger Collection)*

rhenius, who suggested that life on the Earth was introduced billions of years ago from space, originally in the form of microscopic spores that float through the cosmos, landing here and there to act as seeds for new biological systems. This idea, called the *panspermia* hypothesis, cannot be ruled out, but several arguments make it seem unlikely. It would take a very long time for such spores to permeate the galaxy, and there would have to be a very high density of them in space for one or more of them to reach the Earth by chance. More importantly, it seems very unlikely that the spores could survive the hazards of space, such as ultraviolet light and cosmic rays. Even if the panspermia concept is correct, the question of the ultimate origin of life remains, although it is transferred to some other location. In view of what is known today about the evolution of life and the early conditions on the Earth, scientists generally agree that life arose through natural processes occurring here and was not introduced from elsewhere.

It is believed that the early atmosphere of the Earth was composed chiefly of hydrogen and hydrogen-bearing molecules such as ammonia (NH_3) and methane (CH_4),

Figure 32.2 Simulating the early Earth. *This is an apparatus constructed by Urey and Miller to reproduce conditions on the primitive Earth, in hopes of learning how life forms could have developed. (© West Publishing Co.)*

as well as water (H_2O). Therefore, the first organisms must have developed in the presence of these ingredients.

In the 1950s, scientists began to perform experiments in which they attempted to reproduce the conditions of the early Earth. The starting point of these experiments was to place water in containers filled with the type of atmosphere just described. Water was introduced because it is apparent that the Earth had oceans from very early times, and because it is thought that life started in the oceans, where the liquid environment provided a medium in which complex chemical reactions could take place. Reactions occur much more slowly in solids and in gases—in solids because the atoms are not free to move about easily and interact, and in gases because the density of gas is low, and particles are relatively unlikely to encounter each other. Water is the most stable and abundant liquid that can form from the common elements thought to be present when the Earth was young.

The first of these experiments (Fig. 32.2) was performed in 1953 by the American scientists H. Urey and S. Miller, who concocted a mixture of methane, ammonia, water, and hydrogen, and exposed it to elec-

trical discharges, a possible source of energy on the primitive Earth. (Ultraviolet light from the Sun is another, but it was more difficult to work with in the lab.) After a week, the mixture turned dark brown, and Urey and Miller analyzed its composition. What they found were large quantities of amino acids, complex molecules that form the basis of proteins, which are the fundamental substance of living matter. Other experimenters later showed that exposure to ultraviolet light produced the same results. These experiments demonstrated that at least some of the precursors of life probably existed in the primitive oceans almost immediately after the Earth had cooled enough to support liquid water. Other similar experiments have produced more complex molecules, including sugars and larger fragments of proteins.

As noted in Chapter 15, it may be that amino acids were present in the solar system even before the Earth formed, because traces of them have been found in some meteorites (Fig. 32.3), and we know that meteorites are very old, representing the first solid material in the solar system. We also know that several kinds of complex molecules, including organic (carbon-

Figure 32.3 Primordial amino acids.   *This is a section of the Murchison meteorite, which fell in Australia. Amino acids found in this carbonaceous chondrite were apparently present in it when it fell. (Photo by John Fields, the Trustees, the Australian Museum.)*

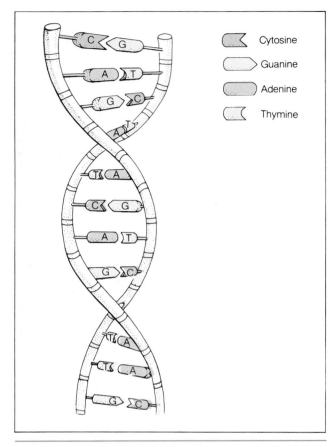

Figure 32.4 The DNA molecule.   *This is a schematic diagram of a section of a DNA molecule. DNA carries the genetic code that allows organisms to reproduce themselves. A critical question in understanding the development of life on Earth is to learn how DNA arose. (© West Publishing Co.)*

bearing) molecules, exist inside dense interstellar clouds (see Chapter 25), and we may speculate that perhaps amino acids may also have formed in these regions. To have survived on the Earth, these primordial amino acids would have to reach our planet sometime *after* its molten period.

It is not so clear what direction things took once amino acids and other organic molecules existed. Somehow these building blocks had to combine to form **ribonucleic acid (RNA)** and **deoxyribonucleic acid (DNA;** Fig. 32.4). These very complex molecules carry the genetic codes that allow living creatures to reproduce. Experiments have successfully produced molecules that are fragments of RNA and DNA from conditions like those that prevailed on the early Earth, but not the complete forms required. Maybe it is a simple matter of time; if such experiments could be performed for years or millennia, perhaps the vital forms of these proteins would appear. This is one of the areas of greatest uncertainty in our present knowledge of how life began.

Fossil records (Fig. 32.5) tell us that the first microorganisms appeared some 3 to 3.5 billion years ago, when the Earth was barely 1 billion years old. Following their appearance, the evolution of increasingly complex species seems to have followed naturally

(Table 32.1). At first the development was very slow, only reaching the level of simple plants such as algae 1 billion years later. Increasingly elaborate multicellular plant forms followed, and gradually altered the Earth's atmosphere by introducing free oxygen. Meanwhile, the gases hydrogen and helium, light and fast-moving enough to escape the Earth's gravity, essentially disappeared. Nitrogen, always present from outgassing and volcanic activity, became more predominant through the decay of dead organisms. By about 1 or 2 billion years ago, the Earth's atmosphere had reached its present composition.

The first broad proliferation of animal life occurred about 600 million years ago, and the great reptiles

Table 32.1 Steps in the evolution of life on Earth

Era	Period	Age (yr)	Biological Developments
Archeozoic		3.5×10^9	
	Archean		No life forms
Proteozoic		$1.5–3.5 \times 10^9$	
	Algonkin		Radiolaria; marine algae
Paleozoic		$0.25–1.5 \times 10^9$	
	Cambrian		Marine faunas, primitive vegetation
	Ordovician		Fishlike vertebrates
	Silurian		Air-breathing invertebrates, first land plants
	Devonian		Fishes and amphibians; primitive ferns
	Mississippian		Ancient sharks, mosses, ferns
	Pennsylvanian		Amphibians, insects
	Permian		Primitive reptiles, mosses ferns
Mesozoic		$70–250 \times 10^6$	
	Triassic		Reptiles, dinosaurs, ferns
	Jurassic		Toothed birds, primitive mammals, palms
	Cretaceous		Decline of dinosaurs, modern insects, birds, snakes
Cenozoic		$1.5–70 \times 10^6$	
	Paleocene		Large land mammals
	Eocene		Primates, first horse, modern plants
	Oligocene		Larger horse, modern plants
	Miocene		Proliferation of mammals, larger horse, modern plants
	Pliocene		Grassland mammals, earliest hominids
Modern		1.5×10^6 to present	
	Pleistocene		Homonids, modern-size horse, modern plants
	Holocene		*Homo sapiens*, present-day vegetation

arose some 350 million years later. The dinosaurs died out after about 200 million years, and mammals came to dominance about 65 million years ago. Our primitive ancestors appeared only in the last 3 or 4 million years. Once the development of intelligence had provided the ability to control the environment, the entire world became our ecological home, and our physical evolution essentially stopped. It remains to be seen whether future ecological pressures will create future evolution of the human species.

COULD LIFE DEVELOP ELSEWHERE?

The scenario just described, if at all accurate, seemingly should occur almost inevitably, given the proper conditions. If this is so, then the question of whether life exists elsewhere amounts to asking whether the conditions that existed on the primitive Earth could have

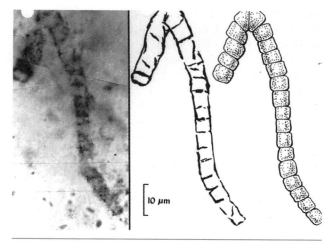

Figure 32.5 A fossil microorganism. *This fossil alga is evidence that primitive life forms existed on the Earth billions of years ago. (Photo courtesy of J. William Schopf, University of California, Los Angeles)*

Figure 32.6 Searching for life in the solar system. *Mars was long thought to be the most likely home in the solar system for extraterrestrial life. Here we see a* Viking *lander in a simulated Martian environment. The* Viking *missions reached Mars in 1976 but found no evidence for life forms. (NASA)*

arisen elsewhere. (For a dissenting view—that life was still improbable—see Astronomical Insight 32.1.)

It is clear that no other planet in the solar system could have provided an environment exactly like that on the early Earth. It therefore seems unlikely that Earth-like organisms will be found anywhere else in the solar system. Mars is the only planet where life forms have been sought so far (Fig. 32.6), but even there the conditions are quite unlike those on the Earth. Given the billions of stars in the galaxy, however, and the vast number that are very similar to the Sun, it seems highly probable that the proper conditions must have been reproduced many times in the history of the galaxy.

So far we have worked from the tacit assumption that life on other planets, if it exists, must be similar to life on the Earth, and we have considered only the question of whether other Earth-like environments may exist. We may question the premise that life could have developed only in the form that we are familiar with, however. Here, obviously, we must indulge in speculation, having no examples of other types of life at hand for examination.

Life as we know it is based on carbon-bearing molecules, and it has been argued that only carbon has the capability of combining chemically in a sufficiently wide variety of ways with other common elements to pro-duce the complexity of molecules thought to be necessary. After all, it is clear that the basic elements available to begin with must be the same everywhere, given the homogeneous composition of the universe. This may seem to rule out life forms based on anything other than carbon. However, it has been pointed out that another common element, silicon, also has a very complex chemistry and therefore might provide a basis for a radically different type of life. If so, we cannot begin to speculate on the conditions necessary for such life forms to arise.

Another assumption that might be subject to question is that water is a necessary medium. As noted earlier, it is the most abundant liquid that can form under the temperature and pressure conditions of the early Earth, and it is thought that only a liquid medium could support the required level of chemical activity. Other liquids can exist under other conditions, however, and it is interesting to consider whether life forms of a wholly different type than we are familiar with might arise in oceans of strange composition. It is interesting to speculate, for example, about what goes on in the lakes of liquid methane or liquid nitrogen that are thought possibly to exist on Titan, the mysterious giant satellite of Saturn.

If we are satisfied that life probably has formed naturally in many places in the universe, we can address a related, and to many, a more important, question: Given the existence of life, how likely is it that intelligence will follow? Here we have no means of answering, except to reiterate that as far as we know, the evolution of our species on Earth was the natural product of environmental pressures.

THE PROBABILITY OF DETECTION

In view of the limitations that prohibit faster-than-light travel, it is exceedingly unlikely that we will ever be able to visit other solar systems, seeking out life forms that may live there. We will continue to explore our own system, so there is a reasonable chance that if life exists on any of the other planets of our Sun, we will someday discover it. It seems, however, that our best hope of finding other intelligent races in the galaxy will be to make long-range contact with them through radio or light signals. Since this requires both a transmitter and a receiver, we can only hope to contact other civilizations as advanced as ours is, with the capability of constructing the necessary devices for interstellar communication.

A mathematical exercise in probabilities has been used for some years as a means of assessing, as objectively as possible, the chances for making contact with an extraterrestrial civilization. The aim is to separate the question into several distinct steps, each of which could then be treated independently. The underlying assumption is that the number of technological civilizations in our galaxy today with the capability for interstellar communication is the product of the number of planets that exist with appropriate conditions, the probability that life developed on those planets, the probability that such life has developed intelligence that gave rise to a technological civilization, and finally, the likelihood that the civilization has not killed itself off, through evolution or catastrophe.

Mathematically, the so-called Drake equation, (after Frank Drake, who has been its best-known advocate) is written

$$N = R_* f_p n_e f_l f_i f_c L,$$

where N is the number of technological civilizations presently in existence in our galaxy, R_* is the number of stars of appropriate spectral type formed per year in the galaxy, f_p is the fraction of these that have planets, n_e is the number of Earth-like planets per star, f_l is the

fraction of these on which life arises, f_i is the fraction of those planets on which intelligence has developed, f_c is the fraction of planets with intelligence on which a technological civilization has evolved to the point at which interstellar communications would be possible, and, finally, L is the average lifetime of such a civilization.

By expressing the number of civilizations in this way, we can isolate the factors about which we can make educated guesses from those about which we are more ignorant. It is an interesting exercise to go through the terms in the equation one by one, to see what conclusions we reach under various assumptions. People who do this have to make sheer guesses for some of the terms, and the result is a variety of answers ranging from very optimistic to the very pessimistic. In the following, we will adopt middle-of-the-road numbers for most of the unknown terms.

The first two factors, R_* and f_p, are, in principle, quantities that can be known with some certainty from observations. The rate of star formation in the equation refers to stars similar to the Sun. A much cooler star would not have a temperate zone around it where a planet could have the moderate temperatures needed for life to begin (we are, throughout this exercise, limiting ourselves to life forms similar to our own), and a very hot star would be short-lived, so that there would be insufficient time for life to develop on its planets, before the parent star blew up in a supernova explosion. Taking these considerations into account, it is estimated by some that up to ten suitable stars form in our galaxy per year; for the sake of discussion, we will be more cautious and adopt $R_* = 1$/year. This may even be a little high for the present epoch in galactic history, but the star-formation rate was surely much higher early in the lifetime of the galaxy, and low-mass stars of solar type formed that long ago are still in their prime. Thus, the adoption of an average formation rate of one Sun-like star per year is probably reasonable.

From what we know of the formation of our solar system, it seems that the formation of planets is almost inevitable, except perhaps in double- or multiple-star systems. For the sake of discussion, let us assume that $f_p = 1$; that is, that all stars of solar type have planets. Observations made with instruments such as the *Hubble Space Telescope* may eventually provide real information on this term.

The number of Earth-like planets, n_e, is highly uncertain and depends on how wide the zone is where the appropriate temperature conditions could exist.

Recent studies show that this zone might be rather small—that is, that the Earth would not have been able to support life if its distance from the Sun were only about 5 percent closer or farther than it is. Estimates of n_e vary from 10^{-6} to 1. Let us be moderate and assume $n_e = 0.1$; that is, that in one out of ten planetary systems around solar type stars, there is a planet within the temperate zone where life can arise.

Now we get to the *really* speculative terms in the equation. We have no way of estimating how likely it is that life should begin, given the right conditions. From the seeming naturalness of its development on Earth, it can be argued that life would always begin if given the chance. Let us be optimistic here and agree with this, adopting $f_l = 1$.

Again, the chance of this life developing intelligence and furthermore advancing to the point of being capable of interstellar communication is completely unknown. All we know is that in the only example that has been observed, both things happened. For the sake of argument, we therefore set both f_i and f_c equal to 1.

At this point, it is instructive to put the values adopted so far into the equation. We find

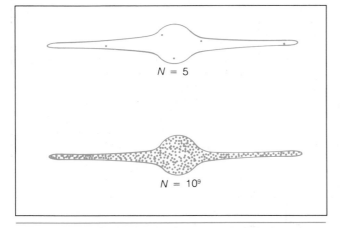

Figure 32.7 Possible values of N. *The number of technological civilizations in the galaxy (**N**) may be very small (upper), in which case the average distance between them is very large; or **N** may be large, so that the distance between civilizations is relatively small (lower). The chances for communication with alien races are much higher in the latter case, where the distances are only a few tens of light years.*

$$N = (1/\text{year})(1)(0.1)(1)(1)(1)L$$
$$= 0.1L$$

Having taken our chances and guessed at the values for all the other terms, we now face a critical question: How long can a technological civilization last? Ours has been sufficiently advanced to send or receive interstellar radio signals for only about 50 years, and there is sufficient instability in our society to lead some pessimists to think that we will not last many more decades. If we take this viewpoint and adopt 50 years as the average lifetime, then we find

$$N = 5,$$

meaning that we should expect the total number of technological civilizations present in the galaxy at one moment to be very small, about five. If this is correct, then the average distance between these outposts of civilization is nearly 20,000 light-years (Fig. 32.7). The time it would take for communications to travel between civilizations would therefore be very much longer than their lifetimes, and there would be no hope of establishing a dialogue with anyone out there. If this estimate is correct, it is no surprise that we haven't heard from anybody yet.

We can be more optimistic, though, and assume that a technological civilization solves its internal problems and lives much longer than 50 years. Extremely optimistic people would argue that a civilization is immortal; that it colonizes star systems other than its own, so that it is immune to any local crises such as planetary wars or suns expanding to become red giants. In that case, allowing a few billion years for the development of such civilizations, we can set $L = 10^{10}$ years (i.e., nearly equal to the age of the galaxy), and we find

$$N = 10^9,$$

where the average distance between civilizations is only about 30 light-years (see Fig. 32.7), coincidentally just a little less than the distance our own radio signals have traveled (Fig. 32.8) since the early days of radio and television. If this estimate of N is correct, we should be hearing from somebody very soon.

We have presented two extreme views of the likelihood that other civilizations exist in our part of the galaxy. As we mentioned earlier, opinions among the scientists who seriously study this question vary throughout this large range. Those who favor the optimistic viewpoint advocate the idea that we should make deliberate attempts to seek out other civilizations.

ASTRONOMICAL INSIGHT 32.1

The Case for a Small Value of N

Although it is an interesting exercise to write down an equation for N, the number of technological civilizations in the galaxy, no two astronomers can agree on the values of the various terms.

There is one very pessimistic argument that concludes that we are alone in the galaxy. This is based on the likely fact that if there were other civilizations, a significant fraction of them would have arisen long before ours. This is a straightforward consequence of the great age of the galaxy, some 5 to 10 billion years greater than the age of the Earth. This argument postulates further that if technological civilizations, once begun, live a long time, then the galaxy must be inhabited by a number of very advanced races, much older and more mature than our own. Up to this point, there is little controversy about any of these assertions.

The next logical step in the argument, however, is debatable; it says that any advanced civilization that has survived for millions or even billions of years will nec-

essarily have done so by perfecting some form of interstellar travel and colonizing other planets. Once this had started, the argument goes, the spread of a given civilization throughout the galaxy would accelerate rapidly, and by this time in galactic history, no habitable planet such as the Earth would remain uncolonized. The first successful galactic civilization, in this view, would quickly rise to complete dominance.

The only logical corollary, if the argument is accepted up to this point, is that no other civilizations exist, because if they did, the Earth would have been colonized long ago. The absence of interstellar visitors on the Earth is taken as proof that there are no other civilizations in the galaxy.

Those who adhere to this line of reasoning believe that $N = 1$. Therefore, at least one of the terms in Drake's equation must be very much smaller than the more optimistic values discussed in the text. One suggestion is that the temperate zone around a Sun-like star where planets can have moderate conditions is really very much narrower than usually supposed, perhaps because of the more ready development of a Venus-like greenhouse effect than is normally thought to be the case. If the temperate zone is very

small, then the term n_e, representing the number of habitable, Earth-like planets per star, would be very much smaller than the value of 0.1 adopted in the text.

Another term that has recently been singled out is f_l, the fraction of Earth-like planets on which life begins. As we learned in the text, somehow the amino acids that were present on the primitive Earth had to arrange themselves into special combinations to produce the long, chainlike protein molecules RNA and DNA. From a purely statistical point of view, the probability that the proper combination would happen to come together by chance is very small. This apparently happened on the Earth, but it may be so unlikely that it has not occurred anywhere else in the galaxy (actually it is thought that formation of these molecules occurred in steps, so that a random encounter of all the requisite atoms, a very unlikely event indeed, was not needed).

Whatever the correct values of the terms in the equation, the debate will rage on until our civilization reaches its life expectancy L and dies, until we discover another civilization, or until we ourselves expand to colonize the galaxy and find no one else out there.

THE STRATEGY FOR SEARCHING

The probability arguments outlined in the previous section are amusing and perhaps somewhat instructive, but obviously not very accurate. There are entirely too many unknowns in the equation for us to develop a reliable estimate of the chances for galactic compan-

ionship. Perhaps we will not know for certain what the answer is until we make contact with another civilization.

The problem of developing an experiment to search for or to send interstellar signals is that we do not know the ground rules. There are an infinite number of ways in which a distant civilization might choose

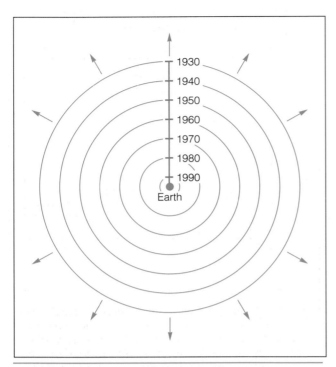

Figure 32.8 Earth's message to the cosmos. *As our entertainment and communications broadcasts travel out into space, they provide a history of our culture for anyone who may be receiving the signals. At the present time, the growing sphere that is filled with our broadcasts has a radius of over 50 light-years.*

to try to communicate, and it is impossible to search for them all. We must try to guess what the most probable technique would be.

In view of the power that is transmitted by radio signals and the relative lack of natural "noise" in the galaxy in that part of the spectrum, it has often been assumed that radio communications are most likely to succeed, although other techniques have been tried (Fig. 32.9). In the early 1960s, the U.S. National Radio Astronomy Observatory (Fig. 32.10) "listened" for transmissions from two nearby Sun-like stars, Tau Ceti and Epsilon Eridani, without success. More recently, larger surveys of neighboring stars have been done in the United States and in the Soviet Union, with similar results. About 40 different searches for extraterrestrial civilizations have been attempted so far.

One of the problems encountered by anyone wanting to search for extraterrestrial radio signals is obtaining observing time on a radio telescope. Most major radio telescopes are already in high demand by astronomers carrying out more standard kinds of research, and it is very difficult to have much time assigned for such a low-probability program as a search for extraterrestrial life. At most, one may obtain a few days a year on a given telescope. Since 1983, however, one radio telescope has been devoted entirely to the search for life. A radio telescope belonging to Harvard University has been dedicated to a program in which a sophisticated computer is used to automatically scan a wide range of wavelengths in as many stars as possible. This search is being funded primarily by private donations (the major donor was movie producer Steven Spielberg).

With the exception of the Harvard program, most searches so far have been confined to a single radio wavelength. The wavelength chosen in most cases was 21 cm, partly for practical reasons (radio telescopes designed for observing at this wavelength already existed), and partly because it was hypothesized that if others out there deliberately wanted to send signals, they might choose to do so at a wavelength that other astronomers around the galaxy would be observing anyway, even if they weren't trying to find distant civ-

Figure 32.9 Another message from Earth. *This recording of a message from Earth is traveling beyond the solar system aboard the* Voyager *spacecraft. Only an advanced race of beings would have the technological skills necessary to learn how to listen and to decode the message of peace that it contains. (NASA)*

Figure 32.10 A radio telescope used in the search for life. *This large dish, at the U.S. National Radio astronomy Observatory, in Green Bank, West Virginia, was used in efforts to detect signals from alien civilizations. Most of these attempts have been made at the 21-centimeter wavelength of atomic hydrogen emission (unfortunately, this particular radio telescope collapsed catestrophically in 1988, and will not be restored). (National Radio Astronomy Observatory)*

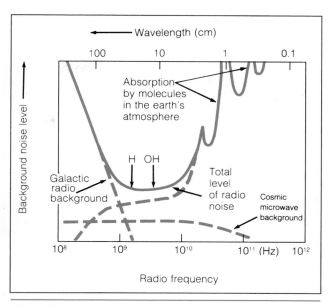

Figure 32.11 The galactic noise spectrum. *This diagram shows the relative intensity of the background noise in the galaxy as a function of wavelength. The quietest portion of the spectrum is in the vicinity of the "water hole" between the natural emission wavelengths of H_2O and OH molecules. (Data from D. Goldsmith and T. Owen 1980, The Search for Life in the Universe [Menlo Park: Benjamin Cummings])*

ilizations. Another wavelength that has been considered, but tried only in a very limited search, lies in the microwave region, between the emission lines of interstellar water vapor and OH. Dubbed the "water hole," this region has the advantages that it, too, might be a target of extraterrestrial astronomers, and it is in the quietest part of the radio spectrum (various objects, ranging from supernova remnants to radio galaxies, create radio noise that fills the galaxy, and the "water hole" lies in a region of the spectrum where this noise is minimized; see Fig 32.11). As in the Harvard search, new programs being planned today will cover wide ranges of wavelengths, avoiding the need to try to outguess the extraterrestrials in trying to decide which specific wavelength to observe.

One pessimistic viewpoint is that we are all listening, and no one is sending. In that case, we can only hope to pick up accidental emissions, such as entertainment broadcasts on radio or television, and these would be much weaker and more difficult to detect. Deliberately sent signals can be detected over far greater distances than accidental transmissions, because more power would be put into a deliberate signal, and it could be directed specifically at candidate stars, whereas our accidental radio and television signals are broadcast indiscriminately in all directions. Therefore, it makes a big difference whether someone out there is trying to send a message or not.

In 1974, a message from Earth was sent from the great Arecibo radio dish (see Fig. 7.25) toward the globular cluster M13, about 25,000 light-years away. The reason for choosing a globular cluster was that there are hundreds of thousands of stars in it, many of which are similar to the Sun in spectral type, and all of them would be within the beam of the transmission. The message consists of a stream of numbers which, when arranged in a two-dimensional array, form a pattern that illustrates schematically such things as the structure of DNA, the form of the human body, Earth's population, and the location of the Earth in the solar system. If this message is received and understood by anyone in M13, it will be some 25,000 years from now, and any answer they may send back will arrive here about 50,000 years from now.

It is not clear what the future will bring in terms of

Figure 32.12 Project Cyclops *The grandest plan seriously considered for the purpose of communicating with extraterrestrial civilizations, Project Cyclops would consist of 1,000 to 2,500 radio antennae, each 100 feet in diameter. This huge array, some 16 kilometers in extent, would be capable of detecting signals over a large portion of the radio spectrum, from planetary systems 1,000 or more parsecs away. (NASA)*

deliberate searches for extraterrestrial intelligence. The most ambitious project yet studied seriously is *Project Cyclops* (Fig. 32.12), but its cost will probably prevent it from ever being built. The plan was to build a veritable forest of large radio antennae, along with a very sophisticated data processing system, so that up to 1 million stars could be scanned for artificial emissions over a broad wavelength region, in a few years' time. By searching over a wide range of wavelengths, the scientists planning this effort hope to eliminate much of the guesswork about which is the most likely portion of the spectrum to be used by alien civilizations. This also enhances the chances of detecting accidental emissions, which, if they are anything like our own radio and television signals, may be spread over much of the spectrum. The *Project Cyclops* antennae could also be used to send powerful radio signals, allowing the Earth to broadcast its own beacon for others to find, and providing for the possibility of two-way communications.

The principal problem in implementing *Cyclops*, or any other large-scale effort, is the enormous cost. The version of *Project Cyclops* just described is currently estimated to require some 10 billion dollars in expenditures. Such sums have never been devoted to astronomy, and if they were, there would still be considerable debate over whether to spend it on this or on some

other research programs that might appear to have greater certainty of success. Presumably, alien astronomers in other solar systems would have to face the same questions, although we can hope that sufficiently advanced civilizations will have overcome the limitations of their planetary resources, and will be willing and able to tackle such a massive task.

A less ambitious project, called the *Search for Extraterrestrial Intelligence (SETI),* has been funded by NASA. The plan is to use existing radio dishes and specially developed, sophisticated computers to carry out a limited search of several stars. Because of advances in computer technology since the time when *Project Cyclops* was first planned, the far less expensive *SETI* program will be capable of a search nearly as thorough. *SETI* will scan some 8 million wavelengths for possible signals and is capable of probing about 1 million nearby Sun-like stars over several years' time. This is a large number of stars, but still only a very small fraction of the total in our galaxy. For *SETI* to succeed in detecting another civilization, the value of *N* will have to be quite high. While the scope of *SETI* is not as grand as that envisioned for *Cyclops*, it is nevertheless noteworthy that a federal government agency has taken the task seriously enough to fund a search for extraterrestrial life. The *SETI* computer system will soon be in full operation.

ASTRONOMICAL INSIGHT 32.2

The Complexity of UFO's.

In this chapter we have discussed the strategies for detecting extraterrestrial civilizations using radio communications, with little or no mention of the possibility of direct physical contact. Yet there is a forty-year history of "sightings" of objects thought to be alien spacecraft visiting Earth that suggests that the first contact might be in the form of face-to-face meetings. Since 1947, there have been countless reports of strange objects or lights in the sky, and some have attributed these "Unidentified Flying Objects" or "UFO's" to alien visitors.

To help decide what to think about UFO's, it is very useful to apply Occam's Razor, the principle that the simplest explanation for an observation is most likely the correct one. By simplest we mean the explanation that requires the fewest unsupported assumptions. Let's see how the alien spacecraft interpretation of UFO's fares under this criterion.

The laws of physics dictate that a spacecraft cannot travel faster than the speed of light. Therefore, allowing for acceleration time, it must take decades or centuries for a ship to travel the distance from even the nearest star to the Earth (recall that the nearest star is over 4 light-years away; calculations based on known propulsion systems indicate that it would take roughly 40 years to get there from here, and changing the propulsion system to some as-yet uninvented type would still require some acceleration time). Furthermore, the energy required for such a journey is immense; it has been estimated that the energy needed to boost a colonizing spaceship from Earth to a nearby star is enough to power the entire United States at its present rate of consumption for one hundred years! Is it likely that Congress would ever approve such an expenditure? This enormous energy requirement would be faced by any civilization planning interstellar travel, and cannot be lessened by advanced technology (unless there are ways of violating the known laws of physics, but there is no evidence that this is possible, and there is plentiful evidence that it is not).

Given the time and energy constraints imposed on any civilization thinking about interstellar travel, it is very difficult indeed to accept the notion that UFO's are sightseeing or spying spacecraft from another star system; that an alien civilization would expend the required time and energy routinely and often. It is far simpler to accept the alternative theory of UFO's, that they are phenomena that occur in the Earth's atmosphere, and are not spacecraft. This explanation is made all the more likely by the fact that virtually every UFO sighting that has been carefully studied has been found to be due to something normal such as an airplane, a balloon of some sort, or a bright planet seen near the horizon.

Many people are tempted to believe that UFO's represent visiting aliens, but objective scientific analysis makes this belief very difficult to accept. Occam's Razor tells us that a much more mundane explanation of UFO's is likely to be correct, and that our best hope of contacting alien civilizations lies in long-range radio communications, which travel at the speed of light and require very little energy. Of course, the fact that no one really *knows* what alien technology and psychology might be like makes the speculation all the more interesting.

PERSPECTIVE

In this chapter we have introduced a bit of speculation, well founded perhaps in the discussion of life on Earth, but less so in the sections on the possibilities of life elsewhere.

We are prepared now to understand and appreciate new advances in astronomy; having completed our study of the present knowledge of the universe, we will be able to put into perspective the major new discoveries that are sure to come as technology and theory continue to improve. The *Hubble Space Telescope,* the large optical telescopes now being designed, the comparable advances in radio, infrared, and X-ray techniques—all of these will inevitably lead to novel and unforeseen breakthroughs in our view of the universe.

The next few decades will be exciting times for astronomy, and we can anticipate their coming with a sense of excitement and wonder.

SUMMARY

1. While there were early beliefs that life started on Earth spontaneously or by primitive spores from space, most scientists today accept the theory that life began through natural, evolutionary processes.

2. Amino acids, fundamental components of living organisms, are formed readily in experiments designed to simulate early conditions on the Earth.

3. The steps that led to the development of the necessary forms of RNA and DNA are not yet fully understood, and probably occurred over a very long period of time.

4. Fossil evidence provides a record of the evolutionary steps leading from the first primitive life forms to modern life.

5. The conditions that prevailed on the early Earth have probably been duplicated on other planets in the galaxy, though probably not on other planets in the solar system.

6. It is often assumed that only Earth-like life could develop, because carbon is nearly unique in the complexity of its chemistry, but there have been suggestions that at least one other element (silicon) may have the necessary properties.

7. Estimates of the number N of technological civilizations now in the galaxy can be made, with great uncertainty, based on what is known of the formation rate of Sun-like stars, and what is guessed for the probability of such stars' having planets with the proper conditions, and the probability that life, leading to intelligence and technology, will develop on these planets. Estimates range from $N = 1$ to $N = 10^9$.

8. Some attempts have been made to search for radio signals at 21 cm and other wavelengths, but with no success.

9. The chances for detecting or being detected by other civilizations depend strongly on whether or not deliberate attempts are made to send signals.

10. Future projects to search for and to send signals have been planned, and one, the *Search for Extraterrestrial Intelligence (SETI)*, has been funded by NASA, and is now undergoing preliminary testing. *SETI* will search about 1 million stars in 8 million radio wavelengths and should result in a detection if the number of technological civilizations in the galaxy is large.

REVIEW QUESTIONS

Thought Questions

1. Why is the natural evolution of life on Earth considered by scientists to be the most likely explanation?

2. To what extent has life on Earth modified its atmosphere? Does this suggest a method for determining, from remote spectroscopic observations, whether a distant planet might have life?

3. Why would a planet only slightly closer to the Sun than the Earth probably not be able to support life? (*Hint:* Recall what you learned in Chapter 10 about the cause of the contrasting conditions on Venus and the Earth.)

4. Summarize the arguments that lead some scientists to expect life elsewhere to be similar in chemistry to life on Earth.

5. Based on what is assumed about how life started on the Earth, do you think it likely that life has developed in the atmosphere of Jupiter? Do you think it could develop in an interstellar cloud?

6. Recall, from Chapter 26, the distinction between Population I and Population II stars, and where these stars are located in the galaxy. Do you think planets orbiting Population II stars are as likely to have life forms as planets orbiting Population I stars like the Sun? Explain. In view of your answer, did it make sense to beam a message to M13?

7. Explain in your own words the meaning of the various terms in the Drake equation, and how this equation represents the number of technological civilizations in the galaxy at any one time.

8. Why are radio communications thought to be the best method for detecting evidence for other civilizations?

9. What are the factors that govern the choice of wavelength at which to search for signals?

10. Explain why signals from a distant civilization would be easier to detect if they were sent deliberately rather than being accidental transmissions (such as our own entertainment broadcasts).

Problems

1. Under the panspermia hypothesis, life propagates through the galaxy in the form of spores that travel randomly about, occasionally landing on planets. If such a spore travels at the speed of a typical interstellar cloud (say 10 km/sec), how long would it take to reach the Earth from the nearest star, 1.3 pc away? How long would it take to travel the distance from the galactic center to the solar system (10,000 pc)? How does your answer compare with the age of the galaxy?

2. Suppose a planet was orbiting a star exactly like the Sun, and the planet's semimajor axis was 0.9 AU. How

would the intensity of sunlight on that planet compare with the intensity of sunlight on the Earth? What effect might this have on the possibility that life could develop on this planet?

3. Repeat the calculation of N, the number of technological civilizations in the galaxy, if the probability of life starting on an Earth-like planet (the term f_l in the Drake equation) is only 10^{-4} instead of 1, as assumed in the text. Assume that the lifetime L is 10^{10} years, then redo the calculation for $L = 10^5$ years.

4. If *I Love Lucy* broadcasts started in 1953, in what year did they first reach Alpha Centauri, 1.3 pc away? When will they reach Canopus, 30 pc away? When might we expect an answer from Canopus, if anyone is there to receive our signals?

5. Using Appendix 8, list the stars among the fifty brightest that might have planets on which life could develop.

ADDITIONAL READINGS

Abt, H. A. 1979. The companions of sun-like stars. *Scientific American* 236(4):96.

Ball, John A. 1980. Extraterrestrial intelligence: Where is everybody? *American Scientist* 68:656.

Barber, V. 1974. Theories of the chemical origin of life on Earth. *Mercury* 3(5):20.

Burt, D. M. 1989. Mining the Moon. *American Scientist* 77:574.

Drake, F. 1988. The *Pioneer* message plaques. *Mercury* 17:88.

Goldsmith, D., ed. 1980. *The quest for extraterrestrial life.* Mill Valley, Calif.: University Science Books.

Guin, Joel. 1980. In the beginning. *Science 80* 1(5):45.

Horowitz, Norman. 1986. *To utopia and back: The search for life in the solar system.* New York: W. H. Freeman.

Olson, Edward C. 1985. Intelligent life in space. *Astronomy* 13(7):6.

O'Neill, G. K. 1974. The colonization of space. *Physics Today* 27(9):32.

Overbye, Dennis. 1982. Is anyone out there? *Discover* 3(3):20.

Papagiannis, M. D. 1982. The search for extraterrestrial civilizations—A new approach. *Griffith Observer* 11(1):112.

———. 1984 Bioastronomy: The search for extraterrestrial life. *Sky and Telescope* 67(6):508.

Pelligrino, C. R. 1979. Organic clues in carbonaceous meteorites. *Sky and Telescope* 57(4):330.

Pollard, W. G. 1979. The prevalence of Earth-like planets. *American Scientist* 67:653.

Rood, R. T., and J. S. Trefil. 1981. *Are we alone?* New York: Scribner.

Sagan, Carl. 1982. The search for who we are. *Discover* 3(3):30.

Sagan, C., and F. Drake. 1975. The search for extraterrestrial intelligence. *Scientific American* 232(5):80.

Sagan, C., and I. S. Shklovskii. *Intelligent life in the universe.* San Francisco: Holden-Day.

Shneartzman, D., and L. J. Rickard, 1988. Being optimistic about the search for extraterrestrial intelligence. *American Scientist* 76:364.

Scheaffer, Robert. 1986. *The UFO verdict.* Buffalo, N.Y.: Prometheus Books.

Special Report on Life in Space. 1983. *Discover* 4(3).

Tipler, F. J. 1982. The most advanced civilization in the galaxy is ours. *Mercury* 11(1):5.

Weissman, P. 1990. Are periodic bombardments real? *Sky and Telescope* 79:266.

Wetherill, C., and W. T. Sullivan. 1979. Eavesdropping on the Earth. *Mercury* 8(2):23.

MATHEMATICAL METHODS

Travel Times for Interstellar Voyagers

Unless our currently accepted laws of physics are very wrong, it is not possible to travel at speeds greater than that of light. This means that travel times to even the nearest stars are very long. The closest Sun-like star is Alpha Centauri, which lies roughly 1.33 pc, or about 4.3 light-years away. This means that if we could build a spaceship that could travel at the speed of light (with no time needed to accelerate to this speed or decelerate upon arrival at the destination) we could travel to this star in 4.3 years. This may seem feasible, but unfortunately it is unrealistic.

The fastest spacecraft yet sent into space are the two *Pioneer* and the two *Voyager* probes, which visited the outer planets and then left the solar system. Upon leaving the solar system, these spacecraft had typical speeds of 15 km/sec (or more than 35,000 miles per hour, about twice the speed of Earth-

(continued on next page)

MATHEMATICAL METHODS

Travel Times for Interstellar Voyagers (continued)

orbiting satellites). These craft will not be significantly accelerated further, until they chance to encounter a massive body such as a star. We can compute the time needed to travel 1.33 pc at a speed of 15 km/sec:

$$t = d/v = (1.3 \text{ pc})(3.09 \times 10^{13} \text{ km/pc})/15 \text{ km/sec}$$
$$= 2.7 \times 10^{12} \text{ sec}$$
$$= 85,000 \text{ years.}$$

This is clearly impractical. But perhaps we can build a spaceship that accelerates to much higher speeds, and thereby shorten the travel time.

One possible technology, yet to be developed, would be to use ion drive engines. These would operate by expelling charged particles out the nozzle, accelerating the particles by a magnetic field. This is feasible, and the fuel (free electrons or protons) would be plentiful in space. It might be possible to keep accelerating all the way to the destination, picking up speed the entire time.

Unfortunately, it is likely that the rate of acceleration would be rather low, since the ship would have a large mass and the acceleration would be created by subatomic particles expelled only by a magnetic field, rather than a violent chemical reaction. An acceleration of 1 km/sec per day may be feasible. Let us see how this would help.

If we start from rest at time zero, then the distance traveled at acceleration a is

$$d = \frac{1}{2}at^2,$$

where d is the distance traveled, and t is the elapsed time since starting. Solving for t, we find

$$t = \sqrt{\frac{2d}{a}}.$$

Now we can substitute 1.4 pc ($= 4.33 \times 10^{13}$ km) for d and for the acceleration $a = 1$ km/sec/day ($= 1.16 \times 10^{-5}$ km/sec/sec). These values yield

$$t = \sqrt{\frac{2(4.33 \times 10^{13} \text{ km})}{(1.16 \times 10^{-5} \text{ km/sec/sec})}}$$
$$= 2.72 \times 10^9 \text{ sec}$$
$$= 86.7 \text{ years.}$$

Thus it might become possible to travel to Alpha Centauri in a bit under 100 years (but don't forget that time will be needed for deceleration: normally it is thought that the ship would be turned around at the midpoint, so that it would decelerate for the second half of the voyage, bringing it to rest when it arrived. To calculate the time more accurately, then, requires finding the time to the halfway point, which is about 61 years, and doubling it, yielding a total time of just over 120 years).

Dinosaur Die-Offs: A Bolt from the Blue, or Death by Earthly Causes

BY ROBERT T. BAKKER

Rarely does one person gain credit for overturning an entire field of research, but Dr. Robert T. Bakker has come very close to doing just that. Largely through his efforts, our view of dinosaurs has changed from one of sluggish, cold-blooded reptiles akin to today's lizards and alligators to one of active, warm-blooded, creatures more like today's birds. These views are presented in convincing fashion in Bakker's book Dinosaur Heresies *(New York: William Morrow, 1986), in which Bakker presents ample paleontological evidence to support his radical views on dinosaur physiology and behavior. Today Dr. Bakker is widely recognized as a leader of a new movement in the understanding of dinosaurs.*

Dr. Bakker was educated at Yale, where he was heavily involved in dinosaur research at the prestigious Peabody Museum, and at Harvard, where he earned a Ph.D. in 1976. At both institutions he won major awards for his excellence in research and in teaching. Since then he has served on the faculty at Johns Hopkins University and is now an adjunct faculty member and museum curator at the University of Colorado. In recent years, he has become a public figure through his numerous educational efforts, par-

ticularly those aimed at children, and through his books and television documentaries, such as his 1985 special on Public Television's "NOVA." While spending considerable time and effort on public education, Dr. Bakker has maintained a high level of research activity through the publication of many technical articles and through frequent field expeditions to hunt for dinosaur remains. (As recently as September 1989, he reported discovering several new specimens in Wyoming, including at least one previously unknown species). He has also discovered new species of dinosaurs in museum collections, where fossilized bones had been either misidentified or unidentified.

In this article, Dr. Bakker refutes the popular theory that the major dinosaur extinctions of the past were triggered by massive meteoroid or asteroid impacts. As recently as November, 1990, he reported a new giant dinosaur discovery in Wyoming that supports the arguments made in this essay. Here Dr. Bakker presents a strong case for more mundane causes of the great extinctions, and in doing so he warns of possible future disasters for life forms as we now know them.

Why did the dinosaurs die? Why did the saber-toothed tigers vanish without a trace? And why did the mammoths disappear? Massive die-offs are the persistent puzzle of fossil history, and the subject is sparking hot debates in evolutionary biology. On one side are astrophysicists and isotope chemists who work with multimillion-dollar spectrometers in clean labs and the most elegant computer modeling. On the other side are field paleontologists whose most complicated piece of equipment usually is a four-wheel-drive pickup splattered up to the windshield with mud.

Mass extinctions were the first theoretical challenges in the early development of paleontological science. As soon as scholars began the systematic excavations of giant, extinct monsters in the late eighteenth century, these scholars began asking some of the grand and terrible questions about the history of life, as they discovered that nearly all the large land animals that had at one time dominated the Earth had been utterly exterminated. In 1820, the great Parisian naturalist Baron Cuvier called these catastrophes "revolutions."

These revolutions aren't just irrelevant dramas of the distant past. Our own species, *Homo sapiens,* owes much of its dominance to mass extinctions. When the dinosaur dynasties fell at the end of the Mesozoic Era 65 million years ago, a new group of animals—the class of furry, warm-blooded mammals to which we belong—seized the opportunity to become the next rulers of the ecosystem. And the final revolution, which ended just 10,000 years ago, made the world safe for human civilization by wiping out nearly all the gigantic predators: huge wolves, hyenas, sabertoothed tigers, most large herbivores, the mammoths, mastodons, 2-ton buffalo, and 1-ton horses. When all the excess megafauna had been removed, neolithic farmers could begin to settle down and clear the land for permanent farms.

What triggers such revolutions? In the nineteenth century environmental explanations offered a popular answer to this question: The dinosaurs died because their swamps had dried up and the winters had gotten too cold—or so it was argued. Mammoths disappeared because the Ice Age turned far too frigid and whole herds of woolly behemoths were frozen in their tracks.

But these theories don't work. Fossil plants and crocodiles are excellent indicators of climate changes, and these environmental gauges simply don't show any dramatic events that coincide with the mass die-offs. As the dinosaurs were vanishing worldwide, big crocodiles, who are very sensitive to temperature fluctuations, went on living and breeding as if nothing at all had happened. Fossil pollen and leaves show that the woolly mammoths flourished through two dozen cycles of glaciers spreading and retreating over a period of 2 million years. But then, after success in such rigorous habitats, all the woolly mammoths expired. There is no good evidence that *any* mass extinction on land was caused by a long-term shift in climate.

In the 1970s, geochemists began to develop an outrageous new theory—that a gigantic meteorite or asteroid had exploded on the Earth and massacred the dinosaurs. The evidence: a thin layer of clay enriched with the element iridium, rare on Earth but concentrated in meteoritelike heavenly bodies, and quartz minerals that had been shocked by an explosion. According to the theory, this cosmic impact sent up a cloud of dust that blacked out the Sun and suddenly chilled the Earth's surface while, at the same time, searing downpours of acid rain polluted streams and lakes. The disaster took only a short time, a few weeks or years at most, and then the climate returned to normal.

Most vertebrate paleontologists who work on dinosaurs don't buy the "Cosmic Zap" hypothesis of death from the heavens. The late Luis Alvarez, principal architect of the meteorite-comet hypothesis, accused the doubting "bone-diggers" of being throwbacks to the eighteenth century who could not grasp the logic of the theory. But the low-tech methods of field paleontology, in fact, have built up a formidable mass of evidence that death came to the dinosaurs from an earthly cause.

Trying to solve the mass-extinction mystery is like searching for a serial killer. We know that some force exists that kills successful creatures every 10 to 20 million years. To catch the culprit, we need to define its modus operandi: how it kills, where it kills, and who is left untouched. Human serial killers almost always choose their victims according to a diabolically precise profile; for example, a serial murderer who preys upon homeless drifters doesn't stalk the halls of college dormitories.

When we take a criminological approach to mass extinction, we see the problems with the cosmic explosion theory. Let's sketch the modus operandi of an exploding meteorite or comet. Who would die out first? Answer: species that are most sensitive to the sudden chill from the dust cloud and the accompanying acid rain. Today, water-loving species like frogs, salamanders, pond and river turtles, and crocodilians are much more vulnerable to temperature drops and acid rain than are big, dry-land species such as elephants and buffalo. When ponds acidify in New England, the first creatures to leave are the salamanders and frogs, because their skin is very permeable to water. And big turtles and crocodilians are very sensitive to chill too; sudden frosts kill alligators in Louisiana and giant softshell turtles rarely live outside the tropics.

If a cosmic projectile exploded 65 million years ago, who, then, would suffer more—the gigantic land-living dinosaurs or the small frogs and salamanders and large turtles and crocodiles? Obviously, the water-bound amphibians and reptiles would be hit harder. However, field work in the badlands of Montana shows exactly the reverse of the predicted murder pattern. You can walk along the Montana outcrops of rock and pick up bones of salamanders and frogs, and turtles and crocodilians that lived before and after the dinosaur extinction—and there is almost no difference whatever. In fact, close relatives of these amphibians and reptiles still swarm in tropical and subtropical habitats.

The main flaw in the meteorite-comet theory is that the wrong animals survived. Dinosaurs were big, active animals like elephants, rhinos, and buffalo. And so dinosaurs would have had far greater resistance to chill and acidification than would the salamander-frog-turtle-croc complex. But the dinosaurs died, and the humble herpetological species survived.

Recent discoveries about dinosaurs make their vulnerability even more puzzling. Dinosaur bone microstructure shows that dinosaurs had high metabolism and fast growth rates, like large birds and mammals of today. And dinosaurs in great variety are turning up in sediments beyond the Arctic Circle in Alaska and Canada, where the climate must have been dark and cold during the four months of arctic winter. Temperatures didn't drop as far during the age of dinosaurs as they do today at those latitudes, but still there must have been snow on the ground and a strong windchill. Arctic dinosaurs either stayed over the winter, as some musk-oxen do today, or migrated far south, as caribou herds do. Either way, the dinosaurs were behaving as warm-blooded animals, not in the cold-blooded manner of modern lizards. And big, warm-blooded species are the most resistant to sudden, short-term cooling of air temperatures.

Whatever killed the dinosaurs had a very precise victim-selection system: Only big, active, warm-blooded species were targeted on the terrestrial ecosystem (deaths in the marine world were more widespread). And exactly the same modus operandi can be seen at all the other mass-extinction events. When the mammoths and giant sloths and saber-tooths died out 10 thousand years ago, the frogs and turtles and other freshwater cold-bloods passed through the crisis without a single mark on their evolutionary histories.

We can ask ourselves "How do we avoid mass extinction?" The answer is simple: "Become a frog." In the entire 350 million year history of nonmarine species, the cold-blooded, freshwater species have been insulated entirely against mass die-offs. Single species of frogs and turtlelike animals have become extinct here and there, but a mass, worldwide extermination has never occurred. The fossil history of frogs, turtles, and alligators is peaceful and crisis-free compared to the story of dinosaurs and big land mammals.

Back to our criminological approach. What agent would kill large, active warm-blooded animals and leave aquatic cold-bloods alone? Human history gives clues. *Homo sapiens* today is the dominant, active, warm-blooded creature on the globe. We are the equivalent of the last dinosaur dynasty. And human beings share one proclivity with other dominant warm-bloods: Our species travels easily from one continent to another. *Homo sapiens* spread rapidly from Eurasia to Australia and the New World in only 30,000 years after the species evolved.

Travel causes epidemiological nightmares. Diseases coevolve with their hosts. Eurasian populations of human beings carried dozens of diseases that had acquired adaptations for host-parasite coexistence. It is in the best interest of a parasite not to exterminate its host, so a long period of coevolution tends to dampen epidemics. However, when a host suddenly meets a new disease with no history of coadaptive accommodation, the disease often runs amuck, producing extraordinary mortality rates. One case in point is smallpox among Native American Indians. Smallpox had a long history in Eurasia, and epidemics rarely threatened the existence of local populations. But when European explorers brought smallpox to this continent, the disease spread almost instantly, killing Native Americans at an alarming rate. Entire tribes disappeared in a season or two.

In the late twentieth century, AIDS is behaving in the same way. It is a tropical disease that only recently made the jump from some nonhuman primate host to *Homo sapiens,* and there is no coevolutionary accommodation.

(continued on next page)

Millions of years ago

80 75 70 65 60 55 50 45

Chasmosaurus ✕
Anchiceratops ✕
Torosaurus ✕

Styracosaurus ✕
Arrhinoceratops ✕
Triceratops ✕

Centrosaurus ✕

Cold-blooded Croc Leidyosuchus
Croc genus lasted 35 million years ✕

Cold-blooded Turtle Aspideretes
Turtle lasted 35 million too ✕

Mass extinctions are selective. *At the end of the Cretaceous Period, the big, active horned dinosaurs suffered three waves of sudden die-off. But the soft shelled turtles and crocodiles continued to evolve slowly without interruption. Why? Perhaps because big mobile animals travel across the globe more easily and spread plagues from continent to continent.*

Nonhuman, warm-blooded mammals share our vulnerability to exotic disease. Rinderpest, an Asian cattle disease, entered Central Africa when the English expeditionary army marched up the Nile, with artillery pulled by Indian oxen. Once released among the dozens of African antelope species, rinderpest began slaughtering local populations, and the disease is still one of the biggest worries of wildlife managers.

The list of new diseases that threaten human beings and their domestic stock is very long indeed. If you arrive at Los Angeles International Airport from Malaya, the customs officials will comb through your luggage very carefully, looking for any insect, seed, or dried fruit that might harbor a dreaded tropical parasite of American food crops. But frogs, turtles, and alligators have far less to fear from exotic pests.

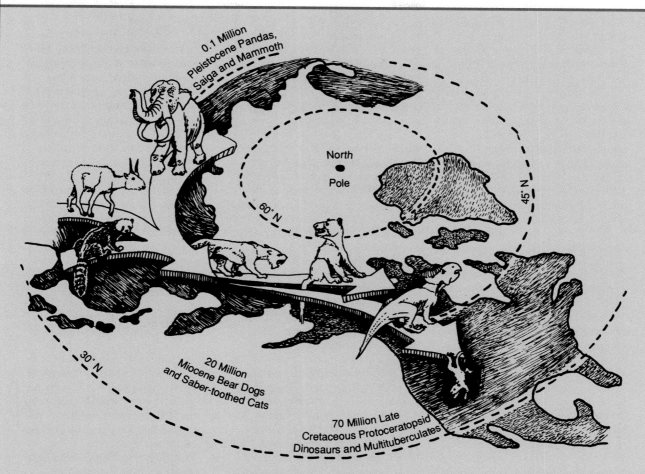

First, a cold-blooded body harbors far fewer diseases than does a warm-blooded one. The guaranteed even temperature of a warm-blooded creature is far more inviting than the fluctuating conditions in cold-blooded organs. And cold-bloods travel far less easily. Many times over the last 2 millions years, horses, mammoths, and camels traveled easily from the Old World to the New World and back again via the Bering Land Bridge. But the aquatic cold-bloods rarely made the trip. Because the frogs and salamanders and other freshwater and cold-bloods usually have very narrow and restrictive habitat requirements—the size of the pond or stream, water pH and temperature, current velocity, and nutrients—these aquatic species cannot often pass from one continent to another.

Mass extinction through exotic disease is not a new idea. In 1900, the great American paleontologist and zoologist H. F. Osborn noted that extinctions of large mammals coincided with times of faunal mixing when

Sudden episodes of immigration flooded North America with new species. *The Bering Land Bridge has opened up on many occasions, each time permitting mobile species to invade the New World from Siberia or vice versa. And every immigrant wave must have carried dozens of new diseases.*

mammal species moved from continent to continent. Such movement, Osborn argued, is bound to mix diseases and cause extinction.

But paleontologists in general have been very reluctant to give disease a chance to be the culprit for mass extinction. Students of human history, on the other hand, have been far quicker to accept the enormous power of epidemiology; the fate of civilizations and entire continents have been changed by invading microorganisms. By conservative estimate, 100 times more Native Americans died of European disease than died of European bullets.

And exotic disease kills native and invader alike. When an expedition led by Europeans went up the Congo River to look for Dr. Livingston in the 1900's, few of the Caucasian invaders survived the Central African pestilences. In exchange for small pox, European explorers carried syphilis back to Spain and England from the New World.

Field paleontologists know that dinosaurs were crossing and re-crossing the Bering Land Bridge late in the age of dinosaurs. Dinosaur species from Mongolia have very close relatives in Alberta and Montana. What would happen during a wholesale exchange of this type between continents? Each big, warm-blooded dinosaur species would carry not just one, but dozens of coadapted parasites, from viruses to tapeworms, just as

each modern rhino or buffalo species have a rich internal fauna of microbial fellow travelers. As the Eurasian and North American dinosaurs mixed and matched their respective germs, multiple epidemics would be unavoidable.

There is no question about whether exotic disease could kill dinosaurs; we know that every sudden invasion carries inevitable plague. No known agent in the natural world could prevent mass die-offs of local dinosaur populations. One disease by itself rarely can wipe out an entire species, but when the dinosaurs were indulging in intercontinental dispersal, hundreds of parasites must have mixed into new hosts. And adding to the devastation of disease would be the direct effects of the invading dinosaurs themselves; new predators and new competitors would put pressure on each native fauna. When Europeans introduced rabbits into Australia, many native species suffered because of the population explosion among the rabbits. Goats, sheep, and dogs have wrought havoc on island faunas.

It is safe to say that zookeepers, customs officials, and wildlife managers agree with historians that an invasion of exotic species is one of the worst events a native fauna can endure. A nearly com-

plete faunal mixing, such as seems to have occurred during the biggest land mass extinctions, could destroy an entire megafauna.

What about the original evidence for a meteorite impact—the layer of iridium-enriched mud with shocked quartz? Perhaps the Earth suffered a cosmic impact about the time of the dinosaur extinction, but the field evidence from the fossils in the sequence of strata shows that most of the dinosaurs were already gone before the clay layer was deposited. And a growing body of evidence suggests that a strange form of vulcanism might have produced the iridium and shocked quartz.

The search for the mass extinction agent is a lesson about how to study complex questions of evolutionary history. A high-tech, simple solution seldom works. Evolution is an intricate business, a web of interacting forces moving over the globe through geological time. We must start with the low-tech science of digging bones and thinking about frogs.

Frog facts and fossil sequences strongly suggest that the dinosaurs did not die out because of a cosmic impact, but through the biological disaster of invasion and counterinvasion.

APPENDIX 1: SYMBOLS COMMONLY USED IN THIS TEXT

Symbol	Meaning	Symbol	Meaning
Å	Angstrom, a unit of length often used to measure wavelengths of light; $1\text{ Å} = 10^{-8}$ cm	$\Delta\lambda$	The Greek letters delta lambda, used to designate a shift in wavelength, as in the Doppler effect
c	Standard symbol for the speed of light	γ	The Greek letter gamma, sometimes used to designate a gamma-ray photon
G	Standard symbol for the gravitational constant		
H	Symbol for the constant in the Hubble expansion law	λ	The Greek letter lambda, usually used to designate wavelength
h	Standard symbol for the Planck constant		
K	Kelvin, the unit of temperature in the absolute scale	ν	The Greek letter nu, the standard symbol for frequency, also used to designate a neutrino in nuclear reactions
z	Symbol commonly used to designate the Doppler shift ($z = \Delta\lambda/\lambda = v/c$ for velocities much less than the speed of light)		
α	The Greek letter alpha, sometimes used to designate an alpha particle	π	The Greek letter pi, usually used to designate the parallax angle; also used for the ratio of the circumference of a circle to its diameter

APPENDIX 2: PHYSICAL AND MATHEMATICAL CONSTANTS

Constant	Symbol	Value
Speed of light	c	2.9979250×10^{10} cm/sec
Gravitation constant	G	6.6726×10^{-8} dyn cm^2/gm^2
Planck constant	h	6.62608×10^{-27} erg/sec
Electron mass	m_e	9.10939×10^{-28} gm
Proton mass	m_p	1.67262×10^{-24} gm
Stefan-Boltzmann constant	σ	5.67051×10^{-5} erg/cm^{2}deg^{4}sec
Wien constant	W	0.289776 cm/deg
Boltzmann constant	k	1.38066×10^{-16} erg/deg
Astronomical unit	AU	1.49599×10^8 km
Parsec	pc	3.085678×10^{13} km = 3.261633 light-years
Light-year	ly	9.460530×10^{12} km
Solar mass	$M_\odot$	1.9891×10^{33} gm
Solar radius	$R_\odot$	6.9600×10^{10} cm
Solar luminosity	$L_\odot$	3.827×10^{33} erg/sec
Earth mass	$M_\oplus$	5.9742×10^{27} gm
Earth radius	$R_\oplus$	6378.140 km
Tropical year (equinox to equinox)		365.241219878 days
Sidereal year (with respect to stars)		365.256366 days = 3.155815×10^7 sec

APPENDIX 3: THE ELEMENTS AND THEIR ABUNDANCES

Element	Symbol	Atomic No.	Atomic Weight[a]	Abundance[b]	Element	Symbol	Atomic No.	Atomic Weight[a]	Abundance[b]
Hydrogen	H	1	1.0080	1.00	Iodine	I	53	126.9045	4.07×10^{-11}
Helium	He	2	4.0026	0.085	Xenon	Xe	54	131.30	1×10^{-10}
Lithium	Li	3	6.941	1.55×10^{-9}	Cesium	Cs	55	132.905	1.26×10^{-11}
Beryllium	Be	4	9.0122	1.41×10^{-11}	Barium	Ba	56	137.34	6.31×10^{-11}
Boron	B	5	10.811	2.00×10^{-10}	Lanthanum	La	57	138.906	6.46×10^{-11}
Carbon	C	6	12.0111	0.000372	Cerium	Ce	58	140.12	4.37×10^{-11}
Nitrogen	N	7	14.0067	0.000115	Praseodymium	Pr	59	140.908	4.27×10^{-11}
Oxygen	O	8	15.9994	0.000676	Neodymium	Nd	60	144.24	6.61×10^{11}
Fluorine	F	9	18.9984	3.63×10^{-8}	Promethium	Pm	61	146	—
Neon	Ne	10	20.179	3.72×10^{-5}	Samarium	Sm	62	150.4	4.57×10^{-11}
Sodium	Na	11	22.9898	1.74×10^{-6}	Europium	Eu	63	151.96	3.09×10^{-12}
Magnesium	Mg	12	24.305	3.47×10^{-5}	Gadolinium	Gd	64	157.25	1.32×10^{-11}
Aluminum	Al	13	26.9815	2.51×10^{-6}	Terbium	Tb	65	158.925	2.63×10^{-12}
Silicon	Si	14	28.086	3.55×10^{-5}	Dysprosium	Dy	66	162.50	1.29×10^{-11}
Phosphorus	P	15	30.9738	3.16×10^{-7}	Holmium	Ho	67	164.930	3.1×10^{-12}
Sulfur	S	16	32.06	1.62×10^{-5}	Erbium	Er	68	167.26	5.75×10^{-12}
Chlorine	Cl	17	35.453	2×10^{-7}	Thulium	Tm	69	168.934	2.69×10^{-11}
Argon	Ar	18	39.948	4.47×10^{-6}	Ytterbium	Yb	70	170.04	6.46×10^{-12}
Potassium	K	19	39.102	1.12×10^{-7}	Lutetium	Lu	71	174.97	6.92×10^{-12}
Calcium	Ca	20	40.08	2.14×10^{-6}	Hafnium	Hf	72	178.49	6.3×10^{-12}
Scandium	Sc	21	44.956	1.17×10^{-9}	Tantalum	Ta	73	180.948	2×10^{-12}
Titanium	Ti	22	47.90	5.50×10^{-8}	Tungsten	W	74	183.85	3.72×10^{-10}
Vanadium	V	23	50.9414	1.26×10^{-8}	Rhenium	Re	75	186.2	1.8×10^{-12}
Chromium	Cr	24	51.996	5.01×10^{-7}	Osmium	Os	76	190.2	5.62×10^{-12}
Manganese	Mn	25	54.9380	2.63×10^{-7}	Iridium	Ir	77	192.2	1.62×10^{-10}
Iron	Fe	26	55.847	2.51×10^{-5}	Platinum	Pt	78	195.09	5.62×10^{-11}
Cobalt	Co	27	58.9332	3.16×10^{-8}	Gold	Au	79	196.967	2.09×10^{-12}
Nickel	Ni	28	58.71	1.91×10^{-6}	Mercury	Hg	80	200.59	1×10^{-9}
Copper	Cu	29	63.546	2.82×10^{-8}	Thallium	Tl	81	204.37	1.6×10^{-12}
Zinc	Zn	30	65.37	2.63×10^{-8}	Lead	Pb	82	207.19	7.41×10^{-11}
Gallium	Ga	31	69.72	6.92×10^{-10}	Bismuth	Bi	83	208.981	6.3×10^{-12}
Germanium	Ge	32	72.59	2.09×10^{-9}	Polonium	Po	84	210	—
Arsenic	As	33	74.9216	2×10^{-10}	Astatine	At	85	210	—
Selenium	Se	34	78.96	3.16×10^{-9}	Radon	Rn	86	222	—
Bromine	Br	35	79.904	6.03×10^{-10}	Francium	Fr	87	223	—
Krypton	Kr	36	83.80	1.6×10^{-9}	Radium	Ra	88	226.025	—
Rubidium	Rb	37	85.4678	4.27×10^{-10}	Actinium	Ac	89	227	—
Strontium	Sr	38	87.62	6.61×10^{-10}	Thorium	Th	90	232.038	6.61×10^{-12}
Yttrium	Y	39	88.9059	4.17×10^{-11}	Protactinium	Pa	91	230.040	—
Zirconium	Zr	40	91.22	2.63×10^{-10}	Uranium	U	92	238.029	4.0×10^{-12}
Niobium	Nb	41	92.906	2.0×10^{-10}	Neptunium	Np	93	237.048	—
Molybdenum	Mo	42	95.94	7.94×10^{-11}	Plutonium	Pu	94	242	—
Technetium	Tc	43	98.906	—	Americium	Am	95	242	—
Ruthenium	Ru	44	101.07	3.72×10^{-11}	Curium	Cm	96	245	—
Rhodium	Rh	45	102.905	3.55×10^{-11}	Berkelium	Bk	97	248	—
Palladium	Pd	46	106.4	3.72×10^{-11}	Californium	Cf	98	252	—
Silver	Ag	47	107.868	4.68×10^{-12}	Einsteinium	Es	99	253	—
Cadmium	Cd	48	112.40	9.33×10^{-12}	Fermium	Fm	100	257	—
Indium	In	49	114.82	5.13×10^{-11}	Mendelevium	Md	101	257	—
Tin	Sn	50	118.69	5.13×10^{-11}	Nobelium	No	102	255	—
Antimony	Sb	51	121.75	5.62×10^{-12}	Lawrencium	Lr	103	256	—
Tellurium	Te	52	127.60	1×10^{-10}					

[a]The atomic weight of an element is its mass in *atomic mass units*. An atomic mass unit is defined as one-twelfth of the mass of the most common isotope of carbon, and has the value 1.660531×10^{-24} grams. In general, the atomic weight of an element is approximately equal to the total number of protons and neutrons in its nucleus.

[b]The abundances are given in terms of the number of atoms of each element compared to hydrogen, and are based on the composition of the Sun. For very rare elements, particularly those toward the end of the list, the abundances can be quite uncertain, and the values given should not be considered exact.

APPENDIX 4: TEMPERATURE SCALES

At the most basic level, temperature can be defined in terms of the motion of particles in a gas (or a solid, or a liquid). We all have an intuitive idea of what heat is, and we are all familiar with at least one scale for measuring temperature.

The most commonly used scales are somewhat arbitrarily defined, with zero points not representing any truly fundamental physical basis. The popular *Fahrenheit* scale, for example, has water freezing at a temperature of 32°F and boiling at 212°F. On this scale absolute zero, the lowest possible temperature (where all molecular motions cease), is −459°F.

The centigrade (or Celsius) scale is perhaps better founded, although it is based on the freezing and boiling points of water, rather than the more fundamental absolute zero. In this system, the freezing point is defined as 0°C, and 100°C is the boiling point. This scale has the advantage over the Fahrenheit scale that there are exactly 100° between the freezing and boiling points, rather than 180°, as in the Fahrenheit system. To con-

vert from Fahrenheit to centigrade, we must subtract 32° first, and then multiply the remainder by 100/180, or 5/9. For example, 50°F is equal to 5/9 × (50 − 32) = 10°C. To convert from centigrade to Fahrenheit, first multiply by 9/5 and then add 32°. Thus, −10°C = 9/5 × (−10 + 32) = 14°F. On the centigrade scale, absolute zero occurs at −273°C.

The temperature scale preferred by scientists is a modification of the centigrade system. In this system, named after its founder, the British physicist Lord Kelvin, the same degree is used as in the centigrade scale; that is, one degree is equal to one one-hundredth of the difference between the freezing and boiling points of water. The zero point is different from the zero point on the centigrade scale, however; it is equal to absolute zero. Hence on this scale water freezes at 273°K and boils at 373°K. Comfortable room temperature is around 300°K. In modern usage the degree symbol (°) is dropped, and we speak simply of temperatures in units of Kelvins (273 K, for example).

APPENDIX 5: RADIATION LAWS

Several laws described in Chapter 6 apply to continuous radiation from hot objects such as stars. In the text these laws were discussed in general terms, and a few simple applications were explained. Here the same laws are given in more precise mathematical form, and their use in that form is illustrated.

Wien's Law

In general terms, Wien's law says that the wavelength of maximum emission from a glowing object is inversely proportional to its temperature. Mathematically this can be written as

$$\lambda_{max} \propto 1/T,$$

where λ_{max} is the wavelength of strongest emission, T is the surface temperature of the object, and $\propto$ is a special mathematical symbol meaning "is proportional to."

Experimentation can determine the *proportionality constant*, specifying the exact relationship between λ_{max} and T, and Wien did this, finding

$$\lambda_{max} = W/T,$$

where W has the value 0.29 if λ_{max} is measured in centimeters and T in degrees absolute. With this equation it is possible to calculate λ_{max}, given T, or vice versa. Thus, if we measure the spectrum of a star and find that it emits most strongly at a wavelength of 2,000 Å = 2×10^{-5} cm, then we can solve for the temperature

$$T = W/\lambda_{max} = 0.29/2 \times 10^{-5} = 14{,}500 \text{ K}.$$

This is a relatively hot star, and it would appear blue-white to the eye. Note that it was necessary to measure the spectrum in ultraviolet wavelengths in order to find λ_{max}.

When solving problems using Wien's law, it is always possible to use the equation form, as we have just done in this example. Often, however, it is more convenient to compare the properties of two objects by considering the ratio of the temperatures or of the wavelengths of maximum emission. In effect this is what we did in the text when we compared two objects of different temperatures in order to determine how their λ_{max} values compared. For example, we said that if one object is twice as hot as another, its value of λ_{max} is half that of the other.

We can see how this ratio technique works by writing the equation for Wien's law separately for object 1 and object 2:

$$\lambda_{max1} = W/T_1, \quad \text{and} \quad \lambda_{max2} = W/T_2.$$

Now we can divide one equation by the other:

$$\frac{\lambda_{max1}}{\lambda_{max2}} = \frac{W/T_1}{W/T_2}$$

or

$$\frac{\lambda_{max1}}{\lambda_{max2}} = \frac{T_2}{T_1}.$$

The numerical factor W has canceled out, and we are left with a simple expression relating the values of λ_{max} and T for the two objects. Now we see that if $T_1 = 2T_2$ (object 1 is twice as hot as object 2), then

$$\frac{\lambda_{max1}}{\lambda_{max2}} = \frac{T_2}{2T_2} = \frac{1}{2},$$

or λ_{max} for object 1 is one-half that for object 2. In this extremely simple example, it probably would have been easier just to work it out in our heads, but what if we have a case where one object is 3.368 times hotter than the other, for example?

A great deal can be learned from making comparisons in this way. Astronomers often use the Sun as the standard for comparison, expressing various quantities in terms of the solar values. Another reason for using comparisons occurs when the numerical constants (such as Wien's constant in the foregoing examples) are not known. If the trick of comparing is kept in mind, it is often possible to work out answers to astronomical questions simply by carrying around in one's head a few numbers describing the Sun.

The Stefan-Boltzmann Law

As discussed in the text, the energy emitted by a glowing object is proportional to T^4, where T is the surface temperature. This can be written mathematically as

$$E = \sigma T^4,$$

where σ stands for a proportionality constant that has the value 5.7×10^{-5}, if the centimeter-gram-second metric units are used.

If we now consider how much surface area a star has, then the total energy it emits, called its *luminosity* and usually denoted L, is

$$L = \text{surface area} \times E$$
$$= 4\pi R^2 \sigma T^4,$$

where R is the radius of the star. This equation is called the Stefan-Boltzmann law.

As in other cases, the law can be used directly in this form, or we can choose to compare properties of stars by writing the equation separately for two stars and then dividing. If we do this, we find

$$\frac{L_1}{L_2} = \frac{R_1^2 T_1^4}{R_2^2 T_2^4} = \left(\frac{R_1}{R_2}\right)^2 \left(\frac{T_1}{T_2}\right)^4$$

The constant factors 4π and σ have canceled out.

As an example of how to use this expression, suppose we determine that a particular star has twice the temperature of the Sun but only one-half the radius, and we wish to know how this star's luminosity compares to that of the Sun. If we designate the star as object 1 and the Sun as object 2, then

$$\frac{L_1}{L_2} = \left(\frac{1}{2}\right)^2 (2)^4 = \frac{1}{4} \times 16 = 4$$

This star has 4 times the luminosity of the Sun.

The Stefan-Boltzmann law is particularly useful because it relates three of the most important properties of stars to each other.

The Planck Function

The radiation laws described above and in the text are actually specific forms of a much more general law, discovered by the great German physicist Max Planck. Wien's law, the Stefan-Boltzmann law, and some others not mentioned in the text bear the same kind of relation to Planck's law as the laws of planetary motion discovered by Kepler do to Newton's mechanics. Kepler's laws were first discovered by observation, but with Newton's laws of motion it is possible to derive Kepler's laws theoretically. In the same fashion the radiation laws discussed so far were found experimentally but can be derived mathematically from the much more general and powerful Planck's law.

Planck's law is usually referred to as the Planck function, a mathematical relationship between intensity and wavelength (or frequency) that describes the spectrum of any glowing object at a given temperature. The Planck function specifically applies only to objects that radiate solely because of their temperature and do not reflect any light or have any spectral lines. The popular term for such an object is *blackbody,* and we often refer to radiation from such an object as *blackbody radiation* or, more commonly, *thermal radiation,* the term used in the text. Stars are not perfect radiators, but to a good approximation can be treated as such; hence in the text we apply Wien's law and the Stefan-Boltzmann law to stars (and even planets in certain circumstances) without pointing out the fact that to do so is only approximately correct.

The form of the Planck function for the radiation intensity B as a function of wavelength is

$$B = \frac{2hc^2}{\lambda^5} \frac{1}{e^{hc/\lambda kT} - 1},$$

where h is the Planck constant, c is the speed of light, and k is the Boltzmann constant (the values of all three are tabulated in Appendix 2), λ is the wavelength (in cm), and T is the temperature of the object (on the absolute scale). The symbol e represents the base of the natural logarithm, something not used elsewhere in this text; for the present purpose, this may be regarded simply as a mathematical constant with the value 2.718.

In terms of frequency v rather than wavelength, the expression is

$$B = \frac{2hv^3}{c^2} \frac{1}{e^{hv/kT} - 1}.$$

Either expression may be used to calculate the spectrum of continuous radiation from a glowing object at a specific temperature. In practice the Planck function is used in a wide assortment of theoretical calculations that call for knowledge of the intensity of radiation so that its effects can be assessed on physical conditions such as ionization.

APPENDIX 6: MAJOR OPTICAL TELESCOPES OF THE WORLD (2 Meters or Larger)

Observatory	Location[a]	Telescope
European Southern Observatory	(Cerro La Silla, Chile)	16-m *Very Large Telescope* (four 8-m instruments)
Keck Observatory (California Institute of Technology and the University of California)	(Mauna Kea, Hawaii)	10.0-m Keck Telescope
National Optical Astronomy Observatories	(Mauna Kea, Hawaii)	8-m
National Optical Astronomy Observatories	(Cerro Tololo, Chile)	8-m
Ohio State University, University of Arizona	(Mt. Graham, Arizona)	*Columbus:* two 8-m instruments
Special Astrophysical Observatory	Mount Pastukhov, USSR	6-m Bol'shoi Teleskop Azimutal'nyi
Hale Observatories	Palomar Mountain, California	5.08-m George Ellery Hale Telescope
Smithsonian Observatory	Mount Hopkins, Arizona	4.5-m Multiple Mirror Telescope
Royal Greenwich Observatory	La Palma, Canary Islands	4.2-m William Herschel Telescope
Cerro Tololo Observatory	Cerro Tololo, Chile	4.0-m
Anglo-Australian Observatory	Siding Spring Mountain, Australia	3.9-m Anglo-Australian Telescope
Kitt Peak National Observatory	Kitt Peak, Arizona	3.8-m Nicholas U. Mayall Telescope
Royal Observatory Edinburgh	Mauna Kea, Hawaii	3.8-m United Kingdom Infrared Telescope
Canada-France-Hawaii Observatory	Mauna Kea, Hawaii	3.6-m Canada-France-Hawaii Telescope
European Southern Observatory	Cerro La Silla, Chile	3.57-m
Max Planck Institute (Bonn)	Calar Alto, Spain	3.5-m
Astronomical Research Corporation	(Sacremento Peak, New Mexico)	3.5-m
European Southern Observatory	Cerro La Silla, Chile	3.5-m New Technology Telescope
Lick Observatory	Mount Hamilton, California	3.05-m C. Donald Shane Telescope
Mauna Kea Observatory	Mauna Kea, Hawaii	3.0-m NASA Infrared Telescope
McDonald Observatory	Mount Locke, Texas	2.7-m
Haute Provence Observatory	Saint Michele, France	2.6-m
Crimean Astrophysical Observatory	Simferopol, USSR	2.6-m
Byurakan Astrophysical Observatory	Byurakan, USSR	2.6-m
Mount Wilson and Las Campanas Observatories	Mount Wilson, California	2.5-m Hooker Telescope
Mount Wilson and Las Campanas Observatories	Las Campanas, Chile	2.5-m Irenée du Pont Telescope
Royal Greenwich Observatory	Canary Islands	2.5-m Isaac Newton Telescope
Dartmouth College and University of Michigan	Kitt Peak, Arizona	2.4-m McGraw-Hill Telescope
Wyoming Infrared Observatory	Mount Jelm, Wyoming	2.3-m Wyoming Infrared Telescope
Steward Observatory	Kitt Peak, Arizona	2.3-m Wyoming Infrared Telescope
Mauna Kea Observatory	Mauna Kea, Hawaii	2.2-m
Max Planck Institute (Bonn)	Calar Alto, Spain	2.2-m
Max Planck Institute	(South West Africa)	2.2-m
University of Mexico Observatory	(Mexico)	2.16-m
La Plata Observatory	La Plata Argentina	2.15-m
Kitt Peak National Observatory	Kitt Peak, Arizona	2.1-m
McDonald Observatory	Mount Locke, Texas	2.1-m Otto Struve Telescope
Karl Schwarzschild Observatory	Tautenberg, East Germany	2.0-m (largest Schmidt telescope)

[a]Telescopes planned or under construction have their locations indicated in parentheses.

APPENDIX 7: PLANETARY AND SATELLITE DATA

Orbital Data for the Planets

Planet	Sidereal Period	Semimajor Axis	Orbital Eccentricity[a]	Inclination of Orbital Plane	Rotation Period	Tilt of Axis
Mercury	0.241 yr	0.387 AU	0.2056	7°0'15"	58^{d}65	28°
Venus	0.615	0.723	0.068	3°23'40"	243	3°
Earth	1.000	1.000	0.0167	0°0'0"	23^{h}56^m	23°27'
Mars	1.881	1.524	0.0934	1°51'0"	24^{h}37^m	23°59'
Jupiter	11.86	5.203	0.0485	1°18'17"	9^{h}55^{m}5	3°5'
Saturn	29.46	9.555	0.0556	2°29'33"	10^{h}39^{m}4	26°44'
Uranus	84.01	19.22	0.0472	0°46'23"	17^{h}14^{m}4	97°52'
Neptune	164.79	30.11	0.0086	1°46'22"	1606.6	29°34'
Pluto	248.5	39.44	0.250	17°10'12"	6^{d}387	122.5°

[a]The eccentricity of an orbit is defined as the ratio of the distance between foci to the semimajor axis. In practice it is related to the perihelion distance P and the semimajor axis a by $P = a(1 - e)$, where e is the eccentricity; and to the aphelion distance A by $A = a(1 + e)$.

Physical Data for the Planets

Planet	Mass[a]	Average, 1 Bar Diameter[a]	Density	Surface Gravity	Escape Speed	1 Bar Temperature	Albedo
Mercury	0.0558	0.381	5.50 g/cc	0.38 g	4.3 km/sec	100–700 K	0.106
Venus	0.815	0.951	5.3	0.90	10.3	730	0.65
Earth	1.000	1.000	5.518	1.00	11.2	200–300	0.37
Mars	0.107	0.531	3.96	0.38	5.0	130–290	0.15
Jupiter	317.89	10.85	1.327	2.64	60.0	165	0.52
Saturn	95.184	8.99	0.688	1.13	36.0	134	0.47
Uranus	14.536	3.96	1.272	0.89	21.2	76	0.50
Neptune	17.148	3.85	1.640	1.13	23.5	74	0.5:
Pluto	0.0022	0.18	1.7	0.06	1.2	40	0.06:
							0.44–0.61

[a]The masses and diameters are given in units of the Earth's mass and diameter, which are 5.974×10^{27} grams and 12,734 km, respectively.

Satellites

Planet	Satellite	Semimajor Axis	Period	Diameter	Mass	Density
Earth	Moon	3.84×10^5 km	27^{d}322	3476 km	7.35×10^{25}	3.34 g/cm^3
Mars	Phobos	9.38×10^3	0.3189	27 × 22 × 19	9.6×10^{18}	2:
	Deimos	2.35×10^4	1.2624	15 × 12 × 11	1.9×10^{18}	2:
Jupiter	Metis	1.280×10^5	0.295	20:	9.5×10^{19}	
	Adrastea	1.290×10^5	0.298	20 × 20 × 15	1.9×19^{19}	
	Amalthea	1.81×10^5	0.498	270 × 170 × 150	7.2×10^{21}	
	Thebe	2.22×10^5	0.675	110 × 90	7.6×10^{20}	
	Io	4.22×10^5	1.769	3630	8.92×10^{25}	3.53
	Europa	6.71×10^5	3.551	3138	4.87×10^{25}	3.03
	Ganymede	1.07×10^6	7.155	5262	1.49×10^{26}	1.93
	Callisto	1.88×10^6	16.689	4800	1.08×10^{26}	1.70
	Leda	1.11×10^7	238.7	16:	5.7×10^{18}	

(continued on next page)

APPENDIX 7: PLANETARY AND SATELLITE DATA, *Continued*

Satellites

Planet	Satellite	Semimajor Axis	Period	Diameter	Mass	Density
Jupiter	Himalia	1.15×10^7	250.6	186:	9.5×10^{21}	
	Lysithea	1.17×10^7	259.2	36:	7.6×10^{19}	
	Elara	1.18×10^7	259.7	76:	7.6×10^{20}	
	Ananke	2.12×10^7	631	30:	3.8×10^{19}	
	Carme	2.26×10^7	692	40:	9.5×10^{19}	
	Pasiphae	2.35×10^7	735	50:	1.9×10^{20}	
	Sinope	2.37×10^7	758	36:	7.6×10^{19}	
Saturn	1981 S13 (Pan)	1.33×10^5	0.573	20:		
	Atlas	1.377×10^5	0.602	40×20		
	Prometheus	1.394×10^5	0.613	$140 \times 100 \times 80$		
	Pandora	1.417×10^5	0.629	$110 \times 90 \times 80$		
	Epimetheus	1.514×10^5	0.694	$140 \times 120 \times 100$		
	Janus	1.514×10^5	0.695	$220 \times 200 \times 160$		
	Mimas	1.855×10^5	0.942	392	4.5×10^{22}	1.43
	Enceladus	2.381×10^5	1.370	500	7.4×10^{22}	1.13
	Telesto	2.947×10^5	1.888	$34 \times 28 \times 26$		
	Calypso	2.947×10^5	1.888	$34 \times 22 \times 22$		
	Tethys	2.947×10^5	1.888	1060	7.4×10^{23}	1.19
	Dione	3.774×10^5	2.737	1120	1.05×10^{24}	1.43
	Helene	3.781×10^5	2.737	$36 \times 32 \times 30$		
	Rhea	5.271×10^5	4.518	1530	2.50×10^{24}	1.33
	Titan	1.222×10^6	15.945	5150	1.35×10^{26}	1.89
	Hyperion	1.481×10^6	21.277	$410 \times 260 \times 220$	1.71×10^{22}	
	Iapetus	3.561×10^6	79.330	1460	1.88×10^{24}	1.15
	Phoebe	1.295×10^7	550.5	220		
Uranus	Cordelia	4.97×10^4	0.336	40		
	Ophelia	5.38×10^4	0.377	50		
	Bianca	5.92×10^4	0.435	50		
	Cressidia	6.18×10^4	0.465	60		
	Nesdimona	6.27×10^4	0.476	60		
	Juliet	6.46×10^4	0.494	80		
	Portia	6.61×10^4	0.515	80		
	Rosalind	6.99×10^4	0.560	60		
	Belinda	7.53×10^4	0.624	60		
	Puck	8.60×10^4	0.764	170		
	Miranda	1.298×10^5	1.413	484	7.5×10^{22}	1.26
	Ariel	1.912×10^5	2.520	1160	1.4×10^{24}	1.65
	Umbriel	2.660×10^5	4.144	1190	1.3×10^{24}	1.44
	Titania	4.358×10^5	8.706	1610	3.5×10^{24}	1.59
	Oberon	5.826×10^5	13.463	1550	2.9×10^{24}	1.50
Neptune	1989N6	4.80×10^4	0.296	54		
	1989N5	5.00×10^4	0.313	80		
	1989N3	5.25×10^4	0.333	150		
	1989N4	6.20×10^4	0.429	180		
	1989N2	7.36×10^4	0.554	190		
	1989N1	1.176×10^5	1.121	400		
	Triton	3.548×10^5	5.875	2705	2.21×10^{25}	2.07
	Nereid	5.5134×10^6	360.129	340	2.1×10^{22}	
Pluto	Charon	1.964×10^4	6.387	1186		

APPENDIX 8: STELLAR DATA

The Fifty Brightest Stars

Star		Spectral Type	Apparent Magnitude	Distance	Position (1980) Right Ascension	Declination
α Eri	Achernar	B3 V	0.51	36 pc	01ʰ37ᵐ.0	− 57°20′
α UMi	Polaris	F8 Ib	1.99	208	02 12.5	+ 89 11
α Per	Mirfak	F5 Ib	1.80	175	03 22.9	+ 49 47
α Tau	Aldebaran	K5 III	0.86	21	04 34.8	+ 16 28
β Ori	Rigel	B8 Ia	0.14	276	05 13.6	− 08 13
α Aur	Capella	G8 III	0.05	14	05 15.2	+ 45 59
γ Ori	Bellatrix	B2 III	1.64	144	05 24.0	+ 06 20
β Tau	Elnath	B7 III	1.65	92	05 25.0	+ 28 36
ε Ori	Alnilam	B0 Ia	1.70	490	05 35.2	− 01 13
ζ Ori	Alnitak	09.5 Ib	1.79	490	05 39.7	− 01 57
α Ori	Betelgeuse	M2 Iab	0.41	159	05 54.0	+ 07 24
β Aur	Menkalinan	A2 V	1.86	27	05 58.0	+ 44 57
β CMa		B1 II-III	1.96	230	06 21.8	− 17 56
α Car	Canopus	F0 Ib-II	− 0.72	30	06 23.5	− 52 41
γ Gem	Alhena	A0 IV	1.93	32	06 36.6	+ 16 25
α CMa	Sirius	A1 V	− 1.47	2.7	06 44.2	− 16 42
ε CMa	Adhara	B2 II	1.48	209	06 57.8	− 28 57
δ CMa		F8 Ia	1.85	644	07 07.6	− 26 22
α Gem	Castor	A1 V	1.97	14	07 33.3	+ 31 56
α CMi	Procyon	F5 IV-V	0.37	3.5	07 38.2	+ 05 17
β Gem	Pollux	K0 III	1.16	11	07 44.1	+ 28 05
γ Vel		WC8	1.83	160	08 08.9	− 47 18
ε Car	Avior	K3 III?	1.90	104	08 22.1	− 59 26
ζ Vel		A2 V	1.95	23	08 44.2	− 54 38
β Car	Miaplacidus	A1 III	1.67	26	09 13.0	− 69 38
α Hya	Alphard	K4 III	1.98	29	09 26.6	− 08 35
α Leo	Regulus	B7 V	1.36	26	10 07.3	+ 12 04
γ Leo		K0 III	1.99	28	10 18.8	+ 19 57
α UMa	Dubhe	K0 III	1.81	32	11 02.5	+ 61 52
α Cru A ⎫	Acrux	B0.5 IV	1.39	114	12 25.4	− 62 59
α Cru B ⎬		B1 V	1.86	114	12 15.4	− 62 59
γ Cru	Gacrux	M4 III	1.69	67	12 30.1	− 57 00
β Cru		B0.5 III	1.28	150	12 46.6	− 59 35
ε UMa	Alioth	A0p	1.79	21	12 53.2	+ 56 04
α Vir	Spica	B1 V	0.91	67	13 24.1	− 11 03
η UMa	Alkaid	B3 V	1.87	64	13 46.8	+ 49 25
β Cen	Hadar	B1 III	0.63	150	14 02.4	− 60 16
α Boo	Arcturus	K2 III	− 0.06	11	14 14.8	+ 19 17
α Cen A ⎫	Rigil Kentaurus	G2 V	0.01	1.3	14 38.4	− 60 46
α Cen B ⎬		K4 V	1.40	1.3	14 38.4	− 60 46
α Sco	Antares	M1 Ib	0.92	160	16 28.2	− 26 23
α TrA	Atria	K2 Ib	1.93	25	16 46.5	− 68 60
λ Sco	Shaula	B1 V	1.60	95	17 32.3	− 37 05
θ Sco		F0 Ib	1.86	199	17 35.9	− 42 59
ε Sgr	Kaus Australis	B9.5 III	1.81	38	18 22.9	− 34 24
α Lyr	Vega	A0 V	0.04	8	18 36.2	+ 38 46
α Aql	Altair	A7 IV-V	0.77	5	19 49.8	+ 08 49
α Pav	Peacock	B2.5 V	1.95	95	20 24.1	− 56 48
α Cyg	Deneb	A2 Ia	1.26	491	20 40.7	+ 45 12
α Gru	Al Na'ir	B7 IV	1.76	20	22 06.9	− 47 04
α PsA	Fomalhaut	A3 V	1.15	7	22 56.5	− 29 44

(continued on next page)

APPENDIX 8: STELLAR DATA, *Continued*

Nearby Stars (Within 5 Parsecs of the Sun)[a]

Star	Spectral Type	Visual Apparent Magnitude	Visual Absolute Magnitude	Parallax	Distance	Proper Motion
α Centauri	G2V	−0.1	4.8	0.753″	1.33 pc	3.68″/yr
Barnard's star	M5V	9.5	13.2	0.544	1.84	10.31
Wolf 359	M8V3	13.5	16.7	0.432	2.31	4.71
BD + 36°2147	M2V	7.5	10.5	0.400	2.50	4.78
Luyten 726-8	M6Ve	12.5	15.4	0.385	2.60	3.36
Sirius	A1V	−1.5	1.4	0.377	2.65	1.33
Ross 154	M5Ve	10.6	13.3	0.345	2.90	0.72
Ross 248	M6Ve	12.3	14.8	0.319	3.13	1.58
ε Eridani	K2V	3.7	6.1	0.305	3.28	0.98
Ross 128	M5V	11.1	13.5	0.302	3.31	1.37
Luyten 789-6	M6V	12.2	14.6	0.302	3.31	3.26
61 Cygni	K5Ve	5.2	7.5	0.292	3.42	5.22
ε Indi	K5Ve	4.7	7.0	0.291	3.44	4.69
τ Ceti	G8V	3.5	5.9	0.289	3.46	1.92
Procyon	F5V	0.4	2.7	0.285	3.51	1.25
Σ 2398	M4V	8.9	11.2	0.284	3.52	2.28
BD + 43°44	M1Ve	8.1	10.4	0.282	3.55	2.89
CD − 36°15693	M2Ve	7.4	9.6	0.279	3.58	6.90
G51-15		14.8	17.0	0.273	3.66	1.26
L725-32	M5Ve	11.5	13.6	0.264	3.79	1.22
BD + 5°1668	M4V	9.8	12.0	0.264	3.79	3.73
CD − 39°14192	M0Ve	6.7	8.8	0.260	3.85	3.46
Kapteyn's star	M0V	8.8	10.8	0.256	3.91	8.89
Kruger 60	M4V	9.7	11.7	0.254	3.94	0.86
Ross 614	M5Ve	11.3	13.3	0.243	4.12	0.99
BD − 12°4523	M5V	10.0	11.9	0.238	4.20	1.18
Wolf 424	M6Ve	13.2	15.0	0.234	4.27	1.75
van Maanen's star	W.D.	12.4	14.2	0.232	4.31	2.95
CD − 37°15492	M3V	8.6	10.4	0.225	4.44	6.08
Luyten 1159-16	M8V	12.3	14.0	0.221	4.52	2.08
BD + 50°1725	K7V	6.6	8.3	0.217	4.61	1.45
CD − 46°11540	M4V	9.4	11.1	0.216	4.63	1.13
CD − 49°13515	M3V	8.7	10.4	0.214	4.67	0.81
CD − 44°11909	M5V	11.2	12.8	0.213	4.69	1.16
BD + 68°946	M3.5V	9.1	10.8	0.213	4.69	1.33
G158-27		13.7	15.5	0.212	4.72	2.06
G208-44/45		13.4	15.0	0.210	4.76	0.75
BD − 15°6290	M5V	10.2	11.8	0.209	4.78	1.16
40 Eridani	K0V	4.4	6.0	0.207	4.83	4.08
L145-141	W.D.	11.4	12.6	0.206	4.85	2.68
BD + 20°2465	M4.5V	9.4	10.9	0.203	4.93	0.49
70 Ophiuchi	K1V	4.2	5.7	0.203	4.93	1.13
BD + 43°4305	M4.5Ve	10.2	11.7	0.200	5.00	0.83

[a]Data from Lippencott, L. S., 1978, *Space Science Reviews* 22:153. Many of the stars listed here are multiple systems; in these cases, the data in the table refer only to the brightest member of the system.

APPENDIX 9: THE CONSTELLATIONS

Name	Genitive	Abbreviation	Position Right Ascension	Declination
Andromeda	Andromedae	And	01^h	+40°
Antlia	Antliae	Ant	10	−35
Apus	Apodis	Aps	16	−75
Aquarius	Aquarii	Aqr	23	−15
Aquila	Aquilae	Aql	20	+05
Ara	Arae	Ara	17	−55
Aries	Arietis	Ari	03	+20
Auriga	Aurigae	Aur	06	+40
Bootes	Bootis	Boo	15	+30
Caelum	Caeli	Cae	05	−40
Camelopardalis	Camelopardalis	Cam	06	−70
Cancer	Cancri	Cnc	09	+20
Canes Venatici	Canum Venaticorum	CVn	13	+40
Canis Major	Canis Majoris	CMa	07	−20
Canis Minor	Canis Minoris	CMi	08	+05
Capricornus	Capricorni	Cap	21	−20
Carina	Carinae	Car	09	−60
Cassiopeia	Cassiopeiae	Cas	01	+60
Centaurus	Centauri	Cen	13	−50
Cepheus	Cephei	Cep	22	+70
Cetus	Ceti	Cet	02	−10
Chamaeleon	Chamaeleonis	Cha	11	−80
Circinis	Circini	Cir	15	−60
Columba	Columbae	Col	06	−35
Coma Berenices	Comae Berenices	Com	13	+20
Corona Australis	Coronae Australis	CrA	19	−40
Coronoa Borealis	Coronae Borealis	CrB	16	+30
Corvus	Corvi	Crv	12	−20
Crater	Crateris	Crt	11	−15
Crux	Crucis	Cru	12	−60
Cygnus	Cygni	Cyg	21	+40
Delphinus	Delphini	Del	21	+10
Dorado	Doradus	Dor	05	−65
Draco	Draconis	Dra	17	+65
Equuleus	Equulei	Equ	21	+10
Eridanus	Eridani	Eri	03	−20
Fornax	Fornacis	For	03	−30
Gemini	Geminorum	Gem	07	+20
Grus	Gruis	Gru	22	−45
Hercules	Herculis	Her	17	+30
Horologium	Horologii	Hor	03	−60
Hydra	Hydrae	Hya	10	−20
Hydrus	Hydri	Hyi	02	−75
Indus	Indi	Ind	21	−55
Lacerta	Lacertae	Lac	22	+45
Leo	Leonis	Leo	11	+15
Leo Minor	Leonis Minoris	LMi	10	+35
Lepus	Leporis	Lep	06	−20
Libra	Librae	Lib	15	−15
Lupus	Lupi	Lup	15	−45
Lynx	Lincis	Lyn	08	+45
Lyra	Lyrae	Lyr	19	+40

(continued on next page)

APPENDIX 9: THE CONSTELLATIONS, *Continued*

Name	Genitive	Abbreviation	Right Ascension	Declination
Mensa	Mensae	Men	05	−80
Microscopium	Microscopii	Mic	21	−35
Monoceros	Monocerotis	Mon	07	−05
Musca	Muscae	Mus	12	−70
Norma	Normae	Nor	16	−50
Octans	Octantis	Oct	22	−85
Ophiuchus	Ophiuchi	Oph	17	00
Orion	Orionis	Ori	05	+05
Pavo	Pavonis	Pav	20	−65
Pegasus	Pegasi	Peg	22	+20
Perseus	Persei	Per	03	+45
Phoenix	Phoenicis	Phe	01	−50
Pictor	Pictoris	Pic	06	−55
Pisces	Piscium	Psc	01	+15
Piscis Austrinus	Piscis Austrini	PsA	22	−30
Puppis	Puppis	Pup	08	−40
Pyxis	Pyxidis	Pyx	09	−30
Reticulum	Reticuli	Ret	04	−60
Sagitta	Sagittae	Sge	20	+10
Sagittarius	Sagittarii	Sgr	19	−25
Scorpius	Scorpii	Sco	17	−40
Sculptor	Sculptoris	Scl	00	−30
Scutum	Scuti	Sct	19	−10
Serpens	Serpentis	Ser	17	00
Sextans	Sextantis	Sex	10	00
Taurus	Tauri	Tau	04	+15
Telescopium	Telescopii	Tel	19	−50
Triangulum	Trianguli	Tri	02	+30
Triangulum Australe	Trianguli Australi	TrA	16	−65
Tucana	Tucanae	Tuc	00	−65
Ursa Major	Ursae Majoris	UMa	11	+50
Ursa Minor	Ursae Minoris	UMi	15	+70
Vela	Velorum	Vel	09	−50
Virgo	Virginis	Vir	13	00
Volans	Volantis	Vol	08	−70
Vulpecula	Vulpeculae	Vul	20	+25

APPENDIX 10: MATHEMATICAL TREATMENT OF STELLAR MAGNITUDES

Logarithmic Representation

In the text, magnitudes are discussed in terms of the brightness ratios between stars of different magnitudes. We generally avoided discussion of cases where two stars differ by a fraction of a magnitude, because in such cases it is no longer simple to calculate the brightness ratio corresponding to the magnitude difference. If star 1 is 0.5 magnitudes brighter than star 2, for example, what is the brightness ratio? Or if star 1 is a factor of 48.76 fainter than star 2, what is the difference in magnitudes?

Astronomers use an exact mathematical relationship between magnitude differences and brightness ratios, written as

$$m_1 - m_2 = 2.5 \log (b_2/b_1),$$

where m_1 and m_2 are the magnitudes of two stars, and

b_2/b_1 is the ratio of their brightnesses. The notation "log (b_2/b_1)" means the **logarithm** of this ratio; a logarithm is the power to which 10 must be raised to give this ratio. Hence, for example, if $b_2/b_1 = 100$, log $(b_2/b_1) = \log (10^2) = 2$, because 10 must be raised to the second power to give 100. The magnitude difference is 2.5 log $(100) = 2.5 \times 2 = 5$. Similarly, if $b_2/b_1 = 0.001$, then log $(b_2/b_1) = \log (0.001) = \log (10^{-3}) = -3$, and in this case the magnitude difference is 2.5 $\times -3 = -7.5$ (the minus sign indicates that star 1 is brighter than star 2 in this example).

The method works equally well in cases where the power of 10 is not a whole number, as in the example where $b_2/b_1 = 48.76$. Here log $(b_2/b_1) = \log (48.76) = 1.69$ (this is usually found by consulting tables of logarithms or by using a scientific calculator). In this example, the magnitude difference is 2.5 log $(b_2/b_1) = 2.5 \log(48.76) = 2.5 \times 1.69 = 4.23$, so star 1 is 4.23 magnitudes fainter than star 2..

The equation can be used in other ways as well; solving for b_2/b_1 yields

$$b_2/b_1 = 10^{(m_1 - m_2)/2.5}$$
$$= 10^{0.4(m_1 - m_2)}.$$

Thus, if we know that the magnitudes of two stars differ by $m_1 - m_2$, we multiply this difference by 0.4 and raise 10 to the power $0.4(m_1 - m_2)$, again using a calculator or tables, to get the brightness ratio b_2/b_1. As a simple example, suppose $m_1 - m_2 = 5$; then $0.4(m_1 - m_2) = 0.4 \times 5 = 2$, and $10^{0.4(m_1 - m_2)} = 10^2 = 100$, as we knew it should. As a more complex example, consider the stars Betelgeuse (magnitude $+0.41$) and Deneb (magnitude $+1.26$). From the equation above, we see that Betelgeuse is $10^{0.4(1.26 - 0.41)} = 10^{0.34} = 2.19$ times brighter than Deneb.

While in most cases it is possible to follow the discussions of magnitudes and brightness ratios in the text without using this exact mathematical technique, it is still useful to be familiar with it.

The Distance Modulus

Whenever we know both the apparent and absolute magnitudes of a star, a comparison of the two will give its distance. In the text we have seen how to make this calculation by the following several steps:

1. Convert the difference $m - M$ between apparent and absolute magnitude into a brightness ratio, that is, a numerical factor indicating how much brighter or fainter the star would appear at 10 parsecs distance than at its actual distance.

2. Using the inverse square law, determine the change in distance required to produce this change in brightness.

3. Multiply this distance factor by 10 parsecs to find the distance to the star.

To do calculations mentally in this way can be laborious, especially in cases where the magnitude difference does not correspond neatly to a simple numerical factor, as it did in the examples given in the text. Hence astronomers use a mathematical equation expressing the relationship between distance and the distance modulus $m - M$. This equation, which works equally well for all cases, is

$$d = 10^{1 + .2(m - M)},$$

where d is the distance in parsecs to a star whose apparent magnitude is m and whose absolute magnitude is M.

In a simple example, where $m = 9$ and $M = -6$, we have

$$d = 10^{1 + .2(15)}$$
$$= 10^{1 + 3}$$
$$= 10^4 \text{ parsecs.}$$

Now let's try a more complex case. Suppose the star is an M2 main-sequence star, so that $M = 13$, as found from the H-R diagram. The apparent magnitude is $m = 16$. Our equation tells us that

$$d = 10^{1 + .2(16 - 13)}$$
$$= 10^{1 + .6}$$
$$= 10^{1.6}$$
$$= 39.8 \text{ parsecs.}$$

It is necessary to use a slide rule, calculator, or mathematical table to do this, of course, but it is still relatively straightforward compared to following the mental steps outlined above.

The Effect of Extinction On the Distance Modulus

When we discussed distance determination techniques (in Chapter 14), we ignored the effects of interstellar extinction. In any method that depends on the apparent brightness of a star, however, extinction can be important, particularly for very distant stars. Because the effect of extinction is to make a star appear fainter than it otherwise would, the tendency is to overestimate distances if no allowance is made for it.

Recall that in the spectroscopic parallax technique, the distance modulus $m - M$ is used to find the distance to a star from the equation given above:

$$d = 10^{1 + .2(m - M)},$$

where m is the apparent magnitude, M is the absolute magnitude, and d is the distance to the star in parsecs.

To correct this equation for extinction, we add a term, A_v, which refers to the extinction (in magnitudes) in visual light. Thus if the extinction toward a particular star makes that star appear 2 magnitudes fainter than it otherwise would, $A_v = 2$. If we insert this into the equation, we find:

$$d = 10^{1 + .2(m - M - A_v)}$$

Let us consider a simple example, a star whose apparent magnitude is $m = 12.4$ and whose absolute magnitude is $M = 2.4$. First let us consider its distance if extinction is ignored:

$$d = 10^{1 + .2(12.4 - 2.4)} = 10^3 = 1,000 \text{ parsecs.}$$

Now suppose it is determined that the extinction in the direction of this star amounts to 1 magnitude. Now the distance is

$$d = 10^{1 + .2(12.4 - 2.4 - 1)} = 10^{2.8} = 631 \text{ parsecs.}$$

One magnitude is only a modest amount of extinction, yet by neglecting it, we overestimated the distance to this star by almost 60 percent. We can see from this example that extinction can have a drastic effect on distance estimates.

It is worthwhile to add a note about how the extinction A_v is determined. It is possible to determine how much redder, in terms of the $B - V$ color index, a star appears because of extinction. To carry that a step further, astronomers define a *color excess* called $E(B - V)$, which is the difference between the observed and intrinsic values of $B - V$:

$$E(B - V) = (B - V)_{observed} - (B - V)_{intrinsic}$$

Studies of the variation of interstellar extinction with wavelength show that the extinction at the visual wavelength is approximately three times the color excess; that is:

$$A_v = 3E(B - V).$$

Hence determination of excess reddening leads to an estimate of A_v, and this in turn can be used in the modified equation for the distance.

APPENDIX 11: NUCLEAR REACTIONS IN STARS

In the text we did not spell out the details of the reactions that occur in stellar cores, although they are quite simple. To do so here, we will use the notation of nuclear physics. This is basically a shorthand in symbols. For example, a helium nucleus, containing two protons and two neutrons, is designated $^4_2$He, the subscript indicating the **atomic number** (the number of protons) and the superscript the **atomic weight** (the total number of protons and neutrons). Similarly a hydrogen nucleus is $^1_1$H, and deuterium, a form of hydrogen with an extra neutron in the nucleus, is $^2_1$H. A special symbol (ν) is used for the **neutrino**, a massless subatomic particle emitted in some reactions, and e^+ indicates a **positron**, which is equivalent to an electron but has a positive electrical charge. The symbol γ indicates a gamma ray, a very short wavelength photon of light emitted in some reactions.

The Proton-Proton Chain

Using this notation system, we can now spell out the proton-proton chain:

$$^1_1\text{H} + {}^1_1\text{H} \rightarrow {}^2_1\text{H} + e^+ + \nu$$
$$^2_1\text{H} + {}^1_1\text{H} \rightarrow {}^3_2\text{He} + \gamma.$$

The $^3_2$He particle is a form of helium, but not the common form. Once we have this particle it will combine with another:

$$^3_2\text{He} + {}^3_2\text{He} \rightarrow {}^4_2\text{He} + {}^1_1\text{H} + {}^1_1\text{H}.$$

We end up with a normal helium nucleus. A total of six hydrogen nuclei went into the reaction (remember, the first two steps had to occur twice, in order to produce two $^3_2$He particles for the final reaction), while there were two left at the end, so the net result is the conversion of four hydrogen nuclei into one helium nucleus.

The CNO Cycle

The CNO cycle, which dominates at higher temperatures, is more complex, involving not only carbon but also nitrogen and oxygen. Each of these elements has more than one form, which differ in their numbers of neutrons. Some of these **isotopes** are unstable and spontaneously emit positrons, decaying into other species in the process. Here is the CNO cycle:

$$^{12}_{6}C + ^{1}_{1}H \rightarrow ^{13}_{7}N + \gamma$$
$$^{13}_{7}N \rightarrow ^{13}_{6}C + e^{+} + \nu$$
$$^{13}_{6}C + ^{1}_{1}H \rightarrow ^{14}_{7}N + \gamma$$
$$^{14}_{7}N + ^{1}_{1}H \rightarrow ^{15}_{8}O + \gamma$$
$$^{15}_{8}O \rightarrow ^{15}_{7}N + e^{+} + \nu$$
$$^{15}_{7}N + ^{1}_{1}H \rightarrow ^{12}_{6}C + ^{4}_{2}He.$$

Here we end up with a helium nucleus and a carbon nucleus, while the particles going into the reaction were four hydrogen nuclei and a carbon nucleus. Along the way, three isotopes of nitrogen and one of oxygen were created and then converted into something else, leaving neither element at the end. As in the proton-proton chain, the net result is the conversion of four hydrogen nuclei into one helium nucleus.

The Triple-Alpha Reaction

Stars that have used up all the available hydrogen in their cores may become hot enough to undergo a new reaction in which helium is converted into carbon. Helium nuclei, consisting of two protons and two neutrons, are called alpha particles, and the reaction is called the triple-alpha reaction, since three of these particles are involved:

$$^{4}_{2}He + ^{4}_{2}He \rightarrow ^{8}_{4}Be$$
$$^{8}_{4}Be + ^{4}_{2}He \rightarrow ^{12}_{6}C + \gamma.$$

In this reaction sequence, $^{8}_{4}Be$ is a form of the element beryllium, and the end product, $^{12}_{6}C$, is the most com-

mon form of carbon. The second step must follow very quickly after the first because $^{8}_{4}Be$ is unstable and will break apart in two $^{4}_{2}He$ particles in a very short time. Therefore the third $^{4}_{2}He$ particle must react with the $^{8}_{4}Be$ particle almost immediately; thus this reaction sequence can be viewed as a three-particle reaction.

Other Reactions

Following helium burning in the triple-alpha reaction, other reactions can take place if the stellar core becomes hot enough. The first of these reactions are **alpha-capture reactions,** in which one form of nucleus adds an alpha particle to become a new form with two more protons and two additional neutrons. One example is the carbon alpha capture:

$$^{12}_{6}C + ^{4}_{2}He \rightarrow ^{16}_{8}O + \gamma,$$

in which $^{16}_{8}O$, the most common form of oxygen, is produced. In another sequence of alpha captures, nitrogen is converted into another isotope of oxygen, which can then be converted into neon:

$$^{14}_{7}N + ^{4}_{2}He \rightarrow ^{18}_{8}O + e^{+} + \nu,$$

and

$$^{18}_{8}O + ^{4}_{2}He \rightarrow ^{22}_{10}Ne + \gamma.$$

Additional alpha-capture reactions can occur, but at high enough temperatures other, more complex, reactions also take place. For example, two carbon nuclei can react, forming a number of different products, including sodium, neon, and magnesium. Two oxygen nuclei can also react, creating such species as sulfur, phosphorus, silicon, and magnesium. As a massive star evolves and its core contracts and heats following each reaction stage, a wide variety of elements is created with generally increasing atomic numbers. As explained in the text, the heaviest and most complex element produced in stable reactions in stellar cores is iron ($^{52}_{26}Fe$). Once a star has an iron core, it cannot undergo any further reaction stages without major disruption by a supernova explosion or collapse to a neutron star or black hole.

APPENDIX 12: DETECTED INTERSTELLAR MOLECULES[a]

Number of Atoms	Symbol	Name	Number of Atoms	Symbol	Name
2	H_2	Molecular hydrogen	4	$NCNH^+$	Protonated hydrogen cyanide
	C_2	Diatomic carbon		HNCS	Isothiocyanic acid
	CH	Methylidyne		C_3N	Cyanoethynyl
	CH^+	Methylidyne ion		C_3O	Tricarbon monoxide
	CN	Cyanogen		H_2CS	Thioformaldehyde
	CO	Carbon monoxide	5	C_4H	Butadiynyl
	CS	Carbon monosulfide		C_3H_2	Cyclopropenylidene
	OH	Hydroxyl		HCO_2H	Formic acid
	NO	Nitric oxide		CH_2CO	Ketene
	NS	Nitrogen sulfide		HC_3N	Cyanoacetylene
	SiC	Silicon carbide			
	SiO	Silicon monoxide		NH_2CN	Cyanamide
	SiS	Silicon sulfide		CH_2NH	Methanimine
	SO	Sulfur monoxide		CH_4	Methane
3	H_2D^+	Protonated hydrogen deuteride		SiH_4	Silane[b]
	C_2H	Ethynyl	6	C_5H	Pentynylidyne[b]
	HCN	Hydrogen cyanide		CH_3OH	Methanol
	HNC	Hydrogen isocyanide		CH_3CN	Methyl cyanide
	HCO	Formyl		CH_3SH	Methyl mercaptan
	HCO^+	Formyl ion		NH_2CHO	Formamide
	N_2H^+	Protonated nitrogen	7	CH_2CHCN	Vinyl cyanide
	HNO	Nitroxyl		CH_3C_2H	Methylacetylene
	H_2O	Water		CH_3CHO	Acetaldehyde
	HCS^+	Thioformyl ion		CH_3NH_2	Methylamine
	H_2S	Hydrogen sulfide		HC_5N	Cyanodiacetylene
	OCS	Carbonyl sulfide	8	$HCOOCH_3$	Methyl formate
	SO_2	Sulfur dioxide		CH_3C_3N	Methylcyanoacetylene
	SiC_2	Silicon dicarbide[b]	9	CH_3C_4H	Methyldiacetylene
4	C_2H_2	Acetylene		CH_3CH_3O	Dimethyl ether
	C_3H	Propynylidyne		CH_3CH_2CN	Ethyl cyanide
	H_2CO	Formaldehyde		CH_3CH_2OH	Ethanol
	NH_3	Ammonia		HC_7N	Cyano-hexa-tri-yne
	HNCO	Isocyanic acid	11	HC_9N	Cyano-octa-tetra-yne
	$HOCO^+$	Protonated carbon monoxide	13	$HC_{11}N$	Cyano-deca-penta-yne

[a]This list does not include isotopic variations (identical molecules except that one or more atoms are in rare isotopic forms, such as deuterium in place of hydrogen, or ^{13}C instead of the much more common ^{12}C).

[b]These species have been detected only in the dense circumstellar clouds surrounding red giant or supergiant stars.

APPENDIX 13: CLUSTERS OF GALAXIES

Galaxies of the local group

Galaxy[a]	Type[b]	Absolute Magnitude	Position Right Ascension	Position Declination
M31 (Andromeda)	Sb	−21.1	00^{h}40.^{m}0	+41° 00′
Milky Way	Sbc	−20.5	17 42.5	−28 59
M33 = NGC 598	Sc	−18.9	01 31.1	−30 24
Large Magellanic Cloud	Irr	−18.5	05 24	−69 50
IC 10	Irr	−17.6	00 17.6	+59 02
Small Magellanic Cloud	Irr	−16.8	00 51	−73 10
M32 = NGC 221	E2	−16.4	00 40.0	+40 36
NGC 205	E6	−16.4	00 37.6	+41 25
NCG 6822	Irr	−15.7	19 42.1	−14 53
NCG 185	Dwarf E	−15.2	00 36.1	+48 04
NGC 147	Dwarf E	−14.9	00 30.4	+48 14
IC 1613	Irr	−14.8	01 02.3	+01 51
WLM	Irr	−14.7	23 59.4	−15 44
Fornax	Dwarf sph	−13.6	02 37.5	−34 44
Leo A	Irr	−13.6	09 56.5	+30 59
IC 5152	Irr	−13.5	21 59.6	−51 32
Pegasus	Irr	−13.4	23 26.1	+14 28
Sculptor	Dwarf sph	−11.7	00 57.5	−33 58
And I	Dwarf sph	−11	00 42.8	+37 46
And II	Dwarf sph	−11	01 13.6	+33 11
And III	Dwarf sph	−11	00 32.7	+36 14
Aquarius	Irr	−11	20 44.1	−13 02
Leo I	Dwarf sph	−11	10 0.58	+12 33
Sagittarius	Irr	−10	19 27.1	−17 47
Leo II	Dwarf sph	−9.4	11 10.8	+22 26
Ursa Minor	Dwarf sph	−8.8	15 08.2	+67 18
Draco	Dwarf sph	−8.6	17 19.4	+57 58
Carina	Dwarf sph		06 40.4	−50 55
Pisces	Irr	−8.5	00 01.2	+21 37

[a]Galaxy names are derived from a variety of sources, including several catalogs (such as those designated M, NGC, and IC) and colloquial names bestowed by discoverers. Many in this list are simply named after the constellation where they are found.

[b]The galaxy types listed here are described in the text, except for the *Dwarf sph* designation, which stands for "dwarf spheroidal" and refers to dwarf galaxies that do not fit easily into the designation of dwarf ellipticals. Note that the absolute magnitudes of some of these are comparable to those of the brightest individual stars in our galaxy.

(continued on next page)

APPENDIX 13: CLUSTERS OF GALAXIES, *Continued*

Clusters of Galaxies within 1,000 Mpc[a]

Cluster	Distance (Mpc)	Radial Velocity (km/sec)	Diameter (Mpc)	Number of Galaxies	Density of Galaxies(Mpc^3)
Virgo	19	1,180	4	2,500	500
Pegasus I	65	3,700	1	100	1,100
Pisces	66	250	12	100	250
Cancer	80	4,800	4	150	500
Perseus	97	5,400	7	500	300
Coma	113	6,700	8	800	40
UMa III	132		2	90	200
Hercules	175	10,300	0.3	300	
Cluster A	240	15,800	4	400	200
Centaurus	250		9	300	10
UMa I	270	15,400	3	300	100
Leo	310	19,500	3	300	200
Cluster B	330		4	300	200
Gemini	350	23,300	3	200	100
CrB	350	21,600	3	400	250
Bootes	650	39,400	3	150	100
UMa II	680	41,000	2		400
Hydra	1,000	60,600			

[a]Data from Allen C. W., 1973, *Astrophysical quantites*, 3d ed. (London: Athlone Press).

APPENDIX 14: THE RELATIVISTIC DOPPLER EFFECT

The simple formula given in the text relating the Doppler shift of an object's spectrum to its velocity is accurate only when the velocity is much less than the speed of light. That simple relation is

$$v = (\Delta\lambda/\lambda)c,$$

where v is the object's velocity, $\Delta\lambda$ is the shift in wavelength of a spectral line whose rest wavelength is λ, and c is the speed of light.

When v is a significant fraction of c, the correct equation must be used. The error in determining v caused by using the approximate formula is 1 percent of v when $\Delta\lambda/\lambda$ is only 0.02; and the error becomes 5 percent when $\Delta\lambda/\lambda$ is 0.1. Thus failure to use the correct, relativistic formula becomes important for speeds of only a few percent of the speed of light.

For simplicity of notation, astronomers usually use the symbol z to represent the Doppler shift; that is,

$$z = \Delta\lambda/\lambda.$$

From Einstein's theory of special relativity, it can be shown that the correct relationship between the shift in wavelength and the speed of the emitting object is

$$z = \Delta\lambda/\lambda = \sqrt{\frac{1 + v/c}{1 - v/c}} - 1,$$

which leads to the following solution for the velocity:

$$v = c\left[\frac{(z + 1)^2 - 1}{(z + 1)^2 + 1}\right]$$

Let us apply this to the redshifts of quasars. One of the first quasars discovered, 3C273, has $z = 0.16$, which would imply a velocity of $0.16c = 48,000$ km/sec, if we used the simple, nonrelativistic equation. Use of the relativistic formula leads instead to $v = 0.147c = 44,200$ km/sec. In this case, use of the approximate formula leads to an error of almost 9 percent. The quasars with the highest known redshifts have z greater than 4.0, which would lead to the nonsensical conclusion that their speeds are more than 4 times that of light, if the nonrelativistic equation were used. Use of the correct equation for $z = 4.0$ gives $v = 0.923c = 277,000$ km/sec, still an enormous velocity.

APPENDIX 15: THE MESSIER CATALOG

Number	Right Ascension	Declination	Magnitude	Description
M1	05^{h}33.^{m}3	+22°01′	11.3	Crab Nebula in Taurus
M2	21 32.4	−00 54	6.3	Globular cluster in Aquarius
M3	13 41.3	+28 29	6.2	Globular cluster in Canes Venatici
M4	16 22.4	−26 27	6.1	Globular cluster in Scorpio
M5	15 17.5	+02 07	6	Globular cluster in Serpens
M6	17 38.9	−32 11	6	Open cluster in Scorpio
M7	17 52.6	−34 48	5	Open cluster in Scorpio
M8	18 02.4	−24 23		Lagoon Nebula in Sagittarius
M9	17 18.1	−18 30	7.6	Globular cluster in Ophiuchus
M10	16 56.0	−04 05	6.4	Globular cluster in Ophiuchus
M11	18 50.0	−06 18	7	Open cluster in Scutum
M12	16 46.1	−01 55	6.7	Globular cluster in Ophiuchus
M13	16 41.0	+36 30	5.8	Globular cluster in Hercules
M14	17 36.5	−03 14	7.8	Globular cluster in Ophiuchus
M15	21 29.1	+12 05	6.3	Globular cluster in Pegasus
M16	18 17.8	−13 48	7	Open cluster in Serpens
M17	18 19.7	−16 12	7	Omega Nebula in Sagittarius
M18	18 18.8	−17 09	7	Open cluster in Sagittarius
M19	17 01.3	−26 14	6.9	Globular cluster in Ophiuchus
M20	18 01.2	−23 02		Trifid Nebula in Sagittarius
M21	18 03.4	−22 30	7	Open cluster in Sagittarius
M22	18 35.2	−23 55	5.2	Globular cluster in Sagittarius
M23	17 55.7	−19 00	6	Open cluster in Sagittarius
M24	18 17.3	−18 27	6	Open cluster in Sagittarius
M25	18 30.5	−19 16	6	Open cluster in Sagittarius
M26	18 44.1	−09 25	9	Open cluster in Scutum
M27	19 58.8	+22 40	8.2	Dumbbell Nebula; planetary nebula in Vulpecula
M28	18 23.2	−24 52	7.1	Globular cluster in Sagittarius
M29	20 23.3	+38 27	8	Open cluster in Cygnus
M30	21 39.2	−23 15	7.6	Globular cluster in Capricornus
M31	00 41.6	+41 09	3.7	Andromeda galaxy
M32	00 41.6	+40 45	8.5	Elliptical galaxy, companion of M31
M33	01 32.8	+30 33	5.9	Spiral galaxy in Triangulum
M34	02 40.7	+42 43	6	Open cluster in Perseus
M35	06 07.6	+24 21	6	Open cluster in Gemini
M36	05 35.0	+34 05	6	Open cluster in Auriga
M37	05 51.5	+32 33	6	Open cluster in Auriga
M38	05 27.3	+35 48	6	Open cluster in Auriga
M39	21 31.5	+48 21	6	Open cluster in Cygnus
M40	12 20	+59		Double star cluster in Ursa Major
M41	06 46.2	−20 43	6	Open cluster in Canis Major
M42	05 34.4	−05 24		Orion Nebula
M43	05 34.6	−05 18		Small extension of the Orion nebula
M44	08 38.8	+20 04	4	Praesepe; open cluster in Cancer
M45	03 46.3	+24 03	2	The Pleiades; open cluster in Taurus
M46	07 40.9	−14 46	7	Open cluster in Puppis
M47	07 35.6	−14 27	5	Open cluster in Puppis
M48	08 12.5	−05 43	6	Open cluster in Hydra
M49	12 28.8	+08 07	8.9	Elliptical galaxy in Virgo
M50	07 02.0	−08 19	7	Open cluster in Monocerotis
M51	13 29.0	+47 18	8.4	Whirlpool galaxy; spiral galaxy in Canes Venatici
M52	23 23.3	+61 29	7	Open cluster in Cassiopeia
M53	13 12.0	+18 17	7.7	Globular cluster in Coma Berenices
M54	18 53.8	−30 30	7.7	Globular cluster in Sagittarius

(continued on next page)

APPENDIX 15: THE MESSIER CATALOG, *Continued*

Number	Right Ascension	Declination	Magnitude	Description
M55	19 38.7	−31 00	6.1	Globular cluster in Sagittarius
M56	19 15.8	+30 08	8.3	Globular cluster in Lyra
M57	18 52.9	+33 01	9.0	Ring Nebula; planetary nebula in Lyra
M58	12 36.7	+11 56	9.9	Spiral galaxy in Virgo
M59	12 41.0	+11 47	10.3	Elliptical galaxy in Virgo
M60	12 42.6	+11 41	9.3	Elliptical galaxy in Virgo
M61	12 20.8	+04 36	9.7	Spiral galaxy in Virgo
M62	16 59.9	−30 05	7.2	Globular cluster in Scorpio
M63	13 14.8	+42 08	8.8	Spiral galaxy in Canes Venatici
M64	12 55.7	+21 48	8.7	Spiral galaxy in Coma Berenices
M65	11 17.8	+13 13	9.6	Spiral galaxy in Leo
M66	11 19.1	+13 07	9.2	Spiral galaxy, companion of M65
M67	08 50.0	+11 54	7	Open cluster in Cancer
M68	12 38.3	−26 38	8	Globulalr cluster in Hydra
M69	18 30.1	−32 23	7.7	Globular cluster in Sagittarius
M70	18 42.0	−32 18	8.2	Globular cluster in Sagittarius
M71	19 52.8	+18 44	6.9	Globular cluster in Sagitta
M72	20 52.3	−12 39	9.2	Globular cluster in Aquarius
M73	20 57.8	−12 44		Open cluster in Aquarius
M74	01 35.6	+15 41	9.5	Spiral galaxy in Pisces
M75	20 04.9	−21 59	8.3	Globular cluster in Sagittarius
M76	01 40.9	+51 28	11.4	Planetary nebula in Perseus
M77	02 41.6	−00 04	9.1	Spiral galaxy in Cetus
M78	05 45.8	+00 02		Emission nebula in Orion
M79	05 23.3	−24 32	7.3	Globular cluster in Lepus
M80	16 15.8	−22 56	7.2	Globular cluster in Scorpio
M81	09 54.2	+69 09	6.9	Spiral galaxy in Ursa Major
M82	09 54.4	+69 47	8.7	Irregular galaxy in Ursa Major
M83	13 35.9	−29 46	7.5	Spiral galaxy in Hydra
M84	12 24.1	+13 00	9.8	Elliptical galaxy in Virgo
M85	12 24.3	+18 18	9.5	Elliptical galaxy in Coma Berenices
M86	12 25.1	+13 03	9.8	Elliptical galaxy in Virgo
M87	12 29.7	+12 30	9.3	Giant elliptical galaxy in Virgo
M88	12 30.9	+14 32	9.7	Spiral galaxy in Coma Berenices
M89	12 34.6	+12 40	10.3	Elliptical galaxy in Virgo
M90	12 35.8	+13 16	9.7	Spiral galaxy in Virgo
M91				Not identified; possibly M58
M92	17 16.5	+43 10	6.3	Globular cluster in Hercules
M93	07 43.6	−23 49	6	Open cluster in Puppis
M94	12 50.1	+41 14	8.1	Spiral galaxy in Canes Venatici
M95	10 42.8	+11 49	9.9	Barred spiral galaxy in Leo
M96	10 45.6	+11 56	9.4	Spiral galaxy in Leo
M97	11 13.7	+55 08	11.1	Owl Nebula; planetary nebula in Ursa Major
M98	12 12.7	+15 01	10.4	Spiral galaxy in Coma Berenices
M99	12 17.8	+14 32	9.9	Spiral galaxy in Coma Berenices
M100	12 21.9	+15 56	9.6	Spiral galaxy in Coma Berenices
M101	14 02.5	+54 27	8.1	Spiral galaxy in Ursa Major
M102				Not identified; possibly M101
M103	01 31.9	+60 35	7	Open cluster in Cassiopeia
M104	12 39.0	−11 35	8	Sombrero galaxy; spiral galaxy in Virgo
M105	10 46.8	+12 51	9.5	Elliptical galaxy in Leo
M106	12 18.0	+47 25	9	Spiral galaxy in Canes Venatici
M107	16 31.8	−13 01	9	Globular cluster in Ophiuchus
M108	11 10.5	+55 47	10.5	Spiral galaxy in Ursa Major
M109	11 56.6	+53 29	10.6	Barred spiral galaxy in Ursa Major

Absolute magnitude The magnitude a star would have if it were precisely 10 parsecs away from the Sun.

Absolute zero The temperature where all molecular or atomic motion stops, equal to $-273°C$ or $-459°F$.

Absorption line A wavelength at which light is absorbed, producing a dark feature in the spectrum.

Abundance gradient A systematic change in the relative abundances of elements over a distance; particularly in a galaxy.

Acceleration Any change—either of speed or direction—in the state of rest or motion of a body.

Accretion disk A rotating disk of gas surrounding a compact object (such as a neutron star or black hole), formed by material falling in.

Aether The substance of which the heavens and all bodies in the sky were thought to be composed, in ancient Greek cosmology.

Albedo The fraction of incident light that is reflected from a surface, such as that of a planet.

Alpha-capture reaction A nuclear fusion reaction in which an alpha particle merges with an atomic nucleus. A typical example is the formation of ^{16}O by the fusion of an alpha particle with ^{12}C.

Alpha particle A nucleus of ordinary helium containing two protons and two neutrons.

Altitude The angular distance of an object above the horizon.

Amino acid A complex organic molecule of the type that forms proteins. Amino acids are fundamental constituents of all living matter.

Andromeda galaxy The large spiral galaxy located some 700,000 parsecs from the Sun; the most distant object visible to the unaided eye.

Angstrom The unit normally used in measuring wavelengths of visible and ultraviolet light; one angstrom is equal to 10^{-8} centimeter.

Angular diameter The diameter of an object as seen on the sky, measured in units of angle.

Angular momentum A measure of the mass, radius, and rotational velocity of a rotating or orbiting body. In the simple case of an object in circular orbit, the angular momentum is equal to the mass of the object, times its distance from the center of the orbit, times its orbital speed.

Annual motions Motions in the sky caused by the Earth's orbital motion about the Sun. These include the seasonal variations of the Sun's declination and the Sun's motion through the zodiac.

Annular eclipse A solar eclipse that occurs when the Moon is near its greatest distance from the Earth, so that its angular diameter is slightly smaller than that of the Sun, and a ring, or annulus, of the Sun's disk is visible surrounding the disk of the Moon.

Anticyclone A rotating wind system around a high-pressure area. On the Earth, an anticyclone rotates clockwise in the Northern Hemisphere and counterclockwise in the Southern Hemisphere.

Antimatter Matter composed of the antiparticles of ordinary matter. For each subatomic particle, there is an antiparticle that is its opposite in such properties as electrical charge, but its equivalent in mass. Matter and antimatter, if combined, annihilate each other, producing energy in the form of gamma rays according to the formula $E = mc^2$.

Aphelion The point in the orbit of a solar system object where it is farthest from the Sun.

Apollo asteroid An asteroid whose orbit brings it closer to the Sun than 1 astronomical unit (AU).

Apparent magnitude The observed magnitude of a star or other celestial object, as seen from the Earth (*see Magnitude*)

Arctic Circle The region extending from the North Pole to 66.5° N latitude, where the Sun stays below the horizon for the full 24-hour day at the time of the winter solstice.

Asteroid (*See Minor planet*)

Asthenosphere The upper portions of the Earth's mantle, below the zone (the lithosphere) where convection currents are thought to operate. The term is also applied to similar zones in the interiors of the Moon and other planets.

Astrology The ancient belief that earthly affairs and human lives are influenced by the positions of the Sun, Moon, and planets with respect to the zodiac.

Astrometric binary A double star recognized as such because the visible star or stars undergo periodic motion that is detected by astrometric measurements.

Astrometry The science of accurately measuring stellar positions.

Astronomical unit (AU) A unit of distance used in astronomy, equal to the average distance between the Sun and the Earth; 1 AU is equal to 1.4959787×10^8 kilometer.

Astronomy The science that deals with the universe beyond the Earth's atmosphere.

Astrophysics The application of physical laws to astronomical phenomena; interchangeable with astronomy in modern usage.

Atomic number The number of protons in the nucleus of an element. The atomic number defines the identity of an element.

Atomic weight The mass of an atomic nucleus in atomic mass units [one atomic mass unit (amu) is defined as the average mass of the protons and neutrons in a nucleus of ordinary carbon, ^{12}C]. For most atoms, the atomic weight is approximately equal to the total number of protons and neutrons in the nucleus.

Aurorae australis   "Southern lights"; the visual emission from the Earth's upper atmosphere, caused by charged particles from space.

Aurorae borealis   "Northern lights"; (*see Aurorae australis*)

Autumnal equinox   The point where the Sun crosses the celestial equator from north to south, around September 21. See also **equinox** and **vernal equinox**.

Bailey's beads   Small glowing regions seen at the edges of the Moon's disk during a solar eclipse, due to irregularities in the shape of the lunar disk, which allows sunlight to pass through to the Earth.

Barred spiral galaxy   A spiral galaxy whose nucleus has linear extensions on opposing sides, giving it a barlike shape. The spiral arms usually appear to emanate from the ends of the bar.

Basalt   An igneous silicate rock common in regions formed by lava flows on the Earth, the Moon, and probably on the other terrestrial planets.

Beta decay   A spontaneous nuclear reaction in which a neutron decays into a proton and an electron, with a neutrino being emitted also. The term has been generalized to mean any spontaneous reaction in which an electron and a neutrino (or their antimatter equivalents) are emitted.

Big bang   A term referring to any theory of cosmogony in which the universe began at a single point, was very hot initially, and has been expanding from that state since.

Binary star   A double star system in which the two stars orbit a common center of mass.

Binary X-ray source   A close binary system containing a compact stellar remnant which emits X rays as it accretes matter from its companion.

BL Lac object   An elliptical galaxy having a very bright, highly concentrated nucleus, presenting a star-like appearance.

Black hole   An object that has collapsed under its own gravitation to such a small radius that its gravitational force traps photons of light.

Bode's law   Also known as the Titius-Bode relation, a simple numerical sequence that approximately represents the relative distances of the inner seven planets from the Sun. The relation is derived by adding 4 to each number in the sequence 0, 3, 6, 12, . . ., and then dividing each sum by 10, resulting in the sequence 0.4, 0.7, 1.0, 1.6, . . ., which is approximately the sequence of planetary distances from the Sun in astronomical units.

Bolide   An extremely bright meteor that explodes in the upper atmosphere.

Bolometric correction   The difference between the bolometric magnitude and the visual magnitude for a star; the bolometric correction is always negative because the star is brighter when all wavelengths are included.

Bolometric magnitude   A magnitude in which all wavelengths of light are included.

Breccia   Lunar rocks consisting of pebbles and soil fused together by meteorite impacts.

Burster   A sporadic source of intense X-rays, probably consisting of a neutron star onto which new matter falls at irregular intervals.

Carbonaceous chondrite   A meteorite containing chondrules having a high abundance of carbon and other volatile elements. Carbonaceous chondrites, thought to be very old, have apparently been unaltered since the formation of the solar system.

Cassegrain focus   A focal arrangement for a reflecting telescope in which a convex secondary mirror reflects the image through a hole in the primary mirror to a focus as the bottom of the telescope tube. This arrangement is commonly used in situations where relatively lightweight instruments for analyzing the light are attached directly to the telescope.

Cataclysmic variable   A binary system in which mass exchange causes irregular outbursts, usually involving a white dwarf which accretes mass from its companion.

Catastrophic theory   Any theory in which observed phenomena are attributed to sudden changes in conditions or to the intervention of an outside force or body.

CCD   Charge-coupled device; an electronic detector that receives photons of light and converts them into electrical charge patterns which are then converted into digital data. (*see Detector*)

Celestial equator   The imaginary circle formed by the intersection of the Earth's equatorial plane with the celestial sphere. The celestial equator is the reference line for north-south (declination) measurements in the standard equatorial coordinate system.

Celestial pole   The point on the celestial sphere directly overhead at either of the Earth's poles.

Celestial sphere   The imaginary sphere formed by the sky. It is a convenient device for discussing and measuring the positions of astronomical objects.

Center of mass   In a binary star system, or any system consisting of several objects, the point about which the mass is "balanced;" that is, the point that moves with a constant velocity through space while the individual bodies in the system move about it.

Cepheid variable   A pulsating variable star, of a class named after the prototype δ Cephei. Cepheid variables obey a period-luminosity relationship and are therefore useful as distance indicators. There are two classes of Cepheid variables, the so-called classical Cepheids, which belong to Population I, and the Population II Cepheids, also known as W Virginis stars.

Chondrite   A stony meteorite containing chondrules.

Chondrule   A spherical inclusion in certain meteorites, usually composed of silicates and always of very great age.

Chromatic aberration   The creation of images that are

dispersed according to wavelength perpendicular to the focal plane, caused by the use of lenses in refracting telescopes; different wavelengths of light are brought to a focus at different positions.

Chromosphere　A thin layer of hot gas just outside the photosphere in the Sun and other cool stars. The temperature in the chromosphere rises from about 4,000 K at its inner edge to 10,000 or 20,000 K at its outer boundary. The chromosphere is characterized by the strong red emission line of hydrogen.

Closed universe　A possible state of the universe in which the expansion will eventually be reversed and which is characterized by positive curvature, being finite in extent but having no boundaries.

Cluster variable　(See *RR Lyrae variable*)

CNO cycle　A nuclear fusion reaction sequence in which hydrogen nuclei are combined to form helium nuclei, and other nuclei, such as isotopes of carbon, oxygen, and nitrogen, appear as catalysts or by-products. The CNO cycle is dominant in the cores of stars on the upper main sequence.

Color-magnitude diagram　The equivalent of an H-R diagram, for a cluster of stars. Because the stars are at a common distance, the apparent magnitude may be plotted on the vertical axis in place of the absolute magnitude; the horizontal axis is usually the color index $(B - V)$, rather than spectral type.

Color excess　The amount, measured in magnitude units, by which a star appears redder in color due to interstellar dust than it would without dust in the line of sight. Technically the definition is $E(B - V) = (B - V)$ observed $- (B - V)$ intrinsic, where $(B - V)$ is the color index.

Color index　The difference $B - V$ between the blue (B) and visual (V) magnitudes of a star. If B is less than V (that is, if the star is brighter in blue than in visual light), the star has a negative color index and is a relatively hot star. If B is greater than V, the color index is positive, and the star is relatively cool.

Coma　The extended glowing region that surrounds the nucleus of a comet.

Comet　An interplanetary body, composed of loosely bound rocky and icy material, which forms a glowing head and extended tail when it enters the inner solar system.

Compressional waves　Waves in which the oscillations are in the direction of wave motion.

Configuration　The position of a planet or the Moon relative to the Sun-Earth line.

Conjunction　The alignment of two celestial bodies on the sky. In connection with the planets, a conjunction is the alignment of a planet with the Sun, an inferior conjunction being the occasion when an inferior planet is directly between the Sun and the Earth, and a superior conjunction being the occasion when any planet is directly behind the Sun as seen from the Earth.

Constellation　A prominent pattern of bright stars, historically associated with mythological figures. In modern usage each constellation incorporates a precisely defined region of the sky.

Constructive interference　The overlapping of waves (including light waves) in which the peaks and valleys of separate waves match each other, adding to their strength (or intensity, in the case of light).

Continental drift　The slow motion of the continental masses over the surface of the Earth, caused by the motions of the Earth's tectonic plates, which in turn are probably caused by convection in the underlying asthenosphere.

Continuous radiation　Electromagnetic radiation that is emitted in a smooth distribution with wavelength, without spectral features such as emission and absorption lines.

Convection　The transport of energy by fluid motions occurring in gases, liquids, or semirigid material such as the Earth's mantle. These motions are usually driven by the buoyancy of heated material, which tends to rise while cooler material descends.

Co-orbital satellites　Satellites that share the same orbit; usually one or more small satellites orbiting 60° ahead or 60° behind a larger one, kept in place by the combined gravitation effects of the large satellite and the parent planet.

Coriolis force　The apparent force felt by a particle moving north or south of the equator on a spinning body such as a planet; the particle veers in the direction of planetary rotation due to the fact the surface rotational speed decreases with distance from the equator.

Coronal holes　Regions of relatively low density in the solar corona, from which the solar wind emanates.

Corona　The very hot, extended outer atmosphere of the Sun and other cool, main-sequence stars. The high temperature in the corona $(1 - 2 \times 10^6$ K) is probably caused by the dissipation of mechanical energy from the convective zone just below the photosphere.

Cosmic background radiation　The primordial radiation field that fills the universe, having been created in the form of gamma rays at the time of the big bang, but having cooled since so that today its temperature is 2.73 K, and its peak wavelength is near 1.1 millimeters, in the microwave portion of the spectrum. Also known as the 3° background radiation.

Cosmic ray　A rapidly moving atomic nucleus from space. Some cosmic rays are produced in the Sun, while others come from interstellar space and probably originate in supernova explosions.

Cosmogony　The study of the origins of the universe.

Cosmological constant　A term added to the field equations by Einstein to allow solutions in which the universe was static (neither expanding nor contracting). Although the need for the term disappeared when it was discovered that the universe is expanding, the cosmological constant is retained in the field equations by modern cosmologists but is usually assigned the value zero.

Cosmological principle　The postulate, made by most cosmologists, that the universe is both homogeneous and isotropic. It is sometimes stated as, "The universe looks the same to all observers everywhere."

Cosmological redshift   A Doppler shift toward longer wavelengths that is caused by a galaxy's motion of recession due to the expansion of the universe.

Cosmology   The study of the universe as a whole.

Coudé focus   A focal arrangement for a reflecting telescope, in which the image is reflected by a series of mirrors to a remote, fixed location where a massive, immovable instrument can be used to analyze it.

Crater   A depression (usually circular) in the surface of a planet, caused by either an impact from space or volcanic activity.

Cyclone   A rotating wind system about a low-pressure center, often associated with storms on the Earth. On the Earth, cyclones rotate counterclockwise in the Northern Hemisphere and clockwise in the Southern Hemisphere.

Dark matter   The invisible material postulated to account for the majority of mass in clusters of galaxies and in the universe as a whole.

Declination   The coordinate in the equatorial system that measures positions in the north-south direction, with the celestial equator as the reference line. Declinations are measured in units of degrees, minutes, and seconds of arc.

Deferent   The large circle centered on or near the Earth on which the epicycle for a given planet moved, in the geocentric theory of the solar system developed by ancient Greek astronomers such as Hipparchus and Ptolemy.

Degenerate gas   A gas in which either free electrons or free neutrons are as densely spaced as allowed by laws of quantum mechanics. Such a gas has extraordinarily high density, and its pressure is not dependent on temperature as it is in an ordinary gas. Degenerate electron gas provides the pressure that supports white dwarfs against collapse, and degenerate neutron gas similarly supports neutron stars.

Density wave   A stable pattern of alternating dense and rarified regions, usually in a rotating fluid disk such as the ring system of Saturn or a spiral galaxy.

Deoxyribonucleic acid (DNA)   A complex protein consisting of a double helix structure composed of amino acid bases, responsible for carrying the genetic code in the nuclei of cells.

Destructive interference   The overlapping of waves (including light waves) in which the peaks of one wave coincide with the valleys of the other, diminishing both (and reducing the intensity, in light).

Detector   The general term for any device that receives and records the intensity of light at the focus of a telescope. Photographic film has been used traditionally in astronomy, but modern telescopes often use electronic devices, such as CCDs, instead.

Deuterium   An isotope of hydrogen containing in its nucleus one proton and one neutron.

Differential gravitational force   A gravitational force acting on an extended object, so that the portions of the object closer to the source of gravitation feel a stronger force than the portions that are farther away. Such a force, also known as a tidal force, acts to deform or disrupt the object and is responsible for many phenomena, ranging from synchronous rotation of Moons or double stars to planetary ring systems and the disruption of galaxies in clusters.

Differentiation   The sinking of relatively heavy elements into the core of a planet or other body. Differentiation can only occur in fluid bodies, so any planet that has undergone this process must once have been at least partially molten.

Diffraction   The process in which waves (including light waves) bend as they pass an obstacle.

Dipole anisotropy   (see *24 hour anisotropy*)

Distance modulus   The difference $m - M$ between the apparent and absolute magnitudes for a given star. This difference, which must be corrected for the effects of interstellar extinction, is a direct measure of the distance to the star.

Diurnal motion   Any motion related to the rotation of the Earth. Diurnal motions include the daily risings and settings of all celestial objects.

DNA   (see *Deoxyribonucleic acid*)

Doppler effect   The shift in wavelength of light caused by relative motion between the source of light and the observer. The Doppler shift, $\Delta\lambda$, is defined as the difference between the observed and rest (laboratory) wavelengths for a given spectral line.

Doppler shift   The observed shift in wavelength (and frequency) of a wave due to relative motion between the source of the wave and the observer.

Dwarf elliptical galaxy   A member of a class of small spheroidal galaxies similar to standard elliptical galaxies except for their small size and low luminosity. Dwarf galaxies are probably the most common in the universe but cannot be detected at distances beyond the Local Group of galaxies.

Dwarf nova   A close binary system containing a white dwarf in which material from the companion star falls onto the other at sporadic intervals, creating brief nuclear outbursts.

Eccentricity   A measure of the degree to which an elliptical orbit is elongated; technically, the eccentricity is equal to the ratio of the distance between the foci to the length of the major axis.

Eclipse   An occurrence in which one object is partially or totally blocked from view by another or passes through the shadow of another.

Eclipsing binary   A double star system in which one or both stars are periodically eclipsed by the other as seen from Earth. This situation can occur only when the orbital plane of the binary is viewed edge-on from the Earth.

Ecliptic   The plane of the Earth's orbit about the Sun, which is approximately the plane of the solar system as a whole. The apparent path of the Sun across the sky is the projection of the ecliptic onto the celestial sphere.

Ejecta Material blasted out of the ground by a meteorite impact.

Electromagnetic force The force created by the interaction of electric and magnetic fields. The electromagnetic force can be either attractive or repulsive and is important in countless situations in astrophysics.

Electromagnetic radiation Waves consisting of alternating electric and magnetic fields. Depending on the wavelength, these waves may be known as gamma rays, X-rays, ultraviolet radiation, visible light, infrared radiation, or radio radiation.

Electromagnetic spectrum The entire array of electromagnetic radiation arranged according to wavelength.

Electron A tiny, negatively charged particle that orbits the nucleus of an atom. The charge is equal and opposite to that of a proton in the nucleus, and in a normal atom the number of electrons and protons is equal so that the overall electrical charge is zero. The electrons emit and absorb electromagnetic radiation by making transitions between fixed energy levels.

Ellipse A geometrical shape such that the sum of the distances from any point on it to two fixed points called foci is constant. In any bound system where two objects orbit a common center of mass, their orbits are ellipses with the center of mass at one focus.

Elliptical galaxy A spherical or spheroidal galaxy having no disk-like component.

Elliptical galaxy One of a class of galaxies characterized by smooth spheroidal forms, few young stars, and little interstellar matter.

Emission line A wavelength at which radiation is emitted, creating a bright line in the spectrum.

Emission nebula A cloud of interstellar gas that glows by the light of emission lines. The source of excitation that causes the gas to emit may be radiation from a nearby star or heating by any of a variety of mechanisms.

Endothermic reaction Any nuclear or chemical reaction that requires more energy to occur than it produces.

Energy The ability to do work. Energy can be in either kinetic form, when it is a measure of the motion of an object, or potential form, when it is stored but capable of being released into kinetic form.

Epicycle A small circle on which a planet revolves, which in turn orbits another, distant body. Epicycles were used in ancient theories of the solar system to devise a cosmology that placed the Earth at the center but accurately accounted for the observed planetary motions.

Equatorial coordinates The astronomical coordinate system in which positions are measured with respect to the celestial equator (in the north–south direction) and a fixed direction (in the east–west dimension). The coordinates used are declination (north–south, in units of angle) and right ascension (east–west, in units of time).

Equinox Either of two points on the sky where the planes of the ecliptic and the Earth's equator intersect. When the Sun is at one of these two points, the lengths of night and day on the Earth are equal. See also **autumnal equinox** and **vernal equinox**.

Erg A unit of energy equal to the kinetic energy of an object of 2 grams mass moving at a speed of 1 centimeter per second, but defined technically as the work required to move a mass of 1 gram through a distance of 1 centimeter at an acceleration of 1 cm/sec^2.

Escape speed The upward speed required for an object to escape the gravitational field of a body such as a planet. In a more technical sense, the escape speed is the speed at which the kinetic energy of the object equals its gravitational potential energy; if the object moves any faster, its kinetic energy exceeds its potential energy, and it can escape the gravitational field.

Event horizon The "surface" of a black hole; the boundary of the region from within which no light can escape.

Evolutionary theory Any theory in which observed phenomena are thought to have arisen as a result of natural processes requiring no outside intervention or sudden changes.

Excitation A state in which one or more electrons of an atom or ion are in energy levels above the lowest possible one.

Excited state A state in which an atom, ion, or molecule has more energy than the lowest possible state (known as the ground state). For an atom or ion, this usually refers to the energy state of one or more electrons; for a molecule, the excited state may refer to an electron energy state, a vibrational energy state, or a rotational energy state.

Field equations A set of equations in general relativity theory that describe the relationship of matter, energy, and gravitation in the universe.

Fission reaction A nuclear reaction in which a large nucleus is split into one or more smaller nuclei.

Flat universe A possible state of the universe in which the momentum of expansion is exactly balanced by self-gravitation, so that the expansion will slow to a stop in an infinite time, but will not reverse itself and become a contraction. This state is predicted to exist by inflationary universe models.

Fluorescence The emission of light at a particular wavelength following excitation of the electron by absorption of light at another, shorter wavelength.

Focus (1) The point at which light collected by a telescope is brought together to form an image; (2) one of two fixed points that define an ellipse (see also **ellipse**).

Force Any agent or phenomenon that produces acceleration of a mass.

Fraunhofer lines The series of prominent solar absorption lines identified and cataloged in the early nineteenth century by Josef Fraunhofer, based on visual observations made with a spectroscope.

Frequency The rate (in units of hertz, or cycles per second) at which electromagnetic waves pass a fixed point.

The frequency, usually designated ν, is related to the wavelength λ and the speed of light c by $\nu = c/\lambda$.

Fusion reaction A nuclear reaction in which atomic nuclei combine to form more massive nuclei.

Galactic cluster A loose cluster of stars located in the disk or spiral arms of the galaxy.

Gamma ray A photon of electromagnetic radiation whose wavelength is very short and whose energy is very high. Radiation whose wavelength is less than one angstrom is usually considered to be gamma-ray radiation.

Gegenschein The diffuse glowing spot, seen on the ecliptic opposite the Sun's direction, created by sunlight reflected off interplanetary dust.

Globular cluster A large spherical cluster of stars located in the halo of the galaxy. These clusters, containing up to several hundred thousand members, are thought to be among the oldest objects in the galaxy.

Gram A unit of mass, equal to the quantity of mass contained in 1 cubic centimeter of water.

Grand Unified Theory A theory being sought by particle physicists in which three of the four fundamental forces (the weak and strong nuclear forces, and the electromagnetic force) are shown to be manifestations of the same phenomenon, and indistinguishable from each other under conditions of sufficiently high temperature and pressure.

Granulation The spotty appearance of the solar surface (the photosphere) caused by convection in the layers just below.

Gravitational redshift A Doppler shift toward long wavelengths caused by the effect of a gravitational field on photons of light. Photons escaping a gravitational field lose energy to the field, which results in the redshift.

Greatest elongation The greatest angular distance from the Sun that an inferior planet can reach, as seen from Earth.

Great Red Spot An oval-shaped, reddish feature on Jupiter's surface, thought to be a long-lived storm system.

Greenhouse effect The trapping of heat near the surface of a planet by atmospheric molecules (such as carbon dioxide) that absorb infrared radiation emitted by the surface.

H-R diagram See **Hertzsprung-Russell diagram.**

H II region A volume of ionized gas surrounding a hot star (see also **emission nebula**).

Half-life The time required for half of the nuclei of an unstable (radioactive) isotope to decay.

Halo (1) The extended outer portions of a galaxy (thought to contain a large fraction of the total mass of the galaxy, mostly in the form of dim stars and interstellar gas); (2) the extensive cloud of gas surrounding the head of a comet.

Helium flash A rapid burst of nuclear reactions in the degenerate core of a moderate mass star in the hydrogen shell-burning phase. The flash occurs when the core temperature reaches a sufficiently high temperature to trigger the triple-alpha reaction.

Herbig-Haro object A bright object often associated with young stars, thought to be a region of ionization caused by the jets of high-speed gas emanating from the polar regions of newly-formed stars.

Hertzsprung-Russell diagram A diagram on which stars are represented according to their absolute magnitudes (on the vertical axis) and spectral types (on the horizontal axis). Because the physical properties of stars are interrelated, they do not fall randomly on such a diagram but lie in well-defined regions according to their state of evolution. Very similar diagrams can be constructed using luminosity instead of absolute magnitude and temperature or color index in place of spectral type.

Hertz A unit of frequency used in describing electromagnetic radiation; 1 hertz (1 Hz) is equal to one cycle or wave per second.

High-velocity star A star whose velocity relative to the solar system is large. As a rule, high-velocity stars are Population II objects following orbital paths that are highly inclined to the plane of the galactic disk.

Horizontal branch A sequence of stars in the H-R diagram of a globular cluster, extending horizontally across the diagram to the left from the red giant region. These are probably stars undergoing helium burning in their cores by the triple-alpha reaction.

Hubble constant The numerical factor, usually denoted H, which describes the rate of expansion of the universe. It is the proportionality constant in the Hubble law $v = Hd$, which relates the speed of recession of a galaxy (v) to its distance (d). The present value of H is not well known, with estimates ranging between 55 and 90 kilometers per second per megaparsec.

Hydrostatic equilibrium The state of balance between gravitational and pressure forces that exists at all points inside any stable object such as a star or planet.

Igneous rock A rock that was formed by cooling and hardening from a molten state.

Impact crater A crater formed on the surface of a terrestrial planet or a satellite by the impact of a meteoroid or planetesimal.

Inertia The tendency of an object to remain in its state of rest or of uniform motion. This tendency is directly related to the mass of the object.

Inferior planet One of the planets whose orbits lie closer to the Sun than that of the Earth (Mercury or Venus).

Inflationary universe A theory of cosmology in which the universe expanded rapidly from a very compact and homogeneous initial state. This early rapid expansion was followed by the big bang expansion, which still continues.

Infrared radiation Electromagnetic radiation in the wavelength region just longer than that of visible light; that is, radiation whose wavelength lies roughly between 7,000 Å and 0.01 centimeter.

Interferometry   The use of interference phenomena in electromagnetic waves to measure positions precisely or to achieve gains in resolution. Interferometry in radio astronomy entails the use of two or more antennae to overcome the normally very coarse resolution of a single radio telescope; in visible-light observations, the object is to eliminate the distorting effects of the Earth's atmosphere.

Interplanetary dust   Tiny solid particles in interplanetary space, concentrated in the ecliptic plane.

Interstellar cloud   A region of relatively high density in the interstellar medium. Interstellar clouds have densities ranging between 1 and 10^6 particles per cubic centimeter and, in aggregate, contain most of the mass in interstellar space.

Interstellar dust   The diffuse medium of tiny solid particles that permeates the space between the stars in a galaxy.

Interstellar extinction   The obscuration of starlight by interstellar dust. Light is scattered by dust grains, so that a distant star appears dimmer than it otherwise would. The scattering process is most effective at short (blue) wavelengths, so that stars seen through interstellar dust are reddened as well as dimmed.

Inverse Compton scattering   Scattering of photons by relativistic electrons, in which photon energies are increased.

Inverse square law   In general, any law describing a force or other phenomenon that decreases in strength as the square of the distance from some central reference point. In particular the term *inverse square law* is often used by itself to mean the law stating that the intensity of light emitted by a source such as a star diminishes as the square of the distance from the source.

Ionization   Any process by which an electron or electrons are freed from an atom or ion. Ionization generally occurs in two ways: by the absorption of a photon with sufficient energy or by collision with another particle.

Ionosphere   The zone of the Earth's upper atmosphere, between 80 and 500 kilometers altitude, where charged subatomic particles (chiefly protons and electrons) are trapped by the Earth's magnetic field. See also **Van Allen belts**.

Ion   Any subatomic particle with a nonzero electrical charge. In standard practice, the term *ion* is usually applied only to positively charged particles, such as atoms missing one or more electrons.

Io torus   A zone of gas particles concentrated in the orbit of Io (the innermost major satellite of Jupiter); the gas originates in volcanic eruptions on Io, and is dispersed throughout the satellite's orbit by magnetic forces, once it becomes ionized.

Isotope   Any form of a given chemical element. Different isotopes of the same element have the same number of protons in their nuclei, but different numbers of neutrons.

Isotropic   Having the property of appearing the same in all directions. In astronomy, this term is often postulated to apply to the universe as a whole.

Jeans criterion   The condition required for a cloud of gas to spontaneously collapse due to its own self-gravitation, usually expressed in terms of the size (or mass) of the cloud relative to its density and temperature.

Kelvin   A unit of temperature, equal to 0.01 of the difference between the freezing and boiling points of water, used in a scale whose zero point is absolute zero. A Kelvin is usually denoted simply by K.

Kiloparsec   A unit of distance, equal to 1,000 parsecs.

Kinetic energy   The energy of motion. The kinetic energy of a moving object is equal to ½ times its mass times the square of its velocity.

Kirkwood's gaps   Narrow gaps in the asteroid belt created by orbital resonance with Jupiter.

Latitude   The distance north or south of the Earth's equator, measured in units of angle.

Lepton   One of the two fundamental classes of elementary particles; includes electrons.

Light curve   A graph showing the intensity of light (or magnitude) of a celestial object versus time. Light curves are often useful in diagnosing the properties of variable stars and other objects.

Light-gathering power   The ability of a telescope to collect light from an astronomical source; the light-gathering power is directly related to the area of the primary mirror or lens.

Limb darkening   The dark region around the edge of the visible disk of the Sun or of a planet caused by a decrease in temperature with height in the atmosphere.

Liquid metallic hydrogen   Hydrogen in a state of semirigidity that can exist only under conditions of extremely high pressure, as in the interiors of Jupiter and Saturn.

Lithosphere   The layer in the Earth, Moon, and terrestrial planets that includes the crust and the outer part of the mantle.

Local Group   The cluster of about thirty galaxies to which the Milky Way belongs.

Logarithm   The logarithm of a number is the power to which 10 must be raised to equal that number. For example, $100 = 10^2$; so the logarithm of 100 is 2.

Luminosity class   One of several classes to which a star can be assigned on the basis of certain luminosity indicators in its spectrum. The classes range from I for supergiants to V for main-sequence stars (also known as dwarfs).

Luminosity   The total energy emitted by an object per second (the power of the object). For stars the luminosity is usually measured in units of ergs per second.

Luminous arcs   Enormous arc-like glowing structures seen in some dense clusters of galaxies. These are thought to be distorted images of background galaxies or quasars, formed by the bending of light due to the cluster gravitation field (*see Gravitational lens*)

Lunar eclipse   An eclipse of the Moon, caused by its passage through the shadow of the earth.

Lunar month The synodic period of the Moon, equal to 29 days, 12 hours, 44 minutes, 11 seconds.

L wave A type of seismic wave that travels only over the surface of the Earth.

Magellanic Clouds The two irregular galaxies that are the nearest neighbors to the Milky Way and are visible to the unaided eye in the Southern Hemisphere.

Magma Molten rock from the Earth's interior.

Magnetic braking The slowing of the spin of a young star such as the early Sun by magnetic forces exerted on the surrounding ionized gas.

Magnetic dynamo A rotating internal zone inside the Sun or a planet, thought to carry the electrical currents that create the solar or planetary magnetic field.

Magnetosphere The region surrounding a star or planet that is permeated by the magnetic field of that body.

Magnitude A measure of the brightness of a star, based on a system established by Hipparchus in which stars were ranked according to how bright they appeared to the unaided eye. In the modern system, a difference of 5 magnitudes corresponds exactly to a brightness ratio of 100, so that a star of a given magnitude has a brightness that is $100^{1/5} = 2.512$ times that of a star 1 magnitude fainter.

Main sequence The strip in the H-R diagram, running from upper left to lower right, where most stars that are converting hydrogen to helium by nuclear reactions in their cores are found.

Main-sequence fitting A distance-determination technique in which an H-R diagram for a cluster of stars is compared with a standard H-R diagram to establish the absolute magnitude scale for the cluster H-R diagram.

Main-sequence turnoff In an H-R diagram for a cluster of stars the point where the main-sequence turns off toward the upper right. The main-sequence turnoff, showing which stars in the cluster have evolved to become red giants, is an indicator of the age of the cluster.

Mantle The semirigid outer portion of the Earth's interior extending from roughly the midway point nearly to the surface and consisting of the mesosphere (the lower portion) and the asthenosphere.

Mare (pl. maria) Any of several extensive, smooth lowland areas on the surface of the Moon or Mercury that were created by extensive lava flows early in the history of the solar system.

Mass A measure of the quantity of matter contained in an object.

Mass-luminosity relation The correspondence between the masses of stars and their luminosities; first found empirically, this relation has since been duplicated by theoretical models of stellar structure.

Mass-to-light ratio The mass of a galaxy, in units of solar masses, divided by its luminosity, in units of the Sun's luminosity. The mass-to-light ratio is an indicator of the relative quantities of Population I and Population II stars in a galaxy.

Maunder minimum An interval during the latter half of the seventeenth century when the number of sunspots was abnormally low.

Maxwell distribution A methematical representation of the distribution of particle speeds in a gas of a certain temperature.

Mean solar day The average length of the solar day as measured throughout the year; the mean solar day is precisely 24 hours.

Megaparsec (Mpc) A unit of distance, equal to 10^6 parsecs.

Meridian The great circle on the celestial sphere that passes through both poles and directly overhead; that is, the north–south line directly overhead.

Mesosphere (1) The layer of the Earth's atmosphere between roughly 50 and 80 kilometers in altitude, where the temperature decreases with height; (2) the layer below the asthenosphere in the Earth's mantle.

Metamorphic rock A rock formed by heat and pressure in the Earth's interior.

Meteor A bright streak or flash of light created when a meteroid enters the Earth's atmosphere from space.

Meteorite The remnant of a meteroid that survives a fall through the Earth's atmosphere and reaches the ground.

Meteoroid A small, interplanetary body.

Meteor shower A period during which meteors are seen with high frequency, occurring when the Earth passes through a swarm of meteoroids.

Micrometeorite A microscopically small meteorite.

Microwave background *See* **cosmic background radiation.**

Milky Way Historically, the diffuse band of light stretching across the sky; our cross-sectional view of the disk of our galaxy. In modern usage, the term *Milky Way* refers to our galaxy as a whole.

Minor planet One of thousands of large (up to 1000 km in diameter) non-planetary bodies orbiting the Sun; most of the orbits lie between Mars and Jupiter.

Moving cluster method A distance-determination technique in which the radial velocities of stars in a cluster are combined with knowledge of the direction of motion gained from observed proper motions, so that the true space velocity can be used in combination with the proper motion to derive the distance.

Neutrino A subatomic particle without mass or electrical charge that is emitted in certain nuclear reactions.

Neutron A subatomic particle with no electrical charge and a mass nearly equal to that of the proton. Neutrons and protons are the chief components of the atomic nucleus.

Neutron-capture reactions Nuclear reactions in which neutrons are captured by nuclei (*see also* **r-process reactions** *and* **s-process reactions**)

Neutron star A very compact, dense stellar remnant whose interior consists entirely of neutrons, and which is supported against collapse by degenerate neutron gas pressure.

Newtonian focus A focal arrangement for reflecting telescopes in which a flat mirror is used to reflect the image through a hole in the side of the telescope tube.

Nonthermal radiation Radiation not due only to the temperature of an object. The term is most often applied to sources of continuous radiation such as synchrotron radiation.

North celestial pole The projection of the Earth's North Pole onto the sky.

Nova A star that temporarily flares up in brightness, most likely as the result of nuclear reactions caused by the deposition of new nuclear fuel on the surface of a white dwarf in a binary system. See also **recurrent nova.**

Nucleus The central, dense concentration in an atom, a comet, or a galaxy.

OB association A group of young stars whose luminosity is dominated by O and B stars.

Obliquity A measure of the tilt of a planet's rotation axis; technically, the angle between the rotation axis and the perpendicular to the orbital plane.

Occam's razor The principle that the simplest explanation of any natural phenomenom is most likely the correct one. *Simple* in this case usually means requiring few assumptions or unverifiable postulates.

Olber's paradox The apparent paradox created by the fact that in a universe that is infinite in extent and in age, every line of sight should intersect a stellar surface, producing a uniformly bright nighttime sky; the fact that the sky is dark demonstrates that the universe is not infinite in both extent and age.

Oort cloud The cloud of bodies, hypothesized to be orbiting the Sun at a great distance, from which comets originate.

Open cluster A loosely-bound association of stars in the galactic disk (*see* **Galactic cluster**)

Open universe A possible state of the universe in which its expansion will never stop, and it is characterized by negative curvature, being infinite in extent and having no boundaries.

Opposition A planetary configuration in which a superior planet is positioned exactly in the opposite direction from the Sun as seen from Earth.

Optical double A pair of stars that happen to appear near each other on the sky but are not in orbit; not a true binary.

Orbital resonance A situation in which the periods of two orbiting bodies are simple multiples of each other so that they are frequently aligned and gravitational forces due to the outer body may move the inner body out of its original orbit. This is one mechanism thought responsible for creating the gaps in the rings of Saturn and Kirkwood's gaps in the asteroid belt.

Organic molecule Any of a large class of carbon-bearing molecules found in living matter.

Outgassing The process by which gases escape from a planetary interior, often through volcanic venting or eruptions.

Ozone A form of oxygen containing three oxygen atoms bonded together.

Paleomagnetism Vestigial traces or artifacts of ancient magnetic fields.

Parallax Any apparent shift in position caused by an actual motion or displacement of the observer. See also **stellar parallax.**

Parsec A unit of distance, equal to the distance to a star whose stellar parallax is 1 arcsecond. A parsec is equal to 206, 265 AU, 3.03×10^{13} kilometers, or 3.26 light-years.

Peculiar velocity The deviation in a star's velocity from perfect circular motion about the galactic center.

Penumbra (1) The light, outer portion of a shadow, such as the portion of the Earth's shadow where the Moon is not totally obscured during a lunar eclipse; (2) the light, outer portion of a sunspot.

Perihelion The point in the orbit of any Sun-orbiting body where it most closely approaches the Sun.

Permafrost A permanent layer of ice just below the surface of certain regions on the Earth and probably on Mars.

Photometer A device, usually using a photoelectric cell, for measuring the brightnesses of astronomical objects.

Photon A particle of light having wave properties but also acting as a discrete unit.

Photosphere The visible surface layer of the Sun and stars; the layer from which continuous radiation escapes and where absorption lines form.

Planck constant The numerical factor h relating the frequency v of a photon to its energy E in the expression $E = hv$. The Planck constant has the value $h = 6.62620 \times 10^{-27}$ erg second.

Planck function (also known as the **Planck law**) The mathematical expression describing the continuous thermal spectrum of a glowing object. For a given temperature, the Planck function specifies the intensity of radiation as a function of either frequency or wavelength.

Planetary nebula A cloud of glowing, ionized gas, usually taking the form of a hollow sphere or shell, ejected by a star in the late stages of its evolution.

Planetesimal A small (diameter up to several hundred kilometers) solar system body of the type that first condensed from the solar nebula. Planetesimals are thought to have been the principal bodies that combined to form the planets.

Plate tectonics A general term referring to the motions of lithospheric plates over the surface of the Earth or other terrestrial planets. See also **continental drift.**

Polarization The preferential alignment of the magnetic and electric fields in electromagnetic radiation.

Population I   The class of stars in a galaxy with relatively high abundances of heavy elements. These stars are generally found in the disk and spiral arms of spiral galaxies and are relatively young. The term *Population I* is also commonly applied to other components of galaxies associated with the star formation, such as the interstellar material.

Population II   The class of stars in a galaxy with relatively low abundances of heavy elements. These stars are generally found in a spheroidal distribution about the galactic center and throughout the halo and are relatively old.

Population III: A population of ancient stars thought to have formed before any enrichment of heavy elements had taken place, so that they contained only hydrogen and helium. No Population III stars have been found.

Positron   A subatomic particle with the same mass as the electron but with a positive electrical charge; the antiparticle of the electron.

Potential energy   Energy that is stored and may be converted into kinetic energy under certain circumstances. In astronomy the most common form of potential energy is gravitational potential energy.

Power   The rate of energy expenditure per second; for stars the world *luminosity* is used instead.

Precession   The slow shifting of star positions on the celestial sphere caused by the 26,000-year periodic wobble of the Earth's rotational axis.

Primary mirror   The principal light-gathering mirror in a reflecting telescope.

Prime focus   The focal arrangement in a reflecting telescope in which the image is allowed to form inside the telescope structure at the focal point of the primary mirror, so that no secondary mirror is needed.

Prograde motion   Orbital or spin motion in the "normal" direction; in the solar system, this is counterclockwise as viewed from above the North Pole.

Prominence   A cloud or column of heated, glowing gas extending from the chromosphere into the corona of the Sun, its structure controlled by magnetic fields.

Proper motion   The motion of a star across the sky, usually measured in units of arcseconds per year.

Proton-proton chain   The sequence of nuclear reactions in which four hydrogen nuclei combine, through intermediate steps involving deuterium and ^{3}He, to form one helium nucleus. The proton-proton chain is responsible for energy production in the cores of stars on the lower main sequence.

Protostar   A star in the process of formation, specifically one that has entered the slow gravitational contraction phase.

Pulsar   A rapidly rotating neutron star that emits periodic pulses of electromagnetic radiation, probably by the emission of beams of radiation from the magnetic poles, which sweep across the sky as the star rotates.

P wave   A seismic wave that is a compressional, or density, wave. P waves can travel through both solid and liquid portions of the Earth and are the first to reach any remote location from an earthquake site.

P wave: (*see Compressional wave*)

QSO   *See quasistellar object.*

Quadrature   The configuration where a superior planet or the Moon is 90 degrees away from the Sun, as seen from the Earth.

Quantum mechanics   The physics of atomic structure and the behavior of subatomic particles, based on the principle of the quantum.

Quantum   The amount of energy associated with a photon, equal to h, where h is the Planck constant, and v is the frequency. The quantum is the smallest amount of energy that can exist at a given frequency.

Quark   One of the two fundamental classes of elementary particles; quarks are the constituent subparticles of protons and neutrons.

Quasar   *See quasistellar object.*

Quasistellar object   Any of a class of extragalactic objects characterized by emission lines with very large redshifts. The quasistellar objects are thought to lie at great distances, in which case they existed only at earlier times in the history of the universe; they may be young galaxies.

Radial velocity   The component of motion of a star or other body along the line of sight; i.e. the portion of the relative velocity that is directed straight toward or away from the observer.

Radiation pressure   Pressure created by the forces exerted by photons of light when they are absorbed or reflected.

Radiative transport   The transport of energy, inside a star or in other situations, by radiation.

Radioactive dating   A technique for estimating the age of material such as rock, based on the known initial isotopic composition and the known rate of radioactive decay for unstable isotopes initially present.

Radioactivity   The spontaneous emission of subatomic particles (alpha rays or beta rays) or high-energy photons (gamma rays) by unstable nuclei. The identity of the emitting substance is changed in the process.

Radio galaxy   Any of a class of galaxies whose luminosity is greatest in radio wavelengths. Radio galaxies are usually large elliptical galaxies, with synchrotron radiation emitted from one or more pairs of lobes located on opposite sides of the visible galaxy.

Ray   A bright streak of ejecta emanating from an impact crater, especially on the Moon or on Mercury.

Recurrent nova   A star known to flare up in nova outbursts more than once. A recurrent nova is thought to be a binary system containing a white dwarf and a mass-losing star, in which the white dwarf sporadically flares up when material falls onto it from the companion.

Red giant A star that has completed its core hydrogen-burning stage and has begun hydrogen shell burning, which causes its outer layers to become very extended and cool.

Reflecting telescope A telescope that brings light to a focus by using mirrors.

Reflection nebula An interstellar cloud containing dust that shines by light reflected from a nearby star.

Refracting telescope A telescope that uses lenses to bring light to a focus.

Refractory The property of being able to exist in solid form under conditions of very high temperature. A refractory element is one that is characterized by a high temperature of vaporization; refractory elements are the first to condense into solid form when a gas cools, as in the solar nebula.

Regolith The layer of debris on the surface of the Moon created by the impact of meteorites; the lunar surface layer.

Resolution In an image the ability to separate closely spaced features, that is, the clarity or fineness of the image. In a spectrum the ability to separate features that are close together in wavelength.

Retrograde motion Orbital or spin motion in the opposite direction from prograde motion; in the solar system retrograde motions are clockwise as seen from above the North Pole.

Ribonucleic acid (RNA) A complex protein consisting of amino acid bases, responsible for transferring the genetic code during reproduction of life forms.

Right ascension The east–west coordinate in the equatorial coordinate system. The right ascension is measured in units of hours, minutes, and seconds to the east from a fixed direction on the sky, which itself is defined as the line of intersection of the ecliptic and the celestial equator.

Rille A type of winding, sinuous valley commonly found on the Moon.

RNA (*see Ribonucleic acid*)

Roche limit The point near a massive body such as a planet or star inside which the tidal forces acting on an orbiting body exceed the gravitational force holding it together. The location of the Roche limit depends on the size of the orbiting body.

Rotation curve A plot showing the orbital velocity of stars in a spiral galaxy versus distance from the galactic center.

r-process reactions Nuclear fusion reactions in which neutrons are captured by nuclei in rapid succession, so that there is insufficient time for beta-decay to take place between captures. These reactions are important only in very high-energy situations, such as in supernova explosions.

RR Lyrae variable A member of a class of pulsational variable stars named after the prototype star, RR Lyrae. These stars are blue-white giants with pulsational periods of less than one day and are Population II objects found primarily in globular clusters.

Russell-Vogt theorem The statement that the properties of a star are fully determined by its mass and its composition.

Saros cycle An eighteen-year, eleven-day repeating pattern of solar and lunar eclipses caused by a combination of the tilt of the lunar orbit with respect to the ecliptic and the precession of the plane of the Moon's orbit.

Scarp A long cliff or series of cliffs, usually created by shrinking or settling of a planetary crust.

Scattering The random reflection of photons by particles such as atoms or ions in a gas or dust particles in interstellar space.

Schwarzschild radius The radius within which an object has collapsed at the point when light can no longer escape the gravitational field as the object becomes a black hole.

Secondary mirror The second mirror in a reflecting telescope (after the primary mirror), usually either convex in shape, to reflect the image out a hole in the bottom of the telescope to the cassegrain focus, or flat, to reflect the image out of the side of the telescope to the Newtonian focus or along the telescope mount axis to the coudé focus.

Sedimentary rock A rock formed by the deposition and hardening of layers of sediment, usually either underwater or in an area subject to flooding.

Seeing The blurring and distortion of point sources of light, such as stars, caused by turbulent motions in the Earth's atmosphere.

Seismic wave A wave created in a planetary or satellite interior, usually caused by an earthquake.

Selection effect The tendency for a conclusion based on observations to be influenced by the method used to select the objects for observation. An example was the early belief that all quasars are radio sources, when the principal method used to discover quasars was to look for radio sources and then see if they had other properties associated with quasars.

Semimajor axis One-half of the major, or long, axis of an ellipse.

Sextant A device consisting of a pair of pointers and a segment (one-sixth) of a circle inscribed with angle markings, for the measurement of altitudes or angular separations between objects on the sky.

Seyfert galaxy Any of a class of spiral galaxies, first recognized by Carl Seyfert, with unusually bright blue nuclei.

Shear wave A wave that consists of transverse motions, that is, motions perpendicular to the direction of wave travel.

Sidereal day The rotation period of the Earth with respect to the stars, or as seen by a distant observer, equal to 23 hours, 56 minutes, 4.091 seconds.

Sidereal period The orbital or rotational period of any object with respect to the fixed stars or as seen by a distant observer.

Singularity A structure that is defined mathematically as a single point, and which therefore cannot be described in physical terms. In astronomy the center of a black hole is described as a singularity.

SNC meteorites   A class of meteorites whose distribution and composition suggests that they are ejecta from one or more impacts that occurred on Mars.

Solar active region   A region on the Sun, usually associated with sunspots, where activity such as solar flares originates.

Solar constant   The intensity of sunlight (in units of ergs per square centimeter) striking the earth above the atmosphere. The value is approximately 1.4×10^6 erg/sec.

Solar day   The synodic rotation period of the Earth with respect to the Sun; that is, the length of time from one local noon, when the Sun is on the meridian, to the next local noon.

Solar eclipse   An eclipse of the Sun, occurring when the Moon's disk blocks some or all of the Sun's disk, as seen from the Earth.

Solar flare   An explosive outburst of ionized gas from the Sun, usually accompanied by X-ray emission and the injection of large quantities of charged particles into the solar wind.

Solar motion   The deviation of the Sun's velocity from perfect circular motion about the center of the galaxy; that is, the Sun's peculiar velocity.

Solar nebula   The primordial gas and dust cloud from which the Sun and the planets condensed.

Solar wind   The stream of charged subatomic particles flowing steadily outward from the Sun.

Solstice   The occasion when the Sun, as viewed from the Earth, reaches its farthest northern (summer solstice) or southern point (winter solstice).

South celestial pole   The projection of the Earth's south pole onto the sky.

s-process reactions   Nuclear reactions in which nuclei capture neutrons at a rate slow enough that beta-decay can occur before an addition neutron is captured. These reactions are responsible for the creation of many heavy elements in massive stars before they explode as supernovae.

Spacetime   The term for four-dimensional space, consisting of the three spatial dimensions plus the time dimension, as described in general relativity theory.

Spectrogram   A photograph of a spectrum.

Spectrograph   An instrument for recording the spectra of astronomical bodies or other sources of light.

Spectroscope   An instrument allowing an observer to view the spectrum of a source of light.

Spectroscopic binary   A binary system recognized as a binary because its spectral lines undergo periodic Doppler shifts as the orbital motions of the two stars cause them to move toward and away from the Earth. If lines of only one star are seen, it is a single-lined spectroscopic binary; if lines of both stars are seen, it is a double-lined spectroscopic binary.

Spectroscopic parallax   The technique of distance determination for stars in which the absolute magnitude is inferred from the H-R diagram and then compared with the observed apparent magnitude to yield the distance.

Spectroscopy   The science of analyzing the spectra of stars or other sources of light.

Spectrum   An arrangement of electromagnetic radiation according to wavelength.

Spectrum binary   A binary system recognized as a binary because its spectrum contains lines of two stars of different spectral types.

Spicules   Narrow, short-lived jets of hot gas extending upward from the solar chromosphere.

Spin-orbit coupling   A simple relationship between the orbital and spin periods of a satellite or planet, caused by tidal forces that have slowed the rate of rotation of the orbiting body. Synchronous rotation is the simplest and most common form of spin-orbit coupling.

Spiral density wave   A spiral wave pattern in a rotating, thin disk, such as the rings of Saturn or the plane of a spiral galaxy like the Milky Way.

Spiral galaxy   Any of a large class of galaxies exhibiting a disk with spiral arms.

Standard candle   A general term for any astronomical object, the absolute magnitude of which can be inferred from its other observed characteristics, and which is therefore useful as a distance indicator.

Starburst galaxy   A galaxy, usually a spiral or an irregular, undergoing a phase of intense star formation. Many are detected only in infrared wavelengths because of obscuration by dense clouds of interstellar dust associated with the star formation.

Stefan-Boltzmann law   The law of continuous radiation stating that for a spherical glowing object such as a star the luminosity is proportional to the square of the radius and the fourth power of the temperature.

Stefan's law   An experimentally derived law of continuous radiation stating that the energy emitted by a glowing body per square centimeter of surface area is proportional to the fourth power of the absolute temperature.

Stellar occultation   The passage of a foreground object in front of a background star. Such occurrences are sometimes useful in analyzing the properties of the occulting body, such as in the *Voyager* spacecraft observations of fine detail in the ring systems of the outer planets, or in the analysis of planetary atmospheres observed as they occult background stars.

Stellar parallax   The apparent annual shifting of position of a nearby star with respect to more distant background stars. The term *stellar parallax* is often assumed to mean the parallax angle, which is one-half of the total angular motion a star undergoes. See also **parallax** and **parsec**.

Stellar wind   Any stream of gas flowing outward from a star, including the very rapid winds from hot, luminous stars; the intermediate-velocity, rarefied winds from stars like the Sun; and the slow, dense winds from cool supergiant stars.

Stratosphere   The layer of the Earth's atmosphere, between 10 and 50 kilometers in altitude, where the temperature increases with height.

Strong nuclear force   The force that holds the nucleons (protons and neutrons) together in the nucleus of an atom. This is the strongest of the four known natural forces, but it acts only over very short (nuclear) distances.

Subduction   The process in which one tectonic plate is submerged below another along a line where two plates collide. A subduction zone is usually characterized by a deep trench and an adjoining mountain range.

Sublimation   Evaporation of a compound directly from the solid to the gaseous state.

Supercluster   A cluster of clusters of galaxies.

Supergiant   A star in its late stages of evolution that is undergoing shell burning and is therefore very extended in size and very cool; an extremely large giant.

Supergranulation   The pattern of large cells seen in the Sun's chromosphere when viewed in the light of the strong emission line of ionized hydrogen.

Superior planet   Any planet whose orbit lies beyond the Earth's orbit around the Sun.

Supernova   The explosive destruction of a massive star that occurs when all sources of nuclear fuel have been consumed, and the star collapses catastrophically.

Synchronous rotation   A situation in which the rotational and orbital periods of an orbiting body are equal so that the same side is always facing the companion object.

Synchrotron emission   Electromagnetic radiation emitted by rapidly-moving charged particles (usually electrons) spiralling around magnetic field lines; the radiation is polarized and is emitted in a conical pattern in the direction of particle motion.

Synchrotron radiation   Continuous radiation produced by rapidly moving electrons traveling along lines of magnetic force.

Synodic period   The orbital or rotational period of an object as seen by an observer on the Earth; for the Moon or a planet the synodic period is the interval between repetitions of the same phase or configuration.

S wave   A type of seismic wave that is a transverse, or shear, wave and can travel only through rigid materials. (*see Shear wave*)

Tectonic activity   Geophysical processes involving motions of tectonic plates and associated volcanic and earthquake activity. *See also* **plate tectonics** and **continental drift.**

Temperature   A measure of the internal energy of a substance. At the atomic level, temperature is related to the motions of atoms and molecules that make up the substance.

Thermal radiation   Continuous radiation emitted by any body whose temperature is above absolute zero.

Thermal spectrum   The spectrum of continuous radiation from a body that glows because it has a temperature above absolute zero. A thermal spectrum is one that is described mathematically by the Planck function.

Thermosphere   The layer of the Earth's atmosphere above the mesosphere, extending upward from a height of 80 kilometers, where the temperature rises with altitude.

Tidal force   A gravitational force that tends to stretch or distort an extended object. See **differential gravitational force.**

Total eclipse   Any eclipse in which the eclipsed body is totally blocked from view or totally immersed in shadow.

Transit telescope   A telescope designed to point straight overhead and accurately measure the times at which stars cross the meridian.

Transverse wave   *See* **shear wave.**

Triple-alpha reaction   A nuclear fusion reaction in which three helium nuclei (or alpha particles) combine to form a carbon nucleus.

Trojan asteroid   Any of several asteroids orbiting the Sun at stable positions in the orbit of Jupiter, either 60 degrees ahead of the planet or 60 degrees behind it.

Tropical year   The year as defined by the length of time required for the Sun to move through the celestial sphere once, as defined by its passage through the vernal equinox.

Troposphere   The lowest temperature zone in the Earth's atmosphere, extending from the surface to a height of about 10 kilometers, in which the temperature decreases with altitude.

Tully-Fisher method   A technique for estimating the distances of spiral galaxies, based on a correlation between the width of the 21-cm emission line and the luminosity of a galaxy.

Tuning fork diagram   A diagram developed by Edwin Hubble in which the morphological types of galaxies are displayed in a sequence having two branches (representing spiral galaxies and barred spiral galaxies), resembling the shape of a tuning fork.

Type II irregular   A galaxy not conforming to any of the standard types, and showing no evidence of spiral structure.

Type I irregular   A galaxy not conforming to any of the standard types, but appearing to have some spiral structure.

Type I Supernova   A violent stellar explosion whose spectrum does not include lines of hydrogen, thought to be caused by the nuclear destruction of a white dwarf star that has gained mass from a companion star.

Type II Supernova   A violent stellar explosion whose spectrum contains lines of hydrogen, thought to be caused by the collapse and rebound of a massive star that has formed an iron core at the end of its nuclear-burning lifetime.

T Tauri star   A young star still associated with the interstellar material from which it formed, typically exhibiting brightness variations and a stellar wind.

Ultraviolet   The portion of the electromagnetic spectrum between roughly 100 and 3,000 angstroms.

Umbra   (1) The dark inner portion of a shadow, such as the part of the Earth's shadow where the Moon is in total eclipse during a lunar eclipse; (2) the dark central portion of a sunspot.

Van Allen belts Zones in the Earth's magnetosphere where charged particles are confined by the Earth's magnetic field. There are two main belts, one centered at an altitude of roughly 1.5 times the Earth's radius, and the other between 4.5 and 6.0 times the Earth's radius.

Velocity dispersion A measure of the average velocity of particles or stars in a group or cluster with random internal motions. In globular clusters and elliptical galaxies, the velocity dispersion can be used to infer the central mass.

Vernal equinox The occasion when the Sun crosses the celestial equator from south to north, usually occurring around March 21. *See also* **autumnal equinox** and **equinox.**

Visual binary Any binary system in which both stars can be seen through a telescope or on photographs.

Volatile The property of being easily vaporized. Volatile elements stay in gaseous form except at very low temperatures and did not condense into solid form during the formation of the solar system.

Wavelength The distance between wavecrests in any type of wave.

White dwarf The compact remnant of a low-mass star, supported against further gravitational collapse by the pressure of the degenerate electron gas that fills its interior.

Wien's law An experimentally discovered law applicable to thermal continuum radiation, which states that the wavelength of maximum emission intensity is inversely proportional to the absolute temperature.

X-ray A photon of electromagnetic radiation in the wavelength interval between about 1 and 100 angstroms.

Zeeman effect The broadening or splitting of spectral lines due to the presence of a magnetic field in the gas where the lines are formed.

Zenith The point directly overhead.

Zero-age main sequence The main sequence in the H-R diagram formed by stars that have just begun their hydrogen-burning lifetimes and have not yet converted any significant fraction of their core mass into helium. The zero-age main sequence forms the lower left boundary of the broader band representing the general main sequence.

Zodiac A band circling the celestial sphere along the ecliptic that is broad enough to encompass the paths of all the planets visible to the naked eye; in some usages the sequence of constellations lying along the ecliptic.

Zodiacal light A diffuse band of light visible along the ecliptic near sunrise and sunset, created by sunlight scattered off interplanetary dust.

21-centimeter line The emission line of atomic hydrogen, whose wavelength is 21.11 centimeters, emitted when the spin of the electron reverses itself with respect to that of the proton. The 21-centimeter line is the most widely used and effective means of tracing the distribution of interstellar gas in the Milky Way and other galaxies.

24-hour anisotropy The daily fluctuation in the observed temperature of the cosmic background radiation caused by a combination of the Earth's motion with respect to the background and its rotation.

3° background The radiation field filling the universe, left over from the big bang. *See also* **cosmic background radiation.**

THE NIGHT SKY IN SPRING

Latitude of chart is 34° N, but it is practical throughout the continental United States.

To use: hold the chart vertically and turn it so the direction you are facing shows at the bottom.

<u>Chart Time (local standard time)</u>	
Mid-March	11 p.m.
Mid-April	9 p.m.
Mid-May	7 p.m.

Star Chart from *Griffith Observer*, Griffith Observatory, Los Angeles

THE NIGHT SKY IN SUMMER

Latitude of chart is 34° N, but it is practical throughout the continental United States.

To use: hold the chart vertically and turn it so the direction you are facing shows at the bottom.

<u>**Chart Time (daylight savings time)**</u>

Mid-June	Midnight
Mid-July	10 p.m.
Mid-August	8 p.m.

Star Chart from *Griffith Observer*, Griffith Observatory, Los Angeles

NORTHERN HORIZON

EASTERN HORIZON

WESTERN HORIZON

SOUTHERN HORIZON

. THE NIGHT SKY IN AUTUMN

Latitude of chart is 34° N, but it is practical throughout the continental United States.

To use: hold the chart vertically and turn it so the direction you are facing shows at the bottom.

<u>Chart Time (local standard time)</u>

Mid-September	11 p.m.
Mid-October	9 p.m.
Mid-November	7 p.m.

Star Chart from *Griffith Observer*, Griffith Observatory, Los Angeles